Interaction Between Neurons and Glia in Aging and Disease

Interaction Between Neurons and Glia in Aging and Disease

Edited by

João O. Malva
Center for Neuroscience and Cell Biology
Faculty of Medicine
University of Coimbra
3004-504 Coimbra, Portugal

Co-Editors

Ana C. Rego, Rodrigo A. Cunha and Catarina R. Oliveira

 Springer

João O. Malva et al.
Center for Neuroscience and Cell Biology
Faculty of Medicine
University of Coimbra
3004-504 Coimbra
Portugal

Library of Congress Control Number: 2007923852

ISBN 978-0-387-70829-4   e-ISBN 978-0-387-70830-0

Printed on acid-free paper.

© 2007 Springer Science+Business Media, LLC
All rights reserved. This work may not be translated or copied in whole or in part without the written permission of the publisher (Springer Science+Business Media, LLC, 233 Spring Street, New York, NY 10013, USA), except for brief excerpts in connection with reviews or scholarly analysis. Use in connection with any form of information storage and retrieval, electronic adaptation, computer software, or by similar or dissimilar methodology now known or hereafter developed is forbidden.
The use in this publication of trade names, trademarks, service marks, and similar terms, even if they are not identified as such, is not to be taken as an expression of opinion as to whether or not they are subject to proprietary rights.

9 8 7 6 5 4 3 2 1

springer.com

Preface

João O. Malva, Ana C. Rego, Rodrigo A. Cunha, and Catarina R. Oliveira

Recent years witness the amazing and enthusiastic growth of research activity dedicated to understand the complex mechanisms of interaction between neurons and glia in brain physiology and the involvement of these processes in brain dysfunction and pathology. Neuroscientists and neuroimmunologists face the major challenge of bringing to light new knowledge about the functional interplay of brain cellular partners. There is now a better understanding of the basic physiopathological processes of many diseases of the nervous system, including the role of inflammation and immune mechanisms. New powerful tools were shown to interfere with the "recital" of cytokines, chemokines and inflammatory molecules and their dynamic effects on neurons, glia, vascular cells and cells of the immune system that invade the brain parenchyma in response to injury. The ultimate goal of this area of research is to develop new therapeutical approaches and to find a cure for the devastating disorders of the nervous system to which inflammation and immune contributions were demonstrated to be key players.

This project was born after joining efforts from expert contributors in neuron–glial cross-talk in health and disease. We are grateful for the efforts of the contributors in this book. The aim of editors is to foster understanding on the mechanisms of brain injury and how they can be modulated to overcome irreversible damage resulting in a better quality of life. There is still much work to be done, but there are reasons to be optimistic.

We hope you will find this book interesting and informative.

Welcome to "Interaction Between Neurons and Glia in Aging and Disease"

The Editors

University of Coimbra, December 1, 2006

Acknowledgements

Special thanks to
All colleagues contributing to the present book.

The colleagues working at "Neuroprotection and Neurogenesis in Brain Repair", "Mitochondrial dysfunction" and "Purines at CNC" groups.

The Center for Neuroscience and Cell Biology, University of Coimbra and to the Portuguese Society for Neuroscience.

Bruno Manadas for his enthusiastic collaboration with figures and animations in the accompanying DVD.

Contents

Section 1 **Neuroimmunity and neuroinflammation** 1
João O. Malva and Catarina R. Oliveira

Chapter 1. Inflammation and Neuronal Susceptibility to Excitotoxic Cell Death 3
Liliana Bernardino and João O. Malva

Chapter 2. Microglial Cell Population Expansion Following Acute Neural Injury 37
M. Wirenfeldt, L. Dissing-Olesen, A.A. Babcock, R. Ladeby,
M.B. Jensen, T. Owens and B. Finsen

Chapter 3. Opposite Modulation of Peripheral Inflammation and Neuroinflammation
by Adenosine A_{2A} Receptors 53
Rodrigo A. Cunha, Jiang-Fan Chen and Michail V. Sitkovsky

Chapter 4. Subventricular Zone Cells as a Tool for Brain Repair 81
Fabienne Agasse, Liliana Bernardino and João O. Malva

Section 2 **Signaling and inflammation in aging** 109
Rodrigo A. Cunha

Chapter 5. Neuron-Glia Interaction in Homeostasis of the Neurotransmitters
Glutamate and GABA 111
Arne Schousboe and Helle S. Waagepetersen

Chapter 6. The Impact of an Imbalance Between Proinflammatory and
Anti-inflammatory Influences on Synaptic Function in the Aged Brain 121
Marina Lynch

Chapter 7. Neurotrophin Signaling and Cell Survival 137
Bruno J. Manadas, Carlos V. Melo, João R. Gomes
and Carlos B. Duarte

Chapter 8. GDNF: a Key Player in Neuron-Glia Crosstalk and Survival
of Nigrostriatal Dopaminergic Neurons 173
Emília P. Duarte, Ana Saavedra and Graça Baltazar

Chapter 9. Molecular Pathways of Mitochondrial Dysfunction in Neurodegeneration: the Paradigms of Parkinson's and Huntington's Diseases 193
Ana Cristina Rego, Sandra Morais Cardoso and Catarina R. Oliveira

Chapter 10. Zinc Homeostasis and Brain Injury 221
Stefano Sensi, Erica Rockabrand and Israel Sekler

Chapter 11. Aging and Cognitive Decline: Neuroprotective Strategies 245
Frederico Simões do Couto and Alexandre de Mendonça

Section 3 Neurodegeneration and inflammation in age related diseases 269
Ana Cristina Rego and Catarina R. Oliveira

Chapter 12. Inflammation and Apoptotic Pathways in the Peripheral Nervous System Related to Protein Misfolding 271
Maria João Saraiva

Chapter 13. Neuroinflammation and Excitotoxicity in Neurobiology of HIV-1 Infection and AIDS: Targets for Neuroprotection 281
Marcus Kaul and Stuart A. Lipton

Chapter 14. Neuroinflammation and Mitochondrial Dysfunction in Alzheimer's and Prion's Diseases 309
Paula Agostinho and Catarina R. Oliveira

Chapter 15. Metal Ions and Alzheimer's Disease 333
Paul A. Adlard and Ashley I. Bush

Chapter 16. Epilepsy, Brain Injury and Cell Death 363
Günther Sperk, Meinrad Drexel, Ramon Tasan and Anna Wieselthaler

Chapter 17. Glia and Hippocampal Neurogenesis in the Normal, Aged and Epileptic Brain 375
William P. Gray and Alexandra Laskowski

Chapter 18. Polyglutamine Expansion Diseases – the Case of Machado-Joseph Disease 391
Sandra Macedo-Ribeiro, Luís Pereira de Almeida, Ana Luísa Carvalho and Ana Cristina Rego

Chapter 19. Remyelination of the Central Nervous System 427
Charlotte C. Bruce, Robin J. M. Franklin and João B. Relvas

Chapter 20. Adult Neurogenesis in Neurodegenerative Diseases 445
Tomas Deierborg, Jia-Yi Li and Patrik Brundin

Chapter 21. ATP Receptors in the Pain Signaling: Glial Contribution in Neuropathic Pain.. 461
Kazuhide Inoue

Chapter 22. Diabetic Retinopathy, Inflammation, and Proteasome.................. 475
António F. Ambrósio, Paulo Pereira and José Cunha-Vaz

Chapter 23. Rejuvenating Neurons and Glia with Microbial Enzymes 503
Aubrey D.N.J. de Grey

Abbreviations 511

Index.. 513

Section 1

NEUROIMMUNITY AND NEUROINFLAMMATION

João O. Malva and Catarina R. Oliveira

Evidence exists showing a clear interaction between neurons, glia, and cells of the immune system, which is highly dynamic and highly regulated, playing a key role in brain homeostasis. There is increasing support for the contribution of the disruption of this interaction to the inflammatory (neuroinflammatory) and immune (neuroimmunity) responses associated to many diseases of the nervous system. In the present section, the major molecular and cellular players in neuroinflammation are introduced and their role in neuronal death/protection and brain repair is discussed.

In Chap. 1, Bernardino and Malva introduce neuroimmunity and neuroinflammation in health and disease, highlighting the close interaction between the innate immune response and excitotoxicity. The key proinflammatory cytokines, chemokines, receptors, and mechanisms involved in IL-1β and TNF-α receptors-mediated signaling transduction are reviewed. The central role of microglia activation in the innate immune response of the brain and its impact in neuronal death/survival is discussed. The authors introduce in this chapter many of the recurrent concepts used in other sections/chapters of the book.

In Chap. 2, by Bente Finsen's group (Wirenfeldt et al.,) the authors revise and discuss the developmental origin and diversity of microglial cells and their response to injury. The highly dynamic plasticity of microglial populations in response to injury is explored and deeply discussed. The readers are invited to understand microglial diversity as a natural resource of brain development and defence against aggression.

Chapter 3 by Cunha, Chen, and Sitkovsky discuss the role of ATP in brain and peripheral response to injury. The shortcut between protection and toxicity and the ambiguous role of adenosine and adenosine receptors is dissected. Clearly, the next years will contribute to recognize the purinergic system as one of the key mediators/modulators of the inflammatory response in periphery and in the central nervous system.

Agasse, Bernardino, and Malva (Chap. 4) discuss the potential of subventricular zone cells as a resource for brain repair. Detailed information about subventricular zone niche of stem/progenitor cells, factors contributing to migration, differentiation, and survival of new neurons is provided. A special attention was given on the impact of brain injury and inflammatory mediators in the migration, differentiation, and survival of subventricular zone cells as endogenous/grafted resources for brain repair. The potential use (at the long-term) of adult brain stem/progenitor cells in the repair of injured central nervous system tissue is envisioned and discussed. Clearly, the stem/progenitor cells research is endowed with enormous potential for the future therapy of the major diseases afflicting the central nervous system. The future years will assist to the enthusiastic expansion in stem/progenitor cell research and hopefully to the outcome of new strategies for brain repair.

In Sect. 2, "Signaling and Inflammation in Aging," and Sect. 3, "Neurodegeneration and Inflammation in Age-Related Diseases," the concepts introduced in Sect. 1 will be used in the context of inflammatory reactions and signaling transduction in aging and disease.

INFLAMMATION AND NEURONAL SUSCEPTIBILITY TO EXCITOTOXIC CELL DEATH

Liliana Bernardino and João O. Malva

1. ABSTRACT

The fast growing research field of Neuroinflammation is challenging the traditional view of the Brain as an immune privileged organ. Our increasing understanding of the mechanisms involved in the close functional interaction between neurons and glia in health and disease is contributing to shed new light into the cellular and molecular mechanisms of neurodegeneration. The response of microglia and astrocytes to injury, releasing signaling molecules (specially, cytokines, chemokines, growth factors, reactive oxygen and nitrogen species), is central in neuroinflammation and in neuron death/survival.

At long-term, our better knowledge of the functional crosstalk between neurons and glia in health and disease and the bidirectional flux of immune cells from the peripheral blood to the brain parenchyma through the blood–brain barrier will allow the build-up of new tools for the treatment of the major diseases afflicting the human brain. In this chapter we review and discuss the immune response of the brain to injury, on the basis of the close interaction between excitotoxicity and inflammation.

2. INFLAMMATION IN THE CENTRAL NERVOUS SYSTEM

Until recently, the brain was considered as an immune privileged organ mostly because it was assumed that blood–brain barrier (BBB) should be completely impermeable to entrance of immune system cells into the brain parenchyma. In this way, the central nervous system (CNS) was considered unable to respond with an immune reaction when challenged by neurodegeneration or by infection. However, in the last decade the immune privileged status of the brain was questioned mostly because in pathological conditions molecules and cells of the immune system enter the brain via the BBB and target the brain parenchyma (Huber et al., 2001; Nguyen et al., 2002). Yet, inflammatory response in the CNS shows some peculiarities when compared with inflammation in other organs. This is most evident in response to acute insults because lymphocyte recruitment is rapid and pronounced in many systemic organs, but modest and delayed in the CNS. Accordingly, the number of patrolling lymphocytes in

---

João O. Malva – jomalva@fmed.uc.pt
Center for Neuroscience and Cell Biology, Institute of Biochemistry, Faculty of Medicine, University of Coimbra, 3004-504 Coimbra, Portugal

the healthy CNS is also lower when compared with other tissues. Moreover, in normal resting conditions, the resident immune effector cells in the brain, the microglia cells (Perry and Gordon, 1991), show a resting phenotype, with a low or absent expression of immunological molecules and their receptors (Neumann and Wekerle, 1998; Neumann, 2001, 2004). Although, besides these peculiarities, following an injury, the brain may initiate and develop by itself a local and rapid immune response resulting in the activation (within minutes or hours after the insult) of brain microglia and the release of inflammatory mediators (Lucas et al., 2006). These reactions induce a rapid and efficient clearance of pathogens and cell debris from the damaged CNS (Schwartz et al., 1999; Becher et al., 2000; Nguyen et al., 2002; Elward and Gasque, 2003). The efficient regulation of the inflammatory cascade results from a fine-tune equilibrium of events that promotes both immune privilege in health and effective responses in injury or disease.

In general, acute inflammation is beneficial to the organism in limiting the survival and proliferation of invading pathogens and promoting regeneration of the tissue. However, prolonged, excessive inflammation is highly detrimental, leading to the onset and/or to the exacerbation of cell damage in neurodegenerative diseases.

2.1. Innate Immunity in the CNS

Among the first line of defense of the body, the innate immunity is responsible for the immediate response to insults and pathogens and plays an essential role in survival of the organism. It is characterized by the rapid and relatively generic recognition of pathogens by immune cells that either kill the invaders directly or activate phagocytic cells to ingest and remove them (Rivest, 2003). Phagocytic cells (including macrophages and microglia cells) are the principal effector cells of innate immunity response in the brain. These cells produce and release substantial amounts of pro-inflammatory cytokines, specially tumor necrosis factor (TNF)-α, interleukin (IL)-1β and IL-6 that facilitate the recruitment and enhance the activity of other immune players, thus improving the overall immune response (Simard and Rivest, 2005).

Pathogen-Associated Molecular Patterns (PAMPs)
Phagocytic cells of the innate immune system express receptors that detect the presence of infectious agents (e.g. bacteria, fungi) by recognizing specific and highly conserved structures produced by these pathogens, which are not expressed by eukaryotic organisms (Anderson, 2000). These elements are called pathogen-associated molecular patterns (PAMPs) and are recognized specially by macrophages and microglial cells that put in motion a rapid response to infection (Medzhitov and Janeway, 2000). The endotoxin lipopolysaccharide (LPS), a component of the outer membrane of Gram-negative bacteria, is one of the most characterized PAMP, able to stimulate the innate immune system phagocytic cells and to trigger a robust innate inflammatory response (Wright, 1999). These immune cells can also recognize disease-associated molecules, such as modified endogenous molecules or cell debris from damaged cells, leading to tissue remodeling and tissue repair (Devitt et al., 1998; Elward and Gasque, 2003).

Toll-Like Receptors (TLR)
The Toll-like receptors (TLRs) are signaling receptors present in immune cells that recognize PAMPs, and represent the first element of contact between pathogens and the host

(Akira et al., 2001; Nguyen et al., 2002). The Toll protein was originally described as a receptor involved in dorsoventral polarization of the *Drosophila* embryo (Anderson et al., 1985; Belvin and Anderson, 1996) with a vital role in the antifungal immunity of adult flies (Lemaitre et al., 1996). The TLRs are mammalian homologues of the Toll protein, both characterized by multiple leucine-rich repeats in extracellular domains and cytoplasmic domains resembling the cytoplasmic portion of the IL-1 receptor (IL-1R), commonly known as Toll/IL-1R motif (TIR) (Wright, 1999; Anderson, 2000). The broad spectrum of ligands recognized by these receptors depends not only from the high diversity of the extracellular domains, but also from the complex pattern of recognition of the ligand by the receptor, including dimerization of some TLRs (Ozinsky et al., 2000; Rivest, 2003).

2.2. Adaptive Immunity in the CNS

Innate immunity is sometimes sufficient to counteract a simple infection, and in these cases the response of the host with an adaptive response is not required. However, sometimes innate immune response cannot eliminate the infectious organisms. In these conditions, following TLRs/PAMPs interaction, glial cells and neurons begin to produce inflammatory cytokines, chemokines, and adhesion molecules, which activate and stimulate the traffic of immune cells with a role in adaptive immune system, such as T and B lymphocytes, to the sites of lesion in the brain (Kielian et al., 2002; Olson and Miller, 2004, Simard and Rivest, 2005). In this sense, TLRs as well as these inflammatory mediators (cytokines and chemokines) play a pivotal role between the innate immune response and the adaptive immunity of the CNS.

One of the characteristic features of acquired immunity in the response to the pathogen is plasticity and development of memory signals. Some viruses and bacterial antigens do not stimulate T lymphocytes directly. In these cases a specific group of highly specialized immune cells, antigen presenting cells (APCs), process and expose the antigens bound to the major histocompatibility complex (MHC) molecules on their cell surface, making these antigens recognizable by specific T lymphocytes (Nguyen et al., 2002). APCs include microglia, macrophages, and dendritic cells. After an injury, T lymphocytes interact with APCs, become activated and develop immune memory. The latter process depends specifically on the recognition of the antigen peptide fragments bound to MHC and on co-stimulatory signals on the APCs, leading ultimately to the clonal selection of lymphocytes. So, after acquiring memory and clonal selection, the lymphocytes better contribute to provide more versatile mechanisms of defense specially when the host is repetitively infected by the same pathogen (Neumann, 2004). Besides T lymphocytes, B lymphocytes are also cellular players with major role in acquired immunity, responsible for the humoral defense. Despite increasing evidences showing that B lymphocytes can enter the CNS in pathological conditions, differentiate into plasma cells and secrete antibodies (Knopf et al., 1998), the ability of B lymphocytes to patrol the neural parenchyma is less understood.

Brain inflammation involves a bidirectional communication between the immune system and the CNS. As indicated before, even in the presence of intact BBB, the CNS is routinely surveyed by blood immune cells (e.g. T lymphocytes) in search for pathogens (Flügel et al., 2001). However, in normal healthy conditions, only few activated T lymphocytes can pass the BBB and most of these cells cannot find antigens and are committed to die by apoptosis (Bauer et al., 1998). Under pathological conditions, T cells accumulate and can reach relatively high density in inflammatory sites in brain parenchyma (Neumann and Wekerle, 1998; Raivich et al., 1998; Yeager et al., 2000; Flügel et al., 2001; Hickey, 2001; Engelhardt,

2006; Man et al., 2006). When a strong inflammatory reaction occurs in the body, even if the CNS is not itself directly involved, patrolling T cells can be found in the CNS. Moreover, systemic inflammation often does not induce evident CNS lesions but contributes to cerebral vulnerability and exacerbate cell damage (Perry, 2004; Cunningham et al., 2005). This crosstalk between periphery immunomodulators of blood and CNS cells occurs directly at regions lacking BBB, which include circumventricular organs (CVOs), the choroid plexus and leptomeninges. These structures are characterized by a rich vascular plexus where the junctions between capillary endothelial cells are absent or modified, allowing the diffusion of molecules such as pro-inflammatory cytokines, but also infectious agents into brain parenchyma (Nguyen et al., 2002; Schulz and Engelhardt, 2005). Another mechanism by which circulating cytokines may communicate with CNS is through direct interaction with BBB endothelial cells. These cells constitutively express receptors for several molecules of the immune system. In the presence of a challenge, the receptor's expression is rapidly up-regulated and, following activation, can induce the release of inflammatory molecules across the brain parenchyma. Moreover, several factors such as the intercellular adhesion molecule 1 (ICAM-1), vascular endothelial growth factor (VEGF), IL-1β, TNF-α, IL-6 and some chemokines such as macrophage inflammatory protein-1α (MIP-1α), CCL19, CCL21 and CCL2, can induce transient alterations in the stability and permeability of the BBB and ultimately facilitate the entry of immunomodulators from the blood into the CNS (Dobrogowska et al., 1998; Mark and Miller, 1999; Huber et al., 2001; Engelhardt, 2006; Mahad et al., 2006; Man et al., 2006).

Therefore, in both models, systemic cytokines interact directly or indirectly with the CNS parenchyma and can trigger a series of events leading to the activation of transduction pathways, generating an inflammatory state in the CNS (Zhang and Rivest, 2003). Furthermore, cytokines amplify the immune response, and lead to the expression of other cytokines and chemokines, contributing to recruit other immune cells such as T and B lymphocytes to the inflammation site and by this way put in motion the onset of the adaptive immune response. So, both innate and acquired immunity are not separate entities, but they are functionally interconnected (Medzhitov and Janeway, 1997; Bailey et al., 2006).

3. MEDIATORS OF INFLAMMATION

The inflammatory response in the CNS comprises a complex and integrated interplay between different cellular types of the immune system (macrophages, T and B lymphocytes, dendritic cells) and resident cells of the CNS (microglia, astrocytes, oligodendrocytes, neurons) as well as a complex orchestra of cytokines, adhesion molecules, chemokines and their receptors. In normal resting conditions, most inflammatory mediators produced by these cells are expressed at very low, or undetectable, levels. However, they are rapidly upregulated in response to infection or tissue injury (Rothwell and Luheshi, 2000; Chavarria and Alcocer-Varela, 2004). In this section we will focus on both the cell component of the BBB and the resident parenchymal immune cells of the CNS, and the interaction between them, in health and disease.

3.1. BBB as a Key Immune Sentinel of the Brain

Immune surveillance at the BBB is ensured by endothelial cells of the brain capillaries and by perivascular cells surrounding cerebral vessels (Thomas, 1999; Williams et al., 2001).

The endothelium is an essential structure of the BBB and thus controls the flow of immune cells from the blood to the CNS and vice-versa. So, endothelial cells are strategically positioned to respond rapidly to circulating endotoxins, bacteria or virus (Huber et al., 2001; Minagar et al., 2006). Vascular endothelial cells are both targets and sources of inflammatory mediators (e.g. nitric oxide (NO) and prostanoids), and adhesion molecules such as ICAM-1, VCAM-1, PECAM-1 and E-selectin, that facilitate the passage of lymphocytes from blood to CNS (Couraud, 1994; Wong et al., 1999; Engelhardt, 2006).

Localized between the vascular endothelial cells and CNS parenchyma, the perivascular space (Virchow-Robins space) contains a specific population of surveillant perivascular cells. These cells are derived from migratory macrophages of the bone marrow and they are regularly exchanged and replaced (Hickey and Kimura, 1988; Lassmann et al., 1993). Perivascular cells show constitutive and inflammatory-induced expression of MHC II and co-stimulatory molecules, cytokines, chemokines, and adhesion molecules (Thomas, 1999; Piehl and Lidman, 2001; Williams et al., 2001). Thus, these cells display all the immunological properties that allow them to be professional and efficient APCs, playing a major role in the first line of the brain immune responses against systemic pathogens (Williams et al., 2001).

3.2. Parenchyma Resident Cells in CNS

Among brain parenchyma resident cells (neurons, astrocytes, oligodendrocytes and microglia), neurons are the most passive cells in the inflammatory response. Nevertheless, they have the ability to express class I MHC molecules, to produce several cytokines like IFN-γ (Neumann et al., 1997) and even to induce apoptosis of T cells (Flügel et al., 2000). Moreover, these cells express several other cytokines such as TNF-α and IL-1β and their receptors (Yabuuchi et al., 1994; Friedman, 2001; Sairanen et al., 2001), that can modulate the outcome of neuronal injury and neuroinflammation (Gruol and Nelson, 2005; Hellstrom et al., 2005; Orellana et al., 2005; Lai et al., 2006). Our knowledge about the function of neurons in immune reactions in brain parenchyma is very restricted but is rapidly growing. New findings about the interplay between neuronal function and inflammation will critically contribute to better understand neuronal dysfunction and neurodegenerative diseases.

Microglia and astrocytes are the two major parenchyma populations of reactive glial cells in CNS that participate in immune reactions. Within the CNS parenchyma, microglia cells are the resident monocyte/macrophage-like population involved in the clearance of invading pathogens and cell debris (Kreutzberg, 1996; Streit, 2002; Kim and de Vellis, 2005). In physiological conditions, microglia display a resting-like down-regulated immune phenotype and a ramified surveillant-like morphology; behave as poor APCs with faint expression of MHC II molecules, cytokines and their receptors, and no co-stimulatory molecules. It has been shown in vitro that substances released during normal neuronal activity can down-regulate the expression of molecules associated with antigen presentation and production of inflammatory molecules by microglia (Frei et al., 1994; O'Keefe et al., 1999). Neumann and collaborators demonstrated that electrically active neurons inhibit IFN-γ-induced expression of MHC class II molecules on astrocytes and microglia (Neumann et al., 1996; Neumann, 2001). Neurotrophins (NGF, BDNF, NT-3) have also been shown to inhibit microglia expression of MHC class II and co-stimulatory (B7-2/CD86 and CD40) molecules involved in antigen presentation (Neumann et al., 1998; Wei and Jonakait, 1999). In response to infection, inflammation or brain injury, microglial cells become rapidly activated and are recruited to the site of injury (Heppner et al., 1998; Bernardino et al., 2005a). Moreover, they strongly up-regulate the expression of immune modulators as well as their receptors. Morphologically, activated microglia are characterized by a round-shape morphology with thickened and short or retracted processes (Kreutzberg, 1996). In this activated condition, microglia are efficient APCs (but less than perivascular microglial cells) (Shrikant and Benveniste, 1996; Aloisi et al., 1998) expressing high levels of MHC II, adhesion molecules (CD11a and CD54), co-stimulatory molecules CD80 and CD86 (Windhagen et al., 1995), receptors involved in phagocytosis (Burudi and Regnier-Vigouroux, 2001) and cytokines and their receptors (Eriksson et al., 2000; Pinteaux et al., 2002; Rappert et al., 2004).

A general assumption is that microglia activation contributes to protect neurons and this is specially important in view of the limited regenerative capacity of these cells. Microglia can be beneficial to CNS because of their phagocytic potential and their ability to eliminate invading pathogens, cell debris and several neurotoxins (Kreutzberg, 1996; Streit, 2002). Moreover, microglia can also produce trophic factors and anti-inflammatory cytokines such as TGF-β, IL-10 and IL-1 receptor antagonist (IL-1ra) (Batchelor et al., 1999; Nakajima and Kohsaka, 2004). It has been assumed that the primary purpose of microglia is to support neuronal function, however, the sustained or excessive activity of these cells may have detrimental consequences and can be associated with the onset and/or exacerbation of neuronal death associated with several neurodegenerative disorders. Accordingly, activated microglia release several pro-inflammatory cytokines, such as TNF-α and IL-1β, chemokines including MIP-1α, reactive oxygen species and NO, that can damage neurons (Aloisi et al., 2000; Piehl and Lidman, 2001; Nakajima and Kohsaka, 2004; Bernardino et al., 2005a). So, microglia can exert both damaging and protective/reparative actions on neighbour cells, causing amplification or suppression of the immune responses (Streit et al., 1999; Hanisch, 2002; Bernardino et al., 2005a; Kim and de Vellis, 2005; Schwartz et al., 2006).

Astrocytes play key physiological roles in metabolic support of neurons, maintenance of ion and neurotransmitter concentration, maintenance of the BBB function and production of trophic and neuroprotective factors (e.g. neurotrophic factors) (Bezzi and Volterra, 2001; Dong and Benveniste, 2001). Astrocytes have a predominant role in limiting the inflammatory reaction in the healthy brain, although, following stress and injury, astrocytes proliferate (astrogliosis), up-regulate the glial fibrillary acidic protein (GFAP) and release inflammatory/potentially neurotoxic molecules like pro-inflammatory cytokines (Piehl and Lidman, 2001; Chen and Swanson, 2003; Moynagh, 2005). Astrogliosis is a key event in the formation of the glial scar, which, depending on the CNS milieu and the disease context, can be beneficial for repair and neuronal survival, by producing neurotrophins, or can be detrimental by inhibiting reinnervation (Hatten et al., 1991; Sofroniew, 2005). Opposite to microglia cells, activated astrocytes are capable of producing and exposing MHC II molecules, but there is much controversy about the expression of important co-stimulatory molecules on these cells. So, overall, astrocytes are not efficient APCs in vivo but depending on the experimental in vitro models they can be efficient in antigen presentation (Aloisi et al., 1998, 2000).

Besides their classic structural and trophic functions, astrocytes can also be involved in more complex functions such as synaptic integration, plasticity and processing of neuronal information (Bezzi and Volterra, 2001; Allen and Barres, 2005; Perea and Araque, 2006). Astrocytes can modulate the function of neurons through chemical synapses and gap-junctions between a presynaptic neuron and surrounding glial cells (Nedergaard, 1994; Alvarez-Maubecin et al., 2000; Bergles et al., 2000; Perea and Araque, 2005). The working hypothesis raised by Bezzi and collaborators was that astrocytes can electrically or chemically communicate with neurons. Thus, during intense neuronal activity some neurotransmitters may diffuse out of the synaptic cleft and induce glial activation and the subsequent increase in the intracellular calcium concentration on glial cells. The raise in intracellular calcium in glial cells can spread in the glial network in the form of calcium waves and can actively induce the release of neurotransmitters, thereby modulating the function of adjacent neurons. On the basis of this functional interaction, inflammatory responses are derived from an integrated signaling among neurons, microglia, and astrocytes (Bezzi and Volterra, 2001; Allen and Barres, 2005; Perea and Araque, 2005, 2006).

3.3. LPS-Induced Inflammation: An Integrated view of an Immune Response in the CNS

LPS is the major inducer of microglia production of pro- and anti-inflammatory cytokines, chemokines, prostaglandins, and nitric oxide. Experimental administration of LPS has been used as a model of inflammatory response to brain infections. It has been described that LPS induces signal transduction by the CD14-mediated activation of TLR4 (Kaisho and Akira, 2000; Kitchens et al., 2000; Dauphinee and Karsan, 2006; Xia et al., 2006). The CD14 and TLR4 receptors are constitutively expressed in the CVOs of the CNS (Nguyen et al., 2002). So, the initial step in the recognition of circulating LPS is the binding of LPS to the LPS-binding protein that guides LPS to the TLR4/CD14 receptor complex present in the CVOs. The interaction within LPS/TLR4/CD14 leads to the nuclear translocation of NFκB and subsequent transcription of inflammatory genes, initially within these organs and thereafter throughout the brain parenchyma mimicking endotoxemia (Laflamme and Rivest, 2001; Xia et al., 2006). It has been proposed that TNF acts as a major mediator of the spread of LPS-induced inflammation throughout the brain parenchyma. TNF, in turn, may diffuse as an autocrine or as a paracrine cytokine, and activate parenchymal microglial cells (Nguyen et al., 2002). Alternatively, LPS may passively diffuse across the BBB, specially under severe endotoxemia (Rivest, 2003) allowing the direct interaction of LPS with its receptors in parenchyma microglia. Lehnardt and colleagues described microglial cells as the major LPS-responsive cell type in the brain parenchyma due to the presence of TLR4 (Lehnardt et al., 2003). All these events may be central to the orchestration of inflammatory responses leading to the activation of the resident phagocytic cells and to the inflammatory status in the brain (Xia et al., 2006). IL-1β and TNF-α are two master pro-inflammatory cytokines with pleiotropic and largely overlapping functions. They are produced by microglia and blood-derived macrophages during CNS inflammation and severe LPS-induced endotoxemia. In this chapter we will focus mainly in the signaling mechanisms and functional implications of these two cytokines.

4. CYTOKINES AND CHEMOKINES

4.1. Cytokines

Cytokines are multifunctional, low molecular weight proteins with important roles in many physiological and pathological processes in the CNS. They interact with specific receptors located on the cell membrane, or with soluble receptors (Wallach et al., 1991; Dinarello, 1996; Aggarwal, 2003) and upon binding to their receptors, several intracellular pathways are usually triggered, which in turn regulate the activity of transcriptional factors such as NFκB and AP-1 (Baeuerle, 1998; MacEwan, 2002; Dunne and O'Neill, 2003). The expression of these molecules and their respective receptors in the CNS are barely detected under normal conditions, but become stimulated in response to pathological conditions, like ischemia (Wang et al., 1997), excitotoxicity (Minami et al., 1991), and in epilepsy (Minami et al., 1990; Eriksson et al., 1998; Vezzani et al., 1999). In the brain, cytokines can be originated from two different sources (1) from infiltrating peripheral immune cells (lymphocytes and macrophages); (2) from CNS resident cells (neurons, microglia, astrocytes, oligodendrocytes, endothelial cells, and perivascular microglia) (Vassalli, 1992; Davies et al., 1999; Rothwell and Luheshi, 2000). Functionally, cytokines have been described as being either pro-inflammatory (e.g. IL-1β, TNF-α) or anti-inflammatory (e.g. IL-10, TGF-β, IL-1ra) depending on the final balance of their effects on the immune system. Although, contributing to the complexity of this interpretation, the functional role of cytokines in CNS inflammation can shift from beneficial to detrimental depending on the type of cells producing and releasing the cytokines, the functional state of neurons, local cytokine concentration and tissue exposure, the presence of specific receptors on target cells, and the presence or absence of modulators of cytokine activity. Moreover, cytokines exhibit pleiotropy and redundancy: an individual cytokine can have multiple functions on different cell types, and different cytokines can act on the same cell population to induce similar or opposite effects. Moreover, many cytokine actions are indirect, acting by stimulating the synthesis and function of other cytokines, resulting in a complex "cytokine cascade" from immune and inflammatory responses (Vilček, 2003).

Cytokines can contribute to initiation, propagation, and regulation of inflammatory reactions in CNS by controlling neuronal and glial activation and plasticity and also by regulating the bidirectional communication between CNS and immune system (Benveniste, 1998). Moreover, they have been described to be involved in the proliferation, differentiation, and survival of new neuronal cells in neurogenic processes (Mason et al., 2001; Monje et al., 2003). Cytokines activity can also be modulated by neurotransmitters, neuropeptides, growth factors, and hormones. Therefore, cytokines can act in the CNS as both immunoregulators and neuromodulators in health and disease (Rothwell, 1999; Wang and Shuaib, 2002; Campbell, 2005).

4.2. Chemokines

Chemokines are a family of chemo-attractant proteins, structurally related to cytokines (CHEMOtactic cytoKINES) that play a major role in chemotaxis in a concentration-dependent manner (Mennicken et al., 1999; Ambrosini and Aloisi, 2004; Ubogu et al., 2006). Chemokines are classified according to the number and the relative position of the first N-terminal cysteine residues in their primary structure. On this basis, there are four distinct subfamilies: alpha (CXC), beta (CC), gamma (XC), and delta (CX3C), where C stands for cysteine and X stands for a separating amino acid residue. These proteins are chemo-attractant for different cell types: CC chemokines are primarily chemo-attractant for monocytes, macrophages, lymphocytes, basophils, eosinophils, dendritic, and natural killer cells; CXC for lymphocytes, monocytes, and neutrophils; XC and CX3C for mononuclear inflammatory cells (Asensio and Campbell, 1999; Zlotnik and Yoshie, 2000).

Chemokines exert their biological activity by binding to cell surface seven-transmembrane domain receptors that signal through coupled heterodimeric G-proteins (Murphy et al., 2000).

The accepted nomenclature for a chemokine receptor includes the identification of its chemokine ligand subclass (CC, CXC, XC, or CX3C), followed by "R" (means receptor), resulting in: CCR, CXCR, XCR or CX3CR (Devalaraja and Richmond, 1999; Mennicken et al., 1999; Murphy et al., 2000; Zlotnik and Yoshie, 2000). In spite of this classification, chemokines functions are very redundant and promiscuous. Thus, multiple chemokines can bind to the same receptor and conversely a single chemokine can bind to several receptors. Nevertheless, chemokine–chemokine receptor interactions are almost always restricted within a single subfamily (Mantovani, 1999) and monogamous chemokine–chemokine receptor interactions exists, being SDF-1-CXCR4 one of the most studied (Biber et al., 2002; Cartier et al., 2005; Vilz et al., 2005). Binding of chemokines to their receptors activates a complex network of intracellular signaling mechanisms involving a variety of second messenger systems, such as calcium, cAMP, and phospholipids, as well as a concerted interplay of kinases cascade (Bajetto et al., 2001; Biber et al., 2006). In the CNS, chemokines and their receptors are constitutively expressed at low levels in astrocytes, microglia, oligodendrocytes, neurons and endothelial cells in the developing and healthy adult brain and are induced by inflammatory mediators (Asensio and Campbell, 1999; Mennicken et al., 1999; Bajetto et al., 2002; Biber et al., 2002, 2006; Minami and Satoh, 2003; Cartier et al., 2005).

Functionally, chemokines and their receptors can be divided into two main categories: homeostatic and inflammatory/secreted. Homeostatic chemokines are secreted constitutively and are generally involved in basal leukocyte trafficking and in development (Rossi and Zlotnik, 2000; Fernandez and Lolis, 2002). Thus, they have been reported to be involved in the developmental organization of the brain and in the maintenance of normal brain homeostasis by regulating the migration, proliferation and differentiation of glial and neuronal precursor cells (Ma et al., 1998; Proudfoot, 2002; Imitola et al., 2004). Moreover, other physiological functions of chemokines have been reported in CNS, such as regulation of synaptic transmission and plasticity, neurodevelopment, regulation of neurotransmitter release, modulation of ion channel activity, and cell death and survival (Bezzi et al., 2001; Mackay, 2001; Cartier et al., 2005; Belmadani et al., 2006). As mentioned above, inflammatory chemokines and their receptors are only produced by cells during infection, injury or inflammation and prompt the migration of leukocytes, microglia, and astrocytes to an injured or infected site. Nevertheless, some chemokines can be both expressed constitutively in normal brain and also up-regulated when an injury or disease happens (Mennicken et al., 1999; Rossi and Zlotnik, 2000; Cartier et al., 2005).

Up-regulation or dysregulation of chemokines expression play a crucial role in neurodegeneration and excitotoxicity, being involved in the communication between damaged neurons and surrounding glial cells and in the regulation of neuronal signaling by a diversity of processes (Mennicken et al., 1999; Murdoch and Finn, 2000; Rossi and Zlotnik, 2000; Bezzi et al., 2001; Minami and Satoh, 2003, 2005; Cartier et al., 2005). Moreover, the expression of chemokines can be regulated by different pro-inflammatory and anti-inflammatory factors, including cytokines, and this has been associated with several acute and chronic inflammatory conditions in CNS (Galasso et al., 1998, 2000; Gourmala et al., 1999; Biber et al., 2002; Sheng et al., 2005; Belmadani et al., 2006).

5. SIGNALING TRANSDUCTION

An essential feature of the early immune response to pathogens and cell debris is secretion of proinflammatory cytokines. So, understanding how cytokine signaling works can provide strategies to look for new therapeutic targets in several inflammatory and auto–immune disorders and also to improve mechanisms of defense against several pathogens. In a simplistic description, both TNF-α and IL-1β signaling activate similar pathways which culminate with the increased expression of target genes, most of which encode proteins involved in immunity and inflammation. Both IL-1β and TNF-α receptors lack metabolic or enzymatic activity and recruit several intracellular adaptor proteins responsible for transducing the signal downstream receptor level, amplifying a cascade of signal transduction pathways for both cytokines (MacEwan, 2002).

It is important to note that the majority of experimental work aiming to unravel the signal pathways activated by these two cytokines has been done in diverse experimental models including blood cells and different cell lines. The obtained data lead people to think that similar mechanisms are responsible for most IL-1β and TNF-α actions in the CNS. However, we must keep in mind that similar pathways, in different experimental models, do not necessarily result in the same toxic or protective effects partially because some signaling proteins and their interactions can slightly differ.

5.1. IL-1β Signaling

IL-1β is a pleiotropic cytokine that has been implicated in a number of neurodegenerative/neurotoxic conditions and is generally assumed to result in neurotoxic processes (Bernardino et al., 2005a, b), although it has been also reported that IL-1β can play a neuroprotective role (Rothwell and Luheshi, 2000; Friedman, 2005; Sayyah et al., 2005). IL-1β can be expressed by astrocytes, oligodendrocytes, neurons, microglia, cerebrovascular cells and circulating immune cells invading the CNS upon injury (Davies et al., 1999; Pearson et al., 1999). The inactive 31 kDa IL-1β is cleaved by a highly specialized IL-1β converting enzyme (ICE or caspase-1) into a 17 kDa IL-1β mature protein (Thornberry et al., 1992). The IL-1 family consists of the two agonists IL-1α and IL-1β and the endogenous IL-1 receptor antagonist (IL-1ra), all of which bind to specific IL-1 receptors such as type I (IL-1RI) and type II (IL-1RII) receptors. The ligands bind both receptors with distinct affinities (Dinarello, 1996). IL-1RI mediates the biological effects of IL-1β (Sims et al., 1988, 1993; Dinarello, 1996), whereas IL-1RII is not capable of transducing the IL-1β mediated signal, but rather acts as a decoy receptor, competing with the IL-1RI for IL-1β (Colotta et al., 1993). This is mainly due to the fact that IL-1RII contains only a short cytoplasmic tail and do not contain a TIR domain (the effector region of the IL-1RI receptor). On the other hand, the endogenous IL-1β antagonist, IL-1ra, also binds to IL-1RI. So, IL-1RII and IL-1ra can act together and cooperate in order to regulate the interaction between IL-1β and its receptor IL-1RI.

Another member of the IL-1 receptor family, IL-1 receptor accessory protein (IL-1RAcP), was also identified. The IL-1RAcP shares limited homology with IL-1RI and IL-1RII and does not bind IL-1β, but appears to increase the affinity of II-1RI for IL-1β (Greenfeder et al., 1995). It has been shown that co-expression of the IL-1RAcP is essential for a fully functional IL-1RI complex (Greenfeder et al., 1995; Hofmeister et al., 1997).

In rat and mouse brain, distribution of the Il-1RI was initially investigated using radiolabeled IL-1α and β. These studies showed high density of binding sites particularly in the granular cell layer of the dentate gyrus of the hippocampus and a weak to moderate signal was obtained over the pyramidal cell layer of the hylus and the CA3 subfield. The signal was also detected in choroids plexus, meninges, and anterior pituitary (Takao et al., 1990; Ban, 1994; Loddick et al., 1998).

IL-1RI has been recognized as part of an interleukin-1 receptor/Toll-like receptor (IL-1R/TLR) superfamily, whose members are involved in host defense and inflammation (Gay and Keith, 1991; O'Neill and Greene, 1998). Moreover, these two receptors share sequence similarity in their cytosolic regions, the TIR domain, and, accordingly, they have similar signaling pathways (Li and Qin, 2005).

Following IL-1RI and IL-1RAcP complex formation induced by IL-1β (Greenfeder et al., 1995; Wesche et al., 1997), the intracellular adaptor protein, termed Myeloid Differentiation factor 88 (MyD88), is recruited and interact with the carboxyl-terminal TIR domain of the IL-1RI (Kawai et al., 1999; Li et al., 2005; Davis et al., 2006). Then, the amplification of signal requires the interaction between IL-1RI/MyD88 and an IL-1 Receptor-Associated Kinase (IRAK). Recruitment of IRAK to IL-1RI/MyD88 complex is mediated by a scaffolding protein, called Toll-interacting protein (Tollip) (Burns et al., 2000). In resting cells, Tollip forms a complex with IRAK and inhibits IL-1RI signaling by binding to and blocking IRAK phosphorylation. Upon activation, Tollip-IRAK complexes are recruited to the activated receptor, allowing the interaction between IRAK and MyD88 (Croston et al., 1995; Cao et al., 1996a; Zhang and Ghosh, 2002). IRAK then undergoes rapid auto-phosphorylation and, in turn, phosphorylates Tollip, resulting in the dissociation of IRAK-Tollip complexes and the activation of downstream signaling components (Burns et al., 2000). IRAK becomes hyperphosphorylated, leaves the receptor complex and interacts with Tumor necrosis factor Receptor-Associated Factor 6 (TRAF6) (Cao et al., 1996b; Yamin and Miller, 1997; Anderson, 2000). TRAF6 then activates NFκB inducing kinase (NIK) (Baeuerle, 1998), a mitogen activated protein kinase kinase kinase (MAPKKK), which in turn phosphorylates and activates the IκB kinase (IKK) complex (Fig. 1). Alternatively, another MAP kinase kinase kinase named TAK1, in association with TAK1-binding proteins 1 and 2 (TAB1 and TAB2), has been identified to interact with TRAF6 (Takaesu et al., 2000, 2001; Qian et al., 2001; Kishida et al., 2005), thereby activating also the IKK complex, leading to NFκB activation (DiDonato et al., 1997; Mercurio et al., 1997; Zandi et al., 1997; Meffert and Baltimore, 2005). Although, it remains unclear which members of the TAK1/TAB1/TAB2 complex are essential for IL-1β signaling (Shim et al., 2005).

In resting cells, NFκB proteins are present in the cytosol as heterodimer of proteins of the Rel family of transcription factors associated with proteins that are known as inhibitory proteins of NFκB (IκBs) (Thanos and Maniatis, 1995; Baeuerle and Baltimore, 1988). The IκBs proteins include IκBε, IκBα, IκBβ, which trap NFκB dimers in the cytoplasm by masking its nuclear localization signal (NLS) and by this process prevent NFκB binding to DNA. In response to a stimulus, such as IL-1RI activation, the degradation of the inhibitory IκBs proteins involves phosphorylation mediated by IκB kinases (IKK) complex. IKK complex is composed by two catalytic subunits, IKKα and IKKβ, and one regulatory subunit, NEMO (IKKγ). Both the IKKγ/NEMO and IKKα subunits are required for activation of NFκB by stimuli such as IL-1β and TNF-α, although only the IKKβ catalytic subunit is essential to trigger NFκB activation (Verma et al., 1995; Rivest, 2003). So, the IKKβ subunit is activated by NIK and phosphorylates IκB proteins leading to the ubiquitin-dependent degradation of IκB by the 26S proteasome subunit (Traenckner et al., 1994, 1995; Delhase et al., 1999). Then, NFκB, free of IκB proteins, translocate to the nucleus and binds to DNA binding sites regulating the transcription of several genes, including inducible cyclo-oxygenase (which leads to the production of prostaglandins, important inflammatory mediators), inducible form of nitric oxide synthase, adhesion molecules, chemokines, cytokines, immune receptors, and growth factors (Moynagh et al., 1994) (Fig. 1). Following its degradation, IκB proteins are rapidly re-synthesized to act as endogenous inhibitors of NFκB.

Besides the NFκB pathway, the recruitment of MyD88/IRAK/TRAF6 complex is also able to activate MAPK kinases (MKKs), in particular the JNK pathway that leads to AP-1 activation (Rivest, 2003) and also the p38 MAP kinases (O'Neill, 2002; Dunne and O'Neill, 2003; Wang et al., 2005b). The specific pathways activated by IL-1β may differ in distinct cell types and may mediate distinct biological consequences of IL-1β (Srinivasan et al., 2004).

5.2. TNF-α SIGNALING

TNF-α is a pleiotropic inflammatory cytokine expressed as a 26 kDa transmembrane precursor from which a soluble 17 kDa polypeptide is released after proteolytic cleavage, mainly by the metalloprotease TNF-α converting enzyme (TACE) (Karkkainen et al., 2000). It was reported that TNF-α is active either as a membrane bound or as a soluble form (Kriegler et al., 1988; Idriss and Naismith, 2000). In general, TNF-α, can be produced by activated microglia and astrocytes, but under pathological conditions TNF-α can also be produced by CNS-infiltrating lymphocytes and macrophages (Vassalli, 1992).

TNF-α can interact with two different receptors (TNFR1 and TNFR2) that are transmembrane proteins with an extracellular carboxy-terminal with cysteine-rich domains and an amino-terminal intracellular domain, both linked by a single transmembrane domain. The C-terminal extracellular domain has approximately 28% amino acid identity between both TNF-α receptors and is responsible for the assembly of receptor trimers and their binding properties (Aggarwal, 2003). The extracellular domain of both TNFR1 and TNFR2 can be proteolytically cleaved, giving rise to soluble receptors for TNF (sTNFR), which can neutralize TNF-α (Wallach et al., 1991). Moreover, unlike IL-1/Toll like receptors, TNF-α receptors show almost no homology in their intracellular sequences and suggest that TNFR1 and TNFR2 activate distinct signaling pathways, contributing to a variety of different biological responses mediated by TNF-α (Grell et al., 1994).

TNF-α-induced signal transduction starts with the pre-assembling of both TNFR1 and TNFR2 on the cell membrane before TNF-α binding. The formation of this complex requires the extracellular pre-ligand-binding assembly domain (PLAD), a highly conserved domain containing cysteine resides in the extracellular region of these receptors. PLAD mediates ligand-independent assembly of receptor trimers and facilitates the interaction between TNF-α and its receptors (Chan, 2000; Chan et al., 2000; Jones, 2000; Locksley et al., 2001). Nevertheless, TNF-α signaling is only transduced following physical binding of TNF-α to TNF receptors.

5.2.1. TNFR1-Mediated Signaling

Among the two TNFR, TNFR1 has been the mostly investigated receptor, partially due to its potential role in the control of cell death and consequently they are looked as good potential therapeutic targets in several diseases. The main difference between TNFR1 and TNFR2 is the presence of intracellular regions so-called death domains (DD) in the TNFR1 (Itoh and Nagata, 1993; Tartaglia et al., 1993). DD are also present on other associating proteins that are primarily involved in cell death signaling. Moreover, deletion of DD region abolishes ligand-induced apoptosis, indicating that this region is required for TNF-α-induced apoptosis. Moreover, silencer of death domain (SODD) proteins associate constitutively with DD region of TNFR1 and so inhibits the recruitment of several DD-interacting proteins, and prevent ligand-independent activation of TNFR1 and spontaneous cell death signaling (Jiang et al., 1999; Takada et al., 2003). After a stimulus, TNF-α binds to the TNFR complex and the SODD is released from the DD of TNFR1, allowing the interaction between TNFR1 and adaptor cytoplasmic molecules containing DD domains (MacEwan, 2002).

TRADD-Mediated Pathways

After binding of TNF-α, TNFR1 recruits a cytoplasmic protein called TNFR associated death domain (TRADD) acting as a platform adaptor for the downstream TNF-α signaling. TRADD interacts with the TNFR1 cytoplasmic domain by DD regions in both molecules. This TRADD–DD complex recruits the downstream signaling adaptor molecules fas-associated death domain (FADD) and receptor interacting protein (RIP) (Aggarwal, 2000; Park et al., 2005). FADD is the major adaptor protein involved in the TNFR1 mediated-cell death. Besides a DD at the carboxyl terminus able to interact with TRADD–DD complex, FADD contains a death effector domain (DED) motif allowing the interaction and activation of caspases and other DED-containing molecules involved in execution of cell death (Chinnaiyan et al., 1995; Hsu et al., 1996b; Park et al., 2005). So, FADD can activate caspase-8 known as an apoptotic initiator caspase, that in turn activates the executioner caspases such as caspase-3 and -6, causing cell death (Kischkel et al., 2000; Denecker et al., 2001; Chen and Goeddel, 2002) (Fig. 2). This death pathway mediated by TRADD, FADD, and caspases is dependent on the internalization of activated TNF/TNFR1 complexes (TNF receptosomes) (Higuchi and Aggarwal, 1994). During the endocytic process, TRADD, FADD and caspase-8 form the death-inducing signaling complex, ultimately resulting in apoptosis (Aggarwal, 2000). This endocytic mechanism can also regulate the activation of TNFR1 and contributes to stop TNF-α signaling.

Another adaptor protein able to interact with the DD of TRADD is RIP. Upon TNF-α binding, TNFR1 is translocated to cholesterol- and sphingolipid enriched membrane microdomains, known as lipid rafts, where it associates with TRADD and RIP, forming a signaling complex. The interaction between TNFR1/TRADD and RIP in lipid rafts is a potent inducer of apoptosis (Legler et al., 2003). Moreover, besides the indirect role on apoptosis, RIP can also play a role in gene expression and proliferation (Blonska et al., 2005; Thakar et al., 2006). RIP can activate NFκB, mediating the activation of IKK, and can also activate JNK (Hsu et al., 1996a) (Fig. 2). Deletion of the gene encoding RIP, however, abolished TNF-induced activation of NFκB, but had minimal effect on JNK activation, indicating that RIP plays a central role in the activation of NFκB (Hsu et al., 1996a; Kelliher et al., 1998). The activation of p38 MAPK by TNFR1 is less well understood and involves the recruitment of RIP by TRADD, which in turn recruits mitogen-activated protein kinase kinase (MKK)3 leading to p38 MAPK activation (Wajant et al., 2003).

Besides FADD and RIP, TRADD can also interact with TNF receptor associated factor (TRAF) proteins, specially TRAF2, that plays a central role in NFκB activation (Hsu et al., 1995, 1996a, b). TNFR1 indirectly interacts with TRAF2 by using the TRADD protein (Fig. 2). TRAF proteins can be involved in death receptor signaling with the indirect recruitment of cytoplasmic adaptor proteins. However, the predominant role attributed to TRAF2 is cell survival through the NIK protein. As previously mentioned in the IL-1β signaling section, NIK is a member of the serine/threonine mitogen-activated protein kinase kinase kinase (MAPKKK) (Malinin et al., 1997) and has been implicated in TNF-α induced NFκB activation by the phosphorylation of IκBα mediated by the IKK complex (Ling et al., 1998). TNFR1 can also activate JNK through the sequential recruitment of TRAF2, MAP/ERK kinase kinase 1 (MEKK1) and MAPK kinase 7 (MKK7). Deletion of the TRAF2 gene has a minimal effect on TNF-induced NFκB activation, but results in deficient JNK signaling

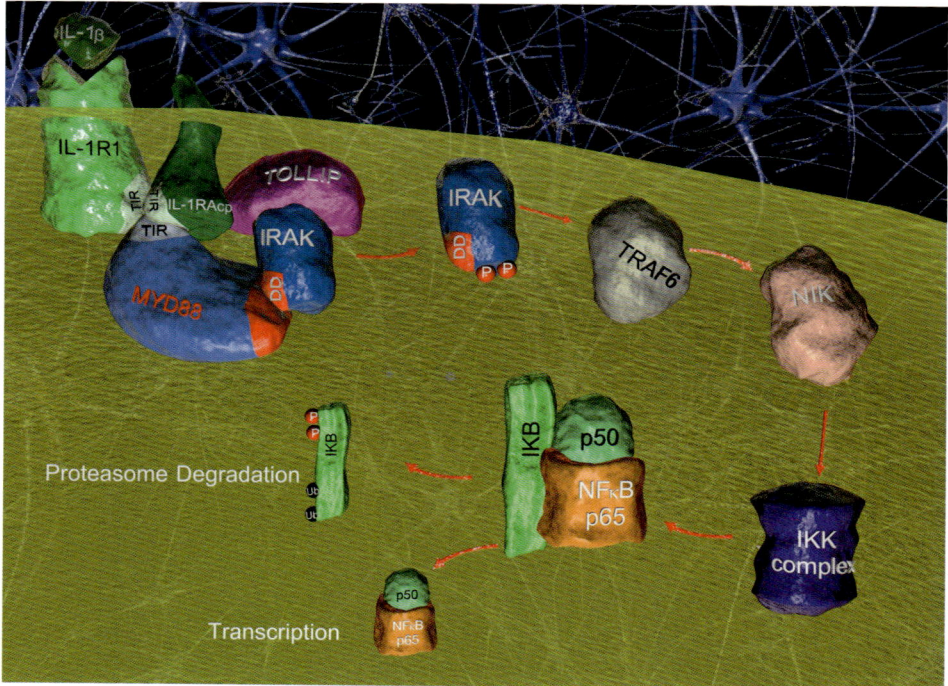

Fig. 1. Schematic representation of the major signal transduction pathways activated by IL-1β. Abbreviations: IL-1R1, IL-1 receptor type I; IL-1RAcP, IL-1 receptor accessory protein; MyD88, myeloid differentiation factor 88; IRAK, IL-1 receptor-associated kinase; Tollip, Toll-interacting protein; TRAF6, Tumor necrosis factor Receptor-Associated Factor 6; NIK, NFκB inducing kinase; IKK, inhibitors of κB (IκB) kinases. Further details are described in the text (see Sect. 5.1) and in the supplementary movies accompanying this book

(Yeh et al., 1997), indicating that TRAF2 is essential for JNK activation but not absolutely required for NFκB activation. NFκB, JNK, and p38 MAPK ultimately lead to the synthesis of immune and inflammatory proteins with several different functions. The activation of NFκB leads to the expression of inflammatory genes such as vascular adhesion molecule-1 (VCAM-1), MIP-1β, MIP-2, and several other cytokines and chemokines (Ghosh and Karin, 2002). Moreover, several NFκB-regulated proteins able to suppress apoptosis have been identified. These include TRAF2, inhibitor of apoptosis 1 (IAP1) and IAP2, survivin, FAS-associated death domain-like interleukin-1 converting enzyme inhibitory protein (FLIP), decoy receptor (DCR) and Bcl-xl. These proteins inhibit different steps of the apoptotic pathways. For example, DCR sequesters the receptors, whereas FLIP blocks the activation of caspase-8, survivin inhibits caspase-3 activation, IAP inactivate effector caspases, and members of the bcl-2 family inhibit cytochrome c release (Wang et al., 1998) (Fig. 2).

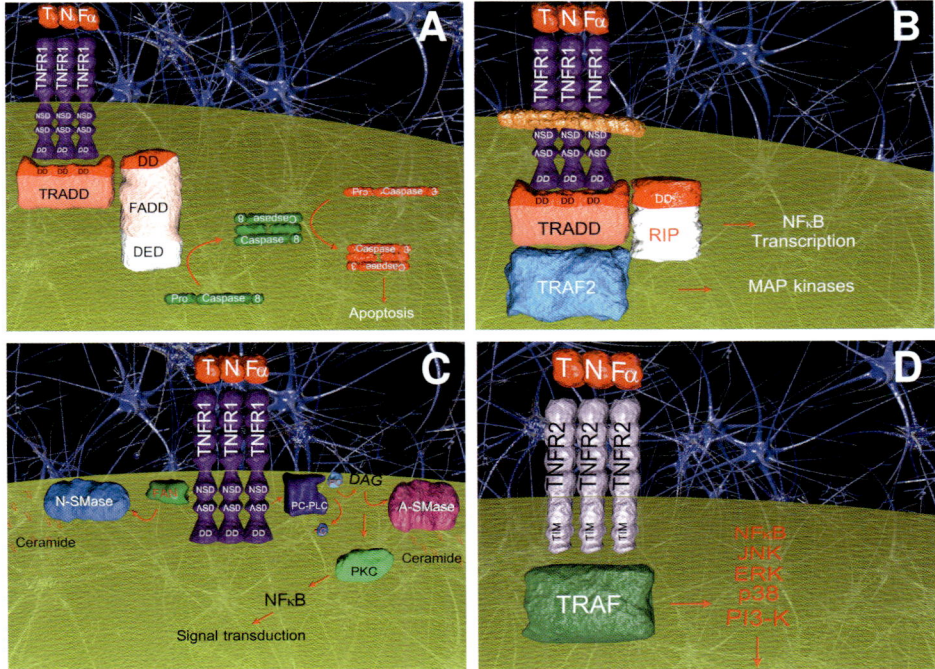

Fig. 2. Schematic representation of the major signal transduction pathways activated by the two TNF receptors subtypes (TNFR1 and TNFR2). TNF Receptor 1 triggers different signaling pathways via three different functional domains: the death domain (A and B) and the NSD domain or the ASD domain (C). Instead, TNF receptor 2-mediating signaling transduction pathways are mediated by a TIM domain (D). Further details are described in the text (see Sect. 5.2) and in the supplementary movies accompanying this book

So, while recruitment of DD containing proteins such as FADD and RIP can lead predominantly to apoptosis or necrosis, recruitment of TRAF2 can lead to the activation of transcription factors such as NFκB and AP-1 and therefore promoting inflammatory reactions, cell survival and differentiation (Wajant et al., 2003). So, TNFR1 can simultaneously activate both apoptotic and antiapoptotic signals (Zheng et al., 2006). The apoptotic/necrotic signals do not require protein synthesis, whereas anti-apoptotic signals involve the activation of several transcription factors, including NFκB (Karin and Lin, 2002). The balance between these two signals is regulated at several levels such as regulation of receptor/ligand expression, soluble decoy receptor expression and regulation of protein synthesis.

Sphingomyelinase-Mediated Pathways

Besides the DD in its intracellular terminal, TNFR1 has also other intracellular adjacent regions named neutral-sphingomyelinase activating domain (NSD) and acidic sphingomyelinase activating domain (ASD). Much interest has also been paid to the ability of TNF to stimulate these SMase enzymes (Kolesnick and Kronke, 1998). The proximal half of TNFR1 was found to be responsible for the stimulation of membrane-associated N-SMase partially via ceramide interconverted from diacylglycerol (DAG) (Wiegmann et al., 1994), but also via the Factor Associated with Neutral sphingomyelinase (FAN) (Neumeyer et al., 2006). FAN stimulates N-SMase directly (Adam-Klages et al., 1996; Neumeyer et al., 2006) and therefore catalyses the degradation of sphingolipids into smaller ceramide-containing molecules (Adam-Klages et al., 1998) (Fig. 2). Ceramide production by FAN can induce proliferation or inflammation involving kinase pathways. By decreasing plasma and mitochondrial membrane stability, ceramide can also trigger apoptosis (Malagarie-Cazenave et al., 2002; Pettus et al., 2002; Colombaioni and Garcia-Gil, 2004).

TNF-α can also activate the phosphatidylcholine-specific phospholipase C (PC-PLC) through the TNFR1/A-SMase complex. Stimulation of PC-PLC activity degrades phosphatidylcholine into phosphocholine and DAG causing the activation of protein kinase C (PKC) isoforms (MacEwan, 2002), downstream activating NFκB cascade (Fig 2).

5.2.2. TNFR2-Mediated Signaling

The overall function of TNFR2 is largely unknown. The main difference between TNFR1 and TNFR2 signaling came from the fact that TNFR2 lacks an intracellular DD. Nevertheless, TNFR2 contains the so-called TRAF interacting motifs (TIM) in its cytoplasmic domain. Activation of TIM leads to the direct recruitment of TRAF-family members and the subsequent activation of signaling pathways such as NFκB, JNK, extracellular signal-related kinase (ERK), p38 and phosphoinositide 3-kinase (Pi3K) (Liu et al., 1996; Hehlgans and Mannel, 2002; Dempsey et al., 2003) (Fig. 2). This can result in the expression of cytokines, but also in the expression of regulatory proteins with potential antiapoptotic activity, such as, TRAF2, IAP1, and IAP2 (Wang et al., 1998).

Even if TNFR2 cannot directly induce apoptosis, it was shown that it can potentiate the effects mediated by TNFR1. Accordingly, it was described that TNFR2 can induce ubiquitination and proteasomal degradation of TRAF2 proteins and this can potentiate TNFR1-mediated apoptosis (Li et al., 2002). As mentioned before, the variety of cellular responses driven by different receptors can be due to the different signaling pathways activated by each receptor or by the expression patterns of the two receptors as well as associated adaptor proteins in different cell types and tissues (MacEwan, 2002; Thommesen and Laegreid, 2005). Moreover, different affinity of both receptors for the soluble or membrane associated TNF-α forms and changes in TNFR1:TNFR2 ratio at the plasma membrane work as mechanisms to control the functional outcome of TNF-α stimulus (Grell et al., 1995). Moreover, experimental conditions in in vitro studies such as the biological model used, the methodology or protocol used and the local concentrations of cytokines can also determine the fate of the cells stimulated with TNF-α. Moreover, TNFR1 has broad functional effects due to its ubiquitous expression in different cell types, while TNFR2 is involved in inflammation since it is expressed mainly by immune and endothelial cells (Aggarwal, 2000).

6. CONTRIBUTION OF INFLAMMATION TO EXCITOTOXICITY

Inflammation has an important role in the development and progression of both acute and chronic neuronal injury. Increasing evidences have shown that neurotransmitters released from nerve terminals are involved not only in the communication between neurons, but also between nerve terminals and glial cells including microglia. Most forms of neuronal injury have been associated with excitotoxicity, i.e. excessive release of excitatory amino acids such as glutamate, and subsequent activation of NMDA and AMPA/Kainate receptors (Choi, 1992; Zimmer et al., 2000; Kristensen et al., 2001; Bonde et al., 2005). Besides glutamate, ATP is a co-transmitter also released by injured or dying neurons following brain insults (Dubyak and el-Moatassim, 1993; Cunha and Ribeiro, 2000).

Microglia expresses both ionotropic and metabotropic glutamatergic receptors, that when overactivated in pathological conditions induce microglia activation and subsequent release of pro-inflammatory cytokines such as TNF-α (Noda et al., 2000; Taylor et al., 2005) . ATP can also modulate microglial activation, chemotaxis and release of cytokines, via interaction with both metabotropic (P2YR) and ionotropic (P2XR) receptors (Ferrari et al., 1996, 1997a, b, 2006; Hide et al., 2000; Sanz and Di Virgilio, 2000; Honda et al., 2001; James and Butt, 2002; Boucsein et al., 2003; Nakajima and Kohsaka, 2004; Suzuki et al., 2004; Bernardino, unpublished data). Among P2 receptors, $P2X_7R$ require higher concentrations of ATP to be activated and mediate microglial response to injury (Ferrari et al., 1997b; Di Virgilio et al., 1999; Inoue, 2002). Extracellular ATP acting via $P2X_7$ receptors is a powerful stimulus for the maturation and release of IL-1β leading to activation of ICE/caspase-1, in the presence of an inflammatory environment, as mimicked by cell exposure to LPS (Ferrari et al., 1997a, b; Perregaux and Gabel, 1998; Sanz and Virgilio, 2000; Solle et al., 2001; Bernardino, unpublished data). Moreover, a prolonged activation of $P2X_7$ receptors can induce the formation of a multimeric pore-like structure allowing the influx of large molecules (up to 900 Da) eventually resulting in cell death (Ferrari et al., 1999; Brough et al., 2002; Le Feuvre et al., 2002; Sylte et al., 2005; Bernardino, unpublished data).

Thus, glutamate and ATP, play important roles in modulating microglia activation and in determining the fate and extent of neuronal injury. Depending on the specific injury and inflammatory environment, activated microglia cells can be either detrimental or protective (McCluskey and Lampson, 2000). Both IL-1β and TNF-α can be neuroprotective and/or neurotoxic, contributing directly or indirectly to the initiation, maintenance, and outcome of several forms of neuronal cell death, including acute insults such as ischemia, trauma, and seizures (Barone et al., 1997; Tamatani et al., 1999; Vezzani et al., 1999, 2000; Allan and

Rothwell, 2001; Pringle et al., 2001; Shandra et al., 2002; Balosso et al., 2005; Lu et al., 2005; Patel et al., 2006) and chronic neurodegenerative disorders such as Parkinson's and Alzheimer's disease (Barger et al., 1995; Sriram et al., 2002; Wang et al., 2005a; Griffin et al., 2006).

Several mechanisms have been identified to modulate the neuroprotective and neurotoxic actions of TNF-α and IL-1β. The neuroprotection afforded by TNF-α may depend on the ability of TNF-α to promote maintenance of calcium homeostasis by increasing expression of calbindin (Cheng et al., 1994; Mattson et al., 1995) to stimulate antioxidant pathways (Barger et al., 1995; Tamatani et al., 1999; Wilde et al., 2000) to increase transient outward potassium currents (Houzen et al., 1997) to increase membrane microglia glutamate transporter EAAT2/GLT-1 (Persson et al., 2005) or neuronal glutamate transporter EAAT3/EAAC1 (Pradillo et al., 2006) expression, promoting clearance of glutamate from the extracellular space. Besides, TNF-α may also mediate neuroprotection by a mechanism coupled to the activation of transcription factors such as NFκB (Furukawa and Mattson, 1998). Experiments carried out in knock-out mice for both TNF-α receptors (TNFR1 and TNFR2) exposed to excitotoxic/ischemic injury, showed that these mice had higher susceptibility to cell damage suggesting that the insult-induced production of TNF-α causes the up-regulation of neuronal apoptosis inhibitor protein, and consequent inhibition of apoptosis and neuroprotection (Thompson et al., 2004). Recently, Turrin and Rivest have shown that endogenous TNF-α released in response to acute nitric oxide toxicity, can be neuroprotective by triggering an immediate microglial reactivity, which helps to eliminate cell debris, restricting subsequent damage, and restoring homeostasis (Turrin and Rivest, 2006).

On the other side, there are also evidences for a neurotoxic role mediated by TNF-α (Hermann et al., 2001; Bernardino et al., 2005a; Zou and Crews, 2005). Accordingly, it was described that TNF-α increases neuronal vulnerability to excitotoxic injury by increasing the surface expression of the GluR1 subunit of the AMPA receptor (Beattie et al., 2002; Yu et al., 2002) and by inducing upregulation of calcium permeable AMPA/kainate channels (Ogoshi et al., 2005). Furthermore, it was reported that TNF-α can inhibit astroglial glutamate transporters such as EAAT2/GLT-1, resulting in the increase of extracellular glutamate concentrations and the subsequent predisposition for glutamate excitotoxicity (Hu et al., 2000; Zou and Crews, 2005). Recently, Takeuchi and colleagues had shown that TNF-α induces neurotoxicity via glutamate release from connexin hemichannels of activated microglia. Thus, glutamate released by activated microglia induces excitotoxicity and may contribute to neuronal damage in neurodegenerative diseases (Takeuchi et al., 2006) (Fig. 3).

Both neuroprotective and neurotoxic effects may depend on local microenvironment concentrations and receptors activated. In accordance, it was shown that high concentrations of TNF-α exert protective effects on *Shigella dysenteriae*-induced seizures (Yuhas et al., 2003), whereas lower concentrations were pro-convulsive (Yuhas et al., 1999). We reported that TNF-α exacerbates AMPA-induced neuronal death at relatively high doses, while lower doses of TNF-α had a neuroprotective effect (Bernardino et al., 2005a). Using organotypic cultures of mouse hippocampal slices obtained from TNFR1 −/− and TNFR2 −/− we showed that this duality reflects the balance between multiple signals derived from both receptor subtypes. Changes in the activated receptor subtype (TNFR1 or TNFR2) can affect the overall balance of excitatory to inhibitory inputs and lead to opposite effects. Thus, the mechanisms modulating the expression of these two receptors may determine neuronal/glial response to TNF-α and the consequences of inflammatory reactions in the brain (Beutler and van Huffel, 1994). Several studies showed dissociation between the neurotoxic and neuroprotective effects using knock-out rodents for each receptor or for both receptors. In accordance, it was shown that TNF-α potentiation of glutamate/AMPA excitotoxicity depends on the activation of TNFR1 receptors which contain an intracellular "death domain" and also contributes mainly to cell death on target cells (Fontaine et al., 2002; Yang et al., 2002; Bernardino et al., 2005a; Taylor et al., 2005). Supporting this evidence, Shinoda and colleagues demonstrated that kainate-induced seizures promote the formation of a molecular scaffolding complex that includes TNFR1 receptors (Shinoda et al., 2003). This complex, through death-domains and apoptosis signal-regulating kinase 1 (ASK1), can promote cell death, suggesting that the TNF-α released after the insult can activate this receptor and contribute to neuronal death. Furthermore, others showed that absence of TNFR1 led to a strong reduction of neurodegeneration in a model of retinal ischemia, while lack of TNFR2 led to an enhancement of neurodegeneration clearly indicating that TNFR1 augment neuronal death and TNFR2 promote neuroprotection (Fontaine et al., 2002). Moreover, TNFR2-induced sustained NFκB activation was essential for neuronal survival (Marchetti et al., 2004) (Fig. 3).

Besides TNF-α, also IL-1β can have both neuroprotective and neurotoxic effects in the modulation of brain excitotoxicity. Possible mechanisms by which IL-1β potentiates neuronal excitotoxicity, may involve induction of COX-2 and related prostanoids (Serou et al., 1999), iNOS (Murakami et al., 2002) and increased synthesis of superoxide and peroxynitrite (Fink et al., 1999). Inhibition of astroglial glutamate re-uptake by IL-1β (Ye and Sontheimer, 1996; Hu et al., 2000) and enhanced astrocytic glutamate release either directly or via the production of TNF-α (Bezzi et al., 1999) can result in more extracellular glutamate, thus priming neuronal vulnerability to subsequent insults. An alternative mechanism may involve functional interactions between IL-1RI and AMPA receptors leading to an increase in inward Ca^{2+} currents. Previous evidence has thus shown that IL-1β can increase NMDA-receptor function in hippocampal neurons by activation of tyrosine kinases, resulting in an increased susceptibility of neurons to glutamate mediated cell loss (Viviani et al., 2003). Furthermore, Lai and colleagues showed that IL-1β impairs memory functions and long-term potentiation (LTP) by selectively down-regulating the surface expression and Ser831 phosphorylation of the AMPA receptor subunit GluR1 (Lai et al., 2006). In vivo studies indicated that intra-hippocampal injection of IL-1β activates kainate-induced pro-convulsant mechanisms that contribute to the increase in the duration of electroencephalographic recorded seizures (Vezzani et al., 1999), an effect blocked by IL-1ra (Vezzani et al., 2000). On the other hand, other studies also involved the GABAergic system and $GABA_A$ receptor in the pro-convulsant effect of IL-1β (Miller et al., 1991; Luk et al., 1999).

Similarly to TNF-α, IL-1β can play also a potential neuroprotective role depending on the dosage and the experimental conditions used. In mice organotypic hippocampal slice cultures we found that IL-1β exacerbated AMPA-induced neuronal death at relatively low doses, while higher doses of IL-1β had a neuroprotective role (Bernardino et al., 2005a). Important functional crosstalk was identified between IL-1β and neurotrophic factors (e.g. NGF, CNTF) and this interaction may, in part, account for the neuroprotective role attributed to IL-1β (Strijbos and Rothwell, 1995; Herx et al., 2000; Miyachi et al., 2001). Furthermore, recently it has been shown that IL-1β can also be neuroprotective by down-regulating L-type Ca^{2+} channel activity in neurons, thus preventing excessive Ca^{2+} entry (Zhou et al., 2006). Recent studies showed that IL-1β suppresses development of focal and generalized seizures in the amygdala kindling model of epilepsy, mediated, in part, by nitric oxide and prostaglandins (Sayyah et al., 2005). Most of the reports discussed above show that the effects of IL-1β were mediated mainly through IL-1RI, since they were inhibited in the presence of IL-1ra. Accordingly, it was reported that endogenous IL-1ra, mainly released by microglial cells in response to an excitotoxic insult, regulates and inhibits the neurotoxic effects mediated by IL-1β on neurons (Pinteaux et al., 2006) (Fig. 3).

Since both TNF-α and IL-1β affect differentially the outcome of various brain injuries by interfering either in neuronal and glial cell functions, pharmacological approaches specifically targeted to manipulate the diverse effects TNF-α and IL-1β and its signaling mechanisms in diseased conditions may reveal new targets for therapeutic strategies in several CNS diseases.

7. ACKNOWLEDGMENTS

Supported by FCT-Portugal, SFRH/BD/12731/2003, POCI/NSE/58492/2004 and FEDER.

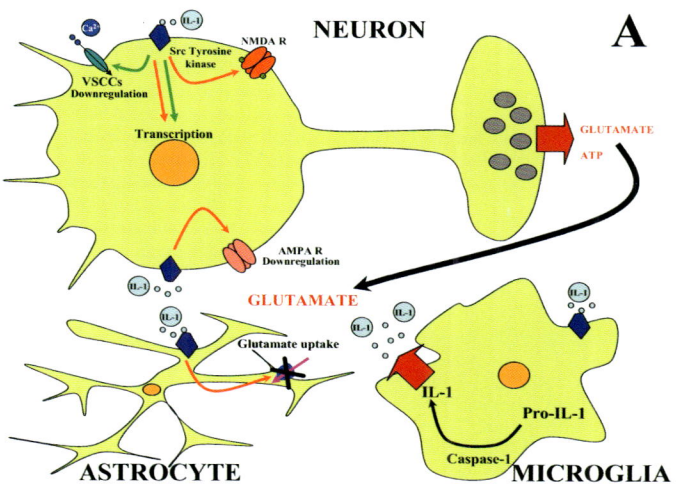

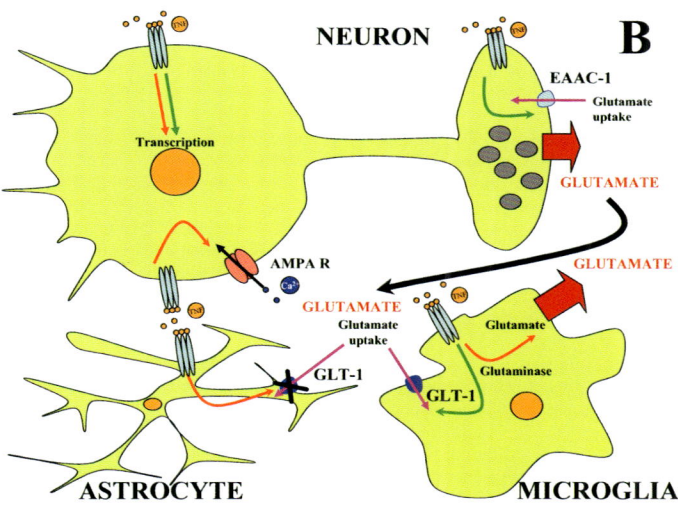

Fig. 3. Simplified representation of the complex network of cellular interaction between neurons, microglia and astrocytes, following stimulation of IL-1β receptors (A) or TNF-α receptors (B), in response to injury. *Green arrows* represent the main neuroprotective pathways, whereas *red arrows* illustrate the main neurotoxic processes, mediated by IL-1β (A) or TNF-α (B). Abbreviations: IL-1, Interleukin-1; IL-1R, Interleukin-1 receptor; VSCC, voltage sensitive calcium channels; NMDA R, NMDA receptors; TNF, Tumor Necrosis Factor, TNFR, TNF receptors; EAAC-1, Excitatory Amino Acid Carrier-1; GLT-1, Glutamate Transporter-1; AMPA R, AMPA receptors. See text for further details (Sect. 6)

8. REFERENCES

Adam-Klages, S., Adam, D., Wiegmann, K., Struve, S., Kolanus, W., Schneider-Mergener, J. and Kronke, M., 1996, FAN, a novel WD-repeat protein, couples the p55 TNF-receptor to neutral sphingomyelinase. *Cell* **86**: 937.
Adam-Klages, S., Schwandner, R., Adam, D., Kreder, D., Bernardo, K. and Kronke, M., 1998, Distinct adapter proteins mediate acid versus neutral sphingomyelinase activation through the p55 receptor for tumor necrosis factor. *J. Leukoc. Biol.* **63**: 678.
Aggarwal, B.B., 2000, Tumour necrosis factors receptor associated signaling molecules and their role in activation of apoptosis, JNK and NF-kappaB. *Ann. Rheum. Dis.* **59**(Suppl 1): 6.
Aggarwal, B.B., 2003, Signaling pathways of the TNF superfamily: a double-edged sword. *Nat. Ver. Immunol.* **3**: 745.
Akira, S., Takeda, K. and Kaisho, T., 2001, Toll-like receptors: critical proteins linking innate and acquired immunity. *Nat. Immunol.* **2**: 675.
Allan, S.M. and Rothwell, N.J., 2001, Cytokines and acute neurodegeneration. *Nat. Rev. Neurosci.* **2**: 734.
Allen, N.J. and Barres, B.A., 2005, Signaling between glia and neurons: focus on synaptic plasticity. *Curr. Opin. Neurobiol.* **15**: 542.
Aloisi, F., Ria, F. and Adorini, L., 2000, Regulation of T-cell responses by CNS antigen-presenting cells: different roles for microglia and astrocytes. *Immunol. Today* **21**: 141.
Aloisi, F., Ria, F., Penna, G. and Adorini, L., 1998, Microglia are more efficient than astrocytes in antigen processing and in Th1 but not Th2 cell activation. *J. Immunol.* **160**: 4671.
Alvarez-Maubecin, V., Garcia-Hernandez, F., Williams, J.T. and Van Bockstaele, E.J., 2000, Functional coupling between neurons and glia. *J. Neurosci.* **20**: 4091.
Ambrosini, E. and Aloisi, F., 2004, Chemokines and glial cells: a complex network in the central nervous system. *Neurochem. Res.* **29**: 1017.
Anderson, K.V., 2000, Toll signaling pathways in the innate immune response. *Curr. Opin. Immunol.* **12**: 13.
Anderson, K.V., Bokla, L. and Nusslein-Volhard, C., 1985, Establishment of dorsal-ventral polarity in the Drosophila embryo: the induction of polarity by the Toll gene product. *Cell* **42**: 791.
Asensio, V.C. and Campbell, I.L., 1999, Chemokines in the CNS: plurifunctional mediators in diverse states. *Trends Neurosci.* **22**: 504.
Baeuerle, P.A., 1998, Pro-inflammatory signaling: last pieces in the NF-kappaB puzzle? *Curr. Biol.* **8**: R19.
Baeuerle, P.A. and Baltimore, D., 1988, I kappa B: a specific inhibitor of the NF-kappaB transcription factor. *Science* **242**: 540.
Bailey, S.L., Carpentier, P.A., McMahon, E.J., Begolka, W.S. and Miller, S.D., 2006, Innate and adaptive immune responses of the central nervous system. *Crit. Rev. Immunol.* **26**: 149.
Bajetto, A., Bonavia, R., Barbero, S., Florio, T. and Schettini, G., 2001, Chemokines and their receptors in the central nervous system. *Front. Neuroendocrinol.* **22**: 147.
Bajetto, A., Bonavia, R., Barbero, S. and Schettini, G., 2002, Characterization of chemokines and their receptors in the central nervous system: physiopathological implications. *J. Neurochem.* **82**: 1311.
Balosso, S., Ravizza, T., Perego, C., Peschon, J., Campbell, I.L., De Simoni, M.G. and Vezzani, A., 2005, Tumor necrosis factor-alpha inhibits seizures in mice via p75 receptors. *Ann. Neurol.* **57**: 804.
Ban, E.M., 1994, Interleukin-1 receptors in the brain: characterization by quantitative in situ autoradiography. *Immunomethods* **5**: 31.
Barger, S.W., Horster, D., Furukawa, K., Goodman, Y., Krieglstein, J. and Mattson, M.P., 1995, Tumor necrosis factors alpha and beta protect neurons against amyloid beta-peptide toxicity: evidence for involvement of a kappa B-binding factor and attenuation of peroxide and Ca2+ accumulation. *Proc. Nat. Acad. Sci. USA* **92**: 9328.
Barone, F.C., Arvin, B., White, R.F., Miller, A., Webb, C.L., Willette, R.N., Lysko, P.G and Feuerstein, G.Z., 1997, Tumor necrosis factor-alpha. A mediator of focal ischemic brain injury. *Stroke* **28**: 1233.
Batchelor, P.E., Liberatore, G.T., Wong, J.Y., Porritt, M.J., Frerichs, F., Donnan, G.A. and Howells, D.W., 1999, Activated macrophages and microglia induce dopaminergic sprouting in the injured striatum and express brain-derived neurotrophic factor and glial cell line-derived neurotrophic factor. *J. Neurosci.* **19**: 1708.
Bauer, J., Bradl, M., Hickley, W.F., Forss-Petter, S., Breitschopf, H., Linington, C., Wekerle, H. and Lassmann, H., 1998, T-cell apoptosis in inflammatory brain lesions: destruction of T cells does not depend on antigen recognition. *Am. J. Pathol.* **153**: 715.
Beattie, M.S., Hermann, G.E., Rogers, R.C. and Bresnahan, J.C., 2002, Cell death in models of spinal cord injury. *Prog. Brain Res.* **137**: 37.
Becher, B., Prat, A. and Antel, J.P., 2000, Brain-immune connection: immuno-regulatory properties of CNS-resident cells. *Glia* **29**: 293.

Belmadani, A., Tran, P.B., Ren, D. and Miller, R.J., 2006, Chemokines regulate the migration of neural progenitors to sites of neuroinflammation. *J. Neurosci.* **26**: 3182.
Belvin, M.P. and Anderson, K.V., 1996, A conserved signaling pathway: the Drosophila toll-dorsal pathway. *Annu. Rev. Cell Dev. Biol.* **12**: 393.
Benveniste, E.N., 1998, Cytokine actions in the central nervous system. *Cytokine Growth Factor Rev.* **9**: 259.
Bergles, D.E., Roberts, J.D., Somogyi, P., and Jahr, C.E., 2000, Glutamatergic synapses on oligodendrocyte precursor cells in the hippocampus. *Nature* **405**: 187.
Bernardino, L., Xapelli, S., Silva, A.P., Jakobsen, B., Poulsen, F.R., Oliveira, C.R., Vezzani, A., Malva, J.O. and Zimmer, J., 2005a, Modulator effects of Interleukin-1 beta and Tumor Necrosis Factor–alpha on AMPA-induced excitotoxicity in mouse organotypic hippocampal slice cultures. *J. Neurosci.* **25**: 6734.
Bernardino, L., Ferreira, R., Cristóvão, A.J., Sales, F. and Malva, J.O., 2005b, Inflammation and neurogenesis in temporal lobe epilepsy. *Curr. Drug Targets C.N.S. Neurol. Disord.* **4**: 349.
Beutler, B. and van Huffel, C., 1994, Unravelling function in the TNF ligand and receptor families. *Science* **264**: 667.
Bezzi, P. and Volterra, A., 2001, A neuron-glia signaling network in the active brain. *Curr. Opin. Neurobiol.* **11**: 387.
Bezzi, P., Vesce, S., Panzarasa, P. and Volterra, A., 1999, Astrocytes as active participants of glutamatergic function and regulators of its homeostasis. *Adv. Exp. Med. Biol.* **468**: 69.
Bezzi, P., Domercq, M., Brambilla, L., Galli, R., Schols, D., De Clercq, E., Vescovi, A., Bagetta, G., Kollias, G., Meldolesi, J. and Volterra, A., 2001, CXCR4-activated astrocyte glutamate release via TNFalpha: amplification by microglia triggers neurotoxicity. *Nat. Neurosci.* **4**: 702.
Biber, K., Zuurman, M.W., Dijkstra, I.M. and Boddeke, H.W., 2002, Chemokines in the brain: neuroimmunology and beyond. *Curr. Opin. Pharmacol.* **2**: 63.
Biber, K., de Jong, E.K., van Weering, H.R. and Boddeke, H.W., 2006, Chemokines and their receptors in central nervous system disease. *Curr. Drug Targets* **7**: 29.
Blonska, M., Shambharkar, P.B., Kobayashi, M., Zhang, D., Sakurai, H., Su, B. and Lin, X., 2005, TAK1 is recruited to the tumor necrosis factor-alpha (TNF-alpha) receptor 1 complex in a receptor-interacting protein (RIP)-dependent manner and cooperates with MEKK3 leading to NF-kappaB activation. *J. Biol. Chem.* **280**: 43056.
Bonde, C., Noraberg, J., Noer, H. and Zimmer, J., 2005, Ionotropic glutamate receptors and glutamate transporters are involved in necrotic neuronal cell death induced by oxygen-glucose deprivation of hippocampal slice cultures. *Neuroscience* **136**: 779.
Boucsein, C., Zacharias, R., Farber, K., Pavlovic, S., Hanisch, U.K. and Kettenmann, H., 2003, Purinergic receptors on microglial cells: functional expression in acute brain slices and modulation of microglial activation in vitro. *Eur. J. Neurosci.* **17**: 2267.
Brough, D., Le Feuvre, R.A., Iwakura, Y. and Rothwell, N.J., 2002, Purinergic (P2X7) receptor activation of microglia induces cell death via an interleukin-1-independent mechanism. *Mol. Cell Neurosci.* **19**: 272.
Burns, K., Clatworthy, J., Martin, L., Martinon, F., Plumpton, C., Maschera, B., Lewis, A., Ray, K., Tschopp, J. and Volpe, F., 2000, Tollip, a new component of the IL-1RI pathway, links IRAK to the IL-1 receptor. *Nat. Cell Biol.* **2**: 346.
Burudi, E.M. and Regnier-Vigouroux, A., 2001, Regional and cellular expression of the mannose receptor in the post-natal developing mouse brain. *Cell Tissue Res.* **303**: 307.
Campbell, I.L., 2005, Cytokine-mediated inflammation, tumorigenesis, and disease-associated JAK/STAT/SOCS signaling circuits in the CNS. *Brain Res. Brain Res. Rev.* **48**: 166.
Cao, Z., Henzel, W.J. and Gao, X., 1996a, IRAK: a kinase associated with the interleukin-1 receptor. *Science* **271**: 1128.
Cao, Z., Xiong, J., Takeuchi, M., Kurama, T. and Goeddel, D.V., 1996b, TRAF6 is a signal transducer for interleukin-1. *Nature* **383**: 443.
Cartier, L., Hartley, O., Dubois-Dauphin, M. and Krause K.H., 2005, Chemokine receptors in the central nervous system: role in brain inflammation and neurodegenerative diseases. *Brain Res. Brain Res. Rev.* **48**: 16.
Chan, F.K., 2000, The pre-ligand binding assembly domain: a potential target of inhibition of tumour necrosis factor receptor function. *Ann. Rheum. Dis.* **59**(Suppl. 1): 50.
Chan, F.K., Chun, H.J., Zheng, L., Siegel, R.M., Bui, K.L. and Lenardo, M.J., 2000, A domain in TNF receptors that mediates ligand-independent receptor assembly and signaling. *Science* **288**: 2351.
Chavarria A. and Alcocer-Varela J., 2004, Is damage in central nervous system due to inflammation? *Autoimmun. Rev.* **3**: 251.
Chen, G. and Goeddel, D.V., 2002, TNF-R1 signaling: a beautiful pathway. *Science* **296**: 1634.
Chen, Y. and Swanson, R.A., 2003, Astrocytes and brain injury. *J. Cereb. Blood Flow. Metab.* **23**: 137.
Cheng, B., Christakos, S. and Mattson, M.P., 1994, Tumor necrosis factors protect neurons against metabolic-excitotoxic insults and promote maintenance of calcium homeostasis. *Neuron* **12**: 139.
Chinnaiyan, A.M., O'Rourke, K., Tewari, M. and Dixit, V.M., 1995, FADD, a novel death domain-containing protein, interacts with the death domain of Fas and initiates apoptosis. *Cell* **81**: 505.

Choi, D.W., 1992, Excitotoxic cell death. *J. Neurobiol.* **23**: 1261.
Colombaioni, L. and Garcia-Gil, M., 2004, Sphingolipid metabolites in neural signaling and functions. *Brain Res. Rev.* **46**: 328.
Colotta, F., Re, F., Muzio, M., Bertini, R., Polentarutti, N., Sironi, M., Giri, J.G., Dower, S.K., Sims, J.E. and Mantovani, A., 1993, Interleukin-1 type II receptor: a decoy target for IL-1 that is regulated by IL-4. *Science* **261**: 472.
Couraud, P.O., 1994, Interactions between lymphocytes, macrophages, and central nervous system cells. *J. Leukoc. Biol.* **56**: 407.
Croston, G.E., Cao, Z. and Goeddel, D.V., 1995, NF-kappa B activation by interleukin-1 (IL-1) requires an IL-1 receptor-associated protein kinase activity. *J. Biol. Chem.* **270**: 16514.
Cunha, R.A. and Ribeiro, J.A., 2000, ATP as a presynaptic modulator. *Life Sci.* **68**: 119.
Cunningham, C., Wilcockson, D.C., Campion, S., Lunnon, K. and Perry, V.H., 2005, Central and systemic endotoxin challenges exacerbate the local inflammatory response and increase neuronal death during chronic neurodegeneration. *J. Neurosci.* **25**: 9275.
Dauphinee, S.M. and Karsan, A., 2006, Lipopolysaccharide signaling in endothelial cells. *Lab. Invest.* **86**: 9.
Davies, C.A., Loddick, S.A., Toulmond, S., Stroemer, R.P., Hunt, J. and Rothwell, N.J., 1999, The progression and topographic distribution of interleukin-1beta expression after permanent middle cerebral artery occlusion in the rat. *J. Cereb. Blood Flow Metab.* **19**: 87.
Davis, C.N., Mann, E., Behrens, M.M., Gaidarova, S., Rebek, M., Rebek, J. Jr. and Bartfai, T., 2006, MyD88-dependent and -independent signaling by IL-1 in neurons probed by bifunctional Toll/IL-1 receptor domain/BB-loop mimetics. *Proc. Natl. Acad. Sci. USA* **103**: 2953.
Delhase, M., Hayakawa, M., Chen, Y. and Karin, M., 1999, Positive and negative regulation of IkappaB kinase activity through IKKbeta subunit phosphorylation. *Science* **284**: 309.
Dempsey, P.W., Doyle, S.E., He, J.Q. and Cheng, G., 2003, The signaling adaptors and pathways activated by TNF superfamily. *Cytokine Growth Factor Rev.* **14**: 193.
Denecker, G., Vercammen, D., Declercq, W. and Vandenabeele, P., 2001, Apoptotic and necrotic cell death induced by death domain receptors. *Cell. Mol. Life Sci.* **58**: 356.
Devalaraja, M.N. and Richmond, A., 1999, Multiple chemotactic factors: fine control or redundancy? *Trends Pharmacol. Sci.* **20**: 151.
Devitt, A., Moffatt, O.D., Raykundalia, C., Capra, J.D., Simmons, D.L. and Gregory, C.D., 1998, Human CD14 mediates recognition and phagocytosis of apoptotic cells. *Nature* **392**: 505.
Di Virgilio, F., Sanz, J.M., Chiozzi, P. and Falzoni, S., 1999, The P2Z/P2X7 receptor of microglial cells: a novel immunomodulatory receptor. *Prog. Brain Res.* **120**: 355.
DiDonato, J.A., Hayakawa, M., Rothwarf, D.M., Zandi, E. and Karin, M., 1997, A cytokine-responsive IkappaB kinase that activates the transcription factor NF-kappaB. *Nature* **388**: 548.
Dinarello, C.A., 1996, Biologic basis for interleukin-1 in disease. *Blood* **87**: 2095.
Dobrogowska, D.H., Lossinsky, A.S., Tarnawski, M. and Vorbrodt, A.W., 1998, Increased blood–brain barrier permeability and endothelial abnormalities induced by vascular endothelial growth factor. *J. Neurocytol.* **27**: 163.
Dong, Y. and Benveniste, E.N., 2001, Immune function of astrocytes. *Glia* **36**:180.
Dubyak, G.R. and el-Moatassim, C., 1993, Signal transduction via P2-purinergic receptors for extracellular ATP and other nucleotides. *Am. J. Physiol.* **265**: C577.
Dunne, A. and O'Neill, L.A., 2003, The interleukin-1 receptor/Toll-like receptor superfamily: signal transduction during inflammation and host defense. *Sci. STKE.* **171**: re3.
Elward, K. and Gasque, P., 2003, "Eat me" and "don't eat me" signals govern the innate immune response and tissue repair in the CNS: emphasis on the critical role of the complement system. *Mol. Immunol.* **40**: 85.
Engelhardt, B., 2006, Molecular mechanisms involved in T cell migration across the blood–brain barrier. *J. Neural. Transm.* **113**: 477.
Eriksson, C., Winblad, B. and Schultzberg, M., 1998, Kainic acid induced expression of interleukin-1 receptor antagonist mRNA in the rat brain. *Brain Res. Mol. Brain Res.* **58**: 195.
Eriksson, C., Tehranian, R., Iverfeldt, K., Winblad, B. and Schultzberg, M., 2000, Increased expression of mRNA encoding interleukin-1beta and caspase-1, and the secreted isoform of interleukin-1 receptor antagonist in the rat brain following systemic kainic acid administration. *J. Neurosci. Res.* **60**: 266.
Fernandez, E.J. and Lolis, E., 2002, Structure, function, and inhibition of chemokines. *Annu. Rev. Pharmacol. Toxicol.* **42**: 469.
Ferrari, D., Villalba, M., Chiozzi, P., Falzoni, S., Ricciardi-Castagnoli, P. and Di Virgilio, F., 1996, Mouse microglial cells express a plasma membrane pore gated by extracellular ATP. *J. Immunol.* **156**: 1531.
Ferrari, D., Chiozzi, P., Falzoni, S., Hanau, S. and Di Virgilio, F., 1997a, Purinergic modulation of interleukin-1 beta release from microglial cells stimulated with bacterial endotoxin. *J. Exp. Med.* **185**: 579.

Ferrari, D., Wesselborg, S., Bauer, M.K. and Schulze-Osthoff, K., 1997b, Extracellular ATP activates transcription factor NF-kappaB through the P2Z purinoreceptor by selectively targeting NF-kappaB p65. *J. Cell Biol.* **139**: 1635.
Ferrari, D., Los, M., Bauer, M.K., Vandenabeele, P., Wesselborg, S. and Schulze-Osthoff, K., 1999, P2z purinoreceptor ligation induces activation of caspases with distinct roles in apoptotic and necrotic alterations of cell death. *FEBS Lett.* **447**: 71.
Ferrari, D., Pizzirani, C., Adinolfi, E., Lemoli, R.M., Curti, A., Idzko, M., Panther, E. and Di Virgilio, F., 2006, The P2X7 receptor: a key player in IL-1 processing and release. *J. Immunol.* **176**: 3877.
Fink, K.B., Andrews, L.J., Butler, W.E., Ona, V.O., Li, M., Bogdanov, M., Endres, M., Khan, S.Q., Namura, S., Stieg, P.E., Beal, M.F., Moskowitz, M.A., Yuan, J. and Friedlander, R.M., 1999, Reduction of post-traumatic brain injury and free radical production by inhibition of the caspase-1 cascade. *Neuroscience* **94**: 1213.
Flügel, A., Schwaiger, F.W., Neumann, H., Medana, I., Willem, M., Wekerle, H., Kreutzberg, G.W. and Graeber, M.B., 2000, Neuronal FasL induces cell death of encephalitogenic T lymphocytes. *Brain Pathol.* **10**: 353.
Flügel, A., Berkowicz, T., Ritter, T., Labeur, M., Jenne, D.E., Li, Z., Ellwart, J.W., Willem, M., Lassmann, H. and Wekerle, H., 2001, Migratory activity and functional changes of green fluorescent effector cells before and during experimental autoimmune encephalomyelitis. *Immunity* **14**: 547.
Fontaine, V., Mohand-Said, S., Hanoteau, N., Fuchs, C., Pfizenmaier, K. and Eisel, U., 2002, Neurodegenerative and neuroprotective effects of tumor necrosis factor (TNF) in retinal ischemia: opposite roles of TNF receptor 1 and TNF receptor 2. *J. Neurosci.* **22**: RC216.
Frei, K., Lins, H., Schwerdel, C. and Fontana, A., 1994, Antigen presentation in the central nervous system. The inhibitory effect of IL-10 on MHC class II expression and production of cytokines depends on the inducing signals and the type of cell analyzed. *J. Immunol.* **152**: 2720.
Friedman, W.J., 2001, Cytokines regulate expression of the type 1 interleukin-1 receptor in rat hippocampal neurons and glia. *Exp. Neurol.* **168**: 23.
Friedman, W.J., 2005, Interactions of interleukin-1 with neurotrophic factors in the central nervous system: beneficial or detrimental? *Mol. Neurobiol.* **32**: 133.
Furukawa, K. and Mattson, M.P., 1998, The transcription factor NF-kappaB mediates increases in calcium currents and decreases in NMDA- and AMPA/kainate-induced currents induced by tumor necrosis factor-alpha in hippocampal neurons. *J. Neurochem.* **70**: 1876.
Galasso, J.M., Harrison, J.K. and Silverstein, F.S., 1998, Excitotoxic brain injury stimulates expression of the chemokine receptor CCR5 in neonatal rats. *Am. J. Pathol.* **153**: 1631.
Galasso, J.M., Miller, M.J., Cowell, R.M., Harrison, J.K., Warren, J.S. and Silverstein, FS., 2000, Acute excitotoxic injury induces expression of monocyte chemoattractant protein-1 and its receptor, CCR2, in neonatal rat brain. *Exp. Neurol.*, **165**: 295.
Gay, N.J. and Keith, F.J., 1991, Drosophila Toll and IL-1 receptor. *Nature* **351**: 355.
Ghosh, S. and Karin, M., 2002, Missing pieces in the NF-kappaB puzzle. *Cell* **109**(Suppl): 81.
Gourmala, N.G., Limonta, S., Bochelen, D., Sauter, A. and Boddeke, H.W., 1999, Localization of macrophage inflammatory protein: macrophage inflammatory protein-1 expression in rat brain after peripheral administration of lipopolysaccharide and focal cerebral ischemia. *Neuroscience* **88**: 1255.
Greenfeder, S.A., Nunes, P., Kwee, L., Labow, M., Chizzonite, R.A. and Ju, G., 1995, Molecular cloning and characterization of a second subunit of the interleukin 1 receptor complex. *J. Biol. Chem.* **270**: 13757.
Grell, M., Zimmermann, G., Hulser, D., Pfizenmaier, K. and Scheurich, P., 1994, TNF receptors TR60 and TR80 can mediate apoptosis via induction of distinct signal pathways. *J. Immunol.* **153**: 1963.
Grell, M., Douni, E., Wajant, H., Lohden, M., Clauss, M., Maxeiner, B., Georgopoulos, S., Lesslauer, W., Kollias, G., Pfizenmaier, K. and Scheurich, P., 1995, The transmembrane form of tumor necrosis factor is the prime activating ligand of the 80 kDa tumor necrosis factor receptor. *Cell* **83**: 793.
Griffin, W.S., Liu, L., Li, Y., Mrak, R.E. and Barger, S.W., 2006, Interleukin-1 mediates Alzheimer and Lewy body pathologies. *J. Neuroinflammation* **3**: 5.
Gruol, D.L. and Nelson, T.E., 2005, Purkinje neuron physiology is altered by the inflammatory factor interleukin-6. *Cerebellum* **4**: 198.
Hanisch, U.K., 2002, Microglia as a source and target of cytokines. *Glia* **40**: 140.
Hatten, M.E., Liem, R.K., Shelanski, M.L. and Mason, C.A., 1991, Astroglia in CNS injury. *Glia* **4**: 233.
Hehlgans, T. and Mannel, D.N., 2002, The TNF-TNF receptor system. *Biol. Chem.* **383**: 1581.
Hellstrom, I.C., Danik, M., Luheshi, G.N. and Williams, S., 2005, Chronic LPS exposure produces changes in intrinsic membrane properties and a sustained IL-beta-dependent increase in GABAergic inhibition in hippocampal CA1 pyramidal neurons. *Hippocampus* **15**: 656.
Heppner, F.L., Skutella, T., Hailer, N.P., Haas, D. and Nitsch, R., 1998, Activated microglial cells migrate towards sites of excitotoxic neuronal injury inside organotypic hippocampal slice cultures. *Eur. J. Neurosci.* **10**: 3284.
Hermann, G.E., Rogers, R.C., Bresnahan, J.C. and Beattie, M.S., 2001, Tumor necrosis factor-alpha induces cFOS and strongly potentiates glutamate-mediated cell death in the rat spinal cord. *Neurobiol. Dis.* **8**: 590.

Herx, L.M., Rivest, S. and Yong, V.W., 2000, Central nervous system-initiated inflammation and neurotrophism in trauma: IL-1 beta is required for the production of ciliary neurotrophic factor. *J. Immunol.* **165**: 2232.
Hickey, W.F., 2001, Basic principles of immunological surveillance of the normal central nervous system. *Glia* **36**: 118.
Hickey, W.F. and Kimura, H., 1988, Perivascular microglial cells of the CNS are bone marrow-derived and present antigen in vivo. *Science* **239**: 290.
Hide, I., Tanaka, M., Inoue, A., Nakajima, K., Kohsaka, S., Inoue, K. and Nakata, Y., 2000, Extracellular ATP triggers tumor necrosis factor-alpha release from rat microglia. *J. Neurochem.* **75**: 965.
Higuchi, M. and Aggarwal, B.B., 1994, TNF induces internalization of the p60 receptor and shedding of the p80 receptor. *J. Immunol.* **152**: 3550.
Hofmeister, R., Wiegmann, K., Korherr, C., Bernardo, K., Kronke, M. and Falk, W., 1997, Activation of acid sphingomyelinase by interleukin-1 (IL-1) requires the IL-1 receptor accessory protein. *J. Biol. Chem.* **272**: 27730.
Honda, S., Sasaki, Y., Ohsawa, K., Imai, Y., Nakamura, Y., Inoue, K. and Kohsaka, S., 2001, Extracellular ATP or ADP induce chemotaxis of cultured microglia through Gi/o-coupled P2Y receptors. *J. Neurosci.* **21**: 1975.
Houzen, H., Kikuchi, S., Kanno, M., Shinpo K. and Tashiro, K., 1997, Tumor necrosis factor enhancement of transient outward potassium currents in cultured rat cortical neurons. *J. Neurosci. Res.* **50**: 990.
Hsu, H., Xiong, J. and Goeddel, D.V., 1995, The TNF receptor 1-associated protein TRADD signals cell death and NF-kappa B activation. *Cell* **81**: 495.
Hsu, H., Huang, J., Shu, H.B., Baichwal, V. and Goeddel, D.V., 1996a, TNF-dependent recruitment of the protein kinase RIP to the TNF receptor-1 signaling complex. *Immunity* **4**: 387.
Hsu, H., Shu, H.B., Pan, M.G. and Goeddel, D.V., 1996b, TRADD-TRAF2 and TRADD-FADD interactions define two distinct TNF receptor 1 signal transduction pathways. *Cell* **84**: 299.
Hu, S., Sheng, W.S., Ehrlich, L.C., Peterson, P.K. and Chao, C.C., 2000, Cytokine effects on glutamate uptake by human astrocytes. *Neuroimmunomodulation* **7**: 153.
Huber, J.D., Egleton, R.D. and Davis, T.P., 2001, Molecular physiology and pathophysiology of tight junctions in the blood–brain barrier. *Trends Neurosci.* **24**: 719.
Idriss, H.T. and Naismith, J.H., 2000, TNF alpha and the TNF receptor superfamily: structure-function relationship(s). *Microsc. Res. Tech.* **50**: 184.
Imitola, J., Raddassi, K., Park, K.I., Mueller, F.J., Nieto, M., Teng, Y.D., Frenkel, D., Li, J., Sidman, R.L., Walsh, C.A., Snyder, E.Y. and Khoury, S.J., 2004, Directed migration of neural stem cells to sites of CNS injury by the stromal cell-derived factor 1alpha/CXC chemokine receptor 4 pathway. *Proc. Natl. Acad. Sci. USA* **101**: 18117.
Inoue, K., 2002, Microglial activation by purines and pyrimidines. *Glia* **40**: 156.
Itoh, N. and Nagata, S., 1993, A novel protein domain required for apoptosis. Mutational analysis of human Fas antigen. *J. Biol. Chem.* **268**: 10932.
James, G. and Butt, M., 2002, P2Y and P2X purinoceptor mediated Ca2+ signaling in glial cell pathology in the central nervous system. *Eur. J. Pharmacol.* **447**: 247.
Jiang, Y., Woronicz, J.D., Liu, W. and Goeddel, D.V., 1999, Prevention of constitutive TNF receptor 1 signaling by silencer of death domains. *Science* **283**: 543.
Jones, E.Y., 2000, The tumour necrosis factor receptor family: life or death choices. *Curr. Opin. Struct. Biol.* **10**: 644.
Kaisho T. and Akira S., 2000, Critical roles of Toll-like receptors in host defense. *Crit. Ver. Immunol.* **20**: 393.
Karin, M. and Lin, A., 2002, NF-kappaB at the crossroads of life and death. *Nat. Immunol.* **3**: 221.
Karkkainen, I., Rybnikova, E., Pelto-Huikko, M. and Huovila, A.P., 2000, Metalloprotease-disintegrin (ADAM) genes are widely and differentially expressed in the adult CNS. *Mol. Cell Neurosci.* **15**: 547.
Kawai, T., Adachi, O., Ogawa, T., Takeda, K. and Akira, S., 1999, Unresponsiveness of MyD88-deficient mice to endotoxin. *Immunity* **11**: 115.
Kelliher, M.A., Grimm, S., Ishida, Y., Kuo, F., Stanger, B.Z. and Leder, P., 1998, The death domain kinase RIP mediates the TNF-induced NF-kappaB signal. *Immunity* **8**: 297.
Kielian, T., Mayes, P. and Kielian, M., 2002, Characterization of microglial responses to Staphylococcus aureus: effects on cytokine, costimulatory molecule, and Toll-like receptor expression. *J. Neuroimmunol.* **130**: 86.
Kim, S.U. and de Vellis, J., 2005, Microglia in health and disease. *J. Neurosci. Res.* **81**: 302.
Kischkel, F.C., Lawrence, D.A., Chuntharapai, A., Schow, P., Kim, K.J. and Ashkenazi, A., 2000, Apo2L/TRAIL-dependent recruitment of endogenous FADD and caspase-8 to death receptors 4 and 5. *Immunity* **12**: 611.
Kishida, S., Sanjo, H., Akira, S., Matsumoto, K. and Ninomiya-Tsuji, J., 2005, TAK-1 binding protein 2 facilitates ubiquitination of TRAF6 and assembly of TRAF6 with IKK in the signaling pathway. *Genes Cell* **10**: 447.
Kitchens, R.L., Thompson, P.A., O'Keefe, G.E. and Munford, R.S., 2000, Plasma constituents regulate LPS binding to, and release from, the monocyte cell surface. *J. Endotoxin Res.* **6**: 477.

Knopf, P.M., Harling-Berg, C.J., Cserr, H.F., Basu, D., Sirulnick, E.J., Nolan, S.C., Park, J.T., Keir, G., Thompson, E.J. and Hickey, W.F., 1998, Antigen-dependent intrathecal antibody synthesis in the normal rat brain: tissue entry and local retention of antigen-specific B cells. *J. Immunol.* **161**: 692.

Kolesnick, R.N. and Kronke, M., 1998, Regulation of ceramide production and apoptosis. *Annu. Rev. Physiol.* **60**: 643.

Kreutzberg, G.W., 1996, Microglia: a sensor for pathological events in the CNS. *Trends Neurosci.* **19**: 312.

Kriegler, M., Perez, C., DeFay, K., Albert, I. and Lu, S.D., 1988, A novel form of TNF/cachectin is a cell surface cytotoxic transmembrane protein: ramifications for the complex physiology of TNF. *Cell* **53**: 45.

Kristensen, B.W., Noraberg, J. and Zimmer, J., 2001, Comparison of excitotoxic profiles of ATPA, AMPA, KA and NMDA in organotypic hippocampal slice cultures. *Brain Res.* **917**: 21.

Laflamme, N. and Rivest, S., 2001, Toll-like receptor 4: the missing link of the cerebral innate immune response triggered by circulating gram-negative bacterial cell wall components. *FASEB J.* **15**: 155.

Lai, A.Y., Swayze, R.D., El-Husseini, A. and Song, C., 2006, Interleukin-1 beta modulates AMPA receptor expression and phosphorylation in hippocampal neurons. *J. Neuroimmunol.* **175**: 97.

Lassmann, H., Schmied, M., Vass, K. and Hickey, W.F., 1993, Bone marrow derived elements and resident microglia in brain inflammation. *Glia* **7**: 19.

Le Feuvre, R.A., Brough, D., Iwakura, Y., Takeda, K. and Rothwell, N.J., 2002, Priming of macrophages with lipopolysaccharide potentiates P2X7-mediated cell death via a caspase-1-dependent mechanism, independently of cytokine production. *J. Biol. Chem.* **277**: 3210.

Legler, D.F., Micheau, O., Doucey, M.A., Tschopp, J. and Bron, C., 2003, Recruitment of TNF receptor 1 to lipid rafts is essential for TNFalpha-mediated NF-kappaB activation. *Immunity* **18**: 655.

Lehnardt, S., Massillon, L., Follett, P., Jensen, F.E., Ratan, R., Rosenberg, P.A., Volpe, J.J. and Vartanian, T., 2003, Activation of innate immunity in the CNS triggers neurodegeneration through a Toll-like receptor 4-dependent pathway. *Proc. Natl. Acad. Sci. USA* **100**: 8514.

Lemaitre, B., Nicolas, E., Michaut, L., Reichhart, J.M. and Hoffmann, J.A., 1996, The dorsoventral regulatory gene cassette spatzle/Toll/cactus controls the potent antifungal response in Drosophila adults. *Cell* **86**: 973.

Li, X. and Qin, J., 2005, Modulation of Toll-interleukin 1 receptor mediated signaling. *J. Mol. Med.* **83**: 258.

Li, X., Yang, Y. and Ashwell, J.D., 2002, TNF-RII and c-IAP1 mediate ubiquitination and degradation of TRAF2. *Nature* **416**: 345.

Li, C., Zienkiewicz, J. and Hawiger, J., 2005, Interactive sites in the MyD88 Toll/interleukin (IL) 1 receptor domain responsible for coupling to the IL1beta signaling pathway. *J. Biol. Chem.* **280**: 26152.

Ling, L., Cao, Z. and Goeddel, D.V., 1998, NF-kappaB-inducing kinase activates IKK-alpha by phosphorylation of Ser-176. *Proc. Natl. Acad. Sci. USA* **95**: 3792.

Liu, Z.G., Hsu, H., Goeddel, D.V. and Karin, M., 1996, Dissection of TNF receptor 1 effector functions: JNK activation is not linked to apoptosis while NF-kappaB activation prevents cell death. *Cell* **87**: 565.

Locksley, R.M., Killeen, N. and Lenardo, M.J., 2001, The TNF and TNF receptor superfamilies: integrating mammalian biology. *Cell* **104**: 487.

Loddick, S.A., Liu, C., Takao, T., Hashimoto, K. and De Souza, E.B., 1998, Interleukin-1 receptors: cloning studies and role in central nervous system disorders. *Brain Res. Brain Res. Rev.* **26**: 306.

Lu, K.T., Wang, Y.W., Wo, Y.Y. and Yang, Y.L., 2005, Extracellular signal-regulated kinase-mediated IL-1-induced cortical neuron damage during traumatic brain injury. *Neurosci. Lett.* **386**: 40.

Lucas, S.M., Rothwell, N.J. and Gibson, R.M., 2006, The role of inflammation in CNS injury and disease. *Br. J. Pharmacol.* **147**: S232.

Luk, W.P., Zhang, Y., White, T.D., Lue, F.A., Wu, C., Jiang, C.G., Zhang, L. and Moldofsky, H., 1999, Adenosine: a mediator of interleukin-1beta-induced hippocampal synaptic inhibition. *J. Neurosci.* **19**: 4238.

Ma, Q., Jones, D., Borghesani, P.R., Segal, R.A., Nagasawa, T., Kishimoto, T., Bronson, R.T. and Springer, T.A., 1998, Impaired B-lymphopoiesis, myelopoiesis, and derailed cerebellar neuron migration in CXCR4- and SDF-1-deficient mice. *Proc. Natl. Acad. Sci. USA* **95**: 9448.

MacEwan, D.J., 2002, TNF receptor subtype signaling: differences and cellular consequences. *Cell Signal* **14**: 477.

Mackay, C.R., 2001, Chemokines: immunology's high impact factors. *Nat. Immunol.* **2**: 95.

Mahad, D., Callahan, M.K., Williams, K.A., Ubogu, E.E., Kivisakk, P., Tucky, B., Kidd, G., Kingsbury, G.A., Chang, K., Fox, R.J., Mack, M., Sniderman, M.B., Ravid, R., Staugaitis, S.M., Stins, M.F. and Ransohoff, R.M., 2006, Modulating CCR2 and CCL2 at the blood–brain barrier: relevance for multiple sclerosis pathogenesis. *Brain* **129**: 212.

Malagarie-Cazenave, S., Andrieu-Abadie, N., Sgui, B., Gouaz, V., Tardy, C., Cuvillier, O. and Levade, T., 2002, Sphingolipid signaling: molecular basis and role in TNF-induced cell death. *Exp. Rev. Mol. Med.* **515**: 20.

Malinin, N.L., Boldin, M.P., Kovalenko, A.V. and Wallach, D., 1997, MAP3K-related kinase involved in NF-kappaB induction by TNF, CD95 and IL-1. *Nature* **385**: 540.

Man, S.M., Ma, Y.R., Shang, D.S., Zhao, W.D., Li B., Guo, D.W., Fang, W.G., Zhu, L. and Chen, Y.H., 2006, Peripheral T cells overexpress MIP-1alpha to enhance its transendothelial migration in Alzheimer's disease. *Neurobiol. Aging* [Epub ahead of print].

Mantovani, A., 1999, The chemokine system: redundancy for robust outputs. *Immunol. Today* **20**: 254.
Marchetti, L., Klein, M., Schlett, K., Pfizenmaier, K. and Eisel, U.L., 2004, Tumor necrosis factor (TNF)-mediated neuroprotection against glutamate-induced excitotoxicity is enhanced by N-methyl-D-aspartate receptor activation. Essential role of a TNF receptor 2-mediated phosphatidylinositol 3-kinase-dependent NF-kappa B pathway. *J. Biol. Chem.* **30**: 32869.
Mark, K.S. and Miller, D.W., 1999, Increased permeability of primary cultured brain microvessel endothelial cell monolayers following TNF-alpha exposure. *Life Sci.* **64**: 1941.
Mason, J.L., Suzuki, K., Chaplin, D.D. and Matsushima, G.K., 2001, Interleukin-1beta promotes repair of the CNS. *J. Neurosci.* **21**: 7046.
Mattson, M.P., Cheng, B., Baldwin, S.A., Smith-Swintosky, V.L., Keller, J., Geddes, J.W., Scheff, S.W. and Christakos, S., 1995, Brain injury and tumor necrosis factors induce calbindin D28k in astrocytes: evidence for a cytoprotective response. *J. Neurosci. Res.* **42**: 357.
McCluskey, L.P. and Lampson, L.A., 2000, Local neurochemicals and site-specific immune regulation in the CNS. *J. Neuropathol. Exp. Neurol.* **59**: 177.
Medzhitov, R. and Janeway, C.A. Jr., 1997, Innate immunity: impact on the adaptive immune response. *Curr. Opin. Immunol.* **9**: 4.
Medzhitov, R. and Janeway, C. Jr., 2000, Innate immune recognition: mechanisms and pathways. *Immunol. Rev.* **173**: 89.
Meffert, M.K. and Baltimore, D., 2005, Physiological functions for brain NF-kappaB. *Trends Neurosci.* **28**: 37.
Mennicken, F., Maki, R., de Souza, E.B. and Quirion, R., 1999, Chemokines and chemokine receptors in the CNS: a possible role in neuroinflammation and patterning. *Trends Pharmacol. Sci.* **20**: 73.
Mercurio, F., Zhu, H., Murray, B.W., Shevchenko, A., Bennett, B.L., Li, J., Young, D.B., Barbosa, M., Mann, M., Manning, A. and Rao, A., 1997, IKK-1 and IKK-2: cytokine-activated IkappaB kinases essential for NF-kappaB activation. *Science* **278**: 860.
Miller, L.G., Galpern, W.R., Dunlap, K., Dinarello, C.A. and Turner, T.J., 1991, Interleukin-1 augments gamma-aminobutyric acid A receptor function in brain. *Mol. Pharmacol.* **39**: 105.
Minagar, A., Jy, W., Jimenez, J.J. and Alexander, J.S., 2006, Multiple sclerosis as a vascular disease. *Neurol. Res.* **28**: 230.
Minami, M. and Satoh, M., 2003, Chemokines and their receptors in the brain: pathophysiological roles in ischemic brain injury, *Life Sci.* **74**: 321.
Minami, M. and Satoh, M., 2005, Role of chemokines in ischemic neuronal stress. *Neuromolecular Med.* **7**: 149.
Minami, M., Kuraishi, Y., Yamaguchi, T., Nakai, S., Hirai, Y. and Satoh, M., 1990, Convulsants induce interleukin-1 beta messenger RNA in rat brain. *Biochem. Biophys. Res. Commun.* **171**: 832.
Minami, M., Kuraishi, Y. and Satoh, M., 1991, Effects of kainic acid on messenger RNA levels of IL-1 beta, IL-6, TNF alpha and LIF in the rat brain. *Biochem. Biophys. Res. Commun.* **176**: 593.
Miyachi, T., Asai, K., Tsuiki, H., Mizuno, H., Yamamoto, N., Yokoi, T., Aoyama, M., Togari, H., Wada, Y., Miura, Y. and Kato, T., 2001, Interleukin-1beta induces the expression of lipocortin 1 mRNA in cultured rat cortical astrocytes. *Neurosci. Res.* **40**: 53.
Monje, M.L., Toda, H. and Palmer, T.D., 2003, Inflammatory blockade restores adult hippocampal neurogenesis. *Science* **302**: 1760.
Moynagh, P.N., 2005, The interleukin-1 signaling pathway in astrocytes: a key contributor to inflammation in the brain. *J. Anat.* **207**: 265.
Moynagh, P.N., Williams, D.C., and O'Neill, L.A., 1994, Activation of NF-kappa B and induction of vascular cell adhesion molecule-1 and intracellular adhesion molecule-1 expression in human glial cells by IL-1. Modulation by antioxidants. *J. Immunol.* **153**: 2681.
Murakami, Y., Okada, S. and Yokotani, K., 2002, Brain inducible nitric oxide synthase is involved in interleukin-1beta-induced activation of the central sympathetic outflow in rats. *Eur. J. Pharmacol.* **455**: 73.
Murdoch, C. and Finn, A., 2000, Chemokine receptors and their role in inflammation and infectious diseases. *Blood* **95**: 3032.
Murphy, P.M., Baggiolini, M., Charo, I.F., Hebert, C.A., Horuk, R., Matsushima, K., Miller, L.H., Oppenheim, J.J. and Power, C.A., 2000, International union of pharmacology. XXII. Nomenclature for chemokine receptors. *Pharmacol. Rev.* **52**: 145.
Nakajima, K. and Kohsaka, S., 2004, Microglia: neuroprotective and neurotrophic cells in the central nervous system. *Curr. Drug Targets Cardiovasc. Haematol. Disord.* **4**: 65.
Nedergaard, M., 1994, Direct signaling from astrocytes to neurons in cultures of mammalian brain cells. *Science* **263**: 1768.
Neumann, H., 2001, Control of glial immune function by neurons. *Glia* **36**: 191.
Neumann, H., 2004, Inflammation, Neuroprotection. Models, Mechanisms and Therapies. Mathias Bahr, ed., Wiley-VGH, pp. 173–190.

Neumann, H. and Wekerle, H., 1998, Neuronal control of the immune response in the central nervous system: linking brain immunity to neurodegeneration. *J. Neuropathol. Exp. Neurol.* **57**: 1.

Neumann, H., Boucraut, J., Hahnel, C., Misgeld, T. and Wekerle, H., 1996, Neuronal control of MHC class II inducibility in rat astrocytes and microglia. *Eur. J. Neurosci.* **8**: 2582.

Neumann, H., Schmidt, H., Wilharm, E., Behrens, L. and Wekerle, H., 1997, Interferon gamma gene expression in sensory neurons: evidence for autocrine gene regulation. *J. Exp. Med.* **186**: 2023.

Neumann, H., Misgeld, T., Matsumuro, K. and Wekerle, H., 1998, Neurotrophins inhibit major histocompatibility class II inducibility of microglia: involvement of the p75 neurotrophin receptor. *Proc. Natl. Acad. Sci. USA* **95**: 5779.

Neumeyer, J., Hallas, C., Merkel, O., Winoto-Morbach, S., Jakob, M., Thon, L., Adam, D., Schneider-Brachert, W. and Schutze, S., 2006, TNF-receptor I defective in internalization allows for cell death through activation of neutral sphingomyelinase. *Exp. Cell Res.* [Epub ahead of print].

Nguyen, M.D., Julien, J.P. and Rivest, S., 2002, Innate immunity: the missing link in neuroprotection and neurodegeneration? *Nat. Rev. Neurosci.* **3**: 216.

Noda, M., Nakanishi, H., Nabekura, J. and Akaike, N., 2000, AMPA-kainate subtypes of glutamate receptor in rat cerebral microglia. *J. Neurosci.* **20**: 251.

Ogoshi, F., Yin, H.Z., Kuppumbatti, Y., Song, B., Amindari, S. and Weiss, J.H., 2005, Tumor necrosis-factor-alpha (TNF-alpha) induces rapid insertion of Ca2+-permeable alpha-amino-3-hydroxyl-5-methyl-4-isoxazole-propionate (AMPA)/kainate (Ca-A/K) channels in a subset of hippocampal pyramidal neurons. *Exp. Neurol.* **193**: 384.

O'Keefe, G.M., Nguyen, V.T. and Benveniste, E.N., 1999, Class II transactivator and class II MHC gene expression in microglia: modulation by the cytokines TGF-beta, IL-4, IL-13 and IL-10. *Eur. J. Immunol.* **29**: 1275.

Olson, J.K. and Miller, S.D., 2004, Microglia initiate central nervous system innate and adaptive immune responses through multiple TLRs. *J. Immunol.* **173**: 3916.

O'Neill, L.A., 2002, Signal transduction pathways activated by the IL-1 receptor/Toll-like receptor superfamily. *Curr. Top. Microbiol. Immunol.* **270**: 47.

O'Neill, L.A. and Greene, C., 1998, Signal transduction pathways activated by the IL-1 receptor family: ancient signaling machinery in mammals, insects, and plants. *J. Leukoc. Biol.* **63**: 650.

Orellana, D.I., Quintanilla, R.A., Gonzalez-Billault, C. and Maccioni, R.B., 2005, Role of the JAKs/STATs pathway in the intracellular calcium changes induced by interleukin-6 in hippocampal neurons. *Neurotox. Res.* **8**: 295.

Ozinsky, A., Underhill, D.M., Fontenot, J.D., Hajjar, A.M., Smith, K.D., Wilson, C.B., Schroeder, L. and Aderem, A., 2000, The repertoire for pattern recognition of pathogens by the innate immune system is defined by cooperation between toll-like receptors. *Proc. Natl. Acad. Sci. USA* **97**: 13766.

Park, S.M., Schickel, R. and Peter, M.E., 2005, Nonapoptotic functions of FADD-binding death receptors and their signaling molecules. *Curr. Opin. Cell Biol.* **17**: 610.

Patel, H.C., Ross, F.M., Heenan, L.E., Davies, R.E., Rothwell, N.J. and Allan, S.M., 2006, Neurodegenerative actions of interleukin-1 in the rat brain are mediated through increases in seizure activity. *J. Neurosci. Res.* **83**: 385.

Pearson, V.L., Rothwell, N.J. and Toulmond, S., 1999, Excitotoxic brain damage in the rat induces interleukin-1beta protein in microglia and astrocytes: correlation with the progression of cell death. *Glia* **25**: 311.

Perea, G. and Araque, A., 2005, Glial calcium signaling and neuron-glia communication. *Cell Calcium* **38**: 375.

Perea, G. and Araque, A., 2006, Synaptic information processing by astrocytes. *J. Physiol. Paris* **99**: 92.

Perregaux, D.G. and Gabel, C.A., 1998, Post-translational processing of murine IL-1: evidence that ATP-induced release of IL-1 alpha and IL-1 beta occurs via a similar mechanism. *J. Immunol.* **160**: 2469.

Perry, V.H., 2004, The influence of systemic inflammation on inflammation in the brain: implications for chronic neurodegenerative disease. *Brain Behav. Immun.* **18**: 407.

Perry, V.H. and Gordon, S., 1991, Macrophages and the nervous system. *Int. Rev. Cytol.* **125**: 203.

Persson, M., Brantefjord, M., Hansson, E. and Ronnback, L., 2005, Lipopolysaccharide increases microglial GLT-1 expression and glutamate uptake capacity in vitro by a mechanism dependent on TNF-alpha. *Glia* **51**: 111.

Pettus, B.J., Chalfant, C.E. and Hannun, Y.A., 2002, Ceramide in apoptosis: an overview and current perspectives. *Biochim. Biophy. Acta* **1585**: 114.

Piehl, F. and Lidman, O., 2001, Neuroinflammation in the rat-CNS cells and their role in the regulation of immune reactions. *Immunol. Rev.* **184**: 212.

Pinteaux, E., Parker, L.C., Rothwell, N.J. and Luheshi, G.N., 2002, Expression of interleukin-1 receptors and their role in interleukin-1 actions in murine microglial cells. *J. Neurochem.* **83**: 754.

Pinteaux, E., Rothwell, N.J. and Boutin, H., 2006, Neuroprotective actions of endogenous interleukin-1 receptor antagonist (IL-1ra) are mediated by glia. *Glia* **53**: 551.

Pradillo, J.M., Hurtado, O., Romera, C., Cardenas, A., Fernandez-Tome, P., Alonso-Escolano, D., Lorenzo, P., Moro, M.A. and Lizasoain, I., 2006, TNFR1 mediates increased neuronal membrane EAAT3 expression after in vivo cerebral ischemic preconditioning. *Neuroscience* **138**: 1171.

Pringle, A.K., Niyadurupola, N., Johns, P., Anthoni, D.C. and Iannotti, F., 2001, Interleukin-1 beta exacerbates hypoxia-induced neuronal damage, but attenuates toxicity produced by simulated ischaemia and excitotoxicity in rat organotypic hippocampal slice cultures. *Neurosci. Lett.* **305**: 29.
Proudfoot, A.E., 2002, Chemokine receptors: multifaceted therapeutic targets. *Nat. Rev. Immunol.* **2**: 106.
Qian, Y., Commane, M., Ninomiya-Tsuji, J., Matsumoto, K. and Li, X., 2001, IRAK-mediated translocation of TRAF6 and TAB2 in the interleukin-1-induced activation of NFkappa B. *J. Biol. Chem.* **276**: 41661.
Raivich, G., Jones, L.L., Kloss, C.U., Werner, A., Neumann, H. and Kreutzberg, G.W., 1998, Immune surveillance in the injured nervous system: T-lymphocytes invade the axotomized mouse facial motor nucleus and aggregate around sites of neuronal degeneration. *J. Neurosci.* **18**: 5804.
Rappert, A., Bechmann, I., Pivneva, T., Mahlo, J., Biber, K., Nolte, C., Kovac, A.D., Gerard, C., Boddeke, H.W., Nitsch, R. and Kettenmann, H., 2004, CXCR3-dependent microglial recruitment is essential for dendrite loss after brain lesion. *J. Neurosci.* **24**: 8500.
Rivest, S., 2003, Molecular insights on the cerebral innate immune system. *Brain Behav. Immun.* **17**: 13.
Rossi, D. and Zlotnik, A., 2000, The biology of chemokines and their receptors. *Annu. Rev. Immunol.* **18**: 217.
Rothwell, N.J., 1999, Annual review prize lecture cytokines - killers in the brain? *J. Physiol.* **514**: 3.
Rothwell, N.J. and Luheshi, G.N., 2000, Interleukin 1 in the brain: biology, pathology and therapeutic target. *Trends Neurosci.* **23**: 618.
Sairanen, T.R., Lindsberg, P.J., Brenner, M., Carpen, O. and Siren, A., 2001, Differential cellular expression of tumor necrosis factor-alpha and Type I tumor necrosis factor receptor after transient global forebrain ischemia. *J. Neurol. Sci.* **186**: 87.
Sanz, J.M. and Di Virgilio, F., 2000, Kinetics and mechanism of ATP-dependent IL-1 beta release from microglial cells. *J. Immunol.* **164**: 4893.
Sayyah, M., Beheshti, S., Shokrgozar, M.A., Eslami-far, A., Deljoo, Z., Khabiri, A.R. and Haeri Rohani, A., 2005, Antiepileptogenic and anticonvulsant activity of interleukin-1 beta in amygdala-kindled rats. *Exp. Neurol.* **191**: 145.
Schulz, M. and Engelhardt, B., 2005, The circumventricular organs participate in the immunopathogenesis of experimental autoimmune encephalomyelitis. *Cerebrospinal Fluid Res.* **2**: 8.
Schwartz, M., Moalem, G., Leibowitz-Amit, R. and Cohen, I.R., 1999, Innate and adaptive immune responses can be beneficial for CNS repair. *Trends Neurosci.* **22**: 295.
Schwartz, M., Butovsky, O., Bruck, W. and Hanisch, U.K., 2006, Microglial phenotype: is the commitment reversible? *Trends Neurosci.* **29**: 68.
Serou, M.J., DeCoster, M.A. and Bazan, N.G., 1999, Interleukin-1 beta activates expression of cyclooxygenase-2 and inducible nitric oxide synthase in primary hippocampal neuronal culture: platelet-activating factor as a preferential mediator of cyclooxygenase-2 expression. *J. Neurosci. Res.* **58**: 593.
Shandra, A.A., Godlevsky, L.S., Vastyanov, R.S., Oleinik, A.A., Konovalenko, V.L., Rapoport, E.N. and Korobka, N.N., 2002, The role of TNF-alpha in amygdala kindled rats. *Neurosci. Res.* **42**: 147.
Sheng, W.S., Hu, S., Ni, H.T., Rowen, T.N., Lokensgard, J.R. and Peterson, P.K., 2005, TNF-alpha-induced chemokine production and apoptosis in human neural precursor cells. *J. Leukoc. Biol.* **78**: 1233.
Shim, J.H., Xiao, C., Paschal, A.E., Bailey, S.T., Rao, P., Hayden, M.S., Lee, K.Y., Bussey, C., Steckel, M., Tanaka, N., Yamada, G., Akira, S., Matsumoto, K. and Ghosh, S., 2005, TAK1, but not TAB1 or TAB2, plays an essential role in multiple signaling pathways in vivo. *Genes Dev.* **19**: 2668.
Shinoda, S., Skradski, S.L., Araki, T., Schindler, C.K., Meller, R., Lan, J.Q., Taki, W., Simon, R.P. and Henshall, D.C., 2003, Formation of a tumour necrosis factor receptor 1 molecular scaffolding complex and activation of apoptosis signal-regulating kinase 1 during seizure-induced neuronal death. *Eur. J. Neurosci.* **17**: 2065.
Shrikant, P. and Benveniste, E.N., 1996, The central nervous system as an immunocompetent organ: role of glial cells in antigen presentation. *J. Immunol.* **157**: 1819.
Simard, A.R. and Rivest, S., 2005, Do pathogen exposure and innate immunity cause brain diseases? *Neurol. Res.* **27**: 717.
Sims, J.E., March, C.J., Cosman, D., Widmer, M.B., MacDonald, H.R., McMahan, C.J., Grubin, C.E., Wignall, J.M., Jackson, J.L., Call, S.M., et al., 1988, cDNA expression cloning of the IL-1 receptor, a member of the immunoglobulin superfamily. *Science* **241**: 585.
Sims, J.E., Gayle, M.A., Slack, J.L., Alderson, M.R., Bird, T.A., Giri, J.G., Colotta, F., Re, F., Mantovani, A., Shanebeck, K., et al., 1993, Interleukin 1 signaling occurs exclusively via the type I receptor. *Proc. Natl. Acad. Sci. USA* **90**: 6155.
Sofroniew, M.V., 2005, Reactive astrocytes in neural repair and protection. *Neuroscientist* **11**: 400.
Solle, M., Labasi, J., Perregaux, D.G., Stam, E., Petrushova, N., Koller, B.H., Griffiths, R.J and Gabel, C.A., 2001, Altered cytokine production in mice lacking P2X(7) receptors. *J. Biol. Chem.* **276**: 125.
Srinivasan, D., Yen, J.H., Joseph, D.J. and Friedman, W., 2004, Cell type-specific interleukin-1beta signaling in the CNS. *J. Neurosci.* **24**: 6482.

Sriram, K., Matheson, J.M., Benkovic, S.A., Miller, D.B., Luster, M.I. and O'Callaghan, J.P., 2002, Mice deficient in TNF receptors are protected against dopaminergic neurotoxicity: implications for Parkinson's disease. *FASEB J.* **16**: 1474.
Streit, W.J., 2002, Microglia as neuroprotective, immunocompetent cells of the CNS. *Glia* **40**: 133.
Streit, W.J., Walter, S.A. and Pennell, N.A., 1999, Reactive microgliosis. *Prog. Neurobiol.* **57**: 563.
Strijbos, P.J. and Rothwell, N.J., 1995, Interleukin-1 beta attenuates excitatory amino acid-induced neurodegeneration in vitro: involvement of nerve growth factor. *J. Neurosci.* **15**: 3468.
Suzuki, T., Hide, I., Ido, K., Kohsaka, S., Inoue, K. and Nakata, Y., 2004, Production and release of neuroprotective tumor necrosis factor by P2X7 receptor-activated microglia. *J. Neurosci.* **24**: 1.
Sylte, M.J., Kuckleburg, C.J., Inzana, T.J., Bertics, P.J. and Czuprynski, C.J., 2005, Stimulation of P2X receptors enhances lipooligosaccharide-mediated apoptosis of endothelial cells. *J. Leukoc. Biol.* **77**: 958.
Takada, H., Chen, N.J., Mirtsos, C., Suzuki, S., Suzuki, N., Wakeham, A., Mak, T.W. and Yeh, W.C., 2003, Role of SODD in regulation of tumor necrosis factor responses. *Mol. Cell Biol.* **23**: 4026.
Takaesu, G., Kishida, S., Hiyama, A., Yamaguchi, K., Shibuya, H., Irie, K., Ninomiya-Tsuji, J. and Matsumoto, K., 2000, TAB2, a novel adaptor protein, mediates activation of TAK1 MAPKKK by linking TAK1 to TRAF6 in the IL-1 signal transduction pathway. *Mol. Cell.* **5**: 649.
Takaesu, G., Ninomiya-Tsuji, J., Kishida, S., Li, X., Stark, G.R. and Matsumoto, K., 2001, Interleukin-1 (IL-1) receptor-associated kinase leads to activation of TAK1 by inducing TAB2 translocation in the IL-1 signaling pathway. *Mol. Cell Biol.* **21**: 2475.
Takao, T., Tracey, D.E., Mitchell, W.M. and De Souza, E.B., 1990, Interleukin-1 receptors in mouse brain: characterization and neuronal localization. *Endocrinology* **127**: 3070.
Takeuchi, H., Jin, S., Wang, J., Zhang, G., Kawanokuchi, J., Kuno, R., Sonobe, Y., Mizuno, T. and Suzumura, A., 2006, Tumor necrosis factor-alpha induces neurotoxicity via glutamate release from hemichannels of activated microglia in an autocrine manner. *J. Biol. Chem.* **281**: 21362.
Tamatani, M., Che, Y.H., Matsuzaki, H., Ogawa, S., Okado, H., Miyake, S., Mizuno, T. and Tohyama, M., 1999, Tumor necrosis factor induces Bcl-2 and Bcl-x expression through NFkappaB activation in primary hippocampal neurons. *J. Biol. Chem.* **274**: 8531.
Tartaglia, L.A., Ayres, T.M., Wong, G.H. and Goeddel, D.V., 1993, A novel domain within the 55 kd TNF receptor signals cell death. *Cell* **74**: 845.
Taylor, D.L., Jones, F., Kubota, E.S. and Pocock, J.M., 2005, Stimulation of microglial metabotropic glutamate receptor mGlu2 triggers tumor necrosis factor alpha-induced neurotoxicity in concert with microglial-derived Fas ligand. *J. Neurosci.* **25**: 2952.
Thakar, J., Schleinkofer, K., Borner, C. and Dandekar, T., 2006, RIP death domain structural interactions implicated in TNF-mediated proliferation and survival. *Proteins* **63**: 413.
Thanos, D. and Maniatis, T., 1995, NF-kappa B: a lesson in family values. *Cell* **80**: 529.
Thomas, W.E., 1999, Brain macrophages: on the role of pericytes and perivascular cells. *Brain Res. Brain Res. Rev.* **31**: 42.
Thommesen, L. and Laegreid, A., 2005, Distinct differences between TNF receptor 1- and TNF receptor 2-mediated activation of NFkappaB. *J. Biochem. Mol. Biol.* **38**: 281.
Thompson, C., Gary, D., Mattson, M., Mackenzie, A. and Robertson, G.S., 2004, Kainic acid-induced naip expression in the hippocampus is blocked in mice lacking TNF receptors. *Brain Res. Mol. Brain Res.* **123**: 126.
Thornberry, N.A., Bull, H.G., Calaycay, J.R., Chapman, K.T., Howard, A.D., Kostura, M.J., Miller, D.K., Molineaux, S.M., Weidner, J.R., Aunins, J., et al., 1992, A novel heterodimeric cysteine protease is required for interleukin-1 beta processing in monocytes. *Nature* **356**: 768.
Traenckner, E.B., Wilk, S. and Baeuerle, P.A., 1994, A proteasome inhibitor prevents activation of NF-kappa B and stabilizes a newly phosphorylated form of I kappa B-alpha that is still bound to NF-kappa B. *EMBO J.* **13**: 5433.
Traenckner, E.B., Pahl, H.L., Henkel, T., Schmidt, K.N., Wilk, S. and Baeuerle, P.A., 1995, Phosphorylation of human I kappa B-alpha on serines 32 and 36 controls I kappa B-alpha proteolysis and NF-kappa B activation in response to diverse stimuli. *EMBO J.* **14**: 2876.
Turrin, N.P. and Rivest, S., 2006, Tumor necrosis factor alpha but not interleukin 1 beta mediates neuroprotection in response to acute nitric oxide excitotoxicity. *J. Neurosci.* **26**: 143.
Ubogu, E.E., Cossoy, M.B. and Ransohoff, R.M., 2006, The expression and function of chemokines involved in CNS inflammation. *Trends Pharmacol. Sci.* **27**: 48.
Vassalli, P., 1992, The pathophysiology of tumor necrosis factors. *Annu. Rev. Immunol.* **10**: 411.
Verma, I.M., Stevenson, J.K., Schwarz, E.M., Van Antwerp, D. and Miyamoto, S., 1995, Rel/NF-kappa B/I kappa B family: intimate tales of association and dissociation. *Genes Dev.* **9**: 2723.

Vezzani, A., Conti, M., De Luigi, A., Ravizza, T., Moneta, D., Marchesi, F. and De Simoni, M.G., 1999, Interleukin-1beta immunoreactivity and microglia are enhanced in the rat hippocampus by focal kainate application: functional evidence for enhancement of electrographic seizures. *J. Neurosci.* **19**: 5054.

Vezzani, A., Moneta, D., Conti, M., Richichi, C., Ravizza, T., De Luigi, A., De Simoni, M.G., Sperk, G., Andell-Jonsson, S., Lundkvist, J., Iverfeldt, K. and Bartfai, T., 2000, Powerful anticonvulsant action of IL-1 receptor antagonist on intracerebral injection and astrocytic overexpression in mice. *Proc. Natl. Acad. Sci. USA* **97**: 11534.

Vilček, J., 2003, The Cytokine Handbook. Elsevier Science, London, pp. 3–18.

Vilz, T.O., Moepps, B., Engele, J., Molly, S., Littman, D.R. and Schilling, K., 2005, The SDF-1/CXCR4 pathway and the development of the cerebellar system. *Eur. J. Neurosci.* **22**: 1831.

Viviani, B., Bartesaghi, S., Gardoni, F., Vezzani, A., Behrens, M.M., Bartfai, T., Binaglia, M., Corsini, E., Di Luca, M., Galli, C.L. and Marinovich, M., 2003, Interleukin-1beta enhances NMDA receptor-mediated intracellular calcium increase through activation of the Src family of kinases. *J. Neurosci.* **23**: 8692.

Wajant, H., Pfizenmaier, K. and Scheurich, P., 2003, Tumor necrosis factor signaling. *Cell Death Differ.* **10**: 45.

Wallach, D., Engelmann, H., Nophar, Y., Aderka, D., Kemper, O., Hornik, V., Holtmann, H. and Brakebusch, C., 1991, Soluble and cell surface receptors for tumor necrosis factor. *Agents Actions Suppl.* **35**: 51.

Wang, C.X. and Shuaib, A., 2002, Involvement of inflammatory cytokines in central nervous injury. *Prog. Neurobiol.* **67**: 161.

Wang, X., Barone, F.C., Aiyar, N.V. and Feuerstein, G.Z., 1997, Interleukin-1 receptor and receptor antagonist gene expression after focal stroke in rats. *Stroke* **28**: 155.

Wang, C.Y., Mayo, M.W., Korneluk, R.G., Goeddel, D.V. and Baldwin, A.S. Jr., 1998, NF-kappaB antiapoptosis: induction of TRAF1 and TRAF2 and c-IAP1 and c-IAP2 to suppress caspase-8 activation. *Science* **281**: 1680.

Wang, X., Chen, S., Ma, G., Ye, M. and Lu, G., 2005a, Involvement of proinflammatory factors, apoptosis, caspase-3 activation and Ca2+ disturbance in microglia activation-mediated dopaminergic cell degeneration. *Mech. Ageing Dev.* **126**: 1241.

Wang, X.J., Kong, K.M., Qi, W.L., Ye, W.L. and Song, P.S., 2005b, Interleukin-1 beta induction of neuron apoptosis depends on p38 mitogen-activated protein kinase activity after spinal cord injury. *Acta Pharmacol. Sin.* **26**: 934.

Wei, R. and Jonakait, G.M., 1999, Neurotrophins and the anti-inflammatory agents interleukin-4 (IL-4), IL-10, IL-11 and transforming growth factor-beta1 (TGF-beta1) down-regulate T cell costimulatory molecules B7 and CD40 on cultured rat microglia. *J. Neuroimmunol.* **95**: 8.

Wesche, H., Korherr, C., Kracht, M., Falk, W., Resch, K. and Martin, M.U., 1997, The interleukin-1 receptor accessory protein (IL-1RAcP) is essential for IL-1-induced activation of interleukin-1 receptor-associated kinase (IRAK) and stress-activated protein kinases (SAP kinases). *J. Biol. Chem.* **272**: 7727.

Wiegmann, K., Schutze, S., Machleidt, T., Witte, D. and Kronke, M., 1994, Functional dichotomy of neutral and acidic sphingomyelinases in tumor necrosis factor signaling. *Cell* **78**: 1005.

Wilde, G.J., Pringle, A.K., Sundstrom, L.E., Mann, D.A. and Iannoyyi, F., 2000, Attenuation and augmentation of ischaemia-related neuronal damage by tumor necrosis factor-alpha in vitro. *Eur. J. Neurosci.* **12**: 3863.

Williams, K., Alvarez, X. and Lackner, A.A., 2001, Central nervous system perivascular cells are immunoregulatory cells that connect the CNS with the peripheral immune system. *Glia* **36**: 156.

Windhagen, A., Newcombe, J., Dangond, F., Strand, C., Woodroofe, M.N., Cuzner, M.L. and Hafler, D.A., 1995, Expression of costimulatory molecules B7-1 (CD80), B7-2 (CD86), and interleukin 12 cytokine in multiple sclerosis lesions. *J. Exp. Med.* **182**: 1985.

Wong, D., Prameya, R. and Dorovini-Zis, K., 1999, In vitro adhesion and migration of T lymphocytes across monolayers of human brain microvessel endothelial cells: regulation by ICAM-1, VCAM-1, E-selectin and PECAM-1. *J. Neuropathol. Exp. Neurol.* **58**: 138.

Wright, S.D., 1999, Toll, a new piece in the puzzle of innate immunity. *J. Exp. Med.* **189**: 605.

Xia, Y., Yamagata, K. and Krukoff, T.L., 2006, Differential expression of the CD14/TLR4 complex and inflammatory signaling molecules following i.c.v. administration of LPS. *Brain Res.* [Epub ahead of print].

Yabuuchi, K., Minami, M., Katsumata, S. and Satoh, M., 1994, Localization of type I interleukin-1 receptor mRNA in the rat brain. *Brain Res. Mol. Brain Res.* **27**: 2736.

Yamin, T.T. and Miller, D.K., 1997, The interleukin-1 receptor-associated kinase is degraded by proteasomes following its phosphorylation. *J. Biol. Chem.* **272**: 21540.

Yang, L., Lindholm, K., Konishi, Y., Li, R. and Shen, Y., 2002, Target depletion of distinct tumor necrosis factor receptor subtypes reveals hippocampal neuron death and survival through different signal transduction pathways. *J. Neurosci.* **22**: 3025.

Ye, Z.C. and Sontheimer, H., 1996, Cytokine modulation of glial glutamate uptake: a possible involvement of nitric oxide. *Neuroreport* **7**: 2181.

Yeager, M.P., DeLeo, J.A., Hoopes, P.J., Hartov, A., Hildebrandt, L. and Hickey, W.F., 2000, Trauma and inflammation modulate lymphocyte localization in vivo: quantitation of tissue entry and retention using indium-111-labeled lymphocytes. *Crit. Care Med.* **28**: 1477.

Yeh, W.C., Shahinian, A., Speiser, D., Kraunus, J., Billia, F., Wakeham, A., de la Pompa, J.L., Ferrick, D., Hum, B., Iscove, N., Ohashi, P., Rothe, M., Goeddel, D.V. and Mak, T.W., 1997, Early lethality, functional NF-kappaB activation, and increased sensitivity to TNF-induced cell death in TRAF2-deficient mice. *Immunity* **7**: 715.

Yu, Z., Cheng, G., Wen, X., Wu, G.D., Lee, W.T. and Pleasure, D., 2002, Tumor necrosis factor alpha increases neuronal vulnerability to excitotoxic necrosis by inducing expression of the AMPA-glutamate receptor subunit GluR1 via an acid sphingomyelinase- and NF-kappaB-dependent mechanism. *Neurobiol. Dis.* **11**: 199.

Yuhas, Y., Weizman, A. and Ashkenazi, S., 2003, Bidirectional concentration-dependent effects of tumor necrosis factor alpha in Shigella dysenteriae-related seizures. *Infect. Immun.* **71**: 2288.

Yuhas, Y., Shulman, L., Weizman, A., Kaminsky, E., Vanichkin, A. and Ashkenazi, S., 1999, Involvement of tumor necrosis factor alpha and interleukin-1beta in enhancement of pentylenetetrazole-induced seizures caused by Shigella dysenteriae. *Infect. Immun.* **67**: 1455.

Zandi, E., Rothwarf, D.M., Delhase, M., Hayakawa, M. and Karin, M., 1997, The IkappaB kinase complex (IKK) contains two kinase subunits, IKKalpha and IKKbeta, necessary for IkappaB phosphorylation and NF-kappaB activation. *Cell* **91**: 243.

Zhang, G. and Ghosh, S., 2002, Negative regulation of toll-like receptor-mediated signaling by Tollip. *J. Biol. Chem.* **277**: 7059.

Zhang, J. and Rivest, S., 2003, Is survival possible without arachidonate metabolites in the brain during systemic infection? *News Physiol. Sci.* **18**: 137.

Zheng, L., Bidere, N., Staudt, D., Cubre, A., Orenstein, J., Chan, F.K. and Lenardo, M., 2006, Competitive control of independent programs of tumor necrosis factor receptor-induced cell death by TRADD and RIP1. *Mol. Cell Biol.* **26**: 3505.

Zhou, C., Tai, C., Ye, H.H., Ren, X., Chen, J.G., Wang, S.Q. and Chai, Z., 2006, Interleukin-1beta downregulates the L-type Ca2+ channel activity by depressing the expression of channel protein in cortical neurons. *J. Cell Physiol.* **206**: 799.

Zimmer, J., Kristensen, B.W., Jakobsen, B. and Noraberg, J., 2000, Excitatory amino acid neurotoxicity and modulation of glutamate receptor expression in organotypic brain slice cultures. *Amino Acids* **19**: 7.

Zlotnik, A. and Yoshie, O., 2000, Chemokines: a new classification system and their role in immunity. *Immunity* **12**: 121.

Zou, J.Y. and Crews, F.T., 2005, TNF alpha potentiates glutamate neurotoxicity by inhibiting glutamate uptake in organotypic brain slice cultures: neuroprotection by NF kappa B inhibition. *Brain Res.* **1034**: 11.

2

MICROGLIAL CELL POPULATION EXPANSION FOLLOWING ACUTE NEURAL INJURY

M. Wirenfeldt[1], L. Dissing-Olesen[1], A.A. Babcock[1,2],
R. Ladeby[1], M.B. Jensen[1,2], T. Owens[1,2], B. Finsen[1]

1. ABSTRACT

By studying the response of hippocampal microglia to anterograde axonal and terminal degeneration in the dentate gyrus we have identified several subsets of microglia, including immigrant bone marrow (BM)-derived and resident microglia, and resident microglia expressing different levels of CD34, of which the majority were induced to undergo proliferation upon injury. Based on the population kinetics elucidated until now, we have evidence that a subpopulation of microglial cells may undergo repeated proliferation upon injury while some microglial cells may not proliferate at all within the investigated period. Furthermore, resident cells in the mouse may be supplemented by BM-derived cells. A picture emerges where new cells are being born, while other cells, including the newly generated and recruited cells are beginning to be cleared from the site of lesion, at least partly by apoptotic mechanisms. Increased insight into basic microglial cell biology may improve diagnostic and therapeutic possibilities for patients suffering from conditions that are currently without real treatment options.

2. INTRODUCTION

Microglial cells are part of the innate immune system and serve immune surveillance functions in the central nervous system (CNS). Due to their close relationship with other cells of the immune system, mainly macrophages, their role in phagocytosis, antigen presentation, and production of pro- and anti-inflammatory cytokines has been extensively studied (Streit et al., 1999). It has been established that microglia do play major roles in neuroinflammatory and infectious diseases like multiple sclerosis and HIV encephalitis (He et al., 1997; Noseworthy et al., 2000). Recently an important regulatory role for microglia in maintaining homeostasis in the normal CNS has become evident through cases of pre-senile dementia that resulted from specific microglial dysfunctions (Bianchin et al., 2004).

Bente Finse - bfinsen@health.sdu.dk
[1]Medical Biotechnology Center, University of Southern Denmark, Odense C, Denmark
[2]Montreal Neurological Institute, McGill University, Montreal, Canada

In line with their role as immune surveillance cells, microglia are highly sensitive to disturbances in the neural parenchyma and react quickly with morphological and biochemical changes (Finsen et al., 1993a). Microglia display a characteristic graded functional plasticity that depends on and is adapted to the nature and severity of the dysfunction or pathology. This ensures the correct and most appropriate response by microglia in any given situation (Raivich et al., 1999; Streit et al., 1999). If dysregulated, microglial response may facilitate or even cause pathology (Streit, 2004; Takahashi et al., 2005).

Activation of microglia comprises several facets, including changes in morphology, altered surface antigen expression, enhanced production of cytokines, chemokines, and growth factors, and cell population expansion by proliferation and migration, resulting in reactive microgliosis (Raivich et al., 1999; Streit et al., 1999). In addition to proliferation there is recruitment of blood borne cells into the site of injury as shown in BM chimeric mice (Priller et al., 2001; Vallieres and Sawchenko, 2003; Wirenfeldt et al., 2005a). This observation has raised the possibility of using gene manipulated BM-derived cells, hematopoietic stem cells or other types of microglial cell progenitors for site-specific delivery of gene therapy to the injured or diseased CNS (Biffi et al., 2004). Ideally this type of therapy should be based on knowledge about the profile of appearance, and the distribution, differentiation, turnover and the gene expression profile of the recruited progenitor cells.

3. MICROGLIAL MORPHOLOGICAL PLASTICITY

In the normal adult CNS, resting microglial cells can be recognized by their small spherical to slightly elongated nuclei that measure approximately 8–10 µm in diameter and that are surrounded by a tiny rim of cytoplasm with a few thin, branched processes extending into the neural parenchyma (del Rio-Hortega, 1932; Mori and Leblond, 1969). Microglia are territorial cells, with each cell and its processes covering approximately 30–50 µm of neural parenchyma (Raivich, 2005). Imaging of microglia in vivo has shown that microglia dynamically survey the neural parenchyma by continuously extending and retracting their processes (Davalos et al., 2005; Nimmerjahn et al., 2005). Microglia have been estimated to constitute 12% of brain cells (Lawson et al., 1992; Streit et al., 1999). With the close to 1:1 ratio between neocortical neurons and glial cells (Pakkenberg and Gundersen, 1988), it can be calculated that 1 microglial cell in average serves homeostatic functions for 5 neocortical neurons.

A general manifestation of activated microglia, irrespective of the type of lesion, is their transformation into hypertrophied cells with condensed, slightly retracted and hyper-ramified processes. Imaging of microglia in vivo showed that significant changes in microglial morphology occur minutes following stimulation (Davalos et al., 2005; Nimmerjahn et al., 2005), confirming observations from histological studies (Morioka et al., 1991). Activated microglia are evident both in injuries with lethal and sub-lethal neuronal damage. Features displayed by activated microglia depend on the severity of the lesion. Phagocytic activity is only acquired when neuronal death is present and there is a need for phagocytosis of dead or dying neurons and cellular debris. In this situation microglia can turn into brain macrophages (Pennell and Streit, 1998; Streit et al., 1999). It is often assumed that activated microglia can return to a resting quiescent state when pathological stimuli have ceased and neurons survive and recover. This, however, is not thought to be the case for microglia that have transformed into brain macrophages. These cells probably undergo cell death by apoptosis (Jones et al., 1997).

Activated microglia displays distinct morphological features depending on the type of stimulus or lesion. Microglial rod cells are a characteristic form of reactive microglia arising by fusion of several microglias along the degenerating pyramidal cell dendrites (Nissl, 1899; Morioka et al., 1991; Jorgensen et al., 1993; Graeber and Mehraein, 1994). Rod cells are seen in the senile cortex and hippocampus, in regio superior hippocampus in transient global cerebral ischaemia, HIV-encephalitis and subacute sclerosing panencephalitis (Morioka et al., 1991; Jorgensen et al., 1993; Graeber and Mehraein, 1994; Streit et al., 1999; Streit, 2000). Reactive microglia also surround axotomized motor neurons, e.g. after the facial nerve transection where reactive perineuronal microglia congregate around the injured axotomized neurons (Streit et al., 1999). A related phenomenon takes place in the dentate gyrus in transient global cerebral ischaemia, where activated microglia congregate around the early degenerating hilar neurons (Jorgensen et al., 1993).

Activated microglia display a hyperramified "bushy" morphology in areas with anterograde axonal and dense terminal degeneration (Matthews et al., 1976), as it occurs in the outer two thirds of the molecular layer of the dentate gyrus following ablation of the entorhino-dentate perforant pathway (Jensen et al., 1994, 1999) (Fig. 1). Bushy cell transformation can also be observed in regio superior hippocampus following kainic acid-induced Schafferotomy, leading to a dense terminal degeneration in the *stratum radiatum* of regio superior (Finsen et al., 1993a; Jorgensen et al., 1993). In contrast, microglia transform into more ameboid-like cells in areas with classical anterograde axonal (Wallerian) degeneration (Stoll et al., 1989; Finsen et al., 1993b). Taken together, this suggests that the bushy microglia are shaped by the dense terminal degeneration, and not the anterograde axonal degeneration.

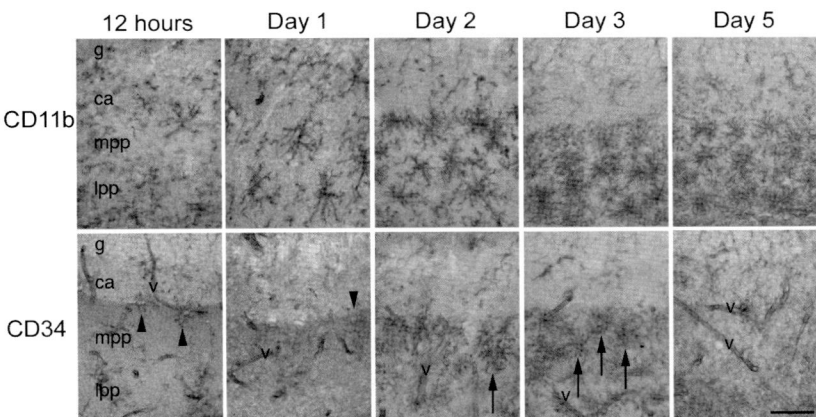

Fig. 1. Microglial CD11b and CD34 expression in the perforant path-denervated dentate gyrus. *Upper panel*, Photomontage illustrating the kinetics of cellular CD11b expression through staining of sections containing the molecular layer from mice with survival times ranging from 12 h to 5 days after perforant path axonal lesion. CD11b$^+$ microglia have attained the typical reactive "bushy" appearance by day 3, at which time there is a markedly increased density of microglia in the denervated mpp and lpp zones in conjunction with a transient clearance of microglia from the inner ca zone. *Lower panel*, Faintly labelled CD34$^+$ cells (arrowheads) are seen from 12 h up till day 2–3 when more strongly labelled cells with microglial-like morphology (arrows) appear in the mpp and lpp. The number of CD34$^+$ cells peaks at day 3 and is reduced towards day 5. Note the stronger staining of the neuropil in the mpp and lpp zones compared to the ca zone. *ca*, commissural associational zone; *g*, granule cell layer; *lpp*, lateral perforant path zone; *mpp*, medial perforant path zone. *v*, vessel. Scale bar 50 μm. Modified from Ladeby et al. (2005a)

4. MICROGLIAL ACTIVATION IS CHARACTERIZED BY CHANGES IN SURFACE ANTIGEN EXPRESSION

Activated microglia alters their surface antigen phenotype. As part of their activation, constitutively expressed antigens are upregulated and a range of novel antigens are induced. CD11b (integrin αM) is a constitutively expressed surface protein on microglial cells. Increased CD11b expression is a hallmark of reactive microgliosis. Increased CD11b expression and morphological changes are among the earliest signs of microglial response, observed within the first 24 h of activation and reaching maximum expression 5–7 days after lesion (Graeber et al., 1988a; Finsen et al., 1993a, b; Jensen et al., 1999). Markers such as the leukocyte common antigen CD45 and the (murine) macrophage marker F4/80 are constitutive antigens on resting microglia and upregulated on activated microglia (Perry et al., 1985; Sedgwick et al., 1991). Microglia express a lower level of CD45 ($CD45^{dim}$) compared to macrophages ($CD45^{high}$), making it possible to separate microglia from macrophages by flow cytometric analysis (Sedgwick et al., 1991; Babcock et al., 2003; Wirenfeldt et al., 2005a). General markers for microglia include certain plant lectins including GSA (I-B4) from griffonia simplicifolia, which reacts with terminal non-reducing glucosamine end groups (Streit et al., 1990), tomato lectin from the plant *Lycopersicon esculentum*, which reacts with P-glycoprotein (Acarin et al., 1994), and ectoenzymes such as nucleoside-diphosphatase (Novikoff and Goldfischer, 1961; Castellano et al., 1991).

In addition to upregulating constitutively expressed markers, activated microglia are induced to express novel antigens, such as MHC class I, and depending on the type of lesion or stimuli, MHC class II. Other novel antigens include CD4 (Jensen et al., 1997), co-stimulatory molecules B7.1 and B7.2 (O'Keefe et al., 2002), and the adhesion molecules ICAM-1, VLA-4 and LFA-1 (Finsen et al., 1993b; Hailer et al., 1997). Characteristically, these antigens reach their maximal level of expression 5–7 days after injury coinciding with peak expression of the constitutively expressed markers. MHC class I expression is a general phenomenon in activated microglia while MHC class II is expressed on a limited number of cells primarily situated in CNS white matter (Rao and Lund, 1989; Stoll et al., 1989; Streit et al., 1999; Finsen et al., 1993a, b). The expression of MHC class II, B7.1, and B7.2 enables microglia to act as antigen presenting cells, although with less efficiency than dendritic cells in other organ systems (Carson et al., 1998, 1999) and perivascular cells in the CNS (Hickey and Kimura, 1988; Greter et al., 2005).

The sialomucin CD34 is an important exception to the graded expression of constitutive markers and novel antigens by activated microglial cells. CD34, originally defined by its expression on early, immature precursor cells of myeloid lineage (Civin et al., 1984; Andrews et al., 1986; Berenson et al., 1988), was recently demonstrated to be transiently expressed at high levels on activated microglial cells in the dentate gyrus 3 days after perforant path axonal lesion (Fig. 1) (Ladeby et al., 2005a). Recently, double immunofluorescence analyses have shown that CD34 is expressed at low levels on most resting microglial cells (Fig. 2A-H), and transiently upregulated by activated microglial cells in the dentate gyrus 3 days after axonal lesion (Fig. 2I–P), at which time the majority of CD34-expressing cells are proliferating (Ladeby et al., 2005a). Furthermore, unlike other activation markers that we have investigated in this lesion paradigm, microglial expression of CD34 declined too close to baseline expression at day 5, when microglial CD11b expression and the expression of other activation markers reached their maximum (Fig. 1) (Ladeby et al., 2005a, b; Wirenfeldt et al., 2005a). As illustrated in Fig. 2, the CD34 molecule is not only associated with hematopoietic cells and microglial cells, it is expressed in varying levels throughout the neuropil, and at high levels on endothelial cells (Brown et al., 1991; Young et al., 1995; Ladeby et al., 2005a). At present the function of microglial CD34 expression is not known.

5. CYTOKINE SYNTHESIS BY ACTIVATED MICROGLIA

Activation of microglia triggers the production of a variety of extracellular signaling molecules. Pro-inflammatory cytokines like IL-1β and TNF are among the molecules produced early in a reactive microglial response (Bartholdi and Schwab, 1997; Streit et al., 1998; Lambertsen et al., 2002). The anti-inflammatory and potentially neurotrophic transforming growth factor-β1 (TGF-β1) is also produced by activated microglia (Kiefer et al., 1993; Lehrmann et al., 1998). Unlike IL-1β and TNF, which are produced early and transiently within hours to 1–2 days following microglial activation, TGF-β1 remains elevated for extended periods, for the entire duration of the microglial reaction (Kiefer et al., 1993; Lehrmann et al., 1998). The same range of cytokines has been shown to be induced at the mRNA or protein level in the perforant path denervated dentate gyrus (Fagan and Gage, 1990; Morgan et al., 1993; Jensen et al., 2000). However, while studies of ischemic and traumatic brain injury suggest that IL-1β has a neurotoxic role (Allan and Rothwell, 2001), whereas TNF has neurotoxic as well as neuroprotective functions (Hallenbeck, 2002), the role of individual cytokines has not yet been defined in the perforant path axonal lesion paradigm.

6. MICROGLIAL CELL POPULATION EXPANSION PREDOMINANTLY OCCURS BY PROLIFERATION OF RESIDENT MICROGLIA

A central element of microglial activation and a prerequisite for reactive microgliosis is microglial cell population expansion. Microglial mitosis is a later event compared to the morphological changes (Cammermeyer, 1965; Kreutzberg, 1966; Graeber et al., 1988b). Proliferation is significant 2–3 days after lesion with the number of microglial cells reaching a maximum after 4–7 days (Graeber et al., 1988b; Streit and Kreutzberg, 1988; Streit et al., 1999). Numerically, a 6.7-fold increase in the number of microglial cells has been reported in regio superior of hippocampus 6–7 days after an episode of transient global cerebral ischaemia in the rat (Kato et al., 2003). This fold increase is comparable to the sixfold increase reported for the perforant path denervated dentate gyrus at day 3 in the rat (Hailer et al., 1999). Similar analysis in the mouse has revealed a somewhat smaller threefold increase at day 3 increasing to a 3.5-fold increase at day 5 after lesion (Ladeby et al., 2005a). As in the rat study (Hailer et al., 1999), the regions counted in Ladeby et al. (2005a) included the non-denervated inner commissural associational zone, indicating that there may be a smaller lesion-induced expansion of the microglial cell population in mouse compared to rat. Importantly, however, data on mice (Ladeby et al., 2005a) supported demonstrations in the rat dentate gyrus of peak proliferative activity at 3 days after lesion (Hailer et al., 1999). At this time we found that up towards 80% of CD34$^+$ microglia had been generated by proliferation. The expansion of the microglial population is not only instituted by the mitotic activity of microglia but also by migration of microglia from the adjoining non-denervated inner commissural associational zone (Rappert et al., 2004). The expansion of the microglial cell population following acute injury is transient. Microglial numbers decline from day

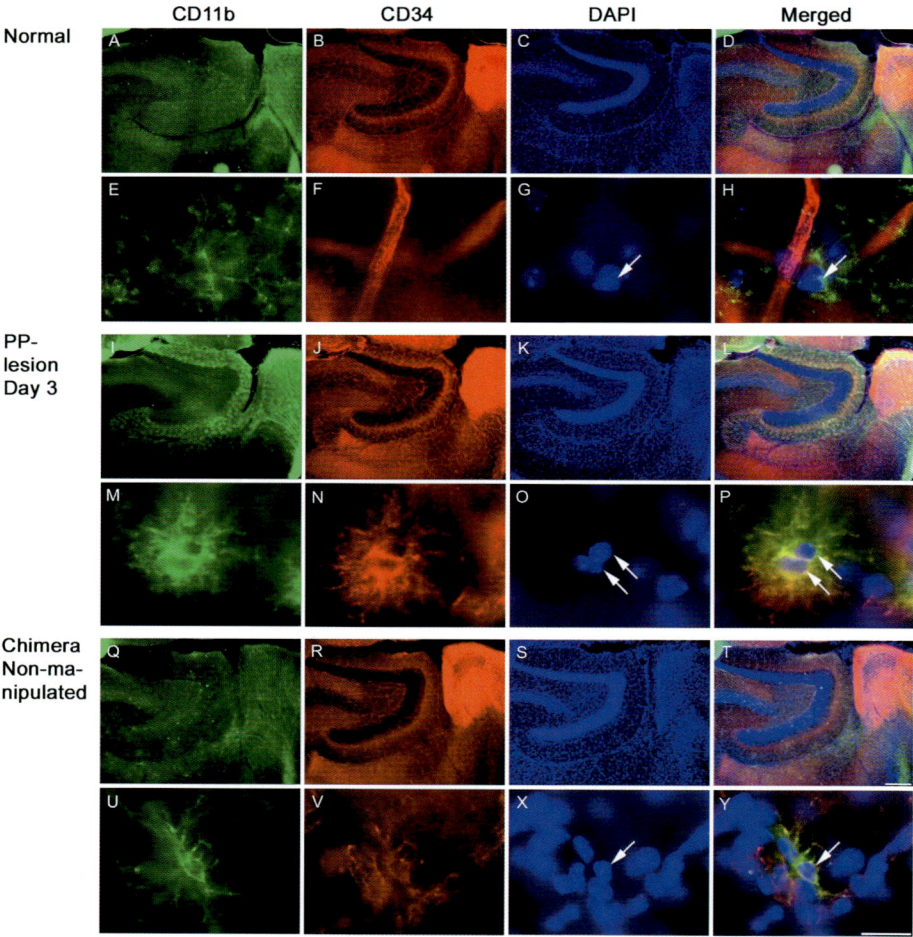

Fig. 2. CD11b$^+$ (green) microglia co-expressing CD34 (red) in three experimental settings. *First column,* CD11b$^+$ microglia visualized using a fluorescein isothiocyanate (FITC)-conjugated primary rat antibody against CD11b and a secondary anti-FITC goat antibody labelled with a green fluorochrome. *Second column,* CD34$^+$ cells visualized using a primary rat antibody against CD34 and a secondary goat anti-rat antibody labelled with a red fluorochrome, this staining was carried out before the staining for CD11b. *Third column,* nuclear stain using 4',6-diamidino-2-phenylindole, DAPI, which is blue. *Fourth column* shows the other three columns merged. (A–H): Dentate gyrus from a normal mouse. At high magnification CD11b (E) and very dim to no CD34 (F) is expressed on resting microglial cells (G, H see arrows). Note the neuropil-staining in (B) and the vascular CD34 staining in (B, F). (I–P): Lesion-reactive dentate gyrus day 3 after perforant path denervation. Both CD11b (M) and CD34 (N) are expressed in high levels here seen on microglial nuclear double profile (O, P see arrows). (Q–Y): Dentate gyrus from a non-manipulated irradiation-BM chimeric mouse. Note the presence of scattered, lightly activated CD11b$^+$ (U) and CD34$^+$ (V) microglial cells in this animal (X, Y see arrows). Scale bars 200 μm (T), 20 μm (Y)

5 to day 10, and have reverted to baseline at 30 days postlesion (Hailer et al., 1999). Similar time profiles have been observed following global cerebral ischaemia in the rat (Morioka et al., 1991; Jorgensen et al., 1993), and facial nerve transection (Graeber et al., 1988b; Streit and Kreutzberg, 1988; Streit et al., 1999).

So far, the elimination of surplus microglia after peak microglial response has been less thoroughly investigated. TUNEL-labelled microglia, however, appear at 4 days and peak at 7 days after facial nerve transection (Jones et al., 1997). In this way reactive microgliosis involves a mitotic factor expanding the microglial population and a presumably apoptotic factor reducing the number of microglia.

7. BONE MARROW DERIVED CELLS CONTRIBUTE TO MICROGLIAL CELL POPULATION EXPANSION IN THE MOUSE

Although it has long been assumed that recruited blood-borne cells supplement proliferating resident microglia in the continuous turnover of microglia in the normal adult CNS (Lawson et al., 1992), it was only recently shown that the microglial cell population is slowly supplemented by exogenous haematogenous microglial precursor cells in normal CNS and at an enhanced rate in injured CNS (Priller et al., 2001; Vallieres and Sawchenko, 2003; Ladeby et al., 2005a; Wirenfeldt et al., 2005a). We have used green-fluorescent protein (GFP) BM chimeric mice to investigate the recruitment of exogenous BM-derived cells into injured CNS. As expected, we observed enhanced lesion-specific recruitment of GFP-positive cells into areas of anterograde axonal and terminal degeneration (Fig. 3A). Using flow cytometry we found that approximately 1% of hippocampal microglial cells were donor BM-derived at postlesion day 3 and 5, in mice with short, post-transplantation reconstitution time (4–7 weeks, Fig. 3B) (Wirenfeldt et al., 2005a). This percentage had increased to 13% at day 7 in mice with longer post-transplantation reconstitution time (7–12 weeks, Wirenfeldt et al., 2005a). While this increase might be due to post-lesion infiltration of cells of BM origin, it might also reflect proliferation of BM-derived cells that had taken up residence in the hippocampus prior to lesion. Since the flow cytometry data were obtained from isolated whole hippocampi which includes non-denervated hippocampal tissue, they are not directly comparable to the previously discussed 3–3.5-fold increase in microglial numbers (Ladeby et al., 2005a).

A frequently raised criticism against experiments performed on BM chimeric mice is that cell recruitment into the CNS may be altered in whole body irradiated animals. Indeed, microscopic analyses of CD11b immunostained brain sections from non-manipulated irradiation-BM chimeric mice show signs of above-baseline microglial activity, visualized both in CD11b and CD34 immunostained sections (Fig. 2Q–Y). Since microglial cells renew themselves by proliferation in normal CNS, irradiation might kill actively proliferating cells, which theoretically might facilitate cellular recruitment. It has also been reported that irradiated endothelial cells enhances hematopoietic transmigration and proliferation after radiation exposure (Gaugler et al., 2001), and that irradiation not only kills but also mobilizes BM cells (Furuya et al., 2003). In line with these observations it has been shown that the percentage of BM-derived cells in irradiated CNS correlated with the amount of radiation given to the skull (Furuya et al., 2003). Importantly, the demonstration of recruitment and microglial transformation of BM-derived cells has recently been supported by studies showing transformation of 6-carboxylfluorescein diacetate labelled spleen cells into ameboid

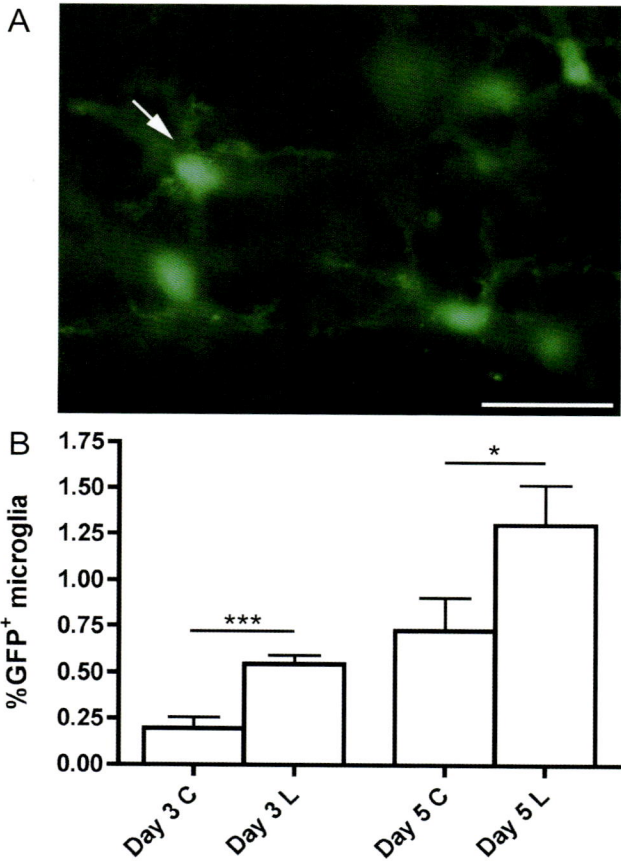

Fig. 3. Recruitment of blood borne GFP^+ cells into perforant path denervated hippocampus in BM chimeric mice. (A): Ramified GFP^+ microglial-like cells in denervated areas of the dentate gyrus. Modified from Wirenfeldt et al. (2005b). (B): Identification of BM-derived immigrant microglia in injured and control hippocampi by flow cytometry. The diagram shows the proportions of GFP^+ microglia in contralateral (C) and lesion-reactive (L) hippocampi of chimeric mice with 4–7 weeks reconstitution after BM transplantation. Immigrant microglial proportions are larger both 3 and 5 days after axonal lesion in chimeras with 4–7 weeks reconstitution. Day 3 and Day 5–3 and 5 days of post PP-lesion survival, respectively. Data are expressed as mean ± SEM. Modified from Wirenfeldt et al. (2005a)

and ramified microglial cells in perforant path denervated dentate gyrus in non-irradiated mice (Bechmann et al., 2005). In contrast to the mouse, a similar lesion-induced recruitment of precursors of parenchymal microglia has so far not been reported to take place in the rat. Indeed, studies of recruitment of BM-derived cells in the mouse were preceded by studies of BM chimeric rats showing replacement of perivascular cells, and meningeal and choroid plexus macrophages with donor BM-derived cells, but a very limited if any transformation of these cells into microglia (Matsumoto and Fujiwara, 1987; Lassmann and Hickey, 1993; Ford et al., 1995; Popovich and Hickey, 2001; Albini et al., 2005). Similarly, there is limited information as to whether and how microglial turnover and recruitment may take place in the normal, injured or diseased human CNS (Cogle et al., 2004). On the other hand, a natural turnover of microglia in the human may play a role in the pathophysiology of infectious diseases like HIV infection. Infected monocytes have been suggested to act like Trojan horses, delivering the virus into the perivascular spaces along CNS vessels, from where virus has easy access to infect microglia in the neural parenchyma (Garden, 2002).

8. SIGNALS THAT INDUCE MICROGLIAL ACTIVATION

The fact that microglial activation is a quick and almost universal event ensuing dysfunction or pathology in the CNS implies a very effective and close communication between neurons and microglia. It is emerging that neuron-microglial signaling comprises direct cell to cell signaling (Hoek et al., 2000) as well as a paracrine humoral component (Davalos et al., 2005). Free extracellular potassium ions (K^+) in high concentration seem to be able to activate microglia, mainly based on the finding that cortical spreading depression can induce microglial activation throughout the cortex within 24 h. There have also been studies supporting the idea of adenosine tri-phosphate (ATP) as an activator of microglial cells acting through the microglial purinergic receptors, e.g. the P2X ligand-gated ion channels and the G-protein coupled P2Y receptors (Walz et al., 1993; Ferrari et al., 1996; Honda et al., 2001). Studies by Davalos et al. (2005) have suggested that ATP released through connexin channels in astrocytes may be responsible for mediating the fast microglial response to neural damage.

Astrocytes could also act as messengers of neuronal dysfunction through their production of the microglial mitogen macrophage colony stimulating factor (Raivich et al., 1994, 1998; Rogove et al., 2002). Other factors that have been shown to promote microglial proliferation and survival are IL-6 (Klein et al., 1997), IL-1 and TNF (Kloss et al., 1997). In comparison, TGF-β1-3 are strong inhibitors of microglial proliferation (Jones et al., 1998). This indirect pathway of activation is also a possible reason for the delay often seen in microglial mitosis which does not commence until 2–3 days after other signs of microglial reactivity such as cellular hypertrophy, cytokine synthesis or CD11b up-regulation.

Another interesting means of microglial activation is through the CD200–CD200 receptor complex (Hoek et al., 2000). The CD200–CD200R interaction is a regulator of myeloid cell activation (Jenmalm et al., 2006). Within the CNS neurons express CD200 which is a glycoprotein containing two Ig superfamily domains. CD200 interacts with CD200R found on cells of myeloid lineage and in the CNS on microglia. Mice lacking CD200 have been found to have an augmented microglial response with microglia spontaneously showing features of activation as well as an accelerated response to facial nerve transection (Hoek et al., 2000). Microglia are kept in a quiescent state by this direct cell–cell interaction with neurons showing yet another mode of regulation of microglial activation. The diversity of regulatory mechanisms governing microglial activation provides at least in part an explanation for the instant reactivity microglia display in various scenes of injury and pathology in the CNS and is indicative of the importance of reactive microgliosis in these situations.

9. MECHANISMS OF CELLULAR RECRUITMENT INTO AND WITHIN THE DENERVATED HIPPOCAMPUS

While it may be unsurprising that BM-derived cells contribute to the lesion-induced microglial–macrophage reaction following stab wounding and stroke, both of which damage the vasculature, leading to extravasation of blood borne cells and plasma components into the neural parenchyma (Persson et al., 1976; Ting et al., 1986), other mechanisms are likely to be involved in the perforant path denervated dentate gyrus. In this case the denervated area in the dentate gyrus is located approximately 1 mm septal to the site of axonal transection in the entorhinal cortex (Jensen et al., 1999; Ladeby et al., 2005a). Studies of the distribution of intravenously injected horseradish peroxidase and Evans blue have confirmed that there is no junctional breakdown of the blood–brain barrier in the denervated areas (Jensen et al., 1997). Therefore, the lesion-induced infiltration of BM-derived cells must either occur through an undamaged blood–brain barrier or at the site of the postcapillary venules (Gowans and Knight, 1964; Wolburg et al., 2005).

A candidate molecule for cellular recruitment into the denervated hippocampus may be endothelial ICAM-1, which is known to be critical for leukocyte trafficking across the blood–brain barrier in other models (Engelhardt and Ransohoff, 2005). Demonstrations of suppression of EAE and reduced monocyte entry in ischemic lesions by treatment with antibodies against ICAM-1 or its ligand LFA-1 (Archelos et al., 1993; Chopp et al., 1996; Greenwood et al., 2003) support the theory that enhanced expression of ICAM-1 on CNS endothelium contributes to leukocyte infiltration. By carefully titrating the histochemical development we were able to show that the endothelial expression of ICAM-1 is upregulated on the microvascular endothelium in the denervated areas within 12h and even more strongly 2 days after lesion (Fig. 4). Although these data remain descriptive, they show that the degenerative changes taking place within the neural parenchyma induce the endothelial cells to upregulate their expression of ICAM-1, which in other models is necessary for leukocyte infiltration.

Axonal lesion also results in the expression of a number of chemokines and cytokines. RNAse protection assays have shown induction of a wide range of chemokines in isolated hippocampus, among which glial-derived CCL2, but not CCL5, was shown to direct leukocyte entry (Babcock et al., 2003). The mRNA expression data by Babcock et al. (2003) have been confirmed in separate analysis of isolated perforant path denervated hippocampi from SJL mice showing induction of CCL2 mRNA and CCL5 mRNA 2 and 5 days after the axonal lesion, respectively (Fig. 5). Another chemokine is CXCL10 that acts through CXCR3 and has been shown to contribute to direct migration within the hippocampus (Rappert et al., 2004). Additional chemokines directing microglial migration within the tissue include fractalkine and CCL21 that like CXCL10 are produced by damaged neurons (Harrison et al., 1998; de Jong et al., 2005).

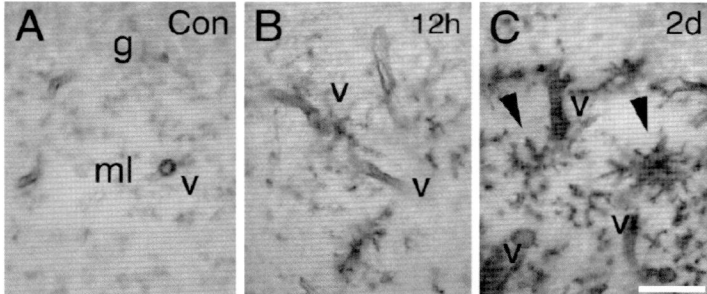

Fig. 4. Axonal lesion-induced ICAM-1 expression on microvascular endothelial cells. (A, B): ICAM-1 was upregulated on the vascular endothelial cells as early as 12 h following surgery (B) compared to control (A) and contralateral control (data not shown). (C): Endothelial ICAM-1 reached a maximum at day 2, when also the microglial cells expressed elevated levels of ICAM-1 (arrowheads). *g*, dentate granule cells; *ml*, dentate molecular layer; *v*, vessels. Bar: 30 μm.

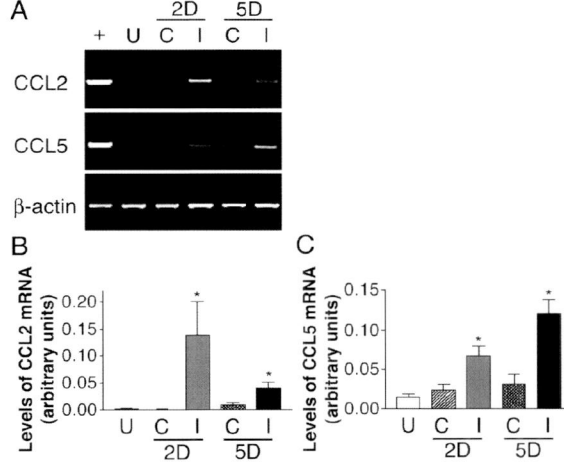

Fig. 5. Chemokine levels in microdissected hippocampi from perforant path-denervated SJL/J mice. (A): Representative ethidium bromide stained gels showing PCR amplified CCL2, CCL5 and β-actin cDNA from hippocampi of PP-lesioned mice ($n = 5–7$). CCL2 message was undetectable or very weakly detected in non-manipulated hippocampi and contralateral control at day 2 and 5 (A, top gel, lanes 2, 3, 5). CCL2 message was upregulated at day 2 and to a lesser extent day 5 in ipsilateral hippocampi (A, top gel, lane 4, 6). Spleen from non-manipulated mice served as positive control (A, top gel, lane 1). CCL5 message was found to be constitutively expressed at low levels in unmanipulated and contralateral control hippocampi (A, middle gel, lane 2, 3, 5). CCL5 message was weakly upregulated at day 2 and markedly upregulated at day 5 in the lesioned hippocampi (A, middle gel, lane 4, 6). Brain from CCL2 transgenic mice served as control (A, middle gel, lane 1). β-actin levels were equivalent in all samples, indicating equal RNA inputs (A, bottom gel). (B, C): CCL2 and CCL5 were not significantly increased in the contralateral hippocampus vs. the unmanipulated hippocampus at 2 or 5 days. CCL2 and CCL5 were significantly increased in the ipsilateral hippocampus vs. contralateral and unmanipulated at 2 and 5 days. This was determined by one-way ANOVA with a positive Bonferroni test. Using the Mann-Whitney *t*-test (two-tailed), the level of CCL5 was significantly elevated at 5 days vs. 2 days in the ipsilateral hippocampus ($p < 0.05$). Although CCL2 levels were decreased at 5 days, this was not significant ($p = 0.0734$). Error bars indicate SD, and asterix indicate $P < 0.05$

10. CONCLUDING REMARKS

We originally studied the expression of CD34 in the anticipation that the exogenous BM-derived cells might express this antigen, and were surprised to find that CD34 was almost exclusively expressed by resident microglia (Ladeby et al. 2005a; Wirenfeldt et al., 2005a). However, since $CD34^+$ murine and human hematopoietic stem cells have been reported to infiltrate the CNS of whole-body irradiated NOD/SCID mice (Asheuer et al., 2004; Biffi et al., 2004) where they transform into microglial-like cells, this demonstrates that $CD34^+$ cells have the potential to infiltrate the murine CNS.

Recent research into the innate immune system triggering receptor expressed on myeloid cells-2 (TREM-2) which is constitutively expressed on microglia has provided new insight into the functional implications of microglia in normal CNS (Cella et al., 2003). TREM-2 deficiency impairs microglial phagocytosis of apoptotic neurons, whereas enhanced TREM-2 expression increases microglial phagocytosis and decreases the expression of tumour necrosis factor (TNF), interleukin-1β (IL-1β) and NOS-2 (Takahashi et al., 2005). TREM-2 on microglia therefore functions as a pro-phagocytic and an anti-inflammatory mediator in the CNS. Human mutations in the TREM-2 and its adaptor protein DAP12 genes lead to the disease Polycystic Lipomembranous Osteodysplasia with Sclerosing Leuko-encephalopathy (PLOSL) where pre-senile dementia is a cardinal symptom (Bianchin et al., 2004).

In conclusion, understanding the regulation of microglial cell population expansion, turnover, clearance and function may lead to improved diagnosis and therapy of some of the disabling and agonizing neurological conditions that currently have few or no treatment options.

11. REFERENCES

Acarin, L., Vela, J.M., Gonzalez, B. and Castellano, B., 1994, Demonstration of poly-N-acetyl lactosamine residues in ameboid and ramified microglial cells in rat-brain by tomato lectin- binding. *J. Histochem. Cytochem.* **42**: 1033.

Albini, T.A., Wang, R.C., Reiser, B., Zamir, E., Wu, G.S. and Rao, N.A., 2005, Microglial stability and repopulation in the retina. *Br. J. Ophthalmol.* **89**: 901.

Allan, S.M. and Rothwell, N.J., 2001, Cytokines and acute neurodegeneration. *Nat. Rev. Neurosci.* **2**: 734.

Andrews, R.G., Singer, J.W. and Bernstein, I.D., 1986, Monoclonal-antibody 12-8 recognizes a 115-kd molecule present on both unipotent and multipotent hematopoietic colony-forming cells and their precursors. *Blood* **67**: 842.

Archelos, J.J., Jung, S., Maurer, M., Schmied, M., Lassmann, H., Tamatani, T., Miyasaka, M., Toyka, K.V. and Hartung, H.P., 1993, Inhibition of experimental autoimmune encephalomyelitis by an antibody to the intercellular adhesion molecule ICAM-1. *Ann. Neurol.* **34**: 145.

Asheuer, M., Pflumio, F.O., Benhamida, S., Dubart-Kupperschmitt, A., Fouquet, F., Imai, Y., Aubourg, P. and Cartier, N., 2004, Human $CD34^+$ cells differentiate into microglia and express recombinant therapeutic protein. *Proc. Natl Acad. Sci. USA* **110**: 3557.

Babcock, A.A., Kuziel, W.A., Rivest, S. and Owens, T., 2003, Chemokine expression by glial cells directs leukocytes to sites of axonal injury in the CNS. *J. Neurosci.* **23**: 7922.

Bartholdi, D. and Schwab, M.E., 1997, Expression of pro-inflammatory cytokine and chemokine mRNA upon experimental spinal cord injury in mouse: An in situ hybridization study. *Eur. J. Neurosci.* **9**: 1422.

Bechmann, I., Goldmann, J., Kovac, A.D., Kwidzinski, E., Simburger, E., Naftolin, F., Dirnagl, U., Nitsch, R., and Priller, J., 2005, Circulating monocytic cells infiltrate layers of anterograde axonal degeneration where they transform into microglia. *FASEB J.* **6**: 647.

Berenson, R.J., Andrews, R.G., Bensinger, W.I., Kalamasz, D., Knitter, G., Buckner, C.D. and Bernstein, I.D., 1988, Antigen Cd34+ marrow-cells engraft lethally irradiated baboons. *J. Clin. Invest.* **81**: 951.

Bianchin, M.M., Capella, H.M., Chaves, D.L., Steindel, M., Grisard, E.C., Ganev, G.G., da Silva, J.P., Neto, E.S., Poffo, M.A., Walz, R., Carlotti, C.G. and Sakamoto, A.C., 2004, Nasu-Hakola disease (polycystic lipomembranous osteodysplasia with sclerosing leukoencephalopathy – PLOSL): a dementia associated with bone cystic lesions. From clinical to genetic and molecular aspects. *Cell. Mol. Neurobiol.* **24**: 1.

Biffi, A., De Palma, M., Quattrini, A., Del Carro, U., Amadio, S., Visigalli, I., Sessa, M., Fasano, S., Brambilla, R., Marchesini, S., Bordignon, C. and Naldini, L., 2004, Correction of metachromatic leukodystrophy in the mouse model by transplantation of genetically modified hematopoietic stem cells. *J. Clin. Invest.* **113**: 1118.

Brown, J., Greaves, M.F. and Molgaard, H.V., 1991, The gene encoding the stem-cell antigen, Cd34, is conserved in mouse and expressed in hematopoietic progenitor-cell lines, brain, and embryonic fibroblasts. *Int. Immunol.* **3**: 175.

Cammermeyer, J., 1965, Juxtavascular karyokinesis and microglia cell proliferation during retrograde reaction in the mouse facial nucleus. *Ergeb. Anat. Entwicklungsgesch.* **38**: 1.

Carson, M.J., Reilly, C.R., Sutcliffe, J.G. and Lo, D., 1998, Mature microglia resemble immature antigen-presenting cells. *Glia* **22**: 72.

Carson, M.J., Sutcliffe, J.G. and Campbell, I.L., 1999, Microglia stimulate naive T-cell differentiation without stimulating T-cell proliferation. *J. Neurosci. Res.* **55**: 127.

Castellano, B., Gonzalez, B., Jensen, M.B., Pedersen, E.B., Finsen, B.R. and Zimmer, J., 1991, A double staining technique for simultaneous demonstration of astrocytes and microglia in brain sections and astroglial cell-cultures. *J. Histochem. Cytochem.* **39**: 561.

Cella, M., Buonsanti, C., Strader, C., Kondo, T., Salmaggi, A. and Colonna, M., 2003, Impaired differentiation of osteoclasts in TREM-2-deficient individuals. *J. Exp. Med.* **198**: 645.

Chopp, M., Li, Y., Jiang, N., Zhang, R.L., and Prostak, J., 1996, Antibodies against adhesion molecules reduce apoptosis after transient middle cerebral artery occlusion in rat brain. *J. Cereb. Blood Flow Metab.* **16**: 578.

Civin, C.I., Strauss, L.C., Brovall, C., Fackler, M.J., Schwartz, J.F. and Shaper, J.H., 1984, Antigenic analysis of hematopoiesis 3: A hematopoietic progenitor-cell surface-antigen defined by a monoclonal-antibody raised against Kg-1A Cells. *J. Immunol.* **133**: 157.

Cogle, C.R., Yachnis, A.T., Laywell, E.D., Zander, D.S., Wingard, J.R., Steindler, D.A. and Scott, E.W., 2004, Bone marrow transdifferentiation in brain after transplantation: a retrospective study. *Lancet* **363**: 1432.

Davalos, D., Grutzendler, J., Yang, G., Kim, J.V., Zuo, Y., Jung, S., Littman, D.R., Dustin, M.L. and Gan, W.B., 2005, ATP mediates rapid microglial response to local brain injury in vivo. *Nat. Neurosci.* **8**: 752.

de Jong, E.K., Dijkstra, I.M., Hensen, M., Brouwer, N., van Amerongen, M., Liem, R.S., Boddeke, H.W. and Biber, K., 2005, Vesicle-mediated transport and release of CCL21 in endangered neurons: a possible explanation for microglia activation remote from a primary lesion. *J. Neurosci.* **25**: 7548.

del Rio-Hortega, P., 1932, Microglia. In: *Cytology and Cellular Pathology of the Nervous System.* W. Penfield, ed., Paul B. Hoeber, New York, pp. 481–534.

Engelhardt, B. and Ransohoff, R.M., 2005, The ins and outs of T-lymphocyte trafficking to the CNS: anatomical sites and molecular mechanisms. *Trends Immunol.* **26**: 485.

Fagan, A.M. and Gage, F.H., 1990, Cholinergic sprouting in the hippocampus: a proposed role for IL-1. *Exp. Neurol.* **110**: 105.

Ferrari, D., Villalba, M., Chiozzi, P., Falzoni, S., Ricciardi-Castagnoli, P. and DiVirgilio, F., 1996, Mouse microglial cells express a plasma membrane pore gated by extracellular ATP. *J. Immunol.* **156**: 1531.

Finsen, B.R., Jorgensen, M.B., Diemer, N.H. and Zimmer, J., 1993a, Microglial MHC antigen expression after ischemic and kainic acid lesions of the adult-rat hippocampus. *Glia* **7**: 41.

Finsen, B.R., Tonder, N., Xavier, G.F., Sorensen, J.C. and Zimmer, J., 1993b, Induction of microglial immunomolecules by anterogradely degenerating mossy fibers in the rat hippocampal-formation *J. Chem. Neuroanat.* **6**: 267.

Ford, A.L., Goodsall, A.L., Hickey, W.F. and Sedgwick, J.D., 1995, Normal adult ramified microglia separated from other central-nervous-system macrophages by flow cytometric sorting – phenotypic differences defined and direct ex-vivo antigen presentation to myelin basic protein-reactive Cd4(+) T-cells compared. *J. Immunol.* **154**: 4309.

Furuya, T., Tanaka, R., Urabe, T., Hayakawa, J., Migita, M., Shimada, T., Mizuno, Y. and Mochizuki, H., 2003, Establishment of modified chimeric mice using GFP bone marrow as a model for neurological disorders. *Neuroreport* **14**: 629.

Garden, G.A., 2002, Microglia in human immunodeficiency virus-associated neurodegeneration. *Glia* **40**: 240.

Gaugler, M.H., Squiban, C., Mouthon, M.A., Gourmelon, P. and van der Meeren, A., 2001, Irradiation enhances the support of haemopoietic cell transmigration, proliferation and differentiation by endothelial cells. *Br. J. Haematol.* **113**: 940.

Gowans, J.L. and Knight, E.J., 1964, Route of re-circulation of lymphocytes in rat. *Proc. R. Soc. Lond. Biol. Sci.* **159**: 257.

Graeber, M.B. and Mehraein, P., 1994, Microglial rod cells. *Neuropathol. Appl. Neurobiol.* **20**: 178.

Graeber, M.B., Streit, W.J. and Kreutzberg, G.W., 1988a, Axotomy of the rat facial-nerve leads to increased Cr3 complement receptor expression by activated microglial cells. *J. Neurosci. Res.* **21**: 18.
Graeber, M.B., Tetzlaff, W., Streit, W.J. and Kreutzberg, G.W., 1988b, Microglial cells but not astrocytes undergo mitosis following rat facial-nerve axotomy. *Neurosci. Lett.* **85**: 317.
Greenwood, J., Amos, C.L., Walters, C.E., Couraud, P.O., Lyck, R., Engelhardt, B. and Adamson, P., 2003, Intracellular domain of brain endothelial intercellular adhesion molecule-1 is essential for T lymphocyte-mediated signaling and migration. *J. Immunol.* **171**: 2099.
Greter, M., Heppner, F.L., Lemos, M.P., Odermatt, B.M., Goebels, N., Laufer, T., Noelle, R.J. and Becher, B., 2005, Dendritic cells permit immune invasion of the CNS in an animal model of multiple sclerosis. *Nat. Med.* **11**: 328.
Hailer, N.P., Bechmann, I., Heizmann, S. and Nitsch, R., 1997, Adhesion molecule expression on phagocytic microglial cells following anterograde degeneration of perforant path axons. *Hippocampus* **7**: 341.
Hailer, N.P., Grampp, A. and Nitsch, R., 1999, Proliferation of microglia and astrocytes in the dentate gyrus following entorhinal cortex lesion: a quantitative bromodeoxyuridine-labelling study. *Eur. J. Neurosci.* **11**: 3359.
Hallenbeck, J.M., 2002, The many faces of tumor necrosis factor in stroke. *Nat. Med.* **8**: 1363.
Harrison, J.K., Jiang, Y., Chen, S., Xia, Y., Maciejewski, D., McNamara, R.K., Streit, W.J., Salafranca, M.N., Adhikari, S., Thompson, D.A., Botti, P., Bacon, K.B. and Feng, L., 1998, Role for neuronally derived fractalkine in mediating interactions between neurons and CXCR1-expressing microglia. *Proc. Natl Acad. Sci. USA* **95**: 10861.
He, J.L., Chen, Y.Z., Farzan, M., Choe, H.Y., Ohagen, A., Gartner, S., Busciglio, J., Yang, X.Y., Hofmann, W., Newman, W., Mackay, C.R., Sodroski, J. and Gabuzda, D., 1997, CCR3 and CCR5 are co-receptors for HIV-1 infection of microglia. *Nature* **385**: 645.
Hickey, W.F. and Kimura, H., 1988, Perivascular microglial cells of the CNS are bone-marrow derived and present antigen in-vivo. *Science* **239**: 290.
Hoek, R.M., Ruuls, S.R., Murphy, C.A., Wright, G.J., Goddard, R., Zurawski, S.M., Blom, B., Homola, M.E., Streit, W.J., Brown, M.H., Barclay, A.N. and Sedgwick, J.D., 2000, Down-regulation of the macrophage lineage through interaction with OX2 (CD200). *Science* **290**: 1768.
Honda, S., Sasaki, Y., Ohsawa, K., Imai, Y., Nakamura, Y., Inoue, K. and Kohsaka, S., 2001, Extracellular ATP or ADP induce chemotaxis of cultured microglia through G(i/o)-coupled P2Y receptors. *J. Neurosci.* **21**: 1975.
Jenmalm, M.C., Cherwinski, H., Bowman, E.P., Phillips, J.H. and Sedgwick, J.D., 2006, Regulation of myeloid cell function through the CD200 Receptor. *J. Immunol.* **176**: 191.
Jensen, M.B., Gonzalez, B., Castellano, B. and Zimmer, J., 1994, Microglial and astroglial reactions to anterograde axonal degeneration – a histochemical and immunocytochemical study of the adult-rat fascia-dentata after entorhinal perforant path lesions. *Exp. Brain Res.* **98**: 245.
Jensen, M.B., Finsen, B. and Zimmer, J., 1997, Morphological and immunophenotypic microglial changes in the denervated fascia dentata of adult rats: correlation with blood–brain barrier damage and astroglial reactions. *Exp. Neurol.* **143**: 103.
Jensen, M.B., Hegelund, I.V., Poulsen, F.R., Owens, T., Zimmer, J. and Finsen, B., 1999, Microglial reactivity correlates to the density and the myelination of the anterogradely degenerating axons and terminals following perforant path denervation of the mouse fascia dentate. *Neuroscience* **93**: 507.
Jensen, M.B., Poulsen, F.R. and Finsen, B., 2000, Axonal sprouting regulates myelin basic protein gene expression in denervated mouse hippocampus. *Int. J. Dev. Neurosci.* **18**: 221.
Jones, L.L., Banati, R.B., Graeber, M.B., Bonfanti, L., Raivich, G. and Kreutzberg, G.W., 1997, Population control of microglia: does apoptosis play a role? *J. Neurocyt.* **26**: 755.
Jones, L.L., Kreutzberg, G.W. and Raivich, G., 1998, Transforming growth factor beta's 1, 2 and 3 inhibit proliferation of ramified microglia on astrocyte monolayer. *Brain Res.* **795**: 301.
Jorgensen, M.B., Finsen, B.R., Jensen, M.B., Castellano, B., Diemer, N.H. and Zimmer, J., 1993, Microglial and astroglial reactions to ischemic and kainic acid-induced lesions of the adult-rat hippocampus. *Exp. Neurol.* **120**: 70.
Kato, H., Takahashi, A. and Itoyama, Y., 2003, Cell cycle protein expression in proliferating microglia and astrocytes following transient global cerebral ischemia in the rat. *Brain Res. Bull.* **60**: 215.
Kiefer, R., Lindholm, D. and Kreutzberg, G.W., 1993, Interleukin-6 and transforming growth-factor-beta-1 messenger-RNAs are induced in rat facial nucleus following motoneuron axotomy. *Eur. J. Neurosci.* **5**: 775.
Klein, M.A., Moller, J.C., Jones, L.L., Bluethmann, H., Kreutzberg, G.W. and Raivich, G., 1997, Impaired neuroglial activation in interleukin-6 deficient mice. *Glia* **19**: 227.
Kloss, C.U.A., Kreutzberg, G.W. and Raivich, G., 1997, Proliferation of ramified microglia on an astrocyte monolayer: characterization of stimulatory and inhibitory cytokines, *J. Neurosci. Res.* **49**: 248.
Kreutzberg, G.W., 1966, Autoradiographische untersuchung uber die beteiligung von gliazellen an der axonalen reaktion im facialiskern der ratte. *Acta Neuropathol. (Berl.)* **7**: 149.

Ladeby, R., Wirenfeldt, M., Dalmau, I., Gregersen, R., Garcia-Ovejero, D., Babcock, A., Owens, T. and Finsen, B., 2005a, Proliferating resident microglia express the stem cell antigen CD34 in response to acute neural injury. *Glia* **50**: 121.

Ladeby, R., Wirenfeldt, M., Garcia-Ovejero, D., Fenger, C., Dissing-Olesen, L., Dalmau, I. and Finsen, B., 2005b, Microglial cell population dynamics in the injured adult central nervous system. *Brain Res. Brain Res. Rev.* **48**: 196.

Lambertsen, K.L., Gregersen, R. and Finsen, B., 2002, Microglial-macrophage synthesis of tumor necrosis factor after focal cerebral ischemia in mice is strain dependent. *J. Cereb. Blood Flow Metab.* **22**: 785.

Lassmann, H. and Hickey, W.F., 1993, Radiation bone-barrow chimeras as a tool to study microglia turnover in normal brain and inflammation. *Clin. Neuropathol.* **12**: 284.

Lawson, L.J., Perry, V.H. and Gordon, S., 1992, Turnover of resident microglia in the normal adult-mouse brain. *Neuroscience* **48**: 405.

Lehrmann, E., Kiefer, R., Christensen, T., Toyka, K.V., Zimmer, J., Diemer, N.H., Hartung, H.P. and Finsen, B., 1998, Microglia and macrophages are major sources of locally produced transforming growth factor beta(1) after transient middle cerebral artery occlusion in rats. *Glia* **24**: 437.

Matsumoto, Y. and Fujiwara, M., 1987, Absence of donor-type major histocompatibility complex class-I antigen-bearing microglia in the rat central-nervous-system of radiation bone-marrow chimeras. *J. Neuroimmunol.* **17**: 71.

Matthews, D.A., Cotman, C. and Lynch, G., 1976, Electron-microscopic study of lesion-induced synaptogenesis in dentate gyrus of adult rat .1. Magnitude and time course of degeneration. *Brain Res.* **115**: 1.

Morgan, T.E., Nichols, N.R., Pasinetti, G.M. and Finch, C.E., 1993, TGF-beta 1 mRNA increases in macrophage/microglial cells of the hippocampus in response to deafferentiation and kainic acid-induced neurodegeneration. *Exp. Neurol.* **120**: 291.

Mori, S. and Leblond, C.P., 1969, Identification of microglia in light and electron microscopy. *J. Comp. Neurol.* **135**: 57.

Morioka, T., Kalehua, A.N. and Streit, W.J., 1991, The microglial reaction in the rat dorsal hippocampus following transient forebrain ischemia. *J. Cereb. Blood Flow Metab.* **11**: 966.

Nimmerjahn, A., Kirchhoff, F. and Helmchen, F., 2005, Resting microglial cells are highly dynamic surveillants of brain parenchyma in vivo. *Science* **308**: 1314.

Nissl, F., 1899, Ueber einige beziehungen zwischen nervenzellenerkrankungen und gliösen erscheinungen bei verschiedenen psychosen. *Arch. Psychiatr.* **32**: 656.

Noseworthy, J.H., Lucchinetti, C., Rodriguez, M. and Weinshenker, B.G., 2000, Medical progress: multiple sclerosis. *New Eng. J. Med.* **343**: 938.

Novikoff, A.B. and Goldfischer, S., 1961, Nucleosidediphosphatase activity in Golgi apparatus and its usefulness for cytological studies. *Proc. Natl Acad. Sci. USA* **47**: 802.

O'Keefe, G.M., Nguyen, V.T. and Benveniste, E.N., 2002, Regulation and function of class II major histocompatibility complex, CD40, and B7 expression in macrophages and microglia: implications in neurological diseases. *J. Neurovir.* **8**: 496.

Pakkenberg, B. and Gundersen, H.J.G., 1988, Total number of neurons and glial-cells in human-brain nuclei estimated by the dissector and the fractionator. *J. Microscopy. Oxford.* **150**: 1.

Pennell, N.A. and Streit, W.J., 1998, Tracing of fluoro-gold prelabeled microglia injected into the adult rat brain. *Glia* **23**: 84.

Perry, V.H., Hume, D.A. and Gordon, S., 1985, Immunohistochemical localization of macrophages and microglia in the adult and developing mouse-brain. *Neuroscience* **15**: 313.

Persson, L., Hansson, H.A. and Sourander, P., 1976, Extravasation, spread and cellular uptake of Evans blue-labelled albumin around a reproducible small stab-wound in the rat brain. *Acta Neuropathol. (Berl.)* **34**: 125.

Popovich, P.G. and Hickey, W.E., 2001, Bone marrow chimeric rats reveal the unique distribution of resident and recruited macrophages in the contused rat spinal cord. *J. Neuropathol. Exp. Neurol.* **60**: 676.

Priller, J., Flugel, A., Wehner, T., Boentert, M., Haas, C.A., Prinz, M., Fernandez-Klett, F., Prass, K., Bechmann, I., de Boer, B.A., Frotscher, M., Kreutzberg, G.W., Persons, D.A. and Dirnagl, U., 2001, Targeting gene-modified hematopoietic cells to the central nervous system: Use of green fluorescent protein uncovers microglial engraftment. *Nat. Med.* **7**: 1356.

Raivich, G., 2005, Like cops on the beat: the active role of resting microglia. *Trends Neurosci.* **28**: 571.

Raivich, G., Morenoflores, M.T., Moller, J.C. and Kreutzberg, G.W., 1994, Inhibition of posttraumatic microglial proliferation in a genetic model of macrophage colony-stimulating factor deficiency in the mouse. *Eur. J. Neurosci.* **6**: 1615.

Raivich, G., Haas, S., Werner, A., Klein, M.A., Kloss, C. and Kreutzberg, G.W., 1998, Regulation of MCSF receptors on microglia in the normal and injured mouse central nervous system: a quantitative immunofluorescence study using confocal laser microscopy. *J. Comp. Neurol.* **395**: 342.

Raivich, G., Bohatschek, M., Kloss, C.U.A., Werner, A., Jones, L.L. and Kreutzberg, G.W., 1999, Neuroglial activation repertoire in the injured brain: graded response, molecular mechanisms and cues to physiological function. *Brain Res. Rev.* **30**: 77.

Rao, K. and Lund, R.D., 1989, Degeneration of optic axons induces the expression of major histocompatibility antigens. *Brain Res.* **488**: 332.

Rappert, A., Bechmann, I., Pivneva, T., Mahlo, J., Biber, K., Nolte, C., Kovac, A.D., Gerard, C., Boddeke, H.W.G.M., Nitsch, R. and Kettenmann, H., 2004, CXCR3-dependent microglial recruitment is essential for dendrite loss after brain lesion. *J. Neurosci.* **24**: 8500.

Rogove, A.D., Lu, W. and Tsirka, S.E., 2002, Microglial activation and recruitment, but not proliferation, suffice to mediate neurodegeneration. *Cell. Death Differ.* **9**: 801.

Sedgwick, J.D., Schwender, S., Imrich, H., Dorries, R., Butcher, G.W. and Termeulen, V., 1991, Isolation and direct characterization of resident microglial cells from the normal and inflamed central-nervous-system. *Proc. Natl Acad. Sci. USA* **88**: 7438.

Stoll, G., Trapp, B.D. and Griffin, J.W., 1989, Macrophage function during Wallerian degeneration of rat optic nerve: clearance of degenerating myelin and Ia expression. *J. Neurosci.* **9**: 2327.

Streit, W.J., 1990, An improved staining method for rat microglial cells using the lectin from griffonia-simplicifolia (GSA I-B4). *J. Histochem. Cytochem.* **38**: 1683.

Streit, W.J., 2000, Microglial response to brain injury: A brief synopsis. *Toxicol. Pathol.* **28**: 28.

Streit, W.J., 2004, Microglia and Alzheimer's disease pathogenesis. *J. Neurosci. Res.* **77**: 1.

Streit, W.J. and Kreutzberg, G.W., 1988, Response of endogenous glial-cells to motor neuron degeneration induced by toxic ricin. *J. Comp. Neurol.* **268**: 248.

Streit, W.J., Semple-Rowland, S.L., Hurley, S.D., Miller, R.C., Popovich, P.G. and Stokes, B.T., 1998, Cytokine mRNA profiles in contused spinal cord and axotomized facial nucleus suggest a beneficial role for inflammation and gliosis. *Exp. Neurol.* **152**: 74.

Streit, W.J., Walter, S.A. and Pennell, N.A., 1999, Reactive microgliosis. *Prog. Neurobiol.* **57**: 563.

Takahashi, K., Rochford, C.D.P. and Neumann, H., 2005, Clearance of apoptotic neurons without inflammation by microglial triggering receptor expressed on myeloid cells-2. *J. Exp. Med.* **201**: 647.

Ting, P., Masaoka, H., Kuroiwa, T., Wagner, H., Fenton, I. and Klatzo, I., 1986, Influence of blood–brain barrier opening to proteins on development of post-ischaemic brain injury. *Neurol. Res.* **8**: 146.

Vallieres, L. and Sawchenko, P.E., 2003, Bone marrow-derived cells that populate the adult mouse brain preserve their hematopoietic identity. *J. Neurosci.* **23**: 5197.

Walz, W., Ilschner, S., Ohlemeyer, C., Banati, R. and Kettenmann, H., 1993, Extracellular ATP activates a cation conductance and a K+ conductance in cultured microglial cells from mouse-brain. *J. Neurosci.* **13**: 4403.

Wirenfeldt, M., Babcock, A.A., Ladeby, R., Lambertsen, K.L., Dagnaes-Hansen, F., Leslie, R.G., Owens, T. and Finsen, B., 2005a, Reactive microgliosis engages distinct responses by microglial subpopulations after minor central nervous system injury. *J. Neurosci. Res.* **82**: 507.

Wirenfeldt, M., Ladeby, R., Dalmau, I., Banati, R.B. and Finsen, B., 2005b, Microglia – biology and relevance to disease. *Ugeskr. Laeger.* **82**: 507.

Wolburg, H., Wolburg-Buchholz, K. and Engelhardt, B., 2005, Diapedesis of mononuclear cells across cerebral venules during experimental autoimmune encephalomyelitis leaves tight junctions intact. *Acta Neuropathol. (Berl.)* **109**: 181.

Young, P.E., Baumhueter, S. and Lasky, L.A., 1995, The sialomucin CD34 is expressed on hematopoietic-cells and blood-vessels during murine development. *Blood* **85**: 96.

OPPOSITE MODULATION OF PERIPHERAL INFLAMMATION AND NEUROINFLAMMATION BY ADENOSINE A_{2A} RECEPTORS

Rodrigo A. Cunha[1], Jiang-Fan Chen[2] and Michail V. Sitkovsky[3]

1. ABSTRACT

Adenosine is a homeostatic modulator in all cells, being produced from intracellular ATP as a consequence of increased workload or noxious stimuli. Extracellular adenosine can then afford tissue protection by a combination of effects operated by inhibitory A_1 receptors, which refrain metabolism, and facilitatory A_{2A} receptors, which cause vasodilatation and act as a potent "Off" signal of immune/inflammatory cells in the periphery. Adenosine also acts as a neuromodulator in the brain through A_1 receptor-mediated inhibition of excitatory transmission. However, adenosine also aggravates brain damage in chronic neurodegenerative conditions by mechanisms that may involve an A_{2A} receptor-mediated potentiation of neuroinflammation. We will now review the evidences suggesting that A_{2A} receptor control neuroinflammation and discuss the possible mechanisms underlying the opposite modulation by A_{2A} receptors of peripheral inflammation and neuroinflammation.

2. ADENOSINE AND ADENOSINE RECEPTORS

Adenosine is a purine nucleoside that behaves as a homeostatic metabolite, apart from its numerous roles in intracellular metabolism (Stone, 1985). Under optimal metabolic conditions and in the absence of increased activity, the intracellular and extracellular concentrations of adenosine are normally kept at a low level, i.e. in the range of low nanomolar (from 20 to 200 nM) concentration (Cunha, 2001; Fredholm et al., 2005). However, in case of metabolic imbalance (i.e. either because of lower glucose and/or oxygen availability or as a result of an increased tissue activity), there is a marked raise both in the intracellular as well as in the extracellular concentration of adenosine (Fredholm, 1997). This is due to the fact that the sequential dephosphorylation of ATP will yield adenosine. Furthermore, the concentration of ATP within cells is in the 3–10 mM range and a minor decrease in ATP levels will cause a disproportional larger relative increase in the concentration of adenosine (i.e. the catabolism of 1% of intracellular ATP will cause a 1,000

Rodrigo A. Cunha – cunharod@gmail.com
[1]Center for Neuroscience of Coimbra, Institute of Biochemistry, Faculty of Medicine, University of Coimbra, Portugal, [2]Department of Neurology, Boston University School of Medicine, Boston, MA, USA and [3]New England Inflammation and Tissue Protection Institute, Northeastern University, Boston, MA, USA.

fold increase in the concentration of adenosine) (Cunha, 2001). Since all cell types are equipped with bidirectional and non-concentrative nucleoside transporters (Baldwin et al., 2004), any modification of the intracellular concentration of adenosine will be translated in an equivalent change in the extracellular amounts of adenosine.

Once present extracellularly, adenosine can act on metabotropic adenosine receptors present at the plasma membrane to influence the activity of neighbouring cells. There are four cloned adenosine receptors named A_1, A_{2A}, A_{2B} and A_3 receptors (Fredholm et al., 2001). We will mainly consider the role of A_1 and A_{2A} receptors since they are the most abundant and the better characterised because of the selectivity of the pharmacological tools and extensive studies in knockout mice and their particular importance in the control of brain function (Fredholm et al., 2005). However, this should not exclude or underscore the possible relevance of the other adenosine receptors (A_{2B} and A_3 receptors), which roles still remain largely unravelled. The most abundant adenosine receptor is the A_1 receptor, which mostly causes an inhibition of cell function. In fact, the activation of A_1 receptors causes a general depression of the oxidative metabolism of cells, which allows cells to cope better with noxious stimuli in general (Newby, 1984). Thus, it is A_1 receptors that are mainly responsible for the ubiquitous homeostatic role of adenosine: once released from a "stressed" cell, adenosine diffuses to neighbouring cells, decreases their metabolic rate through activation of A_1 receptors and allows these neighbouring cells to cope better with noxious conditions.

Historically, the A_{2A} receptor was viewed as an auxiliary receptor aiding in the coordinated homeostatic response of tissues to decreased metabolic availability or increased workload. Thus, the activation of A_{2A} receptors located in endothelial and smooth muscle cells of blood vessels causes a vasodilatation (direct relaxation of the smooth muscle together with an indirect effect resulting from the control of the release of endothelial factors), aimed at increasing oxygen supply (Shryock and Belardinelli, 1997; O'Regan, 2005). Recent years have witnessed a broadening of the roles attributed to A_{2A} receptors both in the periphery and in the brain (Fredholm et al., 2003), which will be the main focus of this review.

3. ADENOSINE AND NEUROMODULATION

Apart from their general homeostatic role, adenosine receptors also fulfil particular roles as neuromodulators in the nervous system (Cunha, 2001), with the ability of controlling neuronal excitability, firing and synaptic plasticity. Thus, it is in the brain that A_1 receptors are most abundant (Dunwiddie and Masino, 2001). They are mostly concentrated in neurons, with a particular density at synapses (Tetzlaff et al., 1987; Rebola et al., 2003). In synapses, A_1 receptors act presynaptically refraining the evoked release of excitatory neurotransmitters (such as glutamate) and postsynaptically controlling calcium entry through inhibition of voltage-sensitive calcium channels as well as by inhibition of NMDA receptors (de Mendonça et al., 2000; Cunha, 2005). Consequently, A_1 receptors have a profound impact on synaptic transmission and plasticity phenomena in excitatory synapses in the brain (de Mendonça and Ribeiro, 1997; Dunwiddie and Masino, 2001). Non-synaptic A_1 receptors also aid in restraining the excitability of excitatory neurons by inhibiting potassium currents, which causes a hyperpolarisation of neurons (Greene and Haas, 1991). Accordingly, adenosine activation of A_1 receptors is an effective endogenous anti-epileptic system (Dunwiddie, 1999).

The role of A_{2A} receptors in neuronal circuits is less well characterised. A_{2A} receptors are most abundantly located in a sub-population of spinny GABAergic neurons of the basal ganglia that forms the indirect pathway (Svenningsson et al., 1999). Here, they oppose the effects of dopamine D2 receptors and have an important role in the control of locomotion, which is best highlighted by the current interest in developing A_{2A} receptor antagonists as novel anti-Parkinsonian drugs (Xu et al., 2005). This high density of A_{2A} receptors in these particular neurons has over-shadowed the recognition of their presence in different other localisation in the brain, where the density of A_{2A} receptors is nearly 20 times lower than in the basal ganglia (Lopes et al., 2004). However, over the years, it has been increasingly recognised that A_{2A} receptors are located in numerous synapses throughout the brain (Cunha, 2005). As for A_1 receptors, A_{2A} receptors also seem to be concentrated in synapses (Rebola et al., 2005a) where they facilitate the evoked release of numerous neurotransmitters (glutamate, acetylcholine, serotonin or GABA) (reviewed in Cunha, 2001) and control postsynaptic responses (Nash and Brotchie, 2000; Wirkner et al., 2000, 2004; Gerevich et al., 2002; Tebano et al., 2005; Ponzio et al., 2006). Interestingly, they seem to be present in the same synapses as A_1 receptors (Rebola et al., 2005b; Ciruela et al., 2006) and they have a major role in controlling A_1 receptor function through intracellular transducing systems (Dixon et al., 1997; Lopes et al., 1999) or through receptor dimerisation (Ciruela et al., 2006). In fact, these synaptic A_{2A} receptors play an important role in the implementation of synaptic plasticity changes, aiding in overcoming the unwanted maximal effect of A_1 receptors to depress synaptic plasticity (reviewed in Ferré et al., 2005).

Thus, adenosine may play a double role in the control of neuronal function: it provides a tonic A_1 receptor-mediated inhibition of excitatory transmission that acts as a hurl avoiding excessive glutamatergic drive of excitotoxicity phenomena (e.g. Fedele et al., 2006). However, at higher frequency and/or intensity of stimulation A_{2A} receptors come into play, to shut down A_1 receptor function and enhancing synaptic efficiency, allowing the implementation of synaptic plasticity phenomena. This ability of A_{2A} receptors to come into play in a selective manner at higher frequencies/intensities of neuronal firing is due to the lower affinity for adenosine of A_{2A} receptors compared to A_1 receptors (Ciruela et al., 2006) as well as by the fact that different sources of adenosine activate A_1 receptors and A_{2A} receptors in synapses: release of adenosine as such activates A_1 receptors whereas A_{2A} receptors are selectively activated by adenosine formed from the extracellular catabolism of ATP that is released in higher amounts at higher frequencies of neuronal firing (Cunha et al., 1996a, b; Pinto-Duarte et al., 2005).

4. ADENOSINE AND NEUROPROTECTION

Because of the potential neurotoxic role associated with excessive release and activation of ionotropic glutamate receptors (Lipton and Rosenberg, 1994), any system able to selectively depress glutamate release and glutamatergic transmission is viewed as a potential neuroprotective system. Since A_1 receptors depress excitatory transmission, refrain calcium entry into neurons and hyperpolarise neurons, they fulfil several of the ideal criteria to act as a neuroprotective system. Accordingly, the acute administration of either A_1 receptor agonists or the use of strategies aimed at enhancing the extracellular levels of adenosine afford a protection against different types of insults either in the in vivo brain or in in vitro brain preparations (de Mendonça et al., 2000). Conversely, the acute administration of A_1 receptor antagonists aggravates brain damage (de Mendonça et al., 2000). This provides a clear confirmation of the role of A_1 receptors in tissue protection against noxious stimulation (developed in the initial section of this review), which is prominent in the brain but is also found in other tissues such as heart or kidney (Osswald et al., 1996; Shryock and Belardinelli, 1997; Modlinger and Welch, 2003).

However, this role of A_1 receptors in neuroprotection (and eventually in other tissues' protection) is time limited. In fact, the A_1 receptor system is prone to a rapid desensitisation (Lee et al., 1986; Coelho et al., 2006) and the effects operated by A_1 receptors undergo desensitisation upon chronic noxious brain conditions (reviewed in Cunha, 2005). Thus, the neuroprotection afforded by A_1 receptors acts as a hurl to avoid the appearance of toxic features, which losses its importance once overtaken. Furthermore, several other systems, normally less effective in refraining glutamatergic transmission may come into play after A_1 receptor blockade (Lucchi et al., 1996; Coelho et al., 2000; Pearson and Frenguelli, 2004), offering a reasonable explanation for the absence of significant modification of brain damage observed in A_1 receptor knockout mice compared to their wild-type littermates (Olsson et al., 2004).

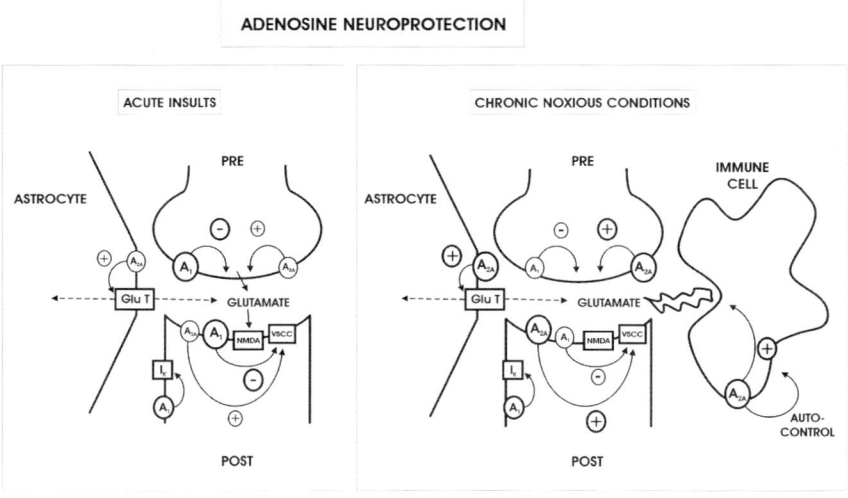

Fig. 1. Different role of adenosine receptors (inhibitory A_1 and facilitatory A_{2A} receptors) in the control of brain tissue damage caused by acute insults (*left*) and chronic noxious brain insults (*right*). In situations of acute insults there is a predominant role of inhibitory A_1 receptors, which are more abundant in brain tissue (represented by their greater size in the left panel). These A_1 receptors can presynaptically inhibit the release of glutamate (a major excitotoxin in the brain) and can post synaptically restrain the over-excitation of NMDA receptors as well as the influx of calcium, two key events leading to neurodegeneration. In parallel, A_1 receptors also promote a hyperpolarisation of neurons, delaying the depolarisation of neurons that contributes to their loss of function. In these situations of acute brain damage, the contribution of adenosine A_{2A} receptors is limited because of their low density in most brain regions (represented by their smaller size in the left panel). In contrast, in situations of chronic noxious brain damage, there is a greater contribution of A_{2A} receptors. This results from an upregulation of A_{2A} receptors together with a downregulation of A_1 receptors in chronic noxious brain conditions (represented by their modified size in the right panel). Thus, the efficient control of glutamatergic transmission by A_1 receptors is considerably reduced whereas the increase density of A_{2A} receptors may lead to greater extracellular levels of glutamate (released from nerve terminals as well as from astrocytes through glutamate transporters) and a greater activation of NMDA receptors. In parallel, A_{2A} receptors also boost neuroinflammation, which impacts on neuronal viability. Any of these effects is likely to contribute to the neuroprotection afforded by blockade of A_{2A} receptors in chronic noxious conditions in adult animals, although recent data indicate that the control of neuroinflammation may be one of the most likely mechanisms

In line with the double and opposite role of A_1 and A_{2A} receptors in the control of synaptic transmission, it could be anticipated that the activation of A_{2A} receptors might aggravate neuronal damage. In fact, several studies carried out using different brain insults and focusing in different brain regions have systematically found that the pharmacological blockade or the genetic inactivation of A_{2A} receptors affords protection against chronic noxious brain insults in adult animals (reviewed in Cunha, 2005). This neuroprotection is particularly effective in more chronic models of brain degeneration since there is an upregulation of the A_{2A} receptor system that is accompanied by a downregulation of A_1 receptor system, as observed upon epilepsy (Rebola et al., 2005c), diabetes (Duarte et al., 2006) or restraint stress (Cunha et al., 2006). Thus, whereas the neuroprotection afforded by the activation of A_1 receptors is effective at the onset of brain damage and has a short time-window of opportunity, the neuroprotection afforded by A_{2A} receptor blockade has a lag-time to become operative but seems not to undergo desensitisation over time (Fig. 1).

5. MECHANISMS OF NEUROPROTECTION AFFORDED BY A_{2A} RECEPTOR BLOCKADE

The mechanism by which A_{2A} receptors impact on neurodegeneration remains to be defined. In fact, the broad neuroprotection afforded by A_{2A} receptor blockade (striatum, cortex, hippocampus) indicate that the mechanism underlying the neuroprotection resulting from A_{2A} receptor blockade should also be a broad mechanism impacting on neuronal viability. Given that A_{2A} receptors control the release of glutamate in different brain regions (Lopes et al., 2002; Marchi et al., 2002; Marcoli et al., 2003, 2004; Rodrigues et al., 2005), it was proposed that the neuroprotection associated with A_{2A} receptors might result from their ability to decrease the evoked release of glutamate (Popoli et al., 2003). This reasoning is probably over-simplistic in essence since most of the extracellular glutamate that builds up upon degeneration of brain tissue is probably not of neuronal origin but rather originates as a consequence of the loss of control by astrocytes of glutamate uptake with the consequent outflow of glutamate upon reversal of flux of astrocytic glutamate transporters (Rossi et al., 2000). In accordance with this contention, we have recently found that selectively knocking out A_{2A} receptors from forebrain neurons abolishes the ability of A_{2A} receptors to modulate the evoked release of glutamate but fails to affect A_{2A} receptor blockade-mediated neuroprotection (unpublished results). Thus, it is now evident that, in adult animals, an alternative general mechanism should be proposed to explain the neuroprotection afforded by A_{2A} receptors against chronic noxious brain insults.

One currently open possibility might be the ability of A_{2A} receptors to control the uptake and release of glutamate from astrocytes. In fact, Chern's group convincingly showed that A_{2A} receptors are located in astrocytes in situ (Lee et al., 2003). Furthermore, the group of Popoli provided in vivo evidence suggesting that A_{2A} receptors might control the uptake of glutamate by GLAST and GLUT-1, the two main putative astrocytic glutamate transporters (Pintor et al., 2004). This confirms in vitro studies showing that A_{2A} receptors could control both the uptake as well as the outflow of glutamate from astrocytes in cultures (Nishizaki et al., 2002). This hypothesis is attractive since it reconciles the ability of A_{2A} receptors to afford neuroprotection with the control of extracellular glutamate levels, a main actor involved in neurodegeneration. The main difference would be that A_{2A} receptors would not act in neurons but rather in astrocytes to control glutamate levels, leaving open the possibility that A_{2A} receptors might also control the effects of glutamate on neurons by way of their ability to control NMDA receptor function (Nash and Brotchie, 2000; Wirkner et al., 2000, 2004; Gerevich et al., 2002; Tebano et al., 2005).

Although attractive, this hypothesis was made less appealing by the observation that the neuroprotection afforded by A_{2A} receptor blockade upon transient forebrain ischaemia predominantly resulted from the ablation of A_{2A} receptors in bone marrow-derived cells rather than from the ablation of A_{2A} receptors in the brain (Yu et al., 2004). In parallel, this work was the first to provide a clear and direct evidence for the hypothesis that it might be the control of neuroinflammation by A_{2A} receptors that may underline the neuroprotection afforded by A_{2A} receptor blockade. However, it should not be immediately concluded that the neuroprotection resulting from A_{2A} receptor blockade is always absolutely dependent on the infiltration of hematogenous cells into the brain. In fact, we have carried out experiment testing if A_{2A} receptor in bone marrow-derived cells were also responsible for the neuroprotection against nigral and striatal dopaminergic toxicity triggered by administration of MPTP, an experimental model of Parkinson's disease. And it was observed that the selective inactivation of A_{2A} receptors in bone marrow-derived cells failed to afford the neuroprotection observed by global blockade/inactivation of A_{2A} receptors (either using a selective A_{2A} receptor antagonist or testing global A_{2A} receptor knockout mice) (Chen et al., 2001). Furthermore, the intracerebroventricular injection of a selective A_{2A} receptor antagonist re-instated neuroprotection against MPTP-induced dopaminergic toxicity in forebrain A_{2A} receptor knockout mice indicating that it is A_{2A} receptors located in the brain but in non-neuronal cells that are responsible for the control of MPTP-induced dopaminergic toxicity. This raises the possibility that the neuroprotection afforded by A_{2A} receptor blockade may result from the control of glia cell-mediated neuroinflammation, since there is increasing evidence in humans and animals models that neuroinflammation contributes to the pathogenesis of Parkinson's disease (Beal, 2003; Hartmann et al., 2003; Hirsch et al., 2003; Liu and Hong, 2003; McGeer and McGeer, 2004; Teismann and Schulz, 2004; Morale et al., 2006; Sriram et al., 2006a, b), which is re-enforced by a recent epidemiological study showing that the chronic treatment with non-steroid anti-inflammatory drugs (NSAIDs) reduces the risk of Parkinson's disease by 45% (Chen et al., 2003).

Altogether, it can be concluded that the control of neuroinflammation by A_{2A} receptors appears to be the most reasonable hypothesis to explain the ability of A_{2A} receptors to afford neuroprotection. However, the cellular localisation of the A_{2A} receptors involved in the control of neuroinflammation might either be in infiltrating cells or in resident microglia cells, probably depending on their relative contribution for the genesis of neuroinflammatory processes, which likely depends on the type of noxious stimuli present.

6. NEUROINFLAMMATION AND NEURODEGENERATION

A burst of interest in the role of neuroinflammation in the demise of brain degeneration has followed the increased recognition that inflammatory features seem to be present in most conditions of brain damage, either acute traumatic conditions (e.g. trauma, stroke) or neurodegenerative disorders (e.g. Alzheimer's disease, Parkinson's disease, epilepsy) (Piehl and Lidman, 2001; Gao et al., 2003; Giovannoni and Baker, 2003; Hartman et al., 2003; Liu and Hong, 2003; Andersen, 2004; Block and Hong, 2005; Herrera et al., 2005; Marchetti and Abbracchio, 2005; Minghetti, 2005; Lucas et al., 2006). However, the cell types responsible for this neuroinflammation associated with brain injury have not been completely clarified. The increased complexity of the brain of higher vertebrates is accompanied by the exclusion of hematogenous cells, the specialised immune cells, probably as a need to restrain auto-immune attack to an increasingly complex and postmitotic structure with little regenerating capacities (see Schwartz et al., 1999; Bechmann, 2003). This is translated into the generalisation that the brain is an "immune privileged" organ, i.e. an organ with restricted local immune responses (reviewed in Matyszak, 1998; Carson and Sutcliffe, 1999; Schwartz and Moalem, 2001; Niederkorn, 2006). This results from the combined existence of a blood–brain barrier, the absence draining lymph nodes, the normal absence of resident antigen-presenting cells (i.e. dendritic cells) and active mechanisms of deviation and suppression of immune responses (reviewed in Schwartz and Moalem, 2001). Since the brain also requires an active system to remove pathogens, toxic cell debris and apoptotic cells, the brain is endowed with a substitute "immune-like" system composed by microglia and astrocytes, which function as a brain-resident innate immune system, albeit comparatively less efficient than hematogenous cells.

Microglia cells are myeloid lineage-derived cells, which infiltrate the brain during the late prenatal and early postnatal period (Ladeby et al., 2005). In the absence of noxious stimuli, microglia cells form a stable cell brain-resident population and comprise near 15% of brain cells (Ladeby et al., 2005). They display a characteristic ramified morphology and their fine branches are highly mobile (Davalos et al., 2005; Nimmerjahn et al., 2005), providing a continuous surveillance of their cellular environment (Raivich, 2005). In their resting state, microglia display a downregulated phenotype characterised by the presence of low amounts of co-stimulatory molecules and undetectable levels of MHC class II molecules (Kreutzberg, 1996; Raivich et al., 1999; O'Keefe et al., 2002). However, in line with their role as the first line of defense of the neural parenchyma, a diversity of stress-like signals (bacterial antigens, lipopolysaccharides, stroke, epileptic seizure activity and neurodegeneration in general) is able to activate microglia cells, changing their morphology and phenotype (Kreutzberg, 1996; Raivich et al., 1999; Farber and Kettenmann, 2005), triggering their movement to the sites of lesion and causing an expansion of microglia cell population (Ladeby et al., 2005). Activated microglia acquires a bushy or amoeboid morphology, which is shaped by the type of underlying neuronal degeneration (Ladeby et al., 2005). There is also an upregulation of different adhesion molecules and co-stimulatory and MHC antigen molecules (Kreutzberg, 1996; Raivich et al., 1999; O'Keefe et al., 2002) and the appearance of microglia endowed with stem cell antigens such as CD34 (Ladeby et al., 2005). This characteristic microglia reactivity in response to different types of brain insults makes it one of the earliest and most sensitive measures of brain damage (Kreutzberg, 1996). Activated microglia may trigger a neuroinflammatory process since they can release a series of pro-inflammatory mediators such as free radicals, cytokines and prostaglandins (Raivich et al., 1999; Block and Hong, 2005; Ladeby et al., 2005; McGeer et al., 2005).

The pathways of activation of microglia after different types of brain injury still remain poorly characterised. However, it is becoming clear that astrocytes play a crucial role in the control of microglia activation (Raivich, 2005; Morale et al., 2006). Astrocytes are far more numerous than microglia in the brain parenchyma, where they fulfil a variety of roles (Fellin and Carmignoto, 2004; Pellerin, 2005). In terms of their possible contribution for an innate immune response, they are far less efficient than microglia (Streit et al., 1999; Gasque et al., 2000; Magnus et al., 2002). Astrocytes can become activated in response to brain injury acquiring a characteristic hypertrophic morphology and producing a series of immune and pro-inflammatory mediators, such as complement proteins, cytokines and prostaglandins (Dong and Benveniste, 2001). However, one of the key roles of astrocytes in the realm of neuroinflammation may reside in their ability to control microglia reactivity. In fact, the extensive astrocytic network is thought to play a key role in maintaining microglia in the resting state and changes in the release of purinergic and pyrimidinergic nucleotides and nucleosides from astrocytes seems to be a main trigger for the activation of microglia (Davalos et al., 2005; Raivich, 2005; Inoue, 2006). Thus, it is probably wiser to discuss the simultaneous participation of an astrocyte–microglia network in neuroinflammation rather than discuss the relative participation of each of these major brain-resident cell types in the development of a neuroinflammatory process. In fact, the role of astrocytes seems central for the proper control of the extent of a neuroinflammatory response, as shown in the case of spinal cord injury where the ablation of reactive astrocytes causes a disruption of the inflammatory response and consequent extent of morphological and functional damage (Faulkner et al., 2004).

Finally, the last set of actors that should be considered when analysing neuroinflammation are peripheral hematogenous cells. As previously discussed when presenting the brain as an immune-privileged organ, the number of leukocytes in the brain is very low (Carson and Sutcliffe, 1999; Aloisi et al., 2001). However, leukocytes have access and are actively traffic in the brain parenchyma in normal conditions and there is an increased accumulation of different populations of monocytes and leukocytes at sites of brain lesions relative to that seen in a healthy brain (Hickey et al., 1991; Perry et al., 1993; Ransohoff and Tani, 1998; Togo et al., 2002; Giraudon et al., 2005). Activated T cells can release pro-inflammatory cytokines such as interferon-γ and TNF-α, which contribute for the activation and antigen-presenting properties of microglia (Gebicke-Haerter, 2001). The role of peripherally derived and brain-invading hematogenous cells for the development of neuroinflammation still remains a matter of discussion and will probably depend on the type of brain damage (Weiner and Selkoe, 2002; Block and Hong, 2005; Schwartz et al., 2006). However, as previously discussed in the case of the intricate interplay between astrocytes and microglia in the build-up of a neuroinflammatory response, there is also a tight interplay between innate and adaptive immune responses regulating the impact of neuroinflammation on brain damage, as will be discussed below.

7. NEUROINFLAMMATION AND NEUROPROTECTION

Probably as important as recognising that neuroinflammatory features are present in different conditions of brain damage, it should be made clear that neuroinflammation is a double-edge sword possibly contributing for brain damage but also for the repair and regeneration of brain tissue (Schwartz and Moalem, 2001; Weiner and Selkoe, 2002; Elward and Gasque, 2003; Kerschensteiner et al., 2003; Schwartz et al., 2003; Marchetti and Abbracchio, 2005). In fact, all cell types potentially involved in mounting a neuroinflammatory response (microglia, astrocytes, macrophages, monocytes, leukocytes) can also directly release or indirectly cause the release of different anti-inflammatory factors, such as IL-4, IL-10 and TGF-β, as well as factors involved in brain tissue repair, such as neurotrophins (Elkabes et al., 1996; Miwa et al., 1997; Heese et al., 1998; Batchelor et al., 1999; Mizuma et al., 1999; Hammarberg et al., 2000; Moalem et al., 2000; Polazzi et al., 2001; Barouch and Schwartz, 2002; Nguyen et al., 2002) These neurotrophins have a broad action on the recruitment, activation and interplay between different types of cells involved in innate immune responses in the brain (reviewed in Kerschensteiner et al., 2003). Thus, neurotrophins can control the migration of hematogenous cells through the blood–brain barrier (Villoslada et al., 2000; Flugel et al., 2001) and prevent the activation of microglia

(Neumann et al., 1998; Wei and Jonakait, 1999), while also controlling the susceptibility of neurons to different insults (reviewed in Castren, 2004). Recent studies also indicate that neurotrophins derived from inflammatory cells also control neurogenesis and, accordingly, NSAIDs or other suppressor of microglia activation can restore neurogenesis (Ekdahl et al., 2003; Monje et al., 2003). This indicates that, apart from their role in promoting an innate immune response in the brain, microglia and astrocytes may switch their function to orchestrate brain repair in the lesioned brain (Hauwel et al., 2005). There is now compelling evidence suggesting that the control of the demise of microglia contribution to neurotoxicity/neuroprotection can have a substantial effect on brain function. For instance, knocking down the intracellular adaptor protein (TYROBP), which controls the activation of type 2 triggering receptor expressed on myeloid cells (TREM2), impairs the phagocytic capacity of microglia and enhances inflammatory gene expression that probably underlies the condition of dementia leukoencephalopathy present in these animals (Nataf et al., 2005; Takahashi et al., 2005).

This potential double role of the inflammatory system in the demise of brain damage is emphasised by the contradictory results obtained in different studies testing the impact of manipulating neuroinflammation on brain damage and/or dysfunction. For instance, it is now evident that there is an inflammatory status in Parkinson's disease patients (Beal, 2003; Chen et al., 2003; Gao et al., 2003; Hartmann et al., 2003; Hirsch et al., 2003; McGeer and McGeer, 2004; Teismann and Schulz, 2004; Morale et al., 2006) and that immunosuppressants, blockade of microglia activation or genetic ablation of TNF-α receptors affords protection in experimental models of Parkinson's disease (Boireau et al., 1997; Matsuura et al., 1997; Wu et al., 2002; Liu and Hong, 2003; Sriram et al., 2006a, b; Wang et al., 2006). Furthermore, chronic treatment with NSAIDs reduces the risk of Parkinson's disease by up to 45% (Chen et al., 2003). In spite of these evidences indicating a deleterious role for neuroinflammation in the course of Parkinson's disease, the injection of IL-1β into the lesioned striatum leads to an increased density of tyrosine hydroxylase reactive fibres and functional recovery of locomotor activity (Herx et al., 2000). Another brain condition where the role of neuroinflammation is also unclear is in Alzheimer's disease (reviewed in Akiyama et al., 2000; McGeer and McGeer, 2002). There is abundant evidence for the presence of a pro-inflammatory status in Alzheimer's disease patients and several neuropathological features characteristic of neuroinflammation are found in the brain of these patients, as well as in patients suffering from other forms of dementia, suggesting the presence of a locally induced chronic inflammatory condition (reviewed in Akiyama et al., 2000; McGeer and McGeer, 2002). A deleterious role of neuroinflammation on the risk of developing Alzheimer's disease was suggested by the observation that patients undergoing NSAIDs therapy to manage heart diseases or rheumatoid arthritis displayed a reduced risk of developing Alzheimer's disease (Zandi et al., 2002; McGeer and McGeer, 2006). However, clinical trials with NSAIDs in Alzheimer's disease patients failed to show consistent benefits (cf. int' Veld et al., 2001; Aisen et al., 2003; van Gool et al., 2003). Also, in the available animal models potentially pertinent for Alzheimer's disease, there is a considerable mismatch between studies indicating that activated microglia can release factors aggravating neuronal viability and studies reaching opposite conclusions, for instance, showing that the complement system or TNF-α actually afford neuroprotection (reviewed in McGeer and McGeer, 2002). Thus, there might be a double and opposite effect of neuroinflammation on Alzheimer's disease: it may be harmful for neurons already under the deleterious influence of fibrillar forms of β-amyloid peptides and it may be beneficial with respect to its ability to remove this deleterious agent. A similar double-edge role for neuroinflammation has been noted in other pathologies, such as stroke or motor neuron diseases (Jones et al., 2002; Sargsyan et al., 2005) where both pro-inflammatory deleterious and repair/regeneration beneficial effects have been associated with the presence of a neuroinflammatory status. The idea emerging from these observations is that neuroinflammation (as inflammation in general) may fulfil a double beneficial/deleterious role according to when and where it is present: at an early stage and under proper control it fulfils a beneficial role affording an effective defensive or clearance mechanism together with a trophic support to allow recovering the damage area; however, if neuroinflammation (or inflammation) evolves to a chronic non-controlled state, then it becomes a deleterious process that may contribute for the amplification of brain damage. Unfortunately, we still lack reliable markers that may allow us to judge the status of neuroinflammation with respect to its dual role.

Another aspect that is relevant to the understanding of the role of neuroinflammation to the development of brain damage is the interplay between the innate and adaptive immune responses in the brain parenchyma. In fact, there is now a burst of interest in the prospect of using therapeutic vaccination as a strategy to manage brain diseases (Weiner and Selkoe, 2002; Angelov et al., 2003; Schwartz et al., 2003; Benner et al., 2004; Schwartz and Kipnis, 2005). This strategy of neuroprotective immunity has been tested in animal models of Parkinson's disease, stroke and Alzheimer's disease, but has so far been translated to humans with promising but still limited success (e.g. Gandy and Heppner, 2005). The use of strategies based on the manipulation of adaptive immune responses to control innate immune responses is based on the particular interplay between these two systems in the brain parenchyma that appears to be central to determining whether neuroinflammation will mainly result in neuroprotection or neurodegeneration. The immune cells infiltrating into the brain are subject to a particular control that includes both suppression and mainly a deviation of their responses. The suppression of T-cell responses is best exemplified by comparative studies showing that the accumulation of T cells in the crush-injured sciatic nerve was significantly greater than in the crush-injured rat optic nerve and, conversely, there was a significantly more extensive death of T cells in the later than in the former model (Hirschberg et al., 1998; Moalem et al., 1999). Probably more relevant for neuroprotection is the evidence showing that microglia and also astrocytes can present leukocytes with cues that determine their particular differentiation towards the more protective T-helper cell type 2/3 (reviewed in Schwartz and Moalem, 2001). These regulatory TH2/3 cells are mostly $CD4^+–CD25^+$ cells that are known to play a key role in the prevention of a number of immune-mediated diseases such as auto-immune disorders or transplant rejection (reviewed in Randolph and Fathman, 2006). They are characterised by the preferential production of anti-inflammatory cytokines (IL-4, IL-10 and TGF-β) in contrast to TH1 cells that mainly release interferon-γ and TNF-α (Randolph and Fathman, 2006). For instance, it is observed that T cells isolated from the brains of mice infected with Sindbis virus encephalitis generated interferon-γ, IL-4 and IL-10 and were defective in the production of IL-2, illustrating the complex regulatory effect of the brain parenchyma on the infiltrating T cells (Irani et al., 1997). This particular effect of the brain milieu on T cells deviation towards regulatory and neuroprotective T-cell helpers translates into a neuroprotective effect of T cells on brain injury. In fact, the local activation of T cells in the brain can induce a particular microglia phenotype and affords a condition of protective auto-immunity (Weiner and Selkoe, 2002; Kerschensteiner et al., 2003; Schwartz and Kipnis, 2005). This concept was first explored in animal models of multiple sclerosis, where it was concluded that glatiramer acetate (a random four amino acid peptide mimicking myelin basic protein antigen presentation) triggers a burst in TH2/3 regulatory T cells which need to be locally activated in the brain to suppress neuroinflammation (Neuhaus et al., 2001). Also, in animal models of Parkinson's disease, the intra-striatal implantation of activated peritoneal leukocytes led to an attenuation of morphological and behavioural deficits characteristic of the disease (Herx et al., 2000; Benner et al., 2004). Likewise, in animal models of Alzheimer's disease, the parental immunisation with synthetic β-amyloid peptides markedly decreases the neurochemical and behavioural deficits associated with this disease (Schenk et al., 1999). Interestingly, a single parental injection with a monoclonal antibody against a β-amyloid peptides produced a near immediate mnemomic benefit in mice models of Alzheimer's disease (Dodart et al., 2002). This participation of T cells in the control of neurotoxicity is further strengthened by observations that brain injury is aggravated in thymectomised rats and can be transferred by splenocytes into naïve rats (Yoles et al., 2001). Another way to confirm the importance of this neuroprotective auto-immune response triggered by brain injury is the observation that there is an aggravation of brain damage by eliminating the normally suppressive effect of $CD4^+$–$CD25^+$ regulatory T cells (Kipnis et al., 2002). All together, these findings suggest that brain injury can trigger a compensatory auto-immune protective response, albeit the control of the magnitude of this response needs to be properly weighted to maximise TH2-mediated neuroprotection and simultaneously avoiding excessive TH1-mediated aggravation of neuronal damage.

Thus, when considering the outcome of a neuroinflammatory status on the control of the viability of injured brain tissue, one needs to consider (1) the time and place of neuroinflammation, and (2) the type of on-going neuroinflammatory reactions, evaluating innate, adaptive and regulatory responses in a time- and space-dependent manner.

8. CONTROL BY ADENOSINE OF PERIPHERAL INFLAMMATION AND NEUROINFLAMMATION

After this general overview on the dual role of neuroinflammation on the outcome of brain injury, we will discuss the possibility that the neuroprotection afforded by adenosine A_{2A} receptors against chronic noxious brain damage in adult animals may be related to their ability to control inflammatory processes.

This hypothesis is mostly based on the crucial role of extracellular adenosine acting through A_{2A} receptors as a "STOP" signal of immune responses, as devised by Sitkovsky's group (Sitkovsky, 2003; Sitkovsky et al., 2004; Sitkovsky and Ohta, 2005). The activation of A_{2A} receptors triggers an increase in the intracellular levels of cAMP, which is long known to vehicle a strong immuno-suppressive effect. In fact, all studies available indicate that the activation of A_{2A} receptors inhibits the release of pro-inflammatory cytokines from different immune cell types, such as macrophages (e.g. Ritchie et al., 1997; Hasko et al., 2000; Leibovich et al., 2002; Pinhal-Enfield et al., 2003), dendritic cells (Panther et al., 2001; Schnurr et al., 2004), monocytes (Bouma et al., 1997; Link et al., 2000; Bshesh et al., 2002; McColl et al., 2006) or T cells (Huang et al., 1997; Apasov et al., 2000; Lappas et al., 2005). Since virtually all types of immune cells are equipped with A_{2A} receptors, no immune cell can escape the inhibitory signaling by adenosine A_{2A} receptors. Thus, the activation of A_{2A} receptors can inhibit the production of pro-inflammatory mediators such as TNF-α, IL-12 and CXCL10 from different immune cell types with the consequent depression of the ability of antigen-presenting cells to initiate and amplify TH1-mediated responses (reviewed in Sitkovsky et al., 2004). In parallel, the activation of A_{2A} receptors can enhance the production of anti-inflammatory factors such as IL-12 or CCL17, indicating a coordinated ability to shift a balance towards an anti-inflammatory state by unbalancing TH2 vs. TH1 responses (reviewed in Sitkovsky and Ohta, 2005). In accordance with their fundamental and non-redundant immuno-suppressive role acting as a second danger system to terminate an immune response, the pharmacological activation of A_{2A} receptors has been shown to confer a robust protection against tissue damage in different organ such as heart (e.g. Maddock et al., 2001; Platts et al., 2003), blood vessels (McPherson et al., 2001), kidney (Okusa et al., 2000; Day et al., 2003), liver (Harada et al., 2000; Day et al., 2004; Odashima et al., 2005a), lung (Khimenko et al., 1995; Ross et al., 1999), intestine (Odashima et al., 2005b), stomach (Odashima et al., 2005c), joints (Cohen et al., 2004) and skin (Peirce et al., 2001; Montesinos et al., 2002). Moreover, A_{2A} receptor blockade actually exacerbates tissue damage involving inflammatory reactions in the periphery (reviewed in Hasko and Cronstein, 2004; Sitkovsky et al., 2004). The group of Joel Linden has been particularly active in identifying the particular A_{2A} receptor-containing cell type that would play a prominent role in this A_{2A} receptor-mediated control of immune response-induced lesion of peripheral tissues (Linden, 2005). In accordance with the ability of A_{2A} receptors to impose a coordinated shift from TH1 into TH2 responses (e.g. Panther et al., 2003), their group recently shown that the key role of A_{2A} receptors in attenuating peripheral tissue damage from ischaemia-reperfusion injury is due to the activation of A_{2A} receptors in CD4$^+$ T cells (Day et al., 2006). This ability of A_{2A} receptors to suppress immune responses not only has a major impact on the management of the recovery of iatrogenic damaged tissue and of tissue damaged by reperfusion injury, but is also anticipated to be of key importance in devising novel promising strategies to burst effective vaccination or to fight tumour growth (Sitkovsky, 2003; Sitkovsky et al., 2004).

In contrast to this clear effect of A_{2A} receptors as the main "OFF" signal of the peripheral inflammatory system, the role of A_{2A} receptors in the control of immune responses in the brain is considerably less explored and certainly less clear. In conditions of acute brain lesion, it appears that the activation of A_{2A} receptors may attenuate neuronal brain damage, possibly through the control of inflammatory processes in the brain. For instance, in situations of haemorrhagic stroke the A_{2A} receptor agonist CGS 21680 inhibits the production of TNF-α mRNA, reduces neutrophil infiltration and decreases the number of TUNEL-positive cells in the area adjacent to the haematoma (Mayne et al., 2001). A similar A_{2A} receptor-mediated control of brain immune responses was reported upon acute brain infection (Sullivan et al., 1999). Also, in model of spinal cord traumatic injury, the administration of selective A_{2A} receptor agonists before or shortly after the injury effectively prevents leukocyte adhesion and consequent tissue damage thus allowing a better functional recovery (Cassada et al., 2001, 2002). However, if A_{2A} receptors are activated long (24 h) after spinal cord injury they now cause a deleterious effect and it is A_{2A} receptor antagonists that are now able to afford tissue protection and functional recovery (communication by Joel Liden at the Eighth International Adenosine and Adenine Nucleotide Symposium).

Interestingly, this deleterious effect of A_{2A} receptors with respect to aggravation of neuroinflammation and contribution for the worsening of brain damage appears to be present in most conditions of chronic noxious insults to the brain (reviewed in Cunha, 2005). For instance in a rodent model of temporal lobe epilepsy, a condition where neuroinflammation is present and is thought to contribute for the amplification of neuronal damage (Vezzani and Granata, 2005), the A_2 receptor antagonist, DMPX, completely prevented the recruitment of activated microglia and tissue damage in the CA3 region of the hippocampus in rats injected with kainate (Lee et al., 2004). However, this might either be due to a direct effect of A_{2A}

receptors known to be present in microglia (Fiebich et al., 1996; Küst et al., 1999) or because A_{2A} receptor antagonists control the evolution of the severity of convulsions, which pre-date microglia recruitment, as observed upon cumulative sub-threshold amygdala kindling (Porciúncula and Cunha, unpublished observations). Thus, we tested in the kainate model of temporal lobe epilepsy the role of A_{2A} receptors, using both a more selective A_{2A} receptor antagonist, SCH 58261, and A_{2A} receptor knockout mice. In this study, we observed that either the pharmacological or genetic blockade of A_{2A} receptors failed to affect the intensity of the kainate-induced convulsions, although it completely abrogated both microglia recruitment and activation in limbic regions as well as the accompanying neuronal damage (communication by Lisiane Porciúncula at the Eighth International Adenosine and Adenine Nucleotide Symposium). A more direct evidence for the ability of A_{2A} receptors to afford neuroprotection by way of controlling neuroinflammation was provided by our (unpublished) observation that a selective A_{2A} receptor antagonist (SCH 58261) can prevent the hippocampal neuronal dysfunction and neurotoxicity triggered by the direct administration of lipopolyssaccharide (LPS), a potent inflammatory trigger and activator of microglia cells (see e.g. Kim et al., 2000; Kloss et al., 2001). Interestingly, SCH 58261 also attenuated the LPS-induced neuroinflammation, as evaluated by the abolishment of LPS-induced increase in IL-1β levels and the recruitment of activated microglia (Rebola, Lynch and Cunha, unpublished observations). This shows that A_{2A} receptor blockade effectively controls neuroinflammation and the neuronal dysfunction and damage resulting from a neuroinflammatory status in the brain. However, these studies do not allow concluding on the involvement of A_{2A} receptors located in microglia since the development of a neuroinflammatory process depends not only on microglia activation, but also on the participation of astrocytes (e.g. Hanisch, 2002) and neurons (e.g. de Simone et al., 2004), as well as on the involvement of infiltrating myeloid cells (see Jensen et al., 1997; Akiyama et al., 2000; Lyons et al., 2000). Jiang-Fan Chen's groups added further evidence supporting the contention that A_{2A} receptor blockade affords neuroprotection by means of a control of neuroinflammation, by showing that brain-resident A_{2A} receptors might only play a minor role in controlling neurodegeneration in an animal model of focal ischaemia, whereas the neuroprotection seem to result from the blockade of A_{2A} receptors located in infiltrating bone marrow-derived cells (Yu et al., 2004). By comparing the infarcted area in the cerebral cortex of γ-irradiated wild-type mice receiving a bone marrow transplant from A_{2A} receptor knockout mice and γ-irradiated A_{2A} receptor knockout mice transplanted with bone marrow from wild-type mice, they found that a superior neuroprotection was observed in the first group (i.e. which possessed A_{2A} receptors in the brain but not in myeloid cells). This proposal that the effect of A_{2A} receptors in myeloid cells is most important for the role of A_{2A} receptors in the control of brain damage is surprising based on the well-established robust role of adenosine in attenuating (rather than exacerbating) inflammation in the periphery (Ohta and Sitkovsky, 2001; reviewed in Hasko and Cronstein, 2004; Sitkovsky et al., 2004). This actually received a direct confirmation in Chen's work, where the authors confirmed that the ablation of A_{2A} receptors actually aggravated (rather than protected) against liver damage (Yu et al., 2004).

However, this ability of A_{2A} receptors in bone marrow cells to control brain damage may not always be so prominent. In fact, as previously mentioned, we have carried out experiment testing if A_{2A} receptors in bone marrow-derived cells were also responsible for the neuroprotection against nigral and striatal dopaminergic toxicity triggered by administration of MPTP, an experimental model of Parkinson's disease. And it was observed that the selective inactivation of A_{2A} receptors in bone marrow-derived cells failed to afford the neuroprotection observed by global blockade of A_{2A} receptors (either using a selective A_{2A} receptor antagonist or global A_{2A} receptor knockout mice) (Chen et al., 2001). Furthermore, the intracerebroventricular injection of a selective A_{2A} receptor antagonist re-instated neuroprotection against MPTP-induced dopaminergic toxicity in forebrain A_{2A} receptor knockout mice indicating that it is A_{2A} receptors located in the brain but in non-neuronal that are responsible for the control of MPTP-induced dopaminergic toxicity. This raises the possibility that the neuroprotection afforded by A_{2A} receptor blockade may result from the control of glial cell-mediated neuroinflammation. Furthermore, since these glial cells are immune competent cells selectively located in the brain parenchyma, it is possible that it is a particular role fulfilled by A_{2A} receptors in these cells that may be responsible for this paradoxical opposite effect of A_{2A} receptors in the control of peripheral and brain tissue damage, an idea that will be developed below.

Both microglia and astrocytes are endowed with A_{2A} receptors (Fredholm et al., 2005; Hasko et al., 2005; van Calker and Biber, 2005). The presence and functional relevance of A_{2A} receptors in microglia has mostly been evaluated in cultured cell models (but see Angulo et al., 2003), whereas the presence and function of A_{2A} receptors in astrocytes is more evident in ex vivo brain preparations rather than in cultures (cf. Lee et al., 2003; Saura et al., 2005). In cultured astrocytes, A_{2A} receptors have been reported to control their proliferation (Hindley et al., 1994) and their reactivity upon stimulation with basic fibroblast growth factor (Brambilla et al., 2003). From the neurochemical point of view, it has also been reported that A_{2A} receptors might enhance the release of NO from cultured astrocytes (Brodie et al., 1998). In cultured microglia cells, initial studies revealed that A_1 and A_{2A} receptors cooperated in controlling calcium transients (Ogata et al., 1996) and proliferation (Gebicke-Haerter et al., 1996). Activation of A_{2A} receptors has also been reported to activate cyclooxygenase-2 (Fiebich et al., 1996), trigger nerve growth factor expression (Heese et al., 1997) and regulate K^+ channel expression, a marker of microglia activation (Küst et al., 1999). Finally, a recent study described the ability of the A_{2A} receptor agonist, CGS 21680 to enhance the LPS-induced expression of iNOS and release of nitrates, without measurable effects on the release of TNF-α (Saura et al., 2005).

Clearly, the genetic inactivation studies carried by the group of Jiang-Fan Chen have provided compelling evidence to consider that the A_{2A} receptor-mediated control of neuro-inflammation might be the strongest candidate mechanism to explain the neuroprotection against brain injury afforded by A_{2A} receptor blockade in adult animals. However, it also seems evident that this conclusion might only be valid for particular conditions of brain injury, where the neuroinflammatory process acquires a chronic profile and neuro-degeneration develops in a gradual rather than catastrophic manner. Finally, it is worth noting that our knowledge on the possible roles of A_{2A} receptors in glial cells is very limited and restricted to cultured cells (obtained from pups or newborns) that still need to prove to be a useful model to understand the role of glial cells in native brain tissue of adult animals (discussed in Melchior et al., 2006).

9. WHY MIGHT THE A_{2A} RECEPTOR-MEDIATED CONTROL OF PERIPHERAL INFLAMMATION AND NEUROINFLAMMATION BE DIFFERENT?

Having clearly stated that our current knowledge of the particular role of A_{2A} receptors in the control of neuroinflammation and neuronal damage is clearly insufficient to provide a coherent mechanism of A_{2A} receptor-mediated neuroprotection, it is nevertheless important to discuss possible reasons that may help to understand why the role of A_{2A} receptors might be opposite in the control of peripheral inflammation and acute inflammatory-mediated brain damage (where it is the activation of A_{2A} receptors that affords protection) and in the control of neuroinflammation-associated chronic brain damage (where the available evidence indicate that it is the blockade of A_{2A} receptors that affords protection). This dichotomy brings us to the concept of the brain as an immune-privileged organ, i.e. to the available evidence suggesting that the processes of neuroinflammation may be different from these occurring in peripheral inflammation. Thus, this pioneering realisation that there is one and the same system (A_{2A} receptors) that acts as controllers of immune responses both in the brain and in the periphery but with exactly opposite effects further pushes us to consider peripheral and central immune responses as two definitively different process involving similar cell types and mediators. We will selectively discuss some possible reasons that may underline these opposite effects of A_{2A} receptors in the control of immune responses and associated tissue damage in the periphery (and in acute brain damage) vs. their role in the control of chronic neuroinflammatory-associated conditions of brain damage.

One obvious difference between central and peripheral inflammation resides in the type of cells involved in these two processes and the existence of cell types in the brain (microglia and astrocytes) that do not participate in peripheral inflammation. As already discussed, it is still largely unclear if the microglia or astrocytic responses constitute an overall beneficial or deleterious process for the viability of brain tissue. In fact, albeit astrogliosis and microgliosis are cardinal features of the chronically injured brain, we have previously alluded to different studies that report either beneficial or deleterious consequences related with microglia and astrocytic activation. Furthermore, too little is yet known on the role of A_{2A} receptors in glial cells to build a coherent hypothesis to explain the particular beneficial role of A_{2A} receptor blockade on brain tissue viability involving a control of brain immune responses. Thus, there is a clear need to detail our knowledge on (1) the role of microgial with different degrees of activation on the control of immune responses; (2) the role of adaptive immune responses on the control of innate immune responses involving microglia and infiltrating macrophages; (3) the role of astrocytes in the control of this interaction; (4) the role of A_{2A} receptors in microglia and in the control by astrocytes of the type and degree of activation of microglia; and (5) the role of A_{2A} receptors in the interaction between innate and adaptive immune responses in the brain parenchyma. Only with this information available will it be possible to further consider the possibility that activation of A_{2A} receptors in glial cells may impose a particular development of neuroinflammatory responses that cannot occur in peripheral tissues where these glial cells are not present.

Another critical parameter to understand neuroinflammatory processes is to realise the important control imposed by neurons on the development of a neuroinflammatory process, an effect that is of minor importance in peripheral inflammation. An excellent example to illustrate this neuronal-induced control of the demise of microglia contribution to neurotoxicity/ neuroprotection was provided by a recently published study based on the knocking down of CX3CR1 (Cardona et al., 2006). This receptor is expressed in microglia cells and responds to CX3CL1 released by neurons (Harrison et al., 1998). CX3CR1 deficiency was associated with increased microglia transformation and increased neuronal cell death in three different in vivo models of brain insults, namely LPS injection, MPTP-model of Parkinson's disease and in the SOD^{G93A} model of amyotrophic lateral sclerosis (Cardona et al., 2006).

Thus, one has to consider the existence of particular characteristics of the brain tissue and in the time-course of the development of immune responses in the brain and in the periphery to seek for possible differences between peripheral and central inflammatory processes. In fact, the brain parenchyma may create a context-dependent contribution of infiltrating bone marrow-derived cells for the development of a brain dysfunction and/or lesion that would be potentially identical or different from that occurring in the periphery. If there is an abrupt rupture of the blood–brain barrier and a massive invasion of haematogenous cells into the brain parenchyma, then the influence of the brain environment would be limited and the inflammatory response would respond to A_{2A} receptors as occurs in peripheral inflammation, i.e. it is the activation of A_{2A} receptors that will limit inflammation and brain damage, as occurs in situations of acute brain infection or haemorrhage. In contrast, when there is a brain insult that causes a progressive disruption of the blood–brain barrier, such as upon ischaemia, then there is an on-going (rather than abrupt) infiltration and fixation of haematogenous cells in the brain parenchyma. These cells would play a major role in the demise of neuroinflammation-related neuronal damage, but would be under a tight control by factors released from the brain parenchyma. This would be the main reason responsible for the shift in effect of A_{2A} receptors, which might now aggravate rather than prevent the release and/or effect of pro-inflammatory mediators. Finally, if there is an insidious and chronic insult to brain tissue without significant perturbation of the blood–brain barrier, then the neuroinflammatory reaction would be mainly dependent on brain-resident microglia cells. Again, these brain-resident glial cells would be under tight control by factors released by the brain parenchyma which would force the A_{2A} receptors to behave as an aggravating factor.

This context-dependent influence of the brain parenchyma on the demise of neuroinflammation may be caused by the presence of a series of transmitter and modulatory molecules in the brain that are not present in the periphery. In fact, microdialysis studies carried out in different brain regions identify the presence of extracellular molecules that are not present (or that are present in much lower amounts) in peripheral tissues. For instance, there is a predominant participation of amino acids (glutamate, GABA, taurine) in processes of information transfer between cells in the brain that is not present in the periphery. Likewise, aminergic molecules (such as noradrenaline, dopamine and acetylcholine) are present in the brain in concentration 10 times higher than in peripheral tissues and their extracellular levels are known to be regulated not only in an activity-dependent manner but most importantly in a context of signaling of noxious conditions. Finally, the repertoire of peptides used as co-transmitters in the brain largely exceeds that of the periphery. Any of these classes of substances are known to exert powerful effects in immune cells, but the extent to which they participate in the control of neuroinflammatory processes is still largely unknown. In fact, one may even conceive a radical situation in which the triggers of neuroinflammation might be different from these operating in peripheral inflammation, i.e. the primary "danger signals" recruiting a neuroinflammatory or a peripheral inflammatory process could be different. And consequently, if the pathways leading to the build-up of an immune response are different, it is quite likely that modulatory pathways (such as those operated by A_{2A} receptors) might also be different. In this context, it is obviously questionable if the golden paradigm to activate brain immune cell responses (use of LPS) has any major significance for the study of neuroinflammation in the context of chronic noxious brain conditions (where bacterial infections are unlikely to be present). It would be interesting to test if increased extracellular levels of glutamate (eventually together with ATP or uric acid) are sufficient to trigger a brain immune response and if the intracellular pathways associated with these eventual alternative triggers of neuroinflammation have any similarity to these involved in triggering an immune response in the periphery.

An alternative reason to help explaining the different role of A_{2A} receptors in the control of neuroinflammation and peripheral inflammation may reside in the intensity of the immune responses in space and in time in the brain and in the periphery. Here, one should consider several possible parameters either individually or in conjunction. Thus, it appears obvious that the levels of pro- and anti-inflammatory mediators found in an inflamed peripheral tissue and in a brain region suffering an insult are different, with a clear trend towards greater levels of inflammatory mediators in the periphery compared to the brain parenchyma. There is therefore the possibility that a modulatory system might cause opposite effects according to the amplitude of the response. And it should be pointed out that the A_{2A} receptor is an excellent example of a modulatory system that has opposite (beneficial/deleterious) effects according to the intensity of a particular recruited response (reviewed in Ferré et al., 2005). For instance, when considering the ability of A_{2A} receptors to modulate NMDA receptors, the facilitation by A_{2A} receptors of NMDA receptor function is of crucial importance to allow the expression of an NMDA-dependent synaptic potentiation when using near-threshold frequency trains (Cunha, 2005); however, if the release of glutamate is too intense, then this A_{2A} receptor-mediated facilitation of NMDA responses becomes deleterious, since it aggravates the NMDA receptor-induced ability to damage neurons (Tebano et al., 2005). In parallel with this aspect of different intensities of immune responses in the brain and in the periphery, one should look at different time courses of the build up of an immune response in the periphery and in the brain (Fig. 2). In fact, one of the characteristics of peripheral immune responses is their rapid kinetics of onset and the relatively short time required to build up a maximally efficient response. This contrasts with immune responses in the brain parenchyma that have a considerably slower onset and especially a considerably longer lag period before a "maximal" (if this has ever been quantified) responses might be reached. Again, according to the time-window of opportunity of a modulatory system, one can easily foresee its possible different impact according to the timing of its recruitment vis-à-vis the evolution of the immune response. Obviously, if one joins these two parameters together, one ends up with potentially different processes in the periphery and in the brain that may be amenable to a different regulation by A_{2A} receptors: in the periphery A_{2A} receptors could stop a rapid and robust immune response (acute) whereas in the brain A_{2A} receptors could actually aggravate a slow and insidious immune response (chronic). Certainly, the efforts currently being made to distinguish markers of acute and chronic inflammation will be of major help to address this hypothesis.

In summary, there is a paradoxical modulation by A_{2A} receptors of peripheral inflammation and of neuroinflammation in chronic noxious brain conditions. This constitutes an opportunity to question the similarities and differences between the processes of peripheral and central inflammation. However, a greater understanding of the processes of brain immune responses, both of its triggers and of the subtle interplay between different cell types, will certainly be required before this question might be solved.

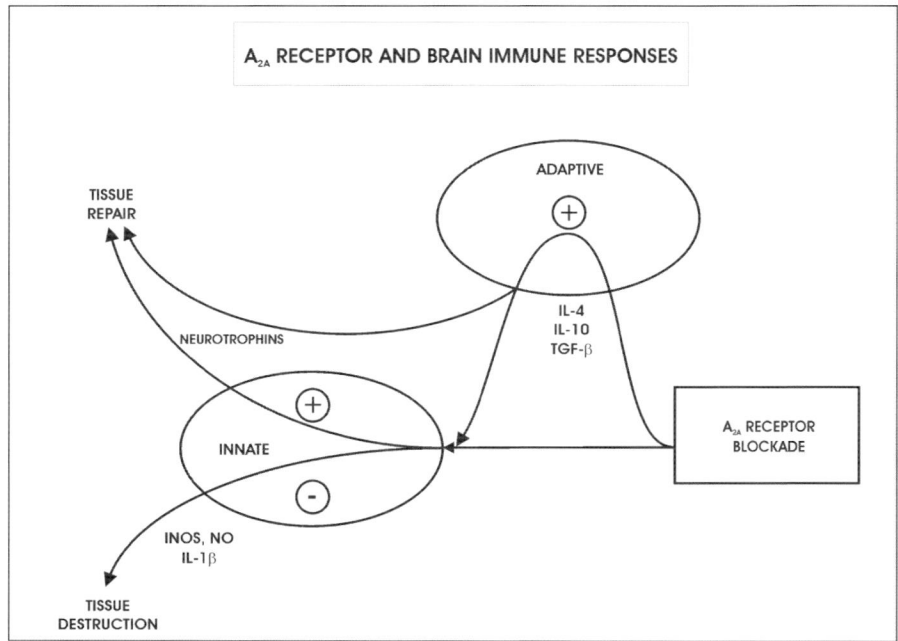

Fig. 2. Potential ability of adenosine A_{2A} receptor blockade to interfere with adaptive and innate immune responses in the brain. Since recent evidence indicates that the neuroprotective effect associated with the blockade of A_{2A} receptors might involve a control of neuroinflammation, it is proposed that the blockade of these receptors may potentiate pathways associated with tissue repair (boosting adaptive immune responses and/or favouring the release of neurotrophins associated with cells responsible for innate immune responses). In parallel, the blockade of A_{2A} receptors might also decrease the production and release of pro-inflammatory substances produced during innate immune responses

10. ACKNOWLEDGMENTS

RAC was supported by a grant from Fundação para a Ciência e a Tecnologia (POCI/BIA-BCM/59980/2004).

11. REFERENCES

Aisen, P.S., Schafer, K.A., Grundman, M., Pfeiffer, E., Sano, M., Davis, K.L., Farlow, M.R., Jin, S., Thomas, R.G. and Thal, L.J., Alzheimer's Disease Cooperative Study, 2003, Effects of rofecoxib or naproxen vs placebo on Alzheimer disease progression: a randomized controlled trial. *JAMA* **289**: 2819.
Akiyama, H., Barger, S., Barnum, S., Bradt, B., Bauer, J., Cole, G.M., Cooper, N.R., Eikelenboom, P., Emmerling, M., Fiebich, B.L., Finch, C.E., Frautschy, S., Griffin, W.S., Hampel, H., Hull, M., Landreth, G., Lue, L., Mrak, R., Mackenzie, I.R., McGeer, P.L., et al., 2000, Inflammation and Alzheimer's disease. *Neurobiol. Aging* **21**: 383.
Aloisi, F., Ambrosini, E., Columba-Cabezas, S., Magliozzi, R. and Serafini, B., 2001, Intracerebral regulation of immune responses. *Ann. Med.* **33**: 510.
Andersen, J.K., 2004, Oxidative stress in neurodegeneration: cause or consequence? *Nat. Med.* **10**: S18.

Angelov, D.N., Waibel, S., Guntinas-Lichius, O., Lenzen, M., Neiss, W.F., Tomov, T.L., Yoles, E., Kipnis, J., Schori, H., Reuter, A., Ludolph, A., and Schwartz, M., 2003, Therapeutic vaccine for acute and chronic motor neuron diseases: implications for amyotrophic lateral sclerosis. *Proc. Natl Acad. Sci. USA* **100**: 4790.

Angulo, E., Casadó, V., Mallol, J., Canela, E.I., Vinals, F., Ferrer, I., Lluis, C. and Franco, R., 2003, A_1 adenosine receptors accumulate in neurodegenerative structures in Alzheimer disease and mediate both amyloid precursor protein processing and tau phosphorylation and translocation. *Brain Pathol.* **13**: 440.

Apasov, S., Chen, J.F., Smith, P. and Sitkovsky, M., 2000, A_{2A} receptor dependent and A_{2A} receptor independent effects of extracellular adenosine on murine thymocytes in conditions of adenosine deaminase deficiency. *Blood* **95**: 3859.

Baldwin, S.A., Beal, P.R., Yao, S.Y., King, A.E., Cass, C.E. and Young, J.D., 2004, The equilibrative nucleoside transporter family, SLC29. *Pflugers Arch.* **447**: 735.

Barouch, R. and Schwartz, M., 2002, Autoreactive T cells induce neurotrophin production by immune and neural cells in injured rat optic nerve: implications for protective autoimmunity. *FASEB J.* **16**: 1304.

Batchelor, P.E., Liberatore, G.T., Wong, J.Y., Porritt, M.J., Frerichs, F., Donnan, G.A. and Howells, D.W., 1999, Activated macrophages and microglia induce dopaminergic sprouting in the injured striatum and express brain-derived neurotrophic factor and glial cell line-derived neurotrophic factor. *J. Neurosci.* **19**: 1708.

Beal, M.F., 2003, Mitochondria, oxidative damage, and inflammation in Parkinson's disease. *Ann. N.Y. Acad. Sci.* **991**: 120.

Bechmann, I., 2003, Failed central nervous system regeneration: a downside of immune privilege? *Neuromolecular Med.* **7**: 217.

Benner, E.J., Mosley, R.L., Destache, C.J., Lewis, T.B., Jackson-Lewis, V., Gorantla, S., Nemachek, C., Green, S.R., Przedborski, S. and Gendelman, H.E., 2004, Therapeutic immunization protects dopaminergic neurons in a mouse model of Parkinson's disease. *Proc. Natl Acad. Sci. USA* **101**: 9435.

Block, M.L. and Hong, J.S., 2005, Microglia and inflammation-mediated neurodegeneration: multiple triggers with a common mechanism. *Prog. Neurobiol.* **76**: 77.

Boireau, A., Bordier, F., Dubedat, P., Peny, C. and Imperato, A., 1997, Thalidomide reduces MPTP-induced decrease in striatal dopamine levels in mice. *Neurosci. Lett.* **234**: 123.

Bouma, M.G., Jeunhomme, T.M., Boyle, D.L., Dentener, M.A., Voitenok, N.N., van den Wildenberg, F.A. and Buurman, W.A., 1997, Adenosine inhibits neutrophil degranulation in activated human whole blood: involvement of adenosine A_2 and A_3 receptors. *J. Immunol.* **158**: 5400.

Brambilla, R., Cottini, L., Fumagalli, M., Ceruti, S. and Abbracchio, M.P., 2003, Blockade of A_{2A} adenosine receptors prevents basic fibroblast growth factor-induced reactive astrogliosis in rat striatal primary astrocytes. *Glia* **43**: 190.

Brodie, C., Blumberg, P.M. and Jacobson, K.A., 1998, Activation of the A_{2A} adenosine receptor inhibits nitric oxide production in glial cells. *FEBS Lett.* **429**: 139.

Bshesh, K., Zhao, B., Spight, D., Biaggioni, I., Feokistov, I., Denenberg, A., Wong, H.R. and Shanley, T.P., 2002, The A_{2A} receptor mediates an endogenous regulatory pathway of cytokine expression in THP-1 cells. *J. Leukoc. Biol.* **72**: 1027.

Cardona, A.E., Pioro, E.P., Sasse, M.E., Kostenko, V., Cardona, S.M., Dijkstra, I.M., Huang, D., Kidd, G., Dombrowski, S., Dutta, R., Lee, J.C., Cook, D.N., Jung, S., Lira, S.A., Littman, D.R. and Ransohoff, R.M., 2006, Control of microglial neurotoxicity by the fractalkine receptor. *Nat. Neurosci.* **9**: 917.

Carson, M.J. and Sutcliffe, J.G., 1999, Balancing function vs. self defense: the CNS as an active regulator of immune responses. *J. Neurosci. Res.* **55**: 1.

Cassada, D.C., Tribble, C.G., Laubach, V.E., Nguyen, B.N., Rieger, J.M., Linden, J., Kaza, A.K., Long, S.M., Kron, I.L. and Kern, J.A., 2001, An adenosine A_{2A} agonist, ATL-146e, reduces paralysis and apoptosis during rabbit spinal cord reperfusion. *J. Vasc. Surg.* **34**: 482.

Cassada, D.C., Tribble, C.G., Long, S.M., Laubach, V.E., Kaza, A.K., Linden, J., Nguyen, B.N., Rieger, J.M., Fiser, S.M., Kron, I.L. and Kern, J.A., 2002, Adenosine A_{2A} analogue ATL-146e reduces systemic tumor necrosing factor-alpha and spinal cord capillary platelet-endothelial cell adhesion molecule-1 expression after spinal cord ischemia. *J. Vasc. Surg.* **35**: 994.

Castren, E., 2004, Neurotrophins as mediators of drug effects on mood, addiction, and neuroprotection. *Mol. Neurobiol.* **29**: 289.

Chen, J.F., Xu, K., Petzer, J.P., Staal, R., Xu, Y.H., Beilstein, M., Sonsalla, P.K., Castagnoli, K., Castagnoli, N. Jr. and Schwarzschild, M.A., 2001, Neuroprotection by caffeine and A_{2A} adenosine receptor inactivation in a model of Parkinson's disease. *J. Neurosci.* **21**: RC143.

Chen, H., Zhang, S.M., Hernan, M.A., Schwarzschild, M.A., Willett, W.C., Colditz, G.A., Speizer, F.E. and Ascherio, A., 2003, Nonsteroidal anti-inflammatory drugs and the risk of Parkinson disease. *Arch. Neurol.* **60**: 1059.

Ciruela, F., Casadó, V., Rodrigues, R.J., Lujan, R., Burgueño, J., Canals, M., Borycz, J., Rebola, N., Goldberg, S.R., Mallol, J., Cortes, A., Canela, E.I., Lopez-Gimenez, J.F., Milligan, G., Lluis, C., Cunha, R.A., Ferré, S. and

Franco, R., 2006, Presynaptic control of striatal glutamatergic neurotransmission by adenosine A_1–A_{2A} receptor heteromers. *J. Neurosci.* **26**: 2080.
Coelho, J.E., de Mendonça, A. and Ribeiro, J.A., 2000, Presynaptic inhibitory receptors mediate the depression of synaptic transmission upon hypoxia in rat hippocampal slices. *Brain Res.* **869**: 158.
Coelho, J.E., Rebola, N., Fragata, I., Ribeiro, J.A., de Mendonça, A. and Cunha, R.A., 2006, Hypoxia-induced desensitization and internalization of adenosine A_1 receptors in the rat hippocampus. *Neuroscience* **138**: 1195.
Cohen, S.B., Gill, S.S., Baer, G.S., Leo, B.M., Scheld, W.M. and Diduch, D.R., 2004, Reducing joint destruction due to septic arthrosis using an adenosine 2A receptor agonist. *J. Orthop. Res.* **22**: 427.
Cunha, R.A., 2001, Adenosine as a neuromodulator and as a homeostatic regulator in the nervous system: different roles, different sources and different receptors. *Neurochem. Int.* **38**: 107.
Cunha, R.A., 2005, Neuroprotection by adenosine in the brain: from A_1 receptor activation to A_{2A} receptor blockade. *Purinergic Signal* **1**: 111.
Cunha, R.A., Correia-de-Sá, P., Sebastião, A.M. and Ribeiro, J.A., 1996a, Preferential activation of excitatory adenosine receptors at rat hippocampal and neuromuscular synapses by adenosine formed from released adenine nucleotides. *Br. J. Pharmacol.* **119**: 253.
Cunha, R.A., Vizi, E.S., Ribeiro, J.A. and Sebastião, A.M., 1996b, Preferential release of ATP and its extracellular catabolism as a source of adenosine upon high- but not low-frequency stimulation of rat hippocampal slices. *J. Neurochem.* **67**: 2180.
Cunha, G.M.A., Canas, P.M., Oliveira, C.R. and Cunha, R.A., 2006, Increased density and synapto-protective effect of adenosine A_{2A} receptors upon sub-chronic restraint stress. *Neuroscience* **141**: 1775.
Davalos, D., Grutzendler, J., Yang, G., Kim, J.V., Zuo, Y., Jung, S., Littman, D.R., Dustin, M.L. and Gan, W.B., 2005, ATP mediates rapid microglial response to local brain injury in vivo. *Nat. Neurosci.* **8**: 752.
Day, Y.J., Huang, L., McDuffie, M.J., Rosin, D.L., Ye, H., Chen, J.F., Schwarzschild, M.A., Fink, J.S., inden, J. and Okusa, M.D., 2003, Renal protection from ischemia mediated by A_{2A} adenosine receptors on bone marrow-derived cells. *J. Clin. Invest.* **112**: 883.
Day, Y.J., Marshall, M.A., Huang, L., McDuffie, M.J., Okusa, M.D. and Linden, J., 2004, Protection from ischemic liver injury by activation of A_{2A} adenosine receptors during reperfusion: inhibition of chemokine induction. *Am. J. Physiol.* **286**: G285.
Day, Y.L., Huang, L., Ye, N., Li, L., Linden, J. and Okusa, M.D., 2006, Renal ischemia-reperfusion injury and adenosine 2A receptor-mediated tissue protection: the role of CD4[+] T cells and INF-γ. *J. Immunol.* **176**: 3108.
de Mendonça, A. and Ribeiro, J.A., 1997, Adenosine and neuronal plasticity. *Life Sci.* **60**: 245.
de Mendonça, A., Sebastião, A.M. and Ribeiro, J.A., 2000, Adenosine: does it have a neuroprotective role after all? *Brain Res. Rev.* **33**: 258.
de Simone, R., Ajmone-Cat, M.A. and Minghetti, L., 2004, Atipical anti-inflammatory activation of microglia induced by apoptotic neurons: possible role of phosphatidylserine–phosphatidylserine receptor interaction. *Mol. Neurobiol.* **29**: 197.
Dixon, A.K., Widdowson, L. and Richardson, P.J., 1997, Desensitisation of the adenosine A_1 receptor by the A_{2A} receptor in the rat striatum. *J. Neurochem.* **69**: 315.
Dodart, J.C., Bales, K.R., Gannon, K.S., Greene, S.J., DeMattos, R.B., Mathis, C., DeLong, C.A., Wu, S., Wu, X., Holtzman, D.M. and Paul, S.M., 2002, Immunization reverses memory deficits without reducing brain Abeta burden in Alzheimer's disease model. *Nat. Neurosci.* **5**: 452.
Dong, Y. and Benveniste, E.N., 2001, Immune function of astrocytes. *Glia* **36**: 180.
Duarte, J.M., Oliveira, C.R., Ambrósio, A.F. and Cunha, R.A., 2006, Modification of adenosine A_1 and A_{2A} receptor density in the hippocampus of streptozotocin-induced diabetic rats. *Neurochem. Int.* **48**: 144.
Dunwiddie, T.V., 1999, Adenosine and suppression of seizures. *Adv. Neurol.* **79**: 1001.
Dunwiddie, T.V. and Masino, S.A., 2001 The role and regulation of adenosine in the central nervous system. *Annu. Rev. Neurosci.* **24**: 31.
Ekdahl, C.T., Claasen, J.H., Bonde, S., Kokaia, Z. and Lindvall, O., 2003, Inflammation is detrimental for neurogenesis in adult brain. *Proc. Natl Acad. Sci. USA* **100**: 13632.
Elkabes, S., DiCicco-Bloom, E.M. and Black, I.B., 1996, Brain microglia/macrophages express neurotrophins that selectively regulate microglial proliferation and function. *J. Neurosci.* **16**: 2508.
Elward, K. and Gasque, P., 2003, "Eat me" and "don't eat me" signals govern the innate immune response and tissue repair in the CNS: emphasis on the critical role of the complement system. *Mol. Immunol.* **40**: 85.
Farber, K. and Kettenmann, H., 2005, Physiology of microglial cells. *Brain Res. Rev.* **48**: 133.
Faulkner, J.R., Herrmann, J.E., Woo, M.J., Tansey, K.E., Doan, N.B. and Sofroniew, M.V., 2004, Reactive astrocytes protect tissue and preserve function after spinal cord injury. *J. Neurosci.* **24**: 2143.
Fedele, D.E., Li, T., Lan, J.Q., Fredholm, B.B. and Boison, D., 2006 Adenosine A_1 receptors are crucial in keeping an epileptic focus localized. *Exp. Neurol.* **200**: 184.

Fellin, T. and Carmignoto, G., 2004, Neurone-to-astrocyte signaling in the brain represents a distinct multifunctional unit. *J. Physiol.* **559**: 3.

Ferré, S., Borycz, J., Goldberg, S.R., Hope, B.T., Morales, M., Lluis, C., Franco, R., Ciruela, F. and Cunha, R.A., 2005, Role of adenosine in the control of homosynaptic plasticity in striatal excitatory synapses. *J. Integr. Neurosci.* **4**: 445.

Fiebich, B.L., Biber, K., Lieb, K., van Calker, D., Berger, M., Bauer, J. and Gebicke-Haerter, P.J., 1996, Cyclooxygenase-2 expression in rat microglia is induced by adenosine A2a-receptors. *Glia* **18**: 152.

Flugel, A., Matsumuro, K., Neumann, H., Klinkert, W.E., Birnbacher, R., Lassmann, H., Otten, U. and Wekerle, H., 2001, Anti-inflammatory activity of nerve growth factor in experimental autoimmune encephalomyelitis: inhibition of monocyte transendothelial migration. *Eur. J. Immunol.* **31**: 11.

Fredholm, B.B., 1997, Adenosine and neuroprotection. *Int. Rev. Neurobiol.* **40**: 259.

Fredholm, B.B., IJzerman, A.P., Jacobson, K.A., Klotz, K.N. and Linden, J., 2001, International Union of Pharmacology. XXV. Nomenclature and classification of adenosine receptors. *Pharmacol. Rev.* **53**: 527.

Fredholm, B.B., Cunha, R.A. and Svenningsson, P., 2003, Pharmacology of adenosine A_{2A} receptors and therapeutic applications. *Curr. Top. Med. Chem.* **3**: 413.

Fredholm, B.B., Chen, J.F., Cunha, R.A., Svenningsson, P. and Vaugeois, J.M., 2005, Adenosine and brain function. *Int. Rev. Neurobiol.* **63**: 191.

Gandy, S. and Heppner, F.L., 2005, Alzheimer's amyloid immunotherapy: quo vadis? *Lancet Neurol.* **4**: 452.

Gao, H.M., Liu, B., Zhang, W. and Hong, J.S., 2003, Novel anti-inflammatory therapy for Parkinson's disease. *Trends Pharmacol. Sci.* **24**: 395.

Gasque, P., Dean, Y.D., McGreal, E.P., VanBeek, J. and Morgan, B.P., 2000, Complement components of the innate immune system in health and disease in the CNS. *Immunopharmacology* **49**: 171.

Gebicke-Haerter, P.J., 2001, Microglia in neurodegeneration: molecular aspects. *Microsc. Res. Tech.* **54**: 47.

Gebicke-Haerter, P.J., Christoffel, F., Timmer, J., Northoff, H., Berger, M. and Van Calker, D., 1996, Both adenosine A_1- and A_2-receptors are required to stimulate microglial proliferation. *Neurochem. Int.* **29**: 37.

Gerevich, Z., Wirkner, K. and Illes, P., 2002, Adenosine A_{2A} receptors inhibit the N-methyl-D-aspartate component of excitatory synaptic currents in rat striatal neurons. *Eur. J. Pharmacol.* **451**: 161.

Giovannoni, G. and Baker, D., 2003, Inflammatory disorders of the central nervous system. *Curr. Opin. Neurol.* **16**: 347.

Giraudon, P., Vincent, P. and Vuaillat, C., 2005, T-cells in neuronal injury and repair: semaphorins and related T-cell signals. *Neuromol. Med.* **7**: 207.

Greene, R.W. and Haas, H.L., 1991, The electrophysiology of adenosine in the mammalian central nervous system. *Prog. Neurobiol.* **36**: 329.

Hammarberg, H., Lidman, O., Lundberg, C., Eltayeb, S.Y., Gielen, A.W., Muhallab, S., Svenningsson, A., Linda, H., van Der Meide, P.H., Cullheim, S., Olsson, T. and Piehl, F., 2000, Neuroprotection by encephalomyelitis: rescue of mechanically injured neurons and neurotrophin production by CNS-infiltrating T and natural killer cells. *J. Neurosci.* **20**: 5283.

Hanisch, U.K., 2002, Microglia as a source and target of cytokines. *Glia* **40**: 140.

Harada, N., Okajima, K., Murakami, K., Usune, S., Sato, C., Ohshima, K. and Katsuragi, T., 2000, Adenosine and selective A_{2A} receptor agonists reduce ischemia/reperfusion injury of rat liver mainly by inhibiting leukocyte activation. *J. Pharmacol. Exp. Ther.* **294**: 1034.

Harrison, J.K., Jiang, Y., Chen, S., Xia, Y., Maciejewski, D., McNamara, R.K., Streit, W.J., Salafranca, M.N., Adhikari, S., Thompson, D.A., Botti, P., Bacon, K.B. and Feng, L., 1998, Role for neuronally derived fractalkine in mediating interactions between neurons and CX3CR1-expressing microglia. *Proc. Natl Acad. Sci. USA* **95**: 10896.

Hartmann, A., Hunot, S. and Hirsch, E.C., 2003, Inflammation and dopaminergic neuronal loss in Parkinson's disease: a complex matter. *Exp. Neurol.* **184**: 561.

Hasko, G. and Cronstein, B.N., 2004, Adenosine: an endogenous regulator of innate immunity. *Trends Immunol.* **25**: 33.

Hasko, G., Kuhel, D.G., Chen, J.F., Schwarzschild, M.A., Deitch, E.A., Mabley, J.G., Marton, A. and Szabo, C., 2000, Adenosine inhibits IL-12 and TNF-α production via adenosine A_{2a} receptor-dependent and independent mechanisms. *FASEB J.* **14**: 2065.

Hasko, G., Pacher, P., Vizi, E.S. and Illes, P., 2005, Adenosine receptor signaling in the brain immune system. *Trends Pharmacol. Sci.* **26**: 511.

Hauwel, M., Furon, E., Canova, C., Griffiths, M., Neal, J. and Gasque, P., 2005, Innate (inherent) control of brain infection, brain inflammation and brain repair: the role of microglia, astrocytes, "protective" glial stem cells and stromal ependymal cells. *Brain Res. Rev.* **48**: 220.

Heese, K., Fiebich, B.L., Bauer, J. and Otten, U., 1997, Nerve growth factor (NGF) expression in rat microglia is induced by adenosine A_{2a}-receptors. *Neurosci. Lett.* **231**: 83.
Heese, K., Hock, C. and Otten, U., 1998, Inflammatory signals induce neurotrophin expression in human microglial cells. *J. Neurochem.* **70**: 699.
Herrera, A.J., Tomas-Camardiel, M., Venero, J.L., Cano, J. and Machado, A., 2005, Inflammatory process as a determinant factor for the degeneration of substantia nigra dopaminergic neurons. *J. Neural Transm.* **112**: 111.
Herx, L.M., Rivest, S. and Yong, V.W., 2000, Central nervous system-initiated inflammation and neurotrophism in trauma: IL-1β is required for the production of ciliary neurotrophic factor. *J. Immunol.* **165**: 2232.
Hickey, W.F., Hsu, B.L. and Kimura, H., 1991, T-lymphocyte entry into the central nervous system. *J. Neurosci. Res.* **28**: 254.
Hindley, S., Herman, M.A. and Rathbone, M.P., 1994, Stimulation of reactive astrogliosis in vivo by extracellular adenosine diphosphate or an adenosine A2 receptor agonist. *J. Neurosci. Res.* **38**: 399.
Hirsch, E.C., Breidert, T., Rousselet, E., Hunot, S., Hartmann, A. and Michel, P.P., 2003, The role of glial reaction and inflammation in Parkinson's disease. *Ann. N.Y. Acad. Sci.* **991**: 214.
Hirschberg, D.L., Moalem, G., He, J., Mor, F., Cohen, I.R. and Schwartz, M., 1998, Accumulation of passively transferred primed T cells independently of their antigen specificity following central nervous system trauma. *J. Neuroimmunol.* **89**: 88.
Huang, S., Apasov, S., Koshiba, M. and Sitkovsky, M., 1997, Role of A_{2a} extracellular adenosine receptor-mediated signaling in adenosine-mediated inhibition of T-cell activation and expansion. *Blood* **90**: 1600.
Inoue, K., 2006, The function of microglia through purinergic receptors: Neuropathic pain and cytokine release. *Pharmacol. Ther.* **109**: 210.
int' Veld, B.A., Ruitenberg, A., Hofman, A., Launer, L.J., van Duijn, C.M., Stijnen, T., Breteler, M.M. and Stricker, B.H., 2001, Nonsteroidal antiinflammatory drugs and the risk of Alzheimer's disease. *N. Engl. J. Med.* **345**: 1515.
Irani, D.N., Lin, K.I. and Griffin, D.E., 1997, Regulation of brain-derived T cells during acute central nervous system inflammation. *J. Immunol.* **158**: 2318.
Jensen, M.B., Finsen, B. and Zimmer, J., 1997, Morphological and immunophenotypic microglial changes in denervated fascia dentate of adult rats: correlation with blood–brain barrier damage and astroglial reactions. *Exp. Neurol.* **143**: 103.
Jones, T.B., Basso, D.M., Sodhi, A., Pan, J.Z., Hart, R.P., MacCallum, R.C., Lee, S., Whitacre, C.C. and Popovich, P.G., 2002, Pathological CNS autoimmune disease triggered by traumatic spinal cord injury: implications for autoimmune vaccine therapy. *J. Neurosci.* **22**: 2690.
Kerschensteiner, M., Stadelmann, C., Dechant, G., Wekerle, H. and Hohlfeld, R., 2003 Neurotrophic cross-talk between the nervous and immune systems: implications for neurological diseases. *Ann. Neurol.* **53**: 292.
Khimenko, P.L., Moore, T.M., Hill, L.W., Wilson, P.S., Coleman, S., Rizzo, A. and Taylor, A.E., 1995, Adenosine A2 receptors reverse ischemia-reperfusion lung injury independent of beta-receptors. *J. Appl. Physiol.* **78**: 990.
Kim, W.G., Mohney, R.P., Wilson, B., Jeohn, G.H., Liu, B. and Hong, J.S., 2000, Regional difference in susceptibility to lipopolysaccharide-induced neurotoxicity: role of microglia. *J. Neurosci.* **20**: 6309.
Kipnis, J., Mizrahi, T., Hauben, E., Shaked, I., Shevach, E. and Schwartz, M., 2002, Neuroprotective autoimmunity: naturally occurring $CD4^+CD25^+$ regulatory T cells suppress the ability to withstand injury to the central nervous system. *Proc. Natl Acad. Sci. USA* **99**: 15620.
Kloss, C.U., Bohatschek, M., Kreutzberg, G.W. and Raivich, G., 2001, Effect of lipopolysaccharide on the morphology and integrin immunoreactivity of ramified microglia in mouse brain and in cell culture. *Exp. Neurol.* **168**: 32.
Kreutzberg, G.W., 1996, Microglia: a sensor for pathological events in the CNS. *Trends Neurosci.* **19**: 312.
Küst, B.M., Biber, K., van Calker, D. and Gebicke-Haerter, P.J., 1999, Regulation of K^+ channel mRNA expression by stimulation of adenosine A_{2a}-receptors in cultured rat microglia. *Glia* **25**: 120.
Ladeby, R., Wirenfeldt, M., Garcia-Ovejero, D., Fenger, C., Dissing-Olesen, L., Dalmau, I. and Finsen, B., 2005, Microglial cell population dynamics in the injured adult central nervous system. *Brain Res. Rev.* **48**: 196.
Lappas, C.M., Rieger, J.M. and Linden, J., 2005, A_{2A} adenosine receptor induction inhibits IFN-γ production in murine $CD4^+$ T cells. *J. Immunol.* **174**: 1073.
Lee, K.S., Tetzlaff, W. and Kreutzberg, G.W., 1986, Rapid down regulation of hippocampal adenosine receptors following brief anoxia. *Brain Res.* **380**: 155.
Lee, Y.C., Chien, C.L., Sun, C.N., Huang, C.L., Huang, N.K., Chiang, M.C., Lai, H.L., Lin, Y.S., Chou, S.Y., Wang, C.K., Tai, M.H., Liao, W.L., Lin, T.N., Liu, F.C. and Chern, Y., 2003, Characterization of the rat A_{2A} adenosine receptor gene: a 4.8-kb promoter-proximal DNA fragment confers selective expression in the central nervous system. *Eur. J. Neurosci.* **18**: 1786.
Lee, H.K., Choi, S.S., Han, K.J., Han, E.J. and Suh, H.W., 2004, Roles of adenosine receptors in the regulation of kainic acid-induced neurotoxic responses in mice. *Mol. Brain Res.* **125**: 76.

Leibovich, S.J., Chen, J.F., Pinhal-Enfield, G., Belem, P.C., Elson, G., Rosania, A., Ramanathan, M., Montesinos, C., Jacobson, M., Schwarzschild, M.A., Fink, J.S. and Cronstein, B., 2002, Synergistic up-regulation of vascular endothelial growth factor expression in murine macrophages by adenosine A_{2A} receptor agonists and endotoxin. *Am. J. Pathol.* **160**: 2231.

Linden, J., 2005, Adenosine in tissue protection and tissue regeneration. *Mol. Pharmacol.* **67**: 1385.

Link, A.A., Kino, T., Worth, J.A., McGuire, J.L., Crane, M.L., Chrousos, G.P., Wilder, R.L. and Elenkov, I.J., 2000, Ligand-activation of the adenosine A_{2a} receptors inhibits IL-12 production by human monocytes. *J. Immunol.* **164**: 436.

Lipton, S.A. and Rosenberg, P.A., 1994, Excitatory amino acids as a final common pathway for neurologic disorders. *N. Engl. J. Med.* **330**: 613.

Liu, B. and Hong, J.S., 2003, Role of microglia in inflammation-mediated neurodegenerative diseases: mechanisms and strategies for therapeutic intervention. *J. Pharmacol. Exp. Ther.* **304**: 1.

Lopes, L.V., Cunha, R.A. and Ribeiro, J.A., 1999, Cross talk between A_1 and A_{2A} adenosine receptors in the hippocampus and cortex of young adult and old rats. *J. Neurophysiol.* **82**: 3196.

Lopes, L.V., Cunha, R.A., Kull, B., Fredholm, B.B. and Ribeiro, J.A., 2002, Adenosine A_{2A} receptor facilitation of hippocampal synaptic transmission is dependent on tonic A_1 receptor inhibition. *Neuroscience* **112**: 319.

Lopes, L.V., Halldner, L., Rebola, N., Johansson, B., Ledent, C., Chen, J.F., Fredholm, B.B. and Cunha, R.A., 2004, Binding of the prototypical adenosine A_{2A} receptor agonist CGS 21680 to the cerebral cortex of adenosine A_1 and A_{2A} receptor knockout mice. *Br. J. Pharmacol.* **141**: 1006.

Lucas, S.M., Rothwell, N.J. and Gibson, R.M., 2006, The role of inflammation in CNS injury and disease. *Br. J. Pharmacol.* **147**: S232.

Lucchi, R., Latini, S., de Mendonça, A., Sebastião, A.M. and Ribeiro, J.A., 1996, Adenosine by activating A1 receptors prevents $GABA_A$-mediated actions during hypoxia in the rat hippocampus. *Brain Res.* **732**: 261.

Lyons, S.A., Pastor, A., Ohlemeyer, C., Kann, O., Wiegand, F., Prass, K., Kettenmann, H. and Dirnagl, U., 2000, Distinct physiologic properties of microglia and blood-borne cells in rat brain slices after permanent middle cerebral artery occlusion. *J. Cereb. Blood Flow Metab.* **20**: 1537.

Maddock, H.L., Broadley, K.J., Bril, A. and Khandoudi, N., 2001, Role of endothelium in ischaemia-induced myocardial dysfunction of isolated working hearts: cardioprotection by activation of adenosine A_{2A} receptors. *J. Auton. Pharmacol.* **21**: 263.

Magnus, T., Chan, A., Linker, R.A., Toyka, K.V. and Gold, R., 2002, Astrocytes are less efficient in the removal of apoptotic lymphocytes than microglia cells: implications for the role of glial cells in the inflamed central nervous system. *J. Neuropathol. Exp. Neurol.* **61**: 760.

Marchetti, B. and Abbracchio, M.P., 2005, To be or not to be (inflamed)-is that the question in anti-inflammatory drug therapy of neurodegenerative disorders? *Trends Pharmacol. Sci.* **26**: 517.

Marchi, M., Raiteri, L., Risso, F., Vallarino, A., Bonfanti, A., Monopoli, A., Ongini, E. and Raiteri, M., 2002, Effects of adenosine A_1 and A_{2A} receptor activation on the evoked release of glutamate from rat cerebrocortical synaptosomes. *Br. J. Pharmacol.* **136**: 434.

Marcoli, M., Raiteri, L., Bonfanti, A., Monopoli, A., Ongini, E., Raiteri, M. and Maura, G., 2003, Sensitivity to selective adenosine A_1 and A_{2A} receptor antagonists of the release of glutamate induced by ischemia in rat cerebrocortical slices. *Neuropharmacology* **45**: 201.

Marcoli, M., Bonfanti, A., Roccatagliata, P., Chiaramonte, G., Ongini, E., Raiteri, M. and Maura, G., 2004, Glutamate efflux from human cerebrocortical slices during ischemia: vesicular-like mode of glutamate release and sensitivity to A_{2A} adenosine receptor blockade. *Neuropharmacology* **47**: 884.

Matsuura, K., Kabuto, H., Makino, H. and Ogawa, N., 1997, Initial cyclosporin A but not glucocorticoid treatment promotes recovery of striatal dopamine concentration in 6-hydroxydopamine lesioned mice. *Neurosci. Lett.* **230**: 191.

Matyszak, M.K., 1998, Inflammation in the CNS: balance between immunological privilege and immune responses. *Prog. Neurobiol.* **56**: 19.

Mayne, M., Fotheringham, J., Yan, H.J., Power, C., Del Bigio, M.R., Peeling, J. and Geiger, J.D., 2001, Adenosine A_{2A} receptor activation reduces proinflammatory events and decreases cell death following intracerebral hemorrhage. *Ann. Neurol.* **49**: 727.

McColl, S.R., St-Onge, M., Dussault, A.A., Laflamme, C., Bouchard, L., Boulanger, J. and Pouliot, M., 2006, Immunomodulatory impact of the A_{2A} adenosine receptor on the profile of chemokines produced by neutrophils. *FASEB J.* **20**: 187.

McGeer, P.L. and McGeer, E.G., 2002, Local neuroinflammation and the progression of Alzheimer's disease. *J. Neurovirol.* **8**: 529.

McGeer, P.L. and McGeer, E.G., 2004, Inflammation and neurodegeneration in Parkinson's disease. *Parkinsonism Relat. Disord.* **10**: S3.

McGeer, P.L. and McGeer, E.G., 2006, NSAIDs and Alzheimer disease: Epidemiological, animal model and clinical studies. *Neurobiol. Aging.* **28**: 639.

McGeer, E.G., Klegeris, A. and McGeer, P.L., 2005, Inflammation, the complement system and the diseases of aging. *Neurobiol. Aging* **26**(Suppl 1): 94.
McPherson, J.A., Barringhaus, K.G., Bishop, G.G., Sanders, J.M., Rieger, J.M., Hesselbacher, S.E., Gimple, L.W., Powers, E.R., Macdonald, T., Sullivan, G., Linden, J. and Sarembock, I.J., 2001, Adenosine A_{2A} receptor stimulation reduces inflammation and neointimal growth in a murine carotid ligation model. *Arterioscler. Thromb. Vasc. Biol.* **21**: 791.
Melchior, B., Puntambekar, S.S. and Carson, M.J., 2006, Microglia and the control of autoreactive T cell responses. *Neurochem. Int.* **49**: 145.
Minghetti, L., 2005, Role of inflammation in neurodegenerative diseases. *Curr. Opin. Neurol.* **18**: 315.
Miwa, T., Furukawa, S., Nakajima, K., Furukawa, Y. and Kohsaka, S., 1997, Lipopolysaccharide enhances synthesis of brain-derived neurotrophic factor in cultured rat microglia. *J. Neurosci. Res.* **50**: 1023.
Mizuma, H., Takagi, K., Miyake, K., Takagi, N., Ishida, K., Takeo, S., Nitta, A., Nomoto, H., Furukawa, Y. and Furukawa, S. 1999, Microsphere embolism-induced elevation of nerve growth factor level and appearance of nerve growth factor immunoreactivity in activated T-lymphocytes in the rat brain. *J. Neurosci. Res.* **55**: 749.
Moalem, G., Leibowitz-Amit, R., Yoles, E., Mor, F., Cohen, I.R. and Schwartz, M., 1999, Autoimmune T cells protect neurons from secondary degeneration after central nervous system axotomy. *Nat. Med.* **5**: 49.
Moalem, G., Gdalyahu, A., Shani, Y., Otten, U., Lazarovici, P., Cohen, I.R. and Schwartz, M., 2000, Production of neurotrophins by activated T cells: implications for neuroprotective autoimmunity. *J. Autoimmun.* **15**: 331.
Modlinger, P.S. and Welch, W.J., 2003, Adenosine A_1 receptor antagonists and the kidney. *Curr. Opin. Nephrol. Hypertens.* **12**: 497.
Monje, M.L., Toda, H. and Palmer, T.D., 2003, Inflammatory blockade restores adult hippocampal neurogenesis. *Science* **302**: 1760.
Montesinos, M.C., Desai, A., Chen, J.F., Yee, H., Schwarzschild, M.A., Fink, J.S. and Cronstein, B.N., 2002, Adenosine promotes wound healing and mediates angiogenesis in response to tissue injury via occupancy of A_{2A} receptors. *Am. J. Pathol.* **160**: 2009.
Morale, M.C., Serra, P.A., L'episcopo, F., Tirolo, C., Caniglia, S., Testa, N., Gennuso, F., Giaquinta, G., Rocchitta, G., Desole, M.S., Miele, E. and Marchetti, B., 2006, Estrogen, neuroinflammation and neuroprotection in Parkinson's disease: glia dictates resistance versus vulnerability to neurodegeneration. *Neuroscience* **138**: 869.
Nash, J.E. and Brotchie, J.M., 2000, A common signaling pathway for striatal NMDA and adenosine A_{2a} receptors: implications for the treatment of Parkinson's disease. *J. Neurosci.* **20**: 7782.
Nataf, S., Anginot, A., Vuaillat, C., Malaval, L., Fodil, N., Chereul, E., Langlois, J.B., Dumontel, C., Cavillon, G., Confavreux, C., Mazzorana, M., Vico, L., Belin, M.F., Vivier, E., Tomasello, E. and Jurdic, P., 2005, Brain and bone damage in KARAP/DAP12 loss-of-function mice correlate with alterations in microglia and osteoclast lineages. *Am. J. Pathol.* **166**: 275.
Neuhaus, O., Farina, C., Wekerle, H. and Hohlfeld, R., 2001, Mechanisms of action of glatiramer acetate in multiple sclerosis. *Neurology* **56**: 702.
Neumann, H., Misgeld, T., Matsumuro, K. and Wekerle, H., 1998, Neurotrophins inhibit major histocompatibility class II inducibility of microglia: involvement of the p75 neurotrophin receptor. *Proc. Natl Acad. Sci. USA* **95**: 5779.
Newby, A.C., 1984, Adenosine and the concept of retaliatory metabolite. *Trends Biochem. Sci.* **9**: 42.
Nguyen, M.D., Julien, J.P. and Rivest, S., 2002, Innate immunity: the missing link in neuroprotection and neurodegeneration? *Nat. Rev. Neurosci.* **3**: 216.
Niederkorn, J.Y., 2006, See no evil, hear no evil, do no evil: the lessons of immune privilege. *Nat. Immunol.* **7**: 354.
Nimmerjahn, A., Kirchhoff, F. and Helmchen, F., 2005, Resting microglial cells are highly dynamic surveillants of brain parenchyma in vivo. *Science* **308**: 1314.
Nishizaki, T., Nagai, K., Nomura, T., Tada, H., Kanno, T., Tozaki, H., Li, X.X., Kondoh, T., Kodama, N., Takahashi, E., Sakai, N., Tanaka, K. and Saito, N., 2002, A new neuromodulatory pathway with a glial contribution mediated via A_{2a} adenosine receptors. *Glia* **39**: 133.
Odashima, M., Otaka, M., Jin, M., Komatsu, K., Wada, I., Matsuhashi, T., Horikawa, Y., Hatakeyama, N., Oyake, J., Ohba, R., Linden, J. and Watanabe, S., 2005a, Selective adenosine A_{2A} receptor agonist, ATL-146e, attenuates stress-induced gastric lesions in rats. *J. Gastroenterol. Hepatol.* **20**: 275.
Odashima, M., Otaka, M., Jin, M., Komatsu, K., Wada, I., Matsuhashi, T., Horikawa, Y., Hatakeyama, N., Oyake, J., Ohba, R., Linden, J. and Watanabe, S., 2005b, Selective A_{2A} adenosine agonist ATL-146e attenuates acute lethal liver injury in mice. *J. Gastroenterol.* **40**: 526.
Odashima, M., Bamias, G., Rivera-Nieves, J., Linden, J., Nast, C.C., Moskaluk, C.A., Marini, M., Sugawara, K., Kozaiwa, K., Otaka, M., Watanabe, S. and Cominelli, F., 2005c, Activation of A_{2A} adenosine receptor attenuates intestinal inflammation in animal models of inflammatory bowel disease. *Gastroenterology* **129**: 26.
Ogata, T., Nakamura, Y. and Schubert, P., 1996, Potentiated cAMP rise in metabotropically stimulated rat cultured astrocytes by a Ca^{2+}-related A1/A2 adenosine receptor cooperation. *Eur. J. Neurosci.* **8**: 1124.

Ohta, A. and Sitkovsky, M., 2001, Role of G-protein-coupled adenosine receptors in downregulation of inflammation and protection from tissue damage. *Nature* **414**: 916.
O'Keefe, G.M., Nguyen, V.T. and Benveniste, E.N., 2002, Regulation and function of class II major histocompatibility complex, CD40, and B7 expression in macrophages and microglia: implications in neurological diseases. *J. Neurovirol.* **8**: 496.
Okusa, M.D., Linden, J., Huang, L., Rieger, J.M., Macdonald, T.L. and Huynh, L.P., 2000, A_{2A} adenosine receptor-mediated inhibition of renal injury and neutrophil adhesion. *Am. J. Physiol.* **279**: F809.
Olsson, T., Cronberg, T., Rytter, A., Asztely, F., Fredholm, B.B., Smith, M.L. and Wieloch, T., 2004, Deletion of the adenosine A_1 receptor gene does not alter neuronal damage following ischaemia in vivo or in vitro. *Eur. J. Neurosci.* **20**: 1197.
O'Regan, M., 2005, Adenosine and the regulation of cerebral blood flow. *Neurol. Res.* **27**: 175.
Osswald, H., Vallon, V. and Muhlbauer, B., 1996, Role of adenosine in tubuloglomerular feedback and acute renal failure. *J. Auton. Pharmacol.* **16**: 377.
Panther, E., Idzko, M., Herouy, Y., Rheinen, H., Gebicke-Haerter, P.J., Mrowietz, U., Dichmann, S. and Norgauer, J., 2001, Expression and function of adenosine receptors in human dendritic cells. *FASEB J.* **15**: 1963.
Panther, E., Corinti, S., Idzko, M., Herouy, Y., Napp, M., la Sala, A., Girolomoni, G. and Norgauer, J., 2003, Adenosine affects expression of membrane molecules, cytokine and chemokine release, and the T-cell stimulatory capacity of human dendritic cells. *Blood* **101**: 3985.
Pearson, T. and Frenguelli, B.G., 2004, Adrenoceptor subtype-specific acceleration of the hypoxic depression of excitatory synaptic transmission in area CA1 of the rat hippocampus. *Eur. J. Neurosci.* **20**: 1555.
Peirce, S.M., Skalak, T.C., Rieger, J.M., Macdonald, T.L. and Linden, J., 2001, Selective A_{2A} adenosine receptor activation reduces skin pressure ulcer formation and inflammation. *Am. J. Physiol.* **281**: H67.
Pellerin, L., 2005 How astrocytes feed hungry neurons. *Mol. Neurobiol.* **32**: 59.
Perry, V.H., Andersson, P.B. and Gordon, S., 1933, Macrophages and inflammation in the central nervous system. *Trends Neurosci.* **16**: 268.
Piehl, F. and Lidman, O., 2001, Neuroinflammation in the rat-CNS cells and their role in the regulation of immune reactions. *Immunol. Rev.* **184**: 212.
Pinhal-Enfield, G., Ramanathan, M., Hasko, G., Vogel, S.N., Salzman, A.L., Boons, G.J. and Leibovich, S.J., 2003, An angiogenic switch in macrophages involving synergy between Toll-like receptors 2, 4, 7, and 9 and adenosine A_{2A} receptors. *Am. J. Pathol.* **163**: 711.
Pinto-Duarte, A., Coelho, J.E., Cunha, R.A., Ribeiro, J.A. and Sebastião, A.M., 2005, Adenosine A_{2A} receptors control the extracellular levels of adenosine through modulation of nucleoside transporters activity in the rat hippocampus. *J. Neurochem.* **93**: 595.
Pintor, A., Galluzzo, M., Grieco, R., Pezzola, A., Reggio, R. and Popoli, P., 2004, Adenosine A_{2A} receptor antagonists prevent the increase in striatal glutamate levels induced by glutamate uptake inhibitors. *J. Neurochem.* **89**: 152.
Platts, S.H., Linden, J. and Duling, B.R., 2003, Rapid modification of the glycocalyx caused by ischemia-reperfusion is inhibited by adenosine A_{2A} receptor activation. *Am. J. Physiol.* **284**: H2360.
Polazzi, E., Gianni, T. and Contestabile, A., 2001, Microglial cells protect cerebellar granule neurons from apoptosis: evidence for reciprocal signaling. *Glia* **36**: 271.
Popoli, P., Frank, C., Tebano, M.T., Potenza, R.L., Pintor, A., Domenici, M.R., Nazzicone, V., Pezzola, A. and Reggio, R., 2003, Modulation of glutamate release and excitotoxicity by adenosine A_{2A} receptors. *Neurology* **61**: S69.
Ponzio, T.A., Wang, Y.F. and Hatton, G.I., 2006, Activation of adenosine A_{2A} receptors alters postsynaptic currents and depolarizes neurons of the supraoptic nucleus. *Am. J. Physiol.* **291**: 359.
Raivich, G., 2005, Like cops on the beat: the active role of resting microglia. *Trends Neurosci.* **28**: 571.
Raivich, G., Bohatschek, M., Kloss, C.U., Werner, A., Jones, L.L. and Kreutzberg, G.W., 1999, Neuroglial activation repertoire in the injured brain: graded response, molecular mechanisms and cues to physiological function. *Brain Res. Rev.* **30**: 77.
Randolph, D.A. and Fathman, C.G., 2006, $Cd4^+Cd25^+$ regulatory T cells and their therapeutic potential. *Annu. Rev. Med.* **57**: 381.
Ransohoff, R.M. and Tani, M., 1998, Do chemokines mediate leukocyte recruitment in post-traumatic CNS inflammation? *Trends Neurosci.* **21**: 154.
Rebola, N., Pinheiro, P.C., Oliveira, C.R., Malva, J.O. and Cunha, R.A., 2003, Subcellular localization of adenosine A_1 receptors in nerve terminals and synapses of the rat hippocampus. *Brain Res.* **987**: 49.
Rebola, N., Canas, P.M., Oliveira, C.R. and Cunha, R.A., 2005a, Different synaptic and subsynaptic localization of adenosine A_{2A} receptors in the hippocampus and striatum of the rat. *Neuroscience* **132**: 893.

Rebola, N., Rodrigues, R.J., Lopes, L.V., Richardson, P.J., Oliveira, C.R. and Cunha, R.A., 2005b, Adenosine A_1 and A_{2A} receptors are co-expressed in pyramidal neurons and co-localized in glutamatergic nerve terminals of the rat hippocampus. *Neuroscience* **133**: 79.

Rebola, N., Porciúncula, L.O., Lopes, L.V., Oliveira, C.R., Soares-da-Silva, P. and Cunha, R.A., 2005c, Long-term effect of convulsive behavior on the density of adenosine A_1 and A_{2A} receptors in the rat cerebral cortex. *Epilepsia* **46**(Suppl 5): 159.

Ritchie, P.K., Spangelo, B.L., Krzymowski, D.K., Rossiter, T.B., Kurth, E. and Judd, A.M., 1997, Adenosine increases interleukin 6 release and decreases tumour necrosis factor release from rat adrenal zona glomerulosa cells, ovarian cells, anterior pituitary cells, and peritoneal macrophages. *Cytokine* **9**: 187.

Rodrigues, R.J., Alfaro, T.M., Rebola, N., Oliveira, C.R. and Cunha, R.A., 2005, Co-localization and functional interaction between adenosine A_{2A} and metabotropic group 5 receptors in glutamatergic nerve terminals of the rat striatum. *J. Neurochem.* **92**: 433.

Ross, S.D., Tribble, C.G., Linden, J., Gangemi, J.J., Lanpher, B.C., Wang, A.Y. and Kron, I.L., 1999, Selective adenosine-A_{2A} activation reduces lung reperfusion injury following transplantation. *J. Heart Lung Transplant.* **18**: 994.

Rossi, D.J., Oshima, T. and Attwell, D., 2000, Glutamate release in severe brain ischaemia is mainly by reversed uptake. *Nature* **403**: 316.

Sargsyan, S.A., Monk, P.N. and Shaw, P.J., 2005, Microglia as potential contributors to motor neuron injury in amyotrophic lateral sclerosis. *Glia* **51**: 241.

Saura, J., Angulo, E., Ejarque, A., Casadó, V., Tusell, J.M., Moratalla, R., Chen, J.F., Schwarzschild, M.A., Lluis, C., Franco, R. and Serratosa, J., 2005, Adenosine A_{2A} receptor stimulation potentiates nitric oxide release by activated microglia. *J. Neurochem.* **95**: 919.

Schenk, D., Barbour, R., Dunn, W., Gordon, G., Grajeda, H., Guido, T., Hu, K., Huang, J., Johnson-Wood, K., Khan, K., Kholodenko, D., Lee, M., Liao, Z., Lieberburg, I., Motter, R., Mutter, L., Soriano, F., Shopp, G., Vasquez, N., Vandevert, C., Walker, S., Wogulis, M., Yednock, T., Games, D. and Seubert, P., 1999, Immunization with amyloid-beta attenuates Alzheimer-disease-like pathology in the PDAPP mouse. *Nature* **400**: 173.

Schnurr, M., Toy, T., Shin, A., Hartmann, E., Rothenfusser, S., Soellner, J., Davis, I.D., Cebon, J. and Maraskovsky, E., 2004, Role of adenosine receptors in regulating chemotaxis and cytokine production of plasmacytoid dendritic cells. *Blood* **103**: 1391.

Schwartz, M. and Kipnis, J., 2005, Therapeutic T cell-based vaccination for neurodegenerative disorders: the role of $CD4^+CD25^+$ regulatory T cells. *Ann. N.Y. Acad. Sci.* **1051**: 701.

Schwartz, M. and Moalem, G., 2001, Beneficial immune activity after CNS injury: prospects for vaccination. *J. Neuroimmunol.* **113**: 185.

Schwartz, M., Moalem, G., Leibowitz-Amit, R. and Cohen, I.R., 1999, Innate and adaptive immune responses can be beneficial for CNS repair. *Trends Neurosci.* **22**: 295.

Schwartz, M., Shaked, I., Fisher, J., Mizrahi, T. and Schori, H., 2003, Protective autoimmunity against the enemy within: fighting glutamate toxicity. *Trends Neurosci.* **26**: 297.

Schwartz, M., Butovsky, O., Bruck, W. and Hanisch, U.K., 2006, Microglial phenotype: is the commitment reversible? *Trends Neurosci.* **29**: 68.

Shryock, J.C. and Belardinelli, L., 1997, Adenosine and adenosine receptors in the cardiovascular system: biochemistry, physiology, and pharmacology. *Am. J. Cardiol.* **79**: 2.

Sitkovsky, M.V., 2003, Use of the A_{2A} adenosine receptor as a physiological immunosuppressor and to engineer inflammation in vivo. *Biochem. Pharmacol.* **65**: 493.

Sitkovsky, M.V. and Ohta, A., 2005, The 'danger' sensors that STOP the immune response: the A2 adenosine receptors? *Trends Immunol.* **26**: 299.

Sitkovsky, M.V., Lukashev, D., Apasov, S., Kojima, H., Koshiba, M., Caldwell, C., Ohta, A. and Thiel, M., 2004, Physiological control of immune response and inflammatory tissue damage by hypoxia-inducible factors and adenosine A_{2A} receptors. *Annu. Rev. Immunol.* **22**: 657.

Sriram, K., Matheson, J.M., Benkovic, S.A., Miller, D.B., Luster, M.I. and O'Callaghan, J.P., 2006a, Deficiency of TNF receptors suppresses microglial activation and alters the susceptibility of brain regions to MPTP-induced neurotoxicity: role of TNF-α. *FASEB J.* **20**: 670.

Sriram, K., Miller, D.B. and O'Callaghan, J.P., 2006b, Minocycline attenuates microglial activation but fails to mitigate striatal dopaminergic neurotoxicity: role of tumor necrosis factor-α. *J. Neurochem.* **96**: 706.

Stone, T.W., 1985, *Purines: Pharmacology and Physiological Roles*. MacMillan, London.

Streit, W.J., Walter, S.A. and Pennell, N.A., 1999, Reactive microgliosis. *Prog. Neurobiol.* **57**: 563.

Sullivan, G.W., Linden, J., Buster, B.L. and Scheld, W.M., 1999, Neutrophil A_{2A} adenosine receptor inhibits inflammation in a rat model of meningitis: synergy with the type IV phosphodiesterase inhibitor, rolipram. *J. Infect. Dis.* **180**: 1550.

Svenningsson, P., Le Moine, C., Fisone, G. and Fredholm, B.B., 1999, Distribution, biochemistry and function of striatal adenosine A_{2A} receptors. *Prog. Neurobiol.* **59**: 355.

Takahashi, K., Rochford, C.D. and Neumann, H., 2005, Clearance of apoptotic neurons without inflammation by microglial triggering receptor expressed on myeloid cells-2. *J. Exp. Med.* **201**: 647.

Tebano, M.T., Martire, A., Rebola, N., Pepponi, R., Domenici, M.R., Gro, M.C., Schwarzschild, M.A., Chen, J.F., Cunha, R.A. and Popoli, P., 2005, Adenosine A_{2A} receptors and metabotropic glutamate 5 receptors are co-localized and functionally interact in the hippocampus: a possible key mechanism in the modulation of N-methyl-D-aspartate effects. *J. Neurochem.* **95**: 1188.

Teismann, P. and Schulz, J.B., 2004, Cellular pathology of Parkinson's disease: astrocytes, microglia and inflammation. *Cell Tissue Res.* **318**: 149.

Tetzlaff, W., Schubert, P. and Kreutzberg, G.W., 1987, Synaptic and extrasynaptic localization of adenosine binding sites in the rat hippocampus. *Neuroscience* **21**: 869.

Togo, T., Akiyama, H., Iseki, E., Kondo, H., Ikeda, K., Kato, M., Oda, T., Tsuchiya, K. and Kosaka, K., 2002, Occurrence of T cells in the brain of Alzheimer's disease and other neurological diseases. *J. Neuroimmunol.* **124**: 83.

van Calker, D. and Biber, K., 2005, The role of glial adenosine receptors in neural resilience and the neurobiology of mood disorders. *Neurochem. Res.* **30**: 1205.

van Gool, W.A., Aisen, P.S. and Eikelenboom, P., 2003, Anti-inflammatory therapy in Alzheimer's disease: is hope still alive? *J. Neurol.* **250**: 788.

Vezzani, A. and Granata, T., 2005, Brain inflammation in epilepsy: experimental and clinical evidence. *Epilepsia* **46**: 1724.

Villoslada, P., Hauser, S.L., Bartke, I., Unger, J., Heald, N., Rosenberg, D., Cheung, S.W., Mobley, W.C., Fisher, S. and Genain, C.P., 2000, Human nerve growth factor protects common marmosets against autoimmune encephalomyelitis by switching the balance of T helper cell type 1 and 2 cytokines within the central nervous system. *J. Exp. Med.* **191**: 1799.

Wang, T., Zhang, W., Pei, Z., Block, M., Wilson, B., Reece, J.M., Miller, D.S. and Hong, J.S., 2006, Reactive microgliosis participates in MPP^+-induced dopaminergic neurodegeneration: role of 67 kDa laminin receptor. *FASEB J.* **20**: 906.

Wei, R. and Jonakait, G.M., 1999, Neurotrophins and the anti-inflammatory agents interleukin-4 (IL-4), IL-10, IL-11 and transforming growth factor-β1 (TGF-β1) down-regulate T cell costimulatory molecules B7 and CD40 on cultured rat microglia. *J. Neuroimmunol.* **95**: 8.

Weiner, H.L. and Selkoe, D.J., 2002, Inflammation and therapeutic vaccination in CNS diseases. *Nature* **420**: 879.

Wirkner, K., Assmann, H., Koles, L., Gerevich, Z., Franke, H., Norenberg, W., Boehm, R. and Illes, P., 2000, Inhibition by adenosine A_{2A} receptors of NMDA but not AMPA currents in rat neostriatal neurons. *Br. J. Pharmacol.* **130**: 259.

Wirkner, K., Gerevich, Z., Krause, T., Gunther, A., Koles, L., Schneider, D., Norenberg, W. and Illes, P., 2004, Adenosine A_{2A} receptor-induced inhibition of NMDA and $GABA_A$ receptor-mediated synaptic currents in a subpopulation of rat striatal neurons. *Neuropharmacology* **46**: 994.

Wu, D.C., Jackson-Lewis, V., Vila, M., Tieu, K., Teismann, P., Vadseth, C., Choi, D.K., Ischiropoulos, H. and Przedborski, S., 2002, Blockade of microglial activation is neuroprotective in the 1-methyl-4-phenyl-1,2,3,6-tetrahydropyridine mouse model of Parkinson disease. *J. Neurosci.* **22**: 1763.

Xu, K., Bastia, E. and Schwarzschild, M., 2005, Therapeutic potential of adenosine A_{2A} receptor antagonists in Parkinson's disease. *Pharmacol. Ther.* **105**: 267.

Yoles, E., Hauben, E., Palgi, O., Agranov, E., Gothilf, A., Cohen, A., Kuchroo, V., Cohen, I.R., Weiner, H. and Schwartz, M., 2001, Protective autoimmunity is a physiological response to CNS trauma. *J. Neurosci.* **21**: 3740.

Yu, L., Huang, Z., Mariani, J., Wang, Y., Moskowitz, M. and Chen, J.F., 2004, Selective inactivation or reconstitution of adenosine A_{2A} receptors in bone marrow cells reveals their significant contribution to the development of ischemic brain injury. *Nat. Med.* **10**: 1081.

Zandi, P.P., Anthony, J.C., Hayden, K.M., Mehta, K., Mayer, L., Breitner, J.C., Cache County Study Investigators, 2002, Reduced incidence of AD with NSAID but not H2 receptor antagonists: the Cache County Study. *Neurology* **59**: 880.

4

SUBVENTRICULAR ZONE CELLS AS A TOOL FOR BRAIN REPAIR

Fabienne Agasse, Liliana Bernardino and João O. Malva

1. ABSTRACT

Neural stem cells of the adult subventricular zone (SVZ) can be easily propagated in culture and give rise to a large number of glial cells and neuronal progenitors, offering promising perspectives for future cell-based therapies. Many factors able to modulate neuronal production have been identified but efficient differentiation into specific neuronal phenotypes is still a challenge. Animal models of brain diseases have been used in neuronal replacement studies in Huntington's and Parkinson's diseases, in stroke and trauma, contributing to highlight the potential of stem cell's therapy. However, inflammatory reaction occurring after injury and the presence of neurogenesis inhibiting factors throughout the brain contribute to limit complete brain repair. Various strategies designed to use SVZ as a source of reparative cells are currently under investigation. Grafting methods are one of the most investigated as it allows ex vivo pre-treatments before grafting. Moreover, stimulation of the SVZ cells through infusion of factors or transfection using viral vectors are also useful to increase grafting efficiency. Furthermore, SVZ cells efficiently migrate in brain tissue under the stimulation of chemoattractive cues making these cells very attractive research tools to target lesioned areas in the brain. Taking advantage of this property, SVZ cells could also be useful as potent drug delivery agents especially for the treatment of brain tumour. However, as stem cells retain the capacity to proliferate and migrate, they also have important potential tumorigenic risk and the efficient control of proliferation/differentiation of these cells is now a major challenge. Nevertheless, the way is paved for the use of SVZ cells in future cell therapies.

2. INTRODUCTION

Neurodegenerative disorders, epilepsy, stroke and head trauma can irreversibly damage the brain. Cells lost as a consequence of these events are not replaced and cause dramatic cognitive and functional deficits. However, the discovery of persistent neurogenesis in defined niches of the adult mammalian brain has challenged the dogma of the irreparable

---

João O. Malva – jomalva@fmed.uc.pt
Center for Neuroscience and Cell Biology, Institute of Biochemistry, Faculty of Medicine University of Coimbra, 3004-504 Coimbra, Portugal

adult brain and rather suggests a weak but promising capacity to self-repair.

Constitutive neurogenesis takes place in the rodent subventricular zone (SVZ) and, at a lower magnitude, in the dentate gyrus (DG) of the hippocampus. Both regions contain stem/progenitor cells that produce new neurons important for olfactory and memory capacities, respectively. In culture, SVZ cells proliferate as neurospheres, which are clonal aggregates of undifferentiated cells. Cultures of SVZ neurospheres can be expanded through dissociation and replating, providing an endless source of cells. In differentiating conditions, both neurons and glial cells are obtained, but SVZ cells can adopt a wider range of cell types depending on the surrounding environment (Vescovi et al., 2001). However, transdifferentiation studies of stem cells are highly controversial and most dismissed on the basis of fusion (Hoofnagle et al., 2004). Nevertheless, SVZ cells have multiple advantages as candidate for cellular therapy (Sayles et al., 2004; Pluchino et al., 2005).

Interestingly, upon brain injury, neurogenesis is promoted in the SVZ (Romanko et al., 2004). Newly born neurons can migrate towards the damaged striatum and differentiate into spiny neurons (Kokaia and Lindvall, 2003) or towards the damaged cortex where they can replace cortical pyramidal neurons, establishing functional connections (Chen et al., 2004). Following demyelination, SVZ newborn cells migrate towards the white matter, and differentiate in myelinating oligodendrocytes (Nait-Oumesmar et al., 1999; Picard-Riera et al., 2002). However, cell replacement is never fully accomplished. Thus, damage-induced neurogenesis/gliogenesis is not a self-repair program by itself but rather an endeavour to repair.

Hence, a better understanding of the mechanisms underlying the SVZ regenerative responses as well as the reasons for unsuccessful cell replacement is essential to the development of brain repair strategies based on endogenous SVZ cell recruitment or grafting. In the present chapter, we propose to review the characteristics of SVZ stem cells in the mammalian brain. We will examine their response upon various brain injuries. A point will be made on the mechanisms involved in this process as well as on the reasons for uncompleted repair. Finally, we will focus our attention on promising studies using SVZ cells as a tool for brain repair.

3. NEUROGENESIS IN THE SVZ OF THE ADULT MAMMALIAN BRAIN

3.1. SVZ and Neurogenesis

At the beginning of the twentieth century, Santiago Ramón y Cajal demonstrated that the cerebral tissue is composed of a dense network of ramified cells making multiple contacts with each other. Because of the complexity of this structure, he concluded that the brain should be devoid of self-repair capacity. However, Joseph Altman reported that new neurons are generated in the hippocampus of the adult rat brain, a structure related with memory, in the SVZ, the telencephalic region lining the lateral ventricles, and in the olfactory bulb (Altman and Das, 1965; Altman, 1969).

In 1992, Reynolds and Weiss showed that stem cells reside in the SVZ of the mouse brain, and in culture these cells proliferate and give rise to neurons and glia. In the mature brain, half of the newborn cells generated in the SVZ die (Morshead and van Der Kooy, 1992; Morshead et al., 1998) and the surviving cells differentiate into neuroblasts that migrate towards the olfactory bulb, both in the rodent and the non-human primate brain (Luskin,

1993; Lois and Alvarez-Buylla, 1994; Lois et al., 1996; Kornack and Rakic, 2001). SVZ is also a source of glial cells during the perinatal period. Astrocytes and oligodendrocytes emerge from SVZ glial cell precursors and then migrate into various regions of the brain (Levison et al., 1993; Kakita et al., 2003). In the rodent hippocampus, neurogenesis observed by Altman and Das (1965) is correlated with the presence of progenitors (Gage et al., 1998). Neurogenesis in the DG of the hippocampus has been shown also in primates, including humans (Eriksson et al., 1998; Gould et al., 2001) as well as in the human olfactory bulb (Bedard and Parent, 2004).

A large number of neurons are generated in the adult rodent brain. In the DG of the rat hippocampus, about 9,000 new neurons are generated daily and the magnitude is even more significant in the SVZ where more than 30,000 neuroblasts are generated daily (Cameron and McKay, 2001; Morshead and Van der Kooy, 2001). Neurogenesis has been studied in other brain structures. In the mouse *substantia nigra*, about 20 new neurons are generated daily by ependymal cells of the third ventricle (Zhao et al., 2003). The *substantia nigra* itself contains cells with properties of progenitor cells (Lie et al., 2002), but their ability to generate new neurons is controversial (Frielingsdorf et al., 2004). Finally, the amygdala and cortex of the primate brain are continuously infused with new neurons derived from the SVZ (Gould et al., 1999; Bernier et al., 2002).

3.2. Functional Role of Adult Neurogenesis

In the mammalian brain, new neurons arising from the SVZ or the DG functionally integrate into the pre-existing neuronal network and participate in specific physiological functions of the tissue.

In the hippocampus, newly generated neurons mature and acquire similar electrophysiological properties of their neighbouring neurons (van Praag et al., 2002). Synaptic connections are established between new neurons and pre-existing ones (Carlen et al., 2002) and new neurons participate in functions such as learning and memory. Accordingly, the pharmacological suppression of rat hippocampal neurogenesis reduces learning capacity (Shors et al., 2001).

Similar to the hippocampus, new olfactory bulb neurons emerging from the SVZ integrate into the pre-existing neuronal network, establish functional synaptic connections and develop similar electrophysiological characteristics of mature neurons (Carlen et al., 2002; Petreanu and Alvarez-Buylla, 2002; Belluzzi et al., 2003). Behavioural studies have shown that bulbar neurogenesis improves odour memory and discrimination (Gheusi et al., 2000; Rochefort et al., 2002), very important for offspring recognition by pregnant mice (Shingo et al., 2003).

3.3. The Adult SVZ Harbours Stem Cells

3.3.1. Stem Cells and Progenitor Cells

Stem cells are found in blastocyst-stage embryos, 5 days after fecundation. At this developmental stage, the embryo is delimited by a layer of cells: the trophoblast that will give rise to the future placenta. In the centre, a cavity or blastocoel is partially filled by cells of the inner cell mass. These totipotent embryonic stem (ES) cells will generate the complete array of the future cells of the organism.

In the adult, stem cells have been primarily identified in the bone marrow (haema-

topoietic stem cells). These cells proliferate and generate two types of proliferating precursors. Precursors of the myeloid lineage produce, among others, erythrocytes, platelets and macrophages. Precursors of the lymphoid lineage give rise to lymphocytes and natural killer cells.
Cardinal features of stem cells have been described by Hall and Watt (1989):
- Proliferate and stay mitotically active throughout life
- Self-renew; the stock of stem cells is maintained during the lifespan of the organism by symmetric divisions of the stem cell into two identical stem daughter cells
- Give rise to progenitors; the asymmetric division of stem cells produces precursors endowed with a limited proliferative capacity and committed to differentiate into defined cell phenotypes

3.3.2. SVZ Stem Cells: Main Characteristics

In 1992, Reynolds and Weiss described a population of stem cells in the adult mouse SVZ. In vitro, cells from the SVZ proliferate in a serum-free medium, supplemented with epidermal growth factor (EGF), and form neurospheres. Neurospheres are made up of stem cells and progenitors cells with limited proliferative capacity. About 67% of the cells dissociated from 6- to 8-day-old neurospheres are able to proliferate and give rise to secondary neurospheres (Reynolds and Weiss, 1992). In the human, cultures of stem cells from the periventricular zone were propagated until 70 passages in the neonatal and 30 passages in the adult or maintained in culture more than 200 days before displaying degenerative signs (Palmer et al., 2001). In vivo, stem cells from the SVZ are maintained throughout life but its number decrease with age (Palmer et al., 2001; Maslov et al., 2004).

After culturing SVZ cells for 21 days, Reynolds and Weiss detected the appearance of neurons at the periphery of neurospheres, which were organized in chains of migrating cells. Astrocytes and oligodendrocytes were also obtained in vitro in rodent and in human cultures (Reynolds and Weiss, 1992; Levison and Goldman, 1997; Luskin et al., 1997; Palmer et al., 2001; Sanai et al., 2004).

Figure 1 provides representative images of SVZ cultures obtained from transgenic mice expressing the green fluorescent protein (GFP). Both neurons and astrocytes can be generated from free-floating neurospheres after platting in polylysine-coated glass coverslips.

Multipotent Cells
It seems that the multipotency of SVZ stem cells allows the differentiation of a wider range of cell phenotypes than previously thought and that the cell fate depends on the surrounding environment. Thus, postnatal mouse SVZ cells injected in the ventricles of a 15-day-old mouse embryo (E15) generate neuroblasts that integrate at various levels of the neuroaxis such as septum, thalamus, hypothalamus and inferior colliculus but not in the cortex. In the colliculi, cells differentiated in neurons presenting dendritic arborization typical of this structure (Lim et al., 1997). Cortico-thalamic projections and corticospinal moto-neurons selectively eliminated by targeted apoptosis were regenerated in the adult mouse cortex from SVZ cells (Magavi et al., 2000; Chen et al., 2004). The corticospinal motoneurons were able to integrate into the pre-existing circuitry and to reform appropriate long-distance connections, demonstrating a true capacity for replacing dead neurons and repair neuronal circuits.

Neural stem cells (NSCs) are also able to generate non-neuronal cells. For instance, hepatocytes, lung, heart and intestine cells were obtained from SVZ cells injected in mice and chicken embryos (Clarke et al., 2000). Systemic injection of stem cells in irradiated mice can regenerate a part of the eliminated blood cells (Bjornson et al., 1999). Also, in co-culture, SVZ cells can differentiate into myocytes (Galli et al., 2000; Rietze et al., 2001). So, in conclusion, adult SVZ stem cells are plastic as they are capable of transdifferentiation, e.g. they have the capacity to differentiate into cells that do not belong to their originating embryonic layer (Vescovi et al., 2001).

3.3.3. SVZ Organization and Stem Cells Identity

The SVZ surrounding the lateral ventricles is divided into rostral and caudal regions. The caudal SVZ is predominantly gliogenic (Law et al., 1999), whereas the rostral SVZ contains mostly stem cells that generate the future olfactory bulb interneurons.

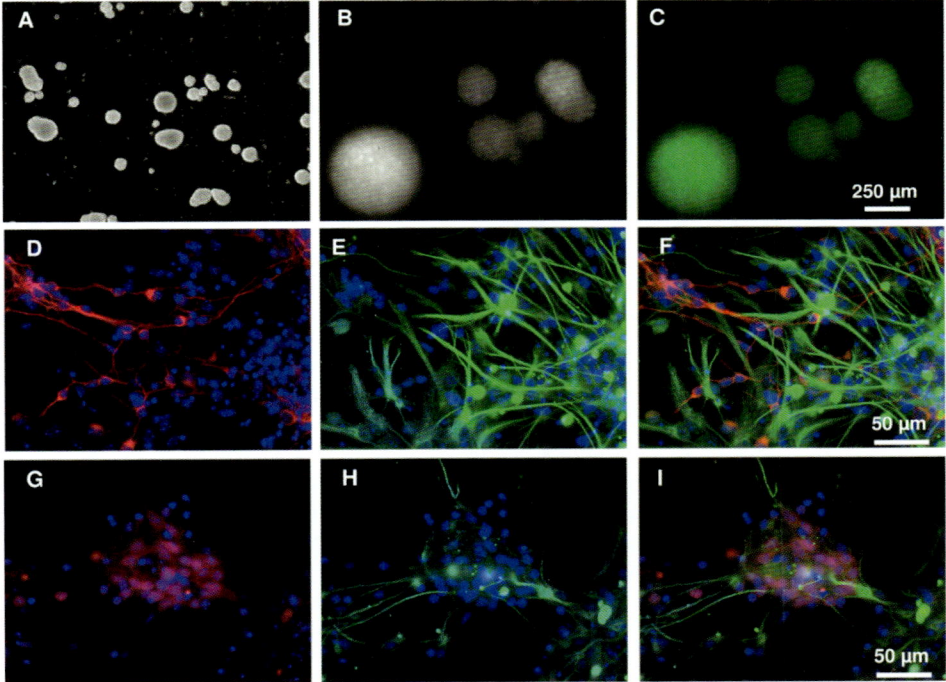

Fig. 1. Representative photographs of SVZ cell cultures from newborn mice in proliferating (A–C) and differentiating (D–I) conditions. **(A–C)**: In serum-free medium with the growth factors EGF and FGF-2, SVZ cells generate clone aggregates of cells called neurospheres (A–B). SVZ neurospheres obtained from GFP mice represent a useful tool for grafting studies (C). (D–I): After plating on poly-D-lysine and in the absence of the growth factors, cells arise at the edge of the neurosphere and develop into neurons expressing MAP-2 (D) and NeuN (G), and glial cells such as astrocytes expressing GFAP (E, H). (F) and (I) photographs are merges of (D, E) and (G, H), respectively. Hoestch nuclear staining is shown in blue fluorescence

Cellular Composition of the SVZ

Three different cell types compose the SVZ (Doetsch et al., 1997): astrocytes (B cells), immature neurons or neuroblasts (A cells) and immature precursors (C cells). The SVZ is separated from the ventricular space by the monolayer of the ventricular zone, composed by ependymal cells (E cells) and tanycytes (D cells), presenting, respectively, motile cilia and microvillosities protruding into the ventricle lumen.

A cells form chains of migrating neuroblasts that associate with aggregates of C cells. A and C cells are ensheathed by a network of B cells that extend from the SVZ towards the olfactory bulb delimitating the rostral migratory stream (RMS) (Lois et al., 1996). The different cell types are characterized by their expression of cellular antigens reported in Table 1.

B and C cells are different according to their antigenic profile and also by the duration of their cell cycle. C cells are rapid-proliferating cells as their cycle spans 12 h, whereas B cells are quiescent with a cell cycle about 15 days (Morshead et al., 1994, 1998). Neuroblasts are also able to divide during their migration towards the bulb (Menezes et al., 1995; Coskun and Luskin, 2001). This ability is correlated with the periodic repression of p19, a cell cycle inhibitor, and, as cells reach the olfactory bulb, neuroblasts stop dividing (Coskun and Luskin, 2001).

Stem Cell Identity

Despite the fact that a unique stem-cell-specific antigen has not been identified, there is a relative consensus that the stem cell in the SVZ has an astrocyte-like phenotype.

In vivo, retroviral infection of astrocyte SVZ cells resulted in the identification of labelled interneurons in the olfactory bulb (Doetsch et al., 1999a). Astrocyte-like cells are the only cell type in the SVZ that is able to proliferate, self-renew and give rise to neurons and glia (Chiasson et al., 1999; Doetsch et al., 1999a; Lim and Alvarez-Buylla, 1999; Morshead and van der Kooy, 2001). Moreover, consistent with this, intracerebroventricular injection of the anti-mitotic cytosine-β-D-arabinofuranoside (ARA-C) in the mouse lateral ventricles selectively suppresses rapidly proliferating cells, i.e. A and C cells. The pool of A and C cells can be regenerated in 10 days from B cells (Doetsch et al., 1999b); the C cells are the first to regenerate and proliferate to give rise to A cells.

SVZ cells derived from a mouse in which SVZ astrocytes have been selectively eliminated fail to form neurospheres (Morshead et al., 2003). Another consistent study also demonstrated that progenitor cells are also astrocyte-like cells in the other main neurogenic area, the DG of the hippocampus (Seri et al., 2001). Furthermore, in the juvenile mouse brain, astrocytes endowed with stem cell's capacities have been identified in non-neurogenic compartments of the adult brain, such as the cerebral cortex, the cerebellum and the spinal cord (Laywell et al., 2000).

Antigenic Profile

Astrocytes represent 20% of SVZ cells (Doetsch et al., 1997). However, only 0.2–0.4% of the SVZ cells are stem cells (Morshead et al., 1998). Several antigens have been identified in the SVZ and contribute to the fine identification of SVZ subtypes (Table 1). Unlike A and C cells, B cells, including stem cells, do not express Dlx2, a transcription factor involved in the development of neurons (Doetsch et al., 2002b). Stem cells express the trisaccharide LeX/ssEA-1, also present on the cell surface of other proliferating cells such as C cells (Capela and Temple, 2002). However, there is no co-expression of Lex and mCD24, an antigen expressed by E cells and neuroblasts, suggesting that stem cells are not E cells (Belvindrah et al., 2002; Capela and Temple, 2002).

Stem cells proliferate in vitro and in vivo in response to the growth factor EGF (Reynolds and Weiss, 1992; Kuhn et al., 1997), and accordingly, C cells and a subpopulation of B cells express the EGF receptor (EGFR) (Doetsch et al., 2002b).

Morphologic Feature

A morphologic feature unambiguously identifies the stem cell. Indeed, some SVZ astrocytes expressing EGFRs extend a cilium through the ventricular zone reaching the ventricular space (Doetsch et al., 1997). Following intracerebroventricular administration of EGF, which promotes proliferation in the SVZ (Kuhn et al., 1997), a significant increase in the number of SVZ astrocytes extending their cilia towards the ventricle is seen (Doetsch et al., 2002b).

The presence of this contact with the ventricle could explain the incorrect identification in 1999 of E cells as stem cells (Johansson et al., 1999). In vivo, the application of the lipophilic fluorescent dye DiI on the ventricular surface resulted in DiI staining of olfactory bulb neurons. Subsequently, in vitro and in vivo studies unambiguously demonstrated that E cells are not stem cells, since they do not form neurospheres, nor self-renew, do not generate neurons (Chiasson et al., 1999; Doetsch et al., 1999a) and remain postmitotic (Spassky et al., 2005). Moreover, transgenic ablation of dividing GFAP cells in the SVZ and in the hippocampus stop neurogenesis, identifying unambiguously the stem cell as GFAP-positive cells (Garcia et al., 2004).

3.3.4. SVZ and Olfactory Neurogenesis

Migration Towards the Olfactory Bulb

SVZ neuroblasts migrate tangentially towards the olfactory bulb in the rodent and non-human primate. This migratory pathway, the RMS, contains a network of astrocytes that guides the neuroblasts to the olfactory bulb (Lois and Alvarez-Buylla, 1994; Doetsch and Alvarez-Buylla, 1996; Lois et al., 1996; Kornack and Rakic, 2001; Pencea et al., 2001b). Once in the olfactory bulb, neuroblasts migrate radially, towards the granule and glomerular cell layers where they differentiate into granular and periglomerular interneurons, expressing dopamine or GABA as main neurotransmitters (Betarbet et al., 1996; Baker et al., 2001).

Migration of SVZ cells is directed towards the olfactory bulb by the action of chemoattractant molecules like astrocyte-derived migration-inducing activity (MIA) factor (Mason et al., 2001b), and chemorepellant factors such as Slit 1 and 2 (Chen et al., 2001) released by cells of the choroid plexus in the adult brain. A recent study showed that the beating of ependymal cilia generates a Slit 2 gradient in the cerebrospinal fluid filling the lateral ventricle that direct the migration of SVZ newborn neurons towards the olfactory bulb (Sawamoto et al., 2006). Interactions between $\alpha 6\beta 1$ integrins of the neuroblast's membrane and proteins from the extracellular matrix, such as laminin, also participate in migration and guidance (Emsley and Hagg, 2003a). Moreover, the olfactory bulb cells secrete glial cell line-derived neurotrophic factor (GDNF), prokineticin2 and yet unidentified molecules that are chemoattractant for SVZ-derived neuroblasts (Liu and Rao, 2003; Ng et al., 2005; Paratcha et al., 2006).

3.4. Regulation of Neurogenesis in the Adult Brain

3.4.1. Modulation by Hormones and Neurotransmitters

In the adult mammalian brain, neurogenesis is stimulated and modulated by hormones. Accordingly, estrogens promote proliferation and survival of DG progenitors in female rat (Tanapat et al., 1999), but despite the expression of the specific receptors in the adult rat SVZ, estrogens do not affect SVZ neurogenesis (Tanapat et al., 1999; Brännvall et al., 2002; Isgor and Watson, 2005). Interestingly, prolactin and thyroid hormones increase the production of neurons in the rodent SVZ (Giardino et al., 2000; Shingo et al., 2003). In the DG and SVZ, proliferation is stimulated by polyamines (Malaterre et al., 2004), whereas stress hormones such as glucocorticoids reduce neurogenesis in the DG of young rats and primates (Gould et al., 1998; Tanapat et al., 1998; Kippin et al., 2004).

Neurogenesis in the adult brain is affected by neurotransmitters. Glutamate reduces proliferation and PSA-NCAM neurons production by hippocampal progenitors (Nacher et al., 2001). On the contrary, glutamate enhances SVZ proliferation in vitro (Brazel et al., 2005). Depletion of serotonin decreases proliferation and emergence of PSA-NCAM expressing neurons in the DG and SVZ (Brezun and Daszuta, 1999). Likely, noradrenaline depletion through selective chemical targeting of noradrenergic neurons decreases proliferation in the rat DG but does not affect SVZ (Kulkarni et al., 2002). However, elimination of nigrostrial dopaminergic projections by the use of the neurotoxins 6-hydroxydopamine (6-OHDA), or 1-methyl-4-phenyl-1,2,3,6-tetrahydropyridine (MPTP), decreases the proliferation of neuronal precursors in the SVZ (Baker et al., 2004; Höglinger et al., 2004), but in vitro, dopamine stimulates proliferation in the SVZ via activation of D3 receptors expressed by SVZ cells (Coronas et al., 2004; van Kampen et al., 2004).

3.4.2. Modulation by Aging and Neuronal Activity

Aging affects neurogenesis. In the hippocampus, the number of new granule cells produced in the DG is reduced in the aged when compared with the young rat (Kuhn et al., 1996) and this can be due to a decrease of IGF-1 production and IGF-1 receptor levels as well as to an increase in glucocorticoids (Cameron and McKay, 1999; Anderson et al., 2002). In the SVZ, there is a good correlation between the decrease of neurogenesis in the aged mouse and the reduction in the number of stem cells as well as a decrease in the expression of the EGFRs (Tropepe et al., 1997; Enwere et al., 2004; Maslov et al., 2004).

Neuronal activity modulates neurogenesis. Neurogenesis is increased in the DG of mice brain, living in an enriched environment (Kempermann et al., 1997, 1998). Indeed, when mice have a free access to a running wheel, the number of newly generated neurons in the DG is increased (van Praag et al., 1999). However the same stimulus does not affect SVZ neurogenesis (Brown et al., 2003). On the other hand, DG neurogenesis is not affected by stimuli that specifically affect the SVZ—olfactory bulb system. In mice exposed to various odours, neurogenesis increases in the olfactory bulb but not in the DG (Rochefort et al., 2002). Moreover, in mice with unilateral odour deprivation, neurogenesis is decreased only in the SVZ ipsilateral to naris closure (Corotto et al., 1994). Molecular mechanisms underlying neuronal activity-induced neurogenesis remain unknown. However, increase in DG neurogenesis in rats undergoing exercise training has been shown to be mediated by the brain uptake of IGF-1 from the blood (Trejo et al., 2001).

4. NEUROGENESIS IN THE DAMAGED BRAIN

4.1. Brain Damage and SVZ Dynamics

4.1.1. Modulation of SVZ Cells Dynamics upon Lesion

Following cerebral damage, neurogenesis/gliogenesis is enhanced in neurogenic niches as well as in regions where no neurogenesis occurs in basal conditions. For instance, in the *substantia nigra*, basal neurogenesis is increased following cytotoxic lesion (Zhao et al., 2003). Neurogenesis also increases in SVZ of patients suffering from neurodegenerative diseases like Huntington's diseases (Curtis et al., 2003) as well as in the hippocampus of Alzheimer's disease patients (Jin et al., 2004). In non-neurogenic areas, such as the striatum and the cortex, ischaemia triggers neurogenesis and new neurons are observed nearby the damaged area (Gu et al., 2000; Jiang et al., 2001; Zhang et al., 2001a; Arvidsson et al., 2002; Jin et al., 2003). In the adult mice cortex subjected to targeted apoptosis of pyramidal neurons, newly generated neurons adopting complex pyramidal phenotype are observed (Magavi et al., 2000; Chen et al., 2004).

Brain insults affect the ability of SVZ cells to proliferate, migrate and differentiate. In vivo, numerous studies report an increased proliferation and neuronal production in the SVZ following various insults (Szele and Chesselet, 1996; Tzeng and Wu, 1999; Zhang et al., 2001a, 2004a, b). Increased proliferation following ischaemic injury is correlated with a decrease of the expression of p27Kip1 (Zhang et al., 2004c) that, in basal conditions, inhibit the G1 phase CDK-cycline complex responsible for slowing down the cell cycle (Doetsch et al., 2002a). Following ischaemia, SVZ proliferating cells divide symmetrically, generating a large number of neuronal precursors (Zhang et al., 2004b). In the SVZ, the number of neuroblasts expressing PSA-NCAM and DCX is increased upon mechanical and ischaemic cortical insult (Szele and Chesselet, 1996; Zhang et al., 2004a).

There is now increasing consensus that immature cells can be a source of reparative cells able to repopulate injured areas. Indeed in the *substantia nigra*, new neurons are originated, following cytotoxic injury, from the ependymal region of the third ventricle (Zhao et al., 2003). In the striatum, the neurons differentiated following ischaemia arise from the adjacent SVZ (Arvidsson et al., 2002). Also, in the cortex, numerous studies showed that the neurons added in injured areas arise from the SVZ. Indeed, upon cortical injury (ischaemia, section, localized aspiration or seizure) proliferation and expression of immature neuronal markers (PSA-NCAM, DCX) are increased in the SVZ (Szele and Chesselet, 1996; Tzeng and Wu, 1999; Jin et al., 2001; Arvidsson et al., 2002; Parent et al., 2002a, b) and neuroblasts migrating out of the SVZ and reaching the damaged cortical territories have been observed (Szele and Chesselet, 1996; Arvidsson et al., 2002; Parent et al., 2002a, b; Jin et al., 2003). Moreover, following experimentally induced demyelination of the *corpus callosum* or in multiple sclerosis (MS), SVZ cells proliferate and produce PSA-NCAM migrating precursors that integrate into the white matter and differentiate in astrocytes and myelinating oligodendrocytes (Nait-Oumesmar et al., 1999; Picard-Riera et al., 2002). The increase in cell proliferation and neuronal/glial cell production, following lesion, strongly support a mechanism of brain repair in the adult brain.

4.1.2. Modifications of the Molecular Microenvironment and Damage-Triggered Neurogenesis

The mechanisms controlling neurogenesis/gliogenesis following lesion have only recently been partially revealed. Lesions and degenerative processes lead to massive apoptosis that change the molecular environment stimulating neurogenesis and gliogenesis in injured areas (Li et al., 1997; Guégan and Sola, 2000; O'Dell et al., 2000). Changes in the extracellular matrix components and secreted factors contribute to the modification of the dynamic status of stem/progenitor cells. Indeed, expression of FGF-2 is enhanced in the DG upon seizures, ischaemia or cortical injury, protecting neurons and stimulating DG neurogenesis (Yoshimura et al., 2001, 2003). Following mouse olfactory bulb ablation, neuronal olfactory receptors enter in apoptosis and secrete LIF enhancing proliferation of progenitor cells and regeneration of the olfactory epithelium (Bauer et al., 2003). In the mouse cortex, following specific degeneration of pyramidal neurons, grafts of neuronal precursors differentiate into pyramidal neurons (Snyder et al., 1997). The authors suggested that molecules promoting neurogenesis are expressed by damaged areas. Indeed, two molecules with properties linked to plasticity and neuronal regeneration, PSA-NCAM and tenascin, are upregulated in the damaged cortex (Brodkey et al., 1995; Hayashi et al., 2001).

Stimulation of SVZ neurogenesis following damage occurring in adjacent or distant regions suggests that diffusible factors mediate signaling between injury and brain repair processes. Indeed, the injured rat cortex secretes factors that stimulate proliferation and neuronal differentiation in SVZ cultures (Agasse et al., 2004).

Pro-neurogenic Factors are Secreted upon Lesion

Neurotrophins are upregulated upon cortical injury (Wang et al., 1998; Hicks et al., 1999; Oyesiku et al., 1999). Accordingly, BDNF stimulates survival and differentiation in SVZ cultures (Ahmed et al., 1995; Kirschenbaum and Goldman, 1995) that express TrkB receptors (Garcia-Verdugo et al., 1998) and in vivo, administration of BDNF promotes production of new neurons in the SVZ (Benraiss et al., 2001; Pencea et al., 2001a). However, the role of BDNF remains controversial because it inhibits ischaemia-induced neurogenesis in the rat DG (Larsson et al., 2002; Kokaia and Lindvall, 2003).

Cytokines involved in haematopoiesis, such as erythropoietin (EPO) and stem cell factor (SCF), are upregulated in the brain following ischaemia or traumatic lesion (Digicaylioglu et al., 1995; Marti et al., 1996; Zhang and Fedoroff, 1999; Sun et al., 2004). These two factors stimulate SVZ neurogenesis after intracerebroventricular administration in adult rodents (Shingo et al., 2001; Jin et al., 2002a) and accordingly, SVZ cells express the EPO receptor (Shingo et al., 2001) as well as c-Kit, the SCF receptor (Jin et al., 2002a). Cytokines such as CNTF (Ip et al., 1993a) and LIF (Banner et al., 1997; Suzuki et al., 2000) are also upregulated upon cortical damage. Following intracerebroventricular administration, CNTF promotes proliferation in the mouse SVZ that expresses receptors for CNTF and LIF (Ip et al., 1993b; Shimazaki et al., 2001; Emsley and Hagg, 2003b).

Growth factors are upregulated following cortical injury. VEGF is overproduced in cortical cells following ischaemia (Jin et al., 2000; Marti et al., 2000). When injected in the rodent lateral ventricle, VEGF stimulates SVZ cell proliferation through VEGF receptor (VEGFR2/Flk-1)-mediated signaling (Jin et al., 2002b). Growth factors FGF-2 and IGF-1 are also upregulated upon injury and stimulate proliferation of neuronal precursors (Frautschy et al., 1991; Yoshimura et al., 2001; Aberg et al., 2003). FGF-2 injected intracerebroventricularly in the rat promotes proliferation and emergence of SVZ neurons (Kuhn et al.,

1997). Eventually, IGF-1 promotes neuronal differentiation in SVZ cultures from adult mice (Brooker et al., 2000).

Migration and Brain Repair
Following brain insult, progenitor cells migrate and invade damaged areas (Parent et al., 2002a, b; Romanko et al., 2004; Salman et al., 2004). Indeed, guidance cues re-routing the migration of SVZ neuroblasts into the site of brain injury are upregulated upon lesion. For instance, SCF expressed by cortical neurons upon ischaemic or mechanical lesion direct the migration of SVZ cells towards the cortex (Jin et al., 2002a, b; Sun et al., 2004). VEGF is upregulated in the ischaemic cortex and has chemoattractant effects on SVZ progenitor in vitro (Zhang et al., 2003c). Extracellular matrix molecules (ECM) might also guide SVZ cells towards the site of injury. Indeed, several ECM that participate in SVZ cell migration have been identified in the brain. Tenascin-R and Reelin guide SVZ neuroblasts in the olfactory bulb (Hack et al., 2004; Saghatelyan et al., 2004). Interactions between laminin and integrins and Eph/ephrin also play an important role in the organization and guidance of SVZ neuroblasts (Conover et al., 2000; Emsley and Hagg, 2003a). Reorganization of ECM proteins that follows the lesion can thus modulate the migration of SVZ cells. Indeed, matrix-metalloproteinases (MMP) and associated inhibitors, involved in invasion of tissues by cancer cells, are expressed in the olfactory epithelium, a site of continuous neurogenesis and regeneration in the adult. Their expression is modified upon lesion and thus may be involved in neuronal regeneration (Tsukatani et al., 2003). Moreover, it has been recently shown that the extracellular MMP-9 promotes migration of the newly generated neuroblasts from the SVZ towards the ischaemic striatum of the mouse brain (Lee et al., 2006).

Inflammation and Brain Repair
Inflammation modulates neurogenesis/gliogenesis and migration/homing into the injured areas. Following cortical stab injury, SVZ cells proliferate and survive in an immune-cell-enriched environment (Tzeng and Wu, 1999). Moreover, it has been shown that the pro-inflammatory cytokine TNF-α can promote proliferation in the SVZ without triggering apoptosis (Wu et al., 2000). Furthermore, injury-activated astrocytes and microglial cells have been shown to produce growth factors known to stimulate stem/progenitor cells dynamic. Accordingly, microglial cells release the pro-neurogenic VEGF (Stoll et al., 2002) and activated astrocytes release FGF-2 that promotes SVZ neurogenesis (Smith et al., 2001). In vitro, microglial cells activated by IL-4 or INF-γ trigger neuronal differentiation in SVZ cultures from adult mice (Butovsky et al., 2006). In experimental demyelination, microglial cells secrete IL-1β (Stoll et al., 2002) that induces the secretion of IGF-1 by astrocytes, promoting the proliferation of oligodendrocyte precursor cells (OPCs) and remyelination (Mason et al., 2001a). Upon ischaemia, IL-1β stimulates the expression of the chemokine SDF-1α (or CXCL12) by activated astrocytes (Zhou et al., 2002; Imitola et al., 2004; Miller et al., 2005). SDF-1α is a chemoattractant for human NSCs and SVZ cells that express its receptor CXCR4 (Imitola et al., 2004).

In vivo, human NSCs grafted in the periventricular areas of mice migrate long distances to home in the ischaemic cortex (Imitola et al., 2004). A very interesting report recently showed that SVZ cells grafted on hippocampal organotypic slice cultures migrate towards the fimbria, the site of experimentally induced inflammation where chemokines are upregulated. The monocyte chemoattractant protein-1 (MCP-1) was shown to mediate the chemoattractive effect on SVZ cells, involving the activation of the chemokine receptor CCR-2 expressed by SVZ cells (Belmadani et al., 2006).

Altogether, these studies shed light on a capacity of the mature brain to self-repair through the stimulation of the endogenous stock of stem/progenitor cells. However, cell replacement is never fully accomplished.

4.2. Failure of Cell Replacement

4.2.1. Inflammation and Glial Scar

The main effect of the inflammatory response is deleterious for neurogenesis and regeneration. Activated microglial cells and astrocytes proliferate at the site of injury and secrete various cytokines such as TNF-α, which can exacerbate tissue damage. Activated astrocytes accumulate in the lesion core, forming a glial scar that limits the propagation of the injury into the surrounding tissue.

Following injury, OPCs proliferate into the glial scar. These OPCs express at their surface the chondroitin sulphate proteoglycan (CSPG) NG2, an inhibitor of the axonal regrowth (for review see Tan et al., 2005). The glial scar is also rich in repulsive axon guidance molecules such as ephrins, semaphorins and Nogo. Hence, the glial scar constitutes a barrier impeding regeneration; targeting components of the glial scar seems to be a good strategy for axonal regeneration. Indeed, treatments that suppress glial scar, in spinal cord-injured rat, promote regeneration of corticospinal motoneurons (Klapka et al., 2005).

In the hippocampus, electrically induced seizures cause neuronal death and microglial activation contributing to inhibit neurogenesis. In these conditions, systemic administration of minocycline, which inhibits microglia activation, restores neurogenesis in the DG (Ekdahl et al., 2003). Part of these effects may involve TNF-α and IL-6 as these cytokines are secreted by activated microglial cells and impair neurogenesis in the DG (Monje et al., 2003). IL-6 triggers the differentiation of neural progenitors cells into astrocytes (Taga and Fukuda, 2005) participating in glial scar formation. On the other hand, the expression of IL-6 is increased after spinal cord injury, and functional recovery has been obtained after blocking IL-6 with a specific antibody (Okano et al., 2005).

Molecules belonging to the TGF-β family mainly inhibit neurogenesis. TGF-β is also produced upon inflammation and promotes the production of extracellular matrix (Smith and Strunz, 2005). In the mammalian olfactory epithelium where new sensory neurons are continuously produced, mature neurons secrete the growth and differentiation factor-11 (GDF11), member of the TGF-β family, that limits the proliferation of neuronal precursors (for review, see Hastings and Gould, 2003).

Nitric oxide (NO) is released by activated microglia (Stoll et al., 2002). NO is a well-known inhibitor of the proliferation of SVZ cells (Packer et al., 2003; Moreno-Lopez et al., 2004). Following treatment with DEtA-NONOate, an NO donor, differentiation of β-III tubulin positive neurons is increased, concomitantly with an increase of cellular adhesion molecules β-catenin and N-cadherin, and the transcription factor promoting neuronal fate neurogenin 1 (Ngn1) (Chen et al., 2006). In summary, NO is a negative regulator of neurogenesis, inhibiting proliferation and promoting differentiation.

In conclusion, inflammatory response and glial scarring appear to hamper brain repair endeavours.

4.2.2. The Non-neurogenic Environment

Mature tissues of the brain, except SVZ and DG, are mainly non-permissive to neurogenesis despite molecular changes in injured tissue that contribute to neural precursor cells homing and differentiation.

Indeed, in the non-injured brain, SVZ cells from adult mouse grafted heterotopically in the cortex or the striatum differentiate into astrocytes or immature neurons but stay aggregated in the graft core (Herrera et al., 1999). Non-neurogenic tissues can modulate the SVZ neurogenic process by releasing soluble factors. Indeed, factors secreted by adult rat cortex explants reduce proliferation and neuronal production in SVZ cultures (Agasse et al., 2004). This secretion of diffusing factors can explain the absence of neurogenesis in the human SVZ in vivo while, in culture, these cells are able to generate neurons (Rakic, 2004; Sanai et al., 2004). A few factors affecting SVZ neurogenesis have been identified. SVZ neurogenesis is inhibited by NO secreted by neurons bordering the SVZ (Packer et al., 2003; Moreno-Lopez et al., 2004). The bone morphogenic protein (BMP), a soluble factor belonging to the TGF-β superfamily of proteins, stops proliferation and promotes differentiation of SVZ immature cells in astrocytes in detriment of neuronal production (Gross et al., 1996; Lim et al., 2000; Coskun et al., 2001; Peretto et al., 2004). BMP is expressed both by cortical and SVZ cells (Ebendal et al., 1998; Mabie et al., 1999; Lim et al., 2000).

Therefore, it seems critical to clearly understand the balance between pro- and anti-neurogenic factors to design new efficient strategies for cell replacement in the different regions of the damaged brain.

5. SVZ CELLS AS THERAPEUTIC TOOLS

5.1. SVZ Cells Grafting

Reconstruction of neuronal circuits by transplantation of neural precursors or stem cells is a promising strategy for brain repair, although multiple questions still remain unsolved and contribute to delayed clinical applications.

What is the cell phenotype appropriate for intracerebral grafting? The candidate cell must be able to proliferate following ex vivo stimulation with mitogens, in order to generate the production of the large number of cells required for transplantation. Then, the cells must be phenotypically plastic, i.e. be able to differentiate into appropriate neurons or glial cells depending on the surrounding host tissue.

Sheen and collaborators (1999) grafted multipotent neural precursor cells, derived from E 16 mouse hippocampus, into the cortex of adult mice following targeted pyramidal neuron degeneration. The grafted cells survived but failed to develop into mature pyramidal cells. By RT-PCR, the authors showed that these immature cells did not express TrkB receptors and so they were not able to sense released BDNF and NT-4/5 by neighbouring injured neurons (Wang et al., 1998). This result shows that in some cases immature cells cannot be fully competent to respond to neurogenic signals released in the damaged environment. However, Sheen and Macklis (1995) grafted immature neurons from E17 neocortex. In this case, transplanted cells differentiated into large pyramidal neurons indicating that, at this developmental stage, they are fully competent to respond to instructive microenvironmental cues.

A large number of studies are based on the use of ES cells obtained from the inner cell mass of human embryos. In spite of the scientific potential, the use of ES cells causes legitimate ethical problems (for review see Ben-Hur, 2006). On the other hand, adult stem cells may provide an ethically accepted source of reparative cells. Promising results have already been obtained with bone marrow stromal cells:

1. It was shown that adult bone marrow cells can transdifferentiate into neuronal cells.
2. In a model of Parkinson's disease where death of dopaminergic neurons of the *substantia nigra* was caused by the intraperitoneal injection of MPTP, adult bone marrow stromal cells grafted in the striatum survived.
3. The same authors also showed that the surviving grafted cells express tyrosine hydroxylase, a key enzyme in the biosynthesis of dopamine, and the mice significantly improved motor performance (Li et al., 2001).

These data clearly contribute to a positive perspective for the future use of bone marrow stromal cells as a tool to regenerate brain tissue.

What about the use of SVZ cells for intracerebral transplantation? SVZ cells fulfil all major criteria to be considered a good candidate for cell-replacement strategies. They are easily expandable in culture, they are plastic, migratory and can efficiently be incorporated in the recipient tissue. Several studies investigated the use of SVZ grafts. In a model of Parkinson's disease with unilateral striatal denervation caused by 6-hydroxy-dopamine (6-OHDA), grafted SVZ cells survived and differentiated in mature neurons expressing MAP-2, β-III tubulin or NeuN and improvement of motor performance was observed (Zigova et al., 1998a; Richardson et al., 2005). In a model of Huntington's disease, quinolinic acid (QA) was injected in the rat striatum to kill GABAergic spiny projection neurons. Degeneration of these neurons is responsible for the symptoms observed in Huntington's disease, such as involuntary movements. Following transplantation in QA-lesioned rats, SVZ adult cells differentiated into neurons expressing NeuN and specific markers for spiny neurons such as DARP-32 (dopamine and cyclic AMP-regulated phosphoprotein, relative molecular mass 32,000) and GAD_{67} (glutamic acid decarboxylase 67) (Vazey et al., 2006) with a concomitant improvement of motor performance.

SVZ cells have also been used in models of MS where chronic inflammation is a major cause of demyelination of the axons. Grafting SVZ cells into the subcortical white matter of myelin-deficient mice results in the migration of PSA-NCAM progenitors cells, differentiation of oligodendrocytes and remyelination (Cayre et al., 2006). In an animal model of MS, SVZ cells injected in the lateral ventricles, or in the blood stream, repopulate the CNS of the mouse, and differentiate into axon-ensheating oligodendrocytes (Pluchino et al., 2003).

However, successful grafting is hampered by the molecular microenvironment. Despite the large body of evidences indicating that damaged areas secrete factors improving progenitor cell dynamics, these factors may not be present at the functional level required to allow full differentiation. For instance, progenitors from the rat cervical spinal cord grafted in the mechanically injured adult spinal cord only differentiate into glial cells and not neurons (Vroemen et al., 2003). Ex vivo pre-treatment with factors promoting differentiation towards desired phenotypes can increase graft efficiency. Indeed, adult rat SVZ cells pre-treated in culture with FGF-2 successfully migrate and mature following graft in the cortex and striatum of adult rats (Zhang et al., 2003b). However, E14 rat spinal cord progenitors primed to a neuronal fate, following treatment with FGF-2 plus retinoic acid and grafted in the injured spinal cord of rats, survived but failed to differentiate into neurons. This result indicates that, despite the lesion, the microenvironment lacks instructive cues adequate to allow neuronal differentiation (Cao et al., 2002).

The host environment plays very important roles in determining the fate of grafted immature cells. Manipulation of the microenvironment in the injured brain seems to be necessary to facilitate neuronal replacement.

5.2. Stimulation of Endogenous Neurogenesis/Gliogenesis

Grafting strategies may become efficient approaches to treat neurological disorders affecting a local population of neurons such as in Parkinson's or Huntington's diseases. However, in multifocal diseases such as inflammatory disorders in which demyelination affect multiple parts of the brain, recruitment of endogenous sources of repairing cells appears to be preferable as compared with grafting strategies.

A variety of studies involving injection of factors in the lateral ventricles of rodents have been conducted in order to increase neurogenesis/gliogenesis in the SVZ. Infusions of EGF and FGF-2 increase proliferation and neurogenesis/gliogenesis in the SVZ (Craig et al., 1996; Kuhn et al., 1997). Likewise, infusion of BDNF or transfection with BDNF gene using adenovirus vectors promotes proliferation and production of neuroblasts that migrate into the olfactory bulb and also into the septum, striatum, thalamus and hypothalamus (Zigova et al., 1998b; Benraiss et al., 2001; Pencea et al., 2001a). Intracerebroventricular administration of VEGF promotes proliferation and production of doublecortin (DCX)-positive neurons (Sun et al., 2006). Orally administered neuroleptics such as olanzapine, used to treat schizophrenia, increases proliferation in the SVZ (Green et al., 2006).

BMP is secreted by SVZ cells and promotes the exit from the cell cycle and differentiation in astroglia. Injection of replication-deficient retroviruses encoding the BMP receptor subtype Ia into the neonatal rat SVZa leads to the expression of the receptor on the cell surface and to a higher expression of the cell cycle inhibitor p19INK4d, showing that BMP signaling promotes cell cycle exit (Coskun et al., 2001). These data suggest that neurogenesis can be increased by inhibiting BMP actions. Noggin, a factor secreted by E cells, binds BMP and thus inhibits its action on SVZ cells (Lim et al., 2000). Indeed, intraventricular injection of adenoviruses encoding Noggin and BDNF potentiate the BDNF-mediated recruitment of new neurons into the adult rat striatum (Chmielnicki et al., 2004). Targeting the interplay between BDNF and Noggin may provide a new elegant strategy for SVZ-driven brain repair.

Administration of NO donors or sildenafil (an inhibitor of the 5-phosphodiesterase) in rats suffering from traumatic brain injury or cerebral ischaemia increases SVZ neurogenesis in the striatum and the DG (Zhang et al., 2001b, 2002; Lu et al., 2003). Rats treated with sildenafil show an improvement in their motor performance as compared with untreated ischaemic rats (Zhang et al., 2002, 2006). The pro-neurogenic effect of these two molecules involves the increase in the concentration of intracellular cGMP that promotes angiogenesis and secretion of VEGF, known to favour neurogenesis (Jin et al., 2002b; Zhang et al., 2003a). Likewise, EPO is critical for SVZ neurogenesis and migration following ischaemia (Tsai et al., 2006). Administration of EPO to rats with brain ischaemia increases SVZ neurogenesis (Wang et al., 2004). Following 6-OHDA injection into the *substantia nigra*, a well-known model of Parkinson's disease, the administration of TGF-α in the rat striatum increases EGFR-mediated proliferation in the SVZ and production of new neurons that migrate towards the striatum and differentiate in new tyrosine hydroxylase positive neurons (Fallon et al., 2000). TGF-α-treated animals show clear improvement in motor performance. However, another study did not report differentiation of SVZ-derived cells into dopaminergic neurons after intrastriatal injection of TGF-α (Cooper and Isacson, 2004).

A major challenge in SVZ cell research is to develop new strategies allowing directed migration of SVZ reparative cells towards the site of injury. For that, injection of chemoattractive factors in the damaged area combined with factors stimulating SVZ neurogenesis can provide optimal conditions for brain repair. Overexpression of SVZ chemoattractant factors such as SCF, VEGF and SDF1-α (or CXCL12) can be good candidates for this purpose (Zhang et al., 2003c; Imitola et al., 2004; Sun et al., 2004).

5.3. SVZ Cells and Drug Delivery

On the basis of the unique ability of progenitor/stem cells to migrate and differentiate in the brain, they have been studied for their capacity to invade brain tumours and to deliver anti-tumoural drugs efficiently and in a site-specific manner. NSCs from hippocampi of 3- to 5-months-old human foetuses were engineered to release IL-12 and injected into rats with C6 tumour cells (Yang et al., 2004). Following IL-12 overproducing NSCs injection into gliomas, a marked decrease of the volume, and in some case the disappearance of the tumour, was observed.

Another study showed that neuronal precursors from the C17.2 lineage, engineered to express cytosine deaminase and injected in the blood stream, are able to migrate and reach intracranial gliomas. The efficient release of the enzyme converts a pro-drug cytotoxic 5-fluorouracil injected in the blood into toxic metabolites targeting rapidly proliferating cells. In this study, a decrease of the volume of the tumour was clearly observed (Aboody et al., 2000).

Stem cells offer a perspective in the development of new strategies for cancer (Yip et al., 2003). Hitherto, no studies reported the successful use of SVZ-engineered cells in cancer therapy, probably because of difficulties in transfection procedures. New avenues will be open for the use of SVZ cells in cancer therapy.

5.4. Stem Cell-Based Therapy and Risk of Tumour Formation

Recently, it has been suggested that adult SVZ stem cells may display a neoplasic potential and that brain tumours could originate from the dysregulation of stem cells' proliferation and self-renewal capacities (Recht et al., 2003). Indeed, tumour-derived cells share many aspects with stem cells. SVZ stem cells present high telomerase activity (Caporaso et al., 2003). Stem cells and tumorigenic cells are both endowed with the capacities to self-renew and proliferate (Fomchenko and Holland, 2005). Moreover, cells derived from human glial tumours express the intermediate filament protein nestin, an immature cell marker that is expressed also by SVZ stem cells (Lendahl et al., 1990; Ignatova et al., 2002). Neurons expressing β-III tubulin and GFAP-positive astrocytes were obtained in a culture of cells derived from human gliomas similar to what is obtained in a classical SVZ cell culture (Ignatova et al., 2002).

A recent study showed a direct link between glial tumours and SVZ cells. The deletion of the tumour-suppressor gene p53 in mice led to an increase in proliferation and neuronal production in the SVZ. Consistently, following exposure to a mutagenic agent, N-ethyl-N-nitrosourea (ENU), $p53^{-/-}$ mice developed tumours, originated from the exacerbate proliferation of B cells (Gil-Perotin et al., 2006).

Altogether, theses studies raise legitimate concerns about the future use of SVZ cells in cell therapy. Indeed, teratocarcinomas were obtained in mice brain following ES cells grafting (Erdö et al., 2003) and hyperplastic areas have been observed after intraventricular grafting of astrocyte-like stem cells (Zheng et al., 2002).

Similar to cancer cells, SVZ cells can migrate towards blood vessels (Zigova et al., 1998a). In the adult mouse, SVZ cells can migrate in the cerebral parenchyma, contact capillaries and are able to differentiate into astrocytes following injection of EGF in the lateral ventricles (Doetsch et al., 2002b). Moreover, after intravenous injection in a model of MS, SVZ cells pass through the vascular endothelium and enter in the brain tissue differentiating into myelinating oligodendrocytes (Pluchino et al., 2003). However, in this case, no tumour formation has been observed. In the study of Aboody and collaborators (2000), SVZ cells were used as drug delivery agent injected in the mice blood and targeting tumours into the brain. The non-neurogenic compartments of the brain release inhibitory factors impeding proliferation (Agasse et al., 2004; Rakic et al., 2004). Hence, "healthy" SVZ cells that do not express oncogenic properties may prove to be a reliable tool for brain repair, as they still respond to environmental cues. Identification of factors limiting proliferation in the non-neurogenic compartments of the brain would provide novel tools for controlling SVZ proliferation following grafting.

6. CONCLUSION

Adult SVZ cells are endowed with proper characteristics to be considered useful in cell replacement brain therapy and provide an important alternative to the ethically controversial use of human ES cells. However, we still need to learn more about how to differentiate SVZ cells in specific neuronal phenotypes.

SVZ cells have proved their efficiency to replace neurons and glial cells and promote functional recovery in model of Parkinson's and Huntington's diseases, injuries and demyelinating disorders. But to guarantee the maintenance and the differentiation of the SVZ-derived progenitors, several targets must be taken into account before efficient therapies can be developed (Fig. 2).

1a- SVZ cells are induced with factors in vitro to promote proliferation and neuronal or glial cell differentiation (Fig. 2).

1b- Endogenous recruitment of SVZ cell is achieved by injection of intracerebral factors or sequence of factors (viral vectors may be useful to increase efficiency). Administration of chemoattractive cues in the damage site will stimulate the correct migration/homing of SVZ-derived cells.

2- Following grafting or endogenous recruitment, survival and adequate differentiation of SVZ-derived newborn cells are assured through long-term administration of neuroprotective and trophic factors.

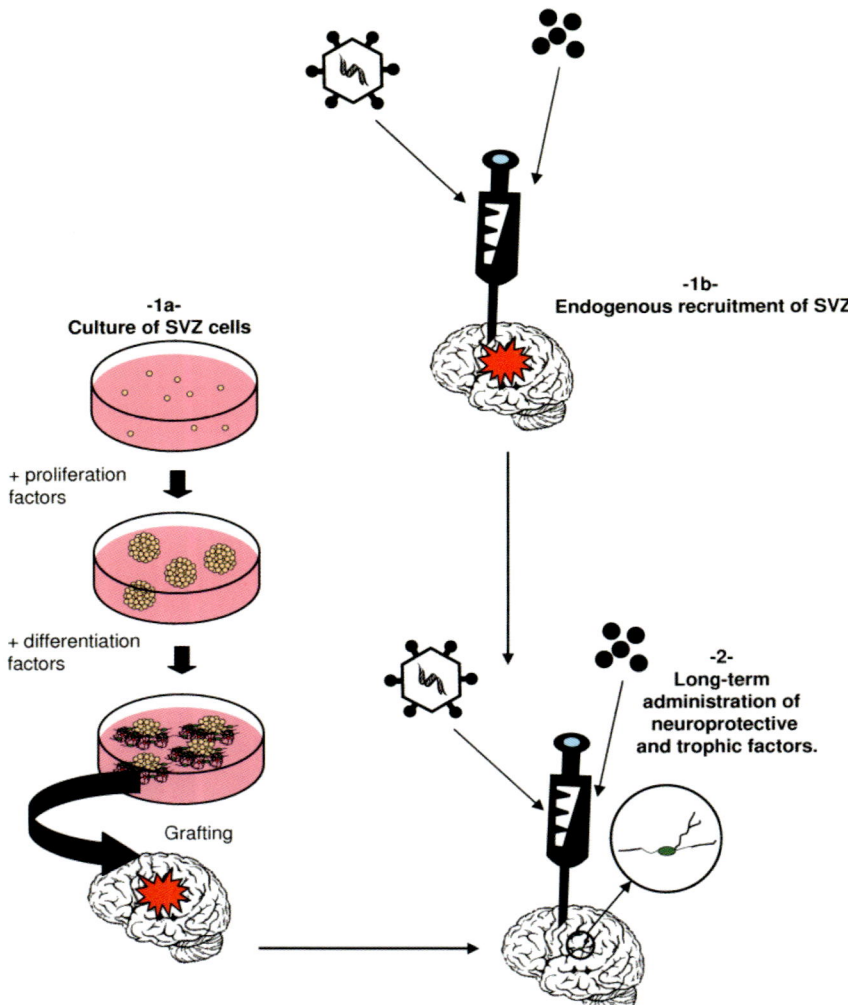

Fig. 2. Schematic view of the different steps needed to achieve cell replacement using SVZ cells. For further comments, see text in the Conclusion section

Table 1. Antigen expressed in the adult mouse SVZ

Expressed antigen	A Cell Neuroblast	B Cell Astrocyte	C cell Progenitor	D Cell Tanycyte	E Cell Ependymal Cell
PSA-NCAM	+	−	−	−	−
β-III tubulin	+	−	−	−	−
GFAP	−	+	−	+	+
Nestin	+	+	+	n.d.	+
Dlx 2	+	−	+	−	−
EGF-R	−	+/−	+	n.d.	n.d.
mCD24	+	−	−	+	+
LeX	−	+	+	−	−

β-III tubulin, cytoskeletal protein; Dlx-2, transcription factor; EGF-R, EGF and TGF-α receptor; GFAP, glial fibrillary acidic protein (cytoskeletal protein); LeX or LeX/ssEA-1 or CD15, glucide expressed at the cell surface; mCD24, mouse cluster of differentiation 24 (transmembrane protein); Nestin, cytoskeletal protein expressed by neuroepithelial cells; PSA-NCAM, polysialylated form of the neural cell adhesion molecule. (+) the antigen is present; (−) the antigen is not present; (+/−) a minority of the cells belonging to this cell type express the antigen; n.d., not determined (Doetsch et al., 1997, 2002b; Belvindrah et al., 2002; Capela and Temple 2002).

7. ACKNOWLEDGEMENTS

This work was supported by FCT-Portugal: grant POCTI/NSE/58492/2004 and by FEDER. We greatly acknowledge important suggestions by Dr. Alexandra Capela.

8. REFERENCES

Aberg, M.A., Aberg, N.D., Palmer, T.D., Alborn, A.M., Carlsson-Skwirut, C., Bang, P., Rosengren, L.E., Olsson, T., Gage, F.H. and Eriksson, P.S., 2003, IGF-I has a direct proliferative effect in adult hippocampal progenitor cells. *Mol. Cell. Neurosci.* **24**: 23.

Aboody, K.S., Brown, A., Rainov, N.G., Bower, K.A., Liu, S., Yang, W., Small, J.E., Herrlinger, U., Ourednik, V., Black, P.M., Breakefield, X.O. and Snyder, E.Y., 2000, Neural stem cells display extensive tropism for pathology in adult brain: evidence from intracranial gliomas. *Proc. Natl. Acad. Sci. USA* **97**: 12846.

Agasse, F., Roger, M. and Coronas, V., 2004, Neurogenic and intact or apoptotic non-neurogenic areas of adult brain release diffusible molecules that differentially modulate the development of subventricular zone cell cultures. *Eur. J. Neurosci.* **19**: 1459.

Ahmed, S., Reynolds, B.A. and Weiss, S., 1995, BDNF enhances the differentiation but not the survival of CNS stem cell-derived neuronal precursors. *J. Neurosci.* **15**: 5765.

Altman, J., 1969, Autoradiographic and histological studies of postnatal neurogenesis. IV. Cell proliferation and migration in the anterior forebrain, with special reference to persisting neurogenesis in the olfactory bulb. *J. Comp. Neurol.* **137**: 433.
Altman, J. and Das, G.D., 1965, Autoradiographic and histological evidence of postnatal hippocampal neurogenesis in rats. *J. Comp. Neurol.* **124**: 319.
Anderson, M.F., Aberg, M.A., Nilsson, M. and Eriksson, P.S., 2002, Insulin-like growth factor-I and neurogenesis in the adult mammalian brain. *Brain Res. Dev. Brain Res.* **134**: 115.
Arvidsson, A., Collin, T., Kirik, D., Kokaia, Z. and Lindvall, O., 2002, Neuronal replacement from endogenous precursors in the adult brain after stroke. *Nat. Med.* **8**: 963.
Baker, H., Liu, N., Chun, H.S., Saino, S., Berlin, R., Volpe, B. and Son, J.H., 2001, Phenotypic differentiation during migration of dopaminergic progenitor cells to the olfactory bulb. *J. Neurosci.* **21**: 8505.
Baker, S.A., Baker, K.A. and Hagg, T., 2004, Dopaminergic nigrostriatal projections regulate neural precursor proliferation in the adult mouse subventricular zone. *Eur. J. Neurosci.* **20**: 575.
Banner, L.R., Moayeri, N.N. and Patterson, P.H., 1997, Leukemia inhibitory factor is expressed in astrocytes following cortical brain injury. *Exp. Neurol.* **147**: 1.
Bauer, S., Rasika, S., Mauduit, C., Raccurt, M., Morel, G., Jourdan, F., Benahmed, M., Moyse, E. and Patterson, P.H., 2003, Leukemia inhibitory factor is a key signal for injury-induced neurogenesis in the adult mouse brain. *J. Neurosci.* **23**: 1792.
Bedard, A. and Parent, A., 2004, Evidence of newly generated neurons in the human olfactory bulb. *Brain Res. Dev. Brain Res.* **151**: 159.
Belluzzi, O., Benedusi, M., Ackman, J. and LoTurco, J.J., 2003, Electrophysiological differentiation of new neurons in the olfactory bulb. *J. Neurosci.* **23**: 10411.
Belmadani, A., Tran, P.B., Ren, D. and Miller, R.J., 2006, Chemokines regulate the migration of neural progenitors to sites of neuroinflammation. *J. Neurosci.* **26**: 3182.
Belvindrah, R., Rougon, G. and Chazal, G., 2002, Increased neurogenesis in adult mCD24-deficient mice. *J. Neurosci.* **22**: 3594.
Ben-Hur, T., 2006, Human embryonic stem cells for neuronal repair. *Isr. Med. Assoc. J.* **8**: 122.
Benraiss, A., Chmielnicki, E., Lerner, K., Roh, D. and Goldman, S.A., 2001, Adenoviral brain-derived neurotrophic factor induces both neostriatal and olfactory neuronal recruitment from endogenous progenitor cells in the adult forebrain. *J. Neurosci.* **21**: 6718.
Bernier, P.J., Bedard, A., Vinet, J., Levesque, M. and Parent, A., 2002, Newly generated neurons in the amygdala and adjoining cortex of adult primates. *Proc. Natl. Acad. Sci. USA* **99**: 11464.
Betarbet, R., Zigova, T., Bakay, R.A. and Luskin, M.B., 1996, Dopaminergic and GABAergic interneurons of the olfactory bulb are derived from the neonatal subventricular zone. *Int. J. Dev. Neurosci.* **14**: 921.
Bjornson, C.C.R., Rietze, R.L., Reynolds, B.A., Magli, M.C. and Vescovi, A.L., 1999, Turning brain into blood: a hematopoietic fate adopted by adult neural stem cells in vivo. *Science* **283**: 534.
Brännvall, K., Korhonen, L. and Lindholm, D., 2002, Estrogen-receptor-dependent regulation of neural stem cell proliferation and differentiation. *Mol. Cell Neurosci.* **21**: 512.
Brazel, C.Y., Nunez, J.L., Yang, Z. and Levison, S.W., 2005, Glutamate enhances survival and proliferation of neural progenitors derived from the subventricular zone. *Neuroscience* **131**: 55.
Brezun, J.M. and Daszuta, A., 1999, Depletion in serotonin decreases neurogenesis in the dentate gyrus and the subventricular zone of adult rats. *Neuroscience* **89**: 999.
Brodkey, J.A., Laywell, E.D., O'Brien, T.F., Faissner, A., Stefansson, K., Dorries, H.U., Schachner, M. and Steindler, D.A., 1995, Focal brain injury and upregulation of a developmentally regulated extracellular matrix protein. *J. Neurosurg.* **82**: 106.
Brooker, G.J., Kalloniatis, M., Russo, V.C., Murphy, M., Werther, G.A. and Bartlett, P.F., 2000, Endogenous IGF-1 regulates the neuronal differentiation of adult stem cells. *J. Neurosci. Res.* **59**: 332.
Brown, J., Cooper-Kuhn, C.M., Kempermann, G., Van Praag, H., Winkler, J., Gage, F.H. and Kuhn, H.G., 2003, Enriched environment and physical activity stimulate hippocampal but not olfactory bulb neurogenesis. *Eur. J. Neurosci.* **17**: 2042.
Butovsky, O., Ziv, Y., Schwartz, A., Landa, G., Talpalar, A.E., Pluchino, S., Martino, G. and Schwartz, M., 2006, Microglia activated by IL-4 or IFN-gamma differentially induce neurogenesis and oligodendrogenesis from adult stem/progenitor cells. *Mol. Cell Neurosci.* **31**: 149.
Cameron, H.A. and McKay, R.D., 1999, Restoring production of hippocampal neurons in old age. *Nat. Neurosci.* **2**: 894.
Cameron, H.A. and McKay, R.D., 2001, Adult neurogenesis produces a large pool of new granule cells in the dentate gyrus. *J. Comp. Neurol.* **435**: 406.
Cao, Q.L., Howard, R.M., Dennison, J.B. and Whittemore, S.R., 2002, Differentiation of engrafted neuronal-restricted precursor cells is inhibited in the traumatically injured spinal cord. *Exp. Neurol.* **177**: 349.

Capela, A. and Temple, S., 2002, LeX/ssea-1 is expressed by adult mouse CNS stem cells, identifying them as nonependymal. *Neuron* **35**: 865.
Caporaso, G.L., Lim, D.A., Alvarez-Buylla, A. and Chao, M.V., 2003, Telomerase activity in the subventricular zone of adult mice. *Mol. Cell. Neurosci.* **23**: 693.
Carlen, M., Cassidy, R.M., Brismar, H., Smith, G.A., Enquist, L.W. and Frisen, J., 2002, Functional integration of adult-born neurons. *Curr. Biol.* **12**: 606.
Cayre, M., Bancila, M., Virard, I., Borges, A. and Durbec, P., 2006, Migrating and myelinating potential of subventricular zone neural progenitor cells in white matter tracts of the adult rodent brain. *Mol. Cell Neurosci.* **31**: 748.
Chen, J.H., Wen, L., Dupuis, S., Wu, J.Y. and Rao, Y., 2001, The N-terminal leucine-rich regions in Slit are sufficient to repel olfactory bulb axons and subventricular zone neurons. *J. Neurosci.* **21**: 1548.
Chen, J., Magavi, S.S. and Macklis, J.D., 2004, Neurogenesis of corticospinal motor neurons extending spinal projections in adult mice. *Proc. Natl. Acad. Sci. USA* **101**: 16357.
Chen, J., Zacharek, A., Li, Y., Li, A., Wang, L., Katakowski, M., Roberts, C., Lu, M. and Chopp, M., 2006, N-cadherin mediates nitric oxide-induced neurogenesis in young and retired breeder neurospheres. *Neuroscience* **140**: 377.
Chiasson, B.J., Tropepe, V., Morshead, C.M. and van der Kooy, D., 1999, Adult mammalian forebrain ependymal and subependymal cells demonstrate proliferating potential, but only subependymal cells have neural stem cell characteristics. *J. Neurosci.* **19**: 4462.
Chmielnicki, E., Benraiss, A., Economides, A.N. and Goldman, S.A., 2004, Adenovirally expressed noggin and brain-derived neurotrophic factor cooperate to induce new medium spiny neurons from resident progenitor cells in the adult striatal ventricular zone. *J. Neurosci.* **24**: 2133.
Clarke, D.L., Johansson, C.B., Wilbertz, J., Veress, B., Nilsson, E., Karlström, H., Lendhal, U. and Frisén, J., 2000, Generalized potential of adult neural stem cells. *Science* **288**: 1660.
Conover, J.C., Doetsch, F., Garcia-Verdugo, J.M., Gale, N.W., Yancopoulos, G.D. and Alvarez-Buylla, A., 2000, Disruption of Eph/ephrin signaling affects migration and proliferation in the adult subventricular zone. *Nat. Neurosci.* **3**: 1091.
Cooper, O. and Isacson, O., 2004, Intrastriatal transforming growth factor alpha delivery to a model of Parkinson's disease induces proliferation and migration of endogenous adult neural progenitor cells without differentiation into dopaminergic neurons. *J. Neurosci.* **24**: 8924.
Coronas, V., Bantubungi, K., Fombonne, J., Krantic, S., Schiffmann, S.N. and Roger, M., 2004, Dopamine D3 receptor stimulation promotes the prolifeartion of cells derived from the postnatal subventricular zone. *J. Neurochem.* **91**: 1292.
Corotto, F.S., Henegar, J.R. and Maruniak, J.A., 1994, Odor deprivation leads to reduced neurogenesis and reduced neuronal survival in the olfactory bulb of the adult mouse. *Neurosci.* **61**: 739.
Coskun, V. and Luskin, M.B., 2001, The expression pattern of the cell cycle inhibitor $p19^{INK4d}$ by progenitor cells of the rat embryonic telencephalon and neonatal anterior subventricular zone. *J. Neurosci.* **21**: 3092.
Coskun, V., Venkatraman, G., Yang, H., Rao, M.S. and Luskin, M.B., 2001, Retroviral manipulation of the expression of bone morphogenetic protein receptor Ia by SVZa progenitor cells leads to changes in their p19(INK4d) expression but not in their neuronal commitment. *Int. J. Dev. Neurosci.* **19**: 219.
Craig, C.G., Tropepe, V., Morshead, C.M., Reynolds, B.A., Weiss, S. and van der Kooy, D., 1996, In vivo growth factor expansion of endogenous subependymal neural precursor cell populations in the adult mouse brain. *J. Neurosci.* **16**: 2649.
Curtis, M.A., Penney, E.B., Pearson, A.G., van Roon-Mom, W.M., Butterworth, N.J., Dragunow, M., Connor, B. and Faull, R.L., 2003, Increased cell proliferation and neurogenesis in the adult human Huntington's disease brain. *Proc. Natl. Acad. Sci. USA* **100**: 9023.
Digicaylioglu, M., Bichet, S., Marti, H.H., Wenger, R.H., Rivas, L.A., Bauer, C. and Gassmann, M., 1995, Localization of specific erythropoietin binding sites in defined areas of the mouse brain. *Proc. Natl. Acad. Sci. USA* **92**: 3717.
Doetsch, F. and Alvarez-Buylla, A., 1996, Network of tangential pathways for neuronal migration in adult mammalian brain. *Proc. Natl. Acad. Sci. USA* **93**: 14895.
Doetsch, F., Garcia-Verdugo, J.M. and Alvarez-Buylla, A., 1997, Cellular composition and three-dimensional organization of the subventricular germinal zone in the adult mammalian brain. *J. Neurosci.* **17**: 5046.
Doetsch, F., Caillé, I., Lim, D.A., Garcia-Verdugo, J.M. and Alvarez-Buylla, A., 1999a, Subventricular zone astrocytes are neural stem cells in the adult mammalian brain. *Cell* **97**: 703.
Doetsch, F., Garcia-Verdugo, J.M. and Alvarez-Buylla, A., 1999b, Regeneration of a germinal layer in the adult mammalian brain. *Proc. Natl. Acad. Sci. USA* **96**: 11619.
Doetsch, F., Verdugo, J.M., Caille, I., Alvarez-Buylla, A., Chao, M.V. and Casaccia-Bonnefil, P., 2002a, Lack of the cell-cycle inhibitor p27Kip1 results in selective increase of transit-amplifying cells for adult neurogenesis. *J. Neurosci.* **22**: 2255.

Doetsch, F., Petreanu, L., Caille, I., Garcia-Verdugo, J.M. and Alvarez-Buylla, A., 2002b, EGF converts transit-amplifying neurogenic precursors in the adult brain into multipotent stem cells. *Neuron* **36**: 1021.

Ebendal, T., Bengtsson, H. and Soderstrom, S., 1998, Bone morphogenetic proteins and their receptors: potential functions in the brain. *J. Neurosci. Res.* **51**: 139.

Ekdahl, C.T., Claasen, J.H., Bonde, S., Kokaia, Z. and Lindvall, O., 2003, Inflammation is detrimental for neurogenesis in adult brain. *Proc. Natl. Acad. Sci. USA* **100**: 13632.

Emsley, J.G. and Hagg, T., 2003a, Alpha6beta1 integrin directs migration of neuronal precursors in adult mouse forebrain. *Exp. Neurol.* **183**: 273.

Emsley, J.G. and Hagg, T., 2003b, Endogenous and exogenous ciliary neurotrophic factor enhances forebrain neurogenesis in adult mice. *Exp. Neurol.* **183**: 298.

Enwere, E., Shingo, T., Gregg, C., Fujikawa, H., Ohta, S. and Weiss, S., 2004, Aging results in reduced epidermal growth factor receptor signaling, diminished olfactory neurogenesis, and deficits in fine olfactory discrimination. *J. Neurosci.* **24**: 8354.

Erdö, F., Buhrle, C., Blunk, J., Hoehn, M., Xia, Y., Fleischmann, B., Focking, M., Kustermann, E., Kolossov, E., Hescheler, J., Hossmann, K.A. and Trapp, T., 2003, Host-dependent tumorigenesis of embryonic stem cell transplantation in experimental stroke. *J. Cereb. Blood Flow Metab.* **23**: 780.

Eriksson, P.S., Perfilieva, E., Bjork-Eriksson, T., Alborn, A.M., Nordborg, C., Peterson, D.A. and Gage, F.H., 1998, Neurogenesis in the adult human hippocampus. *Nat. Med.* **4**: 1313.

Fallon, J., Reid, S., Kinyamu, R., Opole, I., Opole, R., Baratta, J., Korc, M., Endo, T.L., Duong, A., Nguyen, G., Karkehabadhi, M., Twardzik, D., Patel, S. and Loughlin, S., 2000, In vivo induction of massive proliferation, directed migration, and differentiation of neural cells in the adult mammalian brain. *Proc. Natl. Acad. Sci. USA* **97**: 14686.

Fomchenko, E.I. and Holland, E.C., 2005, Stem cells and brain cancer. *Exper. Cell Res.* **306**: 323.

Frautschy, S.A., Walicke, P.A. and Baird, A., 1991, Localization of basic fibroblast growth factor and its mRNA after CNS injury. *Brain Res.* **553**: 291.

Frielingsdorf, H., Schwarz, K., Brundin, P. and Mohapel, P., 2004, No evidence for new dopaminergic neurons in the adult mammalian substantia nigra. *Proc. Natl. Acad. Sci. USA* **101**: 10177.

Gage, F.H., Kempermann, G., Palmer, T.D., Peterson, D.A. and Ray, J., 1998, Multipotent progenitor cells in the adult dentate gyrus. *J. Neurobiol.* **36**: 249.

Galli, R., Borello, U., Gritti, A., Minasi, M.G., Bjornson, C., Coletta, M., Mora, M., De Angelis, M.G., Fiocco, R., Cossu, G. and Vescovi, A.L., 2000, Skeletal myogenic potential of human and mouse neural stem cells. *Nat. Neurosci.* **3**: 986.

Garcia, A.D., Doan, N.B., Imura, T., Bush, T.G. and Sofroniew, M.V., 2004, GFAP-expressing progenitors are the principal source of constitutive neurogenesis in adult mouse forebrain. *Nat. Neurosci.* **7**: 1233.

Garcia-Verdugo, J.M., Doetsch, F., Wichterle, H., Lim, D.A. and Alvarez-Buylla, A., 1998, Architecture and cell types of the adult subventricular zone: in search for the stem cells. *J. Neurobiol.* **36**: 234.

Gheusi, G., Cremer, H., McLean, H., Chazal, G., Vincent, J.D. and Lledo, P.M., 2000, Importance of newly generated neurons in the adult olfactory bulb for odor discrimination. *Proc. Natl. Acad. Sci. USA* **97**: 1823.

Giardino, L., Bettelli, C. and Calza, L., 2000, In vivo regulation of precursor cells in the subventricular zone of adult rat brain by thyroid hormone and retinoids. *Neurosci. Lett.* **295**: 17.

Gil-Perotin, S., Marin-Husstege, M., Li, J., Soriano-Navarro, M., Zindy, F., Roussel, M.F., Garcia-Verdugo, J.M. and Casaccia-Bonnefil, P., 2006, Loss of p53 induces changes in the behavior of subventricular zone cells: implication for the genesis of glial tumors. *J. Neurosci.* **26**: 1107.

Gould, E., Tanapat, P., McEwen, B.S., Flugge, G. and Fuchs, E., 1998, Proliferation of granule cell precursors in the dentate gyrus of adult monkeys is diminished by stress. *Proc. Natl. Acad. Sci. USA* **95**: 3168.

Gould, E., Reeves, A.J., Graziano, M.S. and Gross, C.G., 1999, Neurogenesis in the neocortex of adult primates. *Science* **286**: 548.

Gould, E., Vail, N., Wagers, M. and Gross, C.G., 2001, Adult-generated hippocampal and neocortical neurons in macaques have a transient existence. *Proc. Natl. Acad. Sci. USA* **98**: 10910.

Green, W., Patil, P., Marsden, C.A., Bennett, G.W. and Wigmore, P.M., 2006, Treatment with olanzapine increases cell proliferation in the subventricular zone and prefrontal cortex. *Brain Res.* **1070**: 242.

Gross, R.E., Mehler, M.F., Mabie, P.C., Zang, Z., Santschi, L. and Kessler, J.A., 1996, Bone morphogenetic proteins promote astroglial lineage commitment by mammalian subventricular zone progenitor cells. *Neuron* **17**: 595.

Gu, W., Brannstrom, T. and Wester, P., 2000, Cortical neurogenesis in adult rats after reversible photothrombotic stroke. *J. Cereb. Blood Flow Metab.* **20**: 1166.

Guégan, C. and Sola, B., 2000, Early and sequential recruitment of apoptotic effectors after focal permanent ischemia in mice. *Brain Res.* **856**: 93.

Hack, I., Bancila, M., Loulier, K., Carroll, P. and Cremer, H., 2004, Reelin is a detachment signal in tangential chain-migration during postnatal neurogenesis. *Nat. Neurosci.* **5**: 939.

Hall, P.A. and Watt, F.M., 1989, Stem cells: the generation and maintenance of cellular diversity. *Development* **106**: 619.
Hastings, N.B. and Gould, E., 2003, Neurons inhibit neurogenesis. *Nat. Med.* **9**: 264.
Hayashi, T., Seki, T., Sato, K., Iwai, M., Zhang, W.R., Manabe, Y. and Abe, K., 2001, Expression of polysialylated neural cell adhesion molecule in rat brain after transient middle cerebral artery occlusion. *Brain Res.* **907**: 130.
Herrera, D.G., Garcia-Verdugo, J.M. and Alvarez-Buylla, A., 1999, Adult-derived neural precursors transplanted into multiple regions in the adult brain. *Ann. Neurol.* **46**: 867.
Hicks, R.R., Li, C., Zhang, L., Dhillon, H.S., Prasad, M.R. and Seroogy, K.B., 1999, Alterations in BDNF and trkB mRNA levels in the cerebral cortex following experimental brain trauma in rats. *J. Neurotrauma* **16**: 501.
Höglinger, G.U., Rizk, P., Muriel, M.P., Duyckaerts, C., Oertel, W.H., Caille, I. and Hirsch, E.C., 2004, Dopamine depletion impairs precursor cell proliferation in Parkinson disease. *Nat. Neurosci.* **7**: 726.
Hoofnagle, M.H., Wamhoff, B.R. and Owens, G.K., 2004, Lost in transdifferentiation. *J. Clin. Invest.* **113**: 1249.
Ignatova, T.N., Kukekov, V.G., Laywell, E.D., Suslov, O.N., Vrionis, F.D. and Steindler, D.A., 2002, Human cortical glial tumors contain neural stem-like cells expressing astroglial and neuronal markers in vitro. *Glia* **39**: 193.
Imitola, J., Raddassi, K., Park, K.I., Mueller, F.J., Nieto, M., Teng, Y.D., Frenkel, D., Li, J., Sidman, R.L., Walsh, C.A., Snyder, E.Y. and Khoury, S.J., 2004, Directed migration of neural stem cells to sites of CNS injury by the stromal cell-derived factor 1alpha/CXC chemokine receptor 4 pathway. *Proc. Natl. Acad. Sci. USA* **101**: 18117.
Ip, N.Y., Wiegand, S.J., Morse, J. and Rudge, J.S., 1993a, Injury-induced regulation of ciliary neurotrophic factor mRNA in the adult brain. *Eur. J. Neurosci.* **5**: 25.
Ip, N.Y., McClain, J., Barrezuetta, N.X., Aldrich, T.H., Pan, L., Li, Y., Wiegand, S.J., Friedman, B., Davis, S. and Yancopoulos, G.D., 1993b, The alpha component of CNTF receptor is required for signaling and defines potential CNTF targets in the adult and during development. *Neuron* **10**: 89.
Isgor, C. and Watson, S.J., 2005, Estrogen receptor alpha and beta mRNA expressions by proliferating and differentiating cells in the adult rat dentate gyrus and subventricular zone. *Neuroscience* **134**: 847.
Jiang, W., Gu, W., Brannstrom, T., Rosqvist, R. and Wester, P., 2001, Cortical neurogenesis in adult rats after transient middle cerebral artery occlusion. *Stroke* **32**: 1201.
Jin, K.L., Mao, X.O., Nagayama, T., Goldsmith, P.C. and Greenberg, D.A., 2000, Induction of vascular endothelial growth factor and hypoxia-inducible factor-1alpha by global ischemia in rat brain. *Neuroscience* **99**: 577.
Jin, K., Minami, M., Lan, J.Q., Mao, X.O., Batteur, S., Simon, R.P. and Greenberg, D.A., 2001, Neurogenesis in dentate subgranular zone and rostral subventricular zone after focal cerebral ischemia in the rat. *Proc. Natl. Acad. Sci. USA* **98**: 4710.
Jin, K., Mao, X.O., Sun, Y., Xie, L. and Greenberg, DA., 2002a, Stem cell factor stimulates neurogenesis in vitro and in vivo. *J. Clin. Invest.* **110**: 311.
Jin, K., Zhu, Y., Sun, Y., Mao, X.O., Xie, L. and Greenberg, D.A., 2002b, Vascular endothelial growth factor (VEGF) stimulates neurogenesis in vitro and in vivo. *Proc. Natl. Acad. Sci. USA* **99**: 11946.
Jin, K., Sun, Y., Xie, L., Peel, A., Mao, X.O., Batteur, S. and Greenberg, D.A., 2003, Directed migration of neuronal precursors into the ischemic cerebral cortex and striatum. *Mol. Cell. Neurosci.* **24**: 171.
Jin, K., Peel, A.L., Mao, X.O., Xie, L., Cottrell, B.A., Henshall, D.C. and Greenberg, D.A., 2004, Increased hippocampal neurogenesis in Alzheimer's disease. *Proc. Natl. Acad. Sci. USA* **101**: 343.
Johansson, C.B., Momma, S., Clarke, D.L., Risling, M., Lendahl, U. and Frisén, J., 1999, Identification of a neural stem cell in the adult mammalian central nervous system. *Cell* **96**: 25.
Kakita, A., Zerlin, M., Takahashi, H. and Goldman, J.E., 2003, Some glial progenitors in the neonatal subventricular zone migrate through the corpus callosum to the contralateral cerebral hemisphere. *J. Comp. Neurol.* **458**: 381.
Kempermann, G., Kuhn, H.G. and Gage, F.H., 1997, More hippocampal neurons in adult mice living in an enriched environment. *Nature* **386**: 493.
Kempermann, G., Kuhn, H.G. and Gage, F.H., 1998, Experience-induced neurogenesis in the senescent dentate gyrus. *J. Neurosci.* **18**: 3206.
Kippin, T.E., Cain, S.W., Masum, Z. and Ralph, M.R., 2004, Neural stem cells show bidirectional experience-dependent plasticity in the perinatal mammalian brain. *J. Neurosci.* **24**: 2832.
Kirschenbaum, B. and Goldman, S.A., 1995, Brain-derived neurotrophic factor promotes the survival of neurons arising from the adult rat forebrain subependymal zone. *Proc. Natl. Acad. Sci. USA* **92**: 210.
Klapka, N., Hermanns, S., Straten, G., Masanneck, C., Duis, S., Hamers, F.P., Muller, D., Zuschratter, W. and Muller, H.W., 2005, Suppression of fibrous scarring in spinal cord injury of rat promotes long-distance regeneration of corticospinal tract axons, rescue of primary motoneurons in somatosensory cortex and significant functional recovery. *Eur. J. Neurosci.* **22**: 3047.
Kokaia, Z. and Lindvall, O., 2003, Neurogenesis after ischaemic brain insults. *Curr. Opin. Neurobiol.* **13**: 127.
Kornack, D.R. and Rakic, P., 2001, The generation, migration, and differentiation of olfactory neurons in the adult primate brain. *Proc. Natl. Acad. Sci. USA* **98**: 4752.

Kuhn, H.G., Dickinson-Anson, H. and Gage, F.H., 1996, Neurogenesis in the dentate gyrus of the adult rat: age-related decrease of neuronal progenitor proliferation. *J. Neurosci.* **16**: 2027.

Kuhn, H.G., Winkler, J., Kempermann, G., Thal, L.J. and Gage, F.H., 1997, Epidermal growth factor and fibroblast growth factor-2 have different effects on neural progenitors in the adult rat brain. *J. Neurosci.* **17**: 5820.

Kulkarni, V.A., Jha, S. and Vaidya, V.A., 2002, Depletion of norepinephrine decreases the proliferation, but does not influence the survival and differentiation, of granule cell progenitors in the adult rat hippocampus. *Eur. J. Neurosci.* **16**: 2008.

Larsson, E., Mandel, R.J., Klein, R.L., Muzyczka, N., Lindvall, O. and Kokaia, Z., 2002, Suppression of insult-induced neurogenesis in adult rat brain by brain-derived neurotrophic factor. *Exp. Neurol.* **177**: 1.

Law, A.K., Pencea, V., Buck, C.R. and Luskin, M.B., 1999, Neurogenesis and neuronal migration in the neonatal rat forebrain anterior subventricular zone do not require GFAP-positive astrocytes. *Dev. Biol.* **216**: 622.

Laywell, E.D., Rakic, P., Kukekov, V.G., Holland, E.C. and Steindler, D.A., 2000, Identification of a multipotent astrocytic stem cell in the immature and adult mouse brain. *Proc. Natl. Acad. Sci. USA* **97**: 13883.

Lee, S.R., Kim, H.Y., Rogowska, J., Zhao, B.Q., Bhide, P., Parent, J.M. and Lo, E.H., 2006, Involvement of matrix metalloproteinase in neuroblast cell migration from the subventricular zone after stroke. *J. Neurosci.* **26**: 3491.

Lendahl, U., Zimmerman, L.B. and McKay, R.D., 1990, CNS stem cells express a new class of intermediate filament protein. *Cell* **60**: 585.

Levison, S.W. and Goldman, J.E., 1997, Multipotential and lineage restricted precursors coexist in the mammalian perinatal subventricular zone. *J. Neurosci. Res.* **48**: 83.

Levison, S.W., Chuang, C., Abramson, B.J. and Goldman, J.E., 1993, The migrational patterns and developmental fates of glial precursors in the rat subventricular zone are temporally regulated. *Development* **119**: 611.

Li, Y., Chopp, M., Powers, C. and Jiang, N., 1997, Apoptosis and protein expression after focal cerebral ischemia in rat. *Brain Res.* **765**: 301.

Li, Y., Chen, J., Wang, L., Zhang, L., Lu, M. and Chopp, M., 2001, Intracerebral transplantation of bone marrow stromal cells in a 1-methyl-4-phenyl-1,2,3,6-tetrahydropyridine mouse model of Parkinson's disease. *Neurosci. Lett.* **316**: 67.

Lie, D.C., Dziewczapolski, G., Willhoite, A.R., Kaspar, B.K., Shults, C.W. and Gage, F.H., 2002, The adult substantia nigra contains progenitor cells with neurogenic potential. *J. Neurosci.* **22**: 6639.

Lim, D.A. and Alvarez-Buylla, A., 1999, Interaction between astrocytes and adult subventricular zone precursors stimulates neurogenesis. *Proc. Natl. Acad. Sci. USA* **96**: 7526.

Lim, D.A., Fishell, G.J. and Alvarez-Buylla, A., 1997, Postnatal mouse subventricular zone neuronal precursors can migrate and differentiate within multiple levels of the developing neuraxis. *Proc. Natl. Acad. Sci. USA* **94**: 14832.

Lim, D.A., Tramontin, A.D., Trevejo, J.M., Herrera, D.G., Garcia-Verdugo, J.M. and Alvarez-Buylla, A., 2000, Noggin antagonizes BMP signaling to create a niche for adult neurogenesis. *Neuron* **28**: 713.

Liu, G. and Rao, Y., 2003, Neuronal migration from the forebrain to the olfactory bulb requires a new attractant persistent in the olfactory bulb. *J. Neurosci.* **23**: 6651.

Lois, C. and Alvarez-Buylla, A., 1994, Long-distance neuronal migration in the adult mammalian brain. *Science* **264**: 1145.

Lois, C., Garcia-Verdugo, J.M. and Alvarez-Buylla, A., 1996, Chain migration of neuronal precursors. *Science* **271**: 978.

Lu, D., Mahmood, A., Zhang, R. and Copp, M., 2003, Upregulation of neurogenesis and reduction in functional deficits following administration of DEtA/NONOate, a nitric oxide donor, after traumatic brain injury in rats. *J. Neurosurg.* **99**: 351.

Luskin, M.B., 1993, Restricted proliferation and migration of postnatally generated neurons derived from the forebrain subventricular zone. *Neuron* **11**: 173.

Luskin, M.B., Zigova, T., Soteres, B.J. and Stewart, R.R., 1997, Neuronal progenitor cells derived from the anterior subventricular zone of the neonatal rat forebrain continue to proliferate *in vitro* and express a neuronal phenotype. *Mol. Cell. Neurosci.* **8**: 351.

Mabie, P.C., Mehler, M.F. and Kessler, J.A., 1999, Multiple roles of bone morphogenetic protein signaling in the regulation of cortical cell number and phenotype. *J. Neurosci.* **19**: 7077.

Magavi, S.S., Leavitt, B.R. and Macklis, J.D., 2000, Induction of neurogenesis in the neocortex of adult mice. *Nature* **405**: 951.

Malaterre, J., Strambi, C., Aouane, A., Strambi, A., Rougon, G. and Cayre, M., 2004, A novel role for polyamines in adult neurogenesis in rodent brain. *Eur. J. Neurosci.* **20**: 317.

Marti, H.H., Wenger, R.H., Rivas, L.A., Straumann, U., Digicaylioglu, M., Henn, V., Yonekawa, Y., Bauer, C. and Gassmann, M., 1996, Erythropoietin gene expression in human, monkey and murine brain. *Eur. J. Neurosci.* **8**: 666.

Marti, H.J., Bernaudin, M., Bellail, A., Schoch, H., Euler, M., Petit, E. and Risau, W., 2000, Hypoxia-induced vascular endothelial growth factor expression precedes neovascularization after cerebral ischemia. *Am. J. Pathol.* **156**: 965.

Maslov, A.Y., Barone, T.A., Plunkett, R.J. and Pruitt, S.C., 2004, Neural stem cell detection, characterization, and age-related changes in the subventricular zone of mice. *J. Neurosci.* **24**: 1726.

Mason, J.L., Suzuki, K., Chaplin, D.D. and Matsushima, G.K., 2001a, Interleukin-1beta promotes repair of the CNS. *J. Neurosci.* **21**: 7046.

Mason, H.A., Ito, S. and Corfas, G., 2001b, Extracellular signals that regulate the tangential migration of olfactory bulb neuronal precursors: inducers, inhibitors, and repellents. *J. Neurosci.* **21**: 7654.

Menezes, J.R., Smith, C.M., Nelson, K.C. and Luskin, M.B., 1995, The division of neuronal progenitor cells during migration in the neonatal mammalian forebrain. *Mol. Cell. Neurosci.* **6**: 496.

Miller, J.T., Bartley, J.H., Wimborne, H.J., Walker, A.L., Hess, D.C., Hill, W.D. and Carroll, J.E., 2005, The neuroblast and angioblast chemotaxic factor SDF-1 (CXCL12) expression is briefly up regulated by reactive astrocytes in brain following neonatal hypoxic-ischemic injury. *BMC Neurosci.* **6**: 63.

Monje, M.L., Toda, H. and Palmer, T.D., 2003, Inflammatory blockade restores adult hippocampal neurogenesis. *Science* **302**: 1760.

Moreno-Lopez, B., Romero-Grimaldi, C., Noval, J.A., Murillo-Carretero, M., Matarredona, E.R. and Estrada, C., 2004, Nitric oxide is a physiological inhibitor of neurogenesis in the adult mouse subventricular zone and olfactory bulb. *J. Neurosci.* **24**: 85.

Morshead, C.M. and van der Kooy, D., 1992, Postmitotic death is the fate of constitutively proliferating cells in the subependymal layer of the adult mouse brain. *J. Neurosci.* **12**: 24.

Morshead, C.M. and van der Kooy, D., 2001, A new 'spin' on neural stem cells? *Curr. Opin. Neurobiol.* **11**: 59.

Morshead, C.M., Reynolds, B.A., Craig, C.G., McBurney, M.W., Staines, W.A., Morassutti, D., Weiss, S. and van der Kooy, D., 1994, Neural stem cells in the adult mammalian forebrain: a relatively quiescent subpopulation of subependymal cells. *Neuron* **13**: 1071.

Morshead, C.M., Craig, C.G. and van der Kooy, D., 1998, In vivo clonal analyses reveal the properties of endogenous neural stem cell proliferation in the adult mammalian forebrain. *Development* **125**: 2251.

Morshead, C.M., Garcia, A.D., Sofroniew, M.V. and van Der Kooy, D., 2003, The ablation of glial fibrillary acidic protein-positive cells from the adult central nervous system results in the loss of forebrain neural stem cells but not retinal stem cells. *Eur. J. Neurosci.* **18**: 76.

Nacher, J., Rosell, D.R., Alonso-Llosa, G. and McEwen, B.S., 2001, NMDA receptor antagonist treatment induces a long-lasting increase in the number of proliferating cells, PSA-NCAM-immunoreactive granule neurons and radial glia in the adult rat dentate gyrus. *Eur. J. Neurosci.* **13**: 512.

Nait-Oumesmar, B., Decker, L., Lachapelle, F., Avellana-Adalid, V., Bachelin, C. and Van Evercooren, A.B., 1999, Progenitor cells of the adult mouse subventricular zone proliferate, migrate and differentiate into oligodendrocytes after demyelination. *Eur. J. Neurosci.* **11**: 4357.

Ng, K.L., Li, J.D., Cheng, M.Y., Leslie, F.M., Lee, A.G. and Zhou, Q.Y., 2005, Dependence of olfactory bulb neurogenesis on prokineticin 2 signaling. *Science* **308**: 1923.

O'Dell, D.M., Raghupathi, R., Crino, P.B., Eberwine, J.H. and McIntosh, T.K., 2000, Traumatic brain injury alters the molecular fingerprint of TUNEL-positive cortical neurons in vivo: A single-cell analysis. *J. Neurosci.* **20**: 4821.

Okano, H., Okada, S., Nakamura, M. and Toyama, Y., 2005, Neural stem cells and regeneration of injured spinal cord. *Kidney Int.* **68**: 1927.

Oyesiku, N.M., Evans, C.O., Houston, S., Darrell, R.S., Smith, J.S., Fulop, Z.L., Dixon, C.E. and Stein, D.G., 1999, Regional changes in the expression of neurotrophic factors and their receptors following acute traumatic brain injury in the adult rat brain. *Brain Res.* **833**: 161.

Packer, M.A., Stasiv, Y., Benraiss, A., Chmielnicki, E., Grinberg, A., Westphal, H., Goldman, S.A. and Enikolopov, G., 2003, Nitric oxide negatively regulates mammalian adult neurogenesis. *Proc. Natl. Acad. Sci. USA* **100**: 9566.

Palmer, T.D., Schwartz, P.H., Taupin, P., Kaspar, B., Stein, S.A. and Gage, F.H., 2001, Cell culture. Progenitor cells from human brain after death. *Nature* **411**: 42.

Paratcha, G., Ibanez, C.F. and Ledda, F., 2006, GDNF is a chemoattractant factor for neuronal precursor cells in the rostral migratory stream. *Mol. Cell Neurosci.* **31**: 505.

Parent, J.M., Valentin, V.V. and Lowenstein, D.H., 2002a, Prolonged seizures increase proliferating neuroblasts in the adult rat subventricular zone-olfactory bulb pathway. *J. Neurosci.* **22**: 3174.

Parent, J.M., Vexler, Z.S., Gong, C., Derugin, N. and Ferriero, D.M., 2002b, Rat forebrain neurogenesis and striatal neuron replacement after focal stroke. *Ann. Neurol.* **52**: 802.

Pencea, V., Bingaman, K.D., Wiegand, S.J. and Luskin, M.B., 2001a, Infusion of brain-derived neurotrophic factor into the lateral ventricle of the adult rat leads to new neurons in the parenchyma of the striatum, septum, thalamus, and hypothalamus. *J. Neurosci.* **21**: 6706.

Pencea, V., Bingaman, K.D., Freedman, L.J. and Luskin, M.B., 2001b, Neurogenesis in the subventricular zone and rostral migratory stream of the neonatal and adult primate forebrain. *Exp. Neurol.* **172**: 1.

Peretto, P., Dati, C., De Marchis, S., Kim, H.H., Ukhanova, M., Fasolo, A. and Margolis, F.L., 2004, Expression of the secreted factors noggin and bone morphogenetic proteins in the subependymal layer and olfactory bulb of the adult mouse brain. *Neuroscience* **128**: 685.

Petreanu, L. and Alvarez-Buylla, A., 2002, Maturation and death of adult-born olfactory bulb granule neurons: role of olfaction. *J. Neurosci.* **22**: 6106.

Picard-Riera, N., Decker, L., Delarasse, C., Goude, K., Nait-Oumesmar, B., Liblau, R., Pham-Dinh, D. and Evercooren, A.B., 2002, Experimental autoimmune encephalomyelitis mobilizes neural progenitors from the subventricular zone to undergo oligodendrogenesis in adult mice. *Proc. Natl. Acad. Sci. USA* **99**: 13211.

Pluchino, S., Quattrini, A., Brambilla, E., Gritti, A., Salani, G., Dina, G., Galli, R., Del Carro, U., Amadio, S., Bergami, A., Furlan, R., Comi, G., Vescovi, A.L. and Martino, G., 2003, Injection of adult neurospheres induces recovery in a chronic model of multiple sclerosis. *Nature* **422**: 688.

Pluchino, S., Zanotti, L., Deleidi, M. and Martino, G., 2005, Neural stem cells and their use as therapeutic tool in neurological disorders. *Brain Res. Brain Res. Rev.* **48**: 211.

Rakic, P., 2004, Neuroscience: immigration denied. *Nature* **427**: 685.

Recht, L., Jang, T., Savarese, T. and Litofsky, N.S., 2003, Neural stem cells and neuro-oncology: quo vadis? *J. Cell Biochem.* **88**: 11.

Reynolds, B.A. and Weiss, S., 1992, Generation of neurons and astrocytes from isolated cells of the adult mammalian central nervous system. *Science* **255**: 1707.

Richardson, R.M., Broaddus, W.C., Holloway, K.L. and Fillmore H.L., 2005, Grafts of adult subependymal zone neuronal progenitor cells rescue hemiparkinsonian behavioral decline. *Brain Res.* **1032**: 11.

Rietze, R.L., Valcanis, H., Brooker, G.F., Thomas, T., Voss, A.K. and Bartlett, P.F., 2001, Purification of a pluripotent neural stem cell from the adult mouse brain. *Nature* **412**: 736.

Rochefort, C., Gheusi, G., Vincent, J.D. and Lledo, P.M., 2002, Enriched odor exposure increases the number of newborn neurons in the adult olfactory bulb and improves odor memory. *J. Neurosci.* **22**: 2679.

Romanko, M.J., Rola, R., Fike, J.R., Szele, F.G., Dizon, M.L., Felling, R.J., Brazel, C.Y. and Levison, S.W., 2004, Roles of the mammalian subventricular zone in cell replacement after brain injury. *Prog. Neurobiol.* **74**: 77.

Saghatelyan, A., de Chevigny, A., Schachner, M. and Lledo, P.M., 2004, Tenascin-R mediates activity-dependent recruitment of neuroblasts in the adult mouse forebrain. *Nat. Neurosci.* **7**: 347.

Salman, H., Ghosh, P. and Kernie, S.G., 2004, Subventricular zone neural stem cells remodel the brain following traumatic injury in adult mice. *J. Neurotrauma* **21**: 283.

Sanai, N., Tramontin, A.D., Quinones-Hinojosa, A., Barbaro, N.M., Gupta, N., Kunwar, S., Lawton, M.T., McDermott, M.W., Parsa, A.T., Garcia-Verdugo, J.M., Berger, M.S. and Alvarez-Buylla, A., 2004, Unique astrocyte ribbon in adult human brain contains neural stem cells but lacks chain migration. *Nature* **427**: 740.

Sawamoto, K., Wichterle, H., Gonzalez-Perez, O., Cholfin, J.A., Yamada, M., Spassky, N., Murcia, N.S., Garcia-Verdugo, J.M., Marin, O., Rubenstein, J.L., Tessier-Lavigne, M., Okano, H. and Alvarez-Buylla, A., 2006, New neurons follow the flow of cerebrospinal fluid in the adult brain. *Science* **311**: 629.

Sayles, M., Jain, M. and Barker, R.A., 2004, The cellular repair of the brain in Parkinson's disease-past, present and future. *Transpl. Immunol.* **12**: 321.

Seri, B., Garcia-Verdugo, J.M., McEwen, B.S. and Alvarez-Buylla, A., 2001, Astrocytes give rise to new neurons in the adult mammalian hippocampus. *J. Neurosci.* **21**: 7153.

Sheen, V.L. and Macklis, J.D., 1995, Targeted neocortical cell death in adult mice guides migration and differentiation of transplanted embryonic neurons. *J. Neurosci.* **15**: 8378.

Sheen, V.L., Arnold, M.W., Wang, Y. and Macklis, J.D., 1999, Neural precursor differentiation following transplantation into neocortex is dependent on intrinsic developmental state and receptor competence. *Exp. Neurol.* **158**: 47.

Shimazaki, T., Shingo, T. and Weiss, S., 2001, The ciliary neurotrophic factor/leukemia inhibitory factor/gp130 receptor complex operates in the maintenance of mammalian forebrain neural stem cells. *J. Neurosci.* **21**: 7642.

Shingo, T., Sorokan, S.T., Shimazaki, T. and Weiss, S., 2001, Erythropoietin regulates the in vitro and in vivo production of neuronal progenitors by mammalian forebrain neural stem cells. *J. Neurosci.* **21**: 9733.

Shingo, T., Gregg, C., Enwere, E., Fujikawa, H., Hassam, R., Geary, C., Cross, J.C. and Weiss, S., 2003, Pregnancy-stimulated neurogenesis in the adult female forebrain mediated by prolactin. *Science* **299**: 117.

Shors, T.J., Miesegaes, G., Beylin, A., Zhao, M., Rydel, T. and Gould, E., 2001, Neurogenesis in the adult is involved in the formation of trace memories. *Nature* **410**: 372.

Smith, G.M. and Strunz, C., 2005, Growth factor and cytokine regulation of chondroitin sulfate proteoglycans by astrocytes. *Glia* **52**: 209.

Smith, C., Berry, M., Clarke, W.E. and Logan, A., 2001, Differential expression of fibroblast growth factor-2 and fibroblast growth factor receptor 1 in a scarring and nonscarring model of CNS injury in the rat. *Eur. J. Neurosci.* **13**: 443.

Snyder, E.Y., Yoon, C., Flax, J.D. and Macklis, J.D., 1997, Multipotent neural precursors can differentiate toward replacement of neurons undergoing targeted apoptotic degeneration in adult mouse neocortex. *Proc. Natl. Acad. Sci. USA* **94**: 11663.

Spassky, N., Merkle, F.T., Flames, N., Tramontin, A.D., Garcia-Verdugo, J.M. and Alvarez-Buylla, A., 2005, Adult ependymal cells are postmitotic and are derived from radial glial cells during embryogenesis. *J. Neurosci.* **25**: 10.

Stoll, G., Jander, S. and Schroeter, M., 2002, Detrimental and beneficial effects of injury-induced inflammation and cytokine expression in the nervous system. *Adv. Exp. Med. Biol.* **513**: 87.

Sun, L., Lee, J. and Fine, H.A., 2004, Neuronally expressed stem cell factor induces neural stem cell migration to areas of brain injury. *J. Clin. Invest.* **113**: 1364.

Sun, Y., Jin, K., Childs, J.T., Xie, L., Mao, X.O. and Greenberg, D.A., 2006, Vascular endothelial growth factor-B (VEGFB) stimulates neurogenesis: evidence from knockout mice and growth factor administration. *Dev. Biol.* **289**: 329.

Suzuki, S., Tanaka, K., Nogawa, S., Ito, D., Dembo, T., Kosakai, A. and Fukuuchi, Y., 2000, Immunohistochemical detection of leukemia inhibitory factor after focal cerebral ischemia in rats. *J. Cereb. Blood Flow. Metab.* **20**: 661.

Szele, F.G. and Chesselet, M.F., 1996, Cortical lesion induce an increase in cell number and PSA-NCAM expression in the subventricular zone of adult rats. *J. Comp. Neurol.* **368**: 439.

Taga, T and Fukuda, S., 2005, Role of IL-6 in the neural stem cell differentiation. *Clin. Rev. Allergy Immunol.* **28**: 249.

Tan, A.M., Zhang, W. and Levine, J.M., 2005, NG2: a component of the glial scar that inhibits axon growth. *J. Anat.* **207**: 717.

Tanapat, P., Galea, L.A. and Gould, E., 1998, Stress inhibits the proliferation of granule cell precursors in the developing dentate gyrus. *Int. J. Dev. Neurosci.* **16**: 235.

Tanapat, P., Hastings, N.B., Reeves, A.J. and Gould, E., 1999, Estrogen stimulates a transient increase in the number of new neurons in the dentate gyrus of the adult female rat. *J. Neurosci.* **19**: 5792.

Trejo, J.L., Carro, E. and Torres-Aleman, I., 2001, Circulating insulin-like growth factor I mediates exercise-induced increases in the number of new neurons in the adult hippocampus. *J. Neurosci.* **21**: 162.

Tropepe, V., Craig, C.G., Morshead, C.M. and van der Kooy, D., 1997, Transforming growth factor-alpha null and senescent mice show decreased neural progenitor cell proliferation in the forebrain subependyma. *J. Neurosci.* **17**: 7850.

Tsai, P.T., Ohab, J.J., Kertesz, N., Groszer, M., Matter, C., Gao, J., Liu, X., Wu, H. and Carmichael, S.T., 2006, A critical role of erythropoietin receptor in neurogenesis and post-stroke recovery. *J. Neurosci.* **26**: 1269.

Tsukatani, T., Fillmore, H.L., Hamilton, H.R., Holbrook, E.H. and Costanzo, R.M., 2003, Matrix metalloproteinase expression in the olfactory epithelium. *Neuroreport* **14**: 1135.

Tzeng, S.F. and Wu, J.P., 1999, Responses of microglia and neural progenitors to mechanical brain injury. *Neuroreport* **10**: 2287.

van Kampen, J.M., Hagg, T. and Robertson, H.A., 2004, Induction of neurogenesis in the adult rat subventricular zone and neostriatum following dopamine D3 receptor stimulation. *Eur. J. Neurosci.* **19**: 2377.

van Praag, H., Christie, B.R., Sejnowski, T.J. and Gage, F.H., 1999, Running enhances neurogenesis, learning, and long-term potentiation in mice. *Proc. Natl. Acad. Sci. USA* **96**: 13427.

van Praag, H., Schinder, A.F., Christie, B.R., Toni, N., Palmer, T.D. and Gage, F.H., 2002, Functional neurogenesis in the adult hippocampus. *Nature* **415**: 1030.

Vazey, E.M., Chen, K., Hughes, S.M. and Connor, B., 2006, Transplanted adult neural progenitor cells survive, differentiate and reduce motor function impairment in a rodent model of Huntington's disease. *Exp. Neurol.* **199**: 384.

Vescovi, A.L., Galli, R. and Gritti, A., 2001, The neural stem cells and their transdifferentiation capacity. *Biomed. Pharmacother.* **55**: 201.

Vroemen, M., Aigner, L., Winkler, J. and Weidner, N., 2003, Adult neural progenitor cell grafts survive after acute spinal cord injury and integrate along axonal pathways. *Eur. J. Neurosci.* **18**: 743.

Wang, Y., Sheen, V.L. and Macklis, J.D., 1998, Cortical interneurons upregulate neurotrophins in vivo in response to targeted apoptotic degeneration of neighboring pyramidal neurons. *Exp. Neurol.* **154**: 389.

Wang, L., Zhang, Z., Wang, Y., Zhang, R. and Chopp, M., 2004, Treatment of stroke with erythropoietin enhances neurogenesis and angiogenesis and improves neurological function in rats. *Stroke* **35**: 1732.

Wu, J.P., Kuo, J.S., Liu, Y.L. and Tzeng, S.F., 2000, Tumor necrosis factor-alpha modulates the proliferation of neural progenitors in the subventricular/ventricular zone of adult rat brain. *Neurosci. Lett.* **292**: 203.

Yang, S.Y., Liu, H. and Zhang, J.N., 2004, Gene therapy of rat malignant gliomas using neural stem cells expressing IL-12. *DNA Cell Biol.* **23**: 381.

Yip, S., Aboody, K.S., Burns, M., Imitola, J., Boockvar, J.A., Allport, J., Park, K.I., Teng, Y.D., Lachyankar, M., McIntosh, T., O'Rourke, D.M., Khoury, S., Weissleder, R., Black, P.M., Weiss, W. and Snyder, E.Y., 2003, Neural stem cell biology may be well suited for improving brain tumor therapies. *Cancer J.* **9**: 189.

Yoshimura, S., Takagi, Y., Harada, J., Teramoto, T., Thomas, S.S., Waeber, C., Bakowska, J.C., Breakefield, X.O. and Moskowitz, M.A., 2001, FGF-2 regulation of neurogenesis in adult hippocampus after brain injury. *Proc. Natl. Acad. Sci. USA* **98**: 5874.

Yoshimura, S., Teramoto, T., Whalen, M.J., Irizarry, M.C., Takagi, Y., Qiu, J., Harada, J., Waeber, C., Breakefield, X.O. and Moskowitz, M.A., 2003, FGF-2 regulates neurogenesis and degeneration in the dentate gyrus after traumatic brain injury in mice. *J. Clin. Invest.* **112**: 1202.

Zhang, S.C. and Fedoroff, S., 1999, Expression of stem cell factor and c-kit receptor in neural cells after brain injury. *Acta Neuropathol. (Berl).* **97**: 393.

Zhang, R.L., Zhang, Z.G., Zhang, L. and Chopp, M., 2001a, Proliferation and differentiation of progenitor cells in the cortex and the subventricular zone in the adult rat after focal cerebral ischemia. *Neuroscience* **105**: 33.

Zhang, R., Zhang, L., Zhang, Z., Wang, Y., Lu, M., Lapointe, M. and Chopp. M., 2001b, A nitric oxide donor induces neurogenesis and reduces functional deficits after stroke in rats. *Ann. Neurol.* **50**: 602.

Zhang, R., Wang, Y., Zhang, L., Zhang, Z., Tsang, W., Lu, M., Zhang, L. and Chopp, M., 2002, Sildenafil (Viagra) induces neurogenesis and promotes functional recovery after stroke in rats. *Stroke* **33**: 2675.

Zhang, R., Wang, L., Zhang, L., Chen, J., Zhu, Z., Zhang, Z. and Chopp, M., 2003a, Nitric oxide enhances angiogenesis via the synthesis of vascular endothelial growth factor and cGMP after stroke in the rat. *Circ. Res.* **92**: 308.

Zhang, R.L., Zhang, L., Zhang, Z.G., Morris, D., Jiang, Q., Wang, L., Zhang, L.J. and Chopp, M., 2003b, Migration and differentiation of adult rat subventricular zone progenitor cells transplanted into the adult rat striatum. *Neuroscience* **116**: 373.

Zhang, H, Vutskits, L., Pepper, M.S. and Kiss, J.Z., 2003c, VEGF is a chemoattractant for FGF-2-stimulated neural progenitors. *J. Cell Biol.* **163**: 1375.

Zhang, R., Zhang, Z., Wang, L., Wang, Y., Gousev, A., Zhang, L., Ho, K.L., Morshead, C. and Chopp, M., 2004a, Activated neural stem cells contribute to stroke-induced neurogenesis and neuroblast migration toward the infarct boundary in adult rats. *J. Cereb. Blood Flow. Metab.* **24**: 441.

Zhang, R., Zhang, Z., Zhang, C., Zhang, L., Robin, A., Wang, Y., Lu, M. and Chopp, M., 2004b, Stroke transiently increases subventricular zone cell division from asymmetric to symmetric and increases neuronal differentiation in the adult rat. *J. Neurosci.* **24**: 5810.

Zhang, R., Zhang, Z., Tsang, W., Wang, L. and Chopp, M., 2004c, Down-regulation of p27kip1 increases proliferation of progenitor cells in adult rats. *Neuroreport* **15**: 1797.

Zhang, R.L., Zhang, Z., Zhang, L., Wang, Y., Zhang, C. and Chopp, M., 2006, Delayed treatment with sildenafil enhances neurogenesis and improves functional recovery in aged rats after focal cerebral ischemia. *J. Neurosci. Res.* **83**: 1213.

Zhao, M., Momma, S., Delfani, K., Carlen, M., Cassidy, R.M., Johansson, C.B., Brismar, H., Shupliakov, O., Frisen, J. and Janson, A.M., 2003, Evidence for neurogenesis in the adult mammalian substantia nigra. *Proc. Natl. Acad. Sci. USA* **100**: 7925.

Zheng, T., Steindler, D.A. and Laywell, E.D., 2002, Transplantation of an indigenous neural stem cell population leading to hyperplasia and atypical integration. *Cloning Stem Cells* **4**: 3.

Zhou, Y., Larsen, P.H., Hao, C. and Yong, V.W., 2002, CXCR4 is a major chemokine receptor on glioma cells and mediates their survival. *J. Biol. Chem.* **277**: 49481.

Zigova, T., Pencea, V., Betarbet, R., Wiegand, S.J., Alexander, C., Bakay, R.A. and Luskin, M.B., 1998a, Neuronal progenitor cells of the neonatal subventricular zone differentiate and disperse following transplantation into the adult rat striatum. *Cell Transplant.* **7**: 137.

Zigova, T., Pencea, V., Wiegand, S.J. and Luskin, M.B., 1998b, Intraventricular administration of BDNF increases the number of newly generated neurons in the adult olfactory bulb. *Mol. Cell. Neurosci.* **11**: 234.

Section 2

SIGNALING AND INFLAMMATION IN AGING

Rodrigo A. Cunha

Upon aging, there is a decrease in several brain-related functions and an increased incidence of neurodegenerative diseases. It is presently unclear if this insidious increase of brain susceptibility to dysfunction and damage is mostly caused by modifications intrinsic to neurons or if glia cells might play a prominent role. The first series of chapters of this section explore the possibility that glia-operated functions may be related with brain senescence.

The first chapter, by Schousboe and Waagepetersen, reviews the importance of astrocytes in controlling brain bioenergetics. They emphasize the different importance of astrocytes in the reuptake of glutamate and of GABA and discuss the possibility of manipulating these transporters to differentially affect GABAergic and glutamatergic systems to restore the imbalance between excitation and inhibition present in some pathologic conditions. This opens the perspective of exploring if changes of brain bioenergetics may contribute to the age-associated modifications of brain performance, in particular if modifications of intermediary metabolism, in particular brain components (or between brain components), may be a key underlying factor to determine the appearance of neurological dysfunction. There is an unproved trend to assume that metabolic changes are a consequence rather than a possible cause of disease-modified brain function. Thus, efforts should be made to test if the contribution of astrocytes as an ancillary system to support excitatory and inhibitory transmission may be modified upon aging and whether these putative metabolic modifications would predate overt changes in brain function upon aging.

The chapters by Lynch and by Simões de Couto and de Mendonça present a strong case supporting the role of neuroinflammation in the effect of aging on hippocampal synaptic function and on cognitive status. Lynch describes a decreased ability of hippocampal synapses of the aged rat, in particular in the dentate gyrus, to undergo plastic changes, which correlates with the appearance of an increased neuroinflammatory status. Most importantly, diets aimed at attenuating neuroinflammation also abrogate the aged-associated decrease of synaptic plasticity. Interestingly, this relation between increased neuroinflammation and decreased cognitive abilities is also found in Humans, as elegantly discussed by Simões do Couto and de Mendonça. They also reviewed the trend, still far from conclusive, that drugs with anti-inflammatory properties may prove beneficial to prevent or to manage cognitive impairment, namely in Alzheimer's disease. Provocatively, they argue that the current inability to take therapeutic advantage of the known relation between neuroinflammation and cognitive dysfunction may be related to the difficulties to design robust clinical trials to evaluate neuroprotection and/or disease-modifying abilities of tested drugs to manage Alzheimer's disease. Furthermore, they stress the idea that neuroprotection should be tested in the early phases of the dementia process by selecting patients with mild cognitive impairment. These two perspectives from animal and Human studies clearly warrant the need to investigate if neuroinflammation is indeed an early event during aging and if it precedes or follows the first manifestations of synaptic (in animals) or cognitive (in Humans) impairment.

The chapter by Rego et al., focuses on the role of mitochondria dysfunction in neurodegeneration. It reviews the different molecular pathways that control the demise of cell death, most of which are caused or involve active mitochondria processing, and their participation in motor-related degenerative diseases. Albeit the evidence clearly indicates the involvement of mitochondrial dysfunction in neurodegeneration, it is still unclear if this is a cause or an amplification mechanism in the progression of neurodegenerative diseases. Furthermore, the increased awareness that there are different brain compartments that undergo differential adaptation in the course of the progression of neurodegeneration lead to the anticipation that the time-course of mitochondrial adaptations might be different in different brain compartments. For instance, in view of the particular susceptibility of mitochondria located in nerve terminals to different inhibitors, it might be relevant to test if there is a primary modification of synaptically located mitochondria associated with the early synaptotoxicity that precedes overt neurodegeneration or whether mitochondria dysfunction is mostly associated with the process of neuronal loss in later phases of neurodegeneration. Furthermore, in view of the previously discussed putative roles of astrocytes in supporting brain metabolism and of glia cells in mounting a neuroinflammatory response, it would also be important to distinguish if the mitochondrial dysfunction associated with different neurodegenerative disorders is confined to neurons or if it is particularly relevant in glia cells.

The contribution by Sensi et al. stresses the putative importance of zinc in the control of brain injury. Building on our knowledge of the importance of calcium homeostasis in controlling cell signaling and death, the authors emphasize the particular importance of zinc metabolism in the brain and its likely particular relevance to participate in the demise of neuronal damage. Exploring the role of zinc in different brain compartments may also clarify if zinc is a main modulator of neuronal viability or if it also controls the reactivity and function of astrocytes and/or microglia. Also, further pharmacological development of zinc chelators or inhibitors of particular zinc transporters may help re-enforcing the therapeutic potential of controlling zinc homeostasis as a novel strategy to interfere with brain injury.

Finally, the chapters by Manadas et al., and Duarte et al., develop the role and potential therapeutic possibilities of manipulating two modulation systems operated by growth factors, namely neurotrophins and glia-derived neurotrophic factor (GDNF). Given the key role of these neurotrophic factors in promoting the development of the nervous system, one anticipates that these systems may endow a great potential as restorative strategies in the injured nervous system, which may recapitulate several features of the immature nervous system as an attempt to recover from damage.

Overall, the chapters of this section discuss the possibility that two different types of neuron–glia communication may be relevant for neuronal dysfunction in aging and disease (1) astrocytic control of brain bioenergetics that sustains neuronal function and (2) astrocyte–microglia mounting of an inflammatory response that either triggers or amplifies neuronal damage. They also present evidence that mitochondria and zinc may play a prominent role in this imbalance and discuss the relevance of neuroprotection and signaling by neurotrophic factors.

NEURON–GLIA INTERACTION IN HOMEOSTASIS OF THE NEUROTRANSMITTERS GLUTAMATE AND GABA

Arne Schousboe and Helle S. Waagepetersen

1. ABSTRACT

The functional activity in the brain is primarily composed of interplay between excitation and inhibition. In any given region the output is based upon a complex processing of incoming signals that require both excitatory and inhibitory units. Moreover, these units must be regulated and balanced such that an integrated and finely tuned response is generated. In each of these units or synapses, the activity depends on biosynthesis, release, receptor interaction, and inactivation of the neurotransmitters; thus, it is easily understood that each of these processes needs to be highly regulated and controlled. It is interesting to note that in case of the most prevailing neurotransmitters, glutamate, and GABA, which mediates excitation and inhibition, respectively, the inactivation process is primarily maintained by highly efficient, high-affinity transport systems capable of maintaining transmembrane concentration gradients of these amino acids of 10- to 100-fold. The demonstration of the presence of transporters for glutamate and GABA in both neuronal and astrocytic elements naturally raises the question of the functional importance of the astrocytes in the regulation of the level of the neurotransmitters in the synaptic cleft and hence for the activity of excitatory and inhibitory neurotransmission. Obviously, this discussion has important implications for the understanding of the role of astrocytes in disease states in which imbalances between excitation and inhibition are a triggering factor, for example, epilepsy and neurodegeneration. Thus, glutamate transport in astrocytes is mandatory to maintain extrasynaptic glutamate levels sufficiently low to prevent excitotoxic neuronal damage. In GABA synapses hyperactivity of astroglial GABA uptake may lead to diminished GABAergic inhibitory activity resulting in seizures. The expression and functional activity of astrocytic glutamate and GABA transport is regulated in a number of ways at transcriptional, translational, and post-translational levels. This opens for a number of therapeutic strategies by which the efficacy of excitatory and inhibitory neurotransmission may be manipulated.

Arne Schousboe – as@dfuni.dk
Dept. Pharmacol., Danish Univ. Pharm. Sci., DK-2100 Copenhagen, Denmark

2. INTRODUCTION

Glutamatergic and GABAergic neurotransmission is intimately coupled to astroglial cell function, since these cells that ensheathe the synapses completely (Chao et al., 2002) are of instrumental importance for inactivation of the transmitters as well as for the biosynthesis of transmitter precursors utilized by the neurons to replenish the neurotransmitter pools of the two amino acids (Schousboe and Waagepetersen, 2004; Schousboe et al., 2004a). The present review shall discuss this astrocytic–neuronal relationship with regard to release, uptake, and metabolism of glutamate and GABA in glutamatergic and GABAergic synapses, respectively.

3. GLUTAMATERGIC SYNAPSES

3.1. Release

Classical synaptic neurotransmitter release is vesicular in nature exhibiting Ca^{2+}-dependency, features characteristic also for release of glutamate (McMahon and Nicholls, 1991). Recently, an analogous mechanism was shown to be present in astrocytes that express the entire exocytotic machinery found in presynaptic neurons, i.e. vesicles and synaptic proteins (Parpura et al., 1994; Volterra and Meldolesi, 2005). Membrane depolarization induces vesicular release and may in addition be able to trigger release of glutamate via reversal of glutamate transporters, the inward operation of which is coupled to the transmembrane sodium gradient (Nicholls and Attwell, 1990). Such transporter-coupled release, which involves the cytoplasmic, non-vesicular glutamate pools in both neurons and astrocytes, may also occur during energy failure disrupting the operation of the Na^+, K^+-coupled ATPase (Benveniste et al., 1984; Hagberg et al., 1985; Sandberg et al., 1986; Phillis et al., 2000; Rossi et al., 2000; Bonde et al., 2003). Contrary to this, energy failure will prevent vesicular release from taking place as this process is energy dependent (Nicholls and Attwell, 1990). That reversal of glutamate transporters is of importance for the glutamate overflow noted during energy failure, i.e. depletion of glucose and/or oxygen is underlined by the demonstration that inhibition of the glutamate transporters by the non-transportable inhibitor threo-β-benzyloxyaspartate (TBOA) leads to a reduction of glutamate overflow, which in turn protects neurons against excitotoxic damage (Phillis et al., 2000; Waagepetersen et al., 2001; Bonde et al., 2003).

3.2. Uptake

Kinetic analyses of glutamate uptake in primary cultures of astrocytes and neurons from different brain regions have provided evidence for pronounced capacity for net glutamate transport particularly in astrocytes but nonetheless a substantial capacity for uptake also exists in neurons (Schousboe et al., 1977a; Hertz et al., 1978; Drejer et al., 1982). The advent of cloning of five different glutamate transporters (Gegelashvili and Schousboe, 1997, 1998; Danbolt, 2001) named EAAT 1-5 (see Table 1) has substantiated the view (see Schousboe, 1981) that astrocytic glutamate uptake appears to be the quantitatively most important for inactivation of synaptically released glutamate by binding and subsequent inward transport (Wadiche et al., 1995; Lehre and Danbolt, 1998; Levy, 2002). While there is little doubt that astrocytic glutamate uptake plays a major functional role with regard to maintaining a sufficiently low extracellular glutamate concentration to prevent excitotoxic neuronal damage (Benveniste et al., 1984; Schousboe and Frandsen, 1995; O'Shea et al., 2002; Bonde et al., 2003; Schousboe and Waagepetersen, 2005), the quantitative and functional importance of neuronal and particularly presynaptic glutamate uptake is less clear (Danbolt, 2001). It should, however, be noted that kinetic studies of glutamate uptake in cultured glutamatergic neurons using [^{3}H]D-aspartate as a tracer have demonstrated a very efficient high affinity uptake, which is likely to represent a presynaptic event as [^{3}H]D-aspartate could be released in a manner compatible with exocytosis (Drejer et al., 1982, 1983). Moreover, a recent study of homeostatic mechanisms governing the availability of vesicular glutamate for transmitter release in cultured glutamatergic neurons has shown that uptake of glutamate presynaptically was necessary to maintain transmitter release even in the presence of 0.5 mM glutamine (Waagepetersen et al., 2005). Additionally, the use of an antibody specifically recognizing D-aspartate has provided convincing evidence at the electron microscopic level that D-aspartate is indeed taken up presynaptically (Gundersen et al., 1993, 1996).

Table 1. Nomenclature of cloned glutamate transporters

Original name	Systematic name
GLAST	EAAT1
GLT-1	EAAT2
EAAC1	EAAT3
–	EAAT4
–	EAAT5

3.3. Metabolism

As illustrated in Fig. 1, glutamate metabolism (biosynthesis and degradation) in the brain is highly complex and compartmentalized with individual metabolic reactions preferentially or exclusively expressed in particular cell types. Thus, the five enzymes or groups of enzymes involved, phosphate activated glutaminase (PAG), glutamine synthetase (GS), glutamate decarboxylase (GAD), glutamate dehydrogenase (GDH) and the group of aminotransferases (transaminases) with aspartate and alanine aminotransferases as some of the most prominent members of this group have different cellular expression patterns. While GAD and GS are normally considered specific for GABAergic neurons and astrocytes, respectively (Norenberg and Martinez-Hernandez, 1979; Roberts, 1991; Martin and Rimvall, 1993), the remaining enzymes are less specific with regard to cellular localization. PAG is most prominent in glutamatergic neurons (Laake et al., 1999) but it is expressed in GABAergic neurons and astrocytes as well, albeit with by far the lowest activity in the latter cell type (Schousboe et al., 1979; Kvamme et al., 1982, 2001). GDH has a ubiquitous expression pattern although the activity appears to be higher in astrocytes than in neurons (Plaitakis and Zaganas, 2001). The aminotransferases, the most important of which with regard to glutamate homeostasis appear to be the aspartate-, alanine-, and branched chain amino acid (BCAA)-transaminases, are expressed in all cell types albeit that cytosolic and mitochondrial isoenzymes in case of the BCAA-transaminase exhibit cell specific expression patterns (Bixel et al., 1997; Lieth et al., 2001; Schousboe and Waagepetersen, 2004).

This complexity with regard to expression of the glutamate metabolizing enzymes at the cellular level is, as mentioned above, reflected by the compartmentalized glutamate metabolism as observed more than 50 years ago using radio-isotope labeled glutamate precursors (see Berl and Clarke, 1983). Hence, precursors like acetate and CO_2 labeled with ^{14}C gave rise to higher specific radioactivity of glutamine than that of its precursor glutamate, while ^{14}C-labeled glucose or pyruvate gave higher labeling in glutamate than glutamine, results that can only be interpreted assuming the presence of multiple, distinct pools of glutamate and glutamine having different turnover rates (van den Berg and Garfinkel, 1971; Berl and Clarke, 1983). This obviously is in perfect agreement with the heterogeneous cellular distribution of the key enzymes discussed above.

Table 2. Nomenclature of the cloned GABA transporters from mouse, rat and human

Species	Nomenclature			
Human	GAT-1	BGT-1	NC	GAT-3
Rat	GAT-1	BGT-1	GAT-2	GAT-3
Mouse	GAT1	GAT2	GAT3	GAT4

NC, not cloned.

4. GABAERGIC SYNAPSES

4.1. Release

The first report showing that GABA is released from brain tissue in a manner compatible with classical vesicular, Ca^{2+}-dependent neurotransmitter release was published by Srinivasan et al. (1969). Subsequently, such release was also demonstrated in synaptosomes (Levy et al., 1973; Sihra and Nicholls, 1987). It was almost simultaneously shown that different preparations of glial cells could release GABA upon depolarization but such release did not exhibit the characteristics of vesicular release (Minchin and Iversen, 1974; Roberts, 1974) and thus likely represents reversal of GABA transporters, a mechanism frequently reported to operate in neurons as well (Bernath, 1992; Belhage et al., 1993).

4.2. Uptake

The synaptic action of GABA mediated by activation of GABA receptors is terminated by the concerted action of receptor desensitization, diffusion in the synaptic cleft and subsequent high affinity uptake of GABA into the presynaptic GABAergic nerve endings and surrounding astrocytes (Henn and Hamberger, 1971; Iversen and Kelly, 1975; Schousboe et al., 1977b; Schousboe, 2003; Schousboe and Waagepetersen, 2004). In contrast to inactivation of transmitter glutamate in which astrocytic uptake prevails over that in neurons (see above), most of the released GABA appears to be taken up back into the presynaptic GABAergic nerve endings and hence, it may be reutilized as a neurotransmitter (Schousboe and Waagepetersen, 2004). On the other hand, the fraction of GABA taken up into astrocytes is metabolized (see below) and lost from the neurotransmitter pool. If this process is imbalanced relative to presynaptic GABA uptake, GABA neurotransmission may be impaired resulting in seizure activity. As proof of principle concerning this notion (Schousboe et al.,

1983), it has been shown that a very good correlation exists between protection against seizures and potency of GABA analogs to selectively inhibit astrocytic GABA uptake (White et al., 2002). On the contrary, such a perfect correlation was not found for seizure protection and selective inhibition of neuronal (presynaptic) GABA uptake (White et al., 2002).

Analogous to the glutamate high affinity transporters, multiple GABA transporters have been cloned and characterized (Guastella et al., 1990; Lopéz-Corcuera et al., 1992; Liu et al., 1992, 1993). Since the nomenclature generally used for GABA transporters is somewhat confusing due to different numbering systems for transporters cloned from mouse or rat and human, these nomenclatures are summarized in Table 2. As can be seen, mouse GAT2 is synonymous with the betaine-GABA transporter-1 (BGT-1) from rat and human and thus, mouse GAT3 and GAT4 correspond to rat/human GAT-2 and GAT-3, respectively. GAT1 (GAT-1) is by far the most abundantly expressed transporter with a preferential neuronal localization and less pronounced astroglial expression (Borden, 1996; Gadea and Lopez-Colome, 2001; Schousboe and Kanner, 2002). GAT3 and GAT4 are primarily expressed in glia and also extrasynaptically in neurons (Minelli et al., 1996, 2003; Ribak et al., 1996a, b; De Biasi et al., 1998; Conti et al., 2004). The betaine-GABA carrier (BGT-1/GAT2) also has an extrasynaptic localization and is found both in neurons and astrocytes, although the expression in the latter cell type may be the most prevalent (Borden, 1996; Zhu and Ong, 2004; Olsen et al., 2005). These different cellular expression patterns appear to have profound significance for the functional aspects of GABA transport processes and hence, for the efficacy of GABAergic neurotransmission. This appears to be linked to functional differences between synaptic and extrasynaptic GABA receptors (Mody, 2001; Schousboe et al., 2004b; Clausen et al., 2006a). These aspects are discussed in Sect. 5.2.

5. PHARMACOLOGICAL ASPECTS

5.1. Glutamate Transport and Metabolism

While glutamate receptors have been a target for drug development for decades albeit with limited success (Farber et al., 2002), it has been less attractive to put an emphasis on glutamate metabolism and transport. With regard to metabolism any drug-related manipulation appears problematic, since such interference will disturb brain metabolism at large due to the fact that glutamate is widely distributed and constitutes an important part of general metabolism. So far the available inhibitors of a number of the enzymes involved in the metabolic pathways delineated in Fig. 1 are merely experimental tools to separate the metabolic pathways involved in glutamate turnover (e.g. McKenna et al., 2006). Glutamate transporters have been a target for development of inhibitors as experimental tools to study the function and separate the roles of the different subtypes of the transporters (Bridges et al., 1999; Balcar, 2002; Levy, 2002) but since neurodegeneration most frequently is associated with decreased function of glutamate transport (see above), inhibitors are less interesting as drugs in this context. Since the efficiency of glutamate transport is regulated in a number of ways (e.g. Gegelashvili and Schousboe, 1997), it may well be possible to develop strategies by which glutamate transport capacity can be enhanced, which at least theoretically may be beneficial for treatment of certain neurodegenerative disorders (Schousboe and Waagepetersen, 2005).

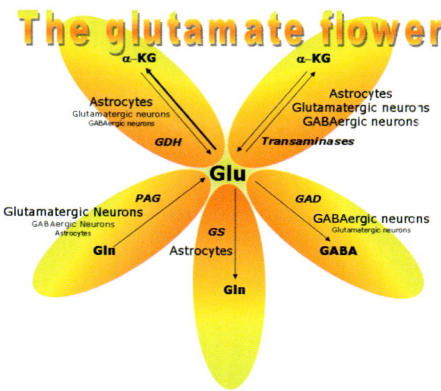

Fig. 1. Schematic representation of the major glutamate metabolizing reactions catalyzed by the enzymes glutamate dehydrogenase (GDH) transaminases, glutamate decarboxylase (GAD), glutamine synthetase (GS), and phosphate activated glutaminase (PAG). The cellular localization of the enzymes is given using letters of different size for some of the enzymes. The size indicates the prevailing cellular localization (largest lettering). In case of the GDH reaction, the bold arrow indicates the prevailing direction of the reaction in situ

5.2. GABA Transport and Metabolism

The GABA metabolizing enzyme GABA-transaminase (GABA-T) has been a successful target for development of drugs that are able to elevate synaptic GABA levels and hence increase the efficacy of GABAergic neurotransmission (Iadarola and Gale, 1980; Wood et al., 1981; Gram et al., 1988). This has resulted in the development of vigabatrin, γ-vinyl GABA (Krämer, 2004), which acts as an active-site-directed suicide inhibitor of GABA-T (Lippert et al., 1977). In keeping with its ability to increase the GABA content of the synaptic pool, vigabatrin is used as a drug to treat certain forms of epilepsy (Krämer, 2004). Recently, analogs of vigabatrin have been synthesized but so far none of these compounds have been developed into clinically useful drugs (Choi and Silverman, 2002).

Since GABA transporters, and in particular GAT1, are responsible for inactivation of synaptically released GABA, these transporters have constituted an interesting target for development of antiepileptic drugs (Dalby, 2003; Schousboe et al., 2004b). Larger series of GABA analogs have during the past 25 years been synthesized as inhibitors of GABA transport (see Clausen et al., 2006b) but only one of these, tiagabine, synthesized by Bræstrup et al. (1990) has been successfully developed into a clinically useful antiepileptic drug (Kälviäinen, 2004). Recent evidence from pharmacological studies of GABA transport inhibitors acting on transporters other than GAT1 has pointed to the possibility that in particular GAT2 (BGT-1) and GAT4 may be interesting drug targets with regard to development of anticonvulsant drugs (Dalby, 2003; Schousboe et al., 2004b, c; Clausen et al., 2005; White et al., 2005). One GABA analog in particular, EF 1502 (N-[4,4-bis(3-methyl-2-thienyl)-3-butenyl]-4-(methylamino)-4,5,6,7-tetrahydrobenzo[d]isoxazol-3-ol), which is a potent inhibitor of both GAT1 and GAT2 (Clausen et al., 2005; White et al., 2005), turned out to act synergistically with tiagabine, the specific GAT1 inhibitor, as an anticonvulsant in several animal models of epilepsy (White et al., 2005). This has led to the hypothesis that GAT2/BGT-1 located extrasynaptically (see above) may be functionally important in the regulation of extrasynaptic GABA levels which, in turn, control the activity of extrasynaptic

GABA receptors that appear to be important drug targets (Krogsgaard-Larsen et al., 2004). Development of novel GABA transport inhibitors with this kind of pharmacological properties may help elucidating this interesting hypothesis and subsequently, this may facilitate development of new antiepileptic drugs having a different therapeutic profile (Clausen et al., 2006a).

6. ACKNOWLEDGMENTS

The expert secretarial assistance of Ms Hanne Danø is highly appreciated. The work has been supported by grants from the Lundbeck, Hørslev, and Benzon Foundations as well as from the Danish Medical Research Council (22-03-0250 and 22-04-0314).

7. REFERENCES

Balcar, V.J., 2002, Molecular pharmacology of the Na^+-dependent transport of acidic amino acids in the mammalian central nervous system. *Biol. Pharm. Bull.* **25**: 291.

Belhage, B., Hansen, G.H. and Schousboe, A., 1993, Depolarization by K^+ and glutamate activates different neurotransmitter release mechanisms in GABAergic neurons: Vesicular versus non-vesicular release of GABA. *Neuroscience* **54**: 1019.

Benveniste, H., Drejer, J., Schousboe, A. and Diemer, N.H., 1984, Elevation of the extracellular concentrations of glutamate and aspartate in rat hippocampus during transient cerebral ischemia monitored by intracerebral microdialysis. *J. Neurochem.* **43**: 1369.

Berl, S. and Clarke, D.D., 1983, The metabolic compartmentation concept. In: *Glutamine, Glutamate and GABA in The Central Nervous System*. L. Hertz, E. Kvamme, E.G. McGeer and A. Schousboe, eds., Alan R. Liss, Inc., New York, pp. 205–217.

Bernath, S., 1992, Calcium-independent release of amino acid neurotransmitters: fact or artifact? *Prog. Neurobiol.* **38**: 57.

Bixel, M.G., Hutson, S.M. and Hamprecht, B., 1997, Cellular distribution of branched-chain amino acid aminotransferase isoenzymes among rat brain glial cells in culture. *J. Histochem. Cytochem.* **45**: 685.

Bonde, C., Sarup, A., Schousboe, A., Gegelashvili, G., Zimmer, J. and Noraberg, J., 2003, Neurotoxic and neuroprotective effects of the glutamate transporter inhibitor DL-threo-beta-benzyloxyaspartate (DL-TBOA) during physiological and ischemia-like conditions. *Neurochem. Int.* **43**: 371.

Borden, L.A., 1996, GABA transporter heterogeneity: pharmacology and cellular localization. *Neurochem. Int.* **29**: 335.

Bræstrup, C., Nielsen, E.B., Sonnewald, U., Knutsen, L.J.S., Andersen, K.E., Jansen, J.A., Frederiksen, K., Andersen, P.H., Mortensen, A. and Suzdak, P.D., 1990, (*R*)-*N*-[4,4-Bis(3-methyl-2-thienyl)but-3-en-1-yl]nipecotic acid binds with high affinity to the brain γ-aminobutyric acid uptake carrier. *J. Neurochem.* **54**: 639.

Bridges, R.J., Kavanaugh, M.P. and Chamberlin, A.R., 1999, A pharmacological review of competitive inhibitors and substrates of high-affinity, sodium-dependent glutamate transport in the central nervous system. *Curr. Pharm. Des.* **5**: 363.

Chao, T.I., Rickmann, M. and Wolff, J.R., 2002, The synapse-astrocyte boundary: anatomical basis for an integrative role of glia in synaptic transmission. In: *Tripartite Synapses: Synaptic Transmission with Glia*. A. Volterra, P. Magistretti and P. Haydon, eds., Oxford University Press, Oxford, New York, pp. 3–23.

Choi, S. and Silverman, R.B., 2002, Inactivation and inhibition of gamma-aminobutyric acid aminotransferase by conformationally restricted vigabatrin analogues. *J. Med. Chem.* **45**: 4531.

Clausen, R.P., Moltzen, E.K., Perregaard, J., Lenz, S.M., Sanchez, C., Falch, E., Frølund, B., Sarup, A., Larsson, O. M., Schousboe, A. and Krogsgaard-Larsen, P., 2005, Selective inhibitors of GABA uptake: synthesis and Molecular Pharmacology of 3-hydroxy-4-*N*-methylamino-4,5,6,7-tetrahydro-1,2-benzo[*d*]isoxazole analogues. *Bioorg. Med. Chem.* **13**: 895.

Clausen, R.P., Frølund, B., Larsson, O.M., Schousboe, A., Krogsgaard-Larsen, P. and White, H.S., 2006a, A novel selective γ-aminobutyric acid transport inhibitor demonstrates a functional role for GABA transporter subtype GAT2/BGT-1 in the CNS. *Neurochem. Int.* **48**: 637.

Clausen, R.P., Madsen, K., Larsson, O.M., Frølund, B., Krogsgaard-Larsen, P. and Schousboe, A., 2006b, Structure-activity relationship and pharmacology of γ-aminobutyric acid (GABA) transport inhibitors. *Adv. Pharmacol.* **54**: 265.
Conti, F., Minelli, A. and Melone, M., 2004, GABA transporters in the mammalian cerebral cortex: localization, development and pathological implications. *Brain Res. Brain Res. Rev.* **45**: 196.
Dalby, N.O., 2003, Inhibition of γ-aminobutyric acid uptake: uptake, physiology and effects against epileptic seizures. *Eur. J. Pharmacol.* **479**: 127.
Danbolt, N.C., 2001, Glutamate uptake. *Progr. Neurobiol.* **65**: 1.
De Biasi, S., Vitellaro-Zuccarello, L. and Brecha, N.C., 1998, Immunoreactivity for the GABA transporter-1 and GABA transporter-3 is restricted to astrocytes in the rat thalamus. A light and electron-microscopic immunolocalization. *Neuroscience* **83**: 815.
Drejer, J., Larsson, O.M. and Schousboe, A., 1982, Characterization of glutamate uptake into and release from astrocytes and neurons cultured from different brain regions. *Exp. Brain Res.* **47**: 259.
Drejer, J., Larsson, O.M. and Schousboe, A., 1983, Characterization of uptake and release processes for D- and L-aspartate in primary cultures of astrocytes and cerebellar granule cells. *Neurochem. Res.* **8**: 231.
Farber, N.B., Newcomer, J.W. and Olney, J.W., 2002, Glutamatergic transmission: therapeutic prospects for schizophrenia and Alzheimer's disease. In: *Glutamate and GABA Receptors and Transporters.* J. Egebjerg, A. Schousboe and P. Krogsgaard-Larsen, eds., Taylor & Francis Publ., London, UK, pp. 385–406.
Gadea, A. and Lopez-Colome, A., 2001, Glial transporters for glutamate, glycine, and GABA: II. GABA transporters. *J. Neurosci. Res.* **63**: 461.
Gegelashvili, G. and Schousboe, A., 1997, High-affinity glutamate transporters: regulation of expression and activity. *Mol. Pharmacol.* **52**: 6.
Gegelashvili, G. and Schousboe, A., 1998, Cellular distribution and kinetic properties of high-affinity glutamate transporters. *Brain Res. Bull.* **45**: 233.
Gram, L., Larsson, O.M., Johnsen, A.H. and Schousboe, A., 1988, Effects of valproate, vigabatrin and aminooxyacetic acid on release of endogenous and exogenous GABA from cultured neurons. *Epilepsy Res.* **2**: 87.
Guastella, J., Nelson, N., Nelson, H., Czyzyk, L., Keynan, S., Miedel, M.C., Davidson, N., Lester, H.A. and Kanner, B.I., 1990, Cloning and expression of a rat brain GABA transporter. *Science* **249**: 1303.
Gundersen, V., Danbolt, N.C., Ottersen, O.P. and Storm-Mathiesen, J., 1993, Demonstration of glutamate/aspartate uptake activity in nerve endings by use of antibodies recognizing exogenous D-aspartate. *Neuroscience* **57**: 97.
Gundersen, V., Ottersen, O.P. and Storm-Mathiesen, J., 1996, Selective excitatory amino acid uptake in glutamatergic nerve terminals and in glia in the rat striatum: quantitative electron microscopic immunocytochemistry of exogenous (D)-aspartate and endogenous glutamate and GABA. *Eur. J. Neurosci.* **8**: 758.
Hagberg, H., Lehmann, A., Sandberg, M., Nyström, B., Jacobsen, I. and Hamberger, A., 1985, Ischemia-induced shift of inhibitory and excitatory amino acids from intra- to extracellular compartments. *J. Cereb. Blood Flow Metab.* **5**: 413.
Henn, F.A. and Hamberger, A., 1971, Glial cell function: uptake of transmitter substances. *Proc. Natl. Acad. Sci. USA* **68**: 2686.
Hertz, L., Schousboe, A., Boechler, N., Mukerji, S. and Fedoroff, S., 1978, Kinetic characteristics of the glutamate uptake into normal astrocytes in cultures. *Neurochem. Res.* **3**: 1.
Iadarola, M.J. and Gale, K., 1980, Evaluation of increases in nerve terminal-dependent vs nerve terminal-independent compartments of GABA in vivo. *Brain Res. Bull.* **5**(Suppl. 2): 13.
Iversen, L.L. and Kelly, J.S., 1975, Uptake and metabolism of γ-aminobutyric acid by neurones and glial cells. *Biochem. Pharmacol.* **24**: 933.
Kälviäinen, R., 2004, Tiagabine. In: *The Treatment of Epilepsy, Sec. Ed.* S. Shorvon, E. Perucca, D. Fish and E. Dodson, eds., Blackwell Science, Oxford, UK, pp. 507–514.
Krämer, G., 2004, Vigabatrin. In: *The Treatment of Epilepsy, Sec. Ed.* S. Shorvon, E. Perucca, D. Fish and E. Dodson, eds., Blackwell Science, Oxford, UK, pp. 540–547.
Krogsgaard-Larsen, P., Frølund, B., Liljefors, T. and Ebert, B., 2004, GABA(A) agonists and partial agonists: THIP (Gaboxadol) as a non-opioid analgesic and a novel type of hypnotic. *Biochem. Pharmacol.* **68**: 1573.
Kvamme, E., Svenneby, G., Hertz, L. and Schousboe, A., 1982, Properties of phosphate activated glutaminase in astrocytes cultured from mouse brain. *Neurochem. Res.* **7**: 761.
Kvamme, E., Torgner, I.A. and Roberg, B., 2001, Kinetics and localization of brain phosphate activated glutaminase. *J. Neurosci. Res.* **66**: 951.
Laake, J.H., Takumi, Y., Eidet, J., Torgner, I.A., Roberg, B., Kvamme, E. and Ottersen, O.P., 1999, Postembedding immunogold labelling reveals subcellular localization and pathway-specific enrichment of phosphate activated glutaminase in rat cerebellum. *Neuroscience* **88**: 1137.

Lehre, K.P. and Danbolt, N.C., 1998, The number of glutamate transporter subtype molecules at glutamatergic synapses: chemical and stereological quantification in young adult rat brain. *J. Neurosci.* **18**: 8751.

Levy, L.M., 2002, Structure, function and regulation of glutamate transporters. In: *Glutamate and GABA Receptors and Transporters. Structure, Function and Pharmacology.* J. Egebjerg, A. Schousboe and P. Krogsgaard-Larsen, eds., Taylor and Francis, London, pp. 307–336.

Levy, W.B., Redburn, D.A. and Cotman, C.W., 1973, Stimulus-coupled secretion of gamma-aminobutyric acid from rat brain synaptosomes. *Science* **181**: 676.

Lieth, E., LaNoue, K.F., Berkich, D.A., Xu, B., Ratz, M., Taylor, C. and Hutson, S.M., 2001, Nitrogen shuttling between neurons and glial cells during glutamate synthesis. *J. Neurochem.* **76**: 1712.

Lippert, B., Metcalf, B.W., Jung, M.J. and Casara, P., 1977, 4-Amino-hex-5-enoic acid, a selective catalytic inhibitor of 4-aminobutyric-acid aminotransferase in mammalian brain. *Eur. J. Biochem.* **74**: 441.

Liu, Q.R., Lopez-Corcuera, B., Nelson, H., Mandiyan, S. and Nelson, N., 1992, Cloning and expression of a cDNA encoding the transporter of taurine and β-alanine in mouse brain. *Proc. Natl. Acad. Sci. USA* **89**: 12145.

Liu, Q.R., Lopez-Corcuera, B., Mandiyan, S., Nelson, H. and Nelson, N., 1993, Molecular characterization of four pharmacologically distinct γ-aminobutyric acid transporters in mouse brain. *J. Biol. Chem.* **268**: 2104.

López-Corcuera, B., Liu, Q.R., Mandiyan, S., Nelson, H. and Nelson, N., 1992, Expression of a mouse brain cDNA encoding novel γ-aminobutyric acid transporter. *J. Biol. Chem.* **267**: 17491.

Martin, D.L. and Rimval, K., 1993, Regulation of γ-aminobutyric acid synthesis in the brain. *J. Neurochem.* **60**: 395.

McKenna, M.C., Waagepetersen, H.S., Schousboe, A. and Sonnewald, U., 2006, Neuronal and astrocytic shuttle mechanisms for cytosolic-mitochondrial transfer of reducing equivalents: Current evidence and pharmacological tools. *Biochem. Pharmacol.* **71**: 399.

McMahon, H.T. and Nicholls, D.G., 1991, Transmitter glutamate release from isolated nerve terminals: evidence for biphasic release and triggering by localized Ca^{2+}. *J. Neurochem.* **56**: 86.

Minchin, M.C.W. and Iversen, L.L., 1974, Release of [^{3}H]gamma-aminobutyric acid from glial cells in rat dorsal root ganglia. *J. Neurochem.* **23**: 535.

Minelli, A., DeBiasi, S., Brecha, N.C., Zuccarello, L.V. and Conti, F., 1996, GAT-3, a high-affinity GABA plasma membrane transporter, is localized to astrocytic processes and is not confined to the vicinity of GABAergic synapses in the cerebral cortex. *J. Neurosci.* **16**: 6255.

Minelli, A., Barbaresi, P. and Conti, F., 2003, Postnatal development of high-affinity plasma membrane GABA transporters GAT-2 and GAT-3 in the rat cerebral cortex. *Dev. Brain Res.* **142**: 7.

Mody, I., 2001, Distinguishing between GABA(A) receptors responsible for tonic and phasic conductances. *Neurochem. Res.* **26**: 907.

Nicholls, D. and Attwell, D., 1990, The release and uptake of excitatory amino acids. *Trends Pharmacol. Sci.* **11**: 462.

Norenberg, M.D. and Martinez-Hernandez, A., 1979, Fine structural localization of glutamine synthetase in astrocytes of rat brain. *Brain Res.* **161**: 303.

Olsen, M., Sarup, A., Larsson, O.M. and Schousboe, A., 2005, Effect of hyperosmotic conditions on the expression of the betaine-GABA-transporter (BGT-1) in cultured mouse astrocytes. *Neurochem. Res.* **30**: 855.

O'Shea, R.D., Fodera, M.V., Aprico, K., Dehnes, Y., Danbolt, C., Crawford, D. and Beart, P.M., 2002, Evaluation of drugs acting at glutamate transporters in organotypic hippocampal cultures: new evidence on substrates and blockers in excitotoxicity. *Neurochem. Res.* **27**: 5.

Parpura, V., Basarsky, T.A., Liu, F., Jeftinija, K., Jeftinija, S. and Haydon, P.G., 1994, Glutamate-mediated astrocyte-neuron signaling. *Nature* **369**: 744.

Phillis, J.W., Ren, J. and O'Regan, M.H., 2000, Transporter reversal as a mechanism of glutamate release from the ischemic rat cerebral cortex: studies with DL-threo-beta-benzyloxyaspartate. *Brain. Res.* **868**: 105.

Plaitakis, A. and Zaganas, I., 2001, Regulation of human glutamate dehydrogenases: implications for glutamate, ammonia and energy metabolism in brain. *J. Neurosci. Res.* **66**: 899.

Ribak, C.E., Tong, W.M. and Brecha, N.C., 1996a, GABA plasma membrane transporters, GAT-1 and GAT-3, display different distributions in the rat hippocampus. *J. Comp. Neurol.* **367**: 595.

Ribak, C.E., Tong, W.M. and Brecha, N.C., 1996b, Astrocytic processes compensate for the apparent lack of GABA transporters in the axon terminals of cerebellar Purkinje cells. *Anat. Embryol.* **193**: 379.

Roberts, P.J., 1974, Amino acid release from isolated rat dorsal root ganglia. *Brain Res.* **74**: 327.

Roberts, E., 1991, Living systems are tonically inhibited, autonomous optimizers, and disinhibition coupled to variability generation is their major organizing principle: Inhibitory command-control at levels of membrane, genome, metabolism, brain, and society. *Neurochem. Res.* **16**: 409.

Rossi, D.J., Oshima, T. and Attwell, D., 2000, Glutamate release in severe brain ischemia is mainly by reversed uptake. *Nature* **403**: 316.

Sandberg, M., Butcher, S.P. and Hagberg, H., 1986, Extracellular overflow of neuroactive amino acids during severe insulin-induced hypoglycemia: In vivo dialysis of rat hippocampus. *J. Neurochem.* **47**: 178.

Schousboe, A., 1981, Transport and metabolism of glutamate and GABA in neurons and glial cells. *Int. Rev. Neurobiol.* **22**: 1.
Schousboe, A., 2003, Role of astrocytes in the maintenance and modulation of glutamatergic and GABAergic neurotransmission. *Neurochem. Res.* **28**: 347.
Schousboe, A. and Frandsen, A., 1995, Glutamate receptors and neurotoxicity. In: *CNS Neurotransmitters and Neuromodulators: Glutamate.* T.W. Stone, ed., CRC Press, Boca Raton, FL, pp. 239–251.
Schousboe, A. and Kanner, B., 2002, GABA transporters: functional and pharmacological properties. In: *Glutamate and GABA Receptors and Transporters.* J. Egebjerg, A. Schousboe and P. Krogsgaard-Larsen, eds., Taylor & Francis Publ., London, UK, pp. 337–349.
Schousboe, A. and Waagepetersen, H.S., 2004, Role of astrocytes in homeostasis of glutamate and GABA during physiological and pathophysiological conditions. In: *Non-Neuronal Cells of the Nervous System: Function and Dysfunction.* L. Hertz, ed., Elsevier Science Publ., Amsterdam, The Netherlands, pp. 461–475.
Schousboe, A. and Waagepetersen, H.S., 2005, Role of astrocytes in glutamate homeostasis: implications for excitotoxicity. *Neurotox. Res.* **8**: 221.
Schousboe, A., Svenneby, G. and Hertz, L., 1977a, Uptake and metabolism of glutamate in astrocytes cultured from dissociated mouse brain hemispheres. *J. Neurochem.* **29**: 999.
Schousboe, A., Hertz, L. and Svenneby, G., 1977b, Uptake and metabolism of GABA in astrocytes cultured from dissociated mouse brain hemispheres. *Neurochem. Res.* **2**: 217.
Schousboe, A., Hertz, L., Svenneby, G. and Kvamme, E., 1979, Phosphate activated glutaminase activity and glutamine uptake in primary cultures of astrocytes. *J. Neurochem.* **32**: 943.
Schousboe, A., Larsson, O.M., Wood, J.D. and Krogsgaard-Larsen, P., 1983, Transport and metabolism of GABA in neurons and glia: implications for epilepsy. *Epilepsia* **24**: 531.
Schousboe, A., Sarup, A., Bak, L.K., Waagepetersen, H.S. and Larsson, O.M., 2004a, Role of astrocytic transport processes in glutamatergic and GABAergic neurotransmission. *Neurochem. Int.* **45**: 512.
Schousboe, A., Sarup, A., Larsson, O.M. and White, H.S., 2004b, GABA transporters as drug targets for modulation of GABAergic activity. *Biochem. Pharmacol.* **68**: 1557.
Schousboe, A., Larsson, O.M., Sarup, A. and White, H.S., 2004c, Role of the betaine/GABA transporter (BGT-1/GAT2) for the control of epilepsy. *Eur. J. Pharmacol.* **500**: 281.
Sihra, T.S. and Nicholls, D.G., 1987, 4-Aminobutyrate can be released exocytotically from guinea-pig cerebral cortical synaptosomes. *J. Neurochem.* **49**: 261.
Srinivasan, V., Neal, M.J. and Mitchell, J.F., 1969, The effect of electrical stimulation and high potassium concentration on efflux of [^{3}H]gamma-aminobutyric acid from brain slices. *J. Neurochem.* **16**: 1235.
van den Berg, C.J. and Garfinkel, D., 1971, A simulation study of brain compartments: Metabolism of glutamate and related substances in mouse brain. *Biochem. J.* **23**: 211.
Volterra, A. and Meldolesi, J., 2005, Quantal release of transmitter: not only from neurons but from astrocytes as well? In: *Neuroglia, Second Edition.* H. Kettenmann and B. Ransom, Oxford University Press, Oxford, UK, pp. 190–201.
Waagepetersen, H.S., Shimamoto, K. and Schousboe, A., 2001, Comparison of effects of DL-threo-β-benzyloxyaspartate (DL-TBOA) and L-trans-pyrrolidine-2,4-dicarboxylate (t-2,4-PDC) on uptake and release of [^{3}H]D-aspartate in astrocytes and glutamatergic neurons. *Neurochem. Res.* **26**: 661.
Waagepetersen, H.S., Qu, H., Sonnewald, U., Shimamoto, K. and Schousboe, A., 2005, Role of glutamine and neuronal glutamate uptake in glutamate homeostasis and synthesis during vesicular release in cultured glutamatergic neurons. *Neurochem. Int.* **47**: 92.
Wadiche, J.I., Arriza, J.L., Amara, S.G. and Kavanaugh, M.P., 1995, Kinetics of a human glutamate transporter. *Neuron* **14**: 1019.
White, H.S., Sarup, A., Bolvig, T., Kristensen, A.S., Petersen, G., Nelson, N., Pickering, D.S., Larsson, O.M., Frølund, B., Krogsgaard-Larsen, P. and Schousboe, A., 2002, Correlation between anticonvulsant activity and inhibitory action on glial GABA uptake of the highly selective mouse GAT1 inhibitor 3-hydroxy-4-amino-4,5,6,7-tetrahydro-1,2-benzisoxazole (*exo*-THPO) and its N-alkylated analogs. *J. Pharmacol. Exp. Therap.* **302**: 636.
White, H.S., Watson, W.P., Hansen, S., Slough, S., Sarup, A., Bolvig, T., Petersen, G., Larsson, O.M., Clausen, R.P., Frølund, B., Krogsgaard-Larsen, P. and Schousboe, A., 2005, First demonstration of a functional role for CNS betaine/GABA transporter (mGAT2) based on synergistic anticonvulsant action among inhibitors of mGAT1 and mGAT2. *J. Pharmacol. Exp. Therap.* **312**: 866.
Wood, J.D., Kurylo, E. and Tsui, S.K., 1981, Interactions of di-n-propylacetate, gabaculine, and aminooxyacetic acid: anticonvulsant activity and the gamma-aminobutyrate system. *J. Neurochem.* **37**: 1440.
Zhu, X.M. and Ong, W.Y., 2004, Changes in GABA transporters in the rat hippocampus after kainate-induced neuronal injury: decrease in GAT-1 and GAT-3 but upregulation of betaine/GABA transporter BGT-1. *J. Neurosci. Res.* **77**: 402.

6

THE IMPACT OF AN IMBALANCE BETWEEN PROINFLAMMATORY AND ANTI-INFLAMMATORY INFLUENCES ON SYNAPTIC FUNCTION IN THE AGED BRAIN

Marina Lynch

1. ABSTRACT

There is now a significant body of evidence indicating that, with age, the brain is exposed to a number of stresses. One of these is a decrease in the concentration of anti-inflammatory cytokines that exert significant neuroprotective effects. On the basis of recent findings, it seems reasonable to propose that among the most critical neuroprotective effects of IL-4 and IL-10 is maintenance of microglia in a quiescent state, preventing excessive release of proinflammatory cytokines like IL-1β, which have been shown to negative impact on hippocampal plasticity. It must therefore be concluded that it is an imbalance between proinflammatory and anti-inflammatory cytokines that leads to age-related deficits in synaptic function and that strategies that restore the balance are likely to be beneficial in reducing the deterioration of function that accompanies age.

2. INTRODUCTION

Aging is an intrinsic property of all aged cells and, in the brain, is characterized by a decline in function. The hippocampal formation is one of the most susceptible regions to the aging process (Hasan and Glees, 1973) and, consistent with this and the knowledge that the hippocampus plays a pivotal part in formation and retrieval of memories, is the finding that hippocampal-dependent learning and memory are impaired with age. Several changes have been described with age, which probably contribute to the deficits in function observed. For example, changes in receptor expression and function have been noted; for example, NMDA receptor binding decreases with age (Bonhaus et al., 1990; Ingram et al., 1992; Shigenaga et al., 1994) and is paralleled by cognitive decline (Barnes, 1988). Deficits observed in aged experimental animals have also been linked with loss of dopamine (Roth and Joseph, 1994; Shigenaga et al., 1994) and 5HT receptors (Shigenaga et al., 1994; Slotkin et al., 2005). Similarly, a reduction in a number of neurotransmitters such as acetylcholine, dopamine and noradrenaline (see Gottfries, 1990; Trollor and Valenzuela, 2001) have been described, while the loss of particular neurons, especially cholinergic neurons, has been linked with functional deficits (Gu et al., 2000). Several additional age-related changes have been documented and have led to the establishment of three main theories of aging, the free radical theory, the membrane theory and the immune theory.

3. THE FREE RADICAL THEORY OF AGING

Harman proposed that accumulation of reactive oxygen species in cells and tissues of the body, which occur during a lifetime, can hinder function and eventually lead to cell death; elaboration and reiteration of his original theory appears in a recent review (Harman, 2003). It has been considered that mitochondrial dysfunction is an important contributor to the accumulation of free radicals and has been suggested as a significant factor in the aging process (Shigenaga et al., 1994; De la Asuncion et al., 1996; Trollor and Valenzuela, 2001), primarily since mitochondria are a major source of reactive oxygen species and are a prime target for oxidative damage (Shigenaga et al., 1994; Schipper, 2004). Mutations in mitochondrial DNA have been documented in the aged brain and this has also been suggested as a contributory factor in the development of age-related neurodegenerative diseases (Shigenaga et al., 1994; Forster et al., 1996; Schipper, 2004). Significantly, a correlation between the impairment in cognitive ability (Pellmar et al., 1991; Forster et al., 1996) and oxidative damage has been reported in aged animals. Consistently, evidence from this laboratory has identified deterioration in synaptic plasticity in aged rats, which is coupled with accumulation of reactive oxygen species, while dietary supplementation with antioxidant vitamins E and C reversed both changes (Murray and Lynch, 1998a).

4. THE MEMBRANE HYPOTHESIS OF AGING

Normal functioning of the brain is dependent on membrane lipid composition, which determines membrane stability and fluidity, and it has been shown that deviations from the optimal lipid content can impact on a host of cellular functions (Zhang et al., 1996). In the aged brain, changes in membrane fluidity occur as a result of alterations in the lipid content of the brain (Zhang et al., 1996); increases in concentrations of membrane cholesterol (Yehuda et al., 2002), dicholol (Zhang et al., 1996) and sphingomyelin (Giusto et al., 1992) together with decreased polyunsaturated fatty acid (McGahon et al., 1999) have been reported. Gabbita and colleagues (1997) have suggested that lipid peroxidation, as a result of oxidative damage, contributes to the decline in membrane fluidity and evidence has indicated that increased lipid peroxidation and reactive oxygen species accumulation are coupled with decreased polyunsaturated fatty acids and deficits in synaptic plasticity, specifically long-term potentiation (LTP) in perforant path-granule cell synapses (Murray and Lynch 1998a; see Lynch, 2004). This is consistent with the observations that the hippocampus is one of the most vulnerable and susceptible areas to changes in fluidity (see Yehuda et al., 2002) and is affirmed by the fact that the cholesterol/phospholipid ratio is highest in the hippocampus (Zhang et al., 1996).

5. THE IMMUNE THEORY OF AGING

In 1962, Walford suggested that normal aging is coupled with faulty immunological processes and this is thought to contribute to the development of age-related diseases such as cancer, autoimmune and cardiovascular disease (see Effros, 2005). It has been proposed that the aging process results from injury to the immune system so that autoantibodies are produced and self attack occurs (Ali, 2005). Fabris (1992) suggested that the stimulus for the conversion of stem cells into T cells decreases with age, while an age-related defect in generation of Th2 responses, which leads to production of anti-inflammatory cytokines, coupled with increased production of Th1 type proinflammatory cytokines has been observed (Smith et al., 2001). A link between the free radical theory of aging and the immune theory of aging is indicated by the finding that reactive oxygen species production triggers an inflammatory response; this is typified by generation of heat shock proteins and the associated recruitment of macrophages and other cells involved in the inflammatory process, which in turn secrete inflammatory cytokines, chemokines and adhesion molecules, resulting in chronic inflammation (Land, 2004).

While the inflammatory response protects the host from invading pathogens as well as tissue insult and injury, inflammation in the brain is a serious and potentially detrimental event, owing to the susceptibility of neurons to injury and insult and due to the restricted limited space that cannot sustain swelling. Prolonged, extensive or non-regulated inflammation contributes to the pathogenesis of diseases and there is an extensive body of evidence indicating that neuroinflammation contributes to the development and progression of age-related neurodegenerative disorders like Alzheimer's disease and Parkinson's disease (Bodles and Barger, 2004; see Block and Hong, 2005). Findings of this type have challenged the originally held view that the brain is an "immune-privileged" site, and the current consensus is that the brain should be regarded as an immunologically specialized site with resident cells like microglia being especially responsible for mediating immunoreactivity.

6. NEUROINFLAMMATION IS A FEATURE OF THE BRAIN IN AGED RATS

In addition to the evidence indicating that inflammatory responses are upregulated in the periphery with age (Smith et al., 2001), there is now a significant body of evidence indicating that proinflammatory cytokine concentration in the brain increases with age. Increases in the concentration of the proinflammatory cytokine, interleukin-1β (IL-1β), in the hippocampus of aged rats has been documented in this laboratory (Lynch and Lynch, 2001; Martin et al., 2002; Maher et al., 2005; Nolan et al., 2005) and similar age-related increase in the concentrations of other inflammatory cytokines like IL-6 (Godbout and Johnson, 2004) and IL-18 (Nally et al., 2004) have also been reported.

6.1. IL-1β and IL-1β-Induced Signaling

Of these inflammatory cytokines, the IL-1 family of cytokines is the most studied. It comprises ten proteins including IL-1α, IL-1β and the endogenous receptor antagonist, IL-1ra, and while it is currently thought that signaling occurs as a consequence of interaction with one receptor, IL-1 receptor type 1 (IL-1RI), a large family of receptors has been described (Sims, 2002). Research has largely focussed on IL-1β, the 17 kDa active form of which is produced by cleavage of the 31 kDa pro-IL-1β by an IL-1β converting enzyme (ICE) or caspase-1. IL-1β has a pleiotropic nature and acts both locally and systemically, and its importance in the central nervous system was spotlighted initially by the finding that it is an endogenous pyrogen, playing a significant role in fever.

The signaling events triggered by the interaction of IL-1β with IL-1RI are well established and involve recruitment of a number of proteins to the ligand–receptor complex to initiate a series of phosphorylation events (see Lynch, 2004). One significant event on which IL-1β relies to stimulate its cellular effect is activation of the mitogen-activated protein kinase (MAPK), c-jun N-terminal kinase (JNK), which responds to diverse extracellular stimuli, particularly stresses (Park et al., 1997). The effects of JNK are mediated by phosphorylation of a variety of nuclear and cytoplasmic substrates and it has been reported that the degenerative changes induced by JNK are likely to require activation of the transcription factor, c-jun (Yang et al., 1997). Work carried out by Xia and colleagues (1995) highlighted the fact that, under various conditions, the stimulation of JNK was a prerequisite for cell death and that blocking JNK activation prevented cell death.

6.2. Proinflammatory Cytokines and LTP

Several reports have indicated that acute administration of IL-1β, IL-6 and IL-18 leads to inhibition of LTP in the hippocampus (Li et al., 1997; Murray et al., 1998b; Curran and O'Connor, 2001; but see Schneider et al. 1998; Avital et al., 2003; Balschun et al., 2004 for alternative interpretations). Significantly, it has been shown that LTP is also attenuated in circumstances in which IL-1β concentration in hippocampus is increased. For example, lipopolysaccharide (LPS), which is a component of the cell wall of gram-negative bacteria, inhibits LTP in perforant path-granule cell synapses and it has been shown that its ability to inhibit LTP is due to an increase in hippocampal IL-1β concentration, since inhibition of caspase-1 prevented both the LPS-induced increase in IL-1β and the inhibition of LTP (Vereker et al., 2000a). Similarly, exposure of animals to irradiation resulted in an increase in IL-1β concentration and inhibition of LTP (Lonergan et al., 2002; Lynch et al., 2003), while similar effects were observed in rats treated with amyloid-β_{1-40} (Aβ; Minogue et al., 2003). In each of these experimental conditions, treatment of rats with the polyunsaturated fatty acid, eicosapentaenoic acid (EPA), reversed the treatment-related increases in hippocampal IL-1β concentration and restored the ability of rats to sustain LTP (Lonergan et al., 2002, 2004). We have also reported that EPA treatment attenuates the age-related increase in hippocampal IL-1β concentration and, in parallel, the age-related attenuation of LTP (Martin et al., 2002).

6.3. Stress-Activated Kinases and LTP

The increase in IL-1β, which occurs in these experimental circumstances, is associated with upregulation of JNK, and the evidence suggests that this is a key event in leading to the impairment in LTP, since inhibiting activation of JNK prevents the deficit in LTP induced by IL-1β (Curran et al., 2003) and by LPS (Barry et al., 2005). The correlation between these parameters was also supported by the finding that intracerebroventricular injection of Aβ, which blocked LTP in CA1 in vivo, induced an increase in IL-1β concentration and an increase in JNK activation in hippocampus, while treatment of rats with the peptide inhibitor of JNK, D-JNKI1, ameliorated all these changes (Minogue et al., 2003). Interestingly, a number of reports have indicated that IL-1β increases activation of another stress-activated protein kinase, p38, in hippocampus (Vereker et al., 2000b) and that the impairment in LTP, which is observed in LPS-treated rats, is attenuated by inhibiting activation of p38 by the inhibitor SB203580 (Kelly et al., 2003).

7. ACTIVATED MICROGLIA ARE THE PRIMARY CELL SOURCE OF IL-1β

Activated microglia secrete proinflammatory cytokines such as IL-1β, IL-6 and tumour necrosis factor-α (TNF-α; Ke and Gibson, 2004; Tuppo and Arias, 2005; Vilhardt, 2005) and indeed are considered to be the primary cell source of IL-1β (Davies et al., 1999; Li et al., 2003; see Hanisch, 2002; Block and Hong, 2005). It is accepted that microglia are ubiquitously distributed in the nervous system (Streit, 2000). During embryogenesis, activated microglia are present at sites of axonal growth after which time they disappear and later reappear in the adult brain during infection or injury (Giulian and Baker, 1986). Thus trauma to the central nervous system triggers a variety of inflammatory responses, which include activation of macrophages at an injury site, release of cytokines and trophic factors, secretion of proteases and, significantly, activation of microglia (reviewed by Mocchetti and Wrathall, 1995).

7.1. Microglial Activation is Upregulated by Age and Stressful Stimuli

Microglia, at rest, exist in a ramified state (Vilhardt, 2005), but undergo morphological changes, involving the contraction of their spines, development of an ameboid shape and proliferation in number in response to insult (von Bernhardi and Eugenin, 2004). Their ability to respond quickly to an injury suggests that their quiescent, resting form represents a state of vigilance to changes in the CNS microenvironment (Kreutsberg, 1996). Thus, microglial activation is a marker of the intracerebral inflammatory response, and while the active state is typified by their ability to secrete proinflammatory cytokines, they can also be identified by increased cell surface expression of markers such as major histocompatibility factor II (MHCII). While constitutive expression of MHCII has been documented (Iglesias et al., 1997), upregulation occurs in diseased states such as Alzheimer's disease (Tuppo and Arias, 2005) and following stimulation in vivo (Maher et al., 2005) and in vitro with INF-γ and LPS (Aloisi et al., 1998; see Aloisi, 2001).

Because activated microglia are probably the major cell source of IL-1β, it might be predicted that evidence of microglial activation will be a feature of the hippocampus in brain in circumstances in which IL-1β concentration is increased. Consistent with this are reports indicating that treatment of rats with LPS markedly increases microglial activation (e.g. Hauss-Wegrzyniak et al., 2002; Clarke and Lynch, 2005). Similarly, it might be expected that the increase in IL-1β concentration in aged rats will be accompanied by evidence of microglial activation; however, although it has been shown to be a feature of neurodegenerative diseases and models of neurodegenerative diseases (Ledeboer et al., 2002; O'Keefe et al., 2002; Tan et al., 2002; McGeer and McGeer, 2004), the evidence indicating a similar change in the aged brain is equivocal. Although an increase in microglial activation has been reported in primates (Sloane et al., 1999), there are conflicting reports relating to changes in aged rodents (Hauss-Wegrzyniak et al., 1999; Felzien et al., 2001; Kullberg et al., 2001). Two studies by Kullberg and colleagues (2001) and Ogura and colleagues (1994) highlighted the fact that age-related changes in microglial activation were region-specific; it is therefore possible that the lack of concordance in reported data reflects the regional specificity. Significantly it has been shown that primary cultures of microglia isolated from aged rats showed a dramatic increase in proliferation, GFAP expression, ameboid morphology and antigen expression (Rozovsky et al., 1998).

Evidence from this laboratory suggests that expression of CD11b, one marker of microglial activation, is increased in the brain of aged, compared with young, rats (Moore et al., 2005) and more recent studies have indicated that other markers of microglial activation are also increased with age including MHCII mRNA expression and an increase in the number of OX6-positive cells in hippocampus (Loane et al., 2005; Lynch et al., 2005). Recent findings showing an age-related increase in hippocampal expression of ICAM-1, CD40 and CD86, which are also markers of microglial activation, support the data that have previously suggested that such activation does indeed increase with age (Nally et al., 2004; Nally et al., unpublished).

7.2. Microglial Activation and LTP

If it is accepted (a) that maintenance of LTP is indirectly correlated with hippocampal IL-1β concentration and (b) that activated microglia are the source of the IL-1β, it follows that if microglial activation is prevented then LTP will be restored. Minocycline, which is a member of the tetracycline class of antibiotics, has been shown in numerous studies to reduce neuroinflammation (see Stirling et al., 2005) and the evidence indicates that it prevents microglial activation and, in some cases, the cell death that results from extensive microglial activation (Zemke and Majid, 2004). Application of minocycline to hippocampal slices has been shown to prevent the Aβ-induced inhibition of LTP (Wang et al., 2004), while data from this laboratory have suggested that oral administration of minocycline to rats attenuates both the Aβ-induced and the age-related inhibition of LTP (Costelloe et al., 2005; Nally et al., 2004; Nally et al., unpublished) and in each of these circumstances the evidence indicated that minocycline decreased MHCII mRNA expression and the accompanying increase in IL-1β concentration in hippocampus.

7.3. What Triggers Activation of Microglia?

It has been established that one of the most potent activators of microglia is INF-γ, and INF-γ receptors, which are almost ubiquitously expressed, have been identified on the surface of microglia (Aloisi, 2001). Unlike IL-1β, which is released primarily from macrophages and microglia, INF-γ is released following appropriate stimulation from T helper 1 (Th1) cells, $CD8^+$ cytotoxic T cells (Pestka et al., 2004), and from natural killer cells, when stimulated by macrophage-derived cytokines such as TNF-α, IL-12 (Boehm et al., 1997) and IL-2 (Borrego et al., 1999). However, it has been revealed that INF-γ can also be released from microglia (Lindberg et al., 2005). The role of INF-γ in the immune system is very broad; one important function is that it modulates polarization of naïve T cells to the active phenotype. The evidence indicates that with IL-12, INF-γ shift the balance towards a Th1 stabilization (see Boehm et al., 1997), while INF-γ also acts by inhibiting Th2 differentiation from naïve T cells (Gajewski and Fitch, 1988); therefore, it propagates inflammatory effects. INF-γ also induces activation of macrophages (Belardelli, 1995), indicated by increased expression of MHCII and it has been shown to exert a similar effect in microglia (Neumann et al., 1996).

Analysis of the effect of INF-γ on microglia has focused to a great extent on assessing changes in vitro (e.g. Delgado, 2003; Kim et al., 2004), with relatively few studies assessing its effect in vivo. However, recent findings from this laboratory have confirmed and extended the observations in vitro and have demonstrated that intracerebroventricular injection of INF-γ results in upregulation of MHCII mRNA expression and IL-1β concentration in hippocampus, while a similar effect occurs following exposure of cultured cortical and/or hippocampal glia to INF-γ (Clarke et al., unpublished). Significantly, a role for INF-γ in activation of caspase-1 has been suggested by the observation that inhibitors of caspase-1 decrease cell death initiated by INF-γ (Kim et al., 2002) and this suggests that IL-1β may mediate the actions of INF-γ in some circumstances. Interestingly, our recent findings have indicated that hippocampal concentration of INF-γ is increased following treatment of rats with LPS or Aβ, while an increase in INF-γ concentration was observed in hippocampus of aged rats (Maher et al., 2005). Significantly in each of these experimental conditions, the increase in INF-γ was accompanied by increased concentration of IL-1β and paralleled by an increase in microglial activation. Consistently, we have shown that there is an inverse correlation between hippocampal concentration of INF-γ and the ability of animals to sustain LTP and that microglial activation is more pronounced in aged rats, which fail to sustain LTP compared with those that exhibit the ability to sustain LTP (Maher et al., 2005). Our evidence has also indicated that intracerebroventricular injection of INF-γ results in inhibition of LTP in perforant path-granule cell synapses. These findings have led us to conclude that one possible sequence of events leading to the age-related deficit in LTP is outlined in Fig. 1. Thus, we propose that there is an age-related increase in INF-γ (the underlying cause of which has not been clarified), which triggers microglial activation, an increase in IL-1β and IL-1β-induced signaling and consequently inhibition of LTP.

8. ANTI-INFLAMMATORY CYTOKINES INHIBIT THE EFFECTS OF IL-1β

While the inflammatory response is driven by proinflammatory cytokines like IL-1 and INF-γ, controlled responses to stress, pathogens and injury are achieved by maintaining a balance between pro- and anti-inflammatory cytokines. Of the anti-inflammatory cytokines, attention has mainly focussed on understanding the mechanisms by which IL-10 and IL-4 act to establish the balance.

8.1. IL-4, IL-10 and Associated Signaling

IL-4 and IL-10 are pleiotropic cytokines that regulate a variety of functions of haematopoietic cells and non-haematopoietic cells and that limit inflammatory responses (Moore et al., 2001). The IL-10 receptor resembles the INF-γ receptor in that at least two subunits of this receptor are members of the INF receptor family. Signal transduction is initiated by binding of dimeric IL-10 with high affinity to the larger receptor chain, IL-10R1, and the subsequent binding of the additional membrane spanning receptor chain, IL-10R2 (Moore et al., 2001; Fickenscher et al., 2002). In the case of IL-4, the receptor comprises an α chain to which IL-4 binds with high affinity and a γc chain, which forms a heterodimer with the α chain. Both IL-10 and IL-4 lack intrinsic kinase activity and signal through the Janus kinases (JAKs), which associate with cytokine receptor subunits and are essential for cytokine signaling (O'Shea, 1997). Four mammalian JAKs have been identified, JAK1, 2, 3 and Tyk 2, the catalytic domain of which facilitates association with a family of proteins known as signal transducers and activators of transcription (STATs; Leonard and O'Shea., 1998); there are seven known mammalian STAT proteins namely STAT1, 2, 3, 4, 5a, 5b and 6. Recruitment of JAK to the receptor complex results in tyrosine phosphorylation of residues in the receptor, the binding of STAT to the phosphorylated receptor and the phosphorylation and dimerization of activated STATs (Horvath and Darnell, 1997; Strehlow et al., 1998; Leonard and O'Shea, 1998). Although the mechanism is unclear, it is known that the dimeric STAT translocates to the nucleus, binds directly to DNA and ultimately modulates the transcription of target genes (Leonard and O'Shea, 1998). IL-4 appears to primarily interact with JAKs 1 and 3 and the target STAT for IL-4 is reported to be STAT6 (Haque et al., 1997; Nelms et al., 1999). Treatment of T cells and monocytes with IL-10 results in the tyrosine phosphorylation of JAK1, but the evidence suggests that IL-10 gene transcription may rely on activation of STAT3 (Finbloom and Winestock, 1995; Kontoyiannis et al., 2001). It is important to recognize that one consequence of tyrosine phosphorylation of the receptor is that pathways that are regulated by signaling through Src-homology 2 domains are stimulated and therefore it has been proposed activation of the Ras-MAPK pathway may contribute to the anti-inflammatory effects of IL-10 and IL-4 (Geng et al., 1994; Nelms et al., 1999). For example, it has been explicitly shown that activation of extracellular signal-activated kinase (ERK) mediates the modulatory effect of IL-10 on LPS-induced changes in monocytes (Geng et al., 1994), although it has also been shown to inhibit LPS-induced phosphorylation of p38 in these cells (Haddad et al., 2003).

Although these pathways are well characterized in circulating cells, there is a paucity of data relating to the mechanisms involved in, and the consequences of, IL-10- and IL-4-activated signaling in the brain. IL-4 receptor expression has been reported, directly or by implication, in various cultured glial cells (Wong et al., 1993; Brodie et al., 1998; Ledeboer et al., 2000; Molina-Holgado et al., 2001; Szczepanik et al., 2001) and in hippocampal neurons (Nolan et al., 2005), while IL-10 receptors are also similarly expressed (Moore et al., unpublished). The evidence indicates that, in hippocampus, as elsewhere, IL-4 and IL-10 receptors signal through activation of JAK/STAT pathways and, consistent with this, we have reported that the age-related decrease in hippocampal IL-4 concentration is coupled with decreased phosphorylation of JAK1 and STAT6 (Nolan et al., 2005), while the age-related decrease in IL-10 concentration is coupled with decreased phosphorylation of JAK1 and STAT3 (Moore et al., 2005).

8.2. IL-4 and IL-10 Downregulate Microglial Activation

The principal effect of both IL-4 and IL-10 is amelioration of inflammation. IL-10 potently inhibits production of proinflammatory cytokines like IL-1 in several cell types (Haddad et al., 2003; Mizuno et al., 1994; Mesples et al., 2003; Sawada et al., 1999). In particular, IL-10 has been shown to play an important role in CNS inflammation by affecting microglial function (Sawada et al., 1999) and by inhibiting cytokine production by microglia (Mizuno et al., 1994; Mesples et al., 2003); consistent with this is the observation that INF-γ-induced upregulation of MHCII mRNA expression is abrogated by IL-10 (Mizuno et al., 1994). The actions of IL-10 in some circumstances have been shown to be mediated by its ability to increase IL-1ra production; for example, this is the mechanism by which IL-10 blocks IL-1-induced effects in leukocytes (Castella et al., 1994). Like IL-10, it has been demonstrated that, in circulating cells at least, IL-4 can inhibit IL-1β synthesis and antagonize the actions of IL-1β (Wong et al., 1993).

8.3. IL-4 and IL-10 Possess Neuroprotective Effects

The evidence indicates that the anti-inflammatory effects of both IL-4 and IL-10 extend to the CNS and that their ability to modulate IL-1β production is a key factor in this. It has been reported that the inhibition of LTP and the activation of signaling cascades induced by intracerebroventricular injection of IL-1β were attenuated by treatment of rats with IL-10 (Kelly et al., 2001). Similarly, the LPS-induced inhibition of LTP (Lynch et al., 2004) and the LPS-induced sickness behaviour (Bluthe et al., 1999) were both ameliorated by IL-10 treatment, and at least in the case of the LTP study, the effect appeared to be mediated by the inhibitory effect of IL-10 on IL-1β production and the consequent IL-1β-induced signaling. However the effects of IL-10 are not confined to its effect on IL-1β, since it has been shown to inhibit LPS-induced production of several other proinflammatory cytokines (Szczepanik et al., 2003). Similarly, IL-4 attenuates the inhibition of LTP induced by intracerebroventricular injection of IL-1β and also by administration of LPS, while the IL-1β-induced cell signaling in both cases was also attenuated by IL-4 (Barry et al., 2005; Nolan et al., 2005). Like IL-10, IL-4 has also been shown to block LPS-induced sickness behaviour.

It is important to note that the neuroprotective effects of IL-10 and IL-4 extend beyond their ability to antagonize IL-1β and LPS-induced changes. For example it has been shown that IL-10 blocks Aβ-induced production of proinflammatory cytokines and cell stress induced by Theiler's murine encephalomyelitis virus (Molina-Holgado et al., 2002), while primary neuronal cortical cultures derived from IL-10$^{-/-}$ animals were found to be more susceptible to excitotoxic insult than cells prepared from wildtype animals (Grilli et al., 2000). Similarly, IL-10 concentration was decreased in mice in which experimental autoimmune encephalomyelitis (EAE) was induced; the severity of the disease was exaggerated in IL-10$^{-/-}$ mice, whereas expansion of Th$_2$-like cells that produce IL-10 was associated with amelioration of symptoms (Moore et al., 2001). A similar effect of IL-4 in attenuating the detrimental changes associated with EAE was suggested by the finding that the beneficial effects of atorvastatin in this condition were mediated by IL-4 (Youssef et al., 2002).

9. THE IMBALANCE IN PRO- AND ANTI-INFLAMMATORY CYTOKINES AND THE CONSEQUENCES ON LTP

Several constituents of the IL-1β signaling pathway are upregulated with age. Thus, the age-related increase in IL-1β concentration in hippocampus is accompanied by an increase in activity of caspase-1 and increases in expression of IL-1RI and phosphorylation of IL-1 receptor-associated kinase (IRAK; Lynch and Lynch 2001). These changes drive the increases in activation of JNK, c-jun and p38 (Martin et al., 2002; Nolan et al., 2005), which lead to inhibition of LTP. Significantly, we have shown that if the age-related increase in IL-1β and the associated signaling events are attenuated, for example by treatment of rats with EPA (Martin et al., 2002), then the age-related deficit in LTP is also attenuated. We have also established that treatment of aged rats with a combination of dexamethasone and vitamin D$_3$ also attenuated both the age-related increases in IL-1β concentration and IL-1β-induced signaling, as well as the deficit in LTP (Moore et al., 2004, 2005). Dexamethasone, a synthetic glucocorticoid, has been shown to block accumulation of IL-1β mRNA in U937 cells following incubation with toxic shock supernatants (Knudsen et al., 1987) and the evidence suggests that this is achieved by inhibiting transcription and secretion of IL-1β ((Kern et al., 1988; but see Amano et al., 1993), who suggest that this occurs because dexamethasone decreases the stability of IL-1β mRNA). Vitamin D_3 has also been shown to possess anti-inflammatory properties, increasing anti-inflammatory cytokines (Cantorna et al., 1998) and decreasing expression of proinflammatory cytokines (DeLuca and Cantorna, 2001).

Recent evidence, which has demonstrated coupled age-related increases in INF-γ concentration, microglial activation and IL-1β concentration in hippocampus, might suggest that the primary age-related change may be an increase in INF-γ (Maher et al., 2005). However, the observation that these changes are associated with decreases in hippocampal concentrations of IL-4 and IL-10, and decreases in the activation of JNK1, STAT3 and STAT6 (Moore and Lynch, 2004; Nolan et al., 2005), bring into sharp focus the fact that age-related deficits in hippocampal function are associated with a myriad of changes that contribute to the observed neuroinflammation. The evidence suggests that IL-4 down-regulates IL-1β expression in hippocampus (Maher et al., 2005) and therefore it must be concluded that the decreased IL-4 concentration in hippocampus of the aged rat relieves the tonic inhibitory effect on IL-1β synthesis. Similarly, Sawada and colleagues (1999) have demonstrated that IL-10 functions as an inhibitory regulator of proinflammatory cytokine activity in the CNS cytokine network and these authors, and others (Moore et al., 2001) reported that the primary target for IL-10 in achieving this was microglia. Consistent with this, our data have revealed that the age-related decreases in IL-4 and IL-10 are associated with evidence of microglial activation (Maher et al., 2005; Nolan et al., 2005).

Recently, we have been attempting to identify the precise mechanism by which EPA, and dexamethasone and vitamin D_3, act to relieve neuroinflammation and restore synaptic plasticity in the hippocampus of aged rats. Our findings indicate that EPA significantly increases IL-4 mRNA and IL-4 protein in hippocampus of aged rats (Lynch et al., 2005) and that treatment of rats with dexamethasone and vitamin D_3 significantly increases hippocampal concentration of IL-10 (Moore et al., 2005; Fig. 1). Both treatments reversed the age-related decreases in the associated signaling cascades and both treatments also attenuated the age-related increases in microglial activation and IL-1β concentration. Perhaps predictably, the age-related increase in JNK activation, which we have shown to be a downstream consequence of the increase in IL-1β, decreased in parallel with the change in IL-1β concentration. These findings add weight to the evidence, which indicates that anti-inflammatory cytokines exert a tonic inhibition on IL-1β synthesis and suggest that the primary defect in the hippocampus of the aged brain might be a decrease in concentration of one or more anti-inflammatory cytokines. At this time, it seems reasonable to propose that the anti-inflammatory effects of IL-10 and/or IL-4 (and perhaps the treatments that increase their expression) may be a consequence of their ability to restore activated microglia to a quiescent state (Fig. 1). In addition to the evidence referred to above, which shows that increasing hippocampal concentration of either or both cytokines decreases microglial activation in the hippocampus of aged rats, there is direct evidence that IL-10 limits microglial activation (Aloisi et al., 1998), and specifically that it inhibits MHCII expression and production of cytokines and chemokines (Iglesias et al., 1997; Aloisi, 2001; Ledeboer et al., 2002). In the case of IL-4, there is compelling evidence that indicates that it potently inhibits the actions of INF-γ, including the INF-γ-induced activation of CD40 in a microglial cell line (Nguyen and Benveniste, 2000), while it completely abrogates the increase in MHCII expression observed in hippocampus of INF-γ-treated rats (Clarke and Lynch, 2005).

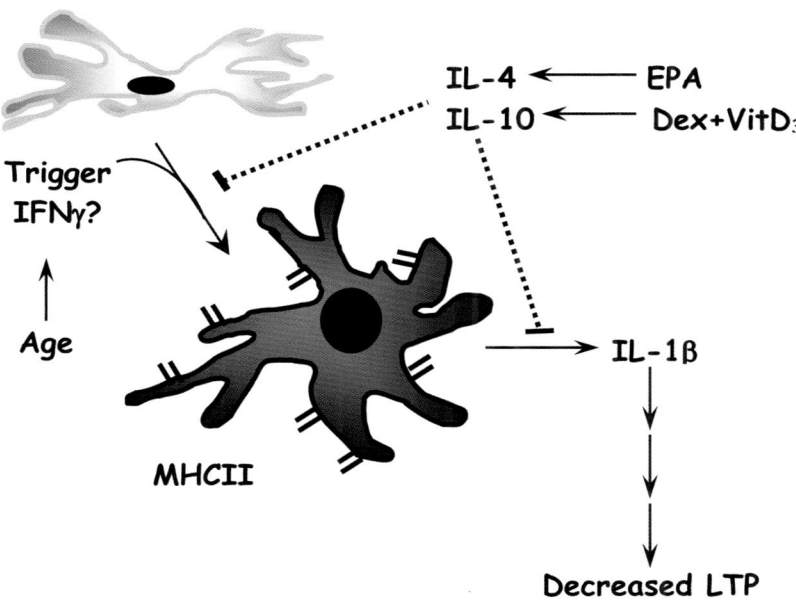

Fig. 1. INF-γ concentration is increased in the hippocampus of aged rats and this induces an increase in microglial activation which, as a consequence, exhibit an increase in cell surface markers for example major histocompatibility factor II (MHCII). Activated microglia are the cell source of IL-1β and therefore there is an age-related increase in the cytokine and a consequent increase in IL-1β-associated signaling. These changes lead to a deficit in LTP. The age-related increase in IL-1β concentration is accompanied by a decrease in hippocampal concentration of anti-inflammatory cytokines, IL-10 and IL-4 and the evidence shows that anti-inflammatory treatments, eicosapentaenoic acid (EPA), and dexamethasone (dex) and vitamin D_3 (vit D_3) increase IL-4 and IL-10, respectively. Both IL-4 and IL-10, by maintaining microglia in a quiescent state, prevent the downstream consequences of the increase in IL-1β and therefore lead to restoration of LTP

10. REFERENCES

Ali, M., 2005, The oxidative-dysoxygenative model of ageing. *J. Integr. Med.* **7**: 1.
Aloisi, F., 2001, Immune function of microglia. *Glia* **36**: 165.
Aloisi, F., Ria, F., Penna, G. and Adorini, L., 1998, Microglia are more efficient than astrocytes in antigen processing and in Th1 but not Th2 cell activation. *J. Immunol.* **160**: 4671.
Amano, Y., Lee, S.W. and Allison, A.C., 1993, Inhibition by glucocorticoids of the formation of interleukin-1α, interleukin-1β, and interleukin-6: mediation by decreased mRNA stability. *Mol. Pharmacol.* **43**: 176.

Avital, A., Goshen, I., Kamsler, A., Segal, M., Iverfeldt, K., Richter-Levin, G. and Yitmiya, R., 2003, Impaired interleukin-1 signaling is associated with deficits in hippocampal memory processes and neural plasticity. *Hippocampus* **13**: 826.

Balschun, D., Wetzel, W., Del Rey, A., Pitossi, F., Schneider, H., Zuschratter, W. and Besedovsky, H.O., 2004, Interleukin-6: a cytokine to forget. *FASEB J.* **18**: 1788.

Barnes, C.A., 1988, Spatial learning and memory processes: the search for their neurobiological mechanisms in the rat. *Trends Neurosci.* **11**: 163.

Barry, C.E., Nolan, Y., Clarke, R.M., Lynch, A. and Lynch, M.A., 2005, Activation of c-Jun-N-terminal kinase is critical in mediating lipopolysaccharide-induced changes in rat hippocampus. *J. Neurochem.* **93**: 221.

Belardelli, F., 1995, Role of interferons and other cytokines in the regulation of the immune response. *Acta Pathol. Microbiol. Immunol. Scand.* **103**: 161.

Block, M.L. and Hong, J.S., 2005, Microglia and inflammation-mediated neurodegeneration: multiple triggers with a common mechanism. *Prog. Neurobiol.* **76**: 77.

Bluthe, R.M., Castanon, N., Pousset, F., Bristow, A., Ball, C., Lestage, J., Michaud, B., Kelley, K.W., Dantzer, R., 1999, Central injection of IL-10 antagonizes the behavioural effects of lipopolysaccharide in rats. *Psychoneuroendocrinol.* **24**: 301.

Bodles, A.M. and Barger, S.W., 2004, Cytokines and the aging brain-what we don't know might help us. *Trends Neurosci.* **27**: 621.

Boehm, U., Klamp, T., Groot, M. and Howard, J.C., 1997, Cellular responses to interferon-γ. *Ann. Rev. Immunol.* **15**: 749.

Bonhaus, D.W., Perry, W.B. and McNamara, J.O., 1990, Decreased density. But not number, of N-methyl-D-aspartate, glycine and phencyclidine binding sites in the hippocampus of senescent rats. *Brain Res.* **532**: 82.

Borrego, F., Alonso, M.C., Galiani, M.D., Carracedo, J., Ramirez, R., Ostos, R., Pena, J. and Solana, R., 1999, NK phenotype markers and IL-2 response in NK cells from elderly people. *Exp. Gerontol.* **34**: 253.

Brodie, C., Goldreich, N., Haiman, T. and Kazimirsky, G., 1998, Functional IL-4 receptors on mouse astrocytes: IL-4 inhibits astrocyte activation and induces NGF secretion. *J. Neuroimmunol.* **81**: 20.

Cantorna, M., Woodward, W., Hayes, C.E. and DeLuca, H.F., 1998, 1, 25-Dihydroxyvitamin D_3 is a positive regulator for the two anti-encephalitogenic cytokines TGF-β1 and IL-4. *J. Immunol.* **160**: 5314.

Castella, M.A., Meda, L., Gasperini, S., Calzetti, F. and Bonora, S., 1994, Interleukin-10 up regulates IL-1 receptor antagonist production from lipopolysaccharide-stimulated human polymorphonuclear leukocytes by delaying mRNA degradation. *J. Exp. Med.* **179**: 1695.

Clarke, R.M. and Lynch, M.A., 2005, Atorvastatin modulates the lipopolysaccharide-induced impairment in long-term potentiation in the rat hippocampus. *Soc. Neurosci. Abstr.* **792**:1.

Costelloe, C.E., Lyons, A. and Lynch, M.A., 2005, Minocycline attenuates the inhibition of LTP induced by amyloid-β. *Soc. Neurosci. Abstr.* **910**:9.

Curran, B. and O'Connor, J.J., 2001, The pro-inflammatory cytokine interleukin-18 impairs long-term potentiation and NMDA receptor-mediated transmission in the rat hippocampus in vitro. *Neuroscience* **108**: 83.

Curran, B.P., Murray, H.J. and O'Connor, J.J., 2003, A role for c-Jun N-terminal kinase in the inhibition of long-term potentiation by interleukin-1beta and long-term depression in the rat dentate gyrus in vitro. *Neuroscience* **118**: 347.

Davies, C.A., Loddickm, S.A., Toulmondm, S., Stroemerm, R.P., Hunt, J. and Rothwell, N.J., 1999, The progression and topographic distribution of interleukin-1beta expression after permanent middle cerebral artery occlusion in the rat. *J. Cereb. Blood Flow Metab.* **19**: 87.

De la Asuncion, J.G., Millan, A., Pla, R., Bruseghini, L., Esteras, A., Pallardo, F.V., Saster, J. and Vina, J., 1996. Mitochondrial glutathione oxidation correlates with age-associated oxidative damage to mitochondrial DNA. *FASEB J.* **10**: 333.

Delgado, M., 2003, Inhibition of interferon (IFN) γ-induced Jak-STAT1 activation in microglia by vasoactive intestinal peptide. *J. Biol. Chem.* **278**: 27620.

DeLuca, H.F. and Cantorna, M.T., 2001, Vitamin D: its role and uses in immunology. *FASEB J.* **15**: 2579.

Effros, R.B., 2005, Roy Walford and the immunologic theory of ageing. *Immunity Ageing* **2**: 7.

Fabris, N., 1992, Biomarkers of ageing in the neuroendocrine-immune domain. Time for a new theory of ageing? *Ann. N.Y. Acad. Sci.* **663**: 335.

Felzien, L.K., McDonald, J.T., Gleason, S.M., Berman, N.E. and Klein, R.M., 2001, Increased chemokine gene expression during aging in the murine brain. *Brain Res.* **26**: 137.

Fickenscher, H., Hor, S., Kupers, H., Knappe, A., Wittman, A. and Sticht, H., 2002, The interleukin-10 family of cytokines. *Trends Immunol.* **23**: 89.

Finbloom, D.S. and Winestock, K.D., 1995, IL-10 induces the tyrosine phosphorylation of tyk2 and Jak1 and the differential assembly of STAT1 and STAT3 complexes in human T cells and monocytes. *J. Immunol.* **155**: 1079.

Forster, M.J., Dubey, A., Dawson, K.M., Stutts, W.A., Lal, H. and Sohal, R.S., 1996, Age-related losses of cognitive function and motor skills in mice are associated with oxidative protein damage in the brain. *Proc. Nat. Acad. Sci. USA* **93**: 4765.

Gabbita, S.P., Butterfield, D.A., Hensley, K., Shaw, W. and Carney, J.M., 1997, Aging and caloric restriction affect mitochondrial respiration and lipid membrane status: an electron paramagnetic resonance investigation. *Free Rad. Biol. Med.* **23**: 191.

Gajewski, T.F. and Fitch, F.W., 1988, Anti-proliferative effect of IFN-γ in immune regulation. I. IFN-γ inhibits the proliferation of Th2 but not Th1 murine helper lymphocyte clones. *J. Immunol.* **140**: 4245.

Geng, Y., Gulbins, E., Altman, A. and Lotz, M., 1994, Monocyte deactivation by interleukin-10 via inhibition of tyrosine kinase activity and the Ras signaling pathway. *Proc. Nat. Acad. Sci. USA* **91**: 8602.

Giulian, D. and Baker, T., 1986, Characterisation of Ameboid microglia isolated from developing mammalian brain. *J. Neurosci.* **6**: 2163.

Giusto, N.M., Roque, M.E. and Ilincheta de Boschero, M.E., 1992, Effects of aging on the content, composition and synthesis of sphingomyelin in the central nervous system. *Lipids* **27**: 835.

Godbout, J.P. and Johnson, R.W., 2004, Interleukin-6 in the aging brain. *J. Immunol.* **147**: 141.

Gottfries, C.G., 1990, Neurochemical aspects on aging and diseases with cognitive impairment. *J. Neurosci. Res.* **27**: 541.

Grilli, M., Barbieri, I., Basudev, H., Brusa, R., Casati, C., Lozza, G. and Ongini, E., 2000, Interleukin-10 modulates neuronal threshold of vulnerability to ischaemic damage. *Eur. J. Neurosci.* **12**: 2265.

Gu, Z., Wortwein, G., Yu, J. and Perez-Polo, J.R., 2000, Model for aging in the basal forebrain cholinergic system. *Antioxid. Redox Signal* **2**: 437.

Haddad, J.J., Saade, N.E. and Safieh-Garabedian, B., 2003, Interleukin-10 and the regulation of mitogen-activated protein kinases: are these signaling modules targets for anti-inflammatory action of this cytokine? *Cell Signal* **15**: 255.

Hanisch, U.K., 2002, Microglia as a source and target of cytokines. *Glia* **40**: 140.

Haque, S.J., Wu, Q., Kammer, W., Friedrich, K., Smith, J.M., Kerr, I.M., Stark, G.R. and Williams, B.R., 1997, Receptor-associated constitutive protein tyrosine phosphatase activity controls the kinase function of JAK1. *Proc. Natl. Acad. Sci. USA* **94**: 8563.

Harman, D., 2003, The free radical theory of aging. *Antioxid. Redox Signal* **5**: 557.

Hasan, M. and Glees, P., 1973, Ultrastructural age changes in hippocampal neurons, synapses and neuroglia. *Exp. Gerontol.* **8**: 75.

Hauss-Wegrzyniak, B., Vraniak, P. and Wenk, G.L., 1999, The effects of a novel NSAID on chronic neuro-inflammation are age dependent. *Neurobiol. Aging* **20**: 305.

Hauss-Wegrzyniak, B., Lynch, M.A., Vraniak, P.D. and Wenk, G.L., 2002, Chronic brain inflammation results in cell loss in the entorhinal cortex and impaired LTP in perforant path-granule cell synapses. *Exp. Neurol.* **176**: 336.

Horvath, C.M. and Darnell, J.E., 1997, The state of Stats: recent developments in the study of signal transduction to the nucleus. *Curr. Opin. Cell Biol.* **9**: 233.

Iglesias, B.M., Cerase, J., Ceracchini, C., Levi, G. and Aloisi, F., 1997, Analysis of B7-1 and B7-2 costimulatory ligands in cultured mouse microglia: upregulation by interferon-γ and lipopolysaccharide and downregulation by interleukin-10, prostaglandin E_2 and cyclic AMP-elevating agents. *J. Neuroimmunol.* **72**: 83.

Ingram, D.K., Garafalo, P., Spangler, E.L., Mantione, C.R., Odano, I. and London, E.D., 1992, Reduced density of NMDA receptors and increased sensitivity to dizocipline-induced learning impairment in aged rats. *Brain Res.* **580**: 273.

Ke, Z. and Gibson, G.E., 2004, Selective response of various brain cell types during neurodegeneration induced by mild impairment of oxidative metabolism. *Neurochem. Int.* **45**: 361.

Kelly, A., Lynch, A., Vereker, E., Nolan, Y., Quennan, P., Whittaker, E., O'Neill, L. and Lynch, M.A., 2001, The anti-inflammatory cytokine, interleukin (IL)-10, blocks the inhibitory effect of IL-1β on long-term potentiation. *J. Biol. Chem.* **276**: 45564.

Kelly, A., Vereker, E., Brady, M., Barry, C., Loscher, C., Mills, K.H.G. and Lynch, M.A., 2003, Activation of p38 plays a pivotal role in the inhibitory effect of lipopolysaccharide and interleukin-1β-induced inhibition of long-term potentiation in rat dentate gyrus. *J. Biol. Chem.* **278**: 19453.

Kern, J.A., Lamb, R.J., Reed, J.C., Daniele, R.P. and Nowell, P.C., 1988, Dexamethasone inhibition of interleukin 1 beta production by human monocytes. *J. Clin. Invest.* **81**: 237.

Kim, E., Lee, J., Namkoong, S., Um, S. and Park, J., 2002, Interferon regulatory factor-1 mediates interferon-γ-induced apoptosis in ovarian carcinoma cells. *J. Cell Biochem.* **85**: 369.

Kim, H., Whang, S., Woo, M., Park, J., Kim, W. and Han, I., 2004, Sodium butyrate suppresses interferon-gamma-, but not lipopolysaccharide-mediated induction of nitric oxide and tumor necrosis factor-alpha in microglia. *J. Neuroimmunol.* **151**: 85.

Knudsen, P.J., Dinarello, C.A. and Strom, T.B., 1987, Glucocorticoids inhibit transcriptional and post-transcriptional expression of Interleukin 1 in U937 cells. *J. Immunol.* **2139**: 4129.

Kontoyiannis, D., Kotlyarov, A., Carballo, E., Alexopoulou, L., Blackshear, P.J., Gaestel, M., Davis, R., Flavell, R. and Kollias, G., 2001, Intereukin-10 targets p38 MAPK to modulate ARE-dependent TNF mRNA translation and limit intestinal pathology. *EMBO J.* **20**: 3760.

Kreutsberg, G.W., 1996, Microglia: a sensor for pathological events in the CNS. *Trends Neurosci.* **19**: 312.

Kullberg, S., Aldskogius, H. and Ulfhake, B., 2001, Microglial activation, emergence of ED1-expressing cells and clusterin upregulation in the aging rat CNS, with special reference to the spinal cord. *Brain Res.* **899**: 169.

Land, W.G., 2004, Ageing and immunosuppression in kidney transplantation. *Exp. Clin. Transplant.* **2**: 229.

Ledeboer, A., Breve, J.J., Poole, S., Tilders, F.J. and Van Dam, A.M., 2000, Interleukin-10, interleukin-4, and transforming growth factor-beta differentially regulate lipopolysaccharide-induced production of pro-inflammatory cytokines and nitric oxide in co-cultures of rat astroglial and microglial cells. *Glia* **30**:134.

Ledeboer, A., Breve, J.J.P., Wierinckx, A., van der Jagt, S., Bristow, A.F., Leysen, J.E., Tilders, F.J.H. and Van Dam A.-M., 2002, Expression and regulation of interleukin-10 and interleukin-10 receptor in rat astroglial and microglial cells. *Eur. J. Neurosci.* **16**: 1175.

Leonard, W.J. and O'Shea, J.J., 1998, Jaks and Stats: biological Implications. *Ann. Rev. Immunol.* **16**: 293.

Li, A.J., Katafuchi, T., Oda, S., Hori, T. and Oomura, Y., 1997, Interleukin-6 inhibits long-term potentiation in rat hippocampal slices. *Brain Res.* **748**: 30.

Li, Y., Liu, L., Barger, S.W. and Griffin, W.S., 2003, Interleukin-1 mediates pathological effects of microglia on tau phosphorylation and on synaptophysin synthesis in cortical neurons through a p38-MAPK pathway. *J. Neurosci.* **23**: 1605.

Lindberg, C., Selenica, M.L., Westlind-Danielsson, A. and Schultzberg, M., 2005, Beta-amyloid protein structure determines the nature of cytokine release from rat microglia. *J. Mol. Neurosci.* **27**: 1.

Loane, D.J., Lynch, A.M., Minogue, A.M., Kilroy, D., Woods, O. and Lynch, M.A., 2005, Eicosapentaenoic acid reduces the age-related increase in microglial activation in the adult hippocampus. *Soc. Neurosci. Abstr.* **910**: 5.

Lonergan, P.E., Martin, D.S.D., Horrobin, D.F. and Lynch, M.A., 2002, Neuroprotective effect of eicosapentanoic acid in hippocampus of rats exposed to γ-irradiation. *J. Biol. Chem.* **277**: 20804.

Lonergan, P.E., Martin, D.S.D., Horrobin, D.F. and Lynch, M.A., 2004, Neuroprotective actions of eicosapentanoic acid on lipopolysaccharide-induced dysfunction in rat hippocampus. *J. Neurochem.* **91**: 20.

Lynch, M.A., 2004, Long-term potentiation and memory. *Physiol. Rev.* **84**: 87.

Lynch, A.M. and Lynch, M.A., 2001, The age-related increase in IL-1 type I receptor in rat hippocampus is coupled with an increase in caspase-3 activation. *Eur. J. Neurosci.* **15**: 1779.

Lynch, A., Moore, M., Craig, S., Lonergan, P., Martin, D. and Lynch, M.A., 2003, Analysis of interleukin-1β-induced cell signaling activation in rat hippocampus following exposure to gamma irradiation. *J. Biol. Chem.* **278**: 51075.

Lynch, A., Walsh, C., Delaney, A., Nolan, Y., Campbell, V. and Lynch, M.A., 2004, Lipopolysaccharide-induced increase in signaling in hippocampus is abrogated by IL-10 – a role for IL-1β? *J. Neurochem.* **88**: 635.

Lynch, A.M., Loane, D.J., Minogue, A.M. and Lynch, M.A., 2005, Cytokine modulation by eicosapentaenoic acid in the aged hippocampus *Soc. Neurosci. Abstr.* **910**: 4.

Maher, F.O., Nolan, Y. and Lynch, M.A., 2005, Downregulation of IL-4-induced signaling in hippocampus contributes to deficits in LTP in the aged rat. *Neurobiol. Aging* **26**: 717.

Martin, D.S., Lonergan, P.E., Boland, B., Fogarty, M.P., Brady, M., Horrobin, D.F., Campbell, V.A. and Lynch, M.A., 2002, Apoptotic changes in the aged brain are triggered by interleukin-1beta-induced activation of p38 and reversed by treatment with eicosapentanoic acid. *J. Biol. Chem.* **277**: 34239.

McGahon, B.M., Martin, D.S.D., Horrobin, D.F. and Lynch, M.A., 1999 Age-related changes in synaptic function: analysis of the effect of dietary supplementation with ω-3 fatty acids. *Neurosci.* **94**: 305.

McGeer, P.L. and McGeer, E.G., 2004, Inflammation and neurodegeneration in Parkinson's disease. *Parkinson Relat. Disord.*, Suppl. 1: S3.

Mesples, B., Plaisant, F. and Gressens, P., 2003, Effects of interleukin-10 on neonatal excitotoxic brain lesions in mice. *Dev. Brain Res.* **141**: 25.

Minogue, A.M., Schmid, A.W., Fogarty, M.P., Moore, A.C., Campbell, V.A., Herron, C.E. and Lynch, M.A., 2003, Activation of the c-Jun N-terminal kinase signaling cascade mediates the effect of amyloid-β on long term potentiation and cell death in hippocampus. *J. Biol. Chem.* **278**: 27971.

Mizuno, T., Sawada, M., Marunouchi, T. and Suzumura, A., 1994, Production of interleukin-10 by mouse glial cells in culture. *Biochem. Biophys. Res. Commun.* **205**: 1907.

Mocchetti, I. and Wrathall, J.R., 1995, Neurotrophic factors in central nervous system trauma. *J. Neurotrauma* **12**: 853.

Molina-Holgado, E., Vela, J.M., Arevalo-Martin, A. and Guaza, C., 2001, LPS/IFN-gamma cytotoxicity in oligodendroglial cells: role of nitric oxide and protection by the anti-inflammatory cytokine IL-10. *Eur. J. Neurosci.* **13**: 493.

Molina-Holgado, E., Arevalo-Martin, A., Ortiz, S., Vela, J.M. and Guaza, C., 2002, Theiler's virus infection induces the expression of cyclooxygenase-2 in murine astrocytes: inhibition by the anti-inflammatory cytokines interleukin-4 and interleukin-10. *Neurosci. Lett.* **324**: 237.

Moore, M. and Lynch, M.A., 2004, Does treatment with dexamethasone prevent age-related changes in hippocampus. *Soc. Neurosci. Abstr.* **565**:14.

Moore, K.W., Malefyt, R., Coffman, R.L. and O'Garra, A., 2001, Interleukin-10 and the interleukin-10 receptor. *Ann. Rev. Immunol.* **19**: 683.

Moore, M.E., Piazza, A., McCartney, Y. and Lynch, M.A., 2005, Evidence that vitamin D(3) reverses age-related inflammatory changes in the rat hippocampus. *Biochem. Soc. Trans.* **33**: 573.

Murray, C.A. and Lynch, M.A., 1998a, Dietary supplementation with vitamin E reverses the age-related deficit in long-term potentiation in dentate gyrus. *J. Biol. Chem.* **273**: 12161.

Murray, C. and Lynch, M.A., 1998b, Evidence that increased hippocampal expression of the cytokine, IL-1β, is a common trigger for age and stress-induced impairments in long-term potentiation. *J. Neurosci.* **18**: 2974.

Nally, R., Walsh, M., Nolan, Y. and Lynch, M.A., 2004, Age-related changes in the rat hippocampus and the effect of exogenously administered interleukin-1β. *Soc. Neurosci. Abstr.* **565**: 2.

Nelms, K., Keegan, A.D., Zamorano, J., Ryan, J.J. and Paul, W.E., 1999, The IL-4 receptor: signaling mechanisms and biologic functions. *Ann. Rev. Immunol.* **17**: 701.

Neumann, H., Boucraut, J., Hahnel, C., Misgeld, T. and Wekerle, H., 1996, Neuronal control of MHC class II inducibility in rat astrocytes and microglia. *Eur. J. Neurosci.* **8**: 2582.

Nguyen, V.T. and Benveniste, E.N., 2000, Involvement of STAT-1 and its family members in interferon-gamma induction of CD40 transcription in microglia/macrophages. *J. Biol. Chem.* **275**: 23674.

Nolan, Y., Maher, F.O., Martin, D.S., Clarke, R.M., Brady, M.T., Bolton, A.E., Mills, K.H. and Lynch, M.A., 2005, Role of interleukin-4 in regulation of age-related inflammatory changes in the hippocampus. *J. Biol. Chem.* **280**: 9354.

Ogura, K., Ogawa, M., Yoshida, M., 1994, Effects of ageing on microglia in the normal rat brain: immunohistochemical observation. *Neuroreport* **5**: 1224.

O'Keefe, G.M., Nguyen, V.T., Benveniste, E.N., 2002 Regulation and function of class II major histocompatibility complex, CD40, and B7 expression in macrophages and microglia: Implications in neurological diseases. *J. Neurovirol.* **8**: 496.

O'Shea, J.J., 1997, Jaks, Stats, cytokine signal transduction and immunoregulation: are we there yet? *Immunity* **7**: 1.

Park, J., Kim, I., Oh, Y.J., Lee, K., Han, P.L. and Choi, E.J., 1997, Activation of c-Jun N-terminal kinase antagonises an anti-apoptotic action of Bcl-2. *J. Biol. Chem.* **272**: 16725.

Pellmar, T.C., Hollinden, G.E. and Sarvey, J.M., 1991, Free radicals accelerate the decay of long-term potentiation in field CA1 of guinea-pig hippocampus, *Neuroscience* **44**: 353.

Pestka, S., Krause, C.D., Sarkar, D., Walter, M.R., Shi, Y. and Fisher, P.B., 2004, Interleukin-10 and related cytokines and receptors. *Ann. Rev. Immunol.* **22**: 929.

Roth, G.S. and Joseph, J.A., 1994, Cellular and molecular mechanisms of impaired dopaminergic function during aging. *Ann. N.Y. Acad. Sci.* **719**: 129.

Rozovsky, I., Finch, C.E. and Morgan, T.E., 1998, Age-related activation of microglia and astrocytes: in vitro studies show persistent phenotypes of aging, increase proliferation and resistance to down-regulation. *Neurobiol. Aging* **19**: 97.

Sawada, M., Suzumura, A., Hosoya, H., Marunouchi, T. and Nagatsu, T., 1999, Interleukin-10 inhibits both production of cytokines and expression of cytokine receptors in microglia. *J. Neurochem.* **72**: 1466.

Schipper, H.M., 2004, Brain iron deposition and the free radical-mitochondrial theory of aging. *Age Res. Rev.* **3**: 265.

Schneider, H., Pitossi, F., Balschun, D., Wagner, A., Del Rey, A. and Besedovsky, H.O., 1998, A neuromodulatory role of interleukin-1beta in the hippocampus. *Proc. Natl. Acad. Sci. USA* **95**: 7778.

Shigenaga, M.K., Hagen, T.M. and Ames, B.N., 1994, Oxidative damage and mitochondrial decay in aging. *Proc. Natl. Acad. Sci. USA* **91**: 10771.

Sims, J.E., 2002, IL-1 and IL-18 receptors, and their extended family. *Curr. Opin. Immunol.* **14**: 117.

Sloane, J.A., Hollander, W., Moss, M.B., Rosene, D.L. and Abraham, C.R., 1999, Increased microglial activation and protein nitration in white matter of the aging monkey. *Neurobiol. Aging* **20**: 395.

Slotkin, T.A., Cousins, M.M., Tate, C.A. and Seidler, F.J., 2005, Serotonergic cell signaling in an animal model of aging and depression: olfactory bulbectomy elicits different adaptations in brain regions of young adult vs aging rats. *Neuropsychopharmacology* **30**: 52.

Smith, P., Dunne D.W. and Fallon, P.G., 2001, Defective in vivo induction of functional type 2 cytokine responses in aged mice. *Eur. J. Immunol.* **31**: 1495.

Stirling, D.P., Koochesfahani, K.M., Steeves, J.D. and Tetzlaff, W., 2005, Minocycline as a neuroprotective agent. *Neuroscientist* **11**: 308.

Strehlow, I. and Schindler, C., 1998, Amino-terminal signal transducer and activators of transcription (STAT) domains regulate nuclear translocation and Stat deactivation. *J. Biol. Chem.* **273**: 28049.

Streit, W.J., 2000, Microglial response to brain injury: a brief synopsis. *Toxicol. Pathol.* **28**: 28.

Szczepanik, A.M. and Ringheim, G.E., 2003, IL-10 and glucocorticoids inhibit Abeta(1-42)- and lipopolysaccharide-induced pro-inflammatory cytokine and chemokine induction in the central nervous system. *J. Alzheimers Dis.* **5**: 105.

Szczepanik, A.M., Funes, S., Petko, W. and Ringheim, G.E., 2001, IL-4, IL-10 and IL-13 modulate A beta(1-42)-induced cytokine and chemokine production in primary murine microglia and a human monocyte cell line. *J. Neuroimmunol.* **113**: 49.

Tan, J., Town, T., Crawford, F., Mori, T., DelleDonne, A., Crescentini, R., Obregon, D., Flavell, R.A. and Mullan, M.J., 2002, Role of CD40 ligand in amyloidosis in transgenic Alzheimer's mice. *Nat. Neurosci.* **5**: 1288.

Trollor, J.N. and Valenzuela, M.J., 2001, Brain ageing in the new millennium. *Aust. NZ. J. Psychiat.* **35**: 788.

Tuppo, E.E. and Arias, H.R., 2005, The role of inflammation in Alzheimer's disease. *Int. J. Biochem. Cell Biol.* **37**: 289.

Vereker, E., Campbell, V., Roche, E., McEntee, E. and Lynch, M.A., 2000a, Lipopolysaccharide inhibits long-term potentiation in the rat dentate gyrus by activating caspase-1. *J. Biol. Chem.* **275**: 26252.

Vereker, E., O'Donnell, E. and Lynch, M.A., 2000b, The inhibitory effect of interleukin-1β on long-term potentiation is coupled with increased activity of stress-activated protein kinases. *J. Neurosci.* **20**: 6811.

Vilhardt, F., 2005, Microglia: phagocyte and glia cell. *Int. J. Biochem. Cell Biol.* **37**: 17.

Von Bernhardi, R. and Eugenin, J., 2004, Microglial reactivity to β-amyloid is modulated by astrocytes and pro-inflammatory factors. *Brain Res.* **1025**: 186.

Wang, Q., Rowan, M.J. and Anwyl, R., 2004, Beta-amyloid-mediated inhibition of NMDA receptor-dependent long-term potentiation induction involves activation of microglia and stimulation of inducible nitric oxide synthase and superoxide. *J. Neurosci.* **24**: 6049.

Wong, H.L., Costa, G.L., Lotze, M.T. and Wahl, S.M., 1993, Interleukin (IL) 4 differentially regulates monocyte IL-1 family gene expression and synthesis in vitro and in vivo. *J. Exp. Med.* **177**: 775.

Xia, Z., Dickens, M., Raingeaud, J., Davis, R.J. and Greenberg, M.E., 1995, Opposing effects of ERK and JNK-p38 MAP kinases on apoptosis. *Science* **270**: 1326.

Yang, D.D., Kuan, C., Whitmarsh, A.J., Rincon, M., Zheng, T.S., Davis, R.J., Rakic, P. and Flavell, R.A., 1997, Absence of excitotoxicity-induced apoptosis in hippocampus of mice lacking the Jnk 3 gene. *Nature* **389**: 865.

Yehuda, S., Rabinovitz, S., Carasso, R.L. and Mostofsky, D.I., 2002, The role of polyunsaturated fatty acids in restoring the aging neuronal membrane. *Neurobiol. Aging* **23**: 843.

Youssef, S., Stuve, O., Patarroyo, J.C., Ruiz, P.J., Radosevich, J.L., Hur, E.M., Bravo, M., Mitchell, D.J., Sobel, R.A., Steinman, L. and Zamvil, S.S., 2002, The HMG-CoA reductase inhibitor, atorvastatin, promotes a Th2 bias and reverses paralysis in central nervous system autoimmune disease. *Nature* **420**: 78.

Zemke, D. and Majid, A., 2004, The potential of minocycline for neuroprotection in human neurological disease. *Clin. Neuropharmacol.* **27**: 293.

Zhang, Y., Appelkvist, E., Kristensson, K. and Dallner, G., 1996, The lipid compositions of different regions of rat brain during development and aging. *Neurobiol. Aging* **17**: 869.

7

NEUROTROPHIN SIGNALING AND CELL SURVIVAL

Bruno J. Manadas, Carlos V. Melo, João R. Gomes and Carlos B. Duarte

1. ABSTRACT

Neurotrophins control survival, differentiation and maintenance of neurons and glial cells in the central and peripheral nervous system. Their biological functions are mediated by two distinct families of receptors, the Trk receptor tyrosine kinases (TrkA, TrkB and TrkC) and the p75 neurotrophin receptors ($p75^{NTR}$), with different signaling activity. Both receptor types convey trophic signals, but $p75^{NTR}$ activation may also cause apoptotic cell death. The abundance of neurotrophins and/or their receptors changes in various pathological conditions, contributing to cell death or cell survival, depending also on the receptors activated. This chapter focuses on the changes in the abundance of neurotrophins and neurotrophin receptors in diseases of the nervous system and how these changes affect neuronal survival.

2. INTRODUCTION

Neurotrophins are a family of highly conserved proteins that play important roles in the regulation of axonal and dendritic growth and guidance, synaptic structure and connections, short- and long-term changes in synaptic activity, and in neuronal survival and neuroprotection (Huang and Reichardt, 2001; Poo, 2001; Lu et al., 2005). Furthermore, neurotrophins contribute to glial cell development and survival (Althaus and Richter-Landsberg, 2000; Syroid et al., 2000; Chan et al., 2004; Husson et al., 2005; Yamauchi et al., 2005). This group of low-molecular weight proteins includes nerve growth factor (NGF), brain-derived neurotrophic factor (BDNF) and neurotrophins-3 and -4/5 (NT-3 and NT-4/5). Neurotrophin-6 and neurotrophin-7 have only been found in fish and probably do not have mammalian orthologues (Gotz et al., 1994; Nilsson et al., 1998). The cellular effects of neurotrophins are mediated by activation of two classes of receptors: the Trk (tropomyosin-related kinase) family, which includes the TrkA, TrkB and TrkC receptors, endowed with tyrosine kinase activity, and the p75 neurotrophin receptors ($p75^{NTR}$), members of the tumour necrosis factor (TNF) receptor family (Chao, 2003; Huang and Reichardt, 2003; Barker, 2004; Teng and Hempstead, 2004). The diversity of the Trk receptors is further enhanced by the existence of TrkB and TrkC receptors lacking the tyrosine kinase domain or containing inserts in the intracellular domain that affect the signaling properties of the receptors (Huang

Carlos B. Duarte – cbduarte@ci.uc.pt
Center for Neuroscience and Cell Biology and Department of Zoology,
University of Coimbra, 3004-517 Coimbra, Portugal

and Reichardt, 2003; see also Sect. 5.4). NGF binds to TrkA receptors, BDNF and NT-4/5 to TrkB receptors, and NT-3 to TrkC; TrkB receptors may also be activated by NT-3. In contrast with the specificity displayed by the Trk family of receptors, the p75^{NTR} bind both the mature form of the neurotrophins and their uncleaved (precursor) forms (proneurotrophins) (Lee et al., 2001). The Trk receptors and p75^{NTR} may be expressed by the same cell, where they coordinate and modulate the neuronal response to neurotrophins. While the former receptors tend to mediate signals leading to cell survival and growth, p75^{NTR} may provide a trophic effect or cause cell death. In this chapter, we review the distribution of neurotrophins and their receptors in the mammalian nervous system, the traffic and secretion, the signaling pathways activated by neurotrophin receptors and the mechanisms whereby neurotrophins regulate cell survival and cell death.

3. NEUROTROPHIN SYNTHESIS AND RELEASE BY NEURONS AND GLIA

3.1. Synthesis and Trafficking of Neurotrophins

Neurotrophins share a high homology, with ~50% amino acid identity, and their three-dimensional structures contain two pairs of anti-parallel β-strands and cysteine residues in a cysteine knot motif (McDonald and Hendrickson, 1993). Like other secreted proteins, neurotrophins are formed from precursors, proneurotrophins, which are proteolytically cleaved, either intracellularly or extracellularly, giving rise to the mature proteins (Seidah et al., 1996a) (Fig. 1). The mature proteins associate non-covalently as biologically active homodimers. BDNF, NT-3 and NT-4/5 are widely distributed in the central nervous system (CNS), whereas NGF expression is limited to specific areas, such as striatal and basal forebrain cholinergic neurons. BDNF and NT-3 are particularly highly expressed in the cerebral cortex, cerebellum and hippocampus (reviewed in: Chao, 1992; Barbacid, 1994; Lindsay et al., 1994). proBDNF is found in many regions of the rat CNS, including the cerebral cortex, hippocampus, cerebellum, *substantia nigra*, amygdala, hypothalamus and the spinal cord (Zhou et al., 2004). The precursor form of BDNF is localized predominantly in nerve terminals in vivo, and its distribution is similar to the mature molecule (Zhou et al., 2004).

The neurotrophin precursors proNGF, proNT3 and proNT4 are primarily packaged into vesicles of the constitutive secretory pathways, whereas proBDNF is preferentially sorted in the trans-Golgi network into vesicles of the regulated secretory pathway, after binding to the lipid-raft-associated sorting receptor carboxypeptidase E (CPE) via the $Ile_{16}Glu_{18}Ile_{105}Asp_{106}$ motif (Goodman et al., 1996; Mowla et al., 1999; Farhadi et al., 2000; Lou et al., 2005). Sorting of the precursor form of BDNF is also influenced by the prodomain, which binds to intracellular sortilin in the Golgi, facilitating the proper folding of the mature domain and thereby the intracellular sorting of BDNF to the regulated secretory pathway (Chen et al., 2005). Mutating the acidic residues of the sorting motifs leads to BDNF mis-sorting to the constitutive pathway, just as a truncated form of sortilin or sortilin small interfering RNA introduced into primary neurons (Chen et al., 2005; Lou et al., 2005). The regulated influx of Ca^{2+} resulting from neuronal activity controls not only the secretion of BDNF but also the expression and transport of the neurotrophin (Shieh and Ghosh, 1999; Du et al., 2000;

Hartmann et al., 2001; Kohara et al., 2001; Balkowiec and Katz, 2002; Tao et al., 2002; Chen et al., 2003; Martinowich et al., 2003).

Under normal conditions, BDNF-containing vesicles may be transported anterogradely and retrogradely along axons and dendrites. However, the pattern of trafficking differs between axons and dendrites, and may depend on the cell type. In cultured cerebrocortical neurons, the anterograde transport of BDNF in axons may constitute the dominant pathway towards the release sites (Adachi et al., 2005). A single nucleotide substitution in the pro-domain region of BDNF (originating a val66met mutation) results in strong impairments in trafficking and regulated secretion of BDNF (Egan et al., 2003; Chen et al., 2004). The human subjects carrying this polymorphism display abnormal hippocampal function and hippocampal-specific short-term plasticity, as well as an increased susceptibility to psychiatric disorders (Egan et al., 2003). This emphasizes the importance of the regulatory mechanisms of BDNF release in restricting to active synapses the action of the neurotrophin, specifically modulating synaptic activity and neuronal connectivity. Cleavage of proBDNF to the mature form by the extracellular serine protease plasmin, at hippocampal synapses, is also critical for protein synthesis-dependent late-phase long-term potentiation (L-LTP), and the application of mature BDNF is sufficient to rescue L-LTP when protein synthesis is inhibited (Pang et al., 2004). In contrast with BDNF, proNGF is cleaved mainly intracellularly, preferentially by furin, and then released in the mature form (Seidah et al., 1996b; Mowla et al., 1999).

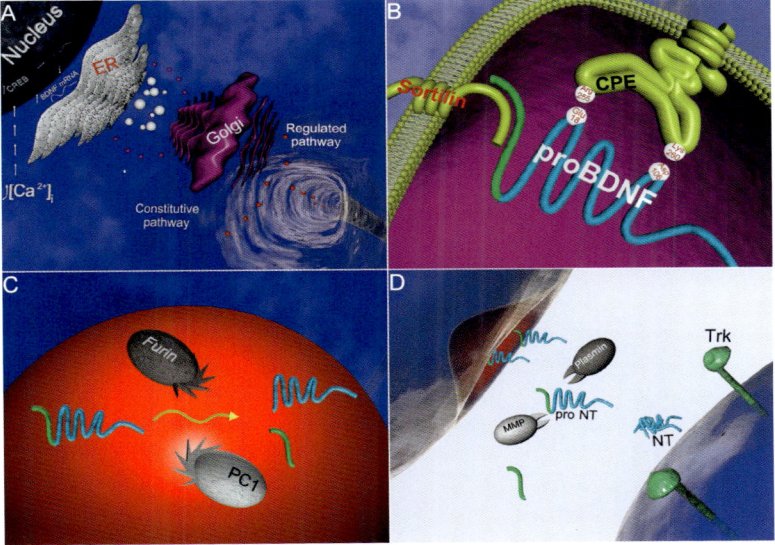

Fig. 1. Synthesis, intracellular trafficking and release of BDNF. BDNF gene expression is regulated by stimuli such as neuronal activity, through activation of CREB and other transcription factors. The neurotrophin is synthesized in the rough endoplasmic reticulum (ER) as proBDNF and is then transported to the Golgi (A), where binding to sortilin contributes to the appropriate folding of the mature region of the protein (B). Interaction of a mature domain of BDNF with carboxypeptidase E (CPE) sorts proBDNF to the regulated secretory pathway. Large dense-core vesicles containing the precursor form of BDNF are transported to the neuritis (A) and their content is released following stimulation of the cells (D). Although proBDNF may be cleaved intracellularly by furin or protein convertase 1 (PC1) (C), a large fraction of the neurotrophin is released in its precursor form and cleaved by extracellular proteases, such as metalloproteinases (MMP) and plasmin (D), giving rise to mature BDNF

In recent years, a significant volume of experimental data has indicated that the release of neurotrophins and other neurotrophic factors in the CNS is not an exclusive feature of neurons, since glial cells are able to produce an array of different trophic factors. In fact, all three major types of glial cells, microglia, astrocytes and oligodendrocytes, have been implicated, in one way or another, in the release of neuron survival factors to the culture media, as indicated in Table 1. The comprehension of trophic factor expression by oligodendrocytes is rather limited so far, especially in comparison with the advances made in regarding their expression by astroglial and microglial cells. Growth factors produced by oligodendrocytes not only influence their own development and survival (Barres et al., 1992) but the neuronal development as well, particularly the maturation of axons (Sanchez et al., 1996) or the assembly of sodium channels (Kaplan et al., 1997).

In most cases, the glial-derived trophic factors are expressed in a region-specific manner (Elkabes et al., 1996) and may be released under control of multiple signaling pathways. For example, protein kinase C (PKC) closely regulates BDNF secretion in rat microglia cultures (Nakajima et al., 2001) and contributes to NGF secretion induced by the non-competitive NMDA receptor antagonist ifenprodil in cultured astrocytes (Toyomoto et al., 2005). The release of NGF by astrocytes is also promoted by extracellular guanosine and GTP (Middlemiss et al., 1995), through inhibition of protein phosphatases 1 and 2A (Pshenichkin and Wise, 1995), or as a result of the elevation of extracellular potassium concentration $[K^+]_e$ during synaptic activity and ischaemia (Abiru et al., 1998). Astrocytes are highly involved in buffering the $[K^+]_e$ under normal physiological stimulation, taking up the excess potassium from the extracellular space. During cerebral ischaemia, the increase in $[K^+]_e$ to 50–90 mM (Hansen, 1977; Hossmann et al., 1977) may lead to neurotrophin release from astrocytes, controlling neuronal survival through activation of TrkA or $p75^{NTR}$ (see below for references).

Table 1. Neurotrophins produced by specific glial cells

Glial cell type	Neurotrophin	References
Microglia	NGF NT-3 NT-4 BDNF	(Elkabes et al., 1996, 1998; Srinivasan et al., 2004) (Elkabes et al., 1996; Srinivasan et al., 2004) (Srinivasan et al., 2004) (Miwa et al., 1997; Elkabes et al., 1998) (Nakajima et al., 2001; Srinivasan et al., 2004; Coull et al., 2005)
Astrocytes	NGF NT-3 BDNF	(Lindsay, 1979; Rudge et al., 1992) (Rudge et al., 1992) (Rudge et al., 1992)
Oligodendrocytes	NGF NT-3 BDNF	(Gonzalez et al., 1990) (Dai et al., 1997) (Acheson et al., 1991; Dai et al., 1997)
Schwann Cells	NT-3 BDNF	(Meier et al., 1999) (Acheson et al., 1991)

3.2. Changes in the Abundance of Neurotrophins in Disorders of the Nervous System

The loss of neurotrophic support from cortical afferents to striatal neurons (unable to produce BDNF), when the transcription of this neurotrophin is decreased (Zuccato et al., 2001) and its vesicular microtubule transport compromised (Zuccato et al., 2001), is responsible for the progressive neurodegeneration observed in Huntington's disease. The activity-dependent secretion of BDNF (Chen et al., 2005; Lou et al., 2005), unique among neurotrophins, is also critical for its specific roles in brain development and cognitive functions (Egan et al., 2003; Chen et al., 2004; Pang et al., 2004), as well as its effect in food intake, energy balance and mood (Xu et al., 2003). The aberrant sorting of BDNF-containing vesicles into the constitutive, instead of the regulated pathway of secretion, leads to a continuous release of the neurotrophin, which may also be associated with psychiatric conditions related with repetitive behaviours, such as bipolar disease (Neves-Pereira et al., 2002; Sklar et al., 2002; Geller et al., 2004), obsessive–compulsion disorder (Hall et al., 2003) or anorexia nervosa (Ribases et al., 2003; 2004). An upregulation of the proNGF is observed in pathological conditions such as mild cognitive impairment (Peng et al., 2004), Alzheimer's disease (Fahnestock et al., 2001; Peng et al., 2004; Pedraza et al., 2005), brain injury (Beattie et al., 2002; Harrington et al., 2004) and retinal dystrophy (Srinivasan et al., 2004). In contrast, proBDNF and mature BDNF are decreased in Alzheimer's disease (Michalski and Fahnestock, 2003; Peng et al., 2005). The changes in abundance of proneurotrophins in the diseased brain may be due, at least in part, to changes in the activity of convertases since the expression of furin and protein convertase 1 in the hippocampus was shown to increase in the hippocampus of mice subjected to kainate-induced seizures (Meyer et al., 1996).

Amyotrophic lateral sclerosis (ALS) is characterized by the progressive degeneration of motor neurons, resulting in part from mutations to the copper- and zinc-containing superoxide dismutase. Neuronal oxidative stress and SOD-1 (superoxide dismutase-1) dysfunction might be associated with FGF-1 (fibroblast growth factor-1) secretion from motor neurons (Cassina et al., 2005). FGF-1 causes an accumulation of the respective receptors (FGFR1) in the nuclei of astrocytes, increases the expression of the inducible form of nitric oxide synthase and promotes the expression and secretion of NGF. The activation of $p75^{NTR}$ by NGF leads to apoptotic death of motor neurons, and this is sensitized by the increased levels of nitric oxide (NO) (Turner et al., 2003; Cassina et al., 2005; Vargas et al., 2006). Still, toxicity to motor neurons is limited to some extent due to activation by FGF-1 of the redox-sensitive transcription nuclear factor erythroid 2-related factor 2 (Nrf2) in astrocytes. This transcription factor was shown to reduce NO-dependent toxicity to motor neurons by increasing GSH biosynthesis (Vargas et al., 2006). Microglial activation may also contribute to motor neuron degeneration in ALS through upregulation of the TNFR1 (tumour necrosis factor receptor 1), amplifying the effect of microglial-secreted TNF-α and thus microglia-induced motor neuron death (Wen et al., 2006).

In general, microglial cells are able to secrete soluble trophic factors upon activation due to inflammation or injury (Nakajima et al., 1998). Activation of microglia precedes or accompanies neuronal and glial demise in several neurodegenerative diseases, including Alzheimer's disease, Parkinson's disease and multiple sclerosis (reviewed in Kim and de Vellis, 2005). The available experimental findings concerning glial neurotrophin secretion raise a fundamental question: what is the purpose of having glial cells, such as activated microglia, producing cytotoxic proinflammatory cytokines, which are involved in progression of injury and disease, and at the same time secreting neurotrophic factors that support the survival of neurons? The answer may be found in studies developed so as to explain the same paradox found in neurons. Most likely, similar to what has been found in neurons, neurotrophins may be capable of bringing about diverse mechanisms of signaling and receptor endocytosis in glial cells, besides inducing a differential release of neurotrophins in their precursor or mature forms. However, so far, there are hardly any equivalent studies performed in glial cells. To mention a few exceptions, in rat astrocytes and Schwann cells, a truncated form of the TrkB receptor mediates the endocytosis and release of BDNF, apparently to facilitate the regulation of this neurotrophic factor for longer periods of time (Alderson et al., 2000). In contrast, the neurotrophin NGF is also secreted by microglia in the precursor form (proNGF) to promote photoreceptor cell death mediated by p75^{NTR} in a rat model of retinal dystrophy (Srinivasan et al., 2004).

The release of neurotrophins by non-neuronal cells may also be relevant in multiple sclerosis, an autoimmune disease where the loss of oligodendrocytes is cumulative with loss of nearby axons (Ferguson et al., 1997; Trapp et al., 1998; Chang et al., 2000). Under normal conditions, neurons, themselves, may signal to oligodendrocytes in their vicinity through the release of neurotransmitters or even neurotrophins since these glial cells express TrkA, TrkB, TrkC and p75^{NTR} (Cohen et al., 1996; Ladiwala et al., 1998; Du et al., 2006). The observed conspicuous BDNF immunoreactivity in infiltrating immune cells in multiple sclerosis lesions, especially in T cells and macrophages, in addition to reactive astrocytes and neurons, might potentiate endogenous neurotrophic support and provide neuronal and glial protection in the affected regions (Kerschensteiner et al., 2003). Nevertheless, it is the lack of trophic factors (Fawcett and Asher, 1999; Logan and Berry, 2002) allied with the inhibitory myelin proteins secreted by oligodendrocytes (Filbin, 2003; Woolf, 2003) that mainly contribute to the failure of the adult CNS to sustain axonal regeneration after an injury or an insult. On the contrary, the peripheral nervous system (PNS) displays regeneration and functional recovery after a nerve injury because Schwann cells are able to secrete survival-inducing neurotrophins (see Table 1) after the lesion (Fawcett and Keynes, 1990). In fact, in transected nerves, Schwann cells can survive even in the absence of axons, by establishing an autocrine circuit of trophic factor production (Meier et al., 1999). This endurance of Schwann cells under degenerative conditions allows them to regulate the survival and differentiation of neurons. For example, behavioural and morphological assays have demonstrated that the grafts of NT-3 or BDNF-transduced Schwann cells are able to promote a fast and strong functional recovery from demyelination of mouse spinal cord neurons by enhancement of oligo-dendrocytic proliferation and neuroprotection against reduced astrogliosis (Girard et al., 2005).

The communication between glial cells and neurons is also crucial for neuropathic pain transmission, which occurs after spinal cord nerve injury with concomitant increase of the synthesis and activity of the P2X$_4$ ATP receptors in the microglia (Tsuda et al., 2003). The damage to spinal lamina I neurons generates chronic pain in response to stimuli that otherwise would not be painful (allodynia) (Torsney and MacDermott, 2005). This arises from the shift in the anion reversal potential (E_{anion}), from negative to positive, in respect to the resting membrane potential (V_{rest}). Consequently, instead of the normal influx of chloride anions elicited by GABA$_A$ receptor activation, there is an efflux of Cl$^-$, which depolarizes the neurons, i.e., the GABA-activated currents switch from hyperpolarizing (inhibitory) to depolarizing (excitatory). During this process, the glia–neuron interaction is mediated by secretion of BDNF from ATP-stimulated microglia (Coull et al., 2005). Blocking BDNF activity with a BDNF-sequestring fusion protein (TrkB-Fc), or using siRNA directed against BDNF, reversed both allodynia and the depolarizing shift in the E_{anion} (Coull et al., 2005).

Moreover, the authors clearly demonstrated that this neurotrophin, which had previously been reported to cause shifts on neuronal anion gradients (Rivera et al., 2002) and to be involved in chronic pain (Thompson et al., 1999), is not released from primary afferent neurons (Malcangio and Lessmann, 2003) but instead from glial cells. For instance, intrathecal injection of BDNF-deficient ATP-activated microglia, through a lumbar spinal catheter, is unable to cause allodynia or E_{anion} shift in adult rats (Coull et al., 2005).

During early post-natal development, glial cells may also secrete soluble factors that promote the formation and maturation of synapses. Previous studies indicated that a large increase of the astrocytic population supports synaptogenesis in the developing brain (reviewed in Ullian et al., 2004). The generation of functional glutamatergic synapses, in cultures of retinal ganglion cells, spinal motor neurons and hippocampal neurons (Pfrieger and Barres, 1997; Ullian et al., 2001; Zhang et al., 2003a; Hama et al., 2004; Christopherson et al., 2005), as well as of hippocampal GABAergic synapses (Elmariah et al., 2005), can be upregulated by astrocytes. This effect is not dependent on direct contact of astrocytes with the cultured neurons (Liu et al., 1996, 1997; Pfrieger and Barres, 1997; Ullian et al., 2001), which has been demonstrated by the addition of astrocyte-conditioned medium (ACM) to purified neuronal cultures that mimics the effects observed in the co-cultures with astrocyte feeding layers. In developing hippocampal GABAergic synapses, astrocytes enhance synaptogenesis by promoting the formation and post-synaptic localization of $GABA_A$ receptor clusters, through a mechanism involving BDNF and TrkB-mediated signaling in neurons (Elmariah et al., 2005).

In conclusion, the available experimental evidences indicate that neurons and glia interact through neurotrophin secretion and activation of different neurotrophin receptors, in order to regulate CNS homeostasis. These findings also underline the importance of understanding glia–neuron signaling via neurotrophin transport and secretion, for the development of new therapeutic strategies, particularly in regard to the treatment of pain hypersensitivity and several neurodegenerative conditions.

4. NEUROTROPHIN RECEPTORS IN NEURONS AND GLIA

The localization of neurotrophin receptors was first studied mainly at the mRNA level, by in situ hybridization and later by immunohistochemistry. TrkB and TrkC receptors are widely but differentially distributed in the CNS (Merlio et al., 1992; Kokaia et al., 1993; Zhou et al., 1993; Altar et al., 1994; Gibbs and Pfaff, 1994; Muragaki et al., 1995; Shelton et al., 1995; Yan et al., 1997; Aoki et al., 2000), whereas TrkA is mainly restricted to basal forebrain cholinergic neurons and some striatum neurons (Holtzman et al., 1992; 1995; Muragaki et al., 1995; Shelton et al., 1995; Lee et al., 1998b). The TrkB and TrkC receptors are particularly enriched in the hippocampus, cerebellum and in the brain cortex, and their subcellular localization in the adult brain has been investigated in more detail due to their putative role in synaptic plasticity (see below). The Trk family of neurotrophin receptors also includes the truncated isoforms, T1 and T2, resulting from the TrkB and TrkC genes. These receptors possess the same extracelular domain, transmembrane domain and the first 12 intracellular amino acid sequences as the Trk full length receptors; their specificity resides in the C-terminal sequences. The truncated Trk receptors are found mainly in glial cells, being widely expressed during development and in the adult brain (Barbacid, 1994).

The expression of the $p75^{NTR}$ is developmentally regulated (Chao and Hempstead, 1995), similarly to the Trk receptors (Muragaki et al., 1995; Fryer et al., 1996), and in the adult brain is found mainly within the basal forebrain, selected brainstem nuclei and cerebellar Purkinje neurons (Koh et al., 1989; Mufson et al., 1989, 1991, 1992; Lee et al., 1998b). In the PNS, the Trk receptors are differentially expressed in functionally distinct neurons, playing important roles in neuronal development and survival (e.g. Crowley et al., 1994; Klein et al., 1994; Smeyne et al., 1994; Fagan et al., 1996; Weisenhorn et al., 1999; Moqrich et al., 2004). The activation of presynaptic TrkA receptors by NGF contributes to the survival of sympathetic and sensory neurons, but the mechanisms involved remain to be fully elucidated (Zweifel et al., 2005).

The subcellular localization of TrkB receptors was investigated to a large extent in the cerebral cortex and hippocampus, due to the interest in unravelling the role of these receptors in synaptic plasticity. Immunoelectron microscopy studies showed that TrkB receptors are present in the glutamatergic pyramidal and granule cells of the hippocampus, mainly in axons, axon terminals and dendritic spines, and to a lower extent in somata and dendritic shafts. Although a large population of the receptors is intracellular, a significant labelling of the plasma membrane of dendritic spines was observed. TrkB receptors may also be present in some hippocampal GABAergic and neuromodulatory afferents (Drake et al., 1999; Swanwick et al., 2004). Depending on the hippocampal regions, the distribution of TrkB receptors is mainly pre- or post-synaptic, or the receptors may be equally distributed in the synapse (Drake et al., 1999). Immunocytochemistry experiments and fractionation studies using isolated rat hippocampal synaptosomes showed that about 30% of the glutamatergic nerve terminals contain TrkB receptors, but only one-fourth of the total full-length receptors present in the overall population of terminals are located at the plasma membrane. Furthermore, about half of the TrkB-containing hippocampal nerve terminals are non-glutamatergic (Pereira et al., 2006). In the rat hippocampal synapses, about 35% of the total full-length receptors are evenly distributed between the presynaptic active zone and the post-synaptic density, in contrast with the truncated receptors that are absent from the post-synaptic density and only 10% are present at the presynaptic active zone (Xu et al., 2000; Pereira et al., 2006). In the cerebral cortex of the adult rat, the TrkB receptors are mainly post-synaptic (Aoki et al., 2000). The subcellular distribution of the TrkB receptors and, therefore, the cellular response to BDNF may change depending on the recent history of the cell, since plasma membrane depolarization and cAMP elevation rapidly increase the amount of receptors associated with the plasma membrane in CNS neurons (Meyer-Franke et al., 1998). In addition to neurons, astrocytes, oligodendrocytes, microglia and Schwann cells also express receptors for neurotrophins and respond to changes in their concentration in the extracellular space (reviewed in: Althaus and Richter-Landsberg, 2000; see also: Elkabes et al., 1996; Hempstead and Salzer, 2002; Chan et al., 2004).

The abundance of neurotrophin receptors changes under numerous pathological conditions. Transient brain ischaemia upregulates TrkA in selected astrocyte populations and may have a similar effect in neurons, in different brain regions (Lee et al., 1998a; Soltys et al., 2003; Hwang et al., 2005; for conflicting results in neurons see: Lee et al., 1995; Oderfeld-Nowak et al., 2003). The p75^{NTR} is also induced in the hippocampus following transient forebrain or global cerebral ischaemia in rat (Lee et al., 1995; Oderfeld-Nowak et al., 2003). In reactive gliosis induced by an excitotoxic insult to the thalamus, there is an increase in the expression of p75^{NTR} and TrkA in non-astrocytic elements (microglia, Schwann cells and/or oligodendrocytes) and in astrocytes, respectively (Junier et al., 1994). In the human brain, a strong increase in TrkA immunoreactivity was also observed in reactive astrocytes in a number of unrelated pathologies, including Alzheimer's disease, Huntington's disease, progressive supranuclear palsy, multiple sclerosis, Creutzfeldt-Jacob disease, multifocal leukoencephalopathy and residual hypoxic encephalopathy (Connor et al., 1996; Aguado et al., 1998; however for conflicting results see: Allen et al., 1999; Ginsberg et al., 2006). Similarly, an upregulation in TrkA protein levels in astrocytes is found in mesial temporal lobe epilepsy (Ozbas-Gerceker et al., 2004) and was observed in the astroglia of grey and white matter of the spinal cord in experimental autoimmune encephalomyelitis (Oderfeld-Nowak et al., 2001). Changes in TrkA protein levels were also reported in focal cortical dysplasia and ganglioglioma (Aronica et al., 2004). Since astrocytes may also produce NGF (see above), the neurotrophin may function as an autocrine or paracrine factor in TrkA-expressing reactive and neoplastic glial cells. The upregulation of TrkA in astrocytes in response to injury and neurological disorders may limit cell damage.

The TrkB and TrkC protein levels and the truncated form of these receptors are also changed in several neurological disorders. Ischaemia after middle cerebral artery occlusion in the rat decreased TrkB in the infarcted core, increased the truncated TrkB protein levels in astrocytes surrounding the area of the infarction, and upregulated the full-length TrkB in distant neurons (Ferrer et al., 2001), and similar findings were reported in young rat brains subjected to hypoxic/ischaemic injury (Narumiya et al., 1998). Chronic injury to the CNS results in the expression of both truncated and full-length TrkB receptor transcripts by reactive astrocytes, suggesting that neurotrophins may regulate astrocytic response under these conditions (McKeon et al., 1997). Spinal cord injury also leads to a chronic upregulation of truncated TrkB receptors in ependymal cells and astrocytes surrounding the lesion cavity, but the full length TrkA-C receptors are acutely downregulated in the damaged area (Liebl et al., 2001). This acute reduction in Trk protein levels may contribute to the extensive cell loss associated with contusion injury of the spinal cord. The full-length TrkB receptor is also found in CNS neurons, in the immediate vicinity of multiple sclerosis plaques and in reactive astrocytes within the lesion, and activation of these receptors by BDNF released by immune cells (T cells and macrophages/microglia) and reactive astrocytes may play a neurotrophic role under these conditions (Kerschensteiner et al., 1999; Stadelmann et al., 2002). Furthermore, due to the effect of neurotrophins on the maturation of oligodendrocytes (Heinrich et al., 1999), they may actively promote remyelination (Jean et al., 2003; Bruck, 2005).

5. NEUROTROPHIN SIGNALING

Neurotrophin receptors belonging to the Trk family, including the TrkA, TrkB and TrkC receptors, are activated by the mature form of neurotrophins, which exist in solution as non-covalent dimers (Narhi et al., 1993). Stimulation of metabotropic receptors for PACAP (pituitary adenylate cyclase-activating polypeptide) and adenosine, coupled to adenylate cyclase, also activates Trk receptors. This mechanism is particularly relevant for the activation of an intracellular pool of Trk receptors and does not require the presence of neurotrophins (Lee and Chao, 2001; Rajagopal et al., 2004).

Binding of neurotrophins to the Trk receptors promotes their dimerization (Jing et al., 1992) and, in this form, the receptors phosphorylate each other on specific tyrosine residues, in the cytoplasmic domain. This, in turn, creates docking sites for different adaptor proteins and enzymes, leading to the activation of various parallel signal transduction cascades, with distinct functions (Atwal et al., 2000). Given the high homology in the intracellular region of Trk receptors, it is not surprising that their phosphorylation sites and the nature of the pathways activated are highly conserved (Atwal et al., 2000). Three main cascades are activated by the neurotrophin Trk receptors: the Ras/ERK pathway, the PI3K/Akt pathway and the phospholipase C-γ (PLC-γ) pathway (reviewed by Chao, 2003; Huang and Reichardt, 2003) (Fig. 2). Interestingly, the former pathway appears to be preferentially activated by the TrkA receptors, whereas TrkC receptors are preferentially coupled to the activation of Akt (Markus et al., 2002).

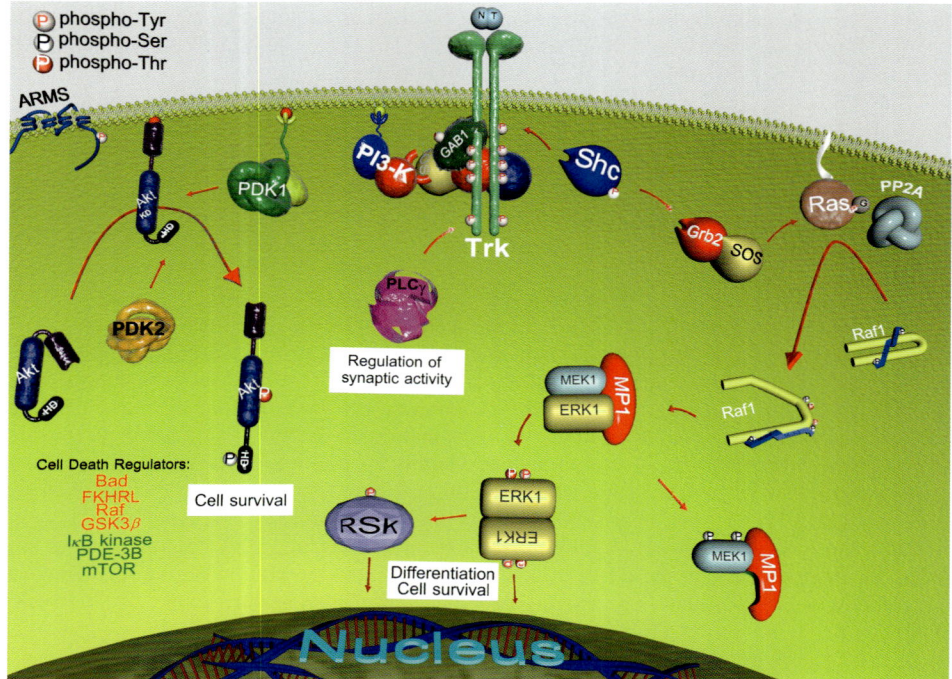

Fig. 2. Intracellular signaling mechanisms activated by Trk neurotrophin receptors. Neurotrophin binding to Trk receptors induces transphosphorylation of intracellular tyrosine residues, which constitute binding sites for adaptor proteins, such as Shc and Frs-2 (not shown), and enzymes (PLC-γ). Phosphorylation of Shc leads to the activation of PI3K and ERK (see text for further details), whereas PLC-γ is activated directly by tyrosine phosphorylation. The latter signaling pathway is involved in the regulation of synaptic transmission. The PI3K pathway plays a major role in neurotrophin-induced cell survival and these effects are mediated through the regulation of the activity of various enzymes by Akt phosphorylation [phosphorylation may increase (green) or decrease (red) enzyme activity]. Activity of the Ras/ERK pathway accounts for the effects of neurotrophins on cell differentiation and also contributes to cell survival under conditions of neuronal injury or toxicity (see Sect. 6)

5.1. PI3Kinase Pathway

Production of P3-phosphorylated phosphoinositides is critical in mediating survival and neuroprotection of many populations of neurons (reviewed by Brunet et al., 2001; see Sect. 6.1), and in the regulation of vesicular trafficking (Chen and Wang, 2001). Binding of protein pleckstrin homology (PH) domains to these inositol phospholipids, generated by PI3 kinase, results in translocation of PKB/Akt to the membrane, where it is activated through phosphorylation by upstream kinases, including PDK1 (phosphoinositide-dependent protein kinase 1) and PDK2 (possibly the rictor-mTOR complex; Sarbassov et al., 2005). PKB/Akt phosphorylates several proteins that are important in the promotion of cell survival (Brunet et al., 2001), most of them are regulators of apoptosis (Yuan and Yankner, 2000). Bad, for example, is a Bcl-2 family member that promotes apoptosis through sequestration of Bcl-xl, which otherwise would inhibit Bax, a proapoptotic protein (Datta et al., 1997). Phosphorylation of Bad by PKB/Akt sequesters the protein in the cytosol, bound to the protein 14-3-3, thereby avoiding its interaction with Bcl-xl. Many other proteins in the apoptotic pathway have consensus sequences for Akt phosphorylation but, up to now, no data have proved that they are substrates of Akt. Glycogen synthase kinase 3-β (GSK3β) is also phosphorylated by Akt, preventing its pro apoptotic actions (Hetman et al., 2000). Akt also interferes with the nuclear factor-κB (NFκB) pathway by phosphorylating and thus promoting the degradation of IκB, the inhibitory partner of this pathway. This results in the liberation of active NFκB that promotes gene transcription, associated with neuronal survival (Foehr et al., 2000; Wooten et al., 2001). The forkhead transcription factor (FKHRL1) is also the target of Akt and upon phosphorylation interacts with 14-3-3, becoming sequestered in the cytoplasm. The phosphorylated FKHRL1 is therefore kept away from the nucleus, preventing its pro-apoptotic activity (Brunet et al., 2002).

5.2. Ras/ERK Pathway

Several pathways lead from Trk receptors to activation of Ras and most of them appear to involve phosphorylation on Y490 (in TrkA). Upon initial phosphorylation of the receptor, at least two possible adaptor molecules compete for the direct binding to the phosphorylated tyrosine residue: Shc and Frs-2 (fibroblast receptor substrate-2) (Meakin et al., 1999). The Shc pathway contributes to a transient activation of ERK signaling (Grewal et al., 1999), and this may involve the participation of Shc isoforms, A, B and C, with different expressions and functions (Segal, 2003). Neurotrophins can induce each one of these isoforms, but in mature neurons ShcC is preferred (Conti et al., 2001). Shc binds through its PTB (phosphotyrosine binding) domain to the tyrosine phosphorylated receptor and is itself phosphorylated on tyrosine. Phosphorylated Shc recruits the adaptor protein Grb2, complexed with SOS (son of Sevenless), a guanine nucleotide-exchange factor (GEF) for Ras (Nimnual et al., 1998). Once Ras becomes activated, it stimulates intracellular signaling through several downstream kinases, including the PI3K and c-Raf (Xing et al., 1998). c-Raf phosphorylates MEK1/2 (MAPK and ERK kinase) which, in turn, phosphorylates ERK1 and ERK2. The Trk receptor-induced activation of Ras, through Shc and Grb2/SOS, promotes a transient activation of ERK signaling (Grewal et al., 1999). A Trk-induced sustained activation of ERK1 is carried out through recruitment of Frs-2 in detriment of Shc, creating an alternative pathway. Active Frs-2 binds several adaptor proteins, including Grb2 and Crk, the protein phosphatase Shp2 and the Src kinase (Meakin et al., 1999; Huang and Reichardt, 2003). Crk binds and activates the exchange factor C3G, which in turn stimulates a small G-protein called Rap-1, followed by activation of the downstream kinase B-Raf. This kinase initiates the ERK1/2 signaling cascade by phosphorylating MEK1/2.

ARMS/Kidins220 (ankyrin-rich membrane spanning protein) is a membrane associated protein that is also rapidly phosphorylated on tyrosine following activation of the Trk receptors (Kong et al., 2001). Furthermore, ARMS/Kidins220 is also phosphorylated on serine residues by protein kinase D. Phosphorylation of ARMS on Tyr1096 plays a critical role in neurotrophin signaling, by producing a switch in the motif of recognition of the adaptor molecule CrkL. Tyrosine phosphorylated ARMS interacts with the SH2 domain of CrkL, freeing its SH3 domain to recruit C3G, which triggers a sustained activation of the Ras/ERK pathway. Mutation of this amino acid impairs prolonged ERK activation and differentiation of PC12 cells in response to NGF treatment (Arevalo et al., 2006).

In the Ras/ERK pathway, a small G-protein (Ras) initiates a signaling cascade where different kinases phophorylate each other until they reach and activate ERKs, a group of kinases belonging to the MAPK (mitogen-activated protein kinases) family (Huang and Reichardt, 2003). Of the various ERKs activated, four are known to respond to neurotrophin/Trk signaling: ERK1, 2, 4 and 5 (Segal, 2003). These kinases regulate gene expression directly, by phosphorylating transcription factors, or indirectly, through the modulation of other kinases. ERK1, 2 and 5 activate the RSK (ribosomal S6 kinase) family of protein kinases, which in turn phosphorylate the transcription factor CREB. However, phosphorylation of CREB on Ser133, which is critical for the expression of CREB-regulated genes, is mainly mediated by the ERK substrate MSK following Trk receptor activation (Arthur et al., 2004). Once in the nucleus, ERKs may also increase the activity of several different transcription factors, including CREB [through regulation of the CREB-binding protein, CBP]; e.g. (Liu et al., 1998], MEF2, Egr-1 and Elk-1, among others (Hazzalin and Mahadevan, 2002). These effects of ERK are isoform specific since, for example, ERK5 activates MEF2 directly, whereas ERK1/2 activates Elk1 (Grewal et al., 1999).

5.3. PLC-γ Pathway

Phosphorylation of TrkA on Y785 [and corresponding sites on TrkB (Y816) and TrkC] recruits the cytoplasmatic enzyme PLC-γ that binds directly to the Trk receptor through an SH2 domain. PLC-γ (mainly the isoform γ1; Middlemas et al., 1994) is activated by tyrosine phosphorylation (e.g. Pereira et al., 2006) and hydrolyses phosphatidylinositol 4,5-bisphosphate (PtdIns(4,5)P$_2$) to generate inositol 1,4,5-trisphosphate (Ins(1,4,5)P$_3$) and DAG (diacylglycerol). Ins(1,4,5)P$_3$ promotes the release of Ca^{2+} from internal stores, thus increasing the cytosolic Ca^{2+} concentration and activating enzymes regulated by the intracellular Ca^{2+} concentration, such as Ca^{2+}- and calmodulin-dependent protein kinases and the Ca^{2+}-regulated isoforms of PKC (Ouyang et al., 1997). At the same time, DAG stimulates DAG-regulated PKC isoforms (e.g. PKCδ; Huang and Reichardt, 2003). The physiological roles of TrkB-mediated PLC-γ signaling have been investigated in vivo, by mutating the recruitment site, Y816, and revealed a key role in the mechanisms of synaptic plasticity (Minichiello et al., 2002).

5.4. Intracellular Signaling Activity of Truncated Trk Receptors

The truncated Trk receptors, found in developing and mature neurons and astrocytes, can inhibit signaling by full-length receptors (Eide et al., 1996), an effect that may be mediated, at least in part, by a reduction in the cell surface expression of the full-length receptor (Haapasalo et al., 2002). Although the truncated Trk receptors lack tyrosine kinase activity, they are still capable of inducing multiple signaling responses. Activation of the truncated TrkB receptor induces the release of Ca^{2+} from Ins(1,4,5)P$_3$-sensitive intracellular stores in glial cells (Rose et al., 2003), which can be interpreted as messages sent from neurons to glia, regulating the release of glutamate from these cells. This regulatory process may be propagated through many astroglia as calcium waves, which are able to pass through gap junctions. Truncated TrkB receptors also regulate glial cell morphology via Rho GDP dissociation inhibitor (Ohira et al., 2005), and a similar role may be played by truncated TrkC receptors through a mechanism involving the activation of ADP-ribosylation factor 6 (Arf6) and the Rac1 GTPase (Esteban et al., 2006).

5.5. $p75^{NTR}$ Signaling Pathways

The $p75^{NTR}$ is a member of the TNF receptor superfamily (Bibel and Barde, 2000) and binds each of the neurotrophins with similar affinity. The unprocessed neurotrophins (proNGF and proBDNF) also bind to the $p75^{NTR}$, with higher affinity than the processed NGF and BDNF (Lee et al., 2001; Teng et al., 2005). In contrast with the Trk receptors, $p75^{NTR}$ plays a dual role in cell survival, since they can be protective under certain conditions and favour apoptosis in distinct settings. These actions originate from signaling pathways quite distinct from those activated by the Trk receptors [reviewed in Schor, 2005]. The pro-survival role of $p75^{NTR}$ is thought to rely on Akt and the NFκB pathway. In the latter case, receptor activation promotes the association of the adaptor protein TRAF6 (Khursigara et al., 1999) with the intracellular portion of $p75^{NTR}$, activating several intermediate proteins that ultimately cause phosphorylation and degradation of the NFκB inhibitory protein IκB (Wooten et al., 2001). This allows the activation of the NFκB-dependent transcription activity, which promotes cell survival synergistically with the Trk signaling (Hamanoue et al., 1999).

The proapoptotic pathways of $p75^{NTR}$ are much more numerous and include the Jun N-terminal kinase (JNK) signaling cascade, increase in sphingolipid turnover and interaction of the receptor with several protein adaptors that directly promote cell cycle arrest and apoptosis (reviewed in: Nykjaer et al., 2005; Schor, 2005). G proteins like Rac (a known activator of JNK) and RhoA are also involved (Harrington et al., 2004). The cytosolic proteins that interact with $p75^{NTR}$ present in a certain cell determine, to some extent, the type of response observed. Also, the presence of Trk receptors may suppress the proapoptotic effects of $p75^{NTR}$. Thus, the effect of $p75^{NTR}$ on sphingolipid turnover is antagonized by active Trk receptors, which prevent the interference of the former receptors in cell survival and differentiation induced by neurotrophins (Dobrowsky et al., 1995). The ceramide produced by the activated sphingomyelinase has been shown to promote apoptosis and mitogenic responses, depending on the experimental settings, through the control of various signaling pathways, including ERK, PI3K and atypical PKC isoforms (Muller et al., 1998; Zhou et al., 1998). Activation of Trk receptors suppresses the activation of acidic sphingomyelinase by $p75^{NTR}$ through association of PI3K with acidic sphingomyelinase in caveoli-related domains (Dobrowsky et al., 1995; Bilderback et al., 2001).

6. NEUROTROPHINS IN CELL SURVIVAL AND CELL DEATH

6.1. Trk-Mediated Cell Survival

Neurotrophins were first identified as promoters of neuronal survival, particularly in peripheral sympathetic ganglia (Levi-Montalcini, 1987), and subsequent studies in mice expressing reduced levels of neurotrophins extended these findings to other neuronal populations of the PNS. $NGF^{+/-}$ mice also displayed decreased cholinergic innervation of the hippocampus (reviewed in Chao, 2003). Neuronal losses were also observed in the PNS of Trk-deficient mice and the reduction in the expression of TrkB/TrkC increase apoptosis in hippocampal and cerebellar granule neurons (reviewed in Huang and Reichardt, 2001). The use of cultured neurons has also provided multiple examples of protection by neurotrophins from trophic support deprivation (e.g. Koh et al., 1995; Kubo et al., 1995; Bonni et al., 1999; Hetman et al., 1999; Yamada et al., 2001; Lee et al., 2002a; Poser et al., 2003; Liot et al., 2004; see Tables 1 and 3 for additional references). Neuroprotective neurotrophins may be produced by glial cells, as demonstrated for dorsal root ganglion neurons subjected to axotomy or NGF deprivation. In this case, the injured neurons produce NO that stimulate the production of neuroprotective NGF and NT-3 by satellite glia cells, through a cGMP-dependent mechanism (Thippeswamy et al., 2005).

Administration of mature neurotrophins to specific brain regions and to cultured neurons has also been shown to rescue cell loss associated with aging or induced by chemical or mechanical insults. Because of the very short half-life of neurotrophins in the plasma (Hetman et al., 1999) and their poor blood–brain barrier permeability, modified neurotrophins have been used in some in vivo studies, in order to overcome those limitations (Gozes, 2001; Wu, 2005). A BDNF chimeric peptide is neuroprotective in rats subjected to transient forebrain ischaemia (Wu and Pardridge, 1999), permanent focal ischaemia (Zhang and Pardridge, 2001a, b) or transient focal ischaemia (Zhang and Pardridge, 2001b). In these studies, protection by BDNF was effective in the reduction of the cortical infarct, as observed in studies where the neurotrophin was administered intravenously (Schabitz et al., 2000). Application of BDNF, NT-3 and NT-4 in the brain all decreased infarct volume in models of cerebral ischaemia (Beck et al., 1994; Chan et al., 1996; Yamamoto et al., 1997) and, interestingly, mice lacking NT-4 or deficient in BDNF develop significantly bigger infarcts than their wild-type littermates (Larsson et al., 1999; Endres et al., 2000), suggesting that endogenous neurotrophins also play a neuroprotective role under these conditions. The expression of BDNF increases in the ischaemic penumbra (Kinoshita et al., 2001; Lee et al., 2002b; Miyake et al., 2002; Rickhag et al., 2006), due to glutamate receptor activation (Zafra et al., 1990), constituting a feed-forward mechanism that allows sustaining the neuroprotective BDNF signaling activity.

Neuronal injury in brain ischaemia is primarily due to metabolic deprivation and excessive release of glutamate, leading to overactivation of glutamate receptors (excitotoxicity) (Choi, 1988). This can be mimicked, to some extent, by stimulating cultured neurons with toxic concentrations of glutamate or glutamate receptor agonists, which cause cell death by an apoptotic-like mechanism or by necrosis, depending on the magnitude of the insult (Ankarcrona et al., 1995). Neurotrophins, particularly BDNF and NT-3, have been shown to protect cultured hippocampal and cerebrocortical neurons from cell death evoked by glutamate (Mattson et al., 1995; Wu et al., 2004; Almeida et al., 2005; Jiang et al., 2005) or metabolic insults (Cheng and Mattson, 1991; 1994; Kim et al., 2004), in agreement with the in vivo findings. BDNF also reduced cortical and white matter lesions induced by ibotenate, an NMDA and metabotropic glutamate receptor agonist, injected at the neopallium at postnatal day 5 (Husson et al., 2005). The toxin produces neuronal cell death and white matter cysts resembling those seen in periventricular leukomalacia, a lesion found in numerous human premature newborns.

Ischaemic retinal injury is implicated in several pathological conditions, including retinal artery occlusion, glaucoma and diabetic retinopathy (Lipton, 2001). Similarly to the brain, retinal ischaemia leads to neuronal damage, partly due to activation of glutamate receptors, being the inner part of the retina particularly vulnerable (Osborne and Herrera, 1994; Duarte et al., 1998). NT-4/5 affords neural protection against ischaemic retinal injury in vivo (Harada et al., 2003) and BDNF is also neuroprotective against excitotoxic damage to the retina (Kido et al., 2000; Kano et al., 2002; Schuettauf et al., 2004). Glutamate upregulates the release of BDNF, NT-3, NT-4/5 and glial-derived neurotrophic factor (GDNF) by Müller glial cells (Taylor et al., 2003) and this is likely to upregulate the endogenous neuroprotective defences in the retina. Although BDNF has been consistently reported as neuroprotective to photoreceptors, in models of light-induced degeneration (Harada et al., 2000; 2002) and primary rod photoreceptor degeneration (Okoye et al., 2003), these cells do not express their receptors. This leads to the suggestion that microglia-derived NGF, BDNF and CNTF influence photoreceptor survival by modulating the production and release of basic fibroblast growth factor (bFGF) and GDNF by Müller glia (Harada et al., 2002). The trophic effects of BDNF in retinal ganglion cells (RGC) have also been extensively studied in relation to glaucoma. BDNF produced at the superior colliculus (in rodents) or at the lateral geniculate body (in primates) binds to TrkB receptors on the RGC axons and is transported to the cell body by retrograde transport. The retrograde transport of BDNF and TrkB receptors is impaired upon increase of the intraocular pressure, suggesting that the optic nerve damage observed in patients with glaucoma may be due to the impairment of BDNF transport (Pease et al., 2000; Quigley et al., 2000). Accordingly, BDNF has been shown to promote RGC survival in a rat model of glaucoma (Ko et al., 2000; Martin et al., 2003) and after transection of the optic nerve in the rat (Cheng et al., 2002; Krueger-Naug et al., 2003).

Disruption in the trophic activity of neurotrophins is also involved in cell death in chronic neurodegenerative disorders such as Parkinson's disease, Huntington's disease and Alzheimer's disease (Choi, 1988; Alexi et al., 2000). The Huntington's disease is characterized by a massive and progressive loss of striatal output neurons, without substantial loss of striatal interneurons and afferents, and many of the neurochemical, histological and behavioral features of this disease can be replicated by intrastriatal injection of quinolinate, an NMDA receptor agonist. Quinolinate-induced loss of striatal projection neurons was prevented by grafting of a BDNF-secreting cell line into the striatum or by intrastriatal injection of adenovirus encoding BDNF, and a similar effect, but with less efficiency, was observed in animals grafted with NT-3 or NT-4/5 secreting cells (Bemelmans et al., 1999; Perez-Navarro et al., 2000; 2005). The neuroprotective effects of BDNF in this model of Huntington's disease were attributed to the inhibition of caspase-3 activation and Akt dephosphorylation, and to counteraction of the changes in Bcl-2 family proteins. Thus, grafting of BDNF-secreting cells into the striatum prevented the quinolinate-induced upregulation of the pro-apoptotic protein Bax and the down-regulation of the anti-apoptotic proteins Bcl-2 and Bcl-xl (Perez-Navarro et al., 2005). In patients with Huntington's disease, the downregulation of BDNF by mutant Huntingtin, a protein involved in the vesicular transport of the neurotrophin, is thought to play a key role in motor dysfunction through the control of the survival of striatal enkephalinergic neurons (Canals et al., 2004; Gauthier et al., 2004).

Parkinson's disease is characterized by a selective loss of dopaminergic neurons from the *substantia nigra*, resulting in a significant depletion of dopamine from the striatum. These neurons synthesize BDNF (Seroogy et al., 1994; Schmidt-Kastner et al., 1996) and inhibition of its expression causes loss of nigral dopaminergic neurons (Porritt et al., 2005). A decrease in BDNF protein levels was also found in the *substantia nigra* in Parkinson's disease (reviewed in Siegel and Chauhan, 2000), suggesting that the loss of dopaminergic neurons may be due to a downregulation of BDNF expression. Pathogenic mutations in α-synuclein may contribute to the loss in BDNF production in autosomal dominant Parkinson's disease (Kohno et al., 2004).

A decrease in BDNF, proBDNF and TrkB expression is also found in selected post-mortem brain regions from Alzheimer's disease patients, together with an increase in the precursor form of NGF, which may reflect disturbances in the processing of proNGF to mature NGF (Siegel and Chauhan, 2000; Fahnestock et al., 2001; Peng et al., 2004; 2005). In addition to basal forebrain cholinergic deterioration, Alzheimer's disease is associated with pathological changes in the entorhinal cortex and hippocampus, regions critical for learning and memory. It has been proposed that the degeneration of the forebrain cholinergic neurons is due to the lack of trophic support and both NGF and BDNF may promote survival of these neurons (Tuszynski et al., 1990; Siegel and Chauhan, 2000; Fahnestock et al., 2002; Zassler and Humpel, 2006). Noteworthy, in a phase 1 clinical trial genetically modified NGF expressing fibroblasts were implanted into the forebrain and slowed the clinical decline of patients with Alzheimer's disease (Tuszynski et al., 2005). However, the actions of NGF are restricted to the cholinergic systems in the forebrain and these neurons are just one of several systems degenerating in Alzheimer's disease. NT-3 protects cultured cerebrocortical neurons from the toxicity of amyloid-β, which is thought to play a central role in neuronal cell death in Alzheimer's disease (Lesne et al., 2005). However, neurotrophins may also play an active role in neuronal cell death in Alzheimer's disease, since the accumulation of proNGF observed in the parietal brain cortex of Alzheimer's disease patients may affect the ratio of proNGF/matureNGF, a critical regulatory event in the maintenance of survival and death balance (see below).

Neurotrophins were also shown to protect neurons from other types of aggression. Pre-incubation of cultured cerebellar granule neurons with BDNF partly prevents apoptosis induced by the HIV-1 exterior envelope glycoprotein gp120, which is shed from the virus and may be one of the agents responsible for acquired immune deficiency syndrome-associated dementia complex (Mocchetti and Bachis, 2004; Bachis and Mocchetti, 2005). BDNF also reduced apoptotic and necrotic neuronal death in bacterial meningitis (Bifrare et al., 2005), and apoptotic cell death caused by DNA topoisomerase-I and -II inhibition (Hetman et al., 1999; Leeds et al., 2005), endoplasmic reticulum stress (Shimoke et al., 2004) and methamphetamine (Matsuzaki et al., 2004). Furthermore, activation of TrkB receptors by BDNF protects neuroblastoma cells from chemotherapy-induced cell death (Jaboin et al., 2003; Li et al., 2005). NGF also protects oligodendrocytes from the toxic effects of TNF-α. This cytokine is one of the major contributors to the damage of myelin sheaths and/or oligodendrocytes observed in multiple sclerosis, affecting the conduction of action potentials (Takano et al., 2000). Since apoptotic cell death induced by many toxic insults and by withdrawal of trophic factors involves many common effector pathways, it is likely that the mechanisms responsible by neuronal and glial protection by neurotrophins also overlap, at least in part.

6.1.1. Intracellular Signaling Mechanisms in Neuroprotection by BDNF

Activation of neurotrophin Trk receptors triggers parallel signaling pathways, including the PI3K pathway, the Ras/ERK pathway and PLCγ (see Sect. 5 and Fig. 2). The PI3K pathway, and to some extent the Ras/ERK pathway, plays a major role in neuroprotection by neurotrophins and other neurotrophic factors, but their relative contribution has been shown to depend on the cell type and the survival factor. Neurotrophic signaling may result from the activation of receptors located in the cell body or in the neurites. The retrograde neurotrophic signaling, originated in distal axons, has been studied in neurons of the PNS and is triggered by ligand-gated and Pincher-mediated internalization of the Trk receptors (Valdez et al., 2005). According to one of the current models, the internalized receptors are incorporated into "signaling endosomes" that convey the trophic effect to the cell body, along the microtubules, using dynein as a motor protein (reviewed by Zweifel et al., 2005).

Studies using chemical inhibitors or dominant negative forms of PI3K pathway intermediates showed a role for this signaling cascade in neurotrophin-mediated survival of CNS and PNS neurons, as well as neuronal cell lines, when subjected to trophic factor deprivation or chemical injuries (Table 2). This pathway suppresses cell death mainly by inhibiting the apoptotic effects of the forkhead transcription factor and Bcl-2-associated death protein (Bad) (Brunet et al., 2001; Downward, 2004). Furthermore, NT-3 was shown to protect cerebrocortical neurons from amyloid-β toxicity through the PI3K pathway by upregulating the expression of NAIP-1, a protein that directly inhibits caspase-3 and -7 (Lesne et al., 2005). In contrast with the PI3K pathway, the major role of the Ras/ERK pathway is to protect neurons from death due to injury or toxicity (e.g. excitotoxicity, calcium overload, oxidative injury, hypoxia), rather than from growth factor withdrawal (Table 3). This pathway acts mainly by stimulating the activity and/or expression of anti-apoptotic proteins (e.g. Bcl-2) and the transcription factor CREB (Watson et al., 2001; Hetman and Gozdz, 2004), but a downregulation of pro-apoptotic proteins may also be involved (Biswas and Greene, 2002). Thus, in cortical neurons, inhibition of MEK prevented BDNF-induced protection from campothecin-induced apoptosis, but it was without effect on the trophic effects of the neurotrophin in serum-free medium (Hetman et al., 1999). Interestingly, the PI3K pathway mediates the trophic effect of Trk receptors under trophic factor deprivation, following activation of PACAP and adenosine receptors coupled to adenylate cyclase (Lee and Chao, 2001). Stimulation of these receptors activates preferentially a population of intracellular Trk receptors in the absence of neurotrophins (Rajagopal et al., 2004).

The signaling pathway(s) involved in the trophic effects of neurotrophins under serum deprivation may also depend on the development stage of the cells, since in developing cerebellar granule neurons, but not in differentiated cells, BDNF-induced survival is mediated by the ERK5-MEF2-NT-3 signaling pathway (Liu et al., 2003; Shalizi et al., 2003). Furthermore, the trophic signals responsible for neuroprotection by neurotrophins under serum deprivation may also depend on the location of the receptors. Thus, although survival of dorsal root ganglion neurons in the absence of serum can be ensured by addition of NGF to distal axons or to the cell bodies, ERK5 is involved in the retrograde survival response, originated in distal axons, whereas ERK1/2 mediates the effect resulting from direct cell body stimulation (Watson et al., 2001). Similarly, there is a more strict dependence on the PI3K pathway for survival of sympathetic neurons supported by NGF applied to the distal axons as compared to neurons supported by NGF added to the cell bodies (Kuruvilla et al., 2000). The unique properties of each neurotrophin may also contribute to the complexity of cellular responses, contributing to their role in cell survival. Accordingly, although the axons of sympathetic neurons possess TrkA receptors sensitive to NGF and NT-3, cell survival and differentiation can only be induced by NGF (Kuruvilla et al., 2004). This apparent discrepancy was attributed to the ability of NGF, but not NT-3, to induce TrkA receptor internalization and retrograde transport of the active receptor (Kuruvilla et al., 2004).

The overlap in the contribution of the PI3K and Ras/ERK pathways in neuroprotection by neurotrophins observed in some experimental paradigms (e.g. Nakazawa et al., 2002; Almeida et al., 2005) may be due to cross-talk between the PI3K and the Ras/ERK pathways. Depending on the cellular background, the PI3K has been shown to stimulate or inhibit the

Table 2. Role of the PI3K pathway in protection by neurotrophins

Neurotrophin	Cell type	Toxic insult	References
NGF	PC12 cells	Serum starvation	Yao and Cooper, 1995; Klesse et al., 1999
	Superior cervical ganglion neurons	Serum starvation	Philpott et al., 1997; Crowder and Freeman, 1998; Meyer-Franke et al., 1998; Mazzoni et al., 1999; Kuruvilla et al., 2000, 2004
	Dorsal root ganglia neurons	Serum starvation	Bartlett et al., 1997; Klesse and Parada, 1998
	Oligodendrocytes	TNF-α	Takano et al., 2000
NGF+ KCl depolarization	Superior cervical ganglion neurons	Serum starvation	Vaillant et al., 1999
BDNF	Retinal ganglion cells	Axotomy	Nakazawa et al., 2002
	Differentiated neuroblastoma SH-SY5Y cells	Serum starvation	Encinas et al., 1999
	HT22 immortalized hippocampal neurons	Serum starvation	Rossler et al., 2004
	Spinal cord motoneuron	Trophic factor deprivation	Dolcet et al., 1999
	Cerebellar granule neurons	Trophic factor deprivation	Shimoke et al., 1997; Skaper et al., 1998
		Cytosine arabinoside (inhibitor of DNA topoisomerase-II)	Leeds et al., 2005
	Hippocampal neurons	Glutamate	Almeida et al., 2005
	Cortical neurons	Ionomycin (Ca^{2+} ionophore)	Takei et al., 1999
		Tunicamycin (inhibitor of protein glycosylation, causing ER stress)	Shimoke et al., 2004
		Serum starvation	Hetman et al., 1999; Yamada et al., 2001; Cheng et al., 2003
	Parkinson's disease cybrids	H_2O_2	Onyango et al., 2005
	TrkB-expressing SY5Y neuroblastoma	Chemotherapy	Jaboin et al., 2002; Li et al., 2005
NT-4/5	Cerebellar granule neurons	Trophic factor deprivation	Skaper et al., 1998
NT-3	Cerebellar granule neurons	Trophic factor deprivation	Skaper et al., 1998
	Cortical neurons	Serum starvation	Liot et al., 2004
		Ionomycin (Ca^{2+} ionophore)	Takei et al., 1999
		Amyloid β	Lesne et al., 2005

Table 3. Role of the Ras/ERK pathway in protection by neurotrophins

Neurotrophin	Cell type	Toxic insult	References
NGF	Superior cervical ganglion neurons	Serum starvation	Meyer-Franke et al., 1998; Mazzoni et al., 1999; Xue et al., 2000
		Cytosine arabinoside (inhibitor of DNA topoisomerase-II)	Anderson and Tolkovsky, 1999
	Dorsal root ganglia neurons	Serum starvation	Watson et al., 2001
BDNF	Cerebellar granule neurons	Trophic factor deprivation	Skaper et al., 1998; Bonni et al., 1999
		Glycoprotein gp120	Mocchetti and Bachis, 2004
	Cortical neurons	Ionomycin (Ca^{2+} ionophore)	Takei et al., 1999
		Camptothecin (inhibitor of DNA topoisomerase-I)	Leeds et al., 2005
	Hippocampal neurons	Glutamate	Almeida et al., 2005
	Retinal ganglion cells	Axotomy	Nakazawa et al., 2002
	Striatum, cortex and hippocampus	Hypoxic-ischemic injury	Han and Holtzman, 2000
	Parkinson's disease cybrids	H_2O_2	Onyango et al., 2005
BDNF+cAMP	Retinal ganglion cells	Serum starvation	Meyer-Franke et al., 1998
NT-3	Cortical neurons	Ionomycin (Ca^{2+} ionophore)	Takei et al., 1999

activity of ERK (Moelling et al., 2002; Sato et al., 2004; Almeida et al., 2005). In cultured hippocampal neurons, where ERK stimulation by BDNF requires PI3K activity, the neurotrophin has a neuroprotective effect under excitotoxic conditions that is mediated by the PI3K and the Ras/ERK pathways (Almeida et al., 2005).

The mechanisms acting in neuroprotection by neurotrophins downstream of the ERK and PI3K pathways may also differ depending on the insult. While the upregulation of anti-apoptotic proteins (e.g. Bcl-2) by neurotrophins and inhibition of caspase activity accounts for neuroprotection under conditions where cell death is typically apoptotic (Michaelidis et al., 1996; Aloyz et al., 1998; Madeddu et al., 2004; Lesne et al., 2005), other mechanisms may be involved in protection from insults where caspase activation plays a minor role in the demise process. Although calpains rather than caspases play a major role in excitotoxic cell death (Higuchi et al., 2005), neurotrophins have been shown to play a protective role under these conditions, both in cultured neurons and in vivo (see references above). Therefore, the activation of a complex program of changes in gene expression by Trk receptors (Schulte et al., 2005) is expected to cause multiple changes in the proteome, rendering the cells better prepared against a wide range of insults.

Neuroprotection by neurotrophins may depend on the metabolic state of the cell, since, for example, the protection of cerebrocortical neurons from serum-withdrawal-induced apoptosis is inhibited by cAMP (Poser et al., 2003). Although in some cases the Trk-activated signaling activity may not be enough to promote survival, due to limited range of changes induced in the cell and/or to the magnitude of the response induced, neuroprotection may still be induced in combination with other trophic factors. Accordingly, the platelet-derived growth factor-BB supports the survival of cultured rat Schwann cell precursors in synergy with NT-3 (Lobsiger et al., 2000), and BDNF and glial cell line-derived neurotrophic factor (GDNF) act as target-derived trophic factors and are required together for survival of some primary sensory neurons (Erickson et al., 2001).

6.2. Trk-Induced Cell Death

Although activation of Trk receptors is generally associated with cell survival, these receptors may also be linked to increased vulnerability to death signals, usually causing necrotic cell death (activation of glutamate receptors, oxidants, oxygen/glucose deprivation) (Fryer et al., 2000; Ishikawa et al., 2000; Kim et al., 2003). The signaling cascades that have been implicated in these effects are PI3K (Fryer et al., 2000), ERK (Ishikawa et al., 2000; Kim et al., 2003) and the p38 MAPK pathway (Ishikawa et al., 2000). These effects of neurotrophins may be mediated, at least in some systems, by induction of the NMDA receptor NR2A subunits and increase in NR2A and NR2B phosphorylation, thereby upregulating NMDA receptor activity, as described in cerebrocortical cultures stimulated with NT-4/5 (Choi et al., 2004). In these cells, BDNF increases the activity of NADPH oxidase, causing necrotic cell death (Kim et al., 2002). The increase in neuronal excitability upon TrkB-induced activation of Na^+ currents, mediated by $Na_v1.9$, may also render neurons more vulnerable to certain toxic insults (Blum et al., 2002).

6.3. Role of $p75^{NTR}$ in Cell Survival and Cell Death

The coexpression of Trk and $p75^{NTR}$ enhances the ability of the tyrosine kinase receptors to bind and respond to the presence of neurotrophins and favours discrimination of their preferential neurotrophin ligands. This is particularly relevant in peripheral tissues, where limited amounts of neurotrophins are produced and yet maintain neuronal survival and enervation (Lee et al., 1992; von Schack et al., 2001). Furthermore, $p75^{NTR}$ can bind Shc, an adaptor molecule in the Trk signaling, stimulating its phosphorylation and upregulating the response to activation of Trk receptors (Epa et al., 2004). In contrast, in sympathetic neurons, NGF increases $p75^{NTR}$ protein levels and these receptors sequester the neurotrophin that becomes less available to bind the Trk receptors (Hannila et al., 2004).

The $p75^{NTR}$ interacts with various proteins in its intracellular domain and similar to other members of the TNF receptor superfamily, it comprises a "death domain". Therefore, it is not surprising that these receptors are able to signal independently and promote death of various cell types, including developing retinal and hippocampal neurons (Frade et al., 1996; Friedman, 2000; Troy et al., 2002; Harada et al., 2006), oligodendrocytes (Casaccia-Bonnefil et al., 1996; Beattie et al., 2002), Schwann cells (Boyle et al., 2005; Teng et al., 2005), corticospinal neurons (Harrington et al., 2004), photoreceptors (Srinivasan et al., 2004), smooth muscle cells (Lee et al., 2001; Nykjaer et al., 2004) and superior cervical ganglion cells (Lee et al., 2001; Nykjaer et al., 2004; Teng et al., 2005; Kenchappa et al., 2006). A furin-resistant form of proNGF was instrumental in showing that the precursor form of the neurotrophin binds $p75^{NTR}$ with high affinity and induces apoptotic cell death in various cell types (Lee et al., 2001; Nykjaer et al., 2004). proBDNF also induces neuronal apoptosis by activating p75^{NTR} (Teng et al., 2005), and the pro-apoptotic effects of the precursor forms of the neurotrophins require the transmembrane protein sortilin, which acts as a co-receptor. It remains to be determined whether the precursor form of the other neurotrophins also induces cell death through the p75^{NTR}.

The mechanisms of cell death following activation of p75^{NTR} are less well understood than the signaling pathways accounting for the survival-promoting effects of Trk receptors. A number of p75^{NTR} interacting proteins have been implicated in apoptosis, including NRIF (neurotrophin-receptor interacting factor), TRAF6 (tumour necrosis factor receptor-associated factor 6), Rac, NADE (neurotrophin-associated cell death executor) and the MAGE (melanoma-associated antigen) family members NRAGE (neurotrophin-receptor interacting MAGE homologue) and Necdin (reviewed in Nykjaer et al., 2005; see also: Linggi et al., 2005; Kenchappa et al., 2006). Ligand binding to p75^{NTR} leads to the cleavage of the intracellular domain of the receptor by γ-secretase, facilitating ubiquitination and nuclear translocation of NRIF (Kenchappa et al., 2006). The p75^{NTR}-dependent apoptotic response is also believed to involve the activation of JNK, c-jun phosphorylation, activation of p53, Bad and Bim, mitochondrial translocation of Bax, cytochrome C release form mitochondria and activation of caspases-9, -3 and -6 (Troy et al., 2002; Bhakar et al., 2003; Becker et al., 2004; Linggi et al., 2005).

The p75^{NTR} is abundantly expressed during development, but is downregulated in the adult organism. However, the receptor can be dynamically regulated upon injury or insult to the nervous system, suggesting that this may constitute a mechanism to eliminate damaged cells. Thus, upregulation of p75^{NTR} was found in different types of neurons following injury by axotomy (Ernfors et al., 1989; Koliatsos et al., 1991; Hayes et al., 1992), trauma (Martinez-Murillo et al., 1998), seizures (Beattie et al., 2002; Troy et al., 2002), ischaemia (Kokaia et al., 1998; Park et al., 2000), zinc (Park et al., 2000), ALS (Kerkhoff et al., 1991; Seeburger et al., 1993) and in Alzheimer's disease (Mufson and Kordower, 1992). An increase in the abundance of p75^{NTR} was also observed in Schwann cells after axotomy (Lemke and Chao, 1988; Taniuchi et al., 1988) and in oligodendrocytes following spinal cord injury (Beattie et al., 2002), as well as in multiple sclerosis lesions (Dowling et al., 1999; Chang et al., 2000).

In patients with Alzheimer's disease the upregulation of p75^{NTR} in the cerebral cortex, together with the decrease in Trk protein levels, may account for an imbalance of survival vs. death signals, leading to cell death (Mufson et al., 2003b). The observation that amyloid-β (Aβ) peptides induce neurotoxicity by binding to p75^{NTR} and activating JNK and caspases-9 and -3 suggested that this could account for cell death in Alzheimer's disease (Tsukamoto et al., 2003; Hashimoto et al., 2004; Costantini et al., 2005). However, this may not be the case, since in a different study, p75^{NTR} expression protected cultured human neurons from Aβ-induced toxicity (Zhang et al., 2003b) and the levels of Aβ necessary to induce cell death are much higher than the levels found in the cerebrospinal fluid and in the plasma of Alzheimer's disease patients. Since proNGF is upregulated in Alzheimer's disease, it is a more likely mediator of p75^{NTR}-induced cell death in these patients. The release of proNGF by activated retinal microglia may also account for photoreceptor cell death observed in retinal dystrophy (Sheedlo et al., 2002; Srinivasan et al., 2004). Similarly, the activation of p75^{NTR} by proNGF upon spinal cord injury is responsible for apoptotic death of oligodendrocytes (Beattie et al., 2002) and the same mechanism is involved in death of corticospinal neurons following axotomy (Harrington et al., 2004). The expression of the p75^{NTR} receptor has also been observed in motoneurons in post-mortem spinal cord of patients with ALS, a neurodegenerative disease characterized by motoneuron loss in the cortex, brainstem and spinal cord, resulting in progressive paralysis (Kerkhoff et al., 1991; Seeburger et al., 1993). Motor neuron death may be due to activation of p75^{NTR} by neurotrophins released by reactive astrocytes and require NO and peroxynitrite (Pehar et al., 2004).

7. ACKNOWLEDGEMENTS

The work in the authors laboratory is supported by research grants from Fundação para a Ciência and Tecnologia (POCTI/NSE/46441/2002 and POCTI/BCI/46466/2002) and FEDER (Portugal).

8. REFERENCES

Abiru, Y., Katoh-Semba, R., Nishio, C. and Hatanaka, H., 1998, High potassium enhances secretion of neurotrophic factors from cultured astrocytes. *Brain Res.* **809**: 115.

Acheson, A., Barker, P.A., Alderson, R.F., Miller, F.D. and Murphy, R.A., 1991, Detection of brain-derived neurotrophic factor-like activity in fibroblasts and Schwann cells: inhibition by antibodies to NGF. *Neuron* **7**: 265.

Adachi, N., Kohara, K. and Tsumoto, T., 2005, Difference in trafficking of brain-derived neurotrophic factor between axons and dendrites of cortical neurons, revealed by live-cell imaging. *BMC Neurosci.* **6**: 42.

Aguado, F., Ballabriga, J., Pozas, E. and Ferrer, I., 1998, TrkA immunoreactivity in reactive astrocytes in human neurodegenerative diseases and colchicine-treated rats. *Acta Neuropathol. (Berl.)* **96**: 495.

Alderson, R.F., Curtis, R., Alterman, A.L., Lindsay, R.M. and DiStefano, P.S., 2000, Truncated TrkB mediates the endocytosis and release of BDNF and neurotrophin-4/5 by rat astrocytes and Schwann cells in vitro. *Brain Res.* **871**: 210.

Alexi, T., Borlongan, C.V., Faull, R.L., Williams, C.E., Clark, R.G., Gluckman, P.D. and Hughes, P.E., 2000, Neuroprotective strategies for basal ganglia degeneration: Parkinson's and Huntington's diseases. *Prog. Neurobiol.* **60**: 409.

Allen, S.J., Wilcock, G.K. and Dawbarn, D., 1999, Profound and selective loss of catalytic TrkB immunoreactivity in Alzheimer's disease. *Biochem. Biophys. Res. Commun.* **264**: 648.

Almeida, R.D., Manadas, B.J., Melo, C.V., Gomes, J.R., Mendes, C.S., Graos, M.M., Carvalho, R.F., Carvalho, A.P. and Duarte, C.B., 2005, Neuroprotection by BDNF against glutamate-induced apoptotic cell death is mediated by ERK and PI3-kinase pathways. *Cell Death Differ.* **12**: 1329.

Aloyz, R.S., Bamji, S.X., Pozniak, C.D., Toma, J.G., Atwal, J., Kaplan, D.R and Miller, F.D., 1998, p53 is essential for developmental neuron death as regulated by the TrkA and p75 neurotrophin receptors. *J. Cell Biol.* **143**: 1691.

Altar, C.A., Siuciak, J.A., Wright, P., Ip, N.Y., Lindsay, R.M. and Wiegand. S.J., 1994, In situ hybridization of trkB and trkC receptor mRNA in rat forebrain and association with high-affinity binding of [^{125}I]BDNF, [^{125}I]NT-4/5 and [^{125}I]NT-3. *Eur. J. Neurosci.* **6**: 1389.

Althaus, H.H. and Richter-Landsberg, C., 2000, Glial cells as targets and producers of neurotrophins. *Int. Rev. Cytol.* **197**: 203.

Anderson, C.N. and Tolkovsky, A.M., 1999, A role for MAPK/ERK in sympathetic neuron survival: protection against a p53-dependent, JNK-independent induction of apoptosis by cytosine arabinoside. *J. Neurosci.* **19**: 664.

Ankarcrona, M., Dypbukt, J.M., Bonfoco, E., Zhivotovsky, B., Orrenius, S., Lipton, S.A. and Nicotera, P., 1995, Glutamate-induced neuronal death: a succession of necrosis or apoptosis depending on mitochondrial function. *Neuron* **15**: 961.

Aoki, C., Wu, K., Elste, A., Len, G., Lin, S., McAuliffe, G. and Black, I.B., 2000, Localization of brain-derived neurotrophic factor and TrkB receptors to postsynaptic densities of adult rat cerebral cortex. *J. Neurosci. Res.* **59**: 454.

Arevalo, J.C., Pereira, D.B., Yano, H., Teng, K.K. and Chao, M.V., 2006, Identification of a switch in neurotrophin signaling by selective tyrosine phosphorylation. *J. Biol. Chem.* **281**: 1001.

Aronica, E., Ozbas-Gerceker, F., Redeker, S., Ramkema, M., Spliet, W.G., van Rijen, P.C., Leenstra, S., Gorter, J.A. and Troost, D., 2004, Expression and cellular distribution of high- and low-affinity neurotrophin receptors in malformations of cortical development. *Acta Neuropathol. (Berl.)* **108**: 422.

Arthur, J.S., Fong, A.L., Dwyer, J.M., Davare, M., Reese, E., Obrietan, K. and Impey, S., 2004, Mitogen- and stress-activated protein kinase 1 mediates cAMP response element-binding protein phosphorylation and activation by neurotrophins. *J. Neurosci.* **24**: 4324.

Atwal, J.K., Massie, B., Miller, F.D. and Kaplan, D.R., 2000, The TrkB-Shc site signals neuronal survival and local axon growth via MEK and PI3-kinase. *Neuron* **27**: 265.

Bachis, A. and Mocchetti, I., 2005, Brain-derived neurotrophic factor is neuroprotective against human immunodeficiency virus-1 envelope proteins. *Ann. NY Acad. Sci.* **1053**: 247.

Balkowiec, A. and Katz, D.M., 2002, Cellular mechanisms regulating activity-dependent release of native brain-derived neurotrophic factor from hippocampal neurons. *J. Neurosci.* **22**: 10399.

Barbacid, M., 1994, The Trk family of neurotrophin receptors. *J. Neurobiol.* **25**: 1386.

Barker, P.A., 2004, p75NTR is positively promiscuous: novel partners and new insights. *Neuron* **42**: 529.

Barres, B.A., Hart, I.K., Coles, H.S., Burne, J.F., Voyvodic, J.T., Richardson, W.D. and Raff, M.C., 1992, Cell death and control of cell survival in the oligodendrocyte lineage. *Cell* **70**: 31.

Bartlett, S.E., Reynolds, A.J., Weible, M., Heydon, K. and Hendry, I.A., 1997, In sympathetic but not sensory neurones, phosphoinositide-3 kinase is important for NGF-dependent survival and the retrograde transport of ^{125}I-betaNGF. *Brain Res.* **761**: 257.

Beattie, M.S., Harrington, A.W., Lee, R., Kim, J.Y., Boyce, S.L., Longo, F.M., Bresnahan, J.C., Hempstead, B.L. and Yoon, S.O., 2002, ProNGF induces p75-mediated death of oligodendrocytes following spinal cord injury. *Neuron* **36**: 375.

Beck, T., Lindholm, D., Castren, E. and Wree, A., 1994, Brain-derived neurotrophic factor protects against ischemic cell damage in rat hippocampus. *J. Cereb. Blood Flow Metab.* **14**: 689.

Becker, E.B., Howell, J., Kodama, Y., Barker, P.A. and Bonni, A., 2004, Characterization of the c-Jun N-terminal kinase-BimEL signaling pathway in neuronal apoptosis. *J. Neurosci.* **24**: 8762.

Bemelmans, A.P., Horellou, P., Pradier, L., Brunet, I., Colin, P. and Mallet, J., 1999, Brain-derived neurotrophic factor-mediated protection of striatal neurons in an excitotoxic rat model of Huntington's disease, as demonstrated by adenoviral gene transfer. *Hum. Gene Ther.* **10**: 2987.

Bhakar, A.L., Howell, J.L., Paul, C.E., Salehi, A.H., Becker, E.B., Said, F., Bonni, A. and Barker, P.A., 2003, Apoptosis induced by p75NTR overexpression requires Jun kinase-dependent phosphorylation of Bad. *J. Neurosci.* **23**: 11373.

Bibel, M. and Barde, Y.A., 2000, Neurotrophins: key regulators of cell fate and cell shape in the vertebrate nervous system. *Genes Dev.* **14**: 2919.

Bifrare, Y.D., Kummer, J., Joss, P., Tauber, M.G. and Leib, S.L., 2005, Brain-derived neurotrophic factor protects against multiple forms of brain injury in bacterial meningitis. *J. Infect. Dis.* **191**: 40.

Bilderback, T.R., Gazula, V.R. and Dobrowsky, R.T., 2001, Phosphoinositide 3-kinase regulates crosstalk between Trk A tyrosine kinase and p75^{NTR}-dependent sphingolipid signaling pathways. *J. Neurochem.* **76**: 1540.

Biswas, S.C. and Greene, L.A., 2002, Nerve growth factor (NGF) down-regulates the Bcl-2 homology 3 (BH3) domain-only protein Bim and suppresses its proapoptotic activity by phosphorylation. *J. Biol. Chem.* **277**: 49511.

Blum, R., Kafitz, K.W. and Konnerth, A., 2002, Neurotrophin-evoked depolarization requires the sodium channel Na$_V$1.9. *Nature* **419**: 687.

Bonni, A., Brunet, A., West, A.E., Datta, S.R., Takasu, M.A. and Greenberg, M.E., 1999, Cell survival promoted by the Ras-MAPK signaling pathway by transcription-dependent and -independent mechanisms. *Science* **286**: 1358.

Boyle, K., Azari, M.F., Cheema, S.S. and Petratos, S., 2005, TNFalpha mediates Schwann cell death by upregulating p75^{NTR} expression without sustained activation of NFkappaB. *Neurobiol. Dis.* **20**: 412.

Bruck, W., 2005, The pathology of multiple sclerosis is the result of focal inflammatory demyelination with axonal damage. *J. Neurol.* **252**(Suppl 5): v3.

Brunet, A., Datta, S.R. and Greenberg, M.E., 2001, Transcription-dependent and -independent control of neuronal survival by the PI3K-Akt signaling pathway. *Curr. Opin. Neurobiol.* **11**: 297.

Brunet, A., Kanai, F., Stehn, J., Xu, J., Sarbassova, D., Frangioni, J.V., Dalal, S.N., DeCaprio, J.A., Greenberg, M.E. and Yaffe, M.B., 2002, 14-3-3 transits to the nucleus and participates in dynamic nucleocytoplasmic transport. *J. Cell Biol.* **156**: 817.

Canals, J.M., Pineda, J.R., Torres-Peraza, J.F., Bosch, M., Martin-Ibanez, R., Munoz, M.T., Mengod, G., Ernfors, P. and Alberch, J., 2004, Brain-derived neurotrophic factor regulates the onset and severity of motor dysfunction associated with enkephalinergic neuronal degeneration in Huntington's disease. *J. Neurosci.* **24**: 7727.

Casaccia-Bonnefil, P., Carter, B.D., Dobrowsky, R.T. and Chao, M.V., 1996, Death of oligodendrocytes mediated by the interaction of nerve growth factor with its receptor p75. *Nature* **383**: 716.

Cassina, P., Pehar, M., Vargas, M.R., Castellanos, R., Barbeito, A.G., Estevez, A.G., Thompson, J.A., Beckman, J.S. and Barbeito, L., 2005, Astrocyte activation by fibroblast growth factor-1 and motor neuron apoptosis: implications for amyotrophic lateral sclerosis. *J. Neurochem.* **93**: 38.

Chan, K.M., Lam, D.T., Pong, K., Widmer, H.R. and Hefti, F., 1996, Neurotrophin-4/5 treatment reduces infarct size in rats with middle cerebral artery occlusion. *Neurochem. Res.* **21**: 763.

Chan, J.R., Watkins, T.A., Cosgaya, J.M., Zhang, C., Chen, L., Reichardt, L.F., Shooter, E.M. and Barres, B.A., 2004, NGF controls axonal receptivity to myelination by Schwann cells or oligodendrocytes. *Neuron* **43**: 183.

Chang, A., Nishiyama, A., Peterson, J., Prineas, J. and Trapp, B.D., 2000, NG2-positive oligodendrocyte progenitor cells in adult human brain and multiple sclerosis lesions. *J. Neurosci.* **20**: 6404.

Chao, M.V., 1992, Growth factor signaling: where is the specificity? *Cell* **68**: 995.

Chao, M.V., 2003, Neurotrophins and their receptors: a convergence point for many signaling pathways. *Nat. Rev. Neurosci.* **4**: 299.

Chao, M.V. and Hempstead, B.L., 1995, p75 and Trk: a two-receptor system. *Trends Neurosci.* **18**: 321.

Chen, X. and Wang, Z., 2001, Regulation of intracellular trafficking of the EGF receptor by Rab5 in the absence of phosphatidylinositol 3-kinase activity. *EMBO Rep.* **2**: 68.

Chen, W.G., Chang, Q., Lin, Y., Meissner, A., West, A.E., Griffith, E.C., Jaenisch, R. and Greenberg, M.E., 2003, Derepression of BDNF transcription involves calcium-dependent phosphorylation of MeCP2. *Science* **302**: 885.

Chen, Z.Y., Patel, P.D., Sant, G., Meng, C.X., Teng, K.K., Hempstead, B.L. and Lee, F.S., 2004, Variant brain-derived neurotrophic factor (BDNF) (Met66) alters the intracellular trafficking and activity-dependent secretion of wild-type BDNF in neurosecretory cells and cortical neurons. *J. Neurosci.* **24**: 4401.

Chen, Z.Y., Ieraci, A., Teng, H., Dall, H., Meng, C.X., Herrera, D.G., Nykjaer, A., Hempstead, B.L. and Lee, F.S., 2005, Sortilin controls intracellular sorting of brain-derived neurotrophic factor to the regulated secretory pathway. *J. Neurosci.* **25**: 6156.

Cheng, B. and Mattson, M.P., 1991, NGF and bFGF protect rat hippocampal and human cortical neurons against hypoglycemic damage by stabilizing calcium homeostasis. *Neuron* **7**: 1031.

Cheng, B. and Mattson, M.P., 1994, NT-3 and BDNF protect CNS neurons against metabolic/excitotoxic insults. *Brain Res.* **640**: 56.

Cheng, L., Sapieha, P., Kittlerova, P., Hauswirth, W.W. and Di Polo, A., 2002, TrkB gene transfer protects retinal ganglion cells from axotomy-induced death in vivo. *J. Neurosci.* **22**: 3977.

Cheng, A., Wang, S., Yang, D., Xiao, R. and Mattson, M.P., 2003, Calmodulin mediates brain-derived neurotrophic factor cell survival signaling upstream of Akt kinase in embryonic neocortical neurons. *J. Biol. Chem.* **278**: 7591.

Choi, D.W., 1988, Glutamate neurotoxicity and diseases of the nervous system. *Neuron* **1**: 623.

Choi, S.Y., Hwang, J.J. and Koh, J.Y., 2004, NR2A induction and NMDA receptor-dependent neuronal death by neurotrophin-4/5 in cortical cell culture. *J. Neurochem.* **88**: 708.

Christopherson, K.S., Ullian, E.M., Stokes, C.C., Mullowney, C.E., Hell, J.W., Agah, A., Lawler, J., Mosher, D.F., Bornstein, P. and Barres, B.A., 2005, Thrombospondins are astrocyte-secreted proteins that promote CNS synaptogenesis. *Cell* **120**: 421.

Cohen, R.I., Marmur, R., Norton, W.T., Mehler, M.F. and Kessler, J.A., 1996, Nerve growth factor and neurotrophin-3 differentially regulate the proliferation and survival of developing rat brain oligodendrocytes. *J. Neurosci.* **16**: 6433.

Connor, B., Young, D., Lawlor, P., Gai, W., Waldvogel, H., Faull, R.L. and Dragunow, M., 1996, Trk receptor alterations in Alzheimer's disease. *Brain. Res. Mol. Brain Res.* **42**: 1.

Conti, L., Sipione, S., Magrassi, L., Bonfanti, L., Rigamonti, D., Pettirossi, V., Peschanski, M., Haddad, B., Pelicci, P., Milanesi, G., Pelicci, G. and Cattaneo, E., 2001, Shc signaling in differentiating neural progenitor cells. *Nat. Neurosci.* **4**: 579.

Costantini, C., Rossi, F., Formaggio, E., Bernardoni, R., Cecconi, D. and Della-Bianca, V., 2005, Characterization of the signaling pathway downstream p75 neurotrophin receptor involved in beta-amyloid peptide-dependent cell death. *J. Mol. Neurosci.* **25**: 141.

Coull, J.A., Beggs, S., Boudreau, D., Boivin, D., Tsuda, M., Inoue, K., Gravel, C., Salter, M.W. and De Koninck, Y., 2005, BDNF from microglia causes the shift in neuronal anion gradient underlying neuropathic pain. *Nature* **438**: 1017.

Crowder, R.J. and Freeman, R.S., 1998, Phosphatidylinositol 3-kinase and Akt protein kinase are necessary and sufficient for the survival of nerve growth factor-dependent sympathetic neurons. *J. Neurosci.* **18**: 2933.

Crowley, C., Spencer, S.D., Nishimura, M.C., Chen, K.S., Pitts-Meek, S., Armanini, M.P., Ling, L.H., McMahon, S.B., Shelton, D.L., Levinson, A.D. et al., 1994, Mice lacking nerve growth factor display perinatal loss of sensory and sympathetic neurons yet develop basal forebrain cholinergic neurons. *Cell* **76**: 1001.

Dai, X., Lercher, L.D., Yang, L., Shen, M., Black, I.B. and Dreyfus, C.F., 1997, Expression of neurotrophins by basal forebrain (BF) oligodendrocytes. *Soc. Neurosci. Abstr.* **23**: 331.

Datta, S.R., Dudek, H., Tao, X., Masters, S., Fu, H., Gotoh, Y. and Greenberg, M.E., 1997, Akt phosphorylation of BAD couples survival signals to the cell-intrinsic death machinery. *Cell* **91**: 231.

Dobrowsky, R.T., Jenkins, G.M. and Hannun, Y.A., 1995, Neurotrophins induce sphingomyelin hydrolysis. Modulation by co-expression of p75 ^{NTR} with Trk receptors. *J. Biol. Chem.* **270**: 22135.

Dolcet, X., Egea, J., Soler, R.M., Martin-Zanca, D. and Comella, J.X., 1999, Activation of phosphatidylinositol 3-kinase, but not extracellular-regulated kinases, is necessary to mediate brain-derived neurotrophic factor-induced motoneuron survival. *J. Neurochem.* **73**: 521.

Dowling, P., Ming, X., Raval, S., Husar, W., Casaccia-Bonnefil, P., Chao, M., Cook, S. and Blumberg, B., 1999, Up-regulated p75 ^{NTR} neurotrophin receptor on glial cells in MS plaques. *Neurology* **53**: 1676.

Downward, J., 2004, PI 3-kinase, Akt and cell survival. *Semin. Cell Dev. Biol.* **15**: 177.

Drake, C.T., Milner, T.A. and Patterson, S.L., 1999, Ultrastructural localization of full-length trkB immunoreactivity in rat hippocampus suggests multiple roles in modulating activity-dependent synaptic plasticity. *J. Neurosci.* **19**: 8009.

Du, J., Feng, L., Yang, F. and Lu, B., 2000, Activity- and Ca^{2+}-dependent modulation of surface expression of brain-derived neurotrophic factor receptors in hippocampal neurons. *J. Cell Biol.* **150**: 1423.

Du, Y., Fischer, T.Z., Clinton-Luke, P., Lercher, L.D. and Dreyfus, C.F., 2006, Distinct effects of p75 in mediating actions of neurotrophins on basal forebrain oligodendrocytes. *Mol. Cell. Neurosci.* **31**: 366.

Duarte, C.B., Ferreira, I.L., Santos, P.F., Carvalho, A.L., Agostinho, P.M. and Carvalho, A.P., 1998, Glutamate in life and death of retinal amacrine cells. *Gen. Pharmacol.* **30**: 289.

Egan, M.F., Kojima, M., Callicott, J.H., Goldberg, T.E., Kolachana, B.S., Bertolino, A., Zaitsev, E., Gold, B., Goldman, D., Dean, M., Lu, B. and Weinberger, D.R., 2003, The BDNF val66met polymorphism affects activity-dependent secretion of BDNF and human memory and hippocampal function. *Cell* **112**: 257.

Eide, F.F., Vining, E.R., Eide, B.L., Zang, K., Wang, X.Y. and Reichardt, L.F., 1996, Naturally occurring truncated TrkB receptors have dominant inhibitory effects on brain-derived neurotrophic factor signaling. *J. Neurosci.* **16**: 3123.

Elkabes, S., DiCicco-Bloom, E.M. and Black, I.B., 1996, Brain microglia/macrophages express neurotrophins that selectively regulate microglial proliferation and function. *J. Neurosci.* **16**: 2508.

Elkabes, S., Peng, L. and Black, I.B., 1998, Lipopolysaccharide differentially regulates microglial trk receptor and neurotrophin expression. *J. Neurosci. Res.* **54**: 117.

Elmariah, S.B., Oh, E.J., Hughes, E.G. and Balice-Gordon, R.J., 2005, Astrocytes regulate inhibitory synapse formation via Trk-mediated modulation of postsynaptic $GABA_A$ receptors. *J. Neurosci.* **25**: 3638.

Encinas, M., Iglesias, M., Llecha, N. and Comella, J.X., 1999, Extracellular-regulated kinases and phosphatidylinositol 3-kinase are involved in brain-derived neurotrophic factor-mediated survival and neuritogenesis of the neuroblastoma cell line SH-SY5Y. *J. Neurochem.* **73**: 1409.

Endres, M., Fan, G., Hirt, L., Fujii, M., Matsushita, K., Liu, X., Jaenisch, R. and Moskowitz, M.A., 2000, Ischemic brain damage in mice after selectively modifying BDNF or NT4 gene expression. *J. Cereb. Blood Flow Metab.* **20**: 139.

Epa, W.R., Markovska, K. and Barrett, G.L., 2004, The p75 neurotrophin receptor enhances TrkA signaling by binding to Shc and augmenting its phosphorylation. *J. Neurochem.* **89**: 344.

Erickson, J.T., Brosenitsch, T.A. and Katz, D.M., 2001, Brain-derived neurotrophic factor and glial cell line-derived neurotrophic factor are required simultaneously for survival of dopaminergic primary sensory neurons in vivo. *J. Neurosci.* **21**: 581.

Ernfors, P., Henschen, A., Olson, L. and Persson, H., 1989, Expression of nerve growth factor receptor mRNA is developmentally regulated and increased after axotomy in rat spinal cord motoneurons. *Neuron* **2**: 1605.

Esteban, P.F., Yoon, H.Y., Becker, J., Dorsey, S.G., Caprari, P., Palko, M.E., Coppola, V., Saragovi, H.U., Randazzo, P.A. and Tessarollo, L., 2006, A kinase-deficient TrkC receptor isoform activates Arf6-Rac1 signaling through the scaffold protein tamalin. *J. Cell Biol.* **173**: 291.

Fagan, A.M., Zhang, H., Landis, S., Smeyne, R.J., Silos-Santiago, I. and Barbacid, M., 1996, TrkA, but not TrkC, receptors are essential for survival of sympathetic neurons in vivo. *J. Neurosci.* **16**: 6208.

Fahnestock, M., Michalski, B., Xu, B. and Coughlin, M.D., 2001, The precursor pro-nerve growth factor is the predominant form of nerve growth factor in brain and is increased in Alzheimer's disease. *Mol. Cell. Neurosci.* **18**: 210.

Fahnestock, M., Garzon, D., Holsinger, R.M. and Michalski, B., 2002, Neurotrophic factors and Alzheimer's disease: are we focusing on the wrong molecule? *J. Neural Transm. Suppl.* 241.

Farhadi, H.F., Mowla, S.J., Petrecca, K., Morris, S.J., Seidah, N.G. and Murphy, R.A., 2000, Neurotrophin-3 sorts to the constitutive secretory pathway of hippocampal neurons and is diverted to the regulated secretory pathway by coexpression with brain-derived neurotrophic factor. *J. Neurosci.* **20**: 4059.

Fawcett, J.W. and Keynes, R.J., 1990, Peripheral nerve regeneration. *Annu. Rev. Neurosci.* **13**: 43.

Fawcett, J.W. and Asher, R.A., 1999, The glial scar and central nervous system repair. *Brain Res. Bull.* **49**: 377.

Ferguson, B., Matyszak, M.K., Esiri, M.M. and Perry, V.H., 1997, Axonal damage in acute multiple sclerosis lesions. *Brain* **120**: 393.

Ferrer, I., Krupinski, J., Goutan, E., Marti, E., Ambrosio, S. and Arenas, E., 2001, Brain-derived neurotrophic factor reduces cortical cell death by ischemia after middle cerebral artery occlusion in the rat. *Acta Neuropathol. (Berl).* **101**: 229.
Filbin, M.T., 2003, Myelin-associated inhibitors of axonal regeneration in the adult mammalian CNS. *Nat. Rev. Neurosci.* **4**: 703.
Foehr, E.D., Lin, X., O'Mahony, A., Geleziunas, R., Bradshaw, R.A. and Greene, W.C., 2000, NF-kappa B signaling promotes both cell survival and neurite process formation in nerve growth factor-stimulated PC12 cells. *J. Neurosci.* **20**: 7556.
Frade, J.M., Rodriguez-Tebar, A. and Barde, Y.A., 1996, Induction of cell death by endogenous nerve growth factor through its p75 receptor. *Nature* **383**: 166.
Friedman, W.J., 2000, Neurotrophins induce death of hippocampal neurons via the p75 receptor. *J. Neurosci.* **20**: 6340.
Fryer, R.H., Kaplan, D.R., Feinstein, S.C., Radeke, M.J., Grayson, D.R. and Kromer, L.F., 1996, Developmental and mature expression of full-length and truncated TrkB receptors in the rat forebrain. *J. Comp. Neurol.* **374**: 21.
Fryer, H.J., Wolf, D.H., Knox, R.J., Strittmatter, S.M., Pennica, D., O'Leary, R.M., Russell, D.S. and Kalb, R.G., 2000, Brain-derived neurotrophic factor induces excitotoxic sensitivity in cultured embryonic rat spinal motor neurons through activation of the phosphatidylinositol 3-kinase pathway. *J. Neurochem.* **74**: 582.
Gauthier, L.R., Charrin, B.C., Borrell-Pages, M., Dompierre, J.P., Rangone, H., Cordelieres, F.P., De Mey, J., MacDonald, M.E., Lessmann, V., Humbert, S. and Saudou, F., 2004, Huntingtin controls neurotrophic support and survival of neurons by enhancing BDNF vesicular transport along microtubules. *Cell* **118**: 127.
Geller, B., Badner, J.A., Tillman, R., Christian, S.L., Bolhofner, K. and Cook, E.H., Jr., 2004, Linkage disequilibrium of the brain-derived neurotrophic factor Val66Met polymorphism in children with a prepubertal and early adolescent bipolar disorder phenotype. *Am. J. Psychiatry* **161**: 1698.
Gibbs, R.B. and Pfaff, D.W., 1994, In situ hybridization detection of trkA mRNA in brain: distribution, colocalization with p75NGFR and up-regulation by nerve growth factor. *J. Comp. Neurol.* **341**: 324.
Ginsberg, S.D., Che, S., Wuu, J., Counts, S.E. and Mufson, E.J., 2006, Down regulation of trk but not p75[NTR] gene expression in single cholinergic basal forebrain neurons mark the progression of Alzheimer's disease. *J. Neurochem.* **97**: 475.
Girard, C., Bemelmans, A.P., Dufour, N., Mallet, J., Bachelin, C., Nait-Oumesmar, B., Baron-Van Evercooren, A. and Lachapelle, F., 2005, Grafts of brain-derived neurotrophic factor and neurotrophin 3-transduced primate Schwann cells lead to functional recovery of the demyelinated mouse spinal cord. *J. Neurosci.* **25**: 7924.
Gonzalez, D., Dees, W.L., Hiney, J.K., Ojeda, S.R. and Saneto, R.P., 1990, Expression of beta-nerve growth factor in cultured cells derived from the hypothalamus and cerebral cortex. *Brain Res.* **511**: 249.
Goodman, L.J., Valverde, J., Lim, F., Geschwind, M.D., Federoff, H.J., Geller, A.I. and Hefti, F., 1996, Regulated release and polarized localization of brain-derived neurotrophic factor in hippocampal neurons. *Mol. Cell. Neurosci.* **7**: 222.
Gotz, R., Koster, R., Winkler, C., Raulf, F., Lottspeich, F., Schartl, M. and Thoenen, H., 1994, Neurotrophin-6 is a new member of the nerve growth factor family. *Nature* **372**: 266.
Gozes, I., 2001, Neuroprotective peptide drug delivery and development: potential new therapeutics. *Trends Neurosci.* **24**: 700.
Grewal, S.S., York, R.D. and Stork, P.J., 1999, Extracellular-signal-regulated kinase signaling in neurons. *Curr. Opin. Neurobiol.* **9**: 544.
Haapasalo, A., Sipola, I., Larsson, K., Akerman, K.E., Stoilov, P., Stamm, S., Wong, G. and Castren, E., 2002, Regulation of TRKB surface expression by brain-derived neurotrophic factor and truncated TRKB isoforms. *J. Biol. Chem.* **277**: 43160.
Hall, D., Dhilla, A., Charalambous, A., Gogos, J.A. and Karayiorgou, M., 2003, Sequence variants of the brain-derived neurotrophic factor (BDNF) gene are strongly associated with obsessive-compulsive disorder. *Am. J. Hum. Genet.* **73**: 370.
Hama, H., Hara, C., Yamaguchi, K. and Miyawaki, A., 2004, PKC signaling mediates global enhancement of excitatory synaptogenesis in neurons triggered by local contact with astrocytes. *Neuron* **41**: 405.
Hamanoue, M., Middleton, G., Wyatt, S., Jaffray, E., Hay, R.T. and Davies, A.M., 1999, p75-mediated NF-kappaB activation enhances the survival response of developing sensory neurons to nerve growth factor. *Mol. Cell. Neurosci.* **14**: 28.
Han, B.H. and Holtzman, D.M., 2000, BDNF protects the neonatal brain from hypoxic-ischemic injury in vivo via the ERK pathway. *J. Neurosci.* **20**: 5775.
Hannila, S.S., Lawrance, G.M., Ross, G.M. and Kawaja, M.D., 2004, TrkA and mitogen-activated protein kinase phosphorylation are enhanced in sympathetic neurons lacking functional p75 neurotrophin receptor expression. *Eur. J. Neurosci.* **19**: 2903.
Hansen, A.J., 1977, Extracellular potassium concentration in juvenile and adult rat brain cortex during anoxia. *Acta Physiol. Scand.* **99**: 412.

Harada, T., Harada, C., Nakayama, N., Okuyama, S., Yoshida, K., Kohsaka, S., Matsuda, H. and Wada, K., 2000, Modification of glial-neuronal cell interactions prevents photoreceptor apoptosis during light-induced retinal degeneration. *Neuron* **26**: 533.

Harada, T., Harada, C., Kohsaka, S., Wada, E., Yoshida, K., Ohno, S., Mamada, H., Tanaka, K., Parada, L.F. and Wada, K., 2002, Microglia-Müller glia cell interactions control neurotrophic factor production during light-induced retinal degeneration. *J. Neurosci.* **22**: 9228.

Harada, C., Harada, T., Quah, H.M., Maekawa, F., Yoshida, K., Ohno, S., Wada, K., Parada, L.F. and Tanaka, K., 2003, Potential role of glial cell line-derived neurotrophic factor receptors in Müller glial cells during light-induced retinal degeneration. *Neuroscience* **122**: 229.

Harada, C., Harada, T., Nakamura, K., Sakai, Y., Tanaka, K. and Parada, L.F., 2006, Effect of p75^{NTR} on the regulation of naturally occurring cell death and retinal ganglion cell number in the mouse eye. *Dev. Biol.* **290**: 57.

Harrington, A.W., Leiner, B., Blechschmitt, C., Arevalo, J.C., Lee, R., Morl, K., Meyer, M., Hempstead, B.L., Yoon, S.O. and Giehl, K.M., 2004, Secreted proNGF is a pathophysiological death-inducing ligand after adult CNS injury. *Proc. Natl. Acad. Sci. USA* **101**: 6226.

Hartmann, M., Heumann, R. and Lessmann, V., 2001, Synaptic secretion of BDNF after high-frequency stimulation of glutamatergic synapses. *EMBO J.* **20**: 5887.

Hashimoto, Y., Kaneko, Y., Tsukamoto, E., Frankowski, H., Kouyama, K., Kita, Y., Niikura, T., Aiso, S., Bredesen, D.E., Matsuoka, M. and Nishimoto, I., 2004, Molecular characterization of neurohybrid cell death induced by Alzheimer's amyloid-beta peptides via p75NTR/PLAIDD. *J. Neurochem.* **90**: 549.

Hayes, R.C., Wiley, R.G. and Armstrong, D.M., 1992, Induction of nerve growth factor receptor (p75NGFr) mRNA within hypoglossal motoneurons following axonal injury. *Brain Res. Mol. Brain Res.* **15**: 291.

Hazzalin, C.A. and Mahadevan, L.C., 2002, MAPK-regulated transcription: a continuously variable gene switch? *Nat. Rev. Mol. Cell Biol.* **3**: 30.

Heinrich, M., Gorath, M. and Richter-Landsberg, C., 1999, Neurotrophin-3 (NT-3) modulates early differentiation of oligodendrocytes in rat brain cortical cultures. *Glia* **28**: 244.

Hempstead, B.L. and Salzer, J.L., 2002, Neurobiology. A glial spin on neurotrophins. *Science* **298**: 1184.

Hetman, M. and Gozdz, A., 2004, Role of extracellular signal regulated kinases 1 and 2 in neuronal survival. *Eur. J. Biochem.* **271**: 2050.

Hetman, M., Kanning, K., Cavanaugh, J.E. and Xia, Z., 1999, Neuroprotection by brain-derived neurotrophic factor is mediated by extracellular signal-regulated kinase and phosphatidylinositol 3-kinase. *J. Biol. Chem.* **274**: 22569.

Hetman, M., Cavanaugh, J.E., Kimelman, D. and Xia, Z., 2000, Role of glycogen synthase kinase-3beta in neuronal apoptosis induced by trophic withdrawal. *J. Neurosci.* **20**: 2567.

Higuchi, M., Tomioka, M., Takano, J., Shirotani, K., Iwata, N., Masumoto, H., Maki, M., Itohara, S. and Saido, T.C., 2005, Distinct mechanistic roles of calpain and caspase activation in neurodegeneration as revealed in mice overexpressing their specific inhibitors. *J. Biol. Chem.* **280**: 15229.

Holtzman, D.M., Li, Y., Parada, L.F., Kinsman, S., Chen, C.K., Valletta, J.S., Zhou, J., Long, J.B. and Mobley, W.C., 1992, p140trk mRNA marks NGF-responsive forebrain neurons: evidence that trk gene expression is induced by NGF. *Neuron* **9**: 465.

Holtzman, D.M., Kilbridge, J., Li, Y., Cunningham, E.T., Jr., Lenn, N.J., Clary, D.O., Reichardt, L.F. and Mobley, W.C., 1995, TrkA expression in the CNS: evidence for the existence of several novel NGF-responsive CNS neurons. *J. Neurosci.* **15**: 1567.

Hossmann, K.A., Sakaki, S. and Zimmerman, V., 1977, Cation activities in reversible ischemia of the cat brain. *Stroke* **8**: 77.

Huang, E.J. and Reichardt, L.F., 2001, Neurotrophins: roles in neuronal development and function. *Annu. Rev. Neurosci.* **24**: 677.

Huang, E.J. and Reichardt, L.F., 2003, Trk receptors: roles in neuronal signal transduction. *Annu. Rev. Biochem.* **72**: 609.

Husson, I., Rangon, C.M., Lelievre, V., Bemelmans, A.P., Sachs, P., Mallet, J., Kosofsky, B.E. and Gressens, P., 2005, BDNF-induced white matter neuroprotection and stage-dependent neuronal survival following a neonatal excitotoxic challenge. *Cereb. Cortex.* **15**: 250.

Hwang, I.K., Lee, K.Y., Yoo, K.Y., Kim, D.S., Lee, N.S., Jeong, Y.G., Kang, T.C., Han, B.H., Kim, J.S. and Won, M.H., 2005, Tyrosine kinase A but not phosphacan/protein tyrosine phosphatase-zeta/beta immunoreactivity and protein level changes in neurons and astrocytes in the gerbil hippocampus proper after transient forebrain ischemia. *Brain Res.* **1036**: 35.

Ishikawa, Y., Ikeuchi, T. and Hatanaka, H., 2000, Brain-derived neurotrophic factor accelerates nitric oxide donor-induced apoptosis of cultured cortical neurons. *J. Neurochem.* **75**: 494.

Jaboin, J., Kim, C.J., Kaplan, D.R. and Thiele, C.J., 2002, Brain-derived neurotrophic factor activation of TrkB protects neuroblastoma cells from chemotherapy-induced apoptosis via phosphatidylinositol 3'-kinase pathway. *Cancer Res.* **62**: 6756.

Jaboin, J., Hong, A., Kim, C.J. and Thiele, C.J., 2003, Cisplatin-induced cytotoxicity is blocked by brain-derived neurotrophic factor activation of TrkB signal transduction path in neuroblastoma. *Cancer Lett.* **193**: 109.

Jean, I., Lavialle, C., Barthelaix-Pouplard, A. and Fressinaud, C., 2003, Neurotrophin-3 specifically increases mature oligodendrocyte population and enhances remyelination after chemical demyelination of adult rat CNS. *Brain Res.* **972**: 110.

Jiang, X., Tian, F., Mearow, K., Okagaki, P., Lipsky, R.H. and Marini, A.M., 2005, The excitoprotective effect of *N*-methyl-*D*-aspartate receptors is mediated by a brain-derived neurotrophic factor autocrine loop in cultured hippocampal neurons. *J. Neurochem.* **94**: 713.

Jing, S., Tapley, P. and Barbacid, M., 1992, Nerve growth factor mediates signal transduction through Trk homodimer receptors. *Neuron* **9**: 1067.

Junier, M.P., Suzuki, F., Onteniente, B. and Peschanski, M., 1994, Target-deprived CNS neurons express the NGF gene while reactive glia around their axonal terminals contain low and high affinity NGF receptors. *Brain. Res. Mol. Brain Res.* **24**: 247.

Kano, T., Abe, T., Tomita, H., Sakata, T., Ishiguro, S. and Tamai, M., 2002, Protective effect against ischemia and light damage of iris pigment epithelial cells transfected with the BDNF gene. *Invest. Ophthalmol. Vis. Sci.* **43**: 3744.

Kaplan, M.R., Meyer-Franke, A., Lambert, S., Bennett, V., Duncan, I.D., Levinson, S.R. and Barres, B.A., 1997, Induction of sodium channel clustering by oligodendrocytes. *Nature* **386**: 724.

Kenchappa, R.S., Zampieri, N., Chao, M.V., Barker, P.A., Teng, H.K., Hempstead, B.L. and Carter, B.D., 2006, Ligand-dependent cleavage of the p75 neurotrophin receptor is necessary for NRIF nuclear translocation and apoptosis in sympathetic neurons. *Neuron* **50**: 219.

Kerkhoff, H., Jennekens, F.G., Troost, D. and Veldman, H., 1991, Nerve growth factor receptor immunostaining in the spinal cord and peripheral nerves in amyotrophic lateral sclerosis. *Acta Neuropathol. (Berl.)* **81**: 649.

Kerschensteiner, M., Gallmeier, E., Behrens, L., Leal, V.V., Misgeld, T., Klinkert, W.E., Kolbeck, R., Hoppe, E., Oropeza-Wekerle, R.L., Bartke, I., Stadelmann, C., Lassmann, H., Wekerle, H. and Hohlfeld, R., 1999, Activated human T cells, B cells and monocytes produce brain-derived neurotrophic factor in vitro and in inflammatory brain lesions: a neuroprotective role of inflammation? *J. Exp. Med.* **189**: 865.

Kerschensteiner, M., Stadelmann, C., Dechant, G., Wekerle, H. and Hohlfeld, R., 2003, Neurotrophic cross-talk between the nervous and immune systems: implications for neurological diseases. *Ann. Neurol.* **53**: 292.

Khursigara, G., Orlinick, J.R. and Chao, M.V., 1999, Association of the p75 neurotrophin receptor with TRAF6. *J. Biol. Chem.* **274**: 2597.

Kido, N., Tanihara, H., Honjo, M., Inatani, M., Tatsuno, T., Nakayama, C. and Honda, Y., 2000, Neuroprotective effects of brain-derived neurotrophic factor in eyes with NMDA-induced neuronal death. *Brain Res.* **884**: 59.

Kim, S.U. and de Vellis, J., 2005, Microglia in health and disease. *J. Neurosci. Res.* **81**: 302.

Kim, S.H., Won, S.J., Sohn, S., Kwon, H.J., Lee, J.Y., Park, J.H. and Gwag, B.J., 2002, Brain-derived neurotrophic factor can act as a pronecrotic factor through transcriptional and translational activation of NADPH oxidase. *J. Cell Biol.* **159**: 821.

Kim, H.J., Hwang, J.J., Behrens, M.M., Snider, B.J., Choi, D.W. and Koh, J.Y., 2003, TrkB mediates BDNF-induced potentiation of neuronal necrosis in cortical culture. *Neurobiol. Dis.* **14**: 110.

Kim, D.H., Zhao, X., Tu, C.H., Casaccia-Bonnefil, P. and Chao, M.V., 2004, Prevention of apoptotic but not necrotic cell death following neuronal injury by neurotrophins signaling through the tyrosine kinase receptor. *J. Neurosurg.* **100**: 79.

Kinoshita, Y., Ueyama, T., Senba, E., Terada, T., Nakai, K. and Itakura, T., 2001, Expression of c-fos, heat shock protein 70, neurotrophins and cyclooxygenase-2 mRNA in response to focal cerebral ischemia/reperfusion in rats and their modification by magnesium sulfate. *J. Neurotrauma* **18**: 435.

Klein, R., Silos-Santiago, I., Smeyne, R.J., Lira, S.A., Brambilla, R., Bryant, S., Zhang, L., Snider, W.D. and Barbacid, M., 1994, Disruption of the neurotrophin-3 receptor gene trkC eliminates Ia muscle afferents and results in abnormal movements. *Nature* **368**: 249.

Klesse, L.J. and Parada, L.F., 1998, p21 ras and phosphatidylinositol-3 kinase are required for survival of wild-type and NF1 mutant sensory neurons. *J. Neurosci.* **18**: 10420.

Klesse, L.J., Meyers, K.A., Marshall, C.J. and Parada, L.F., 1999, Nerve growth factor induces survival and differentiation through two distinct signaling cascades in PC12 cells. *Oncogene* **18**: 2055.

Ko, M.L., Hu, D.N., Ritch, R. and Sharma, S.C., 2000, The combined effect of brain-derived neurotrophic factor and a free radical scavenger in experimental glaucoma. *Invest. Ophthalmol. Vis. Sci.* **41**: 2967.

Koh, S., Oyler, G.A. and Higgins, G.A., 1989, Localization of nerve growth factor receptor messenger RNA and protein in the adult rat brain. *Exp. Neurol.* **106**: 209.

Koh, J.Y., Gwag, B.J., Lobner, D. and Choi, D.W., 1995, Potentiated necrosis of cultured cortical neurons by neurotrophins. *Science* **268**: 573.

Kohara, K., Kitamura, A., Morishima, M. and Tsumoto, T., 2001, Activity-dependent transfer of brain-derived neurotrophic factor to postsynaptic neurons. *Science* **291**: 2419.

Kohno, R., Sawada, H., Kawamoto, Y., Uemura, K., Shibasaki, H. and Shimohama, S., 2004, BDNF is induced by wild-type alpha-synuclein but not by the two mutants, A30P or A53T, in glioma cell line. *Biochem. Biophys. Res. Commun.* **318**: 113.

Kokaia, Z. andsberg, G., Martinez-Serrano, A. and Lindvall, O., 1998, Focal cerebral ischemia in rats induces expression of p75 neurotrophin receptor in resistant striatal cholinergic neurons. *Neuroscience* **84**: 1113.

Kokaia, Z., Bengzon, J., Metsis, M., Kokaia, M., Persson, H. and Lindvall, O., 1993, Coexpression of neurotrophins and their receptors in neurons of the central nervous system. *Proc. Natl. Acad. Sci. USA* **90**: 6711.

Koliatsos, V.E., Crawford, T.O. and Price, D.L., 1991, Axotomy induces nerve growth factor receptor immunoreactivity in spinal motor neurons. *Brain Res.* **549**: 297.

Kong, H., Boulter, J., Weber, J.L., Lai, C. and Chao, M.V., 2001, An evolutionarily conserved transmembrane protein that is a novel downstream target of neurotrophin and ephrin receptors. *J. Neurosci.* **21**: 176.

Krueger-Naug, A.M., Emsley, J.G., Myers, T.L., Currie, R.W. and Clarke, D.B., 2003, Administration of brain-derived neurotrophic factor suppresses the expression of heat shock protein 27 in rat retinal ganglion cells following axotomy. *Neuroscience* **116**: 49.

Kubo, T., Nonomura, T., Enokido, Y. and Hatanaka, H., 1995, Brain-derived neurotrophic factor (BDNF) can prevent apoptosis of rat cerebellar granule neurons in culture. *Brain Res. Dev. Brain Res.* **85**: 249.

Kuruvilla, R., Ye, H. and Ginty, D.D., 2000, Spatially and functionally distinct roles of the PI3-K effector pathway during NGF signaling in sympathetic neurons. *Neuron* **27**: 499.

Kuruvilla, R., Zweifel, L.S., Glebova, N.O., Lonze, B.E., Valdez, G., Ye, H. and Ginty, D.D., 2004, A neurotrophin signaling cascade coordinates sympathetic neuron development through differential control of TrkA trafficking and retrograde signaling. *Cell* **118**: 243.

Ladiwala, U., Lachance, C., Simoneau, S.J., Bhakar, A., Barker, P.A. and Antel, J.P., 1998, p75 neurotrophin receptor expression on adult human oligodendrocytes: signaling without cell death in response to NGF. *J. Neurosci.* **18**: 1297.

Larsson, E., Nanobashvili, A., Kokaia, Z. and Lindvall, O., 1999, Evidence for neuroprotective effects of endogenous brain-derived neurotrophic factor after global forebrain ischemia in rats. *J. Cereb. Blood Flow Metab.* **19**: 1220.

Lee, F.S. and Chao, M.V., 2001, Activation of Trk neurotrophin receptors in the absence of neurotrophins. *Proc. Natl. Acad. Sci. USA* **98**: 3555.

Lee, K.F., Li, E., Huber, L.J., Landis, S.C., Sharpe, A.H., Chao, M.V. and Jaenisch, R., 1992, Targeted mutation of the gene encoding the low affinity NGF receptor p75 leads to deficits in the peripheral sensory nervous system. *Cell* **69**: 737.

Lee, T.H., Abe, K., Kogure, K. and Itoyama, Y., 1995, Expressions of nerve growth factor and p75 low affinity receptor after transient forebrain ischemia in gerbil hippocampal CA1 neurons. *J. Neurosci. Res.* **41**: 684.

Lee, T.H., Kato, H., Chen, S.T., Kogure, K. and Itoyama, Y., 1998a, Expression of nerve growth factor and TrkA after transient focal cerebral ischemia in rats. *Stroke* **29**: 1687.

Lee, T.H., Kato, H., Pan, L.H., Ryu, J.H., Kogure, K. and Itoyama, Y., 1998b, Localization of nerve growth factor, TrkA and p75 immunoreactivity in the hippocampal formation and basal forebrain of adult rats. *Neuroscience* **83**: 335.

Lee, R., Kermani, P., Teng, K.K. and Hempstead, B.L., 2001, Regulation of cell survival by secreted proneurotrophins. *Science* **294**: 1945.

Lee, T.H., Kato, H., Chen, S.T., Kogure, K. and Itoyama, Y., 2002a, Expression disparity of brain-derived neurotrophic factor immunoreactivity and mRNA in ischemic hippocampal neurons. *Neuroreport* **13**: 2271.

Lee, F.S., Rajagopal, R., Kim, A.H., Chang, P.C. and Chao, M.V., 2002b, Activation of TrkA neurotrophin receptor signaling by pituitary adenylate cyclase-activating polypeptides. *J. Biol. Chem.* **277**: 9096.

Leeds, P., Leng, Y., Chalecka-Franaszek, E. and Chuang, D.M., 2005, Neurotrophins protect against cytosine arabinoside-induced apoptosis of immature rat cerebellar neurons. *Neurochem. Int.* **46**: 61.

Lemke, G. and Chao, M., 1988, Axons regulate Schwann cell expression of the major myelin and NGF receptor genes. *Development* **102**: 499.

Lesne, S., Gabriel, C., Nelson, D.A., White, E., Mackenzie, E.T., Vivien, D. and Buisson, A., 2005, Akt-dependent expression of NAIP-1 protects neurons against amyloid-β toxicity. *J. Biol. Chem.* **280**: 24941.

Levi-Montalcini, R., 1987, The nerve growth factor 35 years later. *Science* **237**: 1154.

Li, Z., Jaboin, J., Dennis, P.A. and Thiele, C.J., 2005, Genetic and pharmacologic identification of Akt as a mediator of brain-derived neurotrophic factor/TrkB rescue of neuroblastoma cells from chemotherapy-induced cell death. *Cancer Res.* **65**: 2070.

Liebl, D.J., Huang, W., Young, W. and Parada, L.F., 2001, Regulation of Trk receptors following contusion of the rat spinal cord. *Exp. Neurol.* **167**: 15.

Lindsay, R.M., 1979, Adult rat brain astrocytes support survival of both NGF-dependent and NGF-insensitive neurones. *Nature* **282**: 80.

Lindsay, R.M., Wiegand, S.J., Altar, C.A. and DiStefano, P.S., 1994, Neurotrophic factors: from molecule to man. *Trends Neurosci.* **17**: 182.
Linggi, M.S., Burke, T.L., Williams, B.B., Harrington, A., Kraemer, R., Hempstead, B.L., Yoon, S.O. and Carter, B.D., 2005, Neurotrophin receptor interacting factor (NRIF) is an essential mediator of apoptotic signaling by the p75 neurotrophin receptor. *J. Biol. Chem.* **280**: 13801.
Liot, G., Gabriel, C., Cacquevel, M., Ali, C., MacKenzie, E.T., Buisson, A. and Vivien, D., 2004, Neurotrophin-3-induced PI-3 kinase/Akt signaling rescues cortical neurons from apoptosis. *Exp. Neurol.* **187**: 38.
Lipton, S.A., 2001, Retinal ganglion cells, glaucoma and neuroprotection. *Prog. Brain Res.* **131**: 712.
Liu, Q.Y., Schaffner, A.E., Li, Y.X., Dunlap, V. and Barker, J.L., 1996, Upregulation of $GABA_A$ current by astrocytes in cultured embryonic rat hippocampal neurons. *J. Neurosci.* **16**: 2912.
Liu, Q.Y., Schaffner, A.E., Chang, Y.H., Vaszil, K. and Barker, J.L., 1997, Astrocytes regulate amino acid receptor current densities in embryonic rat hippocampal neurons. *J. Neurobiol.* **33**: 848.
Liu, Y.Z., Chrivia, J.C. and Latchman, D.S., 1998, Nerve growth factor up-regulates the transcriptional activity of CBP through activation of the p42/p44 MAPK cascade. *J. Biol. Chem.* **273**: 32400.
Liu, L., Cavanaugh, J.E., Wang, Y., Sakagami, H., Mao, Z. and Xia, Z., 2003, ERK5 activation of MEF2-mediated gene expression plays a critical role in BDNF-promoted survival of developing but not mature cortical neurons. *Proc. Natl. Acad. Sci. USA* **100**: 8532.
Lobsiger, C.S., Schweitzer, B., Taylor, V. and Suter, U., 2000, Platelet-derived growth factor-BB supports the survival of cultured rat Schwann cell precursors in synergy with neurotrophin-3. *Glia* **30**: 290.
Logan, A. and Berry, M., 2002, Cellular and molecular determinants of glial scar formation. *Adv. Exp. Med. Biol.* **513**: 115.
Lou, H., Kim, S.K., Zaitsev, E., Snell, C.R., Lu, B. and Loh, Y.P., 2005, Sorting and activity-dependent secretion of BDNF require interaction of a specific motif with the sorting receptor carboxypeptidase E. *Neuron* **45**: 245.
Lu, B., Pang, P.T. and Woo, N.H., 2005, The yin and yang of neurotrophin action. *Nat. Rev. Neurosci.* **6**: 603.
Madeddu, F., Naska, S. and Bozzi, Y., 2004, BDNF down-regulates the caspase 3 pathway in injured geniculo-cortical neurones. *Neuroreport* **15**: 2045.
Malcangio, M. and Lessmann, V., 2003, A common thread for pain and memory synapses? Brain-derived neurotrophic factor and trkB receptors. *Trends Pharmacol. Sci.* **24**: 116.
Markus, A., Zhong, J. and Snider, W.D., 2002, Raf and Akt mediate distinct aspects of sensory axon growth. *Neuron* **35**: 65.
Martin, K.R., Quigley, H.A., Zack, D.J., Levkovitch-Verbin, H., Kielczewski, J., Valenta, D., Baumrind, L., Pease, M.E., Klein, R.L. and Hauswirth, W.W., 2003, Gene therapy with brain-derived neurotrophic factor as a protection: retinal ganglion cells in a rat glaucoma model. *Invest. Ophthalmol. Vis. Sci.* **44**: 4357.
Martinez-Murillo, R., Fernandez, A.P., Bentura, M.L. and Rodrigo, J., 1998, Subcellular localization of low-affinity nerve growth factor receptor-immunoreactive protein in adult rat Purkinje cells following traumatic injury. *Exp. Brain Res.* **119**: 47.
Martinowich, K., Hattori, D., Wu, H., Fouse, S., He, F., Hu, Y., Fan, G. and Sun, Y.E., 2003, DNA methylation-related chromatin remodeling in activity-dependent BDNF gene regulation. *Science* **302**: 890.
Matsuzaki, H., Namikawa, K., Kiyama, H., Mori, N. and Sato, K., 2004, Brain-derived neurotrophic factor rescues neuronal death induced by methamphetamine. *Biol. Psychiatry* **55**: 52.
Mattson, M.P., Lovell, M.A., Furukawa, K. and Markesbery, W.R., 1995, Neurotrophic factors attenuate glutamate-induced accumulation of peroxides, elevation of intracellular Ca^{2+} concentration and neurotoxicity and increase antioxidant enzyme activities in hippocampal neurons. *J. Neurochem.* **65**: 1740.
Mazzoni, I.E., Said, F.A., Aloyz, R., Miller, F.D. and Kaplan, D., 1999, Ras regulates sympathetic neuron survival by suppressing the p53-mediated cell death pathway. *J. Neurosci.* **19**: 9716.
McDonald, N.Q. and Hendrickson, W.A., 1993, A structural superfamily of growth factors containing a cystine knot motif. *Cell* **73**: 421.
McKeon, R.J., Silver, J. and Large, T.H., 1997, Expression of full-length trkB receptors by reactive astrocytes after chronic CNS injury. *Exp. Neurol.* **148**: 558.
Meakin, S.O., MacDonald, J.I., Gryz, E.A., Kubu, C.J. and Verdi, J.M., 1999, The signaling adapter FRS-2 competes with Shc for binding to the nerve growth factor receptor TrkA. A model for discriminating proliferation and differentiation. *J. Biol. Chem.* **274**: 9861.
Meier, C., Parmantier, E., Brennan, A., Mirsky, R. and Jessen, K.R., 1999, Developing Schwann cells acquire the ability to survive without axons by establishing an autocrine circuit involving insulin-like growth factor, neurotrophin-3 and platelet-derived growth factor-BB. *J. Neurosci.* **19**: 3847.
Merlio, J.P., Ernfors, P., Jaber, M. and Persson, H., 1992, Molecular cloning of rat trkC and distribution of cells expressing messenger RNAs for members of the trk family in the rat central nervous system. *Neuroscience* **51**: 513.
Meyer, A., Chretien, P., Massicotte, G., Sargent, C., Chretien, M. and Marcinkiewicz, M., 1996, Kainic acid increases the expression of the prohormone convertases furin and PC1 in the mouse hippocampus. *Brain Res.* **732**: 121.

Meyer-Franke, A., Wilkinson, G.A., Kruttgen, A., Hu, M., Munro, E., Hanson, M.G., Jr., Reichardt, L.F. and Barres, B.A., 1998, Depolarization and cAMP elevation rapidly recruit TrkB to the plasma membrane of CNS neurons. *Neuron* **21**: 681.

Michaelidis, T.M., Sendtner, M., Cooper, J.D., Airaksinen, M.S., Holtmann, B., Meyer, M. and Thoenen, H., 1996, Inactivation of bcl-2 results in progressive degeneration of motoneurons, sympathetic and sensory neurons during early postnatal development. *Neuron* **17**: 75.

Michalski, B. and Fahnestock, M., 2003, Pro-brain-derived neurotrophic factor is decreased in parietal cortex in Alzheimer's disease. *Brain. Res. Mol. Brain Res.* **111**: 148.

Middlemas, D.S., Meisenhelder, J. and Hunter, T., 1994, Identification of TrkB autophosphorylation sites and evidence that phospholipase C-gamma 1 is a substrate of the TrkB receptor. *J. Biol. Chem.* **269**: 5458.

Middlemiss, P.J., Gysbers, J.W. and Rathbone, M.P., 1995, Extracellular guanosine and guanosine-5'-triphosphate increase: NGF synthesis and release from cultured mouse neopallial astrocytes. *Brain Res.* **677**: 152.

Minichiello, L., Calella, A.M., Medina, D.L., Bonhoeffer, T., Klein, R. and Korte, M., 2002, Mechanism of TrkB-mediated hippocampal long-term potentiation. *Neuron* **36**: 121.

Miwa, T., Furukawa, S., Nakajima, K., Furukawa, Y. and Kohsaka, S., 1997, Lipopolysaccharide enhances synthesis of brain-derived neurotrophic factor in cultured rat microglia. *J. Neurosci. Res.* **50**: 1023.

Miyake, K., Yamamoto, W., Tadokoro, M., Takagi, N., Sasakawa, K., Nitta, A., Furukawa, S. and Takeo, S., 2002, Alterations in hippocampal GAP-43, BDNF and L1 following sustained cerebral ischemia. *Brain Res.* **935**: 24.

Mocchetti, I. and Bachis, A., 2004, Brain-derived neurotrophic factor activation of TrkB protects neurons from HIV-1/gp120-induced cell death. *Crit. Rev. Neurobiol.* **16**: 51.

Moelling, K., Schad, K., Bosse, M., Zimmermann, S. and Schweneker, M., 2002, Regulation of Raf-Akt cross-talk. *J. Biol. Chem.* **277**: 31099.

Moqrich, A., Earley, T.J., Watson, J. Andahazy, M., Backus, C., Martin-Zanca, D., Wright, D.E., Reichardt, L.F. and Patapoutian, A., 2004, Expressing TrkC from the TrkA locus causes a subset of dorsal root ganglia neurons to switch fate. *Nat. Neurosci.* **7**: 812.

Mowla, S.J., Pareek, S., Farhadi, H.F., Petrecca, K., Fawcett, J.P., Seidah, N.G., Morris, S.J., Sossin, W.S. and Murphy, R.A., 1999, Differential sorting of nerve growth factor and brain-derived neurotrophic factor in hippocampal neurons. *J. Neurosci.* **19**: 2069.

Mufson, E.J. and Kordower, J.H., 1992, Cortical neurons express nerve growth factor receptors in advanced age and Alzheimer disease. *Proc. Natl. Acad. Sci. USA* **89**: 569.

Mufson, E.J., Bothwell, M., Hersh, L.B. and Kordower, J.H., 1989, Nerve growth factor receptor immunoreactive profiles in the normal, aged human basal forebrain: colocalization with cholinergic neurons. *J. Comp. Neurol.* **285**: 196.

Mufson, E.J., Higgins, G.A. and Kordower, J.H., 1991, Nerve growth factor receptor immunoreactivity in the new world monkey (*Cebus apella*) and human cerebellum. *J. Comp. Neurol.* **308**: 555.

Mufson, E.J., Brashers-Krug, T. and Kordower, J.H., 1992, p75 nerve growth factor receptor immunoreactivity in the human brainstem and spinal cord. *Brain Res.* **589**: 115.

Mufson, E.J., Ginsberg, S.D., Ikonomovic, M.D. and DeKosky, S.T., 2003, Human cholinergic basal forebrain: chemoanatomy and neurologic dysfunction. *J. Chem. Neuroanat.* **26**: 233.

Muller, G., Storz, P., Bourteele, S., Doppler, H., Pfizenmaier, K., Mischak, H., Philipp, A., Kaiser, C. and Kolch, W., 1998, Regulation of Raf-1 kinase by TNF via its second messenger ceramide and cross-talk with mitogenic signaling. *EMBO J.* **17**: 732.

Muragaki, Y., Timothy, N., Leight, S., Hempstead, B.L., Chao, M.V., Trojanowski, J.Q. and Lee, V.M., 1995, Expression of trk receptors in the developing and adult human central and peripheral nervous system. *J. Comp. Neurol.* **356**: 387.

Nakajima, K., Kikuchi, Y., Ikoma, E., Honda, S., Ishikawa, M., Liu, Y. and Kohsaka, S., 1998, Neurotrophins regulate the function of cultured microglia. *Glia* **24**: 272.

Nakajima, K., Honda, S., Tohyama, Y., Imai, Y., Kohsaka, S. and Kurihara, T., 2001, Neurotrophin secretion from cultured microglia. *J. Neurosci. Res.* **65**: 322.

Nakazawa, T., Tamai, M. and Mori, N., 2002, Brain-derived neurotrophic factor prevents axotomized retinal ganglion cell death through MAPK and PI3K signaling pathways. *Invest. Ophthalmol. Vis. Sci.* **43**: 3319.

Narhi, L.O., Rosenfeld, R., Talvenheimo, J., Prestrelski, S.J., Arakawa, T., Lary, J.W., Kolvenbach, C.G., Hecht, R., Boone, T., Miller, J.A. et al., 1993, Comparison of the biophysical characteristics of human brain-derived neurotrophic factor, neurotrophin-3 and nerve growth factor. *J. Biol. Chem.* **268**: 13309.

Narumiya, S., Ohno, M., Tanaka, N., Yamano, T. and Shimada, M., 1998, Enhanced expression of full-length TrkB receptors in young rat brain with hypoxic/ischemic injury. *Brain Res.* **797**: 278.

Neves-Pereira, M., Mundo, E., Muglia, P., King, N., Macciardi, F. and Kennedy, J.L., 2002, The brain-derived neurotrophic factor gene confers susceptibility to bipolar disorder: evidence from a family-based association study. *Am. J. Hum. Genet.* **71**: 651.

Nilsson, A.S., Fainzilber, M., Falck, P. and Ibanez, C.F., 1998, Neurotrophin-7: a novel member of the neurotrophin family from the zebrafish. *FEBS Lett.* **424**: 285.
Nimnual, A.S., Yatsula, B.A. and Bar-Sagi, D., 1998, Coupling of Ras and Rac guanosine triphosphatases through the Ras exchanger Sos. *Science* **279**: 560.
Nykjaer, A., Lee, R., Teng, K.K., Jansen, P., Madsen, P., Nielsen, M.S., Jacobsen, C., Kliemannel, M., Schwarz, E., Willnow, T.E., Hempstead, B.L. and Petersen, C.M., 2004, Sortilin is essential for proNGF-induced neuronal cell death. *Nature* **427**: 843.
Nykjaer, A., Willnow, T.E. and Petersen, C.M., 2005, p75^{NTR}-live or let die. *Curr. Opin. Neurobiol.* **15**: 49.
Oderfeld-Nowak, B., Zaremba, M., Micera, A. and Aloe, L., 2001, The upregulation of nerve growth factor receptors in reactive astrocytes of rat spinal cord during experimental autoimmune encephalomyelitis. *Neurosci. Lett.* **308**: 165-168.
Oderfeld-Nowak, B., Orzylowska-Sliwinska, O., Soltys, Z., Zaremba, M., Januszewski, S., Janeczko, K. and Mossakowski, M., 2003, Concomitant up-regulation of astroglial high and low affinity nerve growth factor receptors in the CA1 hippocampal area following global transient cerebral ischemia in rat. *Neuroscience* **120**: 31.
Ohira, K., Kumanogoh, H., Sahara, Y., Homma, K.J., Hirai, H., Nakamura, S. and Hayashi, M., 2005, A truncated tropomyosin-related kinase B receptor, T1, regulates glial cell morphology via Rho GDP dissociation inhibitor 1. *J. Neurosci.* **25**: 1343.
Okoye, G., Zimmer, J., Sung, J., Gehlbach, P., Deering, T., Nambu, H., Hackett, S., Melia, M., Esumi, N., Zack, D.J. and Campochiaro, P.A., 2003, Increased expression of brain-derived neurotrophic factor preserves retinal function and slows cell death from rhodopsin mutation or oxidative damage. *J. Neurosci.* **23**: 4164.
Onyango, I.G., Tuttle, J.B. and Bennett, J.P., Jr., 2005, Brain-derived growth factor and glial cell line-derived growth factor use distinct intracellular signaling pathways to protect PD cybrids from H_2O_2-induced neuronal death. *Neurobiol. Dis.* **20**: 141.
Osborne, N.N. and Herrera, A.J., 1994, The effect of experimental ischaemia and excitatory amino acid agonists on the GABA and serotonin immunoreactivities in the rabbit retina. *Neuroscience* **59**: 1071.
Ouyang, Y., Kantor, R., Harris, K.M., Schuman, E.M. and Kennedy, M.B., 1997, Visualization of the distribution of autophosphorylated calcium/calmodulin-dependent protein kinase II after tetanic stimulation in the CA1 area of the hippocampus. *J. Neurosci.* **17**: 5416.
Ozbas-Gerceker, F., Gorter, J.A., Redeker, S., Ramkema, M., van der Valk, P., Baayen, J.C., Ozguc, M., Saygi, S., Soylemezoglu, F., Akalin, N., Troost, D. and Aronica, E., 2004, Neurotrophin receptor immunoreactivity in the hippocampus of patients with mesial temporal lobe epilepsy. *Neuropathol. Appl. Neurobiol.* **30**: 651.
Pang, P.T., Teng, H.K., Zaitsev, E., Woo, N.T., Sakata, K., Zhen, S., Teng, K.K., Yung, W.H., Hempstead, B.L. and Lu, B., 2004, Cleavage of proBDNF by tPA/plasmin is essential for long-term hippocampal plasticity. *Science* **306**: 487.
Park, J.A., Lee, J.Y., Sato, T.A. and Koh, J.Y., 2000, Co-induction of p75^{NTR} and p75^{NTR}-associated death executor in neurons after zinc exposure in cortical culture or transient ischemia in the rat. *J. Neurosci.* **20**: 9096.
Pease, M.E., McKinnon, S.J., Quigley, H.A., Kerrigan-Baumrind, L.A. and Zack, D.J., 2000, Obstructed axonal transport of BDNF and its receptor TrkB in experimental glaucoma. *Invest. Ophthalmol. Vis. Sci.* **41**: 764.
Pedraza, C.E., Podlesniy, P., Vidal, N., Arevalo, J.C., Lee, R., Hempstead, B., Ferrer, I., Iglesias, M. and Espinet, C., 2005, Pro-NGF isolated from the human brain affected by Alzheimer's disease induces neuronal apoptosis mediated by p75NTR. *Am. J. Pathol.* **166**: 533.
Pehar, M., Cassina, P., Vargas, M.R., Castellanos, R., Viera, L., Beckman, J.S., Estevez, A.G. and Barbeito, L., 2004, Astrocytic production of nerve growth factor in motor neuron apoptosis: implications for amyotrophic lateral sclerosis. *J. Neurochem.* **89**: 464.
Peng, S., Wuu, J., Mufson, E.J. and Fahnestock, M., 2004, Increased proNGF levels in subjects with mild cognitive impairment and mild Alzheimer disease. *J. Neuropathol. Exp. Neurol.* **63**: 641.
Peng, S., Wuu, J., Mufson, E.J. and Fahnestock, M., 2005, Precursor form of brain-derived neurotrophic factor and mature brain-derived neurotrophic factor are decreased in the pre-clinical stages of Alzheimer's disease. *J. Neurochem.* **93**: 1412.
Pereira, D.B., Rebola, N., Rodrigues, R.J., Cunha, R.A., Carvalho, A.P. and Duarte, C.B., 2006, TrkB receptors modulation of glutamate release is limited to a subset of nerve terminals in the adult rat hippocampus. *J. Neurosci. Res.* **83**: 832.
Perez-Navarro, E., Canudas, A.M., Akerund, P., Alberch, J. and Arenas, E., 2000, Brain-derived neurotrophic factor, neurotrophin-3 and neurotrophin-4/5 prevent the death of striatal projection neurons in a rodent model of Huntington's disease. *J. Neurochem.* **75**: 2190.
Perez-Navarro, E., Gavalda, N., Gratacos, E. and Alberch, J., 2005, Brain-derived neurotrophic factor prevents changes in Bcl-2 family members and caspase-3 activation induced by excitotoxicity in the striatum. *J. Neurochem.* **92**: 678.
Pfrieger, F.W. and Barres, B.A., 1997, Synaptic efficacy enhanced by glial cells in vitro. *Science* **277**: 1684.

Philpott, K.L., McCarthy, M.J., Klippel, A. and Rubin, L.L., 1997, Activated phosphatidylinositol 3-kinase and Akt kinase promote survival of superior cervical neurons. *J. Cell Biol.* **139**: 809.
Poo, M.M., 2001, Neurotrophins as synaptic modulators. *Nat. Rev. Neurosci.* **2**: 24.
Porritt, M.J., Batchelor, P.E. and Howells, D.W., 2005, Inhibiting BDNF expression by antisense oligonucleotide infusion causes loss of nigral dopaminergic neurons. *Exp. Neurol.* **192**: 226.
Poser, S., Impey, S., Xia, Z. and Storm, D.R., 2003, Brain-derived neurotrophic factor protection of cortical neurons from serum withdrawal-induced apoptosis is inhibited by cAMP. *J. Neurosci.* **23**: 4420.
Pshenichkin, S.P. and Wise, B.C., 1995, Okadaic acid increases nerve growth factor secretion, mRNA stability and gene transcription in primary cultures of cortical astrocytes. *J. Biol. Chem.* **270**: 5994.
Quigley, H.A., McKinnon, S.J., Zack, D.J., Pease, M.E., Kerrigan-Baumrind, L.A., Kerrigan, D.F. and Mitchell, R.S., 2000, Retrograde axonal transport of BDNF in retinal ganglion cells is blocked by acute IOP elevation in rats. *Invest. Ophthalmol. Vis. Sci.* **41**: 3460.
Rajagopal, R., Chen, Z.Y., Lee, F.S. and Chao, M.V., 2004, Transactivation of Trk neurotrophin receptors by G-protein-coupled receptor ligands occurs on intracellular membranes. *J. Neurosci.* **24**: 6650.
Ribases, M., Gratacos, M., Armengol, L., de Cid, R., Badia, A., Jimenez, L., Solano, R., Vallejo, J., Fernandez, F. and Estivill, X., 2003, Met66 in the brain-derived neurotrophic factor (BDNF) precursor is associated with anorexia nervosa restrictive type. *Mol. Psychiatry* **8**: 745.
Ribases, M., Gratacos, M., Fernandez-Aranda, F., Bellodi, L., Boni, C. anderluh, M., Cavallini, M.C., Cellini, E., Di Bella, D., Erzegovesi, S., Foulon, C., Gabrovsek, M., Gorwood, P., Hebebrand, J., Hinney, A., Holliday, J., Hu, X., Karwautz, A., Kipman, A., Komel, R., Nacmias, B., Remschmidt, H., Ricca, V., Sorbi, S., Wagner, G., Treasure, J., Collier, D.A. and Estivill, X., 2004, Association of BDNF with anorexia, bulimia and age of onset of weight loss in six European populations. *Hum. Mol. Genet.* **13**: 1205.
Rickhag, M., Wieloch, T., Gido, G., Elmer, E., Krogh, M., Murray, J., Lohr, S., Bitter, H., Chin, D.J., von Schack, D., Shamloo, M. and Nikolich, K., 2006, Comprehensive regional and temporal gene expression profiling of the rat brain during the first 24 h after experimental stroke identifies dynamic ischemia-induced gene expression patterns and reveals a biphasic activation of genes in surviving tissue. *J. Neurochem.* **96**: 14.
Rivera, C., Li, H., Thomas-Crusells, J., Lahtinen, H., Viitanen, T., Nanobashvili, A., Kokaia, Z., Airaksinen, M.S., Voipio, J., Kaila, K. and Saarma, M., 2002, BDNF-induced TrkB activation down-regulates the K^+-Cl^- cotransporter KCC2 and impairs neuronal Cl^- extrusion. *J. Cell Biol.* **159**: 747.
Rose, C.R., Blum, R., Pichler, B., Lepier, A., Kafitz, K.W. and Konnerth, A., 2003, Truncated TrkB-T1 mediates neurotrophin-evoked calcium signaling in glia cells. *Nature* **426**: 74.
Rossler, O.G., Giehl, K.M. and Thiel, G., 2004, Neuroprotection of immortalized hippocampal neurones by brain-derived neurotrophic factor and Raf-1 protein kinase: role of extracellular signal-regulated protein kinase and phosphatidylinositol 3-kinase. *J. Neurochem.* **88**: 1240.
Rudge, J.S., Alderson, R.F., Pasnikowski, E., McClain, J., Ip, N.Y. and Lindsay, R.M., 1992, Expression of ciliary neurotrophic factor and the neurotrophins–nerve growth factor, brain-derived neurotrophic factor and neurotrophin 3–in cultured rat hippocampal astrocytes. *Eur. J. Neurosci.* **4**: 459.
Sanchez, I., Hassinger, L., Paskevich, P.A., Shine, H.D. and Nixon, R.A., 1996, Oligodendroglia regulate the regional expansion of axon caliber and local accumulation of neurofilaments during development independently of myelin formation. *J. Neurosci.* **16**: 5095.
Sarbassov, D.D., Guertin, D.A., Ali, S.M. and Sabatini, D.M., 2005, Phosphorylation and regulation of Akt/PKB by the rictor-mTOR complex. *Science* **307**: 1098.
Sato, S., Fujita, N. and Tsuruo, T., 2004, Involvement of 3-phosphoinositide-dependent protein kinase-1 in the MEK/MAPK signal transduction pathway. *J. Biol. Chem.* **279**: 33759.
Schabitz, W.R., Sommer, C., Zoder, W., Kiessling, M., Schwaninger, M. and Schwab, S., 2000, Intravenous brain-derived neurotrophic factor reduces infarct size and counterregulates Bax and Bcl-2 expression after temporary focal cerebral ischemia. *Stroke* **31**: 2212.
Schmidt-Kastner, R., Wetmore, C. and Olson, L., 1996, Comparative study of brain-derived neurotrophic factor messenger RNA and protein at the cellular level suggests multiple roles in hippocampus, striatum and cortex. *Neuroscience* **74**: 161.
Schor, N.F., 2005, The p75 neurotrophin receptor in human development and disease. *Prog. Neurobiol.* **77**: 201.
Schuettauf, F., Vorwerk, C., Naskar, R., Orlin, A., Quinto, K., Zurakowski, D., Dejneka, N.S., Klein, R.L., Meyer, E.M. and Bennett, J., 2004, Adeno-associated viruses containing bFGF or BDNF are neuroprotective against excitotoxicity. *Curr. Eye Res.* **29**: 379.
Schulte, J.H., Schramm, A., Klein-Hitpass, L., Klenk, M., Wessels, H., Hauffa, B.P., Eils, J., Eils, R., Brodeur, G.M., Schweigerer, L., Havers, W. and Eggert, A., 2005, Microarray analysis reveals differential gene expression patterns and regulation of single target genes contributing to the opposing phenotype of TrkA- and TrkB-expressing neuroblastomas. *Oncogene* **24**: 165.
Seeburger, J.L., Tarras, S., Natter, H. and Springer, J.E., 1993, Spinal cord motoneurons express p75NGFR and p145trkB mRNA in amyotrophic lateral sclerosis. *Brain Res.* **621**: 111.

Segal, R.A., 2003, Selectivity in neurotrophin signaling: theme and variations. *Annu. Rev. Neurosci.* **26**: 299.

Seidah, N.G., Benjannet, S., Pareek, S., Chretien, M. and Murphy, R.A., 1996a, Cellular processing of the neurotrophin precursors of NT3 and BDNF by the mammalian proprotein convertases. *FEBS Lett.* **379**: 247.

Seidah, N.G., Benjannet, S., Pareek, S., Savaria, D., Hamelin, J., Goulet, B., Laliberte, J., Lazure, C., Chretien, M. and Murphy, R.A., 1996b, Cellular processing of the nerve growth factor precursor by the mammalian proprotein convertases. *Biochem. J.* **314**: 951.

Seroogy, K.B., Lundgren, K.H., Tran, T.M., Guthrie, K.M., Isackson, P.J. and Gall, C.M., 1994, Dopaminergic neurons in rat ventral midbrain express brain-derived neurotrophic factor and neurotrophin-3 mRNAs. *J. Comp. Neurol.* **342**: 321.

Shalizi, A., Lehtinen, M., Gaudilliere, B., Donovan, N., Han, J., Konishi, Y. and Bonni, A., 2003, Characterization of a neurotrophin signaling mechanism that mediates neuron survival in a temporally specific pattern. *J. Neurosci.* **23**: 7326.

Sheedlo, H.J., Srinivasan, B., Brun-Zinkernagel, A.M., Roque, C.H., Lambert, W., Wordinger, R.J. and Roque, R.S., 2002, Expression of p75^{NTR} in photoreceptor cells of dystrophic rat retinas. *Brain. Res. Mol. Brain Res.* **103**: 71.

Shelton, D.L., Sutherland, J., Gripp, J., Camerato, T., Armanini, M.P., Phillips, H.S., Carroll, K., Spencer, S.D. and Levinson, A.D., 1995, Human trks: molecular cloning, tissue distribution and expression of extracellular domain immunoadhesins. *J. Neurosci.* **15**: 477.

Shieh, P.B. and Ghosh, A., 1999, Molecular mechanisms underlying activity-dependent regulation of BDNF expression. *J. Neurobiol.* **41**: 127.

Shimoke, K., Utsumi, T., Kishi, S., Nishimura, M., Sasaya, H., Kudo, M. and Ikeuchi, T., 2004, Prevention of endoplasmic reticulum stress-induced cell death by brain-derived neurotrophic factor in cultured cerebral cortical neurons. *Brain Res.* **1028**: 105.

Shimoke, K., Kubo, T., Numakawa, T., Abiru, Y., Enokido, Y., Takei, N., Ikeuchi, T. and Hatanaka, H., 1997, Involvement of phosphatidylinositol-3 kinase in prevention of low K^+-induced apoptosis of cerebellar granule neurons. *Brain Res. Dev. Brain Res.* **101**: 197.

Siegel, G.J. and Chauhan, N.B., 2000, Neurotrophic factors in Alzheimer's and Parkinson's disease brain. *Brain Res. Brain Res. Rev.* **33**: 199.

Skaper, S.D., Floreani, M., Negro, A., Facci, L. and Giusti, P., 1998, Neurotrophins rescue cerebellar granule neurons from oxidative stress-mediated apoptotic death: selective involvement of phosphatidylinositol 3-kinase and the mitogen-activated protein kinase pathway. *J. Neurochem.* **70**: 1859.

Sklar, P., Gabriel, S.B., McInnis, M.G., Bennett, P., Lim, Y.M., Tsan, G., Schaffner, S., Kirov, G., Jones, I., Owen, M., Craddock, N., DePaulo, J.R. and Lander, E.S., 2002, Family-based association study of 76 candidate genes in bipolar disorder: BDNF is a potential risk locus. Brain-derived neutrophic factor. *Mol. Psychiatry* **7**: 579.

Smeyne, R.J., Klein, R., Schnapp, A., Long, L.K., Bryant, S., Lewin, A., Lira, S.A. and Barbacid, M., 1994, Severe sensory and sympathetic neuropathies in mice carrying a disrupted Trk/NGF receptor gene. *Nature* **368**: 246.

Soltys, Z., Janeczko, K., Orzylowska-Sliwinska, O., Zaremba, M., Januszewski, S. and Oderfeld-Nowak, B., 2003, Morphological transformations of cells immunopositive for GFAP, TrkA or p75 in the CA1 hippocampal area following transient global ischemia in the rat. A quantitative study. *Brain Res.* **987**: 186.

Srinivasan, B., Roque, C.H., Hempstead, B.L., Al-Ubaidi, M.R. and Roque, R.S., 2004, Microglia-derived pronerve growth factor promotes photoreceptor cell death via p75 neurotrophin receptor. *J. Biol. Chem.* **279**: 41839.

Stadelmann, C., Kerschensteiner, M., Misgeld, T., Bruck, W., Hohlfeld, R. and Lassmann, H., 2002, BDNF and gp145trkB in multiple sclerosis brain lesions: neuroprotective interactions between immune and neuronal cells? *Brain* **125**: 75.

Swanwick, C.C., Harrison, M.B. and Kapur, J., 2004, Synaptic and extrasynaptic localization of brain-derived neurotrophic factor and the tyrosine kinase B receptor in cultured hippocampal neurons. *J. Comp. Neurol.* **478**: 405.

Syroid, D.E., Maycox, P.J., Soilu-Hanninen, M., Petratos, S., Bucci, T., Burrola, P., Murray, S., Cheema, S., Lee, K.F., Lemke, G. and Kilpatrick, T.J., 2000, Induction of postnatal Schwann cell death by the low-affinity neurotrophin receptor in vitro and after axotomy. *J. Neurosci.* **20**: 5741.

Takano, R., Hisahara, S., Namikawa, K., Kiyama, H., Okano, H. and Miura, M., 2000, Nerve growth factor protects oligodendrocytes from tumor necrosis factor-alpha-induced injury through Akt-mediated signaling mechanisms. *J. Biol. Chem.* **275**: 16360.

Takei, N., Tanaka, O., Endo, Y., Lindholm, D. and Hatanaka, H., 1999, BDNF and NT-3 but not CNTF counteract the Ca^{2+} ionophore-induced apoptosis of cultured cortical neurons: involvement of dual pathways. *Neuropharmacology* **38**: 283.

Taniuchi, M., Clark, H.B., Schweitzer, J.B. and Johnson, E.M., Jr., 1988, Expression of nerve growth factor receptors by Schwann cells of axotomized peripheral nerves: ultrastructural location, suppression by axonal contact and binding properties. *J. Neurosci.* **8**: 664.

Tao, X., West, A.E., Chen, W.G., Corfas, G. and Greenberg, M.E., 2002, A calcium-responsive transcription factor, CaRF, that regulates neuronal activity-dependent expression of BDNF. *Neuron* **33**: 383.

Taylor, S., Srinivasan, B., Wordinger, R.J. and Roque, R.S., 2003, Glutamate stimulates neurotrophin expression in cultured Müller cells. *Brain. Res. Mol. Brain Res.* **111**: 189.

Teng, K.K. and Hempstead, B.L., 2004, Neurotrophins and their receptors: signaling trios in complex biological systems. *Cell. Mol. Life Sci.* **61**: 35.

Teng, H.K., Teng, K.K., Lee, R., Wright, S., Tevar, S., Almeida, R.D., Kermani, P., Torkin, R., Chen, Z.Y., Lee, F.S., Kraemer, R.T., Nykjaer, A. and Hempstead, B.L., 2005, ProBDNF induces neuronal apoptosis via activation of a receptor complex of p75^{NTR} and sortilin. *J. Neurosci.* **25**: 5455.

Thippeswamy, T., McKay, J.S., Morris, R., Quinn, J., Wong, L.F. and Murphy, D., 2005, Glial-mediated neuroprotection: evidence for the protective role of the NO-cGMP pathway via neuron-glial communication in the peripheral nervous system. *Glia* **49**: 197.

Thompson, S.W., Bennett, D.L., Kerr, B.J., Bradbury, E.J. and McMahon, S.B., 1999, Brain-derived neurotrophic factor is an endogenous modulator of nociceptive responses in the spinal cord. *Proc. Natl. Acad. Sci. USA* **96**: 7714.

Torsney, C. and MacDermott, A.B., 2005, Neuroscience: a painful factor. *Nature* **438**: 923.

Toyomoto, M., Inoue, S., Ohta, K., Kuno, S., Ohta, M., Hayashi, K. and Ikeda, K., 2005, Production of NGF, BDNF and GDNF in mouse astrocyte cultures is strongly enhanced by a cerebral vasodilator, ifenprodil. *Neurosci. Lett.* **379**: 185.

Trapp, B.D., Peterson, J., Ransohoff, R.M., Rudick, R., Mork, S. and Bo, L., 1998, Axonal transection in the lesions of multiple sclerosis. *N. Engl. J. Med.* **338**: 278.

Troy, C.M., Friedman, J.E. and Friedman, W.J., 2002, Mechanisms of p75-mediated death of hippocampal neurons. Role of caspases. *J. Biol. Chem.* **277**: 34295.

Tsuda, M., Shigemoto-Mogami, Y., Koizumi, S., Mizokoshi, A., Kohsaka, S., Salter, M.W. and Inoue, K., 2003, P_2X_4 receptors induced in spinal microglia gate tactile allodynia after nerve injury. *Nature* **424**: 778.

Tsukamoto, E., Hashimoto, Y., Kanekura, K., Niikura, T., Aiso, S. and Nishimoto, I., 2003, Characterization of the toxic mechanism triggered by Alzheimer's amyloid-beta peptides via p75 neurotrophin receptor in neuronal hybrid cells. *J. Neurosci. Res.* **73**: 627.

Turner, B.J., Cheah, I.K., Macfarlane, K.J., Lopes, E.C., Petratos, S., Langford, S.J. and Cheema, S.S., 2003, Antisense peptide nucleic acid-mediated knockdown of the p75 neurotrophin receptor delays motor neuron disease in mutant SOD1 transgenic mice. *J. Neurochem.* **87**: 752.

Tuszynski, M.H., U, H.S., Amaral, D.G. and Gage, F.H., 1990, Nerve growth factor infusion in the primate brain reduces lesion-induced cholinergic neuronal degeneration. *J. Neurosci.* **10**: 3604.

Tuszynski, M.H., Thal, L., Pay, M., Salmon, D.P., Hoi Sang, U., Bakay, R., Patel, P., Blesch, A., Vahlsing, H.L., Ho, G., Tong, G., Potkin, S.G., Fallon, J., Hansen, L., Mufson, E.J., Kordower, J.H., Gall, C. and Conner, J., 2005, A phase 1 clinical trial of nerve growth factor gene therapy for Alzheimer disease. *Nat. Med.* **11**: 551.

Ullian, E.M., Sapperstein, S.K., Christopherson, K.S. and Barres, B.A., 2001, Control of synapse number by glia. *Science* **291**: 657.

Ullian, E.M., Christopherson, K.S. and Barres, B.A., 2004, Role for glia in synaptogenesis. *Glia* **47**: 209.

Vaillant, A.R., Mazzoni, I., Tudan, C., Boudreau, M., Kaplan, D.R. and Miller, F.D., 1999, Depolarization and neurotrophins converge on the phosphatidylinositol 3-kinase-Akt pathway to synergistically regulate neuronal survival. *J. Cell Biol.* **146**: 955.

Valdez, G., Akmentin, W., Philippidou, P., Kuruvilla, R., Ginty, D.D. and Halegoua, S., 2005, Pincher-mediated macroendocytosis underlies retrograde signaling by neurotrophin receptors. *J. Neurosci.* **25**: 5236.

Vargas, M.R., Pehar, M., Cassina, P., Beckman, J.S. and Barbeito, L., 2006, Increased glutathione biosynthesis by Nrf2 activation in astrocytes prevents p75^{NTR}-dependent motor neuron apoptosis. *J. Neurochem.* **97**: 687.

von Schack, D., Casademunt, E., Schweigreiter, R., Meyer, M., Bibel, M. and Dechant, G., 2001, Complete ablation of the neurotrophin receptor p75^{NTR} causes defects both in the nervous and the vascular system. *Nat. Neurosci.* **4**: 977.

Watson, F.L., Heerssen, H.M., Bhattacharyya, A., Klesse, L., Lin, M.Z. and Segal, R.A., 2001, Neurotrophins use the Erk5 pathway to mediate a retrograde survival response. *Nat. Neurosci.* **4**: 981.

Weisenhorn, D.M., Roback, J., Young, A.N. and Wainer, B.H., 1999, Cellular aspects of trophic actions in the nervous system. *Int. Rev. Cytol.* **189**: 177.

Wen, W., Sanelli, T., Ge, W., Strong, W. and Strong, M.J., 2006, Activated microglial supernatant induced motor neuron cytotoxicity is associated with upregulation of the TNFR1 receptor. *Neurosci. Res.* **55**: 87.

Woolf, C.J., 2003, No Nogo: now where to go? *Neuron* **38**: 153.

Wooten, M.W., Seibenhener, M.L., Mamidipudi, V., Diaz-Meco, M.T., Barker, P.A. and Moscat, J., 2001, The atypical protein kinase C-interacting protein p62 is a scaffold for NF-kappaB activation by nerve growth factor. *J. Biol. Chem.* **276**: 7709.

Wu, D., 2005, Neuroprotection in experimental stroke with targeted neurotrophins. *NeuroRx* **2**: 120.

Wu, D. and Pardridge, W.M., 1999, Neuroprotection with noninvasive neurotrophin delivery to the brain. *Proc. Natl. Acad. Sci. USA* **96**: 254.

Wu, X., Zhu, D., Jiang, X., Okagaki, P., Mearow, K., Zhu, G., McCall, S., Banaudha, K., Lipsky, R.H. and Marini, A.M., 2004, AMPA protects cultured neurons against glutamate excitotoxicity through a phosphatidylinositol 3-kinase-dependent activation in extracellular signal-regulated kinase to upregulate BDNF gene expression. *J. Neurochem.* **90**: 807.
Xing, J., Kornhauser, J.M., Xia, Z., Thiele, E.A. and Greenberg, M.E., 1998, Nerve growth factor activates extracellular signal-regulated kinase and p38 mitogen-activated protein kinase pathways to stimulate CREB serine 133 phosphorylation. *Mol. Cell Biol.* **18**: 1946.
Xu, B., Gottschalk, W., Chow, A., Wilson, R.I., Schnell, E., Zang, K., Wang, D., Nicoll, R.A., Lu, B. and Reichardt, L.F., 2000, The role of brain-derived neurotrophic factor receptors in the mature hippocampus: modulation of long-term potentiation through a presynaptic mechanism involving TrkB. *J. Neurosci.* **20**: 6888.
Xu, B., Goulding, E.H., Zang, K., Cepoi, D., Cone, R.D., Jones, K.R., Tecott, L.H. and Reichardt, L.F., 2003, Brain-derived neurotrophic factor regulates energy balance downstream of melanocortin-4 receptor. *Nat. Neurosci.* **6**: 736.
Xue, L., Murray, J.H. and Tolkovsky, A.M., 2000, The Ras/phosphatidylinositol 3-kinase and Ras/ERK pathways function as independent survival modules each of which inhibits a distinct apoptotic signaling pathway in sympathetic neurons. *J. Biol. Chem.* **275**: 8817.
Yamada, M., Tanabe, K., Wada, K., Shimoke, K., Ishikawa, Y., Ikeuchi, T., Koizumi, S. and Hatanaka, H., 2001, Differences in survival-promoting effects and intracellular signaling properties of BDNF and IGF-1 in cultured cerebral cortical neurons. *J. Neurochem.* **78**: 940.
Yamamoto, T., Yuki, S., Watanabe, T., Mitsuka, M., Saito, K.I. and Kogure, K., 1997, Delayed neuronal death prevented by inhibition of increased hydroxyl radical formation in a transient cerebral ischemia. *Brain Res.* **762**: 240.
Yamauchi, J., Chan, J.R., Miyamoto, Y., Tsujimoto, G. and Shooter, E.M., 2005, The neurotrophin-3 receptor TrkC directly phosphorylates and activates the nucleotide exchange factor Dbs to enhance Schwann cell migration. *Proc. Natl. Acad. Sci. USA* **102**: 5198.
Yan, Q., Radeke, M.J., Matheson, C.R., Talvenheimo, J., Welcher, A.A. and Feinstein, S.C., 1997, Immunocytochemical localization of TrkB in the central nervous system of the adult rat. *J. Comp. Neurol.* **378**: 135.
Yao, R. and Cooper, G.M., 1995, Requirement for phosphatidylinositol-3 kinase in the prevention of apoptosis by nerve growth factor. *Science* **267**: 2003.
Yuan, J. and Yankner, B.A., 2000, Apoptosis in the nervous system. *Nature* **407**: 802.
Zafra, F., Hengerer, B., Leibrock, J., Thoenen, H. and Lindholm, D., 1990, Activity dependent regulation of BDNF and NGF mRNAs in the rat hippocampus is mediated by non-NMDA glutamate receptors. *EMBO J.* **9**: 3545.
Zassler, B. and Humpel, C., 2006, Transplantation of NGF secreting primary monocytes counteracts NMDA-induced cell death of rat cholinergic neurons in vivo. *Exp. Neurol.* **198**: 391.
Zhang, Y. and Pardridge, W.M., 2001a, Conjugation of brain-derived neurotrophic factor to a blood-brain barrier drug targeting system enables neuroprotection in regional brain ischemia following intravenous injection of the neurotrophin. *Brain Res.* **889**: 49.
Zhang, Y. and Pardridge, W.M., 2001b, Neuroprotection in transient focal brain ischemia after delayed intravenous administration of brain-derived neurotrophic factor conjugated to a blood-brain barrier drug targeting system. *Stroke* **32**: 1378.
Zhang, J.M., Wang, H.K., Ye, C.Q., Ge, W., Chen, Y., Jiang, Z.L., Wu, C.P., Poo, M.M. and Duan, S., 2003a, ATP released by astrocytes mediates glutamatergic activity-dependent heterosynaptic suppression. *Neuron* **40**: 971.
Zhang, Y., Hong, Y., Bounhar, Y., Blacker, M., Roucou, X., Tounekti, O., Vereker, E., Bowers, W.J., Federoff, H.J., Goodyer, C.G. and LeBlanc, A., 2003b, p75 neurotrophin receptor protects primary cultures of human neurons against extracellular amyloid beta peptide cytotoxicity. *J. Neurosci.* **23**: 7385.
Zhou, X.F., Parada, L.F., Soppet, D. and Rush, R.A., 1993, Distribution of TrkB tyrosine kinase immunoreactivity in the rat central nervous system. *Brain Res.* **622**: 63.
Zhou, H., Summers, S.A., Birnbaum, M.J. and Pittman, R.N., 1998, Inhibition of Akt kinase by cell-permeable ceramide and its implications for ceramide-induced apoptosis. *J. Biol. Chem.* **273**: 16568.
Zhou, X.F., Song, X.Y., Zhong, J.H., Barati, S., Zhou, F.H. and Johnson, S.M., 2004, Distribution and localization of pro-brain-derived neurotrophic factor-like immunoreactivity in the peripheral and central nervous system of the adult rat. *J. Neurochem.* **91**: 704.
Zuccato, C., Ciammola, A., Rigamonti, D., Leavitt, B.R., Goffredo, D., Conti, L., MacDonald, M.E., Friedlander, R.M., Silani, V., Hayden, M.R., Timmusk, T., Sipione, S. and Cattaneo, E., 2001, Loss of huntingtin-mediated BDNF gene transcription in Huntington's disease. *Science* **293**: 493.
Zweifel, L.S., Kuruvilla, R. and Ginty, D.D., 2005, Functions and mechanisms of retrograde neurotrophin signaling. *Nat. Rev. Neurosci.* **6**: 615.

8

GDNF: A KEY PLAYER IN NEURON–GLIA CROSSTALK AND SURVIVAL OF NIGROSTRIATAL DOPAMINERGIC NEURONS

Emília P. Duarte[1,3], Ana Saavedra[1] and Graça Baltazar[2]

1. ABSTRACT

Glial cell line-derived neurotrophic factor (GDNF) is a potent survival factor for dopaminergic neurons of the nigrostriatal pathway that degenerate in Parkinson's disease (PD). In animal models of PD, GDNF delivery has been shown to both protect dopaminergic neurons against toxin-induced injury and to rescue damaged neurons, promoting recovery of the motor deficit. GDNF may act both as a target-derived neurotrophic factor in the striatum, and as a local neurotrophic factor at the level of neuronal cell bodies in the *substantia nigra*. The neuroprotective and regenerative effects of GDNF are mediated by increases in the activity of antioxidant enzymes and induction of antiapoptotic proteins and cell adhesion molecules. In addition, GDNF stimulates dopaminergic neurotransmission by increasing dopamine biosynthesis and neuronal excitability. In the normal adult brain, GDNF is expressed mainly by neuronal cells but, upon injury, both astrocytes and microglial cells express GDNF. How neuronal damage signals glial cells to upregulate GDNF is being uncovered. Dopamine agonists, interleukin-1β (IL-1β) and estrogen have been shown to modulate GDNF expression by astrocytes and/or microglial cells. GDNF produced by activated microglial cells is involved in the axonal growth and sprouting of damaged neurons. Moreover, GDNF can modulate the migration, adhesion and phagocytic activity of microglia. The intercellular messengers involved in the crosstalk between dopaminergic neurons, astrocytes and microglia are potential targets for PD therapies aimed at upregulating the endogenous expression of GDNF.

2. INTRODUCTION

Glial cell line-derived neurotrophic factor (GDNF) is widely recognized as a potent survival factor for dopaminergic neurons projecting from the *substantia nigra pars compacta* in the midbrain to the striatum (reviewed by Hurelbrink and Barker, 2004). The progressive

---

Emília P. Duarte – epduarte@cnc.cj.uc.pt
[1]Center for Neuroscience and Cell Biology, University of Coimbra, Coimbra, Portugal. [2]Health Sciences Research Center, University of Beira Interior, Covilhã, Portugal. [3]Department of Zoology, University of Coimbra, Coimbra, Portugal.

degeneration of the nigrostriatal neurons in Parkinson's disease (PD) leads to dopamine depletion in the striatum, or caudate-putamen in primates, causing the slowness of movement, stiffness, resting tremor and postural instability observed in PD.

GDNF was originally identified in conditioned media from a glial cell line based on its ability to promote the survival of embryonic dopaminergic neurons in culture, and to increase cell size, neurite length and the high-affinity uptake of dopamine (Lin et al., 1993, 1994). Subsequently, the neuroprotective effects of GDNF were shown in animal models of PD. The administration of GDNF, either the recombinant protein (reviewed by Grondin et al., 2003) or the delivery of GDNF gene by viral vectors (reviewed by Bjorklund et al., 2000), prevents the degeneration of dopaminergic neurons and the motor impairment induced by a subsequent lesion of the nigrostriatal system. More importantly, GDNF delivery after a toxin-induced lesion, a paradigm that better mimics PD at the time of detection, was shown to increase dopaminergic markers and to restore motor function (Tomac et al., 1995a; Gash et al., 1996, 2005; Mandel et al., 1999; Kordower et al., 2000; Grondin et al., 2002; Wang et al., 2002; Zheng et al., 2005). Therefore, in addition to protecting dopaminergic neurons, GDNF appears to induce neuronal repair, stimulating regenerative growth and axonal sprouting after lesions of the nigrostriatal system. Therefore, GDNF was proposed and tested as a neuroprotective/restorative therapy for PD aimed at slowing down, halting or reversing neurodegeneration. Direct infusion of GDNF into the putamen of PD patients has improved the motor function, decreased the requirements for L-DOPA therapy and induced outgrowth of dopaminergic neurons as assessed by imaging of dopamine transporters (Gill et al., 2003; Love et al., 2005; Patel et al., 2005; Slevin et al., 2005). However, in a double blind, placebo-controlled trial, in a larger number of patients, there were no significant clinical improvements and severe side effects were reported (Lang et al., 2006). The clinical trials were discontinued and several problems remain to be solved before GDNF therapy can be used in PD patients (Sherer et al., 2006).

Nervous tissue responds to injury upregulating protective and repairing mechanisms (reviewed by Benn and Woolf, 2004; Martino, 2004). Damaged neurons change their own gene expression and stimulate nearby astrocytes and microglial cells to provide support. Activated astrocytes are known to upregulate antioxidant molecules, membrane transporters and trophic factors that support neuronal and glial survival and tissue repair (reviewed by Liberto et al., 2004). Microglia, the resident immune cells of the brain, proliferate and become immunocompetent cells, with the ability of secreting a number of cytotoxic and trophic molecules (reviewed by Block and Hong, 2005). Upregulation of GDNF, by both astrocytes and microglial cells, has been shown in several injury models (Liberatore et al., 1997; Batchelor et al., 1999; Satake et al., 2000; Miyazaki et al., 2001; Ikeda et al., 2002). Therefore, the intercellular signals involved in the communication between damaged neurons and glial cells are potential therapeutic targets aimed at upregulating the endogenous expression of neurotrophic factors.

The trophic effects of GDNF are not exclusive of midbrain dopaminergic neurons. GDNF also supports the survival of spinal motoneurons, regulates the differentiation of several peripheral neurons, and has roles outside the nervous system in kidney morphogenesis and in spermatogonial differentiation (reviewed by Sariola and Saarma, 2003). These pleiotropic actions of GDNF are mediated by the activation of a multi-component receptor system. GDNF binds to a glycosylphosphatidylinositol (GPI)-anchored cell surface receptor known as GFRα1 (GDNF family receptor-α1). The GFRα1 acts only as a binding subunit that can activate several signaling receptors. GFRα1 can signal through the transmembrane Ret tyrosine kinase, which in turn activates several intracellular signaling cascades: the mitogen-activated protein (MAP) kinase, the phosphoinositide 3-kinase (PI3K) and the phospholipase Cγ (PLC-γ) pathways. In cells not expressing Ret, GDNF triggers Src family kinase activation and the phosphorylation of the extracellular receptor-activated kinase ERK/MAP kinase, PLC-γ and the transcription factor CREB. More recently, GDNF has been shown to signal through the neural cell adhesion molecule (NCAM; reviewed by Sariola and Saarma, 2003).

The present review will focus on GDNF role in the nigrostriatal system, mainly on data gathered in animal and cell culture models of PD, on its neuroprotective effects and mechanisms, and the control of GDNF expression upon nigrostriatal injury.

3. GDNF SUPPORT OF NIGROSTRIATAL DOPAMINERGIC NEURONS: TARGET-DERIVED AND LOCAL SUPPORT?

According to the neurotrophic theory, neuronal survival during development is controlled by target-derived neurotrophic factors (Purves, 1986; Oppenheim, 1991). Only neurons establishing synaptic contacts have access to sufficient amounts of trophic factors. GDNF is highly expressed by striatal neurons during post-natal development, and it was shown to regulate the natural death of *substantia nigra* dopaminergic neurons (reviewed by Burke, 2004). The levels of GDNF mRNA decrease steeply in the post-natal period (Stromberg et al., 1993; Choi-Lundberg and Bohn, 1995; Oo et al., 2005), and the role of GDNF in the normal adult brain is not well established. However, GDNF (Choi-Lundberg and Bohn, 1995; Pochon et al., 1997; Trupp et al., 1997) and its receptors (Treanor et al., 1996; Trupp et al., 1996, 1997; Kozlowski et al., 2004) are expressed in the adult brain, although some contradictory reports exist (Stromberg et al., 1993). The observation of retrograde axonal transport of GDNF in the adult nigrostriatal neurons also indicates a target-derived function and a neuronal maintenance role for GDNF in the adult nervous system (Tomac et al., 1995b). A deficient neurotrophic support was also suggested as a factor involved in the degeneration of the nigrostriatal neurons in PD (Siegel and Chauhan, 2000).

The effects of GDNF in the adult brain and its potential therapeutic value have been extensively studied in animal models of PD. In a widely used rat model, the selective death of *substantia nigra* dopaminergic neurons is induced by 6-hydroxydopamine (6-OHDA) injected in the *substantia nigra*, in the medial forebrain bundle or in the striatum (reviewed by Deumens et al., 2002). This analogue of dopamine causes the degeneration of neurons having the dopamine transporter by increasing intracellular oxidative stress. In mice and non-human primates, the degeneration of the nigrostriatal pathway is generally induced by the administration of 1-methyl-4-phenyl-1,2,3,6-tetrahydropyridine (MPTP), a hydrophobic compound capable of permeating the blood–brain barrier. This compound is metabolized in astrocytes to 1-methyl-4-phenyl-2,3-dihydropyridinium ($MPDP^+$), and oxidized to 1-methyl-4-phenylpyridinium (MPP^+), the active toxin, which is taken up by dopamine transporters. Once inside neuronal cells, MPP^+ inhibits mitochondrial complex I leading to the release of superoxide, decreased ATP cellular content, dopamine leakage from synaptic vesicles and further oxidative stress (reviewed by Dauer and Przedborski, 2003).

The administration of GDNF into the striatum, or into the putamen of primates, has consistently shown efficacy in preserving and restoring the structure and function of the nigrostriatal pathway. Striatal delivery of viral vectors encoding for GDNF before the 6-OHDA lesion prevents dopaminergic damage and motor impairment (Bilang-Bleuel et al.,

1997; Mandel et al., 1997; Choi-Lundberg et al., 1998; Connor et al., 1999; Kirik et al., 2000b; Eslamboli et al., 2005). When administered after an established lesion, a condition more relevant to the treatment of PD, GDNF delivery to the striatum was shown to increase the number of tyrosine hydroxylase (TH)-positive cell bodies in the *substantia nigra*, the density of TH-positive fibres and dopamine levels in the striatum, and to induce recovery of motor impairments (Grondin et al., 2002; Wang et al., 2002; Zheng et al., 2005; Brizard et al., 2006). The use of TH as a marker of dopaminergic cell survival and repair raises some questions. Using fluorgold retrograde tracing, it was shown that the number of surviving dopaminergic neurons is larger than the number of TH-positive cells, suggesting that loss of dopaminergic phenotype precedes neuronal death (Bowenkamp et al., 1996). Therefore, the observed increase in the number of TH-positive cells or fibres upon GDNF treatment might be due to the recovery of TH expression in damaged neurons, rather than to neuronal regeneration and reinnervation of striatum. On the other hand, it was observed that GDNF-induced structural and behavioural recovery occurs only in partial lesions, probably because a number of surviving neurons and fibres are required to reinnervate the striatum by undergoing axonal sprouting, or to function as guidance cues for axonal growth (Rosenblad et al., 1998; Stanic et al., 2003, 2004; Brizard et al., 2006). It is important to stress that functional regeneration of the nigrostriatal system upon GDNF delivery was observed both in young and aged animals, including models of advanced parkinsonism, which more closely mimic the advanced stage and age of PD patients (Grondin et al., 2002; Zheng et al., 2005).

In addition to being expressed by the targets of the nigrostriatal neurons, GDNF is expressed in the *substantia nigra*, in the perikaria area of dopaminergic neurons (Pochon et al., 1997; Sarabi et al., 2001). The GDNF protein detected in *substantia nigra* might come from the striatum by retrograde transport (Tomac et al., 1995b), but the detection of GDNF mRNA in *substantia nigra* (Choi-Lundberg and Bohn, 1995; Pochon et al., 1997) shows unequivocally that GDNF is also expressed in the soma area of the nigrostriatal pathway. Furthermore, colocalization of GDNF mRNA and TH immunoreactivity showed that dopaminergic neurons express GDNF (Pochon et al., 1997). The presence of GDNF receptors in the *substantia nigra* (Trupp et al., 1997), namely on dopaminergic neurons (Sarabi et al., 2001), suggests that locally produced GDNF might also play a role in dopaminergic function and survival. A microarray study also detected the expression of GDNF mRNA in astrocytes from intact adult *substantia nigra* (Nakagawa and Schwartz, 2004a).

GDNF delivery into the *substantia nigra* before a 6-OHDA lesion in the striatum promotes the survival of nigral dopaminergic neurons (Choi-Lundberg et al., 1997; Connor et al., 1999, 2001; Kirik et al., 2000a, b) but fails to preserve striatal innervation and motor function, whereas the intrastriatal delivery preserves the integrity and function of the nigrostriatal system (Connor et al., 1999, 2001; Kirik et al., 2000a, b). In contrast, most studies on GDNF administration into the *substantia nigra* several weeks after the lesion, both in rodents and in primates, report marked functional recoveries (Hoffer et al., 1994; Bowenkamp et al., 1995, 1996; Hoffman et al., 1997; Lapchak et al., 1997; Kozlowski et al., 2000; Gash et al., 2005; Smith et al., 2005). Surprisingly, motor improvement was not always accompanied by striatal reinnervation, as assessed by TH immunoreactivity or dopamine levels (Hoffman et al., 1997; Lapchak et al., 1997). These observations support the idea that nigral delivery of GDNF does not promote fibre regeneration and striatal reinnervation, despite neuronal survival. It was proposed that the improvement in motor behaviour was due to the increase in somatodendritic release of dopamine in the *substantia nigra* (Hoffman et al., 1997; Gerhardt et al., 1999). However, viral delivery of GDNF into the *substantia nigra*, one week after a striatal 6-OHDA lesion, was shown to increase the number of intact projections from the *substantia nigra* to the striatum as compared to the lesioned animals treated with control vectors (Kozlowski et al., 2000). Moreover, another study, in the same animal model, showed that GDNF treatment increases basal and evoked dopamine release in the striatum as measured by in vivo dialysis (Smith et al., 2005).

In summary, the data on rodent and non-human primates, expressing stable parkinsonian symptoms, show that GDNF delivery into the striatum or the *substantia nigra* can induce marked functional improvements. The regenerative ability of GDNF in the nigrostriatal system depends on the extension of pre-existing damage, as surviving neurons are the substrates for GDNF-induced recovery of dopaminergic phenotype and regenerative sprouting. Moreover, the surviving fibres might produce guidance cues to direct neuronal regeneration towards the targets. It was suggested the possibility that GDNF is activating different mechanisms via axon terminal receptors in the putamen, and somatodendritic receptors in the *substantia nigra* (Gash et al., 2005). If so, the simultaneous GDNF administration to the *substantia nigra* and putamen could have additive effects on regeneration of the nigrostriatal pathway and functional improvements.

4. GDNF AND THE NIGROSTRIATAL RESPONSE TO INJURY

Brain injury appears to change the pattern of trophic factor expression. In the intact adult brain, GDNF expression is believed to be largely neuronal as astroglial cells were not found to express detectable GDNF (Blum and Weickert, 1995; Pochon et al., 1997; Bizon et al., 1999; Bresjanac and Antauer, 2000). However, as mentioned before, a microarray study detected GDNF mRNA in astrocytes from normal adult *substantia nigra* (Nakagawa and Schwartz, 2004a). Upon injury, glial cells appear to become the predominant source of trophic substances. Reactive astrocytes in the nigrostriatal system have been demonstrated to express a number of neurotrophic factors, including GDNF (Bresjanac and Antauer, 2000; Nakagawa and Schwartz, 2004b), basic fibroblast growth factor (bFGF) and nerve growth factor (NGF; Nakagawa and Schwartz, 2004b). The neuronal expression of trophic factors associated with injury is a rapid and transient activity-dependent expression, whereas a delayed and more persistent injury-induced expression is observed in glial cells (Hughes et al., 1999).

The selective damage of nigrostriatal dopaminergic neurons in animal models of PD was shown to alter mRNA and/or protein levels for several trophic factors, both in the striatum and the *substantia nigra*, although some contradictory findings were reported. In the striatum, an increase in GDNF mRNA or protein levels has been observed after 6-OHDA- (Zhou et al., 2000; Nakajima et al., 2001; Yurek and Fletcher-Turner, 2001, 2002) or MPTP-induced injury (Tang et al., 1998). However, other authors did not detect changes in striatal GDNF levels after 6-OHDA lesion (Stromberg et al., 1993; Smith et al., 2003) or MPTP treatment (Inoue et al., 1999; Collier et al., 2005). These conflicting results might arise from differences in the injury models and protocols, or different time points analysed. In general, GDNF protein levels increased in the first 2–3 weeks after the 6-OHDA lesion and decreased thereafter, such that at 6 or 12 weeks no upregulation was detected, or decreased levels were observed (Nakajima et al., 2001; Yurek and Fletcher-Turner, 2002). Moreover, the neurotrophic response to injury was shown to depend on the age of the animals. In general, young animals showed a greater capacity to upregulate neurotrophic factors (Yurek and Fletcher-Turner, 2001; Collier et al., 2005).

Fewer studies focused on nigral GDNF expression after nigrostriatal lesions, yielding also conflicting results. Some studies found no changes in GDNF expression upon MPTP- (Inoue et al., 1999) or 6-OHDA-induced injury (Yurek and Fletcher-Turner, 2001). Later on, the same authors reported a transient increase of GDNF in midbrain three days post-lesion, whereas at 12 weeks a down-regulation was observed (Yurek and Fletcher-Turner, 2002). We have shown that mild damage to dopaminergic neurons in neuron–glia cultures from *substantia nigra*, but not extensive damage, induces GDNF up-regulation in astrocytes (Saavedra et al., 2006a). Since GDNF up-regulation was not observed in cultures previously treated with 6-OHDA, we proposed that GDNF induction in astrocytes requires signaling from injured dopaminergic neurons (Saavedra et al., 2006a).

Gene expression profiles, obtained with cDNA microarrays, also showed GDNF up-regulation in the *substantia nigra* and striatum after MPTP intoxication (Grunblatt et al., 2001; Mandel et al., 2002) and in striatal astrocytes from 6-OHDA-lesioned rats (Nakagawa and Schwartz, 2004b). It should be noted that other trophic factors besides GDNF are up-regulated in the injured nigrostriatal system, namely, BDNF, FGF and NGF (Nakajima et al., 2001; Yurek and Fletcher-Turner, 2002; Nakagawa et al., 2005).

The effect of nigrostriatal injury on the expression of GDNF receptors was also examined since the neuroprotective effects of GDNF depend not only on the available GDNF protein levels but also on the expression of its receptors. Intrastriatal injection of 6-OHDA was reported to transiently increase GFRα1 and Ret mRNA levels in the *substantia nigra* 1 day after lesion. Triple-labelling studies indicated that GFRα1 and Ret are expressed by neurons both in control and in 6-OHDA-lesioned animals (Marco et al., 2002). In contrast, marked decreases in GFRα1 and Ret mRNA levels were found in *substantia nigra* after 6-OHDA lesion of the medial forebrain bundle (Smith et al., 2003). Other study reported no changes in GFRα1 mRNA in the *substantia nigra* or striatum at 4 weeks after an acute 6-OHDA lesion of the medial forebrain bundle, as assessed by real-time quantitative RT-PCR. In contrast, a progressive 6-OHDA lesion induced by striatal injection resulted in a progressive decrease of GFRα1 mRNA in the striatum, whereas in the *substantia nigra* GFRα1 mRNA levels were not significantly affected. Since the changes observed in receptor expression did not always parallel the loss of dopamine neurons, the authors suggested that the expression of the receptors was altered in other cells in addition to the nigral dopaminergic neurons (Kozlowski et al., 2004).

It is clear from the data reviewed above that a systematic analysis of the effects of nigrostriatal injury on the expression of neurotrophic factors and their receptors is needed, both in the striatum and the *substantia nigra*, in order to get a detailed knowledge of the temporal pattern of the neurotrophic response to injury, the dependence on the intensity of neuronal damage, the effectiveness of protection, and the effects of aging. It is not known whether the endogenous production of GDNF might be effective in protecting nigrostriatal dopaminergic neurons, contributing for the very slow progression of PD. It should be noticed that the data showing dopaminergic protection and repair involved the delivery of exogenous GDNF, or the transduction of viral encoded GDNF, as detailed in the preceding section.

It was proposed that changes in the levels of neurotrophic factors, due to alterations in the synthesis, release or activity, associated with aging or genetic factors, might be involved in the neuronal loss observed in neurodegenerative diseases as PD (reviewed by Siegel and Chauhan, 2000). Post-mortem studies investigating GDNF distribution in the human parkinsonian brain have yielded conflicting results. In situ hybridization studies failed to detect GDNF mRNA in the human midbrain (Hunot et al., 1996), and no significant differences were found in GDNF content in the caudate-putamen and *substantia nigra* of control and PD samples (Mogi et al., 2001). Since the levels of other growth factors were decreased in PD brains, it was suggested that the unchanged levels of GDNF in PD might be due to compensatory production by glial cells (Mogi et al., 2001). However, using immunohistochemistry, large reductions in GDNF content were reported in surviving PD *substantia nigra* neurons (Chauhan et al., 2001). More recently, using real-time PCR, modest but significantly increased levels of an isoform of GDNF were found in the putamen of PD patients with marked nigral neuronal loss (Backman et al., 2006). Whatever might be the endogenous changes of neurotrophic factors in PD, it is clear that a therapy with GDNF might prevent the progression of the disease and restore function, at least in early stages of the disease.

5. MICROGLIA IN THE DEATH AND SURVIVAL OF THE NIGROSTRIATAL DOPAMINERGIC NEURONS: A ROLE FOR GDNF?

Reactive microglia were observed in post-mortem samples from the *substantia nigra* of PD patients (McGeer et al., 1988; Langston et al., 1999) and animal models of PD, sometimes long after the initial toxin exposure (Hurley et al., 2003; Barcia et al., 2004). These findings led to the proposal that sustained activation of microglia might perpetuate the initial damage and contribute to the progressive neurodegeneration (reviewed by Block and Hong, 2005). Microglial cells are believed to contribute to dopaminergic cell death due to the production and release of reactive oxygen species (ROS), NO, proteases and inflammatory cytokines (reviewed by Block and Hong, 2005). A large body of evidence supports the role of microglia in the degeneration of dopaminergic neurons. In the MPTP mouse model, inhibition of microglial activation with minocycline decreases dopaminergic death (He et al., 2001; Wu et al., 2002), and neuronal death is greatly diminished in mutant mice deficient in NO synthase (Liberatore et al., 1999; Dehmer et al., 2000), or deficient in NADPH-oxidase that catalyses the production of superoxide (Block and Hong, 2005). Moreover, a model of PD was created by infusing in the *substantia nigra* lipopolysacharide (LPS), a component of the cell wall of gram-negative bacteria, which activates microglial cells and selectively kills dopaminergic neurons (reviewed by Block and Hong, 2005). In cell culture models, dopaminergic cell death induced by MPP^+ is greatly reduced in neuron-enriched cultures, as compared to neuron–glia cultures, and the addition of microglia to neuron-enriched cultures reestablishes MPP^+-induced dopaminergic death (reviewed by Block and Hong, 2005).

Differences in microglial activation were observed in rodent and primate models of PD. While reactive microglial cells are consistently observed both in the *substantia nigra* and the striatum of mice treated with MPTP (Breidert et al., 2002; Wu et al., 2002), microgliosis is not observed in the caudate-putamen of monkeys (Hurley et al., 2003; Barcia et al., 2004) or in humans (McGeer et al., 1988; Langston et al., 1999). Moreover, inhibition of microglia activity prevents or decreases MPTP-induced dopaminergic death in the *substantia nigra* but has no protective effect on dopaminergic terminals (Breidert et al., 2002), suggesting different injury mechanisms in the *substantia nigra* and the striatum. The greater deleterious effect of microglia in the *substantia nigra* might be related to the fact that the density of microglial cells is much higher in the *substantia nigra* compared to other brain areas (Lawson et al., 1990; Kim et al., 2000). Differences were also observed in the time-course of microglial activation in rodents and primates. In mice, the microglial reaction was generally observed soon after MPTP intoxication and was is relatively transient (Wu et al., 2002), whereas in monkeys the density of activated microglia in the *substantia nigra* of

MPTP intoxicated animals was still significantly higher than in control animals 1 year after the last MPTP injection (Hurley et al., 2003; McGeer et al., 2003; Barcia et al., 2004). These data suggest that the glial reaction is associated with the loss of neuronal perikarya, as occurs in the midbrain, rather than with the loss of axon terminals in the striatum (Barcia et al., 2004). In addition, these findings raise the hypothesis that the factors regulating microglial activation in the *substantia nigra* and in the striatum are different.

Despite evidences for the negative impact of microglia on dopaminergic cell survival in cell culture and animal models of PD, activated microglial cells have also been shown to have a neuroprotective role and to be involved in brain repair. In the nigrostriatal system, activated microglia and macrophages were shown to promote axonal growth and sprouting of dopaminergic neurons after a mechanical lesion to the striatum (Batchelor et al., 1999). After striatal injury, sprouting dopaminergic fibres grow towards and surround macrophages expressing GDNF and BDNF mRNA (Batchelor et al., 1999). The dopaminergic sprouting after striatal injury was shown to involve the production of GDNF by macrophages at the wound site, since preventing GDNF expression with antisense oligonucleotides resulted in a marked decrease in the intensity of the periwound sprouting as revealed by immunohistochemistry and activity of the dopamine transporter (Batchelor et al., 2000). Moreover, dopaminergic sprouting was related to a gradient of GDNF (Batchelor et al., 2002). These data clearly show that activated microglia and macrophages induce dopaminergic sprouting through synthesis of neurotrophic factors. Interleukin-1 (IL-1) was also shown to be involved in dopaminergic sprouting since the IL-1 receptor-knockout mice do not show neuronal sprouting after a 6-OHDA lesion (Parish et al., 2002). IL-1, produced by reactive microglia and macrophages, induces astrogliosis. Therefore, activated microglia and macrophages appear to stimulate dopaminergic sprouting both directly, by the secretion of neurotrophic factors, and indirectly by the secretion of IL-1 and the stimulation of reactive astrocytosis (Ho and Blum, 1998; Parish et al., 2002). The production of GDNF by activated microglia and macrophages was observed in the striatum following a mechanical injury (Liberatore et al., 1997; Batchelor et al., 1999) and in other systems: in the injured spinal cord (Satake et al., 2000; Widenfalk et al., 2001), in the cerebral cortex after ischemia (Wei et al., 2001) and in cultured macrophages (Hashimoto et al., 2005a).

GDNF also controls the activity of microglial cells. The presence of GDNF receptors GFRα1 and Ret was shown in primary cultures of rat microglia (Honda et al., 1999). In the brain of elderly normal and PD patients, Ret immunoreactivity was observed in ramified microglia having the appearance of resting microglia (Walker et al., 1998). GDNF was shown to promote survival but not the proliferation of cultured microglia (Salimi et al., 2003), to decrease NO production (Nakajima et al., 1998; Salimi et al., 2003), to increase the phagocytic activity of macrophages and microglial cells (Hashimoto et al., 2005b; Chang et al., 2006) and to increase the activity of superoxide dismutase (SOD) and the expression of some adhesion molecules (intercellular adhesion molecule-1 and the integrin α5), which are involved in microglial migration, adhesion and phagocytosis (Chang et al., 2006). GDNF had no effect on the secretion of the pro-inflammatory cytokines tumour necrosis factor-α (TNF-α) and IL-1β (Chang et al., 2006). The effect of GDNF on the phagocytic activity of microglial cells might increase the clearing of death cell debris, which could delay neuronal regeneration after brain damage.

In summary, GDNF induced upon injury might, in addition to a neuroprotective/ neurotrophic role, participate in the inflammatory response promoting the removal of cellular debris and improving nervous tissue repair. The net effect of microglia on neuronal survival after injury, either neuroprotective or neurotoxic, likely depends on the type and stimulus intensity, on the "cocktail" of molecules secreted by microglia, and the vulnerability of the neuronal populations. Dopaminergic neurons are particularly sensitive to microglial injury due to low levels of antioxidant defences (reviewed by Block and Hong, 2005).

6. NEURON–GLIA CROSSTALK UPON INJURY AND GDNF UPREGULATION

As detailed above, a great deal of evidence suggests that nigrostriatal injury may induce GDNF expression both in astrocytes and microglial cells, but how damaged neurons signal glia is not completely understood. Since the expression of neurotrophic factors is regulated by afferent activity (Hughes et al., 1999), dopamine is a likely candidate to control GDNF expression in the nigrostriatal system. This is consistent with the protective effects reported for dopamine in several models of neuronal lesion. Dopamine agonists were shown to protect mouse striatal neurons against 6-OHDA toxicity (Iida et al., 1999), and to protect neuroblastoma SH-SY5Y cell line and mesencephalic cultures from MPP^+ toxicity (Presgraves et al., 2004). Dopamine D2 agonists are also effective against the excitotoxicity evoked by glutamate agonists in mesencephalic neurons (Sawada et al., 1998).

The neuroprotection afforded by dopamine or dopamine agonists has been associated with their ability to promote the expression of several neurotrophic factors, although other mechanisms have been proposed like the induction of antioxidant defences (Iida et al., 1999). D1 and D2 agonists increase the synthesis of GDNF, BDNF and NGF in mouse astroglial cultures (Ohta et al., 2004), and of GDNF and BDNF in mesencephalic cultures (Guo et al., 2002). The synthesis and release of FGF-2 is also modulated by activation of D1/D2 receptors by apomorphine in striatal astrocytes. In agreement with these data showing the ability of dopamine agonists to promoting the expression of neurotrophic factors, mice deficient in D2 receptors (D2R–/–) showed reduced levels of both GDNF and neurotrophin-4 (NT-4; Bozzi and Borrelli, 1999). Moreover, the protection of mesencephalic neurons against a MPP^+ lesion afforded by D3 receptor preferring agonists was mediated by an increase in the secretion of both GDNF and BDNF (Du et al., 2005).

Trophic actions promoted by dopamine agonists appear to be brain region-specific. While cultured foetal striatal cells respond to apomorphine treatment by increasing BDNF and GDNF, hippocampal cells are unable to respond (Guo et al., 2002). Conditioned media from *substantia nigra* astroglial cultures treated with D3 agonists showed remarkably increased levels of GDNF and BDNF, while no changes were observed in media from striatum or cortex astroglial cultures (Du et al., 2005). Moreover, the DR2 (–/–) knockout mice show a 40%–50% reduction of GDNF mRNA in the major target areas of *substantia nigra*/ventral tegmental area (VTA) dopaminergic neurons, whereas GDNF mRNA levels are unaltered in other brain areas receiving dopaminergic innervation (Bozzi and Borrelli, 1999). Conversely, the knockout mice presented changes in the expression of NT-4 in other brain areas. There were no changes in the NT-4 mRNA levels in the *substantia nigra* but the levels in the parietal cortex, an area that receives innervation from the ventral mesencephalon, were substantially reduced (Bozzi and Borrelli, 1999).

Collectively, these findings support the idea that dopamine regulates the expression of GDNF and other neurotrophic factors, an effect contributing to the neuroprotective role of dopamine. Therefore, depletion of dopamine upon nigrostriatal injury would lead to a decrease of GDNF expression, unless other factors related to the neuronal injury modulate

GDNF expression. In *substantia nigra* neuron–glia cultures, we found that selective injury to dopaminergic neurons induces marked increases in IL-1β, and we have shown that this cytokine mediates GDNF upregulation (Saavedra et al., 2006b). We proposed that IL-1β is involved in the neuron–glia crosstalk upon nigrostriatal injury (Saavedra et al., 2006b). Although IL-1β is mostly considered a pro-inflammatory cytokine having harmful effects on neurons, several studies suggest a neuroprotective role for IL-1β. As already mentioned, IL-1 is involved in dopaminergic sprouting after toxin-induced denervation (Parish et al., 2002). Furthermore, intranigral infusion of IL-1β was shown to activate astrocytes and significantly protected nigral dopaminergic cell bodies from a subsequent 6-OHDA lesion (Saura et al., 2003). Our data suggest that these trophic effects of IL-1β might be mediated, at least partially, by GDNF upregulation (Saavedra et al., 2006b).

Estrogens have also been shown to regulate the expression of neurotrophic factors, which might mediate their neuroprotective effects (reviewed by Morale et al., 2006). In vivo studies showed beneficial effects of estrogens against the toxicity induced by 6-OHDA or MPTP in the nigrostriatal system (Murray et al., 2003; D'Astous et al., 2004). Several studies suggest that the neuroprotection afforded by estrogens is associated with their regulatory action on the expression of neurotrophic factors. Estrogens stimulate GDNF expression in the developing hypothalamic neurons (Ivanova et al., 2002a) and rescue spinal motoneurons from AMPA (ionotropic glutamate receptor agonist) induced toxicity through an increase in the production and release of GDNF by astrocytes (Platania et al., 2005). However, other mechanisms might mediate a direct neuronal protection by estrogens. Studies using cultured mouse midbrain cells demonstrate that estrogen action involves phosphorylation of Akt in midbrain neurons but not astrocytes (Ivanova et al., 2002b).

In addition, estrogens control glial activation and expression of inflammatory mediators, such as cytokines and chemokines implicated in neuroinflammation and neurodegeneration (reviewed by Morale et al., 2006). Estrogens downregulate glial activation promoted by MPTP in the *substantia nigra* and the striatum (Tripanichkul et al., 2006). Estrogens are also able to prevent the LPS-induced inflammatory response in microglia. Activation of estrogen receptors in microglial cells blocks NO production (Vegeto et al., 2001) and prevents toxicity in primary cultures of rat mesencephalic neurons exposed to conditioned medium from LPS-activated microglia (Block and Hong, 2005). Therefore, estrogens were suggested to switch microglia from a neurotoxic to a neuroprotective state (Morale et al., 2006).

7. THE NEUROPROTECTIVE MECHANISMS SET IN MOTION BY GDNF

How GDNF promotes survival and regeneration of dopaminergic neurons is not completely understood. GDNF prevents apoptotic death of midbrain neurons in culture (Clarkson et al., 1997; Burke et al., 1998) and in vivo during development (reviewed by Burke, 2004). In cell cultures, GDNF decreases apoptosis induced by increased oxidative stress (Sawada et al., 2000) or 6-OHDA (Ding et al., 2004). The inhibition of apoptosis might be due to upregulation of anti-apoptotic proteins since GDNF was shown to increase the levels of Bcl-2 and Bcl-xl, and to reduce the subsequent caspase activation in mesencephalic neurons in culture (Sawada et al., 2000) and in other systems (Ghribi et al., 2001; Cheng et al., 2002). Other anti-apoptotic proteins have been shown to mediate the protective effects of GDNF in axotomized motor neurons: the X-linked inhibitor of apoptosis (XIAP) and the neuronal apoptosis inhibitory protein (NAIP; Perrelet et al., 2002). In the mesencephalic cell cultures, neuroprotection by GDNF was shown to involve the PI3K/Akt pathway but the involvement of this signaling pathway in the expression of the anti-apoptotic proteins was not addressed (Sawada et al., 2000).

GDNF also increases the activities of the antioxidant enzymes SOD, catalase and glutathione peroxidase (Chao and Lee, 1999; Cheng et al., 2004), and the levels of reduced glutathione (Onyango et al., 2005). Accordingly, GDNF was shown to suppress the accumulation of oxygen free radicals in mesencephalic cell cultures (Sawada et al., 2000) and in the hippocampus in vivo (Cheng et al., 2004). We have shown that GDNF downregulates an oxidative stress-induced protein, the heme oxygenase-1 (HO-1), in *substantia nigra* cell cultures (Saavedra et al., 2005). HO-1 is involved in heme catabolism, leading to the generation of biliverdin, free iron and CO. HO-1 is strongly induced by oxidant stimuli and confers protection against oxidative stress-mediated cell death by production of the antioxidant bilirubin (Dore et al., 1999; Le et al., 1999; Chen et al., 2000; Yoo et al., 2003). However, HO-1 is considered a "double-edged sword" since the release of haem-derived iron at high levels of HO-1 expression may exacerbate oxidative injury (Suttner and Dennery, 1999). Therefore, we proposed that cells exposed to stimuli mimicking dopamine toxicity upregulate HO-1 in response to increased oxidative stress, as a protective strategy. In a second phase, GDNF, also upregulated in response to injury, downregulates HO-1, keeping its expression at a protective level (Saavedra et al., 2005).

In addition to promoting dopaminergic cell survival, GDNF induces axonal outgrowth and sprouting (Batchelor et al., 1999, 2000). This effect might involve signaling through the NCAM, which was shown to function as an alternative signaling receptor for GDNF (Paratcha et al., 2003). It was shown that NCAM and another adhesion molecule, the integrin αv, mediate the effects of GDNF on dopaminergic neuron survival and outgrowth in culture, and on dopamine turnover and motor function in adult rats (Chao et al., 2003). Subchronic administration of GDNF into the rat *substantia nigra* was found to significantly increase integrin αv and NCAM expression in midbrain dopaminergic neurons, and neutralizing antibodies for NCAM and integrin αv antagonized the effects of GDNF on dopaminergic neuron survival and outgrowth in culture, and on dopamine turnover and locomotor activity in vivo (Chao et al., 2003). It is interesting to note that in the absence of GDNF, GFRα1 interacts with NCAM and decreases cell adhesion, whereas in the presence of GDNF, the complex GFRα–NCAM–GDNF promotes cell adhesion. This signaling pathway has been shown to mediate the effects of GDNF on Schwann cell migration and axonal growth in cortical and hippocampal neurons (Paratcha et al., 2003). On the other hand, GDNF was also implicated in axon guidance. Target-derived GFRα1, in a soluble form after cleavage of the GPI anchor, was suggested to function as a chemoattractant cue (Paratcha et al., 2006) that would promote axonal growth in a process involving the activation of the cyclin-dependent kinase 5 (Cdk5) mediated by the PI3K and the MAPK pathways (Ledda et al., 2002). Thus, this mechanism might be involved in axonal regrowth and target reinnervation upon injury promoted by GDNF.

GDNF also exerts direct effects on dopaminergic neurotransmission, not related to the protective/rescue and regenerative actions on dopaminergic neurons. GDNF has been shown to increase quantal release from post-natal dopaminergic neurons in culture (Pothos et al., 1998), an effect likely related to the enhancement of dopamine synthesis since GDNF has been shown to increase TH phosphorylation leading to increased enzyme activity (Kobori et al., 2004; Salvatore et al., 2004). Furthermore, GDNF induces the activity of GTP-cyclohydrolase I generating BH4 (5,6,7,8-tetrahydrobiopterin), which is a cofactor for TH

(Bauer et al., 2002). In addition to increasing dopamine synthesis, GDNF enhances the excitability of midbrain dopaminergic neurons by inhibiting A-type K^+ channels, a fast action mediated by MAP kinase activation (Yang et al., 2001). GDNF also potentiates the activation of Ca^{2+} channels and excitatory transmission in midbrain neurons (Wang et al., 2003). Moreover, GDNF increases the synaptic efficacy of dopaminergic neurons in culture by promoting the establishment of new functional synaptic terminals (Bourque and Trudeau, 2000). These facilitatory mechanisms of dopaminergic transmission are also likely to underlie the improvement of motor function observed in animal models of PD upon administration of GDNF.

In conclusion, the effects of GDNF on the nigrostriatal system involve dopamine biosynthesis and neurotransmission, neuronal survival, neurite outgrowth and sprouting, depending on a complex web of signaling mechanisms.

8. CONCLUSIONS AND THE MISSING LINKS OF GDNF IN THE NIGROSTRIATAL SYSTEM

The neuroprotective and repairing actions of GDNF in the nigrostriatal dopaminergic pathway are well established, both as a target-derived and a local trophic factor. Whether different mechanisms mediate protection of nerve terminals and cell bodies is not clear. Dopaminergic injury, at least some levels of injury, can trigger GDNF upregulation by astrocytes and microglial cells but the mechanisms involved are just beginning to uncover (Fig. 1). Dopamine might mediate neurotrophic factor expression in a neuronal activity-dependent way. Cytokines like IL-1β, produced upon injury, induce GDNF upregulation. It is not known whether similar or different mechanisms control GDNF expression in the striatum and the *substantia nigra*. Microglial cells are both a source and a target for GDNF. This trophic factor was shown to control the phagocytic activity and adhesion properties of microglia, but it is not known whether GDNF might function as a neurotoxic/neurotrophic switch for microglial cells. Knowledge on the intercellular mediators involved in the crosstalk between damaged dopaminergic neurons, astrocytes and microglial cells (Fig. 1) might help to design new therapeutic strategies for PD aimed at upregulating the endogenous expression of GDNF.

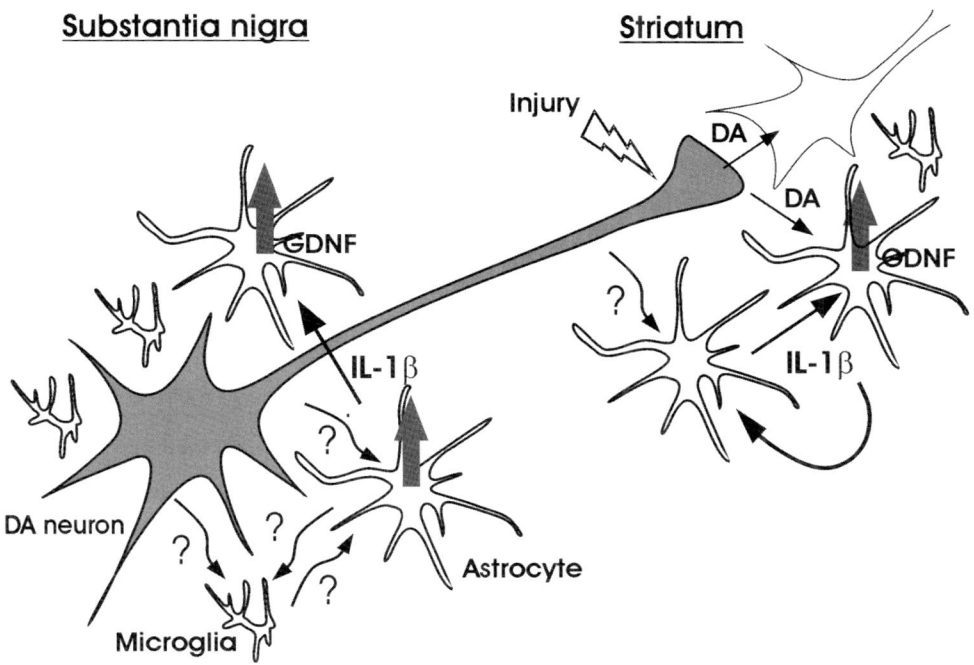

Fig. 1. Neuron–glia crosstalk and GDNF upregulation upon injury in the nigrostriatal pathway. Some studies in animal models of PD show increased expression of GDNF both in the *substantia nigra* and the striatum (Zhou et al., 2000; Nakajima et al., 2001; Yurek and Fletcher-Turner, 2001, 2002). In cell culture models, GDNF upregulation induced by selective damage to dopamine (DA) neurons is mediated by soluble signals since conditioned media from challenged neuron–glia cultures increase GDNF expression in naïve astrocytes cultures (Saavedra et al., 2006a). These putative (?) mediators trigger IL-1β synthesis and release from astrocytes. IL-1β, acting in a paracrine or autocrine way, binds to IL-1 receptors present in astrocytes and stimulates GDNF expression (Saavedra et al., 2006b). DA agonists were shown to upregulate GDNF expression in astroglial cells (Ohta et al., 2004; Du et al., 2005). The signals controlling GDNF expression by microglial cells were not identified in this system. Microglial cells are proposed to have a greater neurotoxic/neuroprotective role in *substantia nigra* than in the striatum, as discussed in the text

9. REFERENCES

Backman, C.M., Shan, L., Zhang, Y.J., Hoffer, B.J., Leonard, S., Troncoso, J.C., Vonsatel, P. and Tomac, A.C., 2006, Gene expression patterns for GDNF and its receptors in the human putamen affected by Parkinson's disease: a real-time PCR study. *Mol. Cell. Endocrinol.* **252**: 160.

Barcia, C., Sanchez, B.A., Fernandez-Villalba, E., Bautista, V., Poza, Y.P., Fernandez-Barreiro, A., Hirsch, E.C. and Herrero, M.T., 2004, Evidence of active microglia in substantia nigra pars compacta of parkinsonian monkeys 1 year after MPTP exposure. *Glia* **46**: 402.

Batchelor, P.E., Liberatore, G.T., Wong, J.Y., Porritt, M.J., Frerichs, F., Donnan, G.A. and Howells, D.W., 1999, Activated macrophages and microglia induce dopaminergic sprouting in the injured striatum and express brain-derived neurotrophic factor and glial cell line-derived neurotrophic factor. *J. Neurosci.* **19**: 1708.

Batchelor, P.E., Liberatore, G.T., Porritt, M.J., Donnan, G.A. and Howells, D.W., 2000, Inhibition of brain-derived neurotrophic factor and glial cell line-derived neurotrophic factor expression reduces dopaminergic sprouting in the injured striatum. *Eur. J. Neurosci.* **12**: 3462.

Batchelor, P.E., Porritt, M.J., Martinello, P., Parish, C.L., Liberatore, G.T., Donnan, G.A. and Howells, D.W., 2002, Macrophages and microglia produce local trophic gradients that stimulate axonal sprouting toward but not beyond the wound edge. *Mol. Cell Neurosci.* **21**: 436.

Bauer, M., Suppmann, S., Meyer, M., Hesslinger, C., Gasser, T., Widmer, H.R. and Ueffing, M., 2002, Glial cell line-derived neurotrophic factor up-regulates GTP-cyclohydrolase I activity and tetrahydrobiopterin levels in primary dopaminergic neurones. *J. Neurochem.* **82**: 1300.

Benn, S.C. and Woolf, C.J., 2004, Adult neuron survival strategies – slamming on the brakes. *Nat. Rev. Neurosci.* **5**: 686.

Bilang-Bleuel, A., Revah, F., Colin, P., Locquet, I., Robert, J.J., Mallet, J. and Horellou, P., 1997, Intrastriatal injection of an adenoviral vector expressing glial-cell-line-derived neurotrophic factor prevents dopaminergic neuron degeneration and behavioral impairment in a rat model of Parkinson disease. *Proc. Natl. Acad. Sci. USA* **94**: 8818.

Bizon, J.L., Lauterborn, J.C. and Gall, C.M., 1999, Subpopulations of striatal interneurons can be distinguished on the basis of neurotrophic factor expression. *J. Comp. Neurol.* **408**: 283.

Bjorklund, A., Kirik, D., Rosenblad, C., Georgievska, B., Lundberg, C. and Mandel, R.J., 2000, Towards a neuroprotective gene therapy for Parkinson's disease: use of adenovirus, AAV and lentivirus vectors for gene transfer of GDNF to the nigrostriatal system in the rat Parkinson model. *Brain Res.* **886**: 82.

Block, M.L. and Hong, J.S., 2005, Microglia and inflammation-mediated neurodegeneration: multiple triggers with a common mechanism. *Prog. Neurobiol.* **76**: 77.

Blum, M. and Weickert, C.S., 1995, GDNF mRNA expression in normal postnatal development, aging, and in Weaver mutant mice. *Neurobiol. Aging* **16**: 925.

Bourque, M.J. and Trudeau, L.E., 2000, GDNF enhances the synaptic efficacy of dopaminergic neurons in culture. *Eur. J. Neurosci.* **12**: 3172.

Bowenkamp, K.E., Hoffman, A.F., Gerhardt, G.A., Henry, M.A., Biddle, P.T., Hoffer, B.J. and Granholm, A.C., 1995, Glial cell line-derived neurotrophic factor supports survival of injured midbrain dopaminergic neurons. *J. Comp. Neurol.* **355**: 479.

Bowenkamp, K.E., David, D., Lapchak, P.L., Henry, M.A., Granholm, A.C., Hoffer, B.J. and Mahalik, T.J., 1996, 6-Hydroxydopamine induces the loss of the dopaminergic phenotype in substantia nigra neurons of the rat. A possible mechanism for restoration of the nigrostriatal circuit mediated by glial cell line-derived neurotrophic factor. *Exp. Brain Res.* **111**: 1.

Bozzi, Y. and Borrelli, E., 1999, Absence of the dopamine D2 receptor leads to a decreased expression of GDNF and NT-4 mRNAs in restricted brain areas. *Eur. J. Neurosci.* **11**: 1275.

Breidert, T., Callebert, J., Heneka, M.T., Landreth, G., Launay, J.M. and Hirsch, E.C., 2002, Protective action of the peroxisome proliferator-activated receptor-gamma agonist pioglitazone in a mouse model of Parkinson's disease. *J. Neurochem.* **82**: 615.

Bresjanac, M. and Antauer, G., 2000, Reactive astrocytes of the quinolinic acid-lesioned rat striatum express GFRalpha1 as well as GDNF in vivo. *Exp. Neurol.* **164**: 53.

Brizard, M., Carcenac, C., Bemelmans, A.P., Feuerstein, C., Mallet, J. and Savasta, M., 2006, Functional reinnervation from remaining DA terminals induced by GDNF lentivirus in a rat model of early Parkinson's disease. *Neurobiol. Dis.* **21**: 90.

Burke, R.E., 2004, Ontogenic cell death in the nigrostriatal system. *Cell Tissue Res.* **318**: 63.

Burke, R.E., Antonelli, M. and Sulzer, D., 1998, Glial cell line-derived neurotrophic growth factor inhibits apoptotic death of postnatal substantia nigra dopamine neurons in primary culture. *J. Neurochem.* **71**: 517.

Chang, Y.P., Fang, K.M., Lee, T.I. and Tzeng, S.F., 2006, Regulation of microglial activities by glial cell line derived neurotrophic factor. *J. Cell Biochem.* **97**: 501.

Chao, C.C. and Lee, E.H., 1999, Neuroprotective mechanism of glial cell line-derived neurotrophic factor on dopamine neurons: role of antioxidation. *Neuropharmacology* **38**: 913.

Chao, C.C., Ma, Y.L., Chu, K.Y. and Lee, E.H., 2003, Integrin alpha and NCAM mediate the effects of GDNF on DA neuron survival, outgrowth, DA turnover and motor activity in rats. *Neurobiol. Aging* **24**: 105.

Chauhan, N.B., Siegel, G.J. and Lee, J.M., 2001, Depletion of glial cell line-derived neurotrophic factor in substantia nigra neurons of Parkinson's disease brain. *J. Chem. Neuroanat.* **21**: 277.

Chen, K., Gunter, K. and Maines, M.D., 2000, Neurons overexpressing heme oxygenase-1 resist oxidative stress-mediated cell death. *J. Neurochem.* **75**: 304.

Cheng, H., Wu, J.P. and Tzeng, S.F. 2002, Neuroprotection of glial cell line-derived neurotrophic factor in damaged spinal cords following contusive injury. *J. Neurosci. Res.* **69**: 397.

Cheng, H., Fu, Y.S. and Guo, J.W., 2004, Ability of GDNF to diminish free radical production leads to protection against kainate-induced excitotoxicity in hippocampus. *Hippocampus* **14**: 77.

Choi-Lundberg, D.L. and Bohn, M.C., 1995, Ontogeny and distribution of glial cell line-derived neurotrophic factor (GDNF) mRNA in rat. *Brain Res. Dev. Brain Res.* **85**: 80.

Choi-Lundberg, D.L., Lin, Q., Chang, Y.N., Chiang, Y.L., Hay, C.M., Mohajeri, H., Davidson, B.L. and Bohn, M.C., 1997, Dopaminergic neurons protected from degeneration by GDNF gene therapy. *Science* **275**: 838.

Choi-Lundberg, D.L., Lin, Q., Schallert, T., Crippens, D., Davidson, B.L., Chang, Y.N., Chiang, Y.L., Qian, J., Bardwaj, L. and Bohn, M.C., 1998, Behavioral and cellular protection of rat dopaminergic neurons by an adenoviral vector encoding glial cell line-derived neurotrophic factor. *Exp. Neurol.* **154**: 261.

Clarkson, E.D., Zawada, W.M. and Freed, C.R., 1997, GDNF improves survival and reduces apoptosis in human embryonic dopaminergic neurons in vitro. *Cell Tissue Res.* **289**: 207.

Collier, T.J., Dung, L.Z., Carvey, P.M., Fletcher-Turner, A., Yurek, D.M., Sladek, J.R., Jr. and Kordower, J.H., 2005, Striatal trophic factor activity in aging monkeys with unilateral MPTP-induced parkinsonism. *Exp. Neurol.* **191**: S60.

Connor, B., Kozlowski, D.A., Schallert, T., Tillerson, J.L., Davidson, B.L. and Bohn, M.C., 1999, Differential effects of glial cell line-derived neurotrophic factor (GDNF) in the striatum and substantia nigra of the aged Parkinsonian rat. *Gene Ther.* **6**: 1936.

Connor, B., Kozlowski, D.A., Unnerstall, J.R., Elsworth, J.D., Tillerson, J.L., Schallert, T. and Bohn, M.C., 2001, Glial cell line-derived neurotrophic factor (GDNF) gene delivery protects dopaminergic terminals from degeneration. *Exp. Neurol.* **169**: 83.

D'Astous, M., Morissette, M. and Di Paolo, T., 2004, Effect of estrogen receptor agonists treatment in MPTP mice: evidence of neuroprotection by an ER alpha agonist. *Neuropharmacology* **47**: 1180.

Dauer, W. and Przedborski, S., 2003, Parkinson's disease: mechanisms and models. *Neuron* **39**: 889.

Dehmer, T., Lindenau, J., Haid, S., Dichgans, J. and Schulz, J.B., 2000, Deficiency of inducible nitric oxide synthase protects against MPTP toxicity in vivo. *J. Neurochem.* **74**: 2213.

Deumens, R., Blokland, A. and Prickaerts, J., 2002, Modeling Parkinson's disease in rats: an evaluation of 6-OHDA lesions of the nigrostriatal pathway. *Exp. Neurol.* **175**: 303.

Ding, Y.M., Jaumotte, J.D., Signore, A.P. and Zigmond, M.J., 2004, Effects of 6-hydroxydopamine on primary cultures of substantia nigra: specific damage to dopamine neurons and the impact of glial cell line-derived neurotrophic factor. *J. Neurochem.* **89**: 776.

Dore, S., Takahashi, M., Ferris, C.D., Zakhary, R., Hester, L.D., Guastella, D. and Snyder, S.H., 1999, Bilirubin, formed by activation of heme oxygenase-2, protects neurons against oxidative stress injury. *Proc. Natl. Acad. Sci. USA* **96**: 2445.

Du, F., Li, R., Huang, Y., Li, X. and Le, W., 2005, Dopamine D3 receptor-preferring agonists induce neurotrophic effects on mesencephalic dopamine neurons. *Eur. J. Neurosci.* **22**: 2422.

Eslamboli, A., Georgievska, B., Ridley, R.M., Baker, H.F., Muzyczka, N., Burger, C., Mandel, R.J., Annett, L. and Kirik, D., 2005, Continuous low-level glial cell line-derived neurotrophic factor delivery using recombinant adeno-associated viral vectors provides neuroprotection and induces behavioral recovery in a primate model of Parkinson's disease. *J. Neurosci.* **25**: 769.

Gash, D.M., Zhang, Z., Ovadia, A., Cass, W.A., Yi, A., Simmerman, L., Russell, D., Martin, D., Lapchak, P.A., Collins, F., Hoffer, B.J. and Gerhardt, G.A., 1996, Functional recovery in parkinsonian monkeys treated with GDNF. *Nature* **380**: 252.

Gash, D.M., Zhang, Z., Ai, Y., Grondin, R., Coffey, R. and Gerhardt, G.A., 2005, Trophic factor distribution predicts functional recovery in parkinsonian monkeys. *Ann. Neurol.* **58**: 224.

Gerhardt, G.A., Cass, W.A., Huettl, P., Brock, S., Zhang, Z. and Gash, D.M., 1999, GDNF improves dopamine function in the substantia nigra but not the putamen of unilateral MPTP-lesioned rhesus monkeys. *Brain Res.* **817**: 163.

Ghribi, O., Herman, M.M., Forbes, M.S., DeWitt, D.A. and Savory, J., 2001, GDNF protects against aluminum-induced apoptosis in rabbits by upregulating Bcl-2 and Bcl-XL and inhibiting mitochondrial Bax translocation. *Neurobiol. Dis.* **8**: 764.

Gill, S.S., Patel, N.K., Hotton, G.R., O'Sullivan, K., McCarter, R., Bunnage, M., Brooks, D.J., Svendsen, C.N. and Heywood, P., 2003, Direct brain infusion of glial cell line-derived neurotrophic factor in Parkinson disease. *Nat. Med.* **9**: 589.

Grondin, R., Zhang, Z., Yi, A., Cass, W.A., Maswood, N., Andersen, A.H., Elsberry, D.D., Klein, M.C., Gerhardt, G.A. and Gash, D.M., 2002, Chronic, controlled GDNF infusion promotes structural and functional recovery in advanced parkinsonian monkeys. *Brain* **125**: 2191.

Grondin, R., Zhang, Z., Ai, Y., Gash, D.M. and Gerhardt, G.A., 2003, Intracranial delivery of proteins and peptides as a therapy for neurodegenerative diseases. *Prog. Drug Res.* **61**: 101.

Grunblatt, E., Mandel, S., Maor, G. and Youdim, M.B., 2001, Gene expression analysis in N-methyl-4-phenyl-1,2,3,6-tetrahydropyridine mice model of Parkinson's disease using cDNA microarray: effect of R-apomorphine. *J. Neurochem.* **78**: 1.

Guo, H., Tang, Z., Yu, Y., Xu, L., Jin, G. and Zhou, J., 2002,. Apomorphine induces trophic factors that support fetal rat mesencephalic dopaminergic neurons in cultures. *Eur. J. Neurosci.* **16**: 1861.

Hashimoto, M., Nitta, A., Fukumitsu, H., Nomoto, H., Shen, L. and Furukawa, S., 2005a, Inflammation-induced GDNF improves locomotor function after spinal cord injury. *Neuroreport* **16**: 99.

Hashimoto, M., Nitta, A., Fukumitsu, H., Nomoto, H., Shen, L. and Furukawa, S., 2005b, Involvement of glial cell line-derived neurotrophic factor in activation processes of rodent macrophages. *J. Neurosci. Res.* **79**: 476.

He, Y., Appel, S. and Le, W., 2001, Minocycline inhibits microglial activation and protects nigral cells after 6-hydroxydopamine injection into mouse striatum. *Brain Res.* **909**: 187.

Ho, A. and Blum, M., 1998, Induction of interleukin-1 associated with compensatory dopaminergic sprouting in the denervated striatum of young mice: model of aging and neurodegenerative disease. *J. Neurosci.* **18**: 5614.

Hoffer, B.J., Hoffman, A., Bowenkamp, K., Huettl, P., Hudson, J., Martin, D., Lin, L.F. and Gerhardt, G.A., 1994, Glial cell line-derived neurotrophic factor reverses toxin-induced injury to midbrain dopaminergic neurons in vivo. *Neurosci. Lett.* **182**: 107.

Hoffman, A.F., van Horne, C.G., Eken, S., Hoffer, B.J. and Gerhardt, G.A., 1997, In vivo microdialysis studies on somatodendritic dopamine release in the rat substantia nigra: effects of unilateral 6-OHDA lesions and GDNF. *Exp. Neurol.* **147**: 130.

Honda, S., Nakajima, K., Nakamura, Y., Imai, Y. and Kohsaka, S., 1999, Rat primary cultured microglia express glial cell line-derived neurotrophic factor receptors. *Neurosci. Lett.* **275**: 203.

Hughes, P.E., Alexi, T., Walton, M., Williams, C.E., Dragunow, M., Clark, R.G. and Gluckman, P.D., 1999, Activity and injury-dependent expression of inducible transcription factors, growth factors and apoptosis-related genes within the central nervous system. *Prog. Neurobiol.* **57**: 421.

Hunot, S., Bernard, V., Faucheux, B., Boissiere, F., Leguern, E., Brana, C., Gautris, P.P., Guerin, J., Bloch, B., Agid, Y. and Hirsch, E.C., 1996, Glial cell line-derived neurotrophic factor (GDNF) gene expression in the human brain: a post mortem in situ hybridization study with special reference to Parkinson's disease. *J. Neural. Transm.* **103**: 1043.

Hurelbrink, C.B. and Barker, R.A., 2004, The potential of GDNF as a treatment for Parkinson's disease. *Exp. Neurol.* **185**: 1.

Hurley, S.D., O'Banion, M.K., Song, D.D., Arana, F.S., Olschowka, J.A. and Haber, S.N., 2003, Microglial response is poorly correlated with neurodegeneration following chronic, low-dose MPTP administration in monkeys. *Exp. Neurol.* **184**: 659.

Ikeda, T., Koo, H., Xia, Y.X., Ikenoue, T. and Choi, B.H., 2002, Bimodal upregulation of glial cell line-derived neurotrophic factor (GDNF) in the neonatal rat brain following ischemic/hypoxic injury. *Int. J. Dev. Neurosci.* **20**: 555.

Iida, M., Miyazaki, I., Tanaka, K., Kabuto, H., Iwata-Ichikawa, E. and Ogawa, N., 1999, Dopamine D2 receptor-mediated antioxidant and neuroprotective effects of ropinirole, a dopamine agonist. *Brain Res.* **838**: 51.

Inoue, T., Tsui, J., Wong, N., Wong, S.Y., Suzuki, F. and Kwok, Y.N., 1999, Expression of glial cell line-derived neurotrophic factor and its mRNA in the nigrostriatal pathway following MPTP treatment. *Brain Res.* **826**: 306.

Ivanova, T., Karolczak, M. and Beyer, C., 2002a, Estradiol stimulates GDNF expression in developing hypothalamic neurons. *Endocrinology* **143**: 3175.

Ivanova, T., Mendez, P., Garcia-Segura, L.M. and Beyer, C. (2002b). Rapid stimulation of the PI3-kinase/Akt signaling pathway in developing midbrain neurones by oestrogen. *J. Neuroendocrinol.* **14**: 73.

Kim, W.G., Mohney, R.P., Wilson, B., Jeohn, G.H., Liu, B. and Hong, J.S., 2000, Regional difference in susceptibility to lipopolysaccharide-induced neurotoxicity in the rat brain: role of microglia. *J. Neurosci.* **20**: 6309.

Kirik, D., Rosenblad, C. and Bjorklund, A., 2000a, Preservation of a functional nigrostriatal dopamine pathway by GDNF in the intrastriatal 6-OHDA lesion model depends on the site of administration of the trophic factor. *Eur. J. Neurosci.* **12**: 3871.

Kirik, D., Rosenblad, C., Bjorklund, A. and Mandel, R.J., 2000b, Long-term rAAV-mediated gene transfer of GDNF in the rat Parkinson's model: intrastriatal but not intranigral transduction promotes functional regeneration in the lesioned nigrostriatal system. *J. Neurosci.* **20**: 4686.

Kobori, N., Waymire, J.C., Haycock, J.W., Clifton, G.L. and Dash, P.K., 2004, Enhancement of tyrosine hydroxylase phosphorylation and activity by glial cell line-derived neurotrophic factor. *J. Biol. Chem.* **279**: 2182.

Kordower, J.H., Emborg, M.E., Bloch, J., Ma, S.Y., Chu, Y., Leventhal, L., McBride, J., Chen, E.Y., Palfi, S., Roitberg, B.Z., Brown, W.D., Holden, J.E., Pyzalski, R., Taylor, M.D., Carvey, P., Ling, Z., Trono, D., Hantraye, P., Deglon, N. and Aebischer, P., 2000, Neurodegeneration prevented by lentiviral vector delivery of GDNF in primate models of Parkinson's disease. *Science* **290**: 767.

Kozlowski, D.A., Connor, B., Tillerson, J.L., Schallert, T. and Bohn, M.C., 2000, Delivery of a GDNF gene into the substantia nigra after a progressive 6-OHDA lesion maintains functional nigrostriatal connections. *Exp. Neurol.* **166**: 1.

Kozlowski, D.A., Miljan, E.A., Bremer, E.G., Harrod, C.G., Gerin, C., Connor, B., George, D., Larson, B. and Bohn, M.C., 2004, Quantitative analyses of GFRalpha-1 and GFRalpha-2 mRNAs and tyrosine hydroxylase protein in the nigrostriatal system reveal bilateral compensatory changes following unilateral 6-OHDA lesions in the rat. *Brain Res.* **1016**: 170.

Lang, A.E., Gill, S., Patel, N.K., Lozano, A., Nutt, J.G., Penn, R., Brooks, D.J., Hotton, G., Moro, E., Heywood, P., Brodsky, M.A., Burchiel, K., Kelly, P., Dalvi, A., Scott, B., Stacy, M., Turner, D., Wooten, V.G., Elias, W.J., Laws, E.R., Dhawan,V., Stoessl, A.J., Matcham, J., Coffey, R.J. and Traub, M., 2006, Randomized controlled trial of intraputamenal glial cell line-derived neurotrophic factor infusion in Parkinson disease. *Ann. Neurol.* **59**: 459.

Langston, J.W., Forno, L.S., Tetrud, J., Reeves, A.G., Kaplan, J.A. and Karluk, D., 1999, Evidence of active nerve cell degeneration in the substantia nigra of humans years after 1-methyl-4-phenyl-1,2,3,6-tetrahydropyridine exposure. *Ann. Neurol.* **46**: 598.

Lapchak, P.A., Araujo, D.M., Hilt, D.C., Sheng, J. and Jiao, S., 1997, Adenoviral vector-mediated GDNF gene therapy in a rodent lesion model of late stage Parkinson's disease. *Brain Res.* **777**: 153.

Lawson, L.J., Perry, V.H., Dri, P. and Gordon, S., 1990, Heterogeneity in the distribution and morphology of microglia in the normal adult mouse brain. *Neuroscience* **39**: 151.

Le, W.D., Xie, W.J. and Appel, S.H., 1999, Protective role of heme oxygenase-1 in oxidative stress-induced neuronal injury. *J. Neurosci. Res.* **56**: 652.

Ledda, F., Paratcha, G. and Ibanez, C.F., 2002, Target-derived GFRalpha1 as an attractive guidance signal for developing sensory and sympathetic axons via activation of Cdk5. *Neuron* **36**: 387.

Liberatore, G.T., Wong, J.Y., Porritt, M.J., Donnan, G.A. and Howells, D.W., 1997, Expression of glial cell line-derived neurotrophic factor (GDNF) mRNA following mechanical injury to mouse striatum. *Neuroreport* **8**: 3097.

Liberatore, G.T., Jackson-Lewis, V., Vukosavic, S., Mandir, A.S., Vila, M., McAuliffe, W.G., Dawson, V.L., Dawson, T.M. and Przedborski, S., 1999, Inducible nitric oxide synthase stimulates dopaminergic neurodegeneration in the MPTP model of Parkinson's disease. *Nat. Med.* **5**: 1403.

Liberto, C.M., Albrecht, P.J., Herx, L.M., Yong, V.W. and Levison, S.W., 2004, Pro-regenerative properties of cytokine-activated astrocytes. *J. Neurochem.* **89**: 1092.

Lin, L.F., Doherty, D.H., Lile, J.D., Bektesh, S. and Collins, F., 1993, GDNF: a glial cell line-derived neurotrophic factor for midbrain dopaminergic neurons. *Science* **260**: 1130.

Lin, L.F., Zhang, T.J., Collins, F. and Armes, L.G., 1994, Purification and initial characterization of rat B49 glial cell line-derived neurotrophic factor. *J. Neurochem.* **63**: 758.

Love, S., Plaha, P., Patel, N.K., Hotton, G.R., Brooks, D.J. and Gill, S.S., 2005, Glial cell line-derived neurotrophic factor induces neuronal sprouting in human brain. *Nat. Med.* **11**: 703.

Mandel, R.J., Spratt, S.K., Snyder, R.O. and Leff, S.E., 1997, Midbrain injection of recombinant adeno-associated virus encoding rat glial cell line-derived neurotrophic factor protects nigral neurons in a progressive 6-hydroxydopamine-induced degeneration model of Parkinson's disease in rats. *Proc. Natl. Acad. Sci. USA* **94**: 14083.

Mandel, R.J., Snyder, R.O. and Leff, S.E., 1999, Recombinant adeno-associated viral vector-mediated glial cell line-derived neurotrophic factor gene transfer protects nigral dopamine neurons after onset of progressive degeneration in a rat model of Parkinson's disease. *Exp. Neurol.* **160**: 205.

Mandel, S., Grunblatt, E., Maor, G. and Youdim, M.B., 2002, Early and late gene changes in MPTP mice model of Parkinson's disease employing cDNA microarray. *Neurochem. Res.* **27**: 1231.

Marco, S., Saura, J., Perez-Navarro, E., Jose, M.M., Tolosa, E. and Alberch, J., 2002, Regulation of c-Ret, GFRalpha1, and GFRalpha2 in the substantia nigra pars compacta in a rat model of Parkinson's disease. *J. Neurobiol.* **52**: 343.

Martino, G., 2004, How the brain repairs itself: new therapeutic strategies in inflammatory and degenerative CNS disorders. *Lancet Neurol.* **3**: 372.

McGeer, P.L., Itagaki, S., Boyes, B.E. and McGeer, E.G., 1988, Reactive microglia are positive for HLA-DR in the substantia nigra of Parkinson's and Alzheimer's disease brains. *Neurology* **38**: 1285.

McGeer, P.L., Schwab, C., Parent, A. and Doudet, D., 2003, Presence of reactive microglia in monkey substantia nigra years after 1-methyl-4-phenyl-1,2,3,6-tetrahydropyridine administration. *Ann. Neurol.* **54**: 599.

Mirza, B., Hadberg, H., Thomsen, P. and Moos, T., 2000, The absence of reactive astrocytosis is indicative of a unique inflammatory process in Parkinson's disease. *Neuroscience* **95**: 425.

Miyazaki, H., Nagashima, K., Okuma, Y. and Nomura, Y., 2001, Expression of glial cell line-derived neurotrophic factor induced by transient forebrain ischemia in rats. *Brain Res.* **922**: 165.

Mogi, M., Togari, A., Kondo, T., Mizuno, Y., Kogure, O., Kuno, S., Ichinose, H. and Nagatsu, T., 2001, Glial cell line-derived neurotrophic factor in the substantia nigra from control and parkinsonian brains. *Neurosci. Lett.* **300**: 179.

Morale, M.C., Serra, P.A., L'Episcopo, F., Tirolo, C., Caniglia, S., Testa, N., Gennuso, F., Giaquinta, G., Rocchitta, G., Desole, M.S., Miele, E. and Marchetti, B., 2006, Estrogen, neuroinflammation and neuroprotection in Parkinson's disease: glia dictates resistance versus vulnerability to neurodegeneration. *Neuroscience* **138**: 869.

Murray, H.E., Pillai, A.V., McArthur, S.R., Razvi, N., Datla, K.P., Dexter, D.T. and Gillies, G.E., 2003, Dose- and sex-dependent effects of the neurotoxin 6-hydroxydopamine on the nigrostriatal dopaminergic pathway of adult rats: differential actions of estrogen in males and females. *Neuroscience* **116**: 213.

Nakagawa, T. and Schwartz, J.P., 2004a, Gene expression patterns in in vivo normal adult astrocytes compared with cultured neonatal and normal adult astrocytes. *Neurochem. Int.* **45**: 203.

Nakagawa, T. and Schwartz, J.P., 2004b, Gene expression profiles of reactive astrocytes in dopamine-depleted striatum. *Brain Pathol.* **14**: 275.

Nakagawa, T., Yabe, T. and Schwartz, J.P., 2005, Gene expression profiles of reactive astrocytes cultured from dopamine-depleted striatum. *Neurobiol. Dis.* **20**: 275.

Nakajima, K., Kikuchi, Y., Ikoma, E., Honda, S., Ishikawa, M., Liu, Y. and Kohsaka, S., 1998, Neurotrophins regulate the function of cultured microglia. *Glia* **24**: 272.

Nakajima, K., Hida, H., Shimano, Y., Fujimoto, I., Hashitani, T., Kumazaki, M., Sakurai, T. and Nishino, H., 2001, GDNF is a major component of trophic activity in DA-depleted striatum for survival and neurite extension of DAergic neurons. *Brain Res.* **916**: 76.

Ohta, K., Fujinami, A., Kuno, S., Sakakimoto, A., Matsui, H., Kawahara, Y. and Ohta, M., 2004, Cabergoline stimulates synthesis and secretion of nerve growth factor, brain-derived neurotrophic factor and glial cell line-derived neurotrophic factor by mouse astrocytes in primary culture. *Pharmacology* **71**: 162.

Onyango, I.G., Tuttle, J.B. and Bennett, J.P., Jr., 2005, Brain-derived growth factor and glial cell line-derived growth factor use distinct intracellular signaling pathways to protect PD cybrids from H2O2-induced neuronal death. *Neurobiol. Dis.* **20**: 141.

Oo, T.F., Ries, V., Cho, J., Kholodilov, N. and Burke, R.E., 2005, Anatomical basis of glial cell line-derived neurotrophic factor expression in the striatum and related basal ganglia during postnatal development of the rat. *J. Comp. Neurol.* **484**: 57.

Oppenheim, R.W., 1991, Cell death during development of the nervous system. *Ann. Rev. Neurosci.* **14**: 453.

Paratcha, G., Ledda, F. and Ibanez, C.F., 2003, The neural cell adhesion molecule NCAM is an alternative signaling receptor for GDNF family ligands. *Cell* **113**: 867.

Paratcha, G., Ibanez, C.F. and Ledda, F., 2006, GDNF is a chemoattractant factor for neuronal precursor cells in the rostral migratory stream. *Mol. Cell Neurosci.* **31**: 505.

Parish, C.L., Finkelstein, D.I., Tripanichkul, W., Satoskar, A.R., Drago, J. and Horne, M.K., 2002, The role of interleukin-1, interleukin-6, and glia in inducing growth of neuronal terminal arbors in mice. *J. Neurosci.* **22**: 8034.

Patel, N.K., Bunnage, M., Plaha, P., Svendsen, C.N., Heywood, P. and Gill, S.S., 2005, Intraputamenal infusion of glial cell line-derived neurotrophic factor in PD: a two-year outcome study. *Ann. Neurol.* **57**: 298.

Perrelet, D., Ferri, A., Liston, P., Muzzin, P., Korneluk, R.G. and Kato, A.C., 2002, IAPs are essential for GDNF-mediated neuroprotective effects in injured motor neurons in vivo. *Nat. Cell Biol.* **4**: 175.

Platania, P., Seminara, G., Aronica, E., Troost, D., Catania, V. and Sortino, A., 2005, 17beta-estradiol rescues spinal motoneurons from AMPA-induced toxicity: a role for glial cells. *Neurobiol. Dis.* **20**: 461.

Pochon, N.A., Menoud, A., Tseng, J.L., Zurn, A.D. and Aebischer, P., 1997, Neuronal GDNF expression in the adult rat nervous system identified by in situ hybridization. *Eur. J. Neurosci.* **9**: 463.

Pothos, E.N., Davila, V. and Sulzer, D., 1998, Presynaptic recording of quanta from midbrain dopamine neurons and modulation of the quantal size. *J. Neurosci.* **18**: 4106.

Presgraves, S.P., Borwege, S., Millan, M.J. and Joyce, J.N., 2004, Involvement of dopamine D(2)/D(3) receptors and BDNF in the neuroprotective effects of S32504 and pramipexole against 1-methyl-4-phenylpyridinium in terminally differentiated SH-SY5Y cells. *Exp. Neurol.* **190**: 157.

Purves, D., 1986, The trophic theory of neural connections. *Trends Neurosci.* **9**: 486.

Rosenblad, C., Martinez-Serrano, A. and Bjorklund, A., 1998, Intrastriatal glial cell line-derived neurotrophic factor promotes sprouting of spared nigrostriatal dopaminergic afferents and induces recovery of function in a rat model of Parkinson's disease. *Neuroscience* **82**: 129.

Saavedra, A., Baltazar, G., Carvalho, C.M. and Duarte, E.P., 2005. GDNF modulates HO-1 expression in substantia nigra postnatal cell cultures. *Free Radic. Biol. Med.* **39**: 1611

Saavedra, A., Baltazar, G., Santos, P., Carvalho, C. and Duarte, E.P., 2006a, Selective injury to dopaminergic neurons up-regulates GDNF in *substantia nigra* postnatal cell cultures: role of neuron-glia crosstalk. *Neurobiol. Dis.* **23**: 533.

Saavedra, A., Baltazar, G. and Duarte, E.P., 2006b, Interleukin-1β mediates GDNF up-regulation upon dopaminergic injury in *substantia nigra* cell cultures. *Neurobiol. Dis.* DOI 10.1016/j.nbd.2006.08.019.

Salimi, K., Moser, K.V., Marksteiner, J., Reindl, M. and Humpel, C., 2003, GDNF and TGF-beta1 promote cell survival in serum-free cultures of primary rat microglia. *Cell Tissue Res.* **312**: 135.

Salvatore, M.F., Zhang, J.L., Large, D.M., Wilson, P.E., Gash, C.R., Thomas, T.C., Haycock, J.W., Bing, G., Stanford, J.A., Gash, D.M. and Gerhardt, G.A., 2004, Striatal GDNF administration increases tyrosine hydroxylase phosphorylation in the rat striatum and substantia nigra. *J. Neurochem.* **90**: 245.

Sarabi, A., Hoffer, B.J., Olson, L. and Morales, M., 2001, GFRalpha-1 mRNA in dopaminergic and nondopaminergic neurons in the substantia nigra and ventral tegmental area. *J. Comp. Neurol.* **441**: 106.

Sariola, H. and Saarma, M., 2003, Novel functions and signaling pathways for GDNF. *J. Cell Sci.* **116**: 3855.

Satake, K., Matsuyama, Y., Kamiya, M., Kawakami, H., Iwata, H., Adachi, K. and Kiuchi, K., 2000, Up-regulation of glial cell line-derived neurotrophic factor (GDNF) following traumatic spinal cord injur. *Neuroreport* **11**: 3877.

Saura, J., Pares, M., Bove, J., Pezzi, S., Alberch, J., Marin, C., Tolosa, E. and Marti, M.J., 2003, Intranigral infusion of interleukin-1beta activates astrocytes and protects from subsequent 6-hydroxydopamine neurotoxicity. *J. Neurochem.* **85**: 651.

Sawada, H., Ibi, M., Kihara, T., Urushitani, M., Akaike, A. and Shimohama, S., 1998, Estradiol protects mesencephalic dopaminergic neurons from oxidative stress-induced neuronal death. *J. Neurosci. Res.* **54**: 707.

Sawada, H., Ibi, M., Kihara, T., Urushitani, M., Nakanishi, M., Akaike, A. and Shimohama, S., 2000, Neuroprotective mechanism of glial cell line-derived neurotrophic factor in mesencephalic neurons. *J. Neurochem.* **74**: 1175

Sherer, T.B., Fiske, B.K., Svendsen, C.N., Lang, A.E. and Langston, J.W., 2006, Crossroads in GDNF therapy for Parkinson's disease. *Mov. Disord.* **21**: 136.

Siegel, G.J. and Chauhan, N.B., 2000, Neurotrophic factors in Alzheimer's and Parkinson's disease brain. *Brain Res. Brain Res. Rev.* **33**: 199.

Slevin, J.T., Gerhardt, G.A., Smith, C.D., Gash, D.M., Kryscio, R. and Young, B., 2005, Improvement of bilateral motor functions in patients with Parkinson disease through the unilateral intraputaminal infusion of glial cell line-derived neurotrophic factor. *J. Neurosurg.* **102**: 216.

Smith, A.D., Antion, M., Zigmond, M.J. and Austin, M.C., 2003, Effect of 6-hydroxydopamine on striatal GDNF and nigral GFRalpha1 and RET mRNAs in the adult rat. *Brain Res. Mol. Brain Res.* **117**: 129.

Smith, A.D., Kozlowski, D.A., Bohn, M.C. and Zigmond, M.J., 2005, Effect of AdGDNF on dopaminergic neurotransmission in the striatum of 6-OHDA-treated rats. *Exp. Neurol.* **193**: 420.

Stanic, D., Finkelstein, D.I., Bourke, D.W., Drago, J. and Horne, M.K., 2003, Timecourse of striatal re-innervation following lesions of dopaminergic SNpc neurons of the rat. *Eur. J. Neurosci.* **18**: 1175.

Stanic, D., Tripanichkul, W., Drago, J., Finkelstein, D.I. and Horne, M.K., 2004, Glial responses associated with dopaminergic striatal reinnervation following lesions of the rat substantia nigra. *Brain Res.* **1023**: 83.

Stromberg, I., Bjorklund, L., Johansson, M., Tomac, A., Collins, F., Olson, L., Hoffer, B. and Humpel, C., 1993, Glial cell line-derived neurotrophic factor is expressed in the developing but not adult striatum and stimulates developing dopamine neurons in vivo. *Exp. Neurol.* **124**: 401.

Suttner, D.M. and Dennery, P.A., 1999, Reversal of HO-1 related cytoprotection with increased expression is due to reactive iron. *FASEB J.* **13**: 1800.

Tang, Y.P., Ma, Y.L., Chao, C.C., Chen, K.Y. and Lee, E.H., 1998, Enhanced glial cell line-derived neurotrophic factor mRNA expression upon (−)-deprenyl and melatonin treatments. *J. Neurosci. Res.* **53**: 593.

Tomac, A., Lindqvist, E., Lin, L.F., Ogren, S.O., Young, D., Hoffer, B.J. and Olson, L., 1995a, Protection and repair of the nigrostriatal dopaminergic system by GDNF in vivo. *Nature* **373**: 335.

Tomac, A., Widenfalk, J., Lin, L.F., Kohno, T., Ebendal, T., Hoffer, B.J. and Olson, L., 1995b, Retrograde axonal transport of glial cell line-derived neurotrophic factor in the adult nigrostriatal system suggests a trophic role in the adult. *Proc. Natl. Acad. Sci. USA* **92**: 8274.

Treanor, J., Goodman, L., Desauvage, F., Stone, D.M., Poulsen, K.T., Beck, C.D., Gray, C., Armanini, M.P., Pollock, R.A., Hefti, F., Phillips, H.S., Goddard, A., Moore, M.W., Bujbello, A., Davies, A.M., Asai, N., Takahashi, M., Vandlen, R., Henderson, C.E. and Rosenthal, A., 1996, Characterization of a multicomponent receptor for GDNF. *Nature* **382**: 80.

Tripanichkul, W., Sripanichkulchai, K. and Finkelstein, D.I., 2006, Estrogen down-regulates glial activation in male mice following 1-methyl-4-phenyl-1,2,3,6-tetrahydropyridine intoxication. *Brain Res.* **108**: 28.

Trupp, M., Arenas, E., Fainzilber, M., Nilsson, A.S., Sieber, B.A., Grigoriou, M., Kilkenny, C., Salazar-Grueso, E., Pachnis, V. and Arumae, U., 1996, Functional receptor for GDNF encoded by the c-ret proto-oncogene. *Nature* **381**: 785.

Trupp, M., Belluardo, N., Funakoshi, H. and Ibanez, C.F., 1997, Complementary and overlapping expression of glial cell line-derived neurotrophic factor (GDNF), c-ret proto-oncogene, and GDNF receptor-alpha indicates multiple mechanisms of trophic actions in the adult rat CNS. *J. Neurosci.* **17**: 3554.

Vegeto, E., Bonincontro, C., Pollio, G., Sala, A., Viappiani, S., Nardi, F., Brusadelli, A., Viviani, B., Ciana, P. and Maggi, A., 2001, Estrogen prevents the lipopolysaccharide-induced inflammatory response in microglia. *J. Neurosci.* **21**: 1809.

Walker, D.G., Beach, T.G., Xu, R., Lile, J., Beck, K.D., McGeer, E.G. and McGeer, P.L., 1998, Expression of the proto-oncogene Ret, a component of the GDNF receptor complex, persists in human substantia nigra neurons in Parkinson's disease. *Brain Res.* **792**: 207.

Wang, L., Muramatsu, S., Lu, Y., Ikeguchi, K., Fujimoto, K., Okada, T, Mizukami, H., Hanazono, Y., Kume, A., Urano, F, Ichinose, H., Nagatsu, T., Nakano, I. and Ozawa, K., 2002, Delayed delivery of AAV-GDNF prevents nigral neurodegeneration and promotes functional recovery in a rat model of Parkinson's disease. *Gene Ther.* **9**: 381.

Wang, J., Chen, G., Lu, B. and Wu, C.P., 2003, GDNF acutely potentiates Ca^{2+} channels and excitatory synaptic transmission in midbrain dopaminergic neurons. *Neurosignals* **12**: 78.

Wei, G., Wu, G. and Cao, X., 2000, Dynamic expression of glial cell line-derived neurotrophic factor after cerebral ischemia. *Neuroreport* **11**: 1177.

Widenfalk, J., Lundstromer, K., Jubran, M., Brene, S. and Olson, L., 2001, Neurotrophic factors and receptors in the immature and adult spinal cord after mechanical injury or kainic acid. *J. Neurosci.* **21**: 3457.

Wu, D.C., Jackson-Lewis, V., Vila, M., Tieu, K., Teismann, P., Vadseth, C., Choi, D.K., Ischiropoulos, H. and Przedborski, S., 2002, Blockade of microglial activation is neuroprotective in the 1-methyl-4-phenyl-1,2,3, 6-tetrahydropyridine mouse model of Parkinson disease. *J. Neurosci.* **22**: 1763.

Yang, F., Feng, L., Zheng, F., Johnson, S.W., Du, J., Shen, L., Wu, C.P. and Lu, B., 2001, GDNF acutely modulates excitability and A-type K^+ channels in midbrain dopaminergic neurons. *Nat. Neurosci.* **4**: 1071.

Yoo, M.S., Chun, H.S., Son, J.J., DeGiorgio, L.A., Kim, D.J., Peng, C. and Son, J.H., 2003, Oxidative stress regulated genes in nigral dopaminergic neuronal cells: correlation with the known pathology in Parkinson's disease. *Brain Res. Mol. Brain Res.* **110**: 76.

Yurek, D.M. and Fletcher-Turner, A., 2001, Differential expression of GDNF, BDNF, and NT-3 in the aging nigrostriatal system following a neurotoxic lesion. *Brain Res.* **891**: 228.

Yurek, D.M. and Fletcher-Turner, A., 2002, Temporal changes in the neurotrophic environment of the denervated striatum as determined by the survival and outgrowth of grafted fetal dopamine neurons. *Brain Res.* **931**: 126.

Zheng, J.S., Tang, L.L., Zheng, S.S., Zhan, R.Y., Zhou ,Y.Q., Goudreau, J., Kaufman, D. and Chen, A.F., 2005, Delayed gene therapy of glial cell line-derived neurotrophic factor is efficacious in a rat model of Parkinson's disease. *Mol. Brain Res.* **134**: 155.

Zhou, J., Yu, Y., Tang, Z., Shen, Y. and Xu, L., 2000, Differential expression of mRNAs of GDNF family in the striatum following 6-OHDA-induced lesion. *Neuroreport* **11**: 3289.

9

MOLECULAR PATHWAYS OF MITOCHONDRIAL DYSFUNCTION IN NEURODEGENERATION: THE PARADIGMS OF PARKINSON'S AND HUNTINGTON'S DISEASES

Ana Cristina Rego[1], Sandra Morais Cardoso[2], and Catarina R. Oliveira[1]

1. ABSTRACT

Mitochondria play an important role as ATP producers through the activity of the citric acid cycle and oxidative phosphorylation, as regulators of intracellular calcium homeostasis, and producers of endogenous reactive oxygen species (ROS). Mitochondria also regulate cell death, marking the point of no return in necrosis and apoptosis. Many evidences have been raised implicating mitochondria defects as crucial mechanisms in the pathogenesis of several neurodegenerative diseases, as well as in aging. This chapter resumes some of the findings that provide evidence for the role of mitochondria in neurodegeneration associated with Parkinson's disease (PD) and Huntington's disease (HD), two neurodegenerative disorders that cause movement disturbances.

2. MITOCHONDRIAL DYSFUNCTION, LOSS OF CALCIUM HOMEOSTASIS, AND FREE RADICAL GENERATION

2.1. The Mitochondrial Respiratory Chain and ATP Production

At the inner mitochondrial membrane, mammalian mitochondria are composed of several carriers of electrons assembled to form the mitochondrial respiratory chain. This chain transfers electrons from mitochondrial components with low to high redox potentials (E_0'), i.e., from highly reducing components with negative E_0' (e.g., NADH), to highly oxidant components with positive E_0' (e.g., O_2). The electron carriers are flavoproteins (containing flavin adenine dinucleotide (FAD) or flavin mononucleotide (FMN) as prosthetic groups), cytochromes, iron–sulfur (nonheme iron) proteins, ubiquinone (UQ or coenzyme Q – a lipid soluble cofactor) and protein-bound Cu^{2+}/Cu^+. Three of four polypeptide chains that contain the electron carriers, namely NADH-UQ oxidoreductase (complex I), UQ-cytochrome c oxidoreductase (complex III) and cytochrome c oxidase (complex IV), act as oxidation–

Ana Cristina Rego – acrego@cnc.cj.uc.pt
Center for Neuroscience and Cell Biology, [1]Institute of Biochemistry and [2]Institute of Biology, Faculty of Medicine, University of Coimbra, 3004-504 Coimbra, Portugal

reduction driven proton pumps (Nicholls and Ferguson, 1992). Although the mechanisms by which the protons are pumped from the mitochondrial matrix (*N*-face of the membrane) to the intermembrane space (*P*-face of the membrane) are not completely understood, a proton-motive force is created along the inner mitochondrial membrane. Likewise, a mitochondrial membrane potential ($\Delta\psi_m$) is originated, with the *N*-face of the membrane being highly negative (−150 to −180 mV). Proton reentry into the matrix completes a proton circuit and helps to generate ATP through the ATP synthase. ATP is mostly produced in the mitochondria by oxidative phosphorylation and reaches the cytosol via the adenine nucleotide translocator (ANT). ATP synthase (also known as F_1/F_0-ATPase or complex V) is a reversible proton-translocating polypeptide. If the respiratory chain is inhibited and ATP is available to mitochondria (cytosolic ATP), this complex functions as an ATPase (Nicholls and Ferguson, 1992). This is possible, for example during anoxia, through the utilization of glycolytic ATP. In cells treated with rotenone, an inhibitor of complex I, generation of glycolytic ATP may help to maintain the $\Delta\psi_m$ despite the reversion of ATP synthase. Experimentally, oligomycin, an ATP synthase inhibitor, can be used to evaluate if mitochondria are synthesizing or hydrolyzing ATP. This "nullpoint" assay can be used to assess the bioenergetic status of in situ mitochondria (Ward et al., 2000). In glutamate-stimulated cerebellar granule cells, oligomycin showed a hyperpolarizing effect indicating ongoing oxidative phosphorylation and mitochondrial ATP synthesis, whereas a depolarizing effect was observed in cells exposed to kainate, indicating the reversal of ATP synthase (Rego et al., 2001b). Oligomycin-induced depolarization was also observed in hypothalamic GT1-7 cells treated with staurosporine, an effect prevented by Bcl-2 overexpression (Rego et al., 2001a). Hence, reversal of ATP synthase is associated with hydrolysis of ATP, most probably resulting from glycolysis, and may be accounted for by the inhibition of components of the mitochondrial respiratory chain. Moreover, ρ0 cells that lack mtDNA cannot produce a functional electron transport chain (ETC), as 13 protein subunits of the mitochondrial respiratory apparatus are mtDNA encoded (seven subunits of NADH-Q oxidoreductase, one subunit of cytochrome c reductase, three subunits of cytochrome c oxidase, and two subunits of ATP synthase) (Attardi and Schatz, 1988; Wallace, 1994). The maintenance of $\Delta\psi_m$ in *ρ0* cells has been documented (Jiang et al., 1999; Cardoso et al., 2001), and probably arises from the transport of protons out of the mitochondrial matrix, a phenomenon that appears to involve the reversal of ATP synthase. In *ρ0* cells nuclear-encoded ATP synthase subunits assemble and can maintain a proton gradient across the mitochondrial membrane (Nijlmans et al., 1995).

2.2. Regulation of Calcium Homeostasis by Mitochondria

Both mitochondrial ATP synthesis and calcium uptake are highly dependent on $\Delta\psi_m$ (Nicholls and Ward, 2000). As cytosolic calcium rises above the "set point" (0.3–1 μM, a concentration exceeding resting cytosolic calcium concentration), mitochondria become net accumulators of calcium (Nicholls and Budd, 2000, for review), helping to maintain low levels of cytosolic calcium. In brain, mitochondria calcium accumulation implicates the transport of calcium to the matrix through the calcium uniport, exceeding matrix calcium efflux occurring through the sodium/calcium exchanger. Under physiological concentrations of phosphate, calcium forms a calcium–phosphate complex in the matrix, which is osmotically inactive. Mitochondrial enzymes such as pyruvate dehydrogenase, isocitrate dehydrogenase and α-ketoglutarate dehydrogenase, the latter two belonging to the citric acid cycle, are activated by matrix calcium (Nicholls and Budd, 2000, for review).

Changes in mitochondrial calcium loading capacity are associated with increased neuronal vulnerability to excitotoxicity, a pathological process resulting from elevations in synaptic glutamate concentration and overactivation of glutamate receptors, namely the N-methyl-D-aspartate (NMDA) receptors. As a result, an increase in intracellular calcium occurs, leading to neuronal death. During NMDA receptor activation, $\Delta\psi_m$ is used not only for ATP synthesis but also for mitochondrial calcium uptake. Mitochondria calcium overload causes mitochondrial swelling, decreased respiratory rate, and mitochondrial depolarization. In fact, mitochondria can only accumulate calcium until the onset of permeability transition (PT), causing the release of cytochrome c and other molecules up to 1.5 kDa, as described in Sect. 3.2.

2.3. Mitochondria and Oxidative Stress

Mitochondria produce ROS, namely the superoxide anion ($O_2^{\bullet-}$), in a constant manner. Mitochondria consume oxygen as part of the normal cellular metabolism and reduce it to water at the terminal step of oxidative phosphorylation. Nevertheless, during this process, the mitochondrial respiratory chain leaks electrons, thereby partially reducing oxygen.

The semiquinone anion ($UQ^{\bullet-}$), a partial reduced form of coenzyme Q, is an important electron leakage site at the inner mitochondrial membrane (Turrens et al., 1985). Ubiquinone is completely reduced by ($2H^+ + 2e^-$) producing ubiquinol (UQH_2). Increased mitochondrial calcium uptake was previously reported to stimulate oxygen radical production from the reduced form of coenzyme Q (Kowaltowski et al., 1995). Besides $O_2^{\bullet-}$, other reactive oxygen species are produced in the mitochondria, namely hydrogen peroxide (H_2O_2), which results from the activity of Mn-superoxide dismutase. Similarly to $O_2^{\bullet-}$, H_2O_2 is not very reactive, unless it encounters Fe^{2+} or Cu^+ to form the highly reactive and short-lived (about 10^{-9} s) hydroxyl radical ($HO^{\bullet}$) by the Fenton–Haber Weiss reaction. $HO^{\bullet}$ potently oxidizes proteins, lipids and DNA, highly contributing to cellular damage. The glutathione redox cycle and particularly the levels of reduced/oxidized glutathione (GSH/GSSG) play an important role as endogenous antioxidants in several neurodegenerative processes by detoxifying the mitochondria from excessive generation of H_2O_2.

Mitochondria are also a source of nitric oxide (NO) produced at the level of the inner mitochondrial membrane by the mitochondrial isoform of nitric oxide synthase (mtNOS) (Ghafourifar and Richter, 1997; Giulivi et al., 1998). NO (half-life ~1 s) is an uncharged radical that readily crosses biological membranes. Because NO reacts very rapidly with $O_2^{\bullet-}$ to give the peroxynitrite anion ($ONOO^-$) at a rate constant of $6.7 \times 10^9 \, M^{-1} \, s^{-1}$, a reaction which is million times faster than the Fenton reaction (Beckman, 1994), $ONOO^-$-induced damage of mitochondrial components is expected. mtNOS activity is stimulated when calcium is taken up by mitochondria (Ghafourifar and Richter, 1997), leading to the release of cytochrome c from isolated mitochondria in a Bcl-2-sensitive manner, most probably due to $ONOO^-$ formation within mitochondria (Ghafourifar et al., 1999). Recent evidences suggest that changes in $\Delta\psi_m$ modulate mtNOS activity: a decrease in membrane potential minimizes NO release, whereas oligomycin, which induces mitochondrial hyperpolarization, generates maximal NO (Valdez et al., 2006). Mitochondrial depolarization has been also associated with higher sensitivity of aged astrocytes in culture to oxidative stress (Lin et al., 2005).

3. NEURONAL CELL DEATH: APOPTOSIS, NECROSIS, AND AUTOPHAGY

3.1. The Mitochondrial-Dependent Apoptotic Pathway

In mammalian cells apoptosis is mainly mediated by caspases, a family of cysteine proteases. Caspases are expressed as inactive procaspase precursors. When initiator caspases, such as caspases-8 and -9, are activated, they cleave the precursor forms of effector caspases, caspases-3, -6, and -7, which then become active. Unlike the effector caspases, activation of initiator caspases such as caspase-9 requires active dimerization by an adaptor molecule. The apoptotic protease activating factor-1 (APAF-1) is the adaptor of caspase-9. Effector caspases are known to cleave several cellular substrates, culminating in DNA fragmentation and condensation.

Two main apoptotic pathways have been identified by which caspases activation is triggered: the extrinsic and the intrinsic apoptotic pathways, the latter also named the mitochondrial-dependent pathway.

Many apoptotic stimuli converge on mitochondria. The intrinsic pathway is triggered by various stresses, such as withdrawal of growth factors, hypoxia, DNA damage, ionizing radiation, or several chemical agents. A key event in mitochondria-dependent apoptosis is the release of molecules such as cytochrome c, the second mitochondrial activator of caspases (Smac), also known as direct IAP binding protein with low pI (Diablo), and the serine protease Omi/HtrA2 to the cytosol. The latter two proteins facilitate caspases activation by inactivating the endogenous inhibitor of caspases, the inhibitor of apoptosis proteins (IAPs), such as the X-linked IAP (XIAP), as previously identified by Verhagen and collaborators and Du and collaborators in 2000 (Du et al., 2000; Verhagen et al., 2000).

The death receptor (extrinsic) pathway links to the mitochondrial pathway through truncated Bid (tBid), which results from the cleavage of Bid, BH3 only member of the B-cell leukemia/lymphoma 2 (Bcl-2) family of proteins, by activated caspase-8. tBid translocates to the mitochondria to activate Bax/Bak, leading to the release of mitochondrial cytochrome c (Luo et al., 1998). The antiapoptotic Bcl-2 family member, Mcl-1, was recently shown to strongly inhibit tBid-induced cytochrome c release through protein–protein interactions (Clohessy et al., 2006). Bcl-2 is one of the antiapoptotic proteins first recognized to prevent the release of mitochondrial cytochrome c (Yang et al., 1997).

Within the cytosol, cytochrome c initiates the formation of the apoptosome, a multimeric protein complex that contains cytochrome c, APAF-1, procaspase-9, and requires ATP. This complex is responsible for activating caspase-9. In fact, once cytochrome c is released the downstream caspases activation is irreversible and activated caspase-9 promotes the cascade of procaspases activation that follows, namely the activation of caspase-3. Effector caspase-3 (and caspase-7) is required for the cleavage of the 45 kDa subunit of the DNA fragmentation factor (DFF) or DFF45 (the inhibitor of caspase activated DNase, ICAD, is the mouse homolog), releasing DFF40 (CAD is the mouse homolog), which has nuclease activity and is responsible for DNA fragmentation during apoptosis (Liu et al., 1997).

Cell death is also modified by the release of two mitochondrial endonucleases, endonuclease G21 (endoG) and the apoptosis-inducing factor (AIF), which induce cell death in a caspase-independent manner. In this form of cell death AIF and EndoG translocate from the mitochondrial intermembrane space to the nucleus, inducing DNA fragmentation and chromatin condensation (Susin et al., 1999; Li et al., 2001). Interestingly, cytochrome c appears to be more easily released from permeabilized mitochondria than AIF (Uren et al., 2005).

While cytochrome c is soluble in the intermembrane space, AIF seems to be partly attached to the mitochondrial inner membrane, requiring more severe treatments (Uren et al., 2005).

Although caspase-9 is generally believed to be the initiator caspase in mitochondria-dependent apoptosis, the ubiquitous caspase-2, one of the first caspases to be identified, was proposed to act upstream of mitochondria (Lassus et al., 2002; Robertson et al., 2002), implicating the mitochondria as amplifiers of caspases activity. Procaspase-2 can be recruited to a high molecular weight complex (similar to the apoptosome in which procaspase-9 is activated) independently of cytochrome c and APAF-1 (Read et al., 2002). The recruitment of caspase-2 to this complex is sufficient to mediate its activation. Caspase-2 may be required for translocation of Bax to the mitochondria, inducing cytochrome c, Smac, and AIF release from mitochondria (Guo et al., 2002; Robertson et al., 2002). Despite permeabilizing the outer mitochondrial membrane and causing cytochrome c release from this organelle, caspase-2 also disrupts the interaction of cytochrome c with cardiolipin and enhances the release of this hemoprotein caused by an apoptotic stimulus, highly suggesting that caspase-2 promotes mitochondrial-dependent intrinsic caspases cascade (Enoksson et al., 2004).

3.2. Permeabilization of Mitochondrial Membrane and the Permeability Transition Pore

Increased permeabilization of the mitochondrial outer membrane is triggered by proapoptotic members of the Bcl-2 family such as Bax or Bak. These proapoptotic proteins can form pores in the mitochondrial outer membrane (without affecting the inner membrane or the matrix), releasing proteins present at the mitochondrial intermembrane space, such as cytochrome c, Smac/DIABLO, and Omi/HtrA2. Bid and Bad, two BH3-only proteins, but not Bax or Bak, were shown to induce cytochrome c release without changes in $\Delta\psi_m$ or the opening of the mitochondrial permeability transition pore (mPTP) and without interacting with the voltage-dependent anion channel (VDAC) (Shimizu and Tsujimoto, 2000). In response to an apoptotic stimulus, Bax was reported to be translocated from the cytosol to the mitochondria, being also responsible for cytochrome c release independently of the calcium-inducible, cyclosporin A-dependent mPTP (Eskes et al., 1998).

Dephosphorylated Bad also acts as a proapoptotic protein in the mitochondria. The calcium–calmodulin activated calcineurin (or protein phosphatase 2B) can dephosphorylate Bad. In this form, Bad is translocated from the cytosol to the mitochondria, where it binds to Bcl-2 or Bcl-xl, inhibiting the antiapoptotic function of these proteins and allowing the release of cytochrome c. In traumatic spinal cord injury, caspase-3 activation was shown to depend upon calcineurin-mediated Bad dephosphorylation (Springer et al., 2000).

A second pathway of mitochondrial-dependent cell death involves increased mitochondrial inner membrane permeability defined by the opening of the mPTP, a nonselective pore that is activated by a variety of signals including mitochondrial calcium overload, oxidative stress, low levels of ATP, and a decrease in $\Delta\psi_m$. The mPTP spans the inner and outer mitochondrial membrane and allows the free diffusion of solutes across the membranes. The mPTP is composed of ANT at the inner membrane, the VDAC and Bax on the outer membrane and cyclophilin D in the matrix. The peripheral benzodiazepine receptor, hexokinase, and creatine kinase might also be associated with the mPTP. The pore has been suggested to be involved in several forms of cell death, namely apoptosis, necrosis, and autophagy. Apart from allowing the passage of several molecules, activation of mitochondrial PT leads to matrix swelling and rupture of the outer membrane (Green, 2005, for review).

Calcium-induced opening of the mPTP in brain mitochondria potentiates the formation of mitochondrial ROS and lipid peroxidation by depleting the levels of NAD(P)H (Maciel et al., 2001). Furthermore, opening of the mPTP results in mitochondrial calcium efflux and mitochondrial swelling, apart from the release of proapoptotic mitochondrial proteins, including cytochrome c and procaspases. Nevertheless, the exact mechanism underlying mPTP opening and its role in the release of proapoptotic proteins is still a matter of ongoing debate. According to Atlante et al. (2006), opening of the mPTP is not required for either cytochrome c release or for the early stages of apoptosis, but it may be involved in the final stages of apoptotic cell death when leaky uncoupled mitochondria release their content.

3.3. Apoptosis and Necrosis in Excitotoxicity

In contrast with apoptosis, necrosis is a passive process, characterized by massive cell destruction, plasma membrane disruption and the consequent disruption of organelle membrane, and release of intracellular components. Mitochondrial swelling is also characteristic of necrotic cell death. Pathophysiological conditions such as infection, inflammation, and ischemia, which involve the activation of NMDA receptors due to excessive glutamate in the synaptic cleft, are normally associated with necrosis, namely secondary necrosis.

The occurrence of necrosis or apoptosis seems to depend on the levels of intracellular energy: if the levels of ATP fall, plasma membrane rupture occurs, along with the deregulation of ion transport and intracellular pH; if ATP levels are maintained, the caspase-dependent apoptotic machinery may proceed. Under pathological conditions, altered calcium extrusion transport (e.g., due to inhibition of calcium-ATPase) leads to the loss of calcium homeostasis, which is responsible for activating several enzymes, namely phospholipases, endonucleases, proteases, nitric oxide synthase, and xanthine dehydrogenase, the latter two mediating the formation of free radicals.

Several reports suggest the occurrence of apoptosis during excitotoxicity. Under these conditions, release of mitochondrial cytochrome c is associated with mitochondrial depolarization and generation of ROS (Atlante et al., 2000; Luetjens et al., 2000). Caspase-3 also plays an important role upon activation of NMDA receptors in cerebrocortical neurons (Tenneti and Lipton, 2000). Moreover, AIF translocation occurs upon NMDA receptor stimulation in a process requiring the activation of poly(ADP-ribose)polymerase (PARP) and the consequent depletion of NAD^+ (Yu et al., 2002). During excitotoxic apoptosis in rat cerebellar granule neurons, a model of glutamate-induced caspase-independent apoptosis, translocation of full-length Bid from the cytosol to the mitochondria was observed, paralleling the collapse of $\Delta\psi_m$ (Ward et al., 2006).

Both caspases and calpains have been demonstrated to be mediators of apoptotic cell death. Calpains are calcium-activated proteases that appear to be important during excitotoxicity. Calpain activation induces the truncation of the C-terminal domain of NR2 subunits of the NMDA receptors, changing its channel binding properties (Bi et al., 1998). mu-Calpain is involved in mediating NMDA-induced excitotoxicity in the rat retina (Chiu et al., 2005). In the hippocampus of epileptic rats, calpain was shown to cause the proteolysis of NR2B subunits of the NMDA receptor (Araújo et al., 2005). Overexpression of calpastatin, the specific endogenous calpain inhibitor, decreases calpain activation and increases caspase-3 activity, accelerating the apoptotic process (Neumar et al., 2003). Caspase activation was also shown to mediate the degradation of calpastatin, suggesting a dual role for calpains during neuronal apoptosis: in early apoptosis calpain decreases caspase-3 activity, which further facilitates the degradation of calpastatin, contributing to secondary necrosis (Neumar et al., 2003).

3.4. Role of Autophagy in Cell Death

Old or damaged proteins are normally degraded by two independent mechanisms, the ubiquitin mediated proteolysis that involves the proteasome and degrades soluble monomeric misfolded proteins and autophagy, a process by which proteins and damaged organelles, namely mitochondria, are degraded within lysosomes (Klionsky and Ohsumi, 1999; Kim and Klionsky, 2000). Autophagy may be activated upon impairment of the ubiquitin proteasome system (UPS), occurring in the presence of non native protein oligomers (Iwata et al., 2005b).

Cells die by autophagy in a nonapoptotic/necrotic manner. Similarly to apoptosis, cells show membrane blebbing, chromatin condensation, although no DNA laddering. This is a caspase-independent process, which is characterized by increased lysosomal activity and increased number of autophagic vesicles (Schwartz et al., 1993; Bursch et al., 2000). According to Xue et al. (1999), apoptotic signals may activate autophagy as an alternative mechanism of cell death. In serum deprived PC12 cells, with recognized apoptotic features, Uchiyama (2001) described the occurrence of autophagy regulated by lysosomal proteinases, cathepsins B and D, in the early stages of apoptosis, before the appearance of nuclear changes. Moreover, apoptosis was prevented in the presence of an autophagy inhibitor (Uchiyama, 2001).

The components of the autophagic machinery are highly conserved (Reggiori and Klionsky, 2002), however their role in mammalian cells is not fully understood. During autophagy, autophagosomes (autophagic vesicles) are formed through the assembly of membranes that originate in the endoplasmic reticulum, which encapsulate organelles or isolated proteins. The autophagosomes then fuse with lysosomes, leading to the degradation of the autophagosomal contents. The signaling pathway that leads to autophagy involves the activity of phosphatidylinositol 3-kinase (PI3K).

Lysosomal degradation of organelles, such as mitochondria, seems to be required for cell remodeling during differentiation, stress or cell damage. In neurodegenerative diseases, namely PD and HD, autophagy has been reported (described in Sects. 4.1.2 and 4.2.2), implicating the degradation of damaged mitochondria, which may occur in the early phases of these diseases as a defense mechanism against injury. In contrast, in accordance to the mitochondrial-lysosomal theory of aging, damaged mitochondria may accumulate during aging as a result of deficient autophagy (Brunk and Terman, 2002).

4. MITOCHONDRIAL DYSFUNCTION IN PARKINSON'S AND HUNTINGTON'S DISEASES

Mitochondrial dysfunction has been shown to be involved in several neurodegenerative diseases. Inhibition of mitochondrial respiratory chain complexes, mitochondrial depolarization, decreased mitochondrial calcium loading capacity, opening of the mPTP and release of cytochrome c triggering the mitochondrial-apoptotic pathway are common features of mitochondrial dysfunction associated with neurodegeneration (see Fig. 1). Mitochondrial toxins have been largely used to model these disorders. These include selective inhibitors of mitochondrial respiratory chain complexes, namely rotenone, an inhibitor of complex I used to model PD, and 3-nitropropionic acid (3-NP), an irreversible inhibitor of succinate dehydrogenase used to model HD. Due to their chronic and progressive nature and the fact that currently there are no available treatments to slow or stop the disease progression, neurodegenerative disorders are attractive targets for drug therapies. In this part of the chapter, we resume some findings evidencing the role of mitochondria in neurodegeneration occurring in two neurodegenerative diseases associated with movement disorders, PD and HD, and mitochondrial neuroprotective compounds with potential therapeutic efficacy.

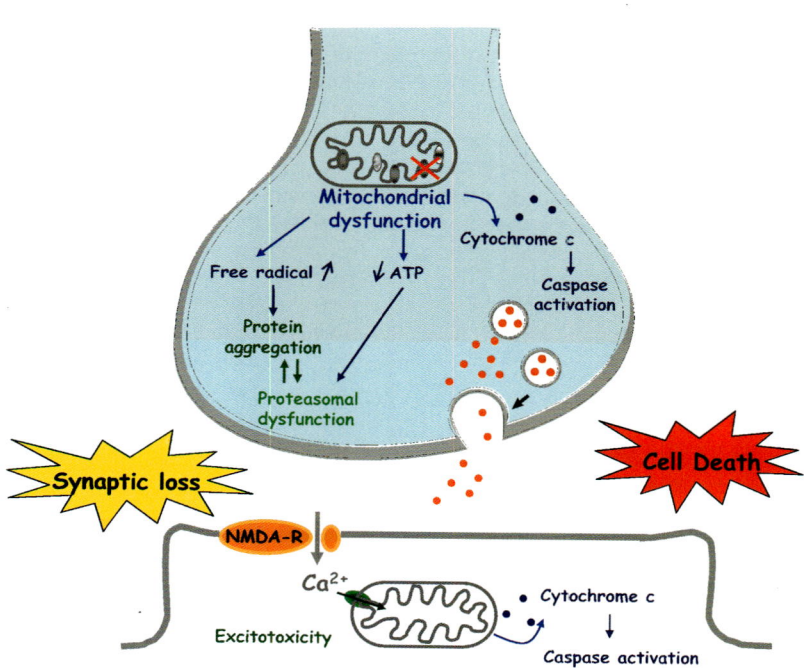

Fig. 1. Schematic representation of the main pathogenic events that may contribute to synaptic and neuronal loss in PD and HD. Mitochondrial dysfunction may trigger oxidative stress, synaptic dysfunction, and proteasome malfunction, promoting protein aggregation. Such impairment contributes to PD and HD pathogenesis through induction of selective neuronal loss (depicted also in Fig. 2). Overactivation of N-methyl-D-aspartate receptors (NMDA-R) leads to loss of calcium homeostasis, contributing to mitochondrial-dependent neuronal death

4.1. Parkinson's Disease

Parkinson's disease (PD) is the most common neurodegenerative movement disorder affecting more than 1% of the population above 60 years of age. Although PD symptomology can be delayed, there is no effective treatment (Roman et al., 1995). Clinically, PD is characterized by rigidity, tremor, slowness, and balance problems. Pathologically, PD is characterized by selective loss of dopaminergic neurons in the *substantia nigra* pars compacta, and formation of Lewy bodies (LB), which are intracytoplasmic inclusions mainly composed of ubiquitin, α-synuclein, synphilin-1, 14-3-3 protein, tubulin and other cytoskeletal proteins (Forno, 1996). Taken together, the data from literature indicate that UPS dysfunction, protein aggregation, mitochondrial dysfunction, and oxidative stress are involved in PD etiopathogenesis.

4.1.1. Evidences of Mitochondrial Dysfunction in PD

Mitochondrial dysfunction has long been implicated in the pathogenesis of PD. Evidence first emerged following the accidental exposure of drug abusers to 1-methyl-4-phenyl-pyridinium (MPP^+), a metabolite of 1-methyl-4-phenyl-1,2,3,6-tetrahydropyridine (MPTP), that induces a parkinsonian syndrome (Langston et al., 1983). MPP^+ accumulates in dopaminergic neurons by the high-affinity dopamine uptake transporter and is a complex I inhibitor of the ETC. A biochemical link between MPTP toxicity and idiopathic PD was established when the deficiency in complex I was identified in the *substantia nigra* of *postmortem* PD brain (Schapira et al., 1989). Over the years, several other groups demonstrated a consistent decrease in complex I activity in PD brains (Schapira et al., 1990a, b; Janetzky et al., 1994; Gu et al., 1998). Hattori et al. (1991) showed a decreased staining of complex I in the *substantia nigra* of PD patients. Moreover, immunoblotting studies on mitochondria isolated from the striatum of PD patients revealed a decrease of complex I subunits (30, 25, and 24 kDa subunits) (Mizuno et al., 1989). More recently, a loss of 8 kDa subunit of the mitochondrial complex I was observed in PD human brains, suggesting disassembly of this protein complex, which further undergoes auto-oxidation of its catalytic subunits, leading to dysfunction of the complex (Keeney et al., 2006). Complex I deficiency has also been described in other PD tissues, namely in peripheral models, like platelets and lymphoblasts (Barroso et al., 1993; Schapira, 1994). Despite some studies failing to show abnormalities of the ETC in platelets of idiopathic PD patients (Mann et al., 1992; Hanagasi et al., 2005), the majority of these studies clearly demonstrate a significant reduction of complex I activity (Parker et al., 1989; Yoshino et al., 1992; Benecke et al., 1993; Haas et al., 1995; Blandini et al., 1998). Furthermore, some of these studies revealed alterations in other ETC complexes, namely reductions in the activities of complexes II/III and IV. Despite the overall reports indicating abnormalities in the ETC in various tissues from PD patients, an increased criticism of the significance of these findings in the etiology of PD has emerged. To rule out that drug treatments or the debilitation of the disease process itself could produce the low ETC activities, Haas et al. (1995) showed a decrease in complex I activity in early untreated PD patients. In addition, the same group excluded that a "home" environmental factor could explain the reduction of complex I activity, since they evaluated ETC complexes' activity in spousal controls. Nonetheless, other environmental factors have not been excluded.

To address the potential causes of complex I defect, namely if it was due to environmental toxins or due to an alteration of mitochondrial or nuclear DNA, the cytoplasmic hybrid (cybrid) technique, first described by King and Attardi (1989), has been applied. A stable decrease in complex I activity, increased ROS production and increased susceptibility to MPP^+ was described in PD cybrids (Swerdlow et al., 1996; Cassarino et al., 1997). These results suggest that the defect in complex I in PD may result from changes in mtDNA. Moreover, Trimmer et al. (2004) reported the generation of fibrillar and vesicular inclusions in a cybrid model of sporadic PD, which replicates the essential antigenic and structural features of LB, occurring independently of exogenous protein expression or inhibition of proteasomal function.

Taken together, these results emphasize the conscious role of mitochondria in age associated neurodegenerative disorders, like sporadic PD. We propose that, similarly to the mitochondrial cascade hypothesis for the late-onset sporadic Alzheimer's disease (Swerdlow and Khan, 2004), mitochondrial dysfunction (due to inherited ETC genes that determine basal ETC efficiency and the consequent production of ROS) is the upstream event leading to neurodegeneration in the late-onset sporadic PD.

4.1.1.1. Evidence Suggesting that Mitochondrial Abnormalities are a Common Event in Familial PD

Although familial forms of PD are relatively rare (<10% of the cases), the identification of single genes linked to the disease has provided crucial insights into possible mechanisms of disease pathogenesis (reviewed in Bossy-Wetzel et al., 2004; Greenamyre and Hastings, 2004). A role for α-synuclein in PD was fuelled by the discovery of two missense mutations in the α-synuclein gene (Ala53Thr, Ala30Pro) as a cause of autosomal dominantly inherited PD (Polymeropoulos et al., 1997; Kruger et al., 1998). Even though mutations in α-synuclein are very rare, the α-synuclein role as the major structural feature of LB has placed this protein at the center in PD pathophysiology (reviewed in Dawson and Dawson, 2003). Furthermore, genomic duplication or triplication of wild-type α-synuclein is the disease-causing mutation in some familial forms of PD (Singleton et al., 2003; Chartier-Harlin et al., 2004). Although α-synuclein function is still unclear, it has been shown to bind synaptic vesicles. Moreover, α-synuclein protofibrils can form pores, leading to membrane vesicle permeabilization and inducing dopamine release to the cytosol (Lashuel et al., 2002), where it undergoes auto-oxidation and increased ROS production (Sulzer et al., 2000). α-Synuclein loss-of-function (by forming protofibrils and/or fibrils) has been shown to indirectly impact on mitochondrial function. Interestingly, α-synuclein knockout mice showed increased resistance to toxicity induced by mitochondrial toxins (Dauer et al., 2002, Klivenyi et al., 2006). Furthermore, overexpression of a α-synuclein mutant form, increased neurons vulnerability to oxidative stress induced by dopamine and MPP^+ (Tabrizi et al., 2000a; Orth and Tabrizi, 2003).

Mutations in *DJ-1* gene (Lys166Pro, Met26Ile, Asp149Ala) cause autosomal recessive PD (ARPD). Structural studies indicate that *DJ-1*, which participates in the oxidative stress response by scavenging H_2O_2, may function as a chaperone to alleviate protein misfolding through the interaction with early-unfolded intermediates (Cookson, 2003; Quigley et al., 2003; Taira et al., 2004). However, recent studies showed that *DJ-1* ability to quench ROS is modest (Junn et al., 2005, Yang et al., 2005) and is more likely to be implicated in apoptosis regulation. *DJ-1* mutation disrupts protein activity by either destabilizing *DJ-1* itself or affecting its subcellular localization. In consequence, a decrease of *DJ-1* nuclear localization and an increase of mitochondrial localization are observed (Junn et al., 2005). Since mitochondrial *DJ-1* function remains to be clarified, we may hypothesize that *DJ-1* loss-of-function can either result from a loss of access to binding patterns on other subcellular compartments or from its interaction with the mitochondria.

A more common causative mutation that affects neuronal viability occurs in a ubiquitin E3 ligase named parkin (Kitada et al., 1998). Parkin mutations (e.g., Lys161Asn, Arg256Cys, and Arg275Trp) cause ARPD possibly impairing its interaction with the proteasome (Sakata et al., 2003). Although parkin mutations have not been associated with LB formation, a common link may exist between ARPD or other familial forms of the disease and sporadic PD. These mutations confer a loss-of-function of parkin E3 ligase activity, therefore a great importance has been placed on protein substrates of parkin. Parkin interacts with α-synuclein interacting protein synphilin-1, and through this interaction it may promote the formation of LBs. It was also proposed that parkin may act specifically on aberrant α-synuclein deposits (protofibrils/fibrils or oligomers). Moreover, parkin rescues unpaired proteasomal function, preventing the toxic effects of mutant α-synuclein (Petrucelli et al., 2002; Feany and Pallanck, 2003). Another possible role for parkin was revealed using a *Drosophila* parkin null mutant that show severe mitochondrial pathology, reduced lifespan, and increased apoptosis (Greene et al., 2003). Mammalian models also support a role for parkin in preserving mitochondrial function and preventing oxidative stress (Palacino et al., 2004). Furthermore, it was demonstrated that complex I is selectively reduced in peripheral leukocytes obtained from parkin-related ARPD patients (Muftuoglu et al., 2004). The mechanism by which parkin regulates mitochondrial function is still unclear. However, the protein has been shown to localize in the outer mitochondrial membrane, where it may be directly involved in maintaining mitochondrial integrity (Darios et al., 2003). Parkin may also have an important role in the elimination of oxidatively damaged proteins formed as a normal consequence of the ETC activity. Hyun et al. (2005) showed that parkin expression was upregulated after exposure to MPP^+ in a neuronal cell model, suggesting a protective role. In addition, it was demonstrated that mitochondrial toxins induce parkin aggregation with the consequent loss of function, suggesting that parkin inactivation might be a crucial step in the pathogenesis of sporadic PD.

Recently, three mutations (Glu240Lys, Leu489Pro, and Leu347Pro) have been identified in a putative mitochondrial protein kinase named PINK1 (PTEN-induced kinase 1) that give rise to a new familial form of PD (Valente et al., 2004). PINK1 contains a highly conserved kinase domain similar to serine/threonine kinases of the calcium–calmodulin family that has been hypothesized to phosphorylate mitochondrial proteins in response to cellular stress, protecting against mitochondrial dysfunction (Valente et al., 2004). Using a cell line that overexpresses a mutant form of PINK1, Valente et al. (2004) showed a significant decrease in $\Delta\psi_m$ and increased cell death following exposure to stress. Consistent with this report, Petit et al. (2005) demonstrated that overexpression of wild-type PINK1 protects cells from staurosporine-induced apoptotic cell death.

Although α-synuclein and parkin mutations indicate that protein misfolding and UPS dysfunction are involved in dopaminergic degeneration, the disclosure of PINK1 and *DJ-1* mutations and, also the new predicted role of parkin, suggests a key role of mitochondria in familial forms of PD.

4.1.2. Evidences of Cell Death Involving the Mitochondria in PD Pathogenesis

The mechanisms by which neurons die in PD are still not clear. Progressive degeneration of dopaminergic neurons can occur through apoptosis, although autophagic cell death has been observed (Anglade et al., 1997). Gómez-Santos et al. (2003) showed that autophagic cell death is induced by dopamine in a human neuroblastoma cell line. It has also been described that PC12 cells carrying the Ala53Thr α-synuclein mutation die by autophagy (Stefanis et al., 2001). Furthermore, Rideout et al. (2004) demonstrated that proteasomal inhibition in primary neurons leads to autophagy, possibly in response to the apoptotic pathway.

Nevertheless, PD postmortem brain analysis reveals nuclear condensation, chromatin fragmentation, and the presence of apoptotic bodies (Mochizuki et al., 1996; Tompkins et al., 1997; Tatton et al., 1998; Hartmann et al., 2000). An increase of apoptotic neuronal nuclei with increased immunoreactivity to caspase-3 and Bax and an upregulation of proapoptotic genes (*c-jun*, *p53*, *gapdh*, *bax*) were observed in the *substantia nigra* of PD patients (Hartmann et al., 2000, 2001; Tatton, 2000). In addition, it was demonstrated that 6-hydroxydopamine (6-OHDA), a dopamine quinone derivative, is present in *postmortem* PD brains (Spencer et al., 1998) and upregulates c-Jun activating caspase-3 in an in vitro study (Holtz and O'Malley, 2003). 6-OHDA also induces apoptosis in rat PC12 cells through the release of cytochrome c and Smac/Diablo from mitochondria, caspase-3 activation, and PARP cleavage (Gorman et al., 2005). Moreover, 6-OHDA causes mitochondrial dysfunction, activation of caspases-3 and -7, nuclear condensation, and apoptosis in a PC12 cell model (Hanrott et al., 2006).

Gomez et al. (2001) correlated caspase-mediated apoptosis with mitochondrial dysfunction in a human neuroblastoma cell line treated with MPP$^+$. They showed that this complex I inhibitor can induce caspase activation at lower concentrations than those needed to trigger mitochondrial dysfunction. In fact, MPP$^+$ is known to induce apoptosis in vitro (Chee et al., 2005) and in vivo (Tatton and Kish, 1997; Spooren et al., 1998). MPP$^+$-induced mitochondrial dependent cell death was shown in MPTP-treated mice, cultured dopaminergic cells, and *postmortem* PD brain tissue (Viswanath et al., 2001). In this work, activation of caspase-9 via cytochrome c release, followed by caspase-8 activation and Bid cleavage was observed. Moreover, MPP$^+$ induced mitochondrial cytochrome c release in a human neuroblastoma cell line (Sharma et al., 2003). Chronic administration of low concentration of rotenone, a highly selective complex I inhibitor, increased basal production of H_2O_2 and induced caspase-dependent cell death in a neuroblastoma cell line (Sherer et al., 2002). Moreover, a study performed by Watabe and Nakaki (2004) showed that rotenone induces neuronal death in SH-SY5Y cells via Bad dephosphorylation and caspase-9 activation.

It has been emphasized that a mitochondrial defect may be associated with mitochondrial dependent apoptotic pathway in vivo. Upon mitochondrial permeabilization with Bax a marked increase in cytochrome c release occurred in complex I defective brain mitochondria (Perier et al., 2005). To assess whether cells activate apoptosis as a consequence of mitochondrial complex I defect, a study using PD cybrids was performed. A decrease in $\Delta\psi_m$ was observed in PD cybrid cells with decreased complex I activity (Trimmer et al., 2000).

α-Synuclein overexpression or loss-of-function may be involved in the induction of apoptosis in vitro (reviewed in Sidhu et al., 2004). Furthermore, expression of mutant α-synuclein (Ala53Thr and Ala30Pro) in the yeast *Saccharomyces cerevisiae* model induces phosphatidylserine externalization, ROS accumulation, and cytochrome c release (Flower et al., 2005). Transgenic mice expressing familial PD-linked Ala53Thr α-synuclein developed intraneuronal inclusions, and revealed a decrease in complex IV activity, mtDNA damage, and caspase-3 cleavage when compared to transgenic mice expressing wild-type human α-synuclein (Martin et al., 2006). In addition, α-synuclein overexpression in rat *substantia nigra* induced a 50% loss of dopaminergic neurons and caspase-9 activation (Yamada et al., 2004).

4.1.3. Mitochondria – a Target for Therapeutic Intervention in PD

Because mitochondrial dysfunction plays a crucial role in the neurodegenerative process in PD, mitochondria have become potential therapeutic targets (see Fig. 2).

Coenzyme Q (CoQ) is an essential cofactor of the ETC (accepts electrons from complexes I and II) and possesses antioxidant properties. Endogenous levels of CoQ10 are significantly lower in mitochondria from PD patients as compared with age-matched controls (Shults et al., 1997). The putative beneficial effects of CoQ10 were evaluated in cultures of skin fibroblasts derived from PD patients. It was demonstrated that in vitro administration of CoQ10 partially ameliorates the activity of ETC complexes I and IV (Winkler-Stuck et al., 2004). Beal et al. (1998) previously reported that the administration of CoQ10 to 24-month-old mice treated with MPTP induced a significant protection against neuronal dopamine depletion and loss of tyrosine hydroxylase. Muller et al. (2003) also tested the symptomatic response of daily oral application of CoQ10 in 28 treated and stable PD patients. They observed a significant mild symptomatic benefit and an improvement of performance compared to placebo. In addition, treatment with the seed powder of the leguminous plant *Mucuna pruriens* increased brain mitochondrial complex I activity in the 6-OHDA lesioned rat model of PD. This treatment significantly restored endogenous levodopa, dopamine, norepinephrine, and serotonin content in the *substantia nigra*. Interestingly, early studies showed that *Mucuna pruriens* cotyledon powder treatment controls PD symptoms, possibly due to increased complex I activity and the presence of NADH and CoQ10 (Manyam et al., 2004).

In 1967, circulating ketones, by-products of incomplete fat oxidation, were discovered to replace glucose as the brain major fuel (Young et al., 1967). Ketone bodies can be efficiently used by the mitochondria to generate ATP and may also help to protect susceptible neurons from free radical damage (Vanitallie and Nufert, 2003). It has been proposed that ketone bodies may bypass the defect in complex I activity implicated in PD. Five PD patients were submitted to a "hyperketogenic" diet for 28 days. The substitution of unsaturated for saturated fatty acids in the diet prevented the increase in cholesterol in four volunteers. Moreover, the Unified Parkinson's Disease Rating Scale scores improved in all five patients during hyperketonemia (Vanitallie et al., 2005).

Despite the clear involvement of mitochondria in PD, no effective therapy approaches have been directed to the mitochondria. Seo et al. (2004) adopted a gene therapy approach to the Ndi1 gene that codes for the single rotenone and MPP^+-insensitive NADH-quinone oxidoreductase subunit (Ndi1) of *Saccharomyces cerevisiae* mitochondria. They showed that Ndi1 can replace or supplement the functionality of complex I in different mammalian cells. Moreover, they demonstrated that Ndi1 was localized in the mitochondria and stimulated complex I activity in rodent brains. This group also tested the possible use of the Ndi1 gene as a therapeutic agent in the MPTP mouse model. They showed that expression of Ndi1 confers resistance to MPTP-induced neuronal injury. The maintenance of tyrosine hydroxylase-positive cells, increased concentrations of dopamine and its metabolites, and significantly less denervation in the striatum was reported (Seo et al., 2006).

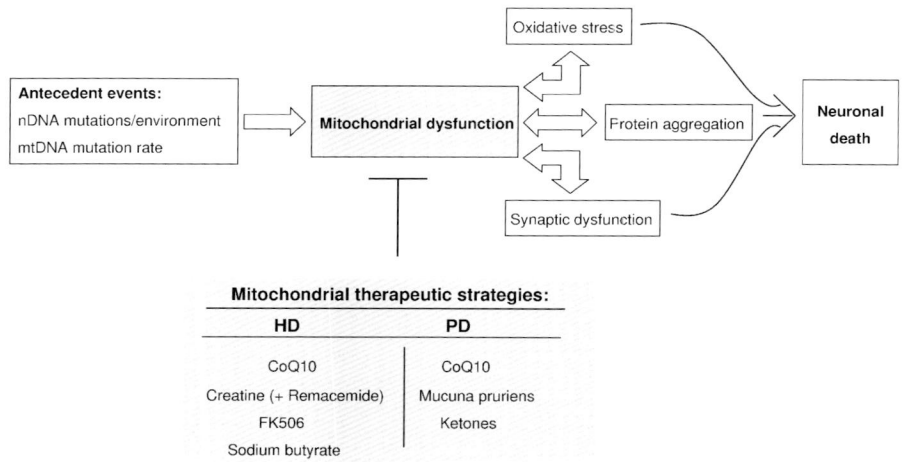

Fig. 2. Mitochondria as a target for therapeutic strategies. Schematic drawing illustrating the current compounds employed in PD and HD in vitro and in vivo. Pharmacological treatment encloses several drugs acting in the mitochondria at different levels (1) mitochondrial energy metabolism and the ETC (such as CoQ10, *Mucuna pruriens*, ketones, and creatine); (2) mitochondrial antioxidants primarily involved in fighting against oxidative stress (CoQ10); and (3) mitochondrial-dependent cell death pathways (such as remacemide, FK506, and sodium butyrate).

4.2. Huntington's Disease

HD is the most common hereditary (autosomal dominant) polyglutamine neurodegenerative disorder (see Chap. 18 for detailed information on polyglutamine expansion diseases; Rego and de Almeida, 2005, for review), caused by a CAG expansion mutation in the exon 1 of the *IT15* gene that codes for huntingtin, a 350 kDa ubiquitously expressed cytoplasmic protein (HDCRG, 1993). Mutant huntingtin containing an expanded polyglutamine stretch in its *N*-terminal is responsible for several deleterious consequences, highly affecting striatal medium spiny γ-aminobutyric acid(GABA)ergic and cortical neurons (Vonsattel and DiFiglia, 1998). Moreover, neuronal loss in HD is accompanied by proliferation of glial cells (Myers et al., 1991). Nuclear and cytosolic protein aggregates are a hallmark of neuropathology in HD. This disorder occurs in the middle-age and is associated with abnormal (choreiform) movements, cognitive decline, psychiatric symptoms, and severe weight loss. Like in many other neurodegenerative diseases, presently there is no effective treatment to prevent the onset or to delay HD progression.

4.2.1. Evidences of Mitochondrial Dysfunction in HD

Mitochondrial dysfunction and metabolic deregulation likely contribute to HD pathogenesis and may underlie the selective neuronal degeneration occurring in HD (Beal, 2005). HD patients show elevated lactate levels in the basal ganglia and cerebral cortex, increased lactate/pyruvate in the cerebrospinal fluid, decreased muscle phosphocreatine/ inorganic phosphate, decreased N-acetylaspartate and weight loss, despite high caloric intake (Jenkins et al., 1993; Koroshetz et al., 1997). A decrease in the activity of mitochondrial complexes II and III (by 53–59%) and a decrease in complex IV activity (by 32–38%) were observed in the caudate nucleus from HD patients (Gu et al., 1996). Recently, a decrease in expression of two constituents of mitochondrial complex II (the 30 kDa iron–sulfur and the 70 kDa FAD subunits) was observed in the striatum of HD patients and in striatal neurons expressing mutant huntingtin with 82 glutamines (Benchoua et al., 2006). In addition, 12-week-old transgenic HD mice expressing the exon 1 of the human huntingtin gene (R6/2 mice) showed a reduction in the activity of mitochondrial complex IV in the striatum and cerebral cortex and a decrease in aconitase activity in the striatum (Tabrizi et al., 2000b). These defects seem relatively selective because mitochondrial complex I is relatively spared in HD, with the exception of HD human platelet mitochondria where a decrease in the activity of complex I was observed (Parker et al., 1990). Increased levels of mitochondrial 4977 nucleotide pair deletion (mDNA4977) were found in the cerebral cortex of HD patients, suggesting damage of mtDNA (Horton et al., 1995). Increased oxidative damage of mtDNA may underlie metabolic defects in HD, as determined in the parietal cortex of HD patients (Polidori et al., 1999).

Impairment of respiratory chain involving decreased intracellular ATP has also been reported in striatal cells from homozygous knock-in HD mice expressing mutant huntingtin (Trettel et al., 2000; Gines et al., 2003). This contrasts with the finding of no significant differences in total intracellular ATP levels or in the activity of mitochondrial respiratory complexes in striatal cells from mutant vs. wild-type mice (Milakovic and Johnson, 2005). However, these studies are in agreement with a decrease in mitochondrial ATP, in addition to a reduction in mitochondrial respiration (Milakovic and Johnson, 2005; Seong et al., 2005). Decreased mitochondrial ADP uptake was proposed to explain the decrease in mitochondrial ATP production (Seong et al., 2005), although more data are required to support these results. Recently, analysis of in situ respiration of mitochondria showed no significant differences between striatal neurons expressing full-length mutant huntingtin from heterozygous knock-in mice and neurons from wild-type mice (Oliveira et al., 2006a).

Mitochondrial dysfunction is believed to contribute to disturbed calcium homeostasis and early neuronal death in HD (Bezprozvanny and Hayden, 2004). In fact, mitochondrial dysfunction in HD may originate from a direct interaction of mutant huntingtin with mitochondria (Panov et al., 2002). Nevertheless, changes in mitochondrial calcium loading capacity in HD are still controversial. Decreased calcium loading was observed in mitochondria isolated from immortalized lymphocytes (lymphoblasts) derived from HD patients, brains of YAC72 HD mice (Panov et al., 2002) or liver of knock-in HD mice (Choo et al., 2004). In contrast, Brustovetsky et al. (2005) reported an age-dependent increase in Ca^{2+}-loading capacity of striatal but not cortical mitochondria isolated from R6/2 and knock-in HD mice. In cells expressing full-length mutant huntingtin and subjected to excitotoxic stimuli a decrease in mitochondrial-dependent calcium handling was observed, supporting cell dysfunction (Oliveira et al., 2006a, b).

Animals (nonhuman primates) exposed to chronic low doses of 3-NP, an inhibitor of succinate dehydrogenase, an enzyme of the tricarboxylic acid cycle that is also the main constituent of the mitochondrial respiratory chain complex II (Alston et al., 1977; Coles et al., 1979) were shown to exhibit motor dysfunction and loss of striatal neurons, replicating HD symptoms observed in humans (Brouillet et al., 1995). Moreover, when administered in rats, 3-NP caused degeneration of medium spiny neurons, while preserving large cholinergic interneurons and NADPH-diaphorase-positive aspiny interneurons (Beal et al., 1993; Brouillet et al., 1993).

4.2.2. Evidences of Cell Death Involving the Mitochondria in HD Pathogenesis

There is substantial evidence that apoptosis and caspases activation contribute to neurodegeneration in HD (Portera-Cailliau et al., 1995; Kiechle et al., 2002).

Striatal cell death in rat models of HD injected with 3-NP were demonstrated to involve caspases plus calpains activation in the acute model and calpains activation in the chronic model (Bizat et al., 2003). Using the same mitochondrial toxic model of HD in vitro, in 3-NP-treated primary cortical and striatal neurons, Galas et al. (2004) showed that activation of caspases were not involved in both cell types, whereas death of striatal neurons was mainly preceded by activation of calpains. In primary cortical neurons, 3-NP induced both caspase-dependent and independent (mitochondrial-dependent) cell death, but no evidences of calpains activation were found (Almeida et al., 2004, 2006). Using immortalized hypothalamic GT1-7 cells, we also showed that Bcl-2 overexpression prevented the activation of caspases-2, -3 and -6 induced by 3-NP (Brito et al., 2003). This apparent discrepancy between the studies concerning the participation of caspases and calpains has been suggested to be due to the different activity of the apoptotic machinery in the cell cultures (Brouillet et al., 2005, for review). In addition, complex II inhibitors, including 3-NP, caused the permeabilization of mitochondrial inner membrane through opening of the mPTP (Maciel et al., 2004).

Activation of caspases-1, -3, -8, and -9 has been demonstrated in end-stage HD transgenic mice models and in *postmortem* human HD brains (Ona et al., 1999; Sanchez et al., 1999; Chen et al., 2000; Kiechle et al., 2002; Zhang et al., 2003). Moreover, caspases-2 and -6 may also play important roles in HD pathogenesis (Hermel et al., 2004). Mutant huntingtin was hypothesized to facilitate mitochondrial PT (Panov et al., 2002; Choo et al., 2004). In addition, the involvement of caspase-dependent and -independent mitochondrial cell death pathways were previously reported in two HD models, namely in ST14A striatal cells stably expressing a mutant huntingtin truncation and in R6/2 mice (Wang et al., 2003). Caspase and calpain cleavage sites have been identified in huntingtin, implicating the generation of potentially more toxic *N*-terminal huntingtin fragments with expanded polyglutamine tract (Gafni and Ellerby, 2002; Wellington et al., 2002). Moreover, the levels of Bax and Bim, two proapoptotic proteins, were shown to increase in the brain of R6/2 mice, particularly at 12 weeks of age, whereas no differences were found in Bcl-2 or Bcl-xl (Zhang et al., 2003). Hence, transgenic expression of Bcl-2 in R6/2 mice very slightly altered animal survival (Zhang et al., 2003).

Excessive activation of NMDA receptors has been reported to be implicated in HD pathogenesis, leading to high levels of intracellular calcium, mitochondrial dysfunction, and cell death (Fig. 1). Glutamate is released by cortical neurons projecting to the striatum, activating ionotropic and metabotropic glutamate receptors. Selective vulnerability of striatal neurons in HD has been attributed to NR1/NR2B subtype of NMDA receptors and to mGluR5, both of which are expressed in striatal medium spiny neurons (Li et al., 2003; Bezprozvanny and Hayden, 2004). Mutant huntingtin also binds to the type 1 inositol 1,4, 5-trisphosphate ($InsP_3$) receptor C-terminus and causes sensitization of this receptor to activation by $InsP_3$ in medium spiny striatal neurons (Bezprozvanny and Hayden, 2004, for review). Expression of mutant huntingtin in glia may also contribute to neuronal excitotoxicity in HD, since decreased expression of glutamate transporters was observed concomitant with glial nuclei accumulation of mutant huntingtin (Shin et al., 2005). In the neostriatum of individuals with HD, TUNEL-positive nuclei were detected in neurons and glia, highly suggesting apoptotic cell death (Portera-Cailliau et al., 1995).

Autophagy also seems to be involved in HD. Dopamine-exposed mutant striatal neurons derived from the R6/2 mice exhibited autophagic granules and electron-dense lysosomes, implicating autophagy as an alternative mode of cell death in HD (Petersen et al., 2001). Recent studies have further demonstrated that cytoplasmic, but not nuclear, aggregate forms of mutant huntingtin are eliminated by autophagy, which may help to explain the toxicity of protein aggregates in the nuclei (Iwata et al., 2005a). Furthermore, degradation of polyglutamine aggregates by autophagy seems to require an intact cytoskeleton and the cytoplasmic histone deacetylase 6 (HDAC6) (Iwata et al., 2005b).

4.2.3. Mitochondria – a Target for Therapeutic Intervention in HD

Several compounds have been tested to ameliorate mitochondrial dysfunction and counteract the bioenergetic defects in HD. These include CoQ10 and creatine, among others (Fig. 2).

Treatment with CoQ10 was shown to decrease the concentration of lactate in the cortex of 18 HD patients (Koroshetz et al., 1997). Furthermore, administration of either CoQ10 or remacemide, a NMDA receptor antagonist, significantly extended survival and delayed the development of motor deficits, weight loss, cerebral atrophy, and neuronal intranuclear inclusions in the R6/2 transgenic mouse model of HD, but the combined therapy was even more efficacious (Ferrante et al., 2002). Moreover, combined minocycline and CoQ10 treatment in the R6/2 mouse ameliorated behavioral and neuropathological alterations. The combined therapy significantly extended survival, improved rotarod performance, attenuated striatal neuron atrophy, and huntingtin aggregation in R6/2 HD mice (Stack et al., 2006).

Dietary creatine supplementation improved survival, slowed the brain atrophy, delayed atrophy of striatal neurons and the formation of huntingtin aggregates, improved body weight and motor performance, attenuated the decrease in *N*-acetylaspartate, and delayed the development of diabetes in both the R6/2 mice and the transgenic mouse model produced by 82 polyglutamine repeats in a 171 amino acid *N*-terminal fragment of huntingtin (N171-82Q) (Ferrante et al., 2000; Andreassen et al., 2001). In a recent study performed in 64 subjects with HD, administration of creatine for 16 weeks was well tolerated and safe, and further decreased the serum levels of 8-hydroxy-2'-deoxyguanosine (8OH2'dG), a marker of DNA oxidative damage (Hersch et al., 2006).

We previously demonstrated that mitochondrial-dependent apoptotic features evoked by exposure of cortical neurons to 3-NP could be prevented by regulating the levels of Bcl-2 family proteins in the mitochondria in the presence of FK506, an inhibitor of calcineurin or protein phosphatase 2B (PP2B), a calcium/calmodulin-dependent serine/threonine phosphatase (Almeida et al., 2004). FK506 was recently shown to favor the phosphorylation of huntingtin at S421, restoring its phosphorylation status in HD cells and preventing cell death (Pardo et al., 2006). The insulin-like growth factor-1 (IGF-1)/Akt pathway and the serum and glucocorticoid-induced kinase (SGK) were also shown to phosphorylate huntingtin at S421, abolishing the toxicity of mutant huntingtin (Humbert et al., 2002; Rangone et al., 2004).

Cystamine (and cysteamine), a transglutaminase competitive inhibitor, was recently shown to prevent 3-NP-induced mitochondrial depolarization in striatal cells (Mao et al.,

2006). Cystamine also showed protective effects in R6/2 mice independently of transglutaminase (Bailey and Johnson, 2006) and also in the YAC128 transgenic model (Van Raamsdonk et al., 2005).

Sodium butyrate, a histone deacetylase (HDAC) inhibitor, was also shown to reduce neuronal death induced by the mitochondrial complex II inhibitor 3-NP (Ferrante et al., 2003), suggesting that HDAC inhibitors may also improve mitochondrial function. In addition, the HDAC inhibitor phenylbutyrate administered after the onset of symptoms in transgenic N171-82Q mice model of HD extended mice survival, attenuated neuronal atrophy, and downregulated caspase-9 implicated in mitochondrial-dependent apoptotic cell death (Gardian et al., 2005). Furthermore, in recent studies Oliveira and collaborators determined that the HDAC inhibitors sodium butyrate and trichostatin A ameliorate the deficits in mitochondrial-dependent calcium handling upon excitotoxic stimuli following expression of mutant huntingtin (Oliveira et al., 2006b).

5. CONCLUDING REMARKS

Mitochondria have a Janus role, since they are both essential ATP producers and integrators and amplifiers of cell death signals in the neuronal system. Mitochondrial dysfunction and impaired energy metabolism underlie neuronal cell death occurring in several neurodegenerative disorders, thus implicating mitochondria at the centre of the neurodegenerative process. Therefore, it is reasonable to propose new therapeutic possibilities such as delivery of specific drugs into the mitochondria, rescuing cells from apoptosis and/or therapies that indirectly influence mitochondrial function through, for example, the regulation/expression of mitochondrial complex subunits or anti-apoptotic proteins. Over the next decades the increase in the elderly population will inevitably lead to a steadily increasing number of sufferers of neurodegenerative disorders associated with aging, namely sporadic PD. Thus, further studies in the mitochondrial field will bring important contributions to the pathology of several neurodegenerative diseases and to the natural process of aging.

6. ACKNOWLEDGMENTS

A.C.R. acknowledges financial support from *Fundação para a Ciência e a Tecnologia*, FCT (POCI/SAL-NEU/57310/2004) and from the *Instituto de Investigação Interdisciplinar*, University of Coimbra (III/BIO/49/2005), Portugal.

7. REFERENCES

Almeida, S., Domingues, A., Rodrigues, L., Oliveira, C.R. and Rego, A.C., 2004, FK506 prevents mitochondrial-dependent apoptotic cell death induced by 3-nitropropionic acid in rat primary cortical cultures. *Neurobiol. Dis.* **17**: 435.
Almeida, S., Brett, A.C., Gois, I.N., Oliveira, C.R. and Rego, A.C., 2006, Caspase-dependent and -independent cell death induced by 3-nitropropionic acid in rat cortical neurons. *J. Cell Biochem.* **98**: 93.
Alston, T.A., Mela, L. and Bright, H.J., 1977, 3-Nitropropionate, the toxic substance of Indigofera, is a suicide inactivator of succinate dehydrogenase. *Proc. Natl. Acad. Sci. USA* **74**: 3767.
Andreassen, O.A., Dedeoglu, A., Ferrante, R.J., Jenkins, B.G., Ferrante, K.L., Thomas, M., Friedlich, A., Browne, S.E., Schilling, G., Borchelt, D.R., Hersch, S.M., Ross, C.A. and Beal, M.F., 2001, Creatine increase survival and delays motor symptoms in a transgenic animal model of Huntington's disease. *Neurobiol. Dis.* **8**: 479.
Anglade, P., Vyas, S., Javoy-Agid, F., Herrero, M.T., Michel, P.P., Marquez, J., Mouatt-Prigent, A., Ruberg, M., Hirsch, E.C. and Agid, Y., 1997, Apoptosis and autophagy in nigral neurons of patients with Parkinson's disease. *Histol. Histopathol.* **12**: 25.
Araújo, I.M., Xapelli, S., Gil, J.M., Mohapel, P., Petersen, A., Pinheiro, P.S., Malva, J.O., Bahr, B.A., Brundin, P. and Carvalho, C.M., 2005, Proteolysis of NR2B by calpain in the hippocampus of epileptic rats. *Neuroreport* **16**: 393.
Atlante, A., Calissano, P., Bobba, A., Azzariti, A., Marra, E. and Passarella, S., 2000, Cytochrome c is released from mitochondria in a reactive oxygen species (ROS)-dependent fashion and can operate as a ROS scavenger and as a respiratory substrate in cerebellar neurons undergoing excitotoxic death. *J. Biol. Chem.* **275**: 37159.
Atlante, A., Bobba, A., de Bari, L., Fontana, F., Calissano, P., Marra, E. and Passarella, S., 2006, Caspase-dependent alteration of the ADP/ATP translocator triggers the mitochondrial permeability transition which is not required for the low-potassium-dependent apoptosis of cerebellar granule cells. *J. Neurochem.* **97**: 1166.
Attardi, G. and Schatz, G., 1988, Biogenesis of mitochondria. *Annu. Rev. Cell Biol.* **4**: 289.
Bailey, C.D. and Johnson, G.V., 2006, The protective effects of cystamine in the R6/2 Huntington's disease mouse involve mechanisms other than the inhibition of tissue transglutaminase. *Neurobiol. Aging* **27**: 871.
Barroso, N., Campos, Y., Huertas, R., Esteban, J., Molina, J.A., Alonso, A., Gutierrez-Rivas, E. and Arenas, J., 1993, Respiratory chain enzyme activities in lymphocytes from untreated patients with Parkinson disease. *Clin. Chem.* **39**: 667.
Beal, M.F., 2005, Mitochondria take center stage in aging and neurodegeneration. *Ann. Neurol.* **58**: 495.
Beal, M.F., Brouillet, E., Jenkins, B.G., Ferrante, R.J., Kowall, N.W., Miller, J.M., Storey, E., Srivastava, R., Rosen, B.R. and Hyman, B.T., 1993, Neurochemical and histologic characterization of striatal excitotoxic lesions produced by the mitochondrial toxin 3-nitropropionic acid. *J. Neurosci.* **13**: 4181.
Beal, M.F., Matthews, R.T., Tieleman, A. and Shults, C.W., 1998, Coenzyme Q10 attenuates the 1-methyl-4-phenyl-1,2,3,tetrahydropyridine (MPTP) induced loss of striatal dopamine and dopaminergic axons in aged mice. *Brain Res.* **783**: 109.
Beckman, J.S., 1994, Peroxynitrite versus hydroxyl radical: the role of nitric oxide in superoxide-dependent cerebral injury. *Ann. N.Y. Acad. Sci.* **738**: 69.
Benchoua, A., Trioulier, Y., Zala, D., Gaillard, M.C., Lefort, N., Dufour, F., Saudou, F., Elalouf, J.M., Hirsch, E., Hantraye, P., Deglon, N. and Brouillet, E., 2006, Involvement of mitochondrial complex II defects in neuronal death produced by N-terminus fragment of mutated huntingtin. *Mol. Biol. Cell.* **17**: 1652.
Benecke, R., Strumper, P. and Weiss, H., 1993, Electron transfer complexes I and IV of platelets are abnormal in Parkinson's disease but normal in Parkinson-plus syndromes. *Brain* **116**: 1451.
Bezprozvanny, I. and Hayden, M.R., 2004, Deranged neuronal calcium signaling and Huntington disease. *Biochem. Biophys. Res. Commun.* **322**: 1310.
Bi, X., Rong, Y., Chen, J., Dang, S., Wang, Z. and Baudry, M., 1998, Calpain-mediated regulation of NMDA receptor structure and function. *Brain Res.* **790**: 245.
Bizat, N., Hermel, J. M., Boyer, F., Jacquard, C., Creminon, C., Ouary, S., Escartin, C., Hantraye, P., Kajewski, S. and Brouillet, E., 2003, Calpain is a major cell death effector in selective striatal degeneration induced in vivo by 3-nitropropionate: implications for Huntington's disease. *J. Neurosci.* **23**: 5020.
Blandini, F., Nappi, G. and Greenamyre, J.T., 1998, Quantitative study of mitochondrial complex I in platelets of parkinsonian patients. *Mov. Disord.* **13**: 11.
Bossy-Wetzel, E., Schwarzenbacher, R. and Lipton, S.A., 2004, Molecular pathways to neurodegeneration. *Nat. Med.* **10**: S2.
Brito, O., Almeida, S., Oliveira, C.R. and Rego, A.C., 2003, Bcl-2 prevents loss of cell viability and caspases activation induced by 3-nitropropionic acid in GT1-7 cells. *Ann. NY Acad. Sci.* **1010**: 148.

Brouillet, E., Jenkins, B.G., Hyman, B.T., Ferrante, R.J., Kowall, N.W., Srivastava, R., Roy, D.S., Rosen, B.R. and Beal, M.F., 1993, Age-dependent vulnerability of the striatum to the mitochondrial toxin 3-nitropropionic acid. *J. Neurochem.* **60**: 356.

Brouillet, E., Hantraye, P., Ferrante, R.J., Dolan, R., Leroy-Willig, A., Kowall, N.W. and Beal, M.F., 1995, Chronic mitochondrial energy impairment produces selective striatal degeneration and abnormal choreiform movements in primates. *Proc. Natl. Acad. Sci. USA* **92**: 7105.

Brouillet, E., Jacquard, C., Bizat, N. and Blum, D., 2005, 3-Nitropropionic acid: a mitochondrial toxin to uncover physiopathological mechanisms underlying striatal degeneration in Huntington's disease. *J. Neurochem.* **95**: 1521.

Brunk, U.T. and Terman, A., 2002, The mitochondrial-lysosomal axis theory of aging: accumulation of damaged mitochondria as a result of imperfect autophagocytosis. *Eur. J. Biochem.* **269**: 1996.

Brustovetsky, N., LaFrance, R., Purl, K.J., Brustovetsky, T., Keene, C.D., Low, W.C. and Dubinsky, J.M. (2005), Age-dependent changes in the calcium sensitivity of striatal mitochondria in mouse models of Huntington's Disease. *J. Neurochem.* **93**: 1361.

Bursch, W., Hochegger, K., Torok, L., Marian, B., Ellinger, A. and Hermann, R.S., 2000, Autophagic and apoptotic types of programmed cell death exhibit different fates of cytoskeletal filaments. *J. Cell Sci.* **113**: 1189.

Cardoso, S.M., Santos, S., Swerdlow, R.H. and Oliveira, C.R., 2001, Functional mitochondria are required for amyloid beta-mediated neurotoxicity. *FASEB J.* **15**: 1439.

Cassarino, D.S., Fall, C.P., Swerdlow, R.H., Smith, T.S., Halvorsen, E.M., Miller, S.W., Parks, J.P., Parker, W.D., Jr. and Bennett, J.P., Jr., 1997, Elevated reactive oxygen species and antioxidant enzyme activities in animal and cellular models of Parkinson's disease. *Biochim. Biophys. Acta.* **1362**: 77.

Chartier-Harlin, M.C., Kachergus, J., Roumier, C., Mouroux, V., Douay, X., Lincoln, S., Levecque, C., Larvor, L., Andrieux, J., Hulihan, M., Waucquier, N., Defebvre, L., Amouyel, P., Farrer, M. and Destee, A., 2004, Alpha-synuclein locus duplication as a cause of familial Parkinson's disease. *Lancet* **364**: 1167.

Chee, J.L., Guan, X.L., Lee, J.Y., Dong, B., Leong, S.M., Ong, E.H., Liou, A.K. and Lim, T.M., 2005, Compensatory caspase activation in MPP+-induced cell death in dopaminergic neurons. *Cell. Mol. Life Sci.* **62**: 227.

Chen, M., Ona, V.O., Li, M., Ferrante, R.J., Fink, K.B., Zhu, S., Bian, J., Guo, L., Farrell, L.A., Hersch, S.M., Hobbs, W., Vonsattel, J.-P., Cha, J.-H. and Friedlander, R.M., 2000, Minocycline inhibits caspase-1 and caspase-3 expression and delays mortality in a transgenic mouse model of Huntington disease. *Nat. Med.* **6**: 797.

Chiu, K., Lam, T.T., Ying Li, W.W., Caprioli, J. and Kwong Kwong, J.M., 2005, Calpain and *N*-methyl-D-aspartate (NMDA)-induced excitotoxicity in rat retinas. *Brain Res.* **1046**: 207.

Choo, Y.S., Johnson, G.V., MacDonald, M., Detloff, P.J. and Lesort, M., 2004, Mutant huntingtin directly increases susceptibility of mitochondria to the calcium-induced permeability transition and cytochrome c release. *Hum. Mol. Genet.* **13**: 1407.

Clohessy, J.G., Zhuang, J., de Boer, J., Gil-Gomez, G. and Brady, H.J., 2006, Mcl-1 interacts with truncated Bid and inhibits its induction of cytochrome c release and its role in receptor-mediated apoptosis. *J. Biol. Chem.* **281**: 5750.

Coles, C.J., Edmondson, D.E. and Singer, T.P., 1979, Inactivation of succinate dehydrogenase by 3-nitropropionate. *J. Biol. Chem.* **254**: 5161.

Cookson, M.R., 2003, Pathways to Parkinsonism. *Neuron* **37**: 7.

Darios, F, Corti, O., Lucking, C.B., Hampe, C., Muriel, M.P., Abbas, N., Gu, W.J., Hirsch, E.C., Rooney, T., Ruberg, M. and Brice, A., 2003, Parkin prevents mitochondrial swelling and cytochrome c release in mitochondria-dependent cell death. *Hum. Mol. Genet.* **12**: 517.

Dauer, W., Kholodilov, N., Vila, M., Trillat, A.C., Goodchild, R., Larsen, K.E, Staal, R., Tieu, K., Schmitz, Y., Yuan, C.A., Rocha, M., Jackson-Lewis, V., Hersch, S., Sulzer, D., Przedborski, S., Burke, R. and Hen, R., 2002, Resistance of alpha-synuclein null mice to the parkinsonian neurotoxin MPTP. *Proc. Natl. Acad. Sci. USA* **99**: 14524.

Dawson, T.M. and Dawson, V.L., 2003, Molecular pathways of neurodegeneration in Parkinson's disease. *Science* **302**: 819.

Du, C., Fang, M., Li, Y., Li, L. and Wang, X., 2000, Smac, a mitochondrial protein that promotes cytochrome c-dependent caspase activation by eliminating IAP inhibition. *Cell* **102**: 33.

Enoksson, M., Robertson, J.D., Gogvadze, V., Bu, P., Kropotov, A., Zhivotovsky, B. and Orrenius, S., 2004, Caspase-2 permeabilizes the outer mitochondrial membrane and disrupts the binding of cytochrome c to anionic phospholipids, *J. Biol. Chem.* **279**: 49575.

Eskes, R., Antonsson, B., Osen-Sand, A., Montessuit, S., Richter, C., Sadoul, R., Mazzei, G., Nichols, A. and Martinou, J.-C., 1998, Bax-induced cytochrome c release from mitochondria is independent of the permeability transition pore but highly dependent on Mg2+ ions. *J. Cell. Biol.* **143**: 217.

Feany, M.B. and Pallanck, L.J., 2003, Parkin: a multipurpose neuroprotective agent? *Neuron* **38**: 13.
Ferrante, R.J., Andreassen, O.A., Jenkins, B.G., Dedeoglu, A., Kuemmerle, S., Kubilus, J.K., Kaddurah-Daouk, R., Hersch, S.M. and Beal, M.F., 2000, Neuroprotective effects of creatine in a transgenic mouse model of Huntington's disease. *J. Neurosci.* **20**: 4389.
Ferrante, R.J., Andreassen, O.A., Dedeoglu, A., Ferrante, K.L., Jenkins, B.G., Hersch, S.M. and Beal, M.F., 2002, Therapeutic effects of coenzyme Q10 and remacemide in transgenic mouse models of Huntington's disease. *J. Neurosci.* **22**: 1592.
Ferrante, R.J., Kubilus, J.K., Lee, J., Ryu, H., Beesen, A., Zucker, B., Smith, K., Kowall, N.W., Ratan, R.R., Luthi-Carter, R. and Hersch, S.M., 2003, Histone deacetylase inhibition by sodium butyrate chemotherapy ameliorates the neurodegenerative phenotype in Huntington's disease mice. *J. Neurosci.* **23**: 9418.
Flower, T.R., Chesnokova, L.S., Froelich, C.A., Dixon, C. and Witt, S.N., 2005, Heat shock prevents alpha-synuclein-induced apoptosis in a yeast model of Parkinson's disease. *J. Mol. Biol.* **351**: 1081.
Forno, L.S., 1996, Neuropathology of Parkinson's disease. *J. Neuropathol. Exp. Neurol.* **55**: 259.
Gafni, J. and Ellerby, L.M., 2002, Calpain activation in Huntington's disease. *J. Neurosci.* **22**: 4842.
Galas, M.C., Bizat, N., Cuvelier, L., Bantubungi, K., Brouillet, E., Schiffmann, S.N. and Blum, D., 2004, Death of cortical and striatal neurons induced by mitochondrial defect involves differential molecular mechanisms. *Neurobiol. Dis.* **15**: 152.
Gardian, G., Browne, S.E., Choi, D.K., Klivenyi, P., Gregorio, J., Kubilus, J.K., Ryu, H., Langley, B., Ratan, R.R., Ferrante, R.J. and Beal, M.F., 2005, Neuroprotective effects of phenylbutyrate in the N171-82Q transgenic mouse model of Huntington's disease. *J. Biol. Chem.* **280**: 556.
Ghafourifar, P. and Richter, C., 1997, Nitric oxide synthase activity in mitochondria. *FEBS Lett.* **418**: 291.
Ghafourifar, P., Schenk, U., Klein, S.D. and Richter, C., 1999, Mitochondrial nitric-oxide synthase stimulation causes cytochrome c release from isolated mitochondria. Evidence for intramitochondrial peroxynitrite formation. *J. Biol. Chem.* **274**: 31185.
Gines, S., Seong, I.S., Fossale, E., Ivanova, E., Trettel, F., Gusella, J.F., Wheeler, V.C., Persichetti, F. and Macdonald, M.E., 2003, Specific progressive cAMP reduction implicates energy deficit in presymptomatic Huntington's disease knock-in mice. *Hum. Mol. Genet.* **12**: 497.
Giulivi, C., Poderoso, J.J. and Boveris, A., 1998, Production of nitric oxide by mitochondria. *J. Biol. Chem.* **273**: 11038.
Gomez, C., Reiriz, J., Pique, M., Gil, J., Ferrer, I. and Ambrosio, S., 2001, Low concentrations of 1-methyl-4-phenylpyridinium ion induce caspase-mediated apoptosis in human SH-SY5Y neuroblastoma cells. *J. Neurosci. Res.* **63**: 421.
Gómez-Santos, C., Ferrer, I., Santidrian, A.F., Barrachina, M., Gil, J. and Ambrosio, S., 2003, Dopamine induces autophagic cell death and alpha-synuclein increase in human neuroblastoma SH-SY5Y cells. *J. Neurosci. Res.* **73**: 341.
Gorman, A.M., Szegezdi, E., Quigney, D.J. and Samali, A., 2005, Hsp27 inhibits 6-hydroxydopamine-induced cytochrome c release and apoptosis in PC12 cells. *Biochem. Biophys. Res. Commun.* **327**: 801.
Green, D.R., 2005, Apoptotic pathways: ten minutes to dead. *Cell* **121**: 671.
Greenamyre, J.T. and Hastings, T.G., 2004, Biomedicine. Parkinson's-divergent causes, convergent mechanisms. *Science* **304**: 1120.
Greene, J.C., Whitworth, A.J., Kuo, I., Andrews, L.A., Feany, M.B. and Pallanck, L.J., 2003, Mitochondrial pathology and apoptotic muscle degeneration in *Drosophila* parkin mutants. *Proc. Natl. Acad. Sci. USA* **100**: 4078.
Gu, M., Gash, M.T., Mann, V.M., Javoy-Agid, F., Cooper, J.M. and Schapira, A.H., 1996, Mitochondrial defect in Huntington's disease caudate nucleus. *Ann. Neurol.* **39**: 385.
Gu, M., Cooper, J.M., Taanman, J.W. and Schapira, A.H., 1998, Mitochondrial DNA transmission of the mitochondrial defect in Parkinson's disease. *Ann. Neurol.* **44**: 177.
Guo, Y., Srinivasula, S.M., Druilhe, A., Fernandes-Alnemri, T. and Alnemri, E.S., 2002, Caspase-2 induces apoptosis by releasing proapoptotic proteins from mitochondria. *J. Biol. Chem.* **277**: 13430.
Haas, R.H., Nasirian, F., Nakano, K., Ward, D., Pay, M., Hill, R. and Shults, C.W., 1995, Low platelet mitochondrial complex I and complex II/III activity in early untreated Parkinson's disease. *Ann. Neurol.* **37**: 714.
Hanagasi, H.A., Ayribas, D., Baysal, K. and Emre, M., 2005, Mitochondrial complex I, II/III, and IV activities in familial and sporadic Parkinson's disease. *Int. J. Neurosci.* **115**: 479.
Hanrott, K., Gudmunsen, L., O'neill, M.J. and Wonnacott, S., 2006, 6-hydroxydopamine-induced apoptosis is mediated via extracellular auto-oxidation and caspase 3-dependent activation of protein kinase C{delta}. *J. Biol. Chem.* **281**: 5373.
Hartmann, A., Hunot, S., Michel, P.P., Muriel, M.P., Vyas, S., Faucheux, B.A., Mouatt-Prigent, A., Turmel, H., Srinivasan, A., Ruberg, M., Evan, G.I., Agid, Y. and Hirsch, E.C., 2000, Caspase-3: a vulnerability factor and final effector in apoptotic death of dopaminergic neurons in Parkinson's disease. *Proc. Natl. Acad. Sci. USA* **97**: 2875.

Hartmann, A., Michel, P.P., Troadec, J.D., Mouatt-Prigent, A., Faucheux, B.A., Ruberg, M., Agid, Y. and Hirsch, E.C., 2001, Is Bax a mitochondrial mediator in apoptotic death of dopaminergic neurons in Parkinson's disease? *J. Neurochem.* **76**: 1785.

Hattori, N., Tanaka, M., Ozawa, T. and Mizuno, Y., 1991, Immunohistochemical studies on complexes I, II, III, and IV of mitochondria in Parkinson's disease. *Ann. Neurol.* **30**: 563.

HDCRG (The Huntington's Disease Collaborative Research Group), 1993, A novel gene containing a trinucleotide repeat that is expanded and unstable on Huntington's disease chromosomes. *Cell* **72**: 971.

Hermel, E., Gafni, J., Propp, S.S., Leavitt, B.R., Wellington, C.L., Young, J.E., Hackam, A.S., Logvinova, A.V., Peel, A.L., Chen, S.F., Hook, V., Singaraja, R., Krajewsk, S., Goldsmith, P.C., Ellerby, H.M., Hayden, M.R., Bredesen, D.E. and Ellerby, L.M., 2004, Specific caspase interactions and amplification are involved in selective neuronal vulnerability in Huntington's disease. *Cell Death Differ.* **11**: 424.

Hersch, S.M., Gevorkian, S., Marder, K., Moskowitz, C., Feigin, A., Cox, M., Como, P., Zimmerman, C., Lin, M., Zhang, L., Ulug, A.M., Beal, M.F., Matson, W., Bogdanov, M., Ebbel, E., Zaleta, A., Kaneko, Y., Jenkins, B., Hevelone, N, Zhang, H., Yu, H., Schoenfeld, D., Ferrante, R. and Rosa, H.D., 2006, Creatine in Huntington disease is safe, tolerable, bioavailable in brain and reduces serum 8OH2'dG. *Neurology* **66**: 250.

Holtz, W.A. and O'Malley, K.L., 2003, Parkinsonian mimetics induce aspects of unfolded protein response in death of dopaminergic neurons. *J. Biol. Chem.* **278**: 19367.

Horton, T.M., Graham, B.H., Corral-Debrinski, M., Shoffner, J.M., Kaufman, A.E., Beal, M.F. and Wallace, D.C., 1995, Marked increase in mitochondrial DNA deletion in the cerebral cortex of Huntington's disease patients. *Neurology* **45**: 1879.

Humbert, S., Bryson, E.A., Cordelieres, F.P., Connors, N.C., Datta, S.R., Finkbeiner, S., Greenberg, M.E. and Saudou, F., 2002, The IGF-1/Akt pathway is neuroprotective in Huntington's disease and involves huntingtin phosphorylation by Akt. *Dev. Cell.* **2**: 831.

Hyun, D.H., Lee, M., Halliwell, B. and Jenner, P., 2005, Effect of overexpression of wild-type or mutant parkin on the cellular response induced by toxic insults. *J. Neurosci. Res.* **82**: 232.

Iwata, A., Christianson, J.C., Bucci, M., Ellerby, L.M., Nukina, N., Forno, L.S. and Kopito, R.R., 2005a, Increased susceptibility of cytoplasmic over nuclear polyglutamine aggregates to autophagic degradation. *Proc. Natl. Acad. Sci. USA* **102**: 13135.

Iwata, A., Riley, B.E., Johnston, J.A. and Kopito, R.R., 2005b, HDAC6 and microtubules are required for autophagic degradation of aggregated huntingtin. *J. Biol. Chem.* **280**: 40282.

Janetzky, B., Hauck, S., Youdim, M.B., Riederer, P., Jellinger, K., Pantucek, F., Zochling, R., Boissl, K.W. and Reichmann, H., 1994, Unaltered aconitase activity, but decreased complex I activity in substantia nigra pars compacta of patients with Parkinson's disease. *Neurosci. Lett.* **169**: 126.

Jenkins, B.G., Koroshetz, W.J., Beal, M.F. and Rosen, B.R., 1993, Evidence for impairment of energy metabolism in vivo in Huntington's disease using localized 1H NMR spectroscopy. *Neurology* **43**: 2689.

Jiang, S., Cai, J., Wallace, D.C. and Jones, D.P., 1999, Cytochrome c-mediated apoptosis in cells lacking mitochondrial DNA. *J. Biol. Chem.* **274**: 29905.

Junn, E., Taniguchi, H., Jeong, B.S., Zhao, X., Ichijo, H. and Mouradian, M.M., 2005, Interaction of *DJ-1* with Daxx inhibits apoptosis signal-regulating kinase 1 activity and cell death. *Proc. Natl. Acad. Sci. USA* **102**: 9691.

Keeney, P.M., Xie, J., Capaldi, R.A. and Bennett, J.P., Jr, 2006, Parkinson's disease brain mitochondrial complex I has oxidatively damaged subunits and is functionally impaired and misassembled. *J. Neurosci.* **26**: 5256.

Kiechle, T., Dedeoglu, A., Kubilus, J., Kowall, N.W., Beal, M.F., Friedlander, R., Hersch, S.M. and Ferrante, R.J., 2002, Cytochrome c and caspase-9 expression in Huntington's disease. *Neuromol. Med.* **1**: 183.

Kim, J. and Klionsky, D.J., 2000, Autophagy, cytoplasm-to-vacuole targeting pathway, and pexophagy in yeast and mammalian cells. *Annu. Rev. Biochem.* **69**: 303.

King, M.P. and Attardi, G., 1989, Human cells lacking mtDNA: repopulation with exogenous mitochondria by complementation. *Science* **246**: 500.

Kitada, T., Asakawa, S., Hattori, N., Matsumine, H., Yamamura, Y., Minoshima, S., Yokochi, M., Mizuno, Y. and Shimizu, N., 1998, Mutations in the parkin gene cause autosomal recessive juvenile parkinsonism. *Nature* **392**: 605.

Klionsky, D.J. and Ohsumi, Y., 1999, Vacuolar import of proteins and organelles from the cytoplasm. *Annu. Rev. Cell Dev. Biol.* **15**: 1.

Klivenyi, P., Siwek, D., Gardian, G., Yang, L., Starkov, A., Cleren, C., Ferrante, R.J., Kowall, N.W., Abeliovich, A. and Beal, M.F., 2006, Mice lacking alpha-synuclein are resistant to mitochondrial toxins. *Neurobiol. Dis.* **21**: 541.

Koroshetz, W.J., Jenkins, B.G., Rosen, B.R. and Beal, M.F., 1997, Energy metabolism defects in Huntington's disease and effects of coenzyme Q10. *Ann. Neurol.* **41**: 160.

Kowaltowski, A.J., Castilho, R.F. and Vercesi, A.E., 1995, Ca2+-induced mitochondrial membrane permeabilization: role of coenzyme Q redox state. *Am. J. Physiol.* **269**: 141.

Kruger, R., Kuhn, W., Muller, T., Woitalla, D., Graeber, M., Kosel, S., Przuntek, H., Epplen, J.T., Schols, L. and Riess, O., 1998, Ala30Pro mutation in the gene encoding alpha-synuclein in Parkinson's disease. *Nature Genet.* **18**: 106.

Langston, J.W., Ballard, P., Tetrud, J.W. and Irwin, I., 1983, Chronic Parkinsonism in humans due to a product of meperidine-analog synthesis. *Science* **219**: 979.

Lashuel, H.A., Hartley, D., Petre, B.M., Walz, T. and Lansbury, P.T., Jr., 2002, Neurodegenerative disease: amyloid pores from pathogenic mutations. *Nature* **418**: 291.

Lassus, P., Opitz-Araya, X. and Lazebnik, Y., 2002, Requirement for caspase-2 in stress-induced apoptosis before mitochondrial permeabilization. *Science* **297**: 1352.

Li, L.Y., Luo, X. and Wang, X., 2001, Endonuclease G is an apoptotic DNase when released from mitochondria. *Nature* **412**: 95.

Li, L., Fan, M., Icton, C.D., Chen, N., Leavitt, B.R., Hayden, M.R., Murphy, T.H. and Raymond, L.A., 2003, Role of NR2B-type NMDA receptors in selective neurodegeneration in Huntington disease. *Neurobiol. Aging* **24**: 1113.

Lin, D.T., Wu, J., Holstein, D., Upadhyay, G., Rourk, W., Muller, E. and Lechleiter, J.D., 2005, Ca(2+) signaling, mitochondria and sensitivity to oxidative stress in aging astrocytes. *Neurobiol. Aging* (Epub ahead of print)

Liu, X., Zou, H., Slaughter, C. and Wang, X., 1997, DFF, a heterodimeric protein that functions downstream of caspase-3 to trigger DNA fragmentation during apoptosis. *Cell* **89**: 175.

Luetjens, C.M., Bui, N.T., Sengpiel, B., Münstermann, G., Poppe, M., Krohn, A.J., Bauerbach, E., Krieglstein, J. and Prehn, J.H.M., 2000, Delayed mitochondrial dysfunction in excitotoxic neuron death: cytochrome c release and a secondary increase in superoxide production. *J. Neurosci.* **20**: 5715.

Luo, X., Budihardjo, I., Zou, H., Slaughter, C. and Wang, X., 1998, Bid, a Bcl2 interacting protein, mediates cytochrome c release from mitochondria in response to activation of cell surface death receptors. *Cell* **94**: 481.

Maciel, E.N., Vercesi, A.E. and Castilho, R.F., 2001, Oxidative stress in Ca(2+)-induced membrane permeability transition in brain mitochondria. *J. Neurochem.* **79**: 1237.

Maciel, E.N., Kowaltowski, A.J., Schwalm, F.D., Rodrigues, J.M., Souza, D.O., Vercesi, A.E., Wajner, M. and Castilho, R.F., 2004, Mitochondrial permeability transition in neuronal damage promoted by Ca2+ and respiratory chain complex II inhibition. *J. Neurochem.* **90**: 1025.

Mann, V.M., Cooper, J.M., Krige, D., Daniel, S.E., Schapira, A.H. and Marsden, C.D., 1992, Brain, skeletal muscle and platelet homogenate mitochondrial function in Parkinson's disease. *Brain* **115**: 333.

Manyam, B.V., Dhanasekaran, M. and Hare, T.A., 2004, Neuroprotective effects of the antiparkinson drug Mucuna pruriens. *Phytother. Res.* **18**: 706.

Mao, Z., Choo, Y.S. and Lesort, M., 2006, Cystamine and cysteamine prevent 3-NP-induced mitochondrial depolarization of Huntington's disease knock-in striatal cells. *Eur. J. Neurosci.* **23**: 1701.

Martin, L.J., Pan, Y., Price, A.C., Sterling, W., Copeland, N.G., Jenkins, N.A., Price, D.L. and Lee, M.K., 2006, Parkinson's disease alpha-synuclein transgenic mice develop neuronal mitochondrial degeneration and cell death. *J. Neurosci.* **26**: 41.

Milakovic, T. and Johnson, G.V., 2005, Mitochondrial respiration and ATP production are significantly impaired in striatal cells expressing mutant huntingtin. *J. Biol. Chem.* **280**: 30773.

Mizuno, Y., Ohta, S., Tanaka, M., Takamiya, S., Suzuki, K., Sato, T., Oya, H., Ozawa, T. and Kagawa, Y., 1989, Deficiencies in complex I subunits of the respiratory chain in Parkinson's disease. *Biochem. Biophys. Res. Commun.* **163**: 1450.

Mochizuki, H., Goto, K., Mori, H. and Mizuno, Y., 1996, Histochemical detection of apoptosis in Parkinson's disease. *J. Neurol. Sci.* **137**: 120.

Muftuoglu, M., Elibol, B., Dalmizrak, O., Ercan, A., Kulaksiz, G., Ogus, H., Dalkara, T. and Ozer, N., 2004, Mitochondrial complex I and IV activities in leukocytes from patients with parkin mutations. *Mov. Disord.* **19**: 544.

Muller, T., Buttner, T., Gholipour, A.F. and Kuhn, W., 2003, Coenzyme Q10 supplementation provides mild symptomatic benefit in patients with Parkinson's disease. *Neurosci. Lett.* **341**: 201.

Myers, R.H., Vonsattel, J.P., Paskevich, P.A., Kiely, D.K., Stevens, T.J., Cupples, L.A., Richardson, E.P., Jr. and Bird, E.D., 1991, Decreased neuronal and increased oligodendroglial densities in Huntington's disease caudate nucleus. *J. Neuropathol. Exp. Neurol.* **50**: 729.

Neumar, R.W., Xu, Y.A., Gada, H., Guttmann, R.P. and Siman, R., 2003, Cross-talk between calpain and caspase proteolytic systems during neuronal apoptosis. *J. Biol. Chem.* **278**: 14162.

Nicholls, D.G. and Ferguson, S.J., 1992, *Bioenergetics 2*. Academic, London, UK.

Nicholls, D.G. and Budd, S.L., 2000, Mitochondria and neuronal survival. *Physiol. Rev.* **80**: 315.

Nicholls, D.G. and Ward, M.W., 2000, Mitochondrial membrane potential and cell death: mortality and millivolts. *Trends Neurosci.* **23**: 166.

Nijlmans, L.G.J., Spelbrink, J.N., Van Galen, M.J.M., Zwaan, M., Klement, P. and Van den Bogert, C., 1995, Expression and fate of the nuclearly encoded subunits of cytochrome c oxidase in cultured human cells depleted of mitochondrial gene products. *Biochem. Biophys. Acta* **1265**: 117.

Oliveira, J.M.A., Jekabsons, M.B., Chen, S., Lin, A., Rego, A.C., Gonçalves, J., Ellerby, L.M. and Nicholls, D.G., 2006a, Mitochondrial dysfunction in Huntington's disease: the bioenergetics of isolated and in situ mitochondria from transgenic mice. *J. Neurochem.* (in press) doi: 10.1111/j.1471-4159.2006.04361.x.

Oliveira, J.M.A., Chen, S., Almeida, S., Riley, R., Gonçalves, J., Oliveira, C.R., Hayden, M.R., Nicholls, D.G., Ellerby, L.M. and Rego, A.C., 2006b, Mitochondrial-dependent Ca^{2+} handling in Huntington's disease striatal cells: effect of histone deacetylase inhibitors. *J. Neurosci.* **26**: 11174.

Ona, V.O., Li, M., Vonsattel, J.-P., Andrews, L.J., Khan, S.Q., Chung, W.M., Frey, A.S., Menon, A.S., Li, X.-J., Stieg, P.E., Yuan, J., Penney, J.B., Young, A.B., Cha, J.-H. and Friedlander, R.M., 1999, Inhibition of caspase-1 slows disease progression in a mouse model of Huntington's disease. *Nature* **399**: 263.

Orth, M. and Tabrizi, S.J., 2003, Models of Parkinson's disease. *Mov. Disord.* **18**: 729.

Palacino, J.J., Sagi, D., Goldberg, M.S., Krauss, S., Motz, C., Wacker, M., Klose, J. and Shen, J., 2004, Mitochondrial dysfunction and oxidative damage in parkin-deficient mice. *J. Biol. Chem.* **279**: 18614.

Panov, A.V., Gutekunst, C.A., Leavitt, B.R., Hayden, M.R., Burke, J.R., Strittmatter, W.J. and Greenamyre, J.T., 2002, Early mitochondrial calcium defects in Huntington's disease are a direct effect of polyglutamines. *Nat. Neurosci.* **5**: 731.

Pardo, R., Colin, E., Regulier, E., Aebischer, P., Deglon, N., Humbert, S. and Saudou F., 2006, Inhibition of calcineurin by FK506 protects against polyglutamine-huntingtin toxicity through an increase of huntingtin phosphorylation at S421. *J. Neurosci.* **26**: 1635.

Parker, W.D., Jr., Boyson, S.J. and Parks, J.K., 1989, Abnormalities of the electron transport chain in idiopathic Parkinson's disease. *Ann. Neurol.* **26**: 719.

Parker, W.D., Jr., Boyson, S.J., Luder, A.S. and Parks, J.K., 1990, Evidence for a defect in NADH: ubiquinone oxidoreductase (complex I) in Huntington's disease. *Neurology* **40**: 1231.

Perier, C., Tieu, K., Guegan, C., Caspersen, C., Jackson-Lewis, V., Carelli, V., Martinuzzi, A., Hirano, M., Przedborski, S. and Vila, M., 2005, Complex I deficiency primes Bax-dependent neuronal apoptosis through mitochondrial oxidative damage. *Proc. Natl. Acad. Sci. USA.* **102**: 19126.

Petersen, A., Larsen, K.E., Behr, G.G., Romero, N., Przedborski, S., Brundin, P. and Sulzer, D., 2001, Expanded CAG repeats in exon 1 of the Huntington's disease gene stimulate dopamine-mediated striatal neuron autophagy and degeneration. *Hum. Mol. Genet.* **10**: 1243.

Petit, A., Kawarai, T., Paitel, E., Sanjo, N., Maj, M., Scheid, M., Chen, F., Gu, Y., Hasegawa, H., Salehi-Rad, S., Wang, L., Rogaeva, E., Fraser, P., Robinson, B., St. George-Hyslop, P. and Tandon, A., 2005, Wild-type PINK1 prevents basal and induced neuronal apoptosis, a protective effect abrogated by Parkinson disease-related mutations. *J. Biol. Chem.* **280**: 34025.

Petrucelli, L., O'Farrell, C., Lockhart, P.J., Baptista, M., Kehoe, K., Vink, L., Choi, P., Wolozin, B., Farrer, M., Hardy, J. and Cookson, M.R., 2002, Parkin protects against the toxicity associated with mutant alpha-synuclein: proteasome dysfunction selectively affects catecholaminergic neurons. *Neuron* **36**: 1007.

Polidori, M.C., Mecocci, P., Browne, S.E., Senin, U. and Beal, M.F., 1999, Oxidative damage to mitochondrial DNA in Huntington's disease parietal cortex. *Neurosci. Lett.* **272**: 53.

Polymeropoulos, M.H., Lavedan, C., Leroy, E., Ide, S.E., Dehejia, A., Dutra, A., Pike, B., Root, H., Rubenstein, J., Boyer, R., Stenroos, E.S., Chandrasekharappa, S., Athanassiadou, A., Papapetropoulos, T., Johnson, W.G., Lazzarini, A.M., Duvoisin, R.C., Di Iorio, G., Golbe, L.I. and Nussbaum, R.L., 1997, Mutation in the alpha-synuclein gene identified in families with Parkinson's disease. *Science* **276**: 2045.

Portera-Cailliau, C., Hedreen, J.C., Price, D.L. and Koliatsos, V.E., 1995, Evidence for apoptotic cell death in Huntington disease and excitotoxic animal models. *J. Neurosci.* **15**: 3775.

Quigley, P.M., Korotkov, K., Baneyx, F. and Hol, W.G., 2003, The 1.6-A crystal structure of the class of chaperones represented by *Escherichia coli* Hsp31 reveals a putative catalytic triad. *Proc. Natl. Acad. Sci. USA* **100**: 3137.

Rangone, H., Poizat, G., Troncoso, J., Ross, C.A., MacDonald, M.E., Saudou, F. and Humbert, S., 2004, The serum- and glucocorticoid-induced kinase SGK inhibits mutant huntingtin-induced toxicity by phosphorylating serine 421 of huntingtin. *Eur. J. Neurosci.* **19**: 273.

Read, S.H., Baliga, B.C., Ekert, P.G., Vaux, D.L. and Kumar, S., 2002, A novel apaf-1-independent putative caspase-2 activation complex. *J. Cell Biol.* **159**: 739.

Reggiori, F. and Klionsky, D.J., 2002, Autophagy in the eukaryotic cell. *Eukaryot. Cell* **1**: 11.

Rego, A.C. and de Almeida, L.P., 2005, Molecular targets and therapeutic strategies in Huntington's disease. *Curr. Drug Targets CNS Neurol. Disord.* **4**: 361.

Rego, A.C., Vesce, S. and Nicholls, D.G., 2001a, The mechanism of mitochondrial membrane potential retention following release of cytochrome c in apoptotic GT1-7 neural cells. *Cell Death Differ.* **8**: 995.

Rego, A.C., Ward, M.W. and Nicholls, D.G., 2001b, Mitochondria control AMPA/kainate receptor-induced cytoplasmic calcium deregulation in rat cerebellar granule cells. *J. Neurosci.* **21**: 1893.

Rideout, H.J., Lang-Rollin, I. and Stefanis, L., 2004, Involvement of macroautophagy in the dissolution of neuronal inclusions. *Int. J. Biochem. Cell Biol.* **36**: 2551.

Robertson, J.D., Enoksson, M., Suomela, M., Zhivotovsky, B. and Orrenius, S., 2002, Caspase-2 acts upstream of mitochondria to promote cytochrome c release during etoposide-induced apoptosis. *J. Biol. Chem.* **277**: 29803.

Roman, G.C., Zhang, Z.X. and Ellenberg, J.H., 1995, The neuroepidemiology of Parkinson's disease. In: *Etiology of Parkinson's disease*. J.H., Ellenberg, W.C., Koller, J.W., Langston, (eds.) Marcel Dekker, New York, pp. 203.

Sakata, E., Yamaguchi, Y., Kurimoto, E., Kikuchi, J., Yokoyama, S., Yamada, S., Kawahara, H., Yokosawa, H., Hattori, N., Mizuno, Y., Tanaka, K. and Kato, K., 2003, Parkin binds the Rpn10 subunit of 26S proteasomes through its ubiquitin-like domain. *EMBO Rep.* **4**: 301.

Sanchez, I., Xu, C.J., Juo, P., Kakizaka, A., Blenis, J. and Yuan, J., 1999, Caspase-8 is required for cell death induced by expanded polyglutamine repeats. *Neuron* **22**: 623.

Schapira, A.H., 1994, Evidence for mitochondrial dysfunction in Parkinson's disease – a critical appraisal. *Mov. Disord.* **9**: 125.

Schapira, A.H., Cooper, J.M., Dexter, D., Jenner, P., Clark, J.B. and Marsden, C.D., 1989, Mitochondrial complex I deficiency in Parkinson's disease. *Lancet* **1**: 1269.

Schapira, A.H., Cooper, J.M., Dexter, D., Clark, J.B., Jenner, P. and Marsden, C.D., 1990a, Mitochondrial complex I deficiency in Parkinson's disease. *J. Neurochem.* **54**: 823.

Schapira, A.H., Mann, V.M., Cooper, J.M., Dexter, D., Daniel, S.E., Jenner, P., Clark, J.B. and Marsden, C.D., 1990b, Anatomic and disease specificity of NADH CoQ1 reductase (complex I) deficiency in Parkinson's disease. *J. Neurochem.* **55**: 2142.

Schwartz, L.M., Smith, S.W., Jones, M.E. and Osborne, B.A., 1993, Do all programmed cell deaths occur via apoptosis? *Proc. Natl. Acad. Sci. USA* **90**: 980.

Seo, B.B., Nakamaru-Ogiso, E., Cruz, P., Flotte, T.R., Yagi, T. and Matsuno-Yagi, A., 2004, Functional expression of the single subunit NADH dehydrogenase in mitochondria in vivo: a potential therapy for complex I deficiencies. *Hum. Gene Ther.* **15**: 887.

Seo, B.B., Nakamaru-Ogiso, E., Flotte, T.R., Matsuno-Yagi, A. and Yagi, T., 2006, In vivo complementation of complex I by the yeast Ndi1 enzyme. Possible application for treatment of Parkinson disease. *J. Biol. Chem.* **281**: 14250.

Seong, I.S., Ivanova, E., Lee, J.M., Choo, Y.S., Fossale, E., Anderson, M., Gusella, J.F., Laramie, J.M., Myers, R.H., Lesort, M. and MacDonald, M.E., 2005, HD CAG repeat implicates a dominant property of huntingtin in mitochondrial energy metabolism. *Hum. Mol. Genet.* **14**: 2871.

Sharma, S.K., Carlson, E.C. and Ebadi, M., 2003, Neuroprotective actions of Selegiline in inhibiting 1-methyl, 4-phenyl, pyridinium ion (MPP+)-induced apoptosis in SK-N-SH neurons. *J. Neurocytol.* **32**: 329.

Sherer, T.B., Betarbet, R., Stout, A.K., Lund, S., Baptista, M., Panov, A.V., Cookson, M.R. and Greenamyre, J.T., 2002, An in vitro model of Parkinson's disease: linking mitochondrial impairment to altered alpha-synuclein metabolism and oxidative damage. *J. Neurosci.* **22**: 7006.

Shimizu, S. and Tsujimoto, Y., 2000, Proapoptotic BH3-only Bcl-2 family members induce cytochrome c release, but not mitochondrial membrane potential loss, and do not directly modulate voltage-dependent anion channel activity. *Proc. Natl. Acad. Sci. USA* **97**: 577.

Shin, J.Y., Fang, Z.H., Yu, Z.X., Wang, C.E., Li, S.H. and Li, X.J., 2005, Expression of mutant huntingtin in glial cells contributes to neuronal excitotoxicity. *J. Cell Biol.* **171**: 1001.

Shults, C.W., Haas, R.H., Passov, D. and Beal, M.F., 1997, Coenzyme Q10 levels correlate with the activities of complexes I and II/III in mitochondria from parkinsonian and nonparkinsonian subjects. *Ann. Neurol.* **42**: 621.

Sidhu, A., Wersinger, C., Moussa, C.E. and Vernier, P., 2004, The Role of {alpha}-Synuclein in Both Neuroprotection and Neurodegeneration. *Ann. NY Acad. Sci.* **1035**: 250.

Singleton, A.B., Farrer, M., Johnson, J., Singleton, A., Hague, S., Kachergus, J., Hulihan, M., Peuralinna, T., Dutra, A., Nussbaum, R., Lincoln, S., Crawley, A., Hanson, M., Maraganore, D., Adler, C., Cookson, M.R., Muenter, M., Baptista, M., Miller, D., Blancato, J., Hardy, J. and Gwinn-Hardy, K., 2003, Alpha-Synuclein locus triplication causes Parkinson's disease. *Science* **302**: 841.

Spencer, J.P., Jenner, P., Daniel, S.E., Lees, A.J., Marsden, D.C. and Halliwell, B.J., 1998, Conjugates of catecholamines with cysteine and GSH in Parkinson's disease: possible mechanisms of formation involving reactive oxygen species. *J. Neurochem.* **71**: 2112.

Spooren, W.P, Gentsch, C. and Wiessner, C., 1998, TUNEL-positive cells in the substantia nigra of C57BL/6 mice after a single bolus of 1-methyl-4-phenyl-1,2,3,6-tetrahydropyridine. *Neuroscience* **85**: 649.

Springer, J.E., Azbill, R.D., Nottingham, S.A. and Kennedy, S.E., 2000, Calcineurin-mediated Bad dephosphorylation activates the caspase-3 apoptotic cascade in traumatic spinal cord injury. *J. Neurosci.* **20**: 7246.

Stack, E.C., Smith, K.M., Ryu, H., Cormier, K., Chen, M., Hagerty, S.W., Del Signore, S.J., Cudkowicz, M.E., Friedlander, R.M. and Ferrante, R.J., 2006, Combination therapy using minocycline and coenzyme Q10 in R6/2 transgenic Huntington's disease mice. *Biochim. Biophys. Acta* **1762**: 373.

Stefanis, L., Larsen, K.E., Rideout, H.J., Sulzer, D. and Greene, L.A., 2001, Expression of A53T mutant but not wild-type alpha-synuclein in PC12 cells induces alterations of the ubiquitin-dependent degradation system, loss of dopamine release, and autophagic cell death. *J. Neurosci.* **21**: 9549.

Sulzer, D., Bogulavsky, J., Larsen, K.E., Behr, G., Karatekin, E., Kleinman, M.H., Turro, N., Krantz, D., Edwards, R.H., Greene, L.A. and Zecca, L., 2000, Neuromelanin biosynthesis is driven by excess cytosolic catecholamines not accumulated by synaptic vesicles. *Proc. Natl. Acad. Sci. USA* **97**: 11869.

Susin, S.A., Lorenzo, H.K., Zamzami, N., Marzo, I., Snow, B.E., Brothers, G.M., Mangion, J., Jacotot, E., Costantini, P., Loeffler, M., Larochette, N., Goodlett, D.R., Aebersold, R., Siderovski, D.P., Penninger, J.M. and Kroemer, G., 1999, Molecular characterization of mitochondrial apoptosis-inducing factor. *Nature* **397**: 441.

Swerdlow, R.H. and Khan, S.M., 2004, A "mitochondrial cascade hypothesis" for sporadic Alzheimer's disease. *Med. Hypotheses* **63**: 8.

Swerdlow, R.H., Parks, J.K., Miller, S.W., Tuttle, J.B., Trimmer, P.A., Sheehan, J.P., Bennett, J.P., Jr., Davis, R.E. and Parker, W.D., Jr., 1996, Origin and functional consequences of the complex I defect in Parkinson's disease. *Ann. Neurol.* **40**: 663.

Tabrizi, S.J., Orth, M., Wilkinson, J.M., Taanman, J.W., Warner, T.T., Cooper, J.M. and Schapira, A.H., 2000a, Expression of mutant alpha-synuclein causes increased susceptibility to dopamine toxicity. *Hum. Mol. Genet.* **9**: 2683.

Tabrizi, S.J., Workman, J., Hart, P.E., Mangiarini, L., Mahal, A., Bates, G., Cooper, J.M. and Schapira, A.H., 2000b, Mitochondrial dysfunction and free radical damage in the Huntington R6/2 transgenic mouse. *Ann. Neurol.* **47**: 80.

Taira, T., Saito, Y., Niki, T., Iguchi-Ariga, S.M., Takahashi, K. and Ariga, H., 2004, *DJ-1* has a role in antioxidative stress to prevent cell death. *EMBO Rep.* **5**: 213.

Tatton, N.A., 2000, Increased caspase 3 and Bax immunoreactivity accompany nuclear GAPDH translocation and neuronal apoptosis in Parkinson's disease. *Exp. Neurol.* **166**: 29.

Tatton, N.A. and Kish, S.J., 1997, In situ detection of apoptotic nuclei in the substantia nigra compacta of 1-methyl-4-phenyl-1,2,3,6-tetrahydropyridine-treated mice using terminal deoxynucleotidyl transferase labeling and acridine orange staining. *Neuroscience* **77**: 1037.

Tatton, N.A., Maclean-Fraser, A., Tatton, W.G., Perl, D.P. and Olanow, C.W., 1998, A fluorescent double-labeling method to detect and confirm apoptotic nuclei in Parkinson's disease. *Ann. Neurol.* **44**: 142.

Tenneti, L. and Lipton, S.A., 2000, Involvement of activated caspase-3-like proteases in N-methyl-D-aspartate-induced apoptosis in cerebrocortical neurons. *J. Neurochem.* **74**: 134.

Tompkins, M.M., Basgall, E.J., Zamrini, E. and Hill, W.D., 1997, Apoptotic-like changes in Lewy-body-associated disorders and normal aging in substantia nigral neurons. *Am. J. Pathol.* **150**: 119.

Trettel, F., Rigamonti, D., Hilditch-Maguire, P., Wheeler, V.C., Sharp, A.H., Persichetti, F., Cattaneo, E. and Macdonald, M.E., 2000, Dominant phenotypes produced by the HD mutation in STHdh(Q111) striatal cells. *Hum. Mol. Genet.* **9**: 2799.

Trimmer, P.A., Swerdlow, R.H., Parks, J.K., Keeney, P., Bennett, J.P., Jr., Miller, S.W., Davis, R.E. and Parker, W.D., Jr., 2000, Abnormal mitochondrial morphology in sporadic Parkinson's and Alzheimer's disease cybrid cell lines. *Exp. Neurol.* **162**: 37.

Trimmer, P.A., Borland, M.K., Keeney, P.M., Bennett, J.P., Jr. and Parker, W.D., Jr., 2004, Parkinson's disease transgenic mitochondrial cybrids generate Lewy inclusion bodies. *J. Neurochem.* **88**: 800.

Turrens, J.F., Alexandre, A. and Lehninger, A.L., 1985, Ubisemiquinone is the electron donor for superoxide formation by complex III of heart mitochondria. *Arch. Biochem. Biophys.* **237**: 408.

Uchiyama, Y., 2001, Autophagy cell death and its execution by lysosomal cathepsins. *Arch. Histol. Cytol.* **64**: 233.

Uren, R.T., Dewson, G., Bonzon, C., Lithgow, T., Newmeyer, D.D. and Kluck, R.M., 2005, Mitochondrial release of proapoptotic proteins. *J. Biol. Chem.* **280**: 2266.

Valdez, L.B., Zaobornyj, T. and Boveris, A., 2006, Mitochondrial metabolic states and membrane potential modulate mtNOS activity. *Biochim. Biophys. Acta* **1757**: 166.

Valente, E.M., Abou-Sleiman, P.M., Caputo, V., Muqit, M.M., Harvey, K., Gispert, S., Ali, Z., Del Turco, D., Bentivoglio, A.R., Healy, D.G., Albanese, A., Nussbaum, R., Gonzalez-Maldonado, R., Deller, T., Salvi, S., Cortelli, P., Gilks, W.P., Latchman, D.S., Harvey, R.J., Dallapiccola, B., Auburger, G. and Wood, N.W., 2004, Hereditary early-onset Parkinson's disease caused by mutations in PINK1. *Science* **304**: 1158.

Van Raamsdonk, J.M., Pearson, J., Bailey, C.D., Rogers, D.A., Johnson, G.V., Hayden, M.R. and Leavitt, B.R., 2005, Cystamine treatment is neuroprotective in the YAC128 mouse model of Huntington disease. *J. Neurochem.* **95**: 210.

Vanitallie, T.B. and Nufert, T.H., 2003, Ketones: metabolism's ugly duckling. *Nutr. Rev.* **61**: 327.

Vanitallie, T.B., Nonas, C., Di Rocco, A., Boyar, K., Hyams, K. and Heymsfield, S.B., 2005, Treatment of Parkinson disease with diet-induced hyperketonemia: a feasibility study. *Neurology* **64**: 728.

Verhagen, A.M., Ekert, P.G., Pakusch, M., Silke, J., Connolly, L.M., Reid, G.E., Moritz, R.L., Simpson, R.J. and Vaux, D.L., 2000, Identification of DIABLO, a mammalian protein that promotes apoptosis by binding to and antagonizing IAP proteins. *Cell* **102**: 43.

Viswanath, V., Wu, Y., Boonplueang, R., Chen, S., Stevenson, F.F., Yantiri, F., Yang, L., Beal, M. F. and Andersen, J.K., 2001, Caspase-9 activation results in downstream caspase-8 activation and bid cleavage in 1-methyl-4-phenyl-1,2,3,6-tetrahydropyridine-induced Parkinson's disease. *J. Neurosci.* **21**: 9519.

Vonsattel, J.P. and DiFiglia, M., 1998, Huntington disease. *J. Neuropathol. Exp. Neurol.* **57**: 369.

Wallace, D.C., 1994, Mitochondrial DNA mutations in disease of energy metabolism. *J. Bioenerg. Biomem.* **26**: 241.

Wang, X., Zhu, S., Drozda, M., Zhang, W., Stavrovskaya, I.G., Cattaneo, E., Ferrante, R.J., Kristal, B.S. and Friedlander, R.M., 2003, Minocycline inhibits caspase-independent and dependent mitochondrial cell death pathways in models of Huntington's disease. *Proc. Natl. Acad. Sci. USA* **100**: 10483.

Ward, M.W., Rego, A.C., Frenguelli, B.G. and Nicholls, D.G., 2000, Mitochondrial membrane potential and glutamate excitotoxicity in cultured cerebellar granule cells. *J. Neurosci.* **20**: 7208.

Ward, M.W., Rehm, M., Duessmann, H., Kacmar, S., Concannon, C.G. and Prehn, J.H., 2006, Real time single cell analysis of Bid cleavage and Bid translocation during caspase-dependent and neuronal caspase-independent apoptosis. *J. Biol. Chem.* **281**: 5837.

Watabe, M. and Nakaki, T., 2004, Rotenone induces apoptosis via activation of bad in human dopaminergic SH-SY5Y cells. *J. Pharmacol. Exp. Ther.* **311**: 948.

Wellington, C.L., Ellerby, L.M., Gutekunst, C.A., Rogers, D., Warby, S., Graham, R.K., Loubser, O., van Raamsdonk, J., Singaraja, R., Yang, Y.Z., Gafni, J., Bredesen, D., Hersch, S.M., Leavitt, B.R., Roy, S., Nicholson, D.W. and Hayden, M.R., 2002, Caspase cleavage of mutant huntingtin precedes neurodegeneration in Huntington's disease. *J. Neurosci.* **22**: 7862.

Winkler-Stuck, K., Wiedemann, F.R., Wallesch, C.W. and Kunz, W.S., 2004, Effect of coenzyme Q10 on the mitochondrial function of skin fibroblasts from Parkinson patients. *J. Neurol. Sci.* **220**: 41.

Xue, L., Fletcher, G.C. and Tolkovsky, A.M., 1999, Autophagy is activated by apoptotic signaling in sympathetic neurons: an alternative mechanism of death execution. *Mol. Cell. Neurosci.* **14**: 180.

Yamada, M., Iwatsubo, T., Mizuno, Y. and Mochizuki, H., 2004, Overexpression of alpha-synuclein in rat substantia nigra results in loss of dopaminergic neurons, phosphorylation of alpha-synuclein and activation of caspase-9: resemblance to pathogenetic changes in Parkinson's disease. *J. Neurochem.* **91**: 451.

Yang, J., Liu, X., Bhalla, K., Kim, C.N., Ibrado, A.M., Cai, J., Peng, T.-I., Jones, D.P. and Wang, X., 1997, Prevention of apoptosis by bcl-2: release of cytochrome c from mitochondria blocked. *Science* **275**: 1129.

Yang, Y., Gehrke, S., Haque, M.E., Imai, Y., Kosek, J., Yang, L., Beal, M.F., Nishimura, I., Wakamatsu, K., Ito, S., Takahashi, R. and Lu, B., 2005, Inactivation of Drosophila *DJ-1* leads to impairments of oxidative stress response and phosphatidylinositol 3-kinase/Akt signaling. *Proc. Natl. Acad. Sci. USA* **102**: 13670.

Yoshino, H., Nakagawa-Hattori, Y., Kondo, T. and Mizuno, Y., 1992, Mitochondrial complex I and II activities of lymphocytes and platelets in Parkinson's disease. *J. Neural Transm. Park. Dis. Dement. Sect.* **4**: 27.

Young, D.R., Pelligra, R., Shapira, J., Adachi, R.R. and Skrettingland, K., 1967, Glucose oxidation and replacement during prolonged exercise in man. *J. Appl. Physiol.* **23**: 734.

Yu, S.W., Wang, H., Poitras, M.F., Coombs, C., Bowers, W.J., Federoff, H.J., Poirier, G.G., Dawson, T.M. and Dawson, V.L., 2002, Mediation of poly(ADP-ribose) polymerase-1-dependent cell death by apoptosis-inducing factor. *Science* **297**: 259.

Zhang, Y., Ona, V.O., Li, M., Drozda, M., Dubois-Dauphin, M., Przedborski, S., Ferrante, R.J. and Friedlander, R.M., 2003, Sequential activation of individual caspases, and of alterations in Bcl-2 proapoptotic signals in a mouse model of Huntington's disease. *J. Neurochem.* **87**: 1184.

10

ZINC HOMEOSTASIS AND BRAIN INJURY

Stefano Sensi[1], Erica Rockabrand[2] and Israel Sekler[3]

1. ABSTRACT

Cumulating evidence suggest that Zn^{2+} dys/homeostasis can play a major role in promoting brain injury in excitotoxic syndromes. Zn^{2+} homeostasis in the brain is regulated through highly dynamic pathways and is deeply connected with other major signaling pathways, such as NO- and MAP kinase-dependent systems. Zn^{2+} signaling in neurons and glia also interplays with proton and Ca^{2+} homeostasis. Zn^{2+} appears to promote injury with greater potency compared to Ca^{2+} and as such the cation may be an underappreciated mediator of excitotoxicity, which for many years has been described mainly as a Ca^{2+}-dependent phenomenon. One intriguing new area of investigation is offered by the injurious interplay between Ca^{2+} and Zn^{2+} dys/homeostasis. In that respect, the harmful actions mediated by $[Ca^{2+}]_i$ increases in many excitotoxic conditions need to be re-evaluated in light of recent data indicating that Zn^{2+} can be mobilized upon Ca^{2+} influx and Ca^{2+}-dependent oxidative stress. This Ca^{2+}-induced mitochondrial ROS generation promotes intracellular Zn^{2+} release and offers the possibility of a more complex injury paradigm than previously thought. In such a model, glutamate-driven $[Ca^{2+}]_i$ rises might be the initiator but not the prime activator of neuronal and glial injury.

Throughout the most recent studies, it is emerging that while Zn^{2+} may play a crucial role in promoting injury in the old and diseased brain, activation of the same Zn^{2+} signaling pathways during brain development can be critically relevant to favor proper brain growth.

The progress made so far in the field of Zn^{2+} neurobiology is remarkable but a more detailed blueprint of the mechanisms involved in Zn^{2+} homeostasis is warranted. Areas that need further investigation include the cross talk between Zn^{2+} pools, the interplay with Ca^{2+}, modulation by Zn^{2+} of the processes controlling neuronal–glial communication, and the molecular and mechanistic aspects of Zn^{2+} transport in the brain. Such knowledge will be crucial to improve our ability to "finely tune" brain Zn^{2+} homeostasis in both development and disease.

Stefano Sensi – ssensi@uci.edu
[1] Molecular Neurology Unit, Center of Excellence on Aging, University"G.d'Annunzio" Chieti, Italy
[2] Dept. of Neurology, University of California-Irvine, Irvine, California, US
[3] Ben Gurion University, Faculty of Health Sciences, Dept. of Physiology and The Slotowski Center of Neuroscience, Ber-Sheva, Israel

2. INTRODUCTION

Cumulating evidence gathered in the past 30 years strongly indicate that Zn^{2+} plays a critical role in many diverse physiological processes. Zn^{2+} can be an important modulator of a plethora of channel and receptor activities (Smart et al., 2004), yet when accumulated in excess amounts, the cation is a potent trigger of neuronal injury. In the brain, Zn^{2+} co-localizes with glutamate at excitatory synapses and acts as a traditional neurotransmitter by being released in a Ca^{2+}-dependent fashion. Synaptically released Zn^{2+} eventually fluxes inside of neurons and glial cells and, under pathological conditions, intracellular accumulation of the cation potently contributes to the neuronal and glial injury observed in cerebral ischemia, epilepsy, and head trauma (Frederickson et al., 2005). Besides acute neuronal and glial injury, Zn^{2+} dysmetabolism also plays a role in Alzheimer's disease as the cation is a critical promoter of β-amyloid aggregation and plaque formation. This chapter will examine toxic pathways involved mainly in neuronal injury. Mechanisms by which Zn^{2+} promotes toxicity in glial cells are still largely unexplored, and we will briefly discuss a few promising areas of research that support a role for Zn^{2+} in the modulation of both physiological and pathological processes in these cells.

To gain entry into neurons, Zn^{2+} can permeate NMDA receptor-mediated channels (NMDAR), voltage-sensitive calcium channels (VSCC), Ca^{2+}-permeable AMPA/kainate (Ca-A/K) receptor-mediated channels, or moved in and out by Zn^{2+}-sensitive membrane transporters. Recent studies suggest that Zn^{2+} can also operate as the "enemy within." The cation is in fact mobilized from intracellular sites of sequestration such as metallothioneins and mitochondria. To trigger neuronal damage, Zn^{2+} activates death signaling pathways that encompass mitochondrial and extra-mitochondrial generation of reactive oxygen species (ROS), disruption of metabolic enzyme activity, and a complex interference in the homeostasis of the ionic milieu, which ultimately leads to the activation of both apoptotic and/or necrotic processes (Sensi and Jeng, 2004) (Fig. 1).

In this chapter, we review the most recent mechanisms of Zn^{2+} dys/homeostasis and the role played by Zn^{2+} in promoting neuronal injury.

3. Zn^{2+} HOMEOSTASIS

Zn^{2+} can regulate both normal and pathological cellular functions, depending on intracellular concentration, and therefore it is important to examine the many systems that control intracellular Zn^{2+} ($[Zn^{2+}]_i$) homeostasis. As with intracellular Ca^{2+} ($[Ca^{2+}]_i$) homeostasis, physiological regulation of $[Zn^{2+}]_i$ levels is the result of a fine balance between ion sequestration, intracellular buffering, and extrusion. Zn^{2+} sequestration and buffering is largely operated by a family of proteins called metallothioneins (MTs), while membrane-associated Zn^{2+} transporters are responsible for Zn^{2+} extrusion (Kägi et al., 1993; Hidalgo et al., 2001; Eide, 2004). Mitochondria also play a critical role in the sequestration and buffering of Zn^{2+}.

3.1. Sequestration and Buffering of Zn^{2+} by Metallothioneins

MTs are low molecular weight, cysteine-rich proteins present in three isoforms (MT-1, -2,-3) in the central nervous system (CNS) (Kägi et al., 1993). In the CNS, MTs show distinct patterns of expression: MT-1 and MT-2 are largely found in astrocytes and spinal glia but largely absent in neurons, while MT-3 is abundant in neurons but poorly expressed in glial cells. MT-3 seems to be particularly relevant to neuronal Zn^{2+} homeostasis in critical brain regions such as the hippocampus because it is abundantly present in the same hippocampal glutamatergic terminals that are also strongly enriched in vesicular Zn^{2+} (Frederickson and Moncrieff, 1994).

The structure of the three MT isoforms is similar and consists of a single polypeptide chain, 61–68 amino acids in size, with a highly conserved sequence of 20 cysteine (cys) residues (Kägi et al., 1993). These cys residues are grouped into two domains for Zn^{2+} binding, resulting in a dumbbell-shaped physical conformation (Arseniev et al., 1988; Robbins et al., 1991). MTs can accommodate a total of seven Zn^{2+} ions by binding the cation to the two cys cluster regions with a very high affinity ($K_d = 2 \times 10^{12}$ M^{-1} at pH 7.0) (Kägi et al., 1993). These two Zn^{2+}/cys cluster regions exert an essential role in the regulation of Zn^{2+} binding, which can be readily modulated by shifts in acid–base equilibrium or, more significantly, by changes in the redox state of the two clusters (Kägi et al., 1993; Jiang et al., 2000).

Oxidative stress for example, has been found to be a key modulator of $[Zn^{2+}]_i$ homeostasis by interfering with Zn^{2+} binding to MTs. Cellular oxidants promote Zn^{2+} release from MTs, while agents that promote a reduction in the intracellular environment facilitate Zn^{2+} binding (Jiang et al., 1998, 2000; Maret and Vallee, 1998). Furthermore, changes in the glutathione redox state [i.e. the ratio between glutathione (GSH) and glutathione disulfide (GSSG)] can also play a major role in regulating Zn^{2+}-MT binding; GSH binds directly to MTs and is thought to "activate" the proteins in order to facilitate GSSG-mediated Zn^{2+} release (Maret, 1994).

Nitrosative stress is another key regulator of Zn^{2+} homeostasis: nitric oxide (NO), a cellular signaling molecule, has also been shown to interact preferentially with MT-3 and to promote Zn^{2+} release from MTs both in vitro and in vivo (Kroncke et al., 1994; Bossy-Wetzel et al., 2004).

3.2. Membrane-Associated Zn^{2+} Transport Proteins

The proteins directly involved in the active movement of Zn^{2+} across cellular membranes generally belong to two families of transporters: the CDF (cation diffusion facilitator) and the ZIP ("Zn^{2+}-regulated metal transporter, Iron-regulated metal transporter-like Protein") families. Those most directly associated with Zn^{2+} transport in humans are the ZnT (Zn^{2+} Transporter) proteins, members of the CDF family, which favor clearance of cytosolic Zn^{2+} either by cation extrusion from the cells, attenuation of cation influx, or promotion of Zn^{2+} sequestration into intracellular compartments (Eide, 2004).

The CDF family consists of nine human CDF genes (also known as SLC30 genes), with ten ZnTs identified (ZnT-1-10) (Palmiter and Huang, 2004). Of these, ZnT-1 and ZnT-3 appear to be the most relevant to Zn^{2+} homeostasis in the brain. The fast changes in extracellular Zn^{2+} and the numerous pathways for permeation of this ion suggest an important role for the ZnT proteins in brain physiology and pathophysiology; however, remarkably little is known about their mechanism of activity or their regulation. Studies on yeast CDF proteins or the bacterial ZitB protein, which is a remotely related homologue of the ZnT proteins, have suggested mechanisms of how these proteins control Zn^{2+} transport (MacDiarmid et al., 2002; Chao and Fu, 2004). Functional analyses have indicated that these proteins might induce an

H^+/Zn^{2+} exchange. Thus, a cytoplasmic drop in pH may lead to the reversal of the activity of the transporter that could promote a rise in cytosolic $[Zn^{2+}]_i$. If a similar mechanism underlies the activity of mammalian ZnT transporters, this mode of action might have important pathophysiological implications when acidosis is produced by metabolic impairment or inflammation like in the context of ischemia, multiple sclerosis or Alzheimer's disease.

ZnT-1, a ubiquitously expressed member of the zinc transporter family SLC30, is found on the plasma membrane of neurons (Palmiter and Huang, 2004). In the mouse brain, ZnT-1 is localized in regions rich in synaptic Zn^{2+} and its expression is developmentally regulated in correlation with the appearance of synaptic Zn^{2+} (Nitzan et al., 2002; Sekler et al., 2002). ZnT-1 has been shown to reduce Zn^{2+} toxicity in neurons and glial cells (Palmiter and Findley, 1995; Nolte et al., 2004; Palmiter and Huang, 2004). Expression of ZnT-1 is highly regulated by Zn^{2+}, via the transcription factor MTF-1 (Langmade et al., 2000), and priming of glial cells with non-toxic Zn^{2+} exposures has been found to promote ZnT-1 expression (Nolte et al., 2004).

The fact that induction of ZnT-1 expression is triggered by synaptically released Zn^{2+} suggests that ZnT-1 might play an important role in the poorly understood experimental phenomena termed "ischemic preconditioning," where "priming" an animal with a sublethal ischemic episode is found to promote neuroprotection against a subsequent and more severe stroke event (Ying et al., 1999; Moncayo et al., 2000). It is conceivable that the neuroprotection observed in ischemic preconditioning can also be linked to Zn^{2+}. Sublethal ischemic events might in fact trigger a synaptic Zn^{2+} release insufficient to produce injury but sufficient to induce a ZnT-1 (and perhaps MT) expression that can counteract otherwise toxic $[Zn^{2+}]_i$ rises in neurons and glial cells. However, it must be emphasized that mechanisms by which ZnT-1 operates are more complex than previously thought.

Previous studies on cell lines have suggested that ZnT-1 may decrease $[Zn^{2+}]_i$ by acting as a Zn^{2+} extruder (Palmiter and Findley, 1995; Palmiter and Huang, 2004). However, Zn^{2+} extrusion operated by ZnT-1 appears to be independent from changes in the concentration of counter ions such as K^+, Na^+, Ca^{2+}, or protons, and it is not affected by depletion of intracellular ATP, a mechanism that poorly fits with the described modus operandi of traditional transporters.

Recent data support the idea that ZnT-1 can also affect Zn^{2+} homeostasis via indirect flux of the cation through L-type VSCC. This model is supported by the fact that expression of ZnT-1 in cells endogenously expressing VSCC, as well as heterologous co-expression of ZnT-1 and VSCC, can result in reduction of $[Zn^{2+}]_i$ (Nolte et al., 2004; Segal et al., 2004).

ZnT-3 is strongly expressed in brain regions that are rich in histochemically reactive Zn^{2+}, such as the entorhinal cortex, the amygdala, and the hippocampus (Palmiter et al., 1996). ZnT-3 is particularly present in the mossy fiber tract and localizes to the membranes of Zn^{2+}-containing vesicles in the mossy fiber synaptic boutons. Further substantiating the idea that this transporter is essential for uploading Zn^{2+} into synaptic vesicles, ZnT-3 knockout mice display a conspicuous lack of Zn^{2+} in their hippocampi, and ultrastructural examination demonstrates the absence of Zn^{2+} in their mossy fiber boutons. The remaining ZnT proteins have variable levels of expression in the brain and much less is known about their function in general.

Another important influx/efflux pathway for Zn^{2+} is through the activation of the Na^+/Zn^{2+} exchanger. Early studies have suggested that the neuronal Na^+/Ca^{2+} exchanger may mediate Zn^{2+} extrusion, yet more recent findings seem to support the existence of a distinct Na^+/Zn^{2+} exchanger (Sensi et al., 1997). These studies have indicated a putative Na^+/Zn^{2+} exchanger operating with a stoichiometry of 3 Na^+ to 1 Zn^{2+}, likely a member of the Na^+/Ca^{2+}

exchanger superfamily, that can promote Zn^{2+} efflux against a 500-fold transmembrane gradient (Ohana et al., 2004). The catalytic similarity between the Na^+/Ca^{2+} exchanger and the Na^+/Zn^{2+} exchanger, both employing the Na^+ gradient and sharing a similar stoichiometry, may suggest that these proteins are molecularly related. However, a distinction between the putative Na^+/Zn^{2+} exchanger and the Na^+/Ca^{2+} exchanger comes from pharmacological studies where the Na^+/Zn^{2+} exchanger shows a unique sensitivity to the well-known inhibitor, KB-R7943. In the context of brain injury, the Na^+/Zn^{2+} exchanger can effectively attenuate excitotoxic $[Zn^{2+}]_i$ accumulation associated with ischemia, epilepsy, and brain trauma.

The activity of the Na^+/Zn^{2+} exchanger might however have a dark side. The steep inward Zn^{2+} gradient in neurons implies that even a modest rise in intracellular Na^+ concentration may reverse the Na^+/Zn^{2+} exchanger, thereby triggering Zn^{2+} influx. A similar scenario has been described for the Na^+/Ca^{2+} exchanger that can mediate injurious Ca^{2+} influx as a consequence of $[Na^+]_i$ rises induced by both cardiac and brain ischemia (Kiedrowski et al., 2004).

3.3. Mitochondria: Key Regulators of Zn^{2+} Homeostasis

As described by many landmark studies in the past thirty years or so, mitochondria play a crucial role in protecting cells from injurious cytosolic Ca^{2+} loads (Nicholls and Budd, 2000). However, the latest studies suggest that these organelles are also critical for cytosolic Zn^{2+} buffering. Mitochondrial Zn^{2+} uptake has been directly visualized in isolated organelles and intact neurons using Zn^{2+}-sensitive mitochondrial fluorophores (Sensi et al., 2000; 2003a). The uptake seems to be largely mediated by the Ca^{2+} uniporter (Saris and Niva, 1994; Jiang et al., 2001); however, more recent data suggest that additional pathways are present (Malaiyandi et al., 2005). Mitochondria have a high Zn^{2+} uptake capacity and therefore they play a key role in the clearance of cytosolic Zn^{2+} loads. In fact blockade of mitochondrial Zn^{2+} sequestration leads to the elevation or prolongation of experimentally induced cytosolic $[Zn^{2+}]_i$ rises (Sensi et al., 2000).

Interestingly, new studies have also indicated that the organelles may also serve as an intracellular store for $[Zn^{2+}]_i$ mobilization. Mitochondria are able to sequester cytosolic $[Zn^{2+}]_i$ and under resting conditions, the sequestered Zn^{2+} may be rereleased into the cytoplasm in a Ca^{2+}-dependent fashion (Sensi et al., 2003b). The physiological purpose of this mitochondrial Zn^{2+} is currently unexplored.

4. TRANS-SYNAPTIC MOVEMENT OF Zn^{2+} IN NEURONAL INJURY

At the excitatory synapses, Zn^{2+} acts as a classical neurotransmitter and is released from terminals upon sustained synaptic activity in a Ca^{2+}-dependent fashion (Assaf and Chung, 1984; Howell et al., 1984; Aniksztejn et al., 1987). The first evidence revealing that Zn^{2+} can act as a trigger for neuronal injury came from an in vivo study showing how epileptic activation of the hippocampus can promote Zn^{2+} loss from the presynaptic mossy fiber terminals together with concurrent $[Zn^{2+}]_i$ rises and neuronal death of hilar interneurons and CA3 pyramidal cells (Frederickson et al., 1988, 1989).

Recently, we have gathered a great deal of data on the ability of Zn^{2+} to trigger neuronal loss (Frederickson et al., 2005). In vitro studies show that exposure to micromolar concentrations of Zn^{2+} promotes both neuronal and glial injury (Choi et al., 1988; Frederickson and Bush, 2001), and in vivo experiments indicate that direct injection of Zn^{2+} into the brain induces strong neurotoxicity (Itoh and Ebadi, 1982).

Injurious cytosolic $[Zn^{2+}]_i$ rises are also observed in ischemia, epilepsy, and brain trauma (Frederickson et al., 1989; Tonder et al., 1990; Koh et al., 1996; Suh et al., 2000a). These $[Zn^{2+}]_i$ elevations precede neuronal degeneration and application of an extracellular Zn^{2+} chelator has been found to be neuroprotective in both transient global and focal ischemia (Koh et al., 1996; Lee et al., 2002a), in epilepsy (Lee et al., 2000; Yi et al., 2003), and brain trauma (Suh et al., 2000a).

Current studies have also elucidated how $[Zn^{2+}]_i$ accumulation and subsequent neuronal injury can be set in motion independently of trans-synaptic movement of the cation. As we will discuss below, a series of very intriguing reports have indicated that neurons possess different pools of releasable intracellular Zn^{2+} that can significantly contribute to "postsynaptic" $[Zn^{2+}]_i$ rises during injury.

4.1. Trans-synaptic Zn^{2+} Movement: Routes of Zn^{2+} Entry

As mentioned above, $[Zn^{2+}]_i$ accumulation is strongly linked to excitotoxicity. Trans-synaptic Zn^{2+} movements are observed upon ischemia and epilepsy, where Zn^{2+} is co-released with glutamate at excitatory synapses and the cation gains entry into neurons by permeating glutamatergic postsynaptic receptors. Among the routes used by Zn^{2+} to flux into neurons, microfluorimetric and electrophysiological studies have revealed that Zn^{2+} may enter neurons through NMDAR and VSCC, both of which are ubiquitously expressed on neurons throughout the brain. However, Zn^{2+} may also flux through Ca-A/K channels, an atypical subtype of AMPAR distinctive for its high Ca^{2+} permeability as well as its selectively increased expression in a subset of forebrain and spinal cord neurons (Weiss et al., 1993; Yin and Weiss, 1995; Sensi et al., 1997; Cheng and Reynolds, 1998; Canzoniero et al., 1999; Sensi et al., 1999a; Jia et al., 2002). Ca-A/K channels lack the GluR2 subunit, whose presence in the typical heterotetrameric AMPAR assembly blocks Ca^{2+} entry. In addition, as discussed above (Sect. 3.2), Zn^{2+} can serve as a substrate for the Na^+/Ca^{2+} exchanger in place of Ca^{2+} or it can use a putative Na^+/Zn^{2+} exchanger to gain entry into neurons (Sensi et al., 1997; Ohana et al., 2004).

It must be pointed out that Zn^{2+} entry routes are not all equally permeable to the cation. On the contrary to Ca^{2+}, which fluxes to the same degree through either NMDAR-associated channels or Ca-A/K channels, in vitro studies indicate that Ca-A/K channels have the greatest permeability to Zn^{2+}, with VSCC and NMDAR having intermediate and minimal permeability, respectively (Yin and Weiss, 1995; Sensi et al., 1999a). Studies in cortical neurons indicate that in fact simultaneous activation of all three entry routes (NMDAR, AMPAR and VSCC) using the endogenous agonist glutamate results in preferential $[Zn^{2+}]_i$ rises in strongly Ca-A/K channel-expressing [Ca-A/K(+)] neurons (Sensi et al., 1999b). Further supporting the idea that Ca-A/K channels provide a key route for injurious $[Zn^{2+}]_i$ accumulation, selective pharmacological inhibition of Ca-A/K channels has been found to be highly neuroprotective against CA1 pyramidal neuron loss in both in vitro and in vivo models of global ischemia, while NMDAR and VSCC blockade were each found to be only marginally beneficial (Yin et al., 2002; Noh et al., 2005).

The greater permeability of Ca-A/K channels to Zn^{2+} may be particularly important in transient global ischemia (TGI), as these channels are concentrated on postsynaptic membranes where the highest levels of synaptically released Zn^{2+} are likely to be achieved, and selectively expressed in subpopulations of neurons such as TGI vulnerable CA1 pyramidal neurons. Although CA1 pyramidal neurons lack Ca-A/K channels at their soma, they appear to express some of these receptors in their dendritic tree, where they would likely play an important role in neurotransmission and injury induction (Lerma et al., 1994; Yin et al., 1999).

Interestingly, not only are these channels present in the dendrites of the most vulnerable neurons, but they may also be subject to dynamic, injury-driven upregulation in the context of TGI. In CA1 pyramidal neurons, several studies have demonstrated that after TGI the number of functional Ca-A/K channels increases through the induction of a selective decrease in the expression of the GluR2 subunit (Gorter et al., 1997; Opitz et al., 2000). On the basis of these observations, Zukin and Bennett have formulated the "GluR2 hypothesis," which postulates that some forms of neuronal insult selectively trigger an increase in the number of Ca-A/K channels present on the plasma membrane of certain neurons, which they propose likely underlies the selective vulnerability of CA1 pyramidal neurons to injury in these conditions (Bennett et al., 1996).

4.2. Sites of Intracellular Zn^{2+} Release

The translocation model described earlier has been recently challenged by new unexpected observations in ZnT-3 knockout mice (ZnT-3 KO). These mice lack histochemically reactive Zn^{2+} in their presynaptic terminals, but nevertheless their hippocampal neurons undergo $[Zn^{2+}]_i$ accumulation and injury following an excitotoxic insult (Lee et al., 2000). The likely explanation for the injurious $[Zn^{2+}]_i$ accumulation seen in ZnT-3 KO mice is that some or all of this Zn^{2+} originates from one or more sites in the postsynaptic neuron itself. Recent studies indicate that intracellular sources of the cation encompass MTs and mitochondria.

In the case of MTs, the latest findings strongly suggest that these Zn^{2+}-binding proteins modulate excitotoxic injury, although their exact role in the process is not entirely clear. For instance, several factors suggest that in some cases, MTs serve a protective purpose in neurons, particularly in focal ischemia. MT-1 and MT-2 mRNA expression is rapidly upregulated following transient focal ischemic insult, and increased expression of MT-1 has been shown to be neuroprotective in focal ischemia (van Lookeren Campagne et al., 1999; Trendelenburg et al., 2002). In addition, MT-1 and MT-2 KO mice were observed to develop infarcts in focal ischemia that were three times larger than in wild-type mice, and a separate study found MT-3 KO mice to be more sensitive to excitotoxic injury as well (Erickson et al., 1997; van Lookeren Campagne et al., 1999; Trendelenburg et al., 2002). Although MTs possess intrinsic antioxidant properties, which might also contribute to the overall beneficial role exerted by these proteins (Ebadi et al., 1996), together these studies seem to support the idea that MTs might act as a "passive" cellular defense mechanism against toxic $[Zn^{2+}]_i$ elevations.

However, emerging evidence suggests that in other circumstances, MTs may serve mainly as a source of injurious Zn^{2+} release. In principle, the ability of MTs to release Zn^{2+} upon changes in the cellular redox state renders these proteins a reservoir of readily available Zn^{2+} under conditions of oxidative stress, which occurs in ischemia. In agreement with this, the additional knockout of MT-3 in ZnT-3 KO mice results in substantial protection from the excitotoxic injury otherwise observed in CA1 neurons (Cole et al., 2000). Furthermore, recent preliminary data indicate that NO-triggered $[Zn^{2+}]_i$ rises and subsequent neuronal loss in the CA1 region are significantly reduced in MT-3 KO mice compared to wild-type animals (Lee et al., 2003). Thus, MTs may ultimately not only act as Zn^{2+} buffers but also as sources for potentially deleterious Zn^{2+} release.

Another site of intracellular Zn^{2+} release is offered by mitochondria (Sensi et al., 2002, 2003b) and the implications of this mitochondrial Zn^{2+} release will be discussed in further detail below. Overall, recent findings suggest that the mechanisms underlying injurious Zn^{2+} accumulation may be more multifaceted than previously thought and encompass presynaptic (vesicular/nonvesicular) as well as postsynaptic sources.

5. INTRACELLULAR MECHANISMS OF Zn^{2+} TOXICITY

Zn^{2+} can promote neuronal demise by activating multiple, intersecting death pathways. Mirroring the cation's multidirectional effects on cellular physiology, Zn^{2+}-dependent injurious pathways are complex. Zn^{2+} has been found to modulate both necrosis and apoptosis. Necrotic and apoptotic processes, once considered mutually exclusive, are actually co-existing, and there is a general consensus that the intensity of the insult and its effect on the status of cellular energy levels determines which process is more prominent in a given cell. Mitochondria are well positioned to be a critical molecular switch between necrosis and apoptosis. For instance, a subacute ischemic insult may trigger levels of mitochondrial dysfunction that allow the organelles to generate sufficient ATP levels to activate an apoptotic demise. In contrast, a fulminant insult results in the abrupt compromise of mitochondria and intracellular energy levels, forcing cells to abandon the apoptotic program in favor of a necrotic exit (Nicotera et al., 1999). As Zn^{2+} can potently disrupt mitochondrial function, it is not surprising that the cation has been implicated in the induction of both necrotic and apoptotic processes (Lobner et al., 2000).

5.1. Zn^{2+} and Necrotic Injury

5.1.1. Zn^{2+}-Induced Generation of Mitochondrial ROS

Mitochondria are important targets for the toxic effects of elevated intracellular Zn^{2+} as well as Ca^{2+} (Nicholls and Budd, 2000; Sensi et al., 1999a, b, 2000, 2003b; Jiang et al., 2001; Dineley et al., 2003). These organelles are the major cellular sources of ROS, which are routinely produced and rapidly utilized as recyclable co-factors in the electron transport chain. Under physiological conditions, ROS are kept within the mitochondrial membranes, where they are unable to adversely affect cellular function. However, under pathological conditions Zn^{2+} enters the cytosol and is sequestrated into mitochondria where it induces a dramatic decrease in the mitochondrial membrane potential ($\Delta\Psi_m$) and an increase in ROS generation (Sensi et al., 1999b, 2000). ROS of mitochondrial origin are released into the cytoplasm where they disrupt plasma membrane lipids, ultimately promoting neuronal injury.

The potent disruptive actions played by Zn^{2+} are largely the result of its inhibition of cellular respiration that leads to mitochondrial ROS production (Skulachev et al., 1967; Nicholls and Malviya, 1968; Kleiner and von Jagow, 1972; Kleiner, 1974). The cation acts upon several sites within the electron transport chain, including cytochrome bc_1 in complex III (Link and von Jagow, 1995) and KGDHG (α-ketoglutarate dehydrogenase) in complex I (Brown et al., 2000). Zn^{2+} also affects multiple activities of LADH (lipoamide dehydrogenase), which is an enzyme that is a component of the KDGHC complex in mitochondria, and also catalyzes NADH oxidation and produces ROS as a by-product. Zn^{2+} strongly inhibits the LADH reaction in mitochondria in both directions: on one hand interfering with respiration, and on the other decreasing the formation of dihydrolipoic acid, the reduced form of lipoic acid, which is a potent ROS scavenger and antioxidant regenerator (Packer et al., 1997; Gazaryan et al., 2002). With respect to NADH oxidation, Zn^{2+} accelerates LADH catalysis of the oxidative reaction fivefold (Gazaryan et al., 2002), resulting in an overall increase in cellular oxidative burden by both direct and indirect means.

5.1.2. Zn^{2+}-Induced Generation of Cytosolic ROS

Milder cytosolic $[Zn^{2+}]_i$ rises may also induce Zn^{2+}-dependent oxidative stress independent of mitochondria, which would also favor necrotic cell death. Zn^{2+} is known to regulate a number of cytosolic enzymes that generate ROS secondary to their main physiological activity. Similar to its effect on LADH in NADH oxidation described earlier, Zn^{2+} appears to activate the enzyme that catalyzes the oxidation of the reduced form of nicotinamide adenine dinucleotide phosphate, (NADPH oxidase; a multisubunit enzyme widely expressed in central neurons) via activation of protein kinase C (Noh et al., 1999; Kim and Koh, 2002), subsequently producing ROS. Another possible inducer of Zn^{2+}-mediated free ROS production in the cytosol is neuronal nitric oxide synthase (nNOS) that leads to nitric oxide (NO) production and peroxynitrite ($^-$ONOO) formation when NO is combined with superoxide. Although a biochemical assay of nNOS activity in vitro demonstrates inhibition by high concentrations of Zn^{2+}, cortical neurons exposed to Zn^{2+} show nNOS activation and increased levels of nitric oxide and nitrites (Persechini et al., 1995). The effect of Zn^{2+} on this enzyme in vivo is of particular interest because the number of neurons expressing nNOS may increase following focal cerebral ischemia (Holtz et al., 2001).

In addition to the lipid membrane degradation described earlier, intracellular oxidative stress can also result in DNA strand breakage. Such DNA damage triggers a cascade of events including the activation of poly(ADP ribose) polymerase (PARP), consumption of NAD^+ during the formation of PAR polymers, and eventually cellular death due to ATP depletion. PARP is an enzyme that regulates its own activity in conjunction with poly(ADP ribose) glycohydrolase (PARG); the PARP/PARG cycle appears to be required for the persistent PARP-dependent activity that leads to NAD^+ and ATP depletion and ultimately cell death (Burkle, 2005). Pharmacological inhibition of both enzymes in cortical neurons results in significant neuroprotection against Zn^{2+} neurotoxicity (Kim et al., 1999a, b).

A final consideration is that oxidative stress may also induce additional intracellular Zn^{2+} release. Recent studies show an intriguing injurious pathway triggered by $^-$ONOO where this free radical species promotes intracellular Zn^{2+} release, leading to 12-lypoxygenase (12-LOX) activation, ROS accumulation, p38 activation, mitochondrial depolarization, caspase-3 activation, and neuronal loss (Zhang et al., 2004). Similar pathways have also been elucidated in myelin producing oligodendrocytes where mobilization of $[Zn^{2+}]_i$ by $^-$ONOO appears to promote the phosphorylation of extracellular signal regulated kinase 42/44 (ERK42/44), activation of 12-LOX, increased ROS production, and oligodendrocyte injury (Zhang et al., 2006). Since cellular oxidation promotes $[Zn^{2+}]_i$ release, and Zn^{2+} can trigger ROS generation, it seems likely that a dangerous feed-forward cycle leading to cellular injury develops as a consequence.

5.1.3. Zn^{2+}-Induced Disruption of Cellular Metabolism

Another injurious pathway initiated by $[Zn^{2+}]_i$ rises appears to be linked to direct modulation of key enzymes involved in neuronal glycolysis. The cation has been shown to inhibit glyceraldehyde-3-phosphate dehydrogenase (GAPDH) (Krotkiewska and Banas, 1992), phosphofructokinase (Ikeda et al., 1980), and NAD^+ glycohydrolase (Kukimoto et al., 1996), in biochemical assays in vitro. In intact cortical neurons, submicromolar $[Zn^{2+}]_i$ rises are sufficient to trigger a powerful inhibition of GAPDH, which leads to ATP depletion and neuronal death. This inhibition by Zn^{2+} involves a reduction in the levels of cytosolic NAD^+ mediated by an unknown mechanism, since restoring NAD^+ by the addition of pyruvate results in strong neuroprotection specifically against Zn^{2+}-dependent toxicity (Sheline et al., 2000). Interestingly, pyruvate has also been shown to dramatically decrease both ischemic $[Zn^{2+}]_i$ rises and injury in an animal model of TGI (Lee et al., 2001).

5.2. Zn^{2+} and the Apoptotic Programme

5.2.1. Zn^{2+}-Induced Release of Proapoptotic Mitochondrial Factors

Robust Ca-A/K channel-mediated $[Zn^{2+}]_i$ rises can promote acute necrotic processes (as seen in selective degeneration of CA1 pyramidal neurons in TGI) by inducing an abrupt decline in intracellular energy levels, mitochondrial dysfunction, and potent oxidative stress. By contrast, milder cytosolic Zn^{2+} loads, resulting from cation entry through less permeable but more ubiquitously expressed routes such as VSCC or ROS-mediated intracellular Zn^{2+} release, may trigger less intense disruption of mitochondrial function, allowing neurons to activate the apoptotic machinery and release proapoptotic mitochondrial factors.

High micromolar levels of $[Zn^{2+}]_i$ inhibit the electron transport chain and decrease cellular respiration, whereas lower levels (submicromolar) of Zn^{2+} have been shown to lead to mitochondrial swelling and increased respiration in isolated mitochondria, effects consistent with the induction of mPTP (mitochondrial permeability transition pore) (Wudarczyk et al., 1999; Jiang et al., 2001; Sensi et al., 2003b; Malaiyandi et al., 2005). This milder Zn^{2+} burden has also been observed to promote the release of proapoptotic mitochondrial proteins such as cytochrome C and apoptosis inducing factor (AIF), both associated with mPTP opening (Jiang et al., 2001). Substantiating the role played by Zn^{2+} in the activation of apoptosis, experiments in intact cortical neurons show that similar submicromolar $[Zn^{2+}]_i$ rises mediated by VSCC are sufficient to induce significant mitochondrial swelling and release of cytochrome C and AIF, while inhibition of mPTP opening attenuates both Zn^{2+}-triggered release of these factors as well as subsequent neuronal loss (Jiang et al., 2001).

5.2.2. Zn^{2+} Modulation of Apoptotic Signaling

Zn^{2+} may also mediate neuronal apoptosis by promoting dyshomeostasis of other ions. For example, mobilization of $[Zn^{2+}]_i$ by oxidative stress can lead to the depletion of intracellular K^+ content, a well-known activator of neuronal apoptosis (Yu et al., 1997). Studies in cultured neurons have shown that oxidizing agents such as 2,2'-dithiodipyridine (DTDP) or nitric oxide (via production of peroxynitrite) promote Zn^{2+} release from MTs, initiating an injury cascade that leads to p38 MAP kinase activation, caspase-independent K^+ efflux, neuronal shrinking, and apoptosis (Aizenman et al., 2000; McLaughlin et al., 2001; Bossy-Wetzel et al., 2004). Zn^{2+} can also influence $[Ca^{2+}]_i$ homeostasis (see Sect. 5.3).

Among the apoptotic signaling molecules modulated by Zn^{2+}, it should be mentioned that the cation can modulate p75 (NTR)/NADE and Egr-1, both of which have been implicated in animal models of cerebral ischemia. In cortical neurons, Zn^{2+} activates the low-affinity neurotrophin receptor p75 (NTR) and its associated death executor protein NADE, a process that leads to caspase-dependent neuronal loss. Upon TGI, both p75 (NTR) and NADE are induced in degenerating CA1 neurons, an effect entirely suppressed by extracellular Zn^{2+} chelation, further supporting a positive correlation between Zn^{2+} accumulation and their induction (Park et al., 2000). Finally Zn^{2+} (but not Ca^{2+}) has been reported to trigger sustained Erk 1/2 activation, a step that is critical in the induction of Egr-1, which is an immediate-early gene transcription factor that is induced after cerebral ischemia (Beckmann and Wilce, 1997). Supporting a Zn^{2+}-dependent injurious cascade that involves Erk1/2 and Egr-1, pharmacological inhibition of Erk 1/2 blocks both Egr-1 activity and Zn^{2+}-dependent neurotoxicity (Park and Koh, 1999).

5.3. Interaction Between Pathways of Injury

As we previously discussed, Zn^{2+} can promote neuronal injury either through interference with mitochondrial functioning or by acting on extramitochondrial death signaling cascades. It is quite possible, however, that these various pathways interact in a synergistic manner. For instance, cytosolic activation of PARP has been shown to lead to decreased NAD^+ levels, an event that prompts the release of AIF from mitochondria. Cytosolic AIF eventually promotes the collapse of $\Delta\Psi_m$, inducing the release of cytochrome C and initiating the apoptotic cascade (Yu et al., 2002; Du et al., 2003).

Another example of mitochondrial and extra-mitochondrial pathway interplay is offered by ROS-dependent $[Zn^{2+}]_i$ mobilization. Recent findings in cultured neurons indicate that MT oxidation can produce $[Zn^{2+}]_i$ rises sufficient to promote partial loss of $\Delta\Psi_m$ (Jiang et al., 2001; Sensi et al., 2003b). Conversely, Zn^{2+}-induced generation of mitochondrial ROS is able to induce more Zn^{2+} release from MTs (Fig. 1). Thus, rather than acting separately, these two processes could together form a self-perpetuating, vicious cycle of injury.

Furthermore, it is also worth noting that PARP and other poly(ADP-ribosyl) transferases are localized within mitochondria as well as in the cytosol, and activation of mitochondrial PARP may be able to trigger loss of $\Delta\Psi_m$ and NAD^+ depletion as well (Du et al., 2003). Zn^{2+}-induced NAD^+ reduction seems to play a key role in glycolysis inhibition and neuronal death (Sheline et al., 2000), and thus the Zn^{2+}-dependent reduction of mitochondrial NAD^+ may further enhance this deleterious cascade.

Finally, Zn^{2+} can also regulate intracellular signaling by interfering with Ca^{2+} homeostasis. The cation has been found to trigger $[Ca^{2+}]_i$ rises by activating a putative Zn^{2+} sensing receptor (ZnR) that appears to be a Gαq-coupled receptor highly specific to Zn^{2+} (Hershfinkel et al., 2001; Maret, 2001). ZnR is implicated in the activation of the MAP kinase and the PI3 kinase pathways and the cation can promote a strong increase in the activity of the Na^+/H^+ exchanger, NHE1 (Azriel-Tamir et al., 2004). ZnR activation might be particularly relevant in brain regions where Zn^{2+} is released during neuronal activity. For instance ZnR-mediated activation of MAP kinase may be crucial for proper brain growth and development, yet in the aging brain, activation of this pathway may elicit neuronal cell death under ischemic conditions (Kaplan and Miller, 2000; Namura et al., 2001). Furthermore, in the brain, Zn^{2+} dependent modulation of the activity of NHE1 might be required for brain development, while decreased activity of the exchanger (as with NHE1 knockout mice) might result in neuronal death and severe epilepsy (Bell et al., 1999; Xia et al., 2003). In addition, regulation of NHE1 activity can also be deleterious in cerebral ischemia when parenchimal acidosis can promote activation of the Na^+/H^+ exchanger and $[Na^+]_i$ overload (Luo et al., 2005). Thus, up-regulation of the activity of this exchanger by translocating Zn^{2+} that can act on the ZnR might further enhance the toxic consequences of the intracellular ionic milieu disruption.

6. NEUROLOGICAL CONDITIONS ASSOCIATED WITH Zn^{2+} DYS-HOMEOSTASIS

6.1. Transient Global Ischemia

TGI is a neurological condition associated with the selective and delayed degeneration of certain hippocampal neurons, particularly in the CA1 subregion (Pulsinelli et al., 1982). One factor that seems to play a critical role in the particular high vulnerability of these neurons is their dendritic expression of Ca-A/K channels (Bennett et al., 1996; Pellegrini-Giampietro et al., 1997). Ca-A/K channel expression in these neurons seems to be TGI-driven. Accordingly, TGI selectively decreases the expression of the GluR2 subunit (and thus increases the number of functional Ca-A/K channels) in CA1 pyramidal neurons 16–24 h after the end of the ischemic insult (Gorter et al., 1997; Opitz et al., 2000).

Recent studies also indicate that selective vulnerability of CA1 pyramidal neurons can be linked to TGI-driven perturbation of Zn^{2+} homeostasis. Trans-synaptic Zn^{2+} movement occurs during the insult, and those neurons that accumulate $[Zn^{2+}]_i$ are in fact the same neurons that eventually die (Tonder et al., 1990; Koh et al., 1996). Furthermore, chelation of extracellular Zn^{2+} with Ca-EDTA, both before and during TGI, prevents $[Zn^{2+}]_i$ rises and is neuroprotective in the CA1 subregion (Koh et al., 1996).

In TGI, increased Ca-A/K channel expression and Zn^{2+} translocation can act synergistically to promote injury as the highly Zn^{2+} permeable Ca-A/K channels are concentrated on postsynaptic membranes where the highest levels of synaptically released Zn^{2+} occur. Providing strong evidence for the pathological role of Zn^{2+} influx through Ca-A/K channels in TGI, selective pharmacological inhibition of these channels has been found to be highly neuroprotective against CA1 pyramidal neuronal loss in both in vitro and in vivo models of TGI (Yin et al., 2002; Noh et al., 2005).

Finally, as predicted by in vitro experiments (Sensi et al., 1999a, b), current findings suggest that $[Zn^{2+}]_i$ rises triggered by TGI might potently target mitochondrial function. Experiments in rats undergoing TGI indicate that the ischemic insult promotes alterations in mitochondrial morphology, mitochondrial Zn^{2+} rises, and activation of large, multiconductance channels. This channel activity is inhibited by pretreatment with Ca-EDTA prior to ischemia or the in vitro application of the membrane-permeable Zn^{2+} chelator tetrakis-(2-pyridylmethyl) ethylenediamine (TPEN), and associated with powerful apoptotic signaling by promoting the proteolytic cleavage of Bcl-xl and generating the proapoptotic N-terminal cleavage fragment of Bcl-xl (ΔN-Bcl-xl) (Bonanni et al., 2004).

6.1.1. Zn^{2+} Influx and Homeostasis under Ischemic Conditions

Among the many parenchymal changes that could affect Zn^{2+} homeostasis, one key element that should not be overlooked is the characteristic acidosis that develops in the ischemic tissue (Lipton, 1999). Acidosis can modulate Zn^{2+} homeostasis in many ways. For instance, acidotic changes can affect Zn^{2+} entry by increasing the Zn^{2+} permeability of Ca-A/K channels and VSCC while decreasing Ca^{2+} permeability of these routes (Traynelis and Cull-Candy, 1990; Kerchner et al., 2000; Jeng et al., 2002). Furthermore, acidic shifts rapidly destabilize the interaction between MTs and Zn^{2+} (Jiang et al., 2000), and promote intracellular Zn^{2+} mobilization (Fig. 1). In neurons, this effect seems to be particularly prominent and can lead to the marked increase in $[Zn^{2+}]_i$ rises generated by MT oxidation (Sensi et al., 2003b). On the other hand, Zn^{2+} can also interfere with the acid–base equilibrium and induce intracellular acidification and/or delay recovery from intracellular acidosis (Dineley et al., 2002). Thus, an acidic environment may influence Zn^{2+} homeostasis through modulation of Zn^{2+} influx and/or intracellular mobilization. Zn^{2+} dyshomeostasis may in turn disrupt neuronal acid–base equilibrium, thereby creating a potentially injurious feed-forward loop.

Finally, it should be mentioned that Zn^{2+} dyshomeostasis occurring in an acidotic environment may also influence excitotoxic $[Ca^{2+}]_i$ rises. Recent studies indicate that Zn^{2+} potently inhibits the activity of the acid sensing ionic channels (ASIC), which are an important route (and alternative to glutamate receptor associated channels) for Ca^{2+} influx during ischemia (Chu et al., 2004; Xiong et al., 2004). In the context of cerebral ischemia, acidosis may therefore exert a complex role. On one hand, it may amplify Zn^{2+} dyshomeostasis and related injury, while on the other hand it can reduce Ca^{2+}-dependent neuronal loss via inhibition of the ASIC channels. However, considering that Zn^{2+} shows far greater potency than Ca^{2+} in promoting ROS generation and mitochondrial dysfunction (Sensi et al., 2000; Jiang et al., 2001), the net effect of acidosis seems to promote neuronal injury.

6.2. Epilepsy

The role of Zn^{2+} in epilepsy is complex. Exogenous Zn^{2+} can act as a powerful convulsant, and intracerebral zinc injection has been used since the 1980s to experimentally induce epilepsy (Itoh and Ebadi, 1982; Pei et al., 1983; Pei and Koyama, 1986). However, more recent studies indicate that endogenous Zn^{2+} released under physiological conditions can inhibit synaptic activity and rather serves instead as an effective anticonvulsant. Indeed, reduction or loss of vesicular Zn^{2+} via dietary zinc deficiency, chelation, or ZnT-3 knockout has consistently been found to enhance seizure susceptibility (Fukahori and Itoh, 1990; Mitchell and Barnes, 1993; Cole et al., 2000; Takeda et al., 2003; Blasco-Ibanez et al., 2004). Another mechanism by which Zn^{2+} can exert its anticonvulsant effect is through the newly described selective inhibition of synaptic GABA transporters (Cohen-Kfir et al., 2005).

In epilepsy, Zn^{2+} can also promote the neuronal loss associated with *status epilepticus*. Zn^{2+} translocation has been described in animal models of *status epilepticus* (Sloviter, 1985; Frederickson et al., 1988, 1989; Suh et al., 2001), and the neuronal loss associated with this condition is attenuated by chelation of extracellular Zn^{2+} (Lee et al., 2000; Yi et al., 2003). The origin of these seizure-driven $[Zn^{2+}]_i$ rises seems to be region specific; a recent study on ZnT-3 and MT3 double-knockout mice indicates that the main source of $[Zn^{2+}]_i$ accumulation in the CA3 hippocampal subregion is translocation, whereas $[Zn^{2+}]_i$ rises in CA1, dentate gyrus, and the thalamus are mainly the result of intracellular mobilization from MT-3 (Lee et al., 2003).

6.3. Brain Trauma

Besides direct mechanical damage to neurons via axonal shearing and membrane rupture, traumatic brain injury (TBI) causes neuronal demise through indirect mechanisms. Such indirect neurotoxic mechanisms include (1) secondary ischemic insult due to brain swelling and increased intracranial pressure; (2) K^+ accumulation with resultant paroxysmal neuronal firing; and (3) increased glutamate release and excitotoxicity (Nilsson et al., 1993; Clark et al., 1994; Bullock et al., 1995). It is therefore not surprising that like TGI and *status epilepticus*, TBI is associated with the preferential loss of hippocampal pyramidal neurons (Lowenstein et al., 1992).

Recent studies indicate that both Zn^{2+} influx and intracellular mobilization can occur in cortical and hippocampal neurons of animals undergoing TBI. Further substantiating an injurious role for Zn^{2+} in TBI, chelation with Ca-EDTA has been found to be neuroprotective (Suh et al., 2000b). Extracellular Zn^{2+} chelation has also been linked to increased expression of neuroprotective genes such as p21, heme oxygenase-1, and heat shock protein 70, as well as several antioxidant enzymes (Hellmich et al., 2004), suggesting that Zn^{2+} might exert at least some of its toxic action in TBI by inhibiting endogenous neuroprotective pathways.

6.4. Alzheimer's Disease

Alzheimer's disease (AD) is a neurodegenerative condition characterized by the accumulation of β-amyloid (Aβ) plaques in the brain. Aβ is a 39-43 amino acid peptide derived from the proteolytic cleavage of the amyloid β protein precursor (APP) through the sequential action of β- and γ-secretases. Aβ can be produced physiologically; the soluble Aβ1-40 form is found in cerebrospinal fluid at low nanomolar concentrations, whereas Aβ1-42 is associated with pathological deposition within plaques. Aβ is distributed in the brain both diffusely in its nonfibrillar form and focally in highly insoluble fibrillary deposits (Selkoe, 2004).

The "amyloid cascade hypothesis" identifies excessive Aβ production as the primary cause of the disease and is supported by clinical data. Autosomal dominant familial forms represent 5–10% of all AD cases and are linked to mutations in APP and presenilin 1 or 2 (PS-1; PS-2) genes, all of which lead to the over-production of total Aβ or Aβ1-42. Although Aβ dysmetabolism has a central role in AD pathogenesis, other factors are likely to be involved. Among these, Zn^{2+} and Cu^{2+} dyshomeostasis have acquired increasing attention in the past two decades, and chelation of these cations offers promising new therapeutic perspectives (Bush, 2003).

6.4.1. Zinc Dyshomeostasis in AD

The role played by Zn^{2+} dys/homeostasis in AD has been discussed for more than twenty years. The well-known physiological actions exerted by the cation have provided the rationale for the idea that perhaps AD could be partially due to Zn^{2+} deficiency. This hypothesis was originally suggested in the early 1980s by Burnet (Burnet, 1981) and supported by evidence of decreased Zn^{2+} levels in the brains of AD patients (Constantinidis, 1990, 1991; Wenstrup et al., 1990). The observation that Zn^{2+} may improve mental alertness in elderly people eventually led researchers to test the potential clinical value of dietary zinc supplementation in AD patients, but the initial trial was suspended within two days due to the marked cognitive deterioration observed in some participants (Bush, 1992; Kaiser, 1994). Other studies lend support to the idea that oral zinc supplements might improve cognitive performance in AD patients (Potocnik et al., 1997), but the hypothesis as a whole remains controversial since brain Zn^{2+} levels appear to be only marginally affected by dietary intake of the cation. Indeed, the central concept of brain Zn^{2+} levels in AD remains uncertain, as it is possible to find several studies supporting either decreased (Constantinidis, 1990, 1991;

Wenstrup et al., 1990; Corrigan et al., 1993; Andrasi et al., 1995), or increased (Thompson et al., 1988; Samudralwar et al., 1995; Deibel et al., 1996; Danscher et al., 1997; Cornett et al., 1998), levels of the cation in the CNS of AD patients. However, it must be noted that earlier studies indicating decreased Zn^{2+} levels in AD brains used fixation methods that may have led to the denaturation of the metal binding sites and artifactual Zn^{2+} deficits. Thus, evidence from recent studies free of methodological artifacts lends support to the idea that Zn^{2+} levels are elevated in AD brains.

6.4.2. Zn^{2+} and Plaque Formation

Biochemical studies have indicated that Zn^{2+} can be actively involved in the amyloid dysmetabolism associated with AD. Amyloid plaques found in AD brains are highly enriched in chelatable zinc (Lovell et al., 1998; Lee et al., 1999; Suh et al., 2000b; Cherny et al., 2001). Zn^{2+} binding to APP has been found to inhibit α-secretase activity and confer resistance to proteolytic Aβ degradation, possibly leading to increased levels of Aβ in the brain. Several studies have also shown that Aβ1-40 binds to Zn^{2+} with both high (K_A 107 nM) and low affinity (K_A 5.2 μM), and revealed that micromolar Zn^{2+} concentrations induce Aβ aggregation and precipitation, a process reversible through chelation (Bush et al., 1994a, b; Brown et al., 1997; Huang et al., 1997).

The importance of synaptic Zn^{2+} in promoting amyloid plaque formation in vivo has clearly been supported by findings in transgenic animal models of AD. Transgenic mice expressing human mutated APP (Tg 2576) have an amyloid plaque burden that is comparable to that found in AD brains, but surprisingly, when these animals are crossed with ZnT-3 KO mice, which have no synaptic Zn^{2+}, they show a dramatic decrease in cerebral plaque formation (Lee et al., 2002b). Furthermore, both AD patients and AD-prone transgenic mice have high Zn^{2+} concentrations in neuritic plaques and cerebrovascular amyloid deposits (Lovell et al., 1998; Lee et al., 1999; Suh et al., 2000b; Friedlich et al., 2004), and postmortem administration of Zn^{2+} chelators promotes Aβ resolubilization in both these models suggesting that Zn^{2+} plays a strong role in AD by promoting amyloid plaque formation in vivo (Cherny et al., 1999, 2001; Lee et al., 2004).

7. Zn^{2+} HOMEOSTASIS IN GLIAL CELLS: POTENTIAL ROLE FOR PHYSIOLOGY AND PATHOLOGY

After an acute brain insult, functional recovery greatly depends on the proper functioning of the surviving astrocytes. Astrocytes exert a critical role in the reuptake of glutamate and also control the balance of extracellular K^+, both of which are key factors involved in the development of necrosis and apoptosis (Yu et al., 2001). Glial cells are sensitive to excitotoxicity; however, these cells show some reduced sensitivity to the toxic effects of glutamate (David et al., 1996). Zn^{2+} is also a less investigated promoter of astrocyte death (Yokoyama et al., 1986). Mechanisms by which Zn^{2+} triggers injury in glia are still largely unknown and, as with excitotoxicity, astrocytes appear to be less sensitive than neurons to the toxic effects of intracellular Zn^{2+} accumulation (Dineley et al., 2000). Recent studies have started to shed some light in the field of Zn^{2+}-mediated glial injury and one the most promising area of interest seems to be offered by the interplay between oxidative stress and Zn^{2+} dyshomeostasis. As mentioned above, ^-ONOO can promote $[Zn^{2+}]_i$ rises in oligodendrocytes that set in motion 12-LOX activation, ROS production, p38 activation, loss of $\Delta\psi_m$, caspase-3 activation, and injury (Zhang et al., 2006).

Data accumulated in the past two decades indicate that like neurons, astrocytes are excitable cells and respond to activation with the release of a plethora of signaling molecules such as glutamate, ATP, D-serine, and eicosanoids. All of these molecules can affect the functioning of neighboring neurons in a process that has been called gliotransmission, and is a key component of the bidirectional process of information exchange between neurons and glia. In glial cells, Ca^{2+} signaling plays an essential role in the activation of gliotransmission and one very promising and fascinating area of future investigation seems to be offered by recent evidence indicating that Zn^{2+} modulates intracellular $[Ca^{2+}]_i$ homeostasis as the cation interferes with capacitative Ca^{2+} entry (Kresse et al., 2005).

8. CONCLUDING REMARKS

Early studies have indicated trans-synaptic Zn^{2+} movement as the major mechanism for Zn^{2+}-dependent brain injury in excitotoxic syndromes. This simplistic view has been radically changed in recent years and the emerging scenario appears to be more complex. Zn^{2+} homeostasis in the brain is regulated through highly dynamic pathways and is deeply connected with other major signaling pathways, such as NO- and MAP kinase-dependent pathways. Zn^{2+} signaling in neurons and glia is also interplaying with proton and Ca^{2+} homeostasis. Moreover, considering the fact that Zn^{2+} seems to promote injury with greater potency compared to Ca^{2+}, Zn^{2+} may be an underappreciated mediator of excitotoxicity, which has for the most part been thought of as a purely Ca^{2+}-dependent phenomenon. The injurious interplay between Ca^{2+} and Zn^{2+} deserves some attention and the harmful actions mediated by $[Ca^{2+}]_i$ increases in many excitotoxic conditions need be re-evaluated in light of recent data indicating that Zn^{2+} can be mobilized upon Ca^{2+} influx and Ca^{2+}-dependent oxidative stress. This Ca^{2+}-induced mitochondrial ROS generation promotes Zn^{2+} release from MTs, and offers the possibility of a more complex injury paradigm than previously thought. In such a model, glutamate-driven $[Ca^{2+}]_i$ rises might be the initiator but not the prime activator of neuronal and glial injury.

Throughout the most recent studies the Janus faces of Zn^{2+} are also emerging. The cation is playing a crucial role in promoting injury in the old and diseased brain. However, the activation of the same Zn^{2+} signaling pathways can be critically relevant to favor proper brain growth in the healthy and young developing brain.

The progress made so far in the field of Zn^{2+} neurobiology is remarkable but a more detailed blueprint of the mechanisms involved in Zn^{2+} homeostasis is warranted. Areas that need further investigation concern the cross talk between Zn^{2+} pools, the interplay with Ca^{2+}, Zn^{2+} modulation of the processes controlling neuronal–glia communication, and the molecular and mechanistic aspects of Zn^{2+} transport in the brain. Such knowledge will be crucial to improve our ability to "finely tune" brain Zn^{2+} homeostasis in both development and disease.

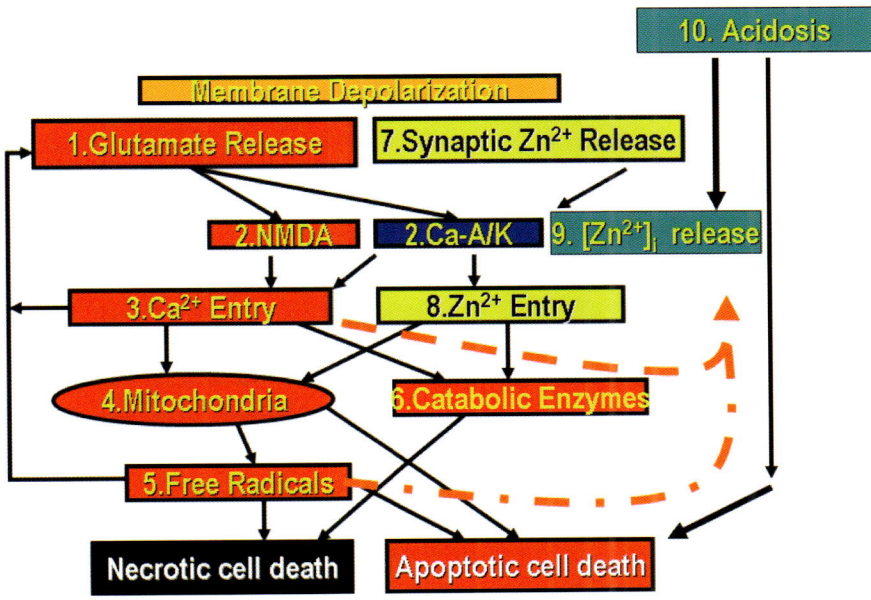

Fig. 1. The excitotoxic cascade. In the past two decades, excitotoxic events have been linked to Ca^{2+} dyshomeostasis triggered by excessive stimulation of glutamate receptors. New evidence indicates that Zn^{2+} acts in synergism with Ca^{2+} to promote brain injury. According with this more complex scenario, excessive release of glutamate (1) activates NMDA and Ca^{2+}-permeable AMPA/kainate (Ca-A/K) receptors (2) and promotes massive $[Ca^{2+}]_i$ rises (3) that leads to mitochondrial dysfunction (4) oxidative stress (5) and induction of catabolic enzymes (6). In some forms of cerebral ischemia, Zn^{2+} is released at synaptic sites (7), which fluxes mainly through Ca-A/K receptors, and increases intracellular free Zn^{2+} ($[Zn^{2+}]_i$) (8). These $[Zn^{2+}]_i$ rises, together with Ca^{2+}, promote mitochondrial dysfunction and oxidative stress. $[Ca^{2+}]_i$ accumulation and oxidative stress can also induce mobilization of Zn^{2+} (9) from intracellular sites such as mitochondria and metallothioneins further enhancing injurious Zn^{2+} dyshomeostasis. Parenchymal acidosis (10), a common feature upon ischemia and inflammation, can inhibit intracellular Zn^{2+} buffering and lead to increased $[Zn^{2+}]_i$ rises (9). Interplaying of all these pathways occur and are induced in both neuronal and glial cells to promote apoptotic and necrotic injury. It is important to note that glial cell death can be critical in promoting a feed-forward injurious cycle. Injured astrocytes are not able to take up glutamate, which leads to increased toxic levels of the neurotransmitter at synapses

9. ACKNOWLEDGMENTS

This work was supported by NIH grant AG00919 and PRIN 2004 (SLS) and ISF, BSF and GIF grants (IS).

10. REFERENCES

Aizenman, E., Stout, A.K., Hartnett, K.A., Dineley, K.E., McLaughlin, B. and Reynolds, I.J., 2000, Induction of neuronal apoptosis by thiol oxidation: putative role of intracellular zinc release. *J. Neurochem.* **75**: 1878.
Andrasi, E., Farkas, E., Scheibler, H., Reffy, A. and Bezur, L., 1995, Al, Zn, Cu, Mn and Fe levels in brain in Alzheimer's disease. *Arch. Gerontol. Geriatr.* **21**: 89.
Aniksztejn, L., Charton, G. and Ben-Ari, Y., 1987, Selective release of endogenous zinc from the hippocampal mossy fibers in situ. *Brain Res.* **404**: 58.
Arseniev, A., Schultze, P., Worgotter, E., Braun, W., Wagner, G., Vasak, M., Kagi, J.H. and Wuthrich, K., 1988, Three-dimensional structure of rabbit liver [Cd7]metallothionein-2a in aqueous solution determined by nuclear magnetic resonance. *J. Mol. Biol.* **201**: 637.
Assaf, S.Y. and Chung, S.H., 1984, Release of endogenous Zn^{2+} from brain tissue during activity. *Nature* **308**: 734.
Azriel-Tamir, H., Sharir, H., Schwartz, B. and Hershfinkel, M., 2004, Extracellular zinc triggers ERK-dependent activation of Na^+/H^+ exchange in colonocytes mediated by the zinc sensing receptor. *J. Biol. Chem.* **279**: 51804.
Beckmann, A.M. and Wilce, P.A., 1997, Egr transcription factors in the nervous system. *Neurochem. Int.* **31**: 477.
Bell, S.M., Schreiner, C.M., Schultheis, P.J., Miller, M.L., Evans, R.L., Vorhees, C.V., Shull, G.E. and Scott, W.J., 1999, Targeted disruption of the murine Nhe1 locus induces ataxia, growth retardation, and seizures. *Am. J. Physiol.* **276**: C788.
Bennett, M.V.L., Pellegrini-Giampietro, D.E., Gorter, J.A., Aronica, E., Connor, J.A. and Zukin, R.S., 1996, The GluR2 hypothesis: Ca^{2+}-permeable AMPA receptors in delayed neurodegeneration. *Cold Spring Harb. Symp. Quant. Biol.* **61**: 373.
Blasco-Ibanez, J.M., Poza-Aznar, J., Crespo, C., Marques-Mari, A.I., Gracia-Llanes, F.J. and Martinez-Guijarro, F.J., 2004, Chelation of synaptic zinc induces overexcitation in the hilar mossy cells of the rat hippocampus. *Neurosci. Lett.* **355**: 101.
Bonanni, L., Li, H., Jover, L., Yokota, H., Sensi, S.L., Zukin, S. and Jonas, E.A., 2004, Zinc-dependent multiconductance channel activity in mitochondria isolated from ischemic brain. *In SFN Annual Meeting. 2004. San Diego, CA: Washington, DC: Society for Neuroscience.*
Bossy-Wetzel, E., Talantova, M.V., Lee, W.D., Scholzke, M.N., Harrop, A., Mathews, E., Gotz, T., Han, J., Ellisman, M.H., Perkins, G.A. and Lipton, S.A., 2004, Crosstalk between nitric oxide and zinc pathways to neuronal cell death involving mitochondrial dysfunction and p38-activated K^+ channels. *Neuron* **41**: 351.
Brown, A.M., Tummolo, D.M., Rhodes, K.J., Hofmann, J.R., Jacobsen, J.S. and Sonnenberg-Reines, J., 1997, Selective aggregation of endogenous beta-amyloid peptide and soluble amyloid precursor protein in cerebrospinal fluid by zinc. *J. Neurochem.* **69**: 1204.
Brown, A.M., Kristal, B.S., Effron, M.S., Shestopalov, A.I., Ullucci, P.A., Sheu, K.-F.R., Blass, J.P. and Cooper, A.J.L., 2000, Zn^{2+} inhibits alpha-ketoglutarate-stimulated mitochondrial respiration and the isolated alpha-ketoglutarate dehydrogenase complex. *J. Biol. Chem.* **275**: 13441.
Bullock, R., Zauner, A., Myseros, J.S., Marmarou, A., Woodward, J.J. and Young, H.F., 1995, Evidence for prolonged release of excitatory amino acids in severe human head trauma. Relationship to clinical events. *Ann. N. Y. Acad. Sci.* **765**: 290.
Burkle, A., 2005, Poly (ADP-ribose). The most elaborate metabolite of $NAD+$. *FEBS J.* **272**: 4576.
Burnet, F.M., 1981, A possible role of zinc in the pathology of dementia. *Lancet* **1**: 186.
Bush, A.I., *Thesis. 1992 The University of Melbourne.*
Bush, A.I., 2003, Copper, zinc, and the metallobiology of Alzheimer disease. *Alzheimer Dis. Assoc. Disord.* **17**: 147.
Bush, A.I., Pettingell, W.H., Multhaup, G., d Paradis, M., Vonsattel, J.P., Gusella, J.F., Beyreuther, K., Masters, C.L. and Tanzi, R.E., 1994a, Rapid induction of Alzheimer A beta amyloid formation by zinc. *Science* **265**: 1464.
Bush, A.I., Pettingell, Jr. W.H., Paradis, M.D. and Tanzi, R.E., 1994b, Modulation of A beta adhesiveness and secretase site cleavage by zinc. *J. Biol. Chem.* **269**: 12152.
Canzoniero, L.M., Turetsky, D.M. and Choi, D.W., 1999, Measurement of intracellular free zinc concentrations accompanying zinc-induced neuronal death. *J. Neurosci.* **19**: RC31.
Chao, Y. and Fu, D., 2004, Kinetic study of the antiport mechanism of an Escherichia coli zinc transporter, ZitB. *J. Biol. Chem.* **279**: 12043.
Cheng, C. and Reynolds, I.J., 1998, Calcium-sensitive fluorescent dyes can report increases in intracellular free zinc concentration in cultured forebrain neurons. *J. Neurochem.* **71**: 2401.
Cherny, R.A., Legg, J.T., McLean, C.A., Fairlie, D.P., Huang, X., Atwood, C.S., Beyreuther, K., Tanzi, R.E., Masters, C.L. and Bush, A.I., 1999, Aqueous dissolution of Alzheimer's disease Abeta amyloid deposits by biometal depletion. *J. Biol. Chem.* **274**: 23223.
Cherny, R.A., Atwood, C.S., Xilinas, M.E., Gray, D.N., Jones, W.D., McLean, C.A., Barnham, K.J., Volitakis, I., Fraser, F.W. and Kim, et al., 2001, Treatment with a copper-zinc chelator markedly and rapidly inhibits beta-amyloid accumulation in Alzheimer's disease transgenic mice. *Neuron* **30**: 665.

Choi, D.W., Yokoyama, M. and Koh, J., 1988, Zinc neurotoxicity in cortical cell culture. *Neuroscience* **24**: 67.
Chu, X.P., Wemmie, J.A., Wang, W.Z., Zhu, X.M., Saugstad, J.A., Price, M.P., Simon, R.P. and Xiong, Z.G., 2004, Subunit-dependent high-affinity zinc inhibition of acid-sensing ion channels. *J. Neurosci.* **24**: 8678.
Clark, R.S., Schiding, J.K., Kaczorowski, S.L., Marion, D.W. and Kochanek, P.M., 1994, Neutrophil accumulation after traumatic brain injury in rats: comparison of weight drop and controlled cortical impact models. *J. Neurotrauma* **11**: 499.
Cohen-Kfir, E., Lee, W., Eskandari, S. and Nelson, N., 2005, Zinc inhibition of gamma-aminobutyric acid transporter 4 (GAT4) reveals a link between excitatory and inhibitory neurotransmission. *Proc. Natl Acad. Sci. USA* **102**: 6154.
Cole, T.B., Robbins, C.A., Wenzel, H.J., Schwartzkroin, P.A. and Palmiter, R.D., 2000, Seizures and neuronal damage in mice lacking vesicular zinc. *Epilepsy Res.* **39**: 153.
Constantinidis, J., 1990, Alzheimer's disease and the zinc theory. *Encephale* **16**: 231.
Constantinidis, J., 1991, The hypothesis of zinc deficiency in the pathogenesis of neurofibrillary tangles. *Med. Hypotheses* **35**: 319.
Cornett, C.R., Markesbery, W.R. and Ehmann, W.D., 1998, Imbalances of trace elements related to oxidative damage in Alzheimer's disease brain. *Neurotoxicology* **19**: 339.
Corrigan, F.M., Reynolds, G.P. and Ward, N.I., 1993, Hippocampal tin, aluminum and zinc in Alzheimer's disease. *Biometals* **6**: 149.
Danscher, G., Jensen, K.B., Frederickson, C.J., Kemp, K., Andreasen, A., Juhl, S., Stoltenberg, M. and Ravid, R., 1997, Increased amount of zinc in the hippocampus and amygdala of Alzheimer's diseased brains: a proton-induced X-ray emission spectroscopic analysis of cryostat sections from autopsy material. *J. Neurosci. Methods* **76**: 53.
David, J.C., Yamada, K.A., Bagwe, M.R. and Goldberg, M.P., 1996, AMPA receptor activation is rapidly toxic to cortical astrocytes when desensitization is blocked. *J. Neurosci.* **16**: 200.
Deibel, M.A., Ehmann, W.D. and Markesbery, W.R., 1996, Copper, iron, and zinc imbalances in severely degenerated brain regions in Alzheimer's disease: possible relation to oxidative stress. *J. Neurol. Sci.* **143**: 137.
Dineley, K.E., Scanlon, J.M., Kress, G.J., Stout, A.K. and Reynolds, I.J., 2000, Astrocytes are more resistant than neurons to the cytotoxic effects of increased [Zn(2+)](i). *Neurobiol. Dis.* **7**: 310.
Dineley, K.E., Brocard, J.B. and Reynolds, I.J., 2002, Elevated intracellular zinc and altered proton homeostasis in forebrain neurons. *Neuroscience* **114**: 439.
Dineley, K.E., Votyakova, T.V. and Reynolds, I.J., 2003, Zinc inhibition of cellular energy production: implications for mitochondria and neurodegeneration. *J. Neurochem.* **85**: 563.
Du, L., Zhang, X., Han, Y., Burke, N.A., Kochanek, P.M., Watkins, S.C., Graham, S.H., Carcillo, J.A., Szabo, C. and Clark, R.S.B., 2003, Intra-mitochondrial poly(adp-ribosylation) contributes to NAD+ depletion and cell death induced by oxidative stress. *J. Biol. Chem.* **278**: 18426.
Ebadi, M., Leuschen, M.P., El Refaey, H., Hamada, F.M. and Rojas, P., 1996, The antioxidant properties of zinc and metallothionein. *Neurochem. Int.* **29**: 159.
Eide, D.J., 2004, The SLC39 family of metal ion transporters. *Pflugers Arch.* **447**: 796.
Erickson, J.C., Hollopeter, G., Thomas, S.A., Froelick, G.J. and Palmiter, R.D., 1997, Disruption of the metallothionein-III gene in mice: analysis of brain zinc, behavior, and neuron vulnerability to metals, aging, and seizures. *J. Neurosci.* **17**: 1271.
Frederickson, C.J. and Bush, A.I., 2001, Synaptically released zinc: physiological functions and pathological effects. *BioMetals* **14**: 353.
Frederickson, C.J. and Moncrieff, D.W., 1994, Zinc-containing neurons. *Biol. Signals* **3**: 127.
Frederickson, C.J., Hernandez, M.D., Goik, S.A., Morton, J.D. and McGinty, J.F., 1988, Loss of zinc staining from hippocampal mossy fibers during kainic acid induced seizures: a histofluorescence study. *Brain Res.* **446**: 383.
Frederickson, C.J., Hernandez, M.D. and McGinty, J.F., 1989, Translocation of zinc may contribute to seizure-induced death of neurons. *Brain Res.* **480**: 317.
Frederickson, C.J., Koh, J.Y. and Bush, A.I., 2005, The neurobiology of zinc in health and disease. *Nat. Rev. Neurosci.* **6**: 449.
Friedlich, A.L., Lee, J.Y., van Groen, T., Cherny, R.A., Volitakis, I., Cole, T.B., Palmiter, R.D., Koh, J.Y. and Bush, A.I., 2004, Neuronal zinc exchange with the blood vessel wall promotes cerebral amyloid angiopathy in an animal model of Alzheimer's disease. *J. Neurosci.* **24**: 3453.
Fukahori, M. and Itoh, M., 1990, Effects of dietary zinc status on seizure susceptibility and hippocampal zinc content in the El (epilepsy) mouse. *Brain Res.* **529**: 16.
Gazaryan, I.G., Krasnikov, B.F., Ashby, G.A., Thorneley, R.N.F., Kristal, B.S. and Brown, A.M., 2002, Zinc is a potent inhibitor of thiol oxidoreductase activity and stimulates reactive oxygen species production by lipoamide dehydrogenase. *J. Biol. Chem.* **277**: 10064.

Gorter, J.A., Petrozzino, J.J., Aronica, E.M., Rosenbaum, D.M., Opitz, T., Bennett, M.V., Connor, J.A. and Zukin, R.S., 1997, Global ischemia induces downregulation of Glur2 mRNA and increases AMPA receptor-mediated Ca2+ influx in hippocampal CA1 neurons of gerbil. *J. Neurosci.* **17**: 6179.

Hellmich, H.L., Frederickson, C.J., DeWitt, D.S., Saban, R., Parsley, M.O., Stephenson, R., Velasco, M., Uchida, T., Shimamura, M. and Prough, D.S., 2004, Protective effects of zinc chelation in traumatic brain injury correlate with upregulation of neuroprotective genes in rat brain. *Neurosci. Lett.* **355**: 221.

Hershfinkel, M., Moran, A., Grossman, N. and Sekler, I., 2001, A zinc-sensing receptor triggers the release of intracellular Ca2+ and regulates ion transport. *Proc. Natl Acad. Sci. USA* **98**:11749.

Hidalgo, J., Aschner, M., Zatta, P. and Vasak, M., 2001, Roles of the metallothionein family of proteins in the central nervous system. *Brain Res. Bull.* **55**: 133.

Holtz, M.L., Craddock, S.D. and Pettigrew, L.C., 2001, Rapid expression of neuronal and inducible nitric oxide synthases during post-ischemic reperfusion in rat brain. *Brain Res.* **898**: 49.

Howell, G.A., Welch, M.G. and Frederickson, C.J., 1984, Stimulation-induced uptake and release of zinc in hippocampal slices. *Nature* **308**: 736.

Huang, X., Atwood, C.S., Moir, R.D., Hartshorn, M.A., Vonsattel, J.P., Tanzi, R.E. and Bush, A.I., 1997, Zinc-induced Alzheimer's Abeta1-40 aggregation is mediated by conformational factors. *J. Biol. Chem.* **272**: 26464.

Ikeda, T., Kimura, K., Morioka, S. and Tamaki, N., 1980, Inhibitory effects of Zn2+ on muscle glycolysis and their reversal by histidine. *J. Nutr. Sci. Vitaminol.* **26**: 357.

Itoh, M. and Ebadi, M., 1982, The selective inhibition of hippocampal glutamic acid decarboxylase in zinc-induced epileptic seizures. *Neurochem. Res.* **7**: 1287.

Jeng, J.-M., Jia, Y., Bonanni, L. and Weiss, J.H., 2002, Divergent effects of pH on Zn2+ and Ca2+ flux through Ca2+-permeable AMPA/kainate channels (CAKR). *In SFN Annual Meeting. 2002. Orlando, FL: Washington, DC: Society for Neuroscience.*

Jia, Y., Jeng, J.M., Sensi, S.L. and Weiss, J.H., 2002, Zn2+ currents are mediated by calcium-permeable AMPA/kainate channels in cultured murine hippocampal neurones, *J. Physiol.* **543**: 35.

Jiang, L.-J., Maret, W. and Vallee, B.L., 1998, The glutathione redox couple modulates zinc transfer from metallothionein to zinc-depleted sorbitol dehydrogenase. *Proc. Natl. Acad. Sci. USA* **95**: 3483.

Jiang, L.J., Vasak, M., Vallee, B.L. and Maret, W., 2000, Zinc transfer potentials of the alpha - and beta-clusters of metallothionein are affected by domain interactions in the whole molecule. *Proc. Natl Acad. Sci. USA* **97**: 2503.

Jiang, D., Sullivan, P.G., Sensi, S.L., Steward, O. and Weiss, J.H., 2001, Zn(2+) induces permeability transition pore opening and release of pro-apoptotic peptides from neuronal mitochondria. *J. Biol. Chem.* **276**: 47524.

Kägi, J.H.R., Suzuki, K.T., Imura, N. and Kimura, M., 1993, Metallothionein III. Birkhäuser, Basel.

Kaiser, J., 1994, Alzheimer's: could there be a zinc link? *Science* **265**: 1365.

Kaplan, D.R. and Miller, F.D., 2000, Neurotrophin signal transduction in the nervous system. *Curr. Opin. Neurobiol.* **10**: 381.

Kerchner, G., Canzoniero, L., Yu, S., Ling, C. and Choi, D.W., 2000, Zn2+ current is mediated by voltage-gated Ca2+ channels and enhanced by extracellular acidity in mouse cortical neurons. *J. Physiol.* **528**: 39.

Kiedrowski, L., Czyz A., Baranauskas, G., Li, X.F. and Lytton, J., 2004, Differential contribution of plasmalemmal Na/Ca exchange isoforms to sodium-dependent calcium influx and NMDA excitotoxicity in depolarized neurons. *J. Neurochem.* **90**: 117.

Kim, Y.H. and Koh, J.Y., 2002, The role of NADPH oxidase and neuronal nitric oxide synthase in zinc-induced poly(ADP-ribose) polymerase activation and cell death in cortical culture. *Exp. Neurol.* **177**: 407.

Kim, E.Y., Koh, J.Y., Kim, Y.H., Sohn, S., Joe, E. and Gwag, B.J., 1999a, Zn2+ entry produces oxidative neuronal necrosis in cortical cell cultures. *Eur. J. Neurosci.* **11**: 327.

Kim, Y.H., Kim, E.Y., Gwag, B.J., Sohn, S. and Koh, J.Y., 1999b, Zinc-induced cortical neuronal death with features of apoptosis and necrosis: mediation by free radicals. *Neuroscience* **89**:175.

Kleiner, D. and von Jagow, G., 1972, On the inhibition of mitochondrial electron transport by Zn(2+) ions. *FEBS Lett.* **20**: 229.

Kleiner, D., 1974, The effect of Zn2+ ions on mitochondrial electron transport. *Arch. Biochem. Biophys.* **165**: 121.

Koh, J.Y., Suh, S.W., Gwag, B.J., He, Y.Y., Hsu, C.Y. and Choi, D.W., 1996, The role of zinc in selective neuronal death after transient global cerebral ischemia. *Science* **272**: 1013.

Kresse, W., Sekler, I., Hoffmann, A., Peters, O., Nolte, C., Moran, A. and Kettenmann, H., 2005, Zinc ions are endogenous modulators of neurotransmitter-stimulated capacitative Ca2+ entry in both cultured and in situ mouse astrocytes. *Eur. J. Neurosci.* **21**: 1626.

Kroncke, K.D., Fehsel, K., Schmidt, T., Zenke, F.T., Dasting, I., Wesener, J.R., Bettermann, H., Breunig, K.D. and Kolb-Bachofen, V., 1994, Nitric oxide destroys zinc-sulfur clusters inducing zinc release from metallothionein and inhibition of the zinc finger-type yeast transcription activator LAC9. *Biochem. Biophys. Res. Commun.* **200**: 1105.

Krotkiewska, B. and Banas, T., 1992, Interaction of Zn2+ and Cu2+ ions with glyceraldehyde-3-phosphate dehydrogenase from bovine heart and rabbit muscle. *Int. J. Biochem.* **24**: 1501.

Kukimoto, I., Hoshino, S., Kontani, K., Inageda, K., Nishina, H., Takahashi, K. and Katada, T., 1996, Stimulation of ADP-ribosyl cyclase activity of the cell surface antigen CD38 by zinc ions resulting from inhibition of its NAD+ glycohydrolase activity. *Eur. J. Biochem.* **239**: 177.

Langmade, S.J., Ravindra, R., Daniels, P.J. and Andrews, G.K., 2000, The transcription factor MTF-1 mediates metal regulation of the mouse ZnT1 gene. *J. Biol. Chem.* **275**: 34803.

Lee, J.Y., Mook-Jung, I. and Koh, J.Y., 1999, Histochemically reactive zinc in plaques of the Swedish mutant beta-amyloid precursor protein transgenic mice. *J. Neurosci.* **19**: RC10.

Lee, J.Y., Cole, T.B., Palmiter, R.D. and Koh, J.Y., 2000, Accumulation of zinc in degenerating hippocampal neurons of ZnT3-null mice after seizures: evidence against synaptic vesicle origin. *J. Neurosci.* **20**: RC79.

Lee, J.Y., Kim, Y.H. and Koh, J.Y., 2001, Protection by pyruvate against transient forebrain ischemia in rats. *J. Neurosci.* **21**: RC171.

Lee, J.M., Zipfel, G.J., Park, K.H., He, Y.Y., Hsu, C.Y. and Choi, D.W., 2002a, Zinc translocation accelerates infarction after mild transient focal ischemia. *Neuroscience* **115**: 871.

Lee, J.Y., Cole, T.B., Palmiter, R.D., Suh, S.W. and Koh, J.Y., 2002b, Contribution by synaptic zinc to the gender-disparate plaque formation in human Swedish mutant APP transgenic mice. *Proc. Natl Acad. Sci. USA* **99**: 7705.

Lee, J.Y., Kim, J.H., Palmiter, R.D. and Koh, J.Y., 2003, Zinc released from metallothionein-iii may contribute to hippocampal CA1 and thalamic neuronal death following acute brain injury. *Exp. Neurol.* **184**: 337.

Lee, J.Y., Friedman, J.E., Angel, I., Kozak, A. and Koh, J.Y., 2004, The lipophilic metal chelator DP-109 reduces amyloid pathology in brains of human beta-amyloid precursor protein transgenic mice. *Neurobiol. Aging* **25**: 1315.

Lerma, J., Morales, M., Ibarz, J.M. and Somohano, F., 1994, Rectification properties and Ca2+ permeability of glutamate receptor channels in hippocampal cells. *Eur. J. Neurosci.* **6**: 1080.

Link, T.A. and von Jagow, G., 1995, Zinc ions inhibit the QP center of bovine heart mitochondrial bc1 complex by blocking a protonatable group. *J. Biol. Chem.* **270**: 25001.

Lipton, P., 1999, Ischemic cell death in brain neurons. *Physiol. Rev.* **79**: 1431.

Lobner, D., Canzoniero, L.M., Manzerra, P., Gottron, F., Ying, H., Knudson, M., Tian, M., Dugan, L.L., Kerchner, G.A., Sheline, C.T., Korsmeyer, S.J. and Choi, D.W., 2000, Zinc-induced neuronal death in cortical neurons. *Cell. Mol. Biol.* **46**: 797.

Lovell, M.A., Robertson, J.D., Teesdale, W.J., Campbell, J.L. and Markesbery, W.R., 1998, Copper, iron and zinc in Alzheimer's disease senile plaques. *J. Neurol. Sci.* **158**: 47.

Lowenstein, D.H., Thomas, M.J., Smith, D.H. and McIntosh, T.K., 1992, Selective vulnerability of dentate hilar neurons following traumatic brain injury: a potential mechanistic link between head trauma and disorders of the hippocampus. *J. Neurosci.* **12**: 4846.

Luo, J., Chen, H., Kintner, D.B., Shull, G.E. and Sun, D., 2005, Decreased neuronal death in Na+/H+ exchanger isoform 1-null mice after in vitro and in vivo ischemia. *J. Neurosci.* **25**: 11256.

MacDiarmid, C.W., Milanick, M.A. and Eide, D.J., 2002, Biochemical properties of vacuolar zinc transport systems of Saccharomyces cerevisiae. *J. Biol. Chem.* 277: 39187.

Malaiyandi, L.M., Vergun, O., Dineley, K.E. and Reynolds, I.J., 2005, Direct visualization of mitochondrial zinc accumulation reveals uniporter-dependent and -independent transport mechanisms. *J. Neurochem.* **93**: 1242.

Maret, W. and Vallee, B.L., 1998, Thiolate ligands in metallothionein confer redox activity on zinc clusters. *Proc. Natl Acad. Sci. USA* **95**: 3478.

Maret, W., 1994, Oxidative metal release from metallothionein via zinc-thiol/disulfide interchange. *Proc. Natl Acad. Sci. USA* **91**: 237.

Maret, W., 2001, Crosstalk of the group IIa and IIb metals calcium and zinc in cellular signaling. *Proc. Natl Acad. Sci. USA* **98**: 12325.

McLaughlin, B., Pal, S., Tran, M.P., Parsons, A.A., Barone, F.C., Erhardt, J.A. and Aizenman, E., 2001, p38 Activation is required upstream of potassium current enhancement and caspase cleavage in thiol oxidant-induced neuronal apoptosis. *J. Neurosci.* **21**: 3303.

Mitchell, C.L. and Barnes, M.I., 1993, Proconvulsant action of diethyldithiocarbamate in stimulation of the perforant path. *Neurotoxicol. Teratol.* **15**: 165.

Moncayo, J., de Freitas, G.R., Bogousslavsky, J., Altieri, M. and van Melle, G., 2000, Do transient ischemic attacks have a neuroprotective effect? *Neurology* **54**: 2089.

Namura, S., Iihara, K., Takami, S., Nagata, I., Kikuchi, H., Matsushita, K., Moskowitz, M.A., Bonventre, J.V. and Alessandrini, A., 2001, Intravenous administration of MEK inhibitor U0126 affords brain protection against forebrain ischemia and focal cerebral ischemia. *Proc. Natl Acad. Sci. USA* **98**: 11569.

Nicholls, D.G. and Budd, S.L., 2000, Mitochondria and neuronal survival. *Physiol. Rev.* **80**: 315.

Nicholls, P. and Malviya, A.N., 1968, Inhibition of nonphosphorylating electron transfer by zinc. The problem of delineating interaction sites. *Biochemistry* **7**: 305.

Nicotera, P., Leist, M. and Ferrando-May, E., 1999, Apoptosis and necrosis: different execution of the same death. *Biochem. Soc. Symp.* **66**: 69.

Nilsson, P., Hillered, L., Olsson, Y., Sheardown, M.J. and Hansen, A.J., 1993, Regional changes in interstitial K+ and Ca2+ levels following cortical compression contusion trauma in rats. *J. Cereb. Blood. Flow. Metab.* **13**: 183.

Nitzan, Y.B., Sekler, I., Hershfinkel, M., Moran, A. and Silverman, W.F., 2002, Postnatal regulation of ZnT-1 expression in the mouse brain. *Brain Res. Dev. Brain Res.* **137**: 149.

Noh, K.M., Kim, Y.H. and Koh, J.Y., 1999, Mediation by membrane protein kinase C of zinc-induced oxidative neuronal injury in mouse cortical cultures. *J. Neurochem.* **72**: 1609.

Noh, K.M., Yokota, H., Mashiko, T., Castillo, P.E., Zukin, R.S. and Bennett, M.V., 2005, Blockade of calcium-permeable AMPA receptors protects hippocampal neurons against global ischemia-induced death. *Proc. Natl Acad. Sci. USA* **102**: 12230.

Nolte, C., Gore, A., Sekler, I., Kresse, W., Hershfinkel, M., Hoffmann, A., Kettenmann, H. and Moran, A., 2004, ZnT-1 expression in astroglial cells protects against zinc toxicity and slows the accumulation of intracellular zinc. *Glia* **48**: 145.

Ohana, E., Segal, D., Palty, R., Ton-That, D., Moran, A., Sensi, S.L., Weiss, J.H., Hershfinkel, M. and Sekler, I., 2004, A sodium zinc exchange mechanism is mediating extrusion of zinc in Mammalian cells. *J. Biol. Chem.* **279**: 4278.

Opitz, T., Grooms, S.Y., Bennett, M.V., Zukin, R.S. and Opitz, T., 2000, Remodeling of alpha-amino-3-hydroxy-5-methyl-4-isoxazole-propionic acid receptor subunit composition in hippocampal neurons after global ischemia. *Proc. Natl Acad. Sci. USA* **97**: 13360.

Packer, L., Tritschler, H.J. and Wessel, K., 1997, Neuroprotection by the metabolic antioxidant -lipoic acid. *Free Radic. Biol. Med.* **22**: 359.

Palmiter, R.D. and Findley, S., 1995, Cloning and functional characterization of a mammalian zinc transporter that confers resistance to zinc. *EMBO J.* **14**: 639.

Palmiter, R.D. and Huang, L., 2004, Efflux and compartmentalization of zinc by members of the SLC30 family of solute carriers. *Pflugers Arch.* 447: 744.

Palmiter, R.D., Cole, T.B., Quaife, C.J. and Findley, S.D., 1996, ZnT-3, a putative transporter of zinc into synaptic vesicles. *Proc. Natl Acad. Sci. USA* **93**: 14934.

Park, J.A. and Koh, J.-Y., 1999, Induction of an immediate early gene egr-1 by zinc through extracellular signal-regulated kinase activation in cortical culture. *J. Neurochem.* **73**: 450.

Park, J.A., Lee, J.Y., Sato, T.A. and Koh, J.Y., 2000, Co-induction of p75NTR and p75NTR-associated death executor in neurons after zinc exposure in cortical culture or transient ischemia in the rat. *J. Neurosci.* **20**: 9096.

Pei, Y.Q. and Koyama, I., 1986, Features of seizures and behavioral changes induced by intrahippocampal injection of zinc sulfate in the rabbit: a new experimental model of epilepsy. *Epilepsia* **27**: 183.

Pei, Y., Zhao, D., Huang, J. and Cao, L., 1983, Zinc-induced seizures: a new experimental model of epilepsy. *Epilepsia* **24**: 169.

Pellegrini-Giampietro, D.E., Gorter, J.A., Bennett, M.V. and Zukin, R.S., 1997, The GluR2 (GluR-B) hypothesis: Ca(2+)-permeable AMPA receptors in neurological disorders. *Trends Neurosci.* **20**: 464.

Persechini, A., McMillan, K. and Masters, B.S., 1995, Inhibition of nitric oxide synthase activity by Zn2+ ion. *Biochemistry* **34**: 15091.

Potocnik, F.C., van Rensburg, S.J., Park, C., Taljaard, J.J. and Emsley, R.A., 1997, Zinc and platelet membrane microviscosity in Alzheimer's disease. The in vivo effect of zinc on platelet membranes and cognition. *S. Afr. Med. J.* **87**: 1116.

Pulsinelli, W.A., Brierley, J.B. and Plum, F., 1982, Temporal profile of neuronal damage in a model of transient forebrain ischemia. *Ann. Neurol.* **11**: 491.

Robbins, A.H., McRee, D.E., Williamson, M., Collett, S.A., Xuong, N.H., Furey, W.F., Wang, B.C. and Stout, C.D., 1991, Refined crystal structure of Cd, Zn metallothionein at 2.0 A resolution. *J. Mol. Biol.* **221**: 1269.

Samudralwar, D.L., Diprete, C.C., Ni, B.F., Ehmann, W.D. and Markesbery, W.R., 1995, Elemental imbalances in the olfactory pathway in Alzheimer's disease. *J. Neurol. Sci.* **130**: 139.

Saris, N.E. and Niva, K., 1994, Is Zn2+ transported by the mitochondrial calcium uniporter? *FEBS Lett.* **356**: 195.

Segal, D., Ohana, E., Besser, L., Hershfinkel, M., Moran, A. and Sekler, I., 2004, A role for ZnT-1 in regulating cellular cation influx. *Biochem. Biophys. Res. Commun.* **323**: 1145.

Sekler, I., Moran, A., Hershfinkel, M., Dori, A., Margulis, A., Birenzweig, N., Nitzan, Y. and Silverman, W.F., 2002, Distribution of the zinc transporter ZnT-1 in comparison with chelatable zinc in the mouse brain. *J. Comp. Neurol.* **447**: 201.

Selkoe, D.J., 2004, Alzheimer disease: mechanistic understanding predicts novel therapies. *Ann. Intern. Med.* **140**: 627.

Sensi, S.L. and Jeng, J.M., 2004, Rethinking the excitotoxic ionic milieu: the emerging role of Zn(2+) in ischemic neuronal injury. *Curr. Mol. Med.* **4**: 87

Sensi, S.L., Canzoniero, L.M., Yu, S.P., Ying, H.S., Koh, J.Y., Kerchner, G.A. and Choi, D.W., 1997, Measurement of intracellular free zinc in living cortical neurons: routes of entry. *J. Neurosci.* **17**: 9554.

Sensi, S.L., Ton-That, D. and Weiss, J.H., 2002, Mitochondrial sequestration and Ca(2+)-dependent release of cytosolic Zn(2+) loads in cortical neurons. *Neurobiol. Dis.* **10**: 100.

Sensi, S.L., Ton-That, D., Sullivan, P.G., Jonas, E.A., Gee, K.R., Kaczmarek, L.K. and Weiss, J.H., 2003b, Modulation of mitochondrial function by endogenous Zn2+ pools. *Proc. Natl. Acad. Sci. USA* **100**: 6157.

Sensi, S.L., Ton-That, D., Weiss, J.H., Rothe, A. and Gee, K.R., 2003a, A new mitochondrial fluorescent zinc sensor. *Cell Calcium* **34**: 281.

Sensi, S.L., Yin, H.Z. and Weiss, J.H., 1999a, Glutamate triggers preferential Zn2+ flux through Ca2+ permeable AMPA channels and consequent ROS production. *Neuroreport* **10**: 1723.

Sensi, S.L., Yin, H.Z. and Weiss, J.H., 2000, AMPA/kainate receptor-triggered Zn2+ entry into cortical neurons induces mitochondrial Zn2+ uptake and persistent mitochondrial dysfunction. *Eur. J. Neurosci.* **12**: 3813.

Sensi, S.L., Yin, H.Z., Carriedo, S.G., Rao, S.S. and Weiss, J.H., 1999b, Preferential Zn2+ influx through Ca2+-permeable AMPA/kainate channels triggers prolonged mitochondrial superoxide production. *Proc. Natl Acad. Sci. USA* **96**: 2414.

Sheline, C.T., Behrens, M.M. and Choi, D.W., 2000, Zinc-induced cortical neuronal death: contribution of energy failure attributable to loss of NAD(+) and inhibition of glycolysis. *J. Neurosci.* **20**: 3139.

Skulachev, V.P., Chistyakov, V.V., Jasaitis, A.A. and Smirnova, E.G., 1967, Inhibition of the respiratory chain by zinc ions. *Biochem. Biophys. Res. Commun.* **26**: 1.

Sloviter, R.S., 1985, A selective loss of hippocampal mossy fiber Timm stain accompanies granule cell seizure activity induced by perforant path stimulation. *Brain Res.* **330**: 150.

Smart, T.G., Hosie, A.M. and Miller P.S., 2004, Zn2+ ions: modulators of excitatory and inhibitory synaptic activity. *Neuroscientist* **10**: 432.

Suh, S.W., Chen, J.W., Motamedi, M., Bell, B., Listiak, K., Pons, N.F., Danscher, G. and Frederickson, C.J., 2000a, Evidence that synaptically-released zinc contributes to neuronal injury after traumatic brain injury. *Brain Res.* **852**: 268.

Suh, S.W., Jensen, K.B., Jensen, M.S., Silva, D.S., Kesslak, P.J., Danscher, G. and Frederickson, C.J., 2000b, Histochemically-reactive zinc in amyloid plaques, angiopathy, and degenerating neurons of Alzheimer's diseased brains. *Brain Res.* **852**: 274.

Suh, S.W., Thompson, R.B. and Frederickson, C.J., 2001, Loss of vesicular zinc and appearance of perikaryal zinc after seizures induced by pilocarpine. *Neuroreport* **12**: 1523.

Takeda, A., Hirate, M., Tamano, H., Nisibaba, D. and Oku, N., 2003, Susceptibility to kainate-induced seizures under dietary zinc deficiency. *J. Neurochem.* **85**: 1575.

Thompson, C.M., Markesbery, W.R., Ehmann, W.D., Mao, Y.X. and Vance, D.E., 1988, Regional brain trace-element studies in Alzheimer's disease. *Neurotoxicology* **9**: 1.

Tonder, N., Johansen, F.F., Frederickson, C.J., Zimmer, J. and Diemer, N.H., 1990, Possible role of zinc in the selective degeneration of dentate hilar neurons after cerebral ischemia in the adult rat. *Neurosci. Lett.* **109**: 247.

Traynelis, S. and Cull-Candy, S., 1990, Proton inhibition of N-methyl-D-aspartate receptors in cerebellar neurons. *Nature* **345**: 347.

Trendelenburg, G., Prass, K., Priller, J., Kapinya, K., Polley, A., Muselmann, C., Ruscher, K., Kannbley, U., A. Schmitt, O., Castell, S., Wiegand, F., Meisel, A., Rosenthal, A. and Dirnagl, U., 2002, Serial analysis of gene expression identifies metallothionein-II as major neuroprotective gene in mouse focal cerebral ischemia. *J. Neurosci.* **22**: 5879.

van Lookeren Campagne, M., Thibodeaux, H., van Bruggen, N., Cairns, B., Gerlai, R., J. Palmer, T., Williams, S.P. and Lowe, D.G., 1999, Evidence for a protective role of metallothionein-1 in focal cerebral ischemia. *Proc. Natl Acad. Sci. USA* **96**: 12870.

Weiss, J.H., Hartley, D.M., Koh, J.Y. and Choi, D.W., 1993, AMPA receptor activation potentiates zinc neurotoxicity. *Neuron* **10**: 43.

Wenstrup, D., Ehmann, W.D. and Markesbery, W.R., 1990, Trace element imbalances in isolated subcellular fractions of Alzheimer's disease brains. *Brain Res.* **533**: 125.

Wudarczyk, J., Debska, G. and Lenartowicz, E., 1999, Zinc as an inducer of the membrane permeability transition in rat liver mitochondria. *Arch. Biochem. Biophys.* **363**: 1.

Xia, Y., Zhao, P., Xue, J., Gu, X.Q., Sun, X., Yao, H. and Haddad, G.G., 2003, Na+ channel expression and neuronal function in the Na+/H+ exchanger 1 null mutant mouse. *J. Neurophysiol.* **89**: 229.

Xiong, Z.G., Zhu, X.M., Chu, X.P., Minami, M., Hey, J., Wei, W.L., MacDonald, J.F., Wemmie, J.A., Price, M.P., Welsh, M.J. and Simon, R.P., 2004, Neuroprotection in ischemia: blocking calcium-permeable acid-sensing ion channels. *Cell* **118**: 687.

Yi, J.S., Lee, S.K., Sato, T.A. and Koh, J.Y., 2003, Co-induction of p75(NTR) and the associated death executor NADE in degenerating hippocampal neurons after kainate-induced seizures in the rat. *Neurosci. Lett.* **347**: 126.

Yin, H.Z. and Weiss, J.H., 1995, Zn(2+) permeates Ca(2+) permeable AMPA/kainate channels and triggers selective neural injury. *Neuroreport* **6**: 2553.

Yin, H.Z., Sensi, S.L., Carriedo, S.G. and Weiss, J.H., 1999, Dendritic localization of Ca(2+)-permeable AMPA/kainate channels in hippocampal pyramidal neurons. *J. Comp. Neurol.* **409**: 250.

Yin, H.Z., Sensi, S.L., Ogoshi, F. and Weiss, J.H., 2002, Blockade of Ca2+-permeable AMPA/kainate channels decreases oxygen-glucose deprivation-induced Zn2+ accumulation and neuronal loss in hippocampal pyramidal neurons. *J. Neurosci.* **22**: 1273.

Ying, W., Han, S.-K., Miller, J.W. and Swanson, R.A., 1999, Acidosis potentiates oxidative neuronal death by multiple mechanisms. *J. Neurochem.* **73**: 1549.

Yokoyama, M., Koh, J. and Choi, D.W., 1986, Brief exposure to zinc is toxic to cortical neurons. *Neurosci. Lett.* **71**: 351.

Yu, S.P., Canzoniero, L.M. and Choi, D.W., 2001, Ion homeostasis and apoptosis. *Curr. Opin. Cell. Biol.* **13**: 405.

Yu, S.P., Yeh, C.H., Sensi, S.L., Gwag, B.J., Canzoniero, L.M., Farhangrazi, Z.S., Ying, H.S., Tian, M., Dugan, L.L. and Choi, D.W., 1997, Mediation of neuronal apoptosis by enhancement of outward potassium current. *Science* **278**: 114.

Yu, S.W., Wang, H., Poitras, M.F., Coombs, C., Bowers, W.J., Federoff, H.J., Poirier, G.G., Dawson, T.M. and Dawson, V.L., 2002, Mediation of poly(ADP-ribose) polymerase-1-dependent cell death by apoptosis-inducing factor. *Science* **297**: 259.

Zhang, Y., Wang, H., Li, J., Jimenez, D.A., Levitan, E.S., Aizenman, E. and Rosenberg, P.A., 2004, Peroxynitrite-induced neuronal apoptosis is mediated by intracellular zinc release and 12-lipoxygenase activation. *J. Neurosci.* **24**: 10616.

Zhang, Y., Wang, H., Li, J., Dong, L., Xu P., Chen, W., Neve, R.L., Volpe, J.J. and Rosenberg, P.A., 2006, Intracellular zinc release and ERK phosphorylation are required upstream of 12-lipoxygenase activation in peroxynitrite toxicity to mature rat oligodendrocytes. *J. Biol. Chem.* **281**: 9460.

11

AGING AND COGNITIVE DECLINE: NEUROPROTECTIVE STRATEGIES

Frederico Simões do Couto and Alexandre de Mendonça

1. ABSTRACT

Neuronal loss occurs with aging or dementia, and an extended concept of neuroprotection assumes that the administration of a drug, or a procedure, is able to reverse or prevent neuronal damage. Neuroprotective strategies have been classified according to the mechanisms of neuronal death, and this chapter will deal mainly with neuroinflammation and related processes that occur in aging, cognitive decline, and dementia. The evidence for inflammatory phenomena in dementia, especially in Alzheimer's disease (AD), is impressive, and the role of inflammatory cells and mediators has been extensively studied. Furthermore, anti-inflammatory drugs (specially the nonsteroidal anti-inflammatory drugs (NSAIDs)) have been shown to attenuate these phenomena in preclinical studies. The evidence from epidemiological studies (with different methodologies) for a protective role of anti-inflammatory drugs is variable, although a favorable trend is suggested, especially if the drugs were started years before the onset of AD. Randomized clinical trials with NSAIDs in AD generally yielded disappointing results, without evidence of protective effects. Several reasons have been proposed, specially the short length of the trials and the fact that patients with dementia already have important degenerative changes and neuronal loss that should limit the clinical usefulness of neuroprotective therapeutic approaches at this stage. In recent years, criteria that detect elderly subjects who are at high risk of progressing to dementia were established. There is evidence that patients with this condition, called Mild Cognitive Impairment already have relevant Alzheimer-like pathological changes in the brain, like β-amyloid deposition and neurofibrillary tangles. Furthermore, inflammatory markers are present in the cerebrospinal fluid, namely proinflammatory cytokines, and evidence for increased oxidative damage in the brain can be found. Patients with MCI thus represent a suitable population for testing the efficacy of proposed neuroprotective therapies.

2. NEUROPROTECTION

The concept of neuroprotection assumes that the administration of a drug, or another intervention, is able to reverse neuronal damage, or prevent further neuronal damage (Levi

---

Alexandre de Mendonça – mendonca@fml.ul.pt
Departments of Neurology and Psychiatry and Institute of Pharmacology and Neurosciences, Institute of Molecular Medicine and Faculty of Medicine of Lisbon, Av. Prof. Egas Moniz, 1649-028 Lisbon, Portugal

and Brimble, 2004). The resistance of neurons is thus enhanced in order to protect them from insults. The evidence for neuroprotection comes from experimental studies in cell cultures, where neuronal damage or death (for instance, evaluated by lactate dehydrogenase release) caused by certain noxious stimuli (for instance, glutamate exposure) can be attenuated by specific interventions, like the presence of a particular compound (for instance, adenosine) (de Mendonça et al., 2000). In experimental studies in brain slices, neuronal damage or death can be attenuated, or neuronal function preserved or recovered (for instance, synaptic activity) by neuroprotective interventions. In studies in vivo, some structural (for instance, infarct area) or functional (for instance, neurological deficit) outcomes may be improved. Extension of the concept of neuroprotection includes the protection of neurons in acute neurological disorders, like stroke, or even chronic neurodegenerative disorders in which there is progressive neuronal cell death, such as Alzheimer's disease (AD).

The strategies proposed to achieve neuroprotection in AD have been classified according to the mechanisms of neural death (Akwa et al., 2005). This chapter will deal mainly with inflammation and related processes, and will give particular emphasis to the approach of neuroprotection from a clinical perspective.

3. INFLAMMATION: A TARGET FOR NEUROPROTECTIVE STRATEGIES IN DEGENERATIVE DISORDERS?

The evidence from an inflammatory component in AD remotes to one of the first histo-pathologists working in this field. In 1910, Oskar Fischer (Fischer, 1910; Stuchbury et al., 2005) hypothesized the presence of inflammation in the brain slices of demented patients, probably related to the presence of the "peculiar substance" described by Alois Alzheimer and others few years ago. However, he did not notice the classic Virchow's signs of inflammation, and his paper was largely ignored in the literature.

In fact, the brain has been considered for years an immunological sanctuary, unable to mount an organized immune response. Actually, this point of view has changed radically, and although with some particularities, the brain is perfectly able to organize an immune response and to suffer the consequences of the inflammatory processes – called neuroinflammation by some (Tuppo and Arias, 2005) (i.e., the inflammation associated with the central nervous system (CNS)).

These neuroinflammatory phenomena have been implicated in AD from the mid-1980s. The role of inflammation in AD has an important place and has been the subject of an intense research, expressed by the hundreds of papers published. Understanding the neuro-inflammatory mechanisms has provided a theoretical framework for the design of new drugs that can block this inflammation and protect the neurons.

4. THE ROLE OF INFLAMMATION IN ALZHEIMER'S DISEASE BRAIN

4.1. Microglia

Microglia cells are composed mostly of mesodermally derived macrophages and are considered to be their representatives in the CNS. Microglia cells are immunocompetent cells that orchestrate the endogenous immune response in the CNS and, obviously, have been extensively studied in understanding inflammation and AD.

Microglial cells possess neuroprotective (Streit et al., 1999) and neurotoxic properties and some authors propose that the deregulation of this equilibrium can lead to AD. Activated microglia are present in clusters interdigitating senile plaques. "Activated" is defined by a specific reactive morphology and by an increase in expression of major histocompatibility complex II (MHC II), cytokines, chemokines, and complement. When compared to the brain of nondemented elderly people, MHC II expression (a marker of microglial activation) is substantially increased in AD (Rogers et al., 1988; Styren et al., 1990).

There are several caveats regarding the characterization and the definition of "activation," because the expression of proteins can be different in cell culture and in vivo (Hurley et al., 1999) and increases with age. Moreover, activation is not synonymous with proliferation.

However, the functional evidence for the central role of microglia in the neuroinflammation of AD is impressive.

4.1.1. Participation in Senile Plaque Formation and Evolution

Microglial cells cluster to senile plaques (SP), probably directed by the chemotactic properties of amyloid β (Aβ) (Davis et al., 1992), although other substances could also be responsible, namely complement activation fragments, cytokines, apolipoprotein E (APOE), α1-antichymotrypsin, and chemokines (Akiyama et al., 2000; Meda et al., 2001). There is evidence that the binding of Aβ to microglia can occur through various receptors (Alarcon et al., 2005), namely the scavenger receptor type A, the scavenger receptor type B, the CD36, the receptor for advanced glycation end-products (RAGE), the low-density lipoprotein-receptor-related protein, and the mannose receptor.

It has been shown that the binding of Aβ to these receptors can trigger the production of several noxious substances, such as more Aβ, macrophage-colony stimulating factor (M-CSF) and reactive oxygen species (ROS) by RAGE activation (Yan et al., 1998; Lue et al., 2005), or ROS and complement by scavenger receptor type A binding (El Khoury et al., 1996, 1998).

It seems that microglia can produce Aβ (Bauer et al., 1991; Bitting et al., 1996), although most authors believe that neurons are the main source of this protein (Akiyama et al., 2000; Tuppo and Arias, 2005).

Microglia also participates in plaque evolution. Based on several morphological findings, it has been proposed that microglia participates in converting nonfibrillar to fibrillar amyloid (Shirahama et al., 1990), and transforming diffuse plaques into neuritic plaques (Cotman et al., 1996).

Microglia can phagocyte exogenous fibrillar Aβ (Kopec and Carroll, 1998). The exact meaning of this phagocytic activity is not clear, as it has been shown that Aβ uptaken by microglia further increases the secretion of neurotoxic substances, namely ROS (Colton et al., 2000). These microglial cells laden with Aβ migrate to the vessels and to the ventricles apparently in an attempt to rid the brain of Aβ, resulting in the deposition of this protein on the vessels (Tuppo and Arias, 2005).

4.1.2. Microglial Activation

Aβ has the property of activating microglia. Exposure of microglial cells to Aβ induces the activation of several complex signal transduction pathways (McDonald et al., 1997, 1998), ultimately leading to the activation of proinflammatory gene expression.

This activation can lead to the production of several inflammatory mediators and neurotoxins (Combs et al., 1999). It has been shown that, once activated in vitro, microglia can produce a variety of inflammatory mediators and neurotoxic substances in response to Aβ, namely interleukin-1 (IL-1), interleukin-6 (IL-6), tumor necrosis factor-α (TNF-α), macrophage inflammatory protein-1α (MIP-1α), monocyte chemo-attractant protein-1 (MCP-1), nitric oxide (NO), prostaglandins, excitatory amino acids, proteases, and ROS (Goodwin et al., 1995; Meda et al., 1995, 2001; Giulian et al., 1996; Klegeris and McGeer, 1997; Akwa et al., 2005).

It has also been shown that microglia from AD brains produce twice more complement C1q than microglia from brains of nondemented people. Moreover, plaque associated microglia from AD patients, when compared to elderly nondemented persons, expresses IL-1 (Griffin, 2006), IL-6 (Dickson et al., 1993), TNF-α (Dickson et al., 1993), MCP-1 (Ishizuka et al., 1997), and other related receptors (Akiyama et al., 2000).

4.2. Astrocytes

Although less clearly than microglia, astrocytes have also been implicated in AD. These data come both from morphological and functional studies.

It has been shown that reactive astrocytes are associated with SPs in AD, clustering at Aβ deposits (Dickson, 1997). Astrocytes produce a wide range of inflammatory mediators, in vivo and in vitro, specifically IL-1, IL-6, transforming growth factor (TGF)-β1, TGF-β2, TGF-β3, α1-antichymotrypsin, cyclo-oxigenase (COX)-2, prostaglandins, leukotrienes, thromboxanes, coagulation factors, complement factors and receptors, proteases, and protease inhibitors, similar to that of the microglia in AD (McGeer et al., 2001). More important, upon response to Aβ, astrocytes produce chemokines, cytokines, and ROS (Johnstone et al., 1999; Smits et al., 2002).

On the other hand, some studies claim a protective role for astrocytes by Aβ uptake (Shaffer et al., 1995), or by modulation of microglial activity (DeWitt et al., 1998).

4.3. Neurons

Neurons of AD patients express high levels of mRNA of the classic complement pathway proteins when compared to healthy controls (Shen et al., 1997; Akiyama et al., 2000). Neurons of AD patients also express pentatraxins, C-reactive protein, and amyloid P (Yasojima et al., 2000).

Neurons are also able to produce several inflammatory mediators, such as IL-1 (Friedman, 2001), IL-6 (Suzuki et al., 1999), TNF-α (Tchelingerian, et al., 1994), complement (Shen et al., 1997), COX (Nakayama et al., 1998; Ho et al., 1999), and others (Akiyama et al., 2000).

Most of the evidence is, however, generated from animal models of acute insults, and the involvement of neurons in the inflammation of chronic degenerative diseases such as AD is less clear.

4.4. Complement

The involvement of complement is one of the most well-characterized phenomena in AD. Both pathways, classical and alternative, had been extensively studied and the mechanisms of activation are reasonably well defined.

Levels of complement mRNA for C1r, C1s, C2, C3, C4, C5, C6, C7, C8, C9, and their resulting protein products have been found to exist in higher levels in AD brains than in the liver of the same patients (Yasojima et al., 1999a). Similar complement proteins, including membrane attack complex (Webster et al., 1997), are expressed in higher levels by neurons of AD when compared to healthy controls (Terai et al., 1997). Also, complement inhibitors such as CD59 and C1q inhibitor occur in lower levels in AD when compared to age matched nondemented controls (Shen and Meri, 2003). However, membrane attack complex has not been found in AD by others (Eikelenboom and Veerhuis, 1996).

The classical complement system in the CNS can be activated by other substances than antibodies (Gewurz et al., 1993). In particular, substances present in the SP have this ability, namely Aβ (Rogers et al., 1992), tau containing neurofibrillary tangles (Shen et al., 2001), serum amyloid P (Gewurz et al., 1993) and exposed cellular by-products of neurodegeneration, namely, naked DNA (Gewurz et al., 1993), neurofilaments (Akiyama et al., 2000), and myelin (Akiyama et al., 2000).

Aβ in a very specific way (Bradt et al., 1998) has been shown to activate the alternative complement pathway, although other substances have this capacity (Akyama et al., 2000). Also in AD, mRNA for complement factors, but not for inhibitors, of the alternative pathway have been found elevated (Strohmeyer et al., 2000), something consistent with a potential disturbance of this pathway (Tuppo and Arias, 2005).

4.5. Chemokines and Cytokines

Several interleukins and chemokines have been found elevated in AD (Akiyama et al., 2000; Wilson et al., 2002).

The mRNA of IL-1 can be induced by Aβ (Lee et al., 2002) and over-expression of this protein occurs in microglia and astrocytes of AD brains (Griffin et al., 1995). The inflammatory actions of IL-1 have been related to AD, namely the promotion of APP synthesis and activation of microglia, further enhancing IL-1 production (Mrak and Griffin, 2001). IL-6 has also been implicated in the inflammatory phenomena related to AD (Hüll et al., 2002).

TNF-α, that have been shown to be elevated in AD brain (Perry et al., 2001), induces Aβ production. However, TNF-α has neuroprotective properties (Tarkowski et al., 2003), so its role in AD is somehow ambiguous.

Other interleukins and cytokines were found in SP (Akiyama et al., 2000; Tuppo and Arias, 2005).

4.6. Pentatraxins

Acute phase proteins, such as C-reactive protein and serum amyloid P are associated with AD (McGeer et al., 2001), and their production is upregulated in AD (Yasojima et al., 2000). These proteins can activate complement (Gewurz et al., 1993) and thus enhance cellular damage.

4.7. Cyclo-Oxigenase and Related Substances

Cyclo-oxigenase is the crucial enzyme of the prostaglandins biosynthetic pathway. There are two COX isoforms: COX-1, constitutive, that exists in almost all organs, and COX-2, inducible in inflammatory situations (although exists constitutively in the kidney and in CNS).

Prostaglandins have several physiological functions, namely protecting the stomach from the acid-pepsin secretion and keeping the osmotic differences responsible for the urine concentration in the kidney. However, in inflammatory situations, prostaglandins are responsible for fever, pain, swelling, and other inflammatory signs.

These inflammatory actions of prostaglandins occur not only due to a direct action on several inflammatory cells (e.g., functioning as a chemotatic substance for macrophages), but also to the extensive interaction with other inflammatory mediators (e.g., complement, acute phase proteins, and interleukins). This promotes a cascade of mutual interactions and potentiations that amplifies the inflammatory response, both in acute and chronic situations (Smyth et al., 2006).

COX protein levels are increased in AD (Kitamura et al., 1999; Yasojima et al., 1999b), more specifically in frontal cortex (Pasinetti and Aisen, 1998) and in subregions of the hippocampal formation, where this elevation could potentiate Aβ toxicity (Ho et al., 1999). However, these findings were not confirmed by others (Lukiw and Bazan, 1997).

Furthermore, prostaglandins have been involved in glutamate toxicity, because they inhibit glutamate reuptake by astrocytes and modulate the postsynaptic stimulation of NMDA glutamate receptors (Kelley et al., 1999).

4.8. Others

Oxidative stress in neurons, resulting in the generation of ROS and other free radicals, has been implicated as an important source of damage to these cells in AD (Butterfield et al., 2001; Moreira et al., 2005). However, free radicals are thought to be involved in inflammation, as inflammatory cells use these mediators to kill opsonized agents. ROS and nitrogen active species have been linked in vivo and in vitro to the inflammatory phenomena of AD (Akiyama et al., 2000).

5. WHAT DRUGS CAN ATTENUATE INFLAMMATION IN THE BRAIN?

NSAIDs are a chemically heterogeneous group of drugs that share some clinical properties (anti-inflammatory, antipyretic, and analgesic) and a common mechanism of action, diminished production of prostaglandins through the inhibition of COX.

NSAIDs are divided into two groups: nonselective COX inhibitors (indomethacin, iboprufen, diclofenac, naproxen, and others) and selective COX-2 inhibitors (the coxibs, namely celecoxib, rofecoxib, and others). Selective COX-2 inhibitors are supposed to have less gastric adverse events, because they relatively spare gastric prostaglandins, essential for stomach protection from the acid secretion (FitzGerald and Patrono, 2001).

The rationale for this eventual anti-AD effect of NSAIDs is not clear, because these drugs have various biologic activities. One obvious mechanism could be the decrease of prostaglandins formation, as these agents have potent proinflammatory properties, and have been involved in AD, as described above.

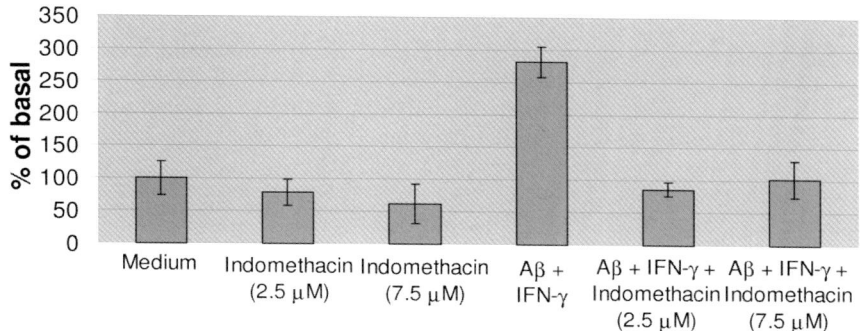

Fig. 1. Effect of indomethacin on the production of NO by murine microglial cells in culture in response to the presence of Aβ1-42 and IFNγ. Both Aβ+IFNγ+indomethacin 2.5 μM and Aβ+IFNγ+indomethacin 7.5 μM were statistically different from Aβ+IFNγ ($P < 0.0001$, Student's t-test)

Besides COX inhibition, NSAIDs modulate gene transcription by agonistic binding to peroxisome proliferator-activated receptor gamma (PPARγ) (Sastre et al., 2006). It has been shown that PPARγ expression is increased in AD (Kitamura et al., 1999) and that PPARγ agonists diminish inflammatory mediators in neuronal cultures, specifically COX-2 expression, IL-6, inducible NO synthase, and thereby attenuate neuronal damage (Combs et al., 2000; Heneka et al., 2000; Ogawa et al., 2000). Potent PPARγ agonists, such as the antidiabetics of thiazolidinedione class, are presently being tested for AD (Watson et al., 2005).

NSAIDs can also diminish the production of NO. We have prepared primary cultures of microglial cells, from one or two days old postnatal mice, showing a high degree of purity (documented immunocytochemically). These cultures were incubated for 24 h in the presence of several compounds, and later the supernatant was assayed for NO determination by the Griess reaction (de Castro et al., 1994). Our results showed a low basal production of NO, slightly stimulated by Aβ, and markedly stimulated synergistically by Aβ plus gamma-interferon (IFNγ), in a way that was reversed in the presence of L-nitro-arginine. Interestingly, the NSAID, indomethacin, markedly inhibited NO production elicited by Aβ plus IFNγ (Fig. 1).

Other authors have shown that NO secretion by microglia leads to neuronal death (Meda et al., 1995). NSAIDs can inhibit the formation of other proinflammatory or neurotoxic substances, such as interleukins (Vane and Botting, 1987; Bour et al., 2000).

Besides interfering with inflammatory phenomena, NSAIDs may have neuroprotective properties through other mechanisms, namely by inhibiting the excitocixity induced by kainate (Kunz and Oliw, 2001) and glutamate (Hewett et al., 2000).

Another mechanism, which is not shared by all NSAIDs, is the inhibition of γ-secretase (Weggen et al., 2001, 2003; Eriksen et al., 2003).

It seems that COX-2 inhibitors cross the blood–brain barrier (BBB) (Dembo et al., 2005), although nonspecific COX inhibitors also seem able to do it (Cohen et al., 1998). Moreover, Aβ can disrupt the BBB integrity (Halliday et al., 2000), thereby making the brain available to the actions of these drugs.

Glucocorticoids are very effective immunosuppressive and anti-inflammatory drugs, used mainly to treat noninfectious inflammatory systemic diseases. These drugs can increase or negatively influence the transcription of some genes, by interacting with specific receptor proteins. Corticosteroids negatively regulate genes for COX-2, inducible nitric oxide synthase (iNOS), and inflammatory cytokines, although multiple actions are involved in the suppression of inflammation by these drugs (Schimmer and Parker, 2006). However, their effect on several organs and systems, namely their deleterious actions on cognition and on the cardiovascular system, makes them less suitable for the use in AD, as discussed later.

6. EPIDEMIOLOGICAL STUDIES ON ANTI-INFLAMMATORY DRUGS AND RISK FOR ALZHEIMER'S DISEASE

The association of anti-inflammatory therapy with Alzheimer's disease has been studied in a number of ways. There are several reviews and meta-analysis on this subject (McGeer et al., 1996; Zandi and Breitner, 2001; in't Veld et al., 2002; Etminan et al., 2003; Szekely et al., 2004).

The first retrospective case-control studies chose a population of AD patients and compared them with matched controls in what concerns to NSAIDs use. Other studies selected a population and looked at the time-dependent association between NSAIDs use and AD.

6.1. Case-Control Studies

Case-control studies evaluated the risk of developing dementia or AD in patients taking NSAIDs. The definition of NSAIDs intake is troublesome, because these drugs are very commonly prescribed, and some of them are over the counter medications. To overcome this problem, many of these studies evaluated arthritis or rheumatoid arthritis patients, since patients with these inflammatory diseases are believed to take NSAIDs in high doses throughout their life. The results of these studies vary widely, with some showing a protective role for NSAIDs (French et al., 1985; Jenkinson et al., 1989), while others failed to confirm this property (Heyman et al., 1984; Broe et al., 1990).

In order to minimize genetic factors, two studies selected patients with a familiar load of AD, defined as having a twin with AD (Breitner et al., 1994) or siblings (Breitner et al., 1995) affected. Both studies yielded odds ratio of the same magnitude of the case-control studies performed before (0.64 and 0.451), supporting a favorable effect of NSAIDs.

A meta-analysis performed by McGeer et al. (1996), with some of these studies, found that subjects suffering from arthritis, and thus presumably treated with anti-inflammatory compounds, had a lower chance of developing AD, with an odds ratio of 0.556 (0.442–0.700, 95% confidence interval (CI), $P < 0.0001$). However, others found these studies largely incomparable (in't Veld et al., 2002).

Other studies directly used NSAIDs as a measure of exposure (Breitner et al., 1994, 1995; The Canadian Study of Health and Aging, 1994; Andersen et al., 1995; Anthony et al., 2000). Although they were quite variable in terms of the methods and results, the meta-analysis of McGeer (1996) found that subjects exposed to anti-inflammatory compounds had a lower chance of developing AD, with an odds ratio of 0.496 (0.343–0.716, 95% CI, $P = 0.0002$) for NSAIDs and 0.559 (0.124–0.738, 95% CI, $P < 0.0001$) for NSAIDs and steroids considered together.

There are several issues regarding these retrospective studies. It is difficult to ascertain if NSAIDs were taken before the onset of dementia, and if so, for how long and with what dose. The studies, with some exceptions, did not differentiate among different NSAIDs.

A different approach was to choose leprosy patients (Endoh et al., 1999), who took dapsone (a drug with anti-inflammatory properties) for long periods of time. A few studies used analgesics as the source of exposure (Henderson et al., 1992), but were inconclusive.

There are a few studies involving corticosteroids that were discrepant. In the twin study (Breitner et al., 1994), the onset of AD was inversely associated with a prior use of corticosteroids and ACTH. However, larger studies (Graves et al., 1990; The Canadian Study of Health and Aging, 1994; Breitner et al., 1995) did not find a significant difference in the risk of AD by the prior use of corticosteroids. The meta-analysis of McGeer et al. (1996), considering these studies, yielded an odds ratio of 0.656 (0.431–0.999, 95% CI, $P = 0.049$).

6.2. Longitudinal Studies

The limitations mentioned above for the retrospective studies can be satisfactorily dealt with in longitudinal (usually prospective) studies. Generally, in these studies, a nondemented population cohort is followed for a long period of time, and both NSAIDs use and incident AD cases are recorded. Mainly these studies gave no support for a protective role of NSAIDs (Fourrier et al., 1996; Henderson et al., 1997), that is, NSAIDs users did not have less AD than nonusers. However, some studies showed a nonsignificant trend toward a reduced risk for AD in people taking NSAIDs (Beard et al., 1998; Cornelius et al., 1998) and, in a stratified analysis, the results favored NSAIDs (Cornelius et al., 2004).

It seems, however, that the relative risk for AD decreases with the increasing duration of NSAIDs use. Important studies, such as the Baltimore Study of Aging (Stewart et al., 1997) and the Rotterdam Study (in't Veld et al., 1998), showed that NSAIDs consumption for more than two years could significantly reduce the risk of AD, while the overall effect of NSAIDs was nonsignificant. More interesting is that the strongest negative association occurred when NSAIDs consumption preceded by at least two years the onset of AD (in't Veld et al., 1998; Zandi and Breitner, 2001). In another meta-analysis (Etminan et al., 2003), the pooled results yielded a relative risk favouring a protective role for NSAIDs with a relative risk of 0.72 (0.56–0.94, 95% CI). A more recent meta-analysis (Szekely et al., 2004) also concluded that NSAIDs exposure was associated with a lower risk of AD.

In line with these results, very recent data from the Cache County Study revealed that the regular use of NSAIDs was associated with less cognitive decline if such use began before the age of 67, and that this relationship was stronger in those with APOE ε4 alleles (Hayden et al., 2006).

These prospective studies are also subject of methodological problems, especially regarding NSAIDs exposure. Repeated assessment can give overestimated values (Stewart et al., 1997) and the pharmacy records, although more reliable, do not contemplate the frequent over-the-counter sale of these drugs. Another issue relates to a possible diminished prescription of NSAIDs to cognitively impaired people due to diminished pain perception or complaining (Farrell et al., 1996).

Other biases, quite common in epidemiological studies, should also be considered (in't Veld et al., 2002). Selection bias relates to the factors that could select a special sample from the population, e.g., people from Memory Clinics are different from people chosen randomly from the community. Information bias depends on how real are the reports, e.g., can we trust the recall of NSAIDs intake by cognitively impaired people? Confounding bias relates to the variables that are an independent risk for AD and for drug use. This last bias is less important in the studies reviewed, although the possibility of a negative association between AD and arthritis has been advanced (Robertson, 2003).

7. RANDOMIZED CLINICAL TRIALS WITH ANTI-INFLAMMATORY DRUGS IN ALZHEIMER'S DISEASE

7.1. Why Randomized Controlled Trials?

Scientific evidence for the effects of drugs in humans must be generated by clinical trials. The best evidence comes from controlled clinical trials in which the tested drug is compared with a standard treatment or a placebo. The process of attribution of the test drug or placebo to the patient should be a random one. Blindness refers to the (non) knowledge of the treatment, double blind meaning that neither the investigator nor the patient knows which drug the patient is taking. Randomized controlled trials (RCTs), preferably double blind, are thus presently considered the gold standard to obtain scientific evidence for the effects of drugs in humans.

Randomized clinical involving NSAIDs in AD generally try to answer the question whether NSAIDs can improve patients with AD as compared to placebo, instead of approaching the issue of AD prevention. Evaluation of patients usually comprises neuropsychological tests to assess cognition, scales of activities of daily living, global scales of dementia evaluation, and sometimes other instruments. Patients are typically assessed at the baseline, then start taking the drug (or the placebo) and are then sequentially evaluated over time.

Some ethical issues have been raised for the design of AD trials. There are presently drugs with some clinical efficacy and approved for the treatment of AD, namely the acetylcholinesterase inhibitors or the *N*-methyl-D-aspartate receptor antagonist, memantine, and placebo-controlled clinical trials may not be considered ethically acceptable (Knopman et al., 1998). Some current studies are now using an *add-on* strategy. In these cases, all patients receive the same approved treatment, in a steady dose, and are then randomly assigned to receive either the test drug or the placebo.

7.2. Randomized Controlled Trials with Anti-Inflammatory Drugs

The RCT assessing the efficacy of anti-inflammatory compounds in AD patients have yielded different results. There is only one positive RCT with indomethacin, which enrolled a small number of patients (Rogers et al., 1993). A favorable trend was observed in the pilot study with diclofenac/misoprostol (Scharf et al., 1999).

Cochrane reviews evaluated iboprufen (Tabet and Feldman, 2003) and indomethacin (Tabet and Feldman, 2002). There are some trials involving iboprufen, but none was a RCT, and they could not generate solid evidence. The only RCT involving indomethacin was referred earlier. The reviewers recommended that both drugs should not be used for AD.

Some negative RCTs can be found in the literature, namely with rofecoxib (Reines et al., 2004), a trial comparing rofecoxib and low-dose naproxen (Aisen et al., 2003), a pilot trial using nimesulid (Aisen et al., 2002), and another pilot study with diclofenac/misoprostol (Scharf et al., 1999). Hydroxychloroquine, an antimalarial drug used to treat inflammatory diseases, can inhibit Aβ neurotoxicity (Giulian, 1999) and an RCT was performed with this compound in AD (van Gool et al., 2001). However, hydroxychloroquine for 18 months did not slow the rate of decline in minimal or mild AD.

Steroidal anti-inflammatory drugs, such as prednisone, were also tried, but the results showed no difference between placebo and the prednisone groups (Aisen et al., 2000). Corticosteroids can be harmful to AD patients because of their effect on cognition and mood, both in acute (Schmidt et al., 1999) and chronic situations (Wolkowitz et al., 1997). Moreover, their metabolic effects on long-term administration can be an additional risk factor for vascular disease.

Cyclophosphamide is an antineoplasic drug, with anti-inflammatory and immunosuppressive properties, especially effective in blocking cytokine-driven processes and complement activation. A pilot trial performed with cyclophosphamide showed improved cognition of 32 AD patients (Leszek and Gasiorowski, 1996). It has been announced that an endovenous trial with this drug was ongoing (Aisen, 2002), but the results could not be found in the literature. A very recent study with endovenous cyclophosphamide was presented. However, it was a dose finding study, and no data were presented on efficacy (Gordon et al., 2006).

A very recent pilot trial with etanercept, an anti-TNFalpha drug, given by perispinal extratecal route in AD, showed a significant improvement in all cognitive scales studied. Although promising, it was an open-label, nonrandomized trial, which clearly reduces the quality of the data generated (Tobinick et al., 2006).

Other drugs proposed for the treatment of AD may also indirectly interfere with inflammatory phenomena in the brain, and will be briefly discussed.

As referred above, ROS have been implicated in AD, and antioxidants have been proposed to have important properties in protecting from AD or delaying its onset. These include selenium, vitamins (vitamin A, [alpha]-tocopherol, vitamin C, natural extracts of papaya, red fruits, etc.), *Ginkgo biloba*, and, more recently, strategies that promote the enhancement of endogenous reducing mechanisms (glutathione, catalase, SOD, melatonin or metallothioneins) (Akwa et al., 2005). There was a controversial trial with vitamin E (Sano et al., 1997) that is discussed later in this chapter. Idebenone, a compound with antioxidant properties, failed to slow cognitive decline in AD (Thal et al., 2003).

Ginkgo biloba has at least one well-designed trial with interesting results (Kanowski and Hoerr, 2003), although other studies were inconclusive (Schneider et al., 2005). Some guidelines do not discourage the use of *G. biloba* (Taylor et al., 2005).

7.3. Special Designs of Clinical Trials to Probe Neuroprotection

Clinical trials in AD have been usually designed to detect some type of amelioration of the clinical symptoms, reflected, as mentioned above, in specific tests or scales. In other words, a symptomatic effect of the drug is searched for. This was typically the case with the clinical trials using the acetylcholinesterase inhibitors. These compounds, by increasing the levels of the neurotransmitter at the damaged cholinergic system, are supposed to improve the synaptic function, with a relatively rapid beneficial effect at the clinical level.

In contrast, the effects of drugs that could attenuate the slow process of neuronal degeneration, what may be called a disease-modifying effect, may not be easily observed using conventional clinical trial designs. Probably, this effect would take longer than 1 year to be statistically significant. This may be one of the reasons why the RCT testing the efficacy of anti-inflammatory compounds in AD patients have been essentially disappointing.

The natural history of AD may be represented, as an oversimplification, by a linear deterioration represented graphically by a descending line (Fig. 2a). Symptomatic improvement can be distinguished graphically from a modification in the course of the disease (slowing down the progression due to neuroprotection). A symptomatic improvement is depicted by a temporary ascending line, followed by a descent that become parallel to the placebo plot (or to eventually reach it, after a long period of time). Neuroprotection is apparent if the slope of deterioration is less steep to the placebo one (Fig. 2a). However, to be statistically different a long period of follow up (>1 year) would be needed.

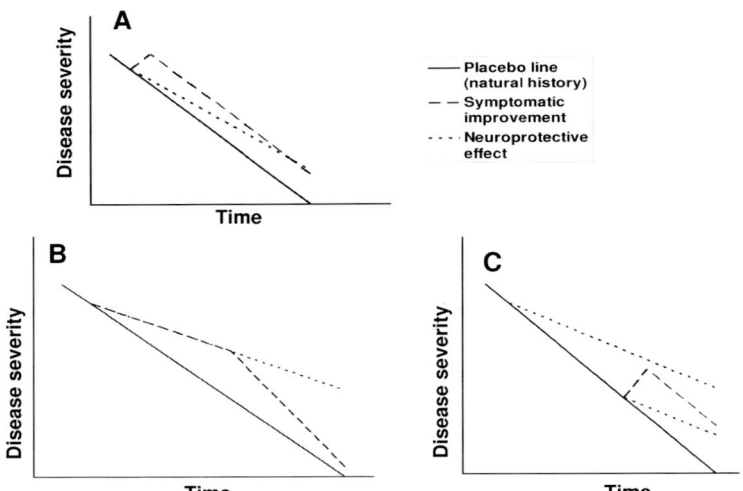

Fig. 2. Natural history of AD and strategies to differentiate between symptomatic improvement and neuroprotection. (**a**) Represents the natural history of AD simplified, showing symptomatic improvement and neuroprotective effect. (**b**) Represents a withdrawal design, in which a symptomatic effect is lost after drug withdrawal. (**c**) Represents a delayed start design, in which the symptomatic effect is similar whether the subject begins the active treatment at the beginning of the trial, or later on

Strategies have been advanced to evaluate more precisely disease-modifying effects of drugs in RCT (Bodick et al., 1997; Whitehouse et al., 1998). One is the drug withdrawal approach. If a neuroprotective effect exists, the patient benefits from attenuation of disease progression while taking the active drug, and after stopping the drug will stay better than the control, i.e., the line for the active drug will not catch with the line of placebo, or will eventually reach it after a very long period of time. In contrast, if the effect of the drug is merely symptomatic, after withdrawal the line for the active drug will become superimposable with the line of placebo (Fig. 2b).

Another possibility is to start the drug at different time-points in separate groups of patients, the so-called delayed-onset design. Again, in case of a neuroprotective effect, the subject that begins the drug later on will stay worse than the subject that started it earlier, because the latter already benefited from attenuation of disease progression. The line for the subject that began the drug later will, thus, not catch with the line of the subject that had it from the beginning. In contrast, if the effect of the drug is merely symptomatic, the patient that began the drug later will benefit from a similar improvement as the one that began it earlier. Both lines will thus essentially converge (Fig. 2c).

It should be emphasized that the attributions of drug withdrawal and delayed-onset must be planned a priori, the patients randomized, and the clinical trial adequately powered to support the pertinent statistical comparisons. It has become commonplace to compare clinical data from patients subjected to a particular drug treatment with a placebo line (e.g., Raskind et al., 2004) extrapolated from Stern equation (Stern et al., 1994). However, the scientific quality of the information generated by this method is clearly limited.

A clinical trial with propentofylline in patients with AD and vascular dementia applied these methods of drug withdrawal and delayed start (Propentofylline Long-term Use Study). At low concentrations, propentofylline is an adenosine uptake blocker, endowed with anti-inflammatory properties (Zhang et al., 1996). Propentofylline inhibits activation of microglial cells, namely proliferation, transformation into macrophages, release of free oxygen radicals, and release of cytokines like TNF-α and IL-1β (Rudolphi et al., 1997). Unfortunately, although beneficial effects for propentofylline were observed, they were mild and did not encourage further development of the drug (Kittner et al., 1997; Rother et al., 1998).

8. MILD COGNITIVE IMPAIRMENT AS A TARGET FOR CLINICAL TRIALS WITH NEUROPROTECTIVE DRUGS

8.1. Mild Cognitive Impairment

It has been emphasized that a neuroprotective treatment should be more efficient if began early in the course of a neurodegenerative disorder. In fact, in advanced stages of the disease a marked neuronal death occurs (Gómez-Isla et al., 1996), which presumably will hamper substantial therapeutic benefits. Selection of mild cognitive impairment (MCI) as the target of new neuroprotective therapies appears reasonable. Patients with MCI are elderly people who suffer from cognitive decline not as severe as to warrant the diagnosis of dementia, but they have the quality of their lives significantly affected and do search for medical help (de Mendonça et al., 2004). Furthermore, those elderly people who meet the criteria for the diagnosis of MCI are at high risk for further cognitive deterioration to the point of dementia, usually AD (Petersen et al., 1999, 2001a). Thus, in a memory clinic setting, as many as 80%

of these patients will actually develop dementia, usually AD, in 6 years (Petersen et al., 2001a). Patients with MCI are also at higher risk for death (Bennett et al., 2002). The prognosis of MCI in a community setting, however, appears to be less consistent and less bleak (Ritchie et al., 2001). It is now accepted that patients with MCI should be identified and monitored for cognitive and functional decline because of their increased risk for subsequent dementia (Petersen et al., 2001b).

Commonly used diagnostic criteria for MCI require (1) memory complaints (preferably corroborated by informant), (2) normal activities of daily living, (3) normal general cognitive function, (4) abnormal memory for age and education, and (5) absence of dementia (Petersen et al., 1999, 2001a). Certainly these criteria need to be operationalized in order to be applied, and there is still no consensus about this operationalization, different groups presenting different studies with distinct criteria. Recruitment of patients for clinical trials in MCI has certainly contributed to this purpose. Several aspects of the operationalization of diagnostic criteria for MCI are discussed below.

8.1.1. Presence of Memory Complaints

Memory complaints are frequent in elderly individuals, and in fact in young and middle-aged adults as well. In a recent study, as much as 64.9% of healthy subjects revealed general forgetfulness with the Subjective Memory Complaints Scale (Schmand et al., 1996), and 67.5% reported that their memory had previously been better (Mendes, 2003).

8.1.2. Abnormal Memory for Age and Education

Many issues remain on the evaluation of memory deficits in patients with MCI. What specific memory domain should be considered, for instance logical memory or verbal learning? What specific test should be used, for instance the Logical Memory subtest of the Wechsler Memory Scale or the New York University (NYU) delayed paragraph recall to probe logical memory? What index is more appropriate, for instance immediate or delayed memory? Should one test be enough or would it be preferable to use more than one? What cut-off values should be preferred, 1 SD or 1.5 SD?

In our studies, we have documented impaired memory function by scores 1 or 1.5 standard deviations (SD) below the normal for age and education, on delayed recall of the two stories from the Logical Memory (LMDelayed) subtest of the Wechsler Memory Scales (WMS; Wechsler, 1969). Certainly, further studies must evaluate the effects of using different memory tests on the characterization and probability of progression of patients with MCI. An interesting issue is that some patients with memory complaints have normal logical memory scores, but display deficits on more demanding memory tasks like the California Verbal Learning Test (CVLT; Delis et al., 1987; Ribeiro et al., 2007). It is not known whether these subjects will behave and progress as typical MCI patients or not. Another interesting topic is the use of so-called ecological tests that might reflect more closely the cognitive demands of everyday life.

8.1.3. Preserved General Cognitive Function

Preserved general cognitive function is required for the diagnosis of MCI. It is not straightforward to define maintained general cognitive function, not as much as memory impinges upon many other cognitive domains. Gauthier and Touchon (2005), in a recent reformulation, mention relatively normal general cognitive function.

We recently compared the neuropsychological profiles of a series of consecutive MCI patients with a control group of healthy elderly subjects (Ribeiro et al., 2006), by using the Battery of Lisbon for the Assessment of Dementia which includes the A's cancellation task, verbal initiative (food products), motor initiative, grapho-motor initiative, object naming, auditory comprehension token test (modified version), orientation, interpretation of proverbs, progressive matrices (Raven), clock draw, calculation, digit span, logical memory (Wechsler), word recall with delay, pairs of words associative learning (Wechsler), information, and visual memory (Wechsler) (BLAD; Garcia, 1984). The presence of a memory deficit on delayed recall was consistent in the MCI sample, as it was an inclusion criterion in the study. But remarkably, patients with MCI frequently had deficits in cognitive domains beyond memory. As much as 30.2% of the patients had deficits in semantic fluency, 33.7% in the Token test, 23.4% in calculation, and 23.9% in motor initiative. If detailed neuropsychological testing is performed, the majority of MCI patients will thus have deficits in cognitive domains other than memory.

These results certainly challenge the notion of preserved general cognitive function in patients with MCI. For practical purposes, and particularly in clinical trials, preserved general cognitive function has been defined as a MMSE score above cut-off.

8.1.4. Normal Activities of Daily Living

It is not clear at present what criteria should be used to define normal activities of daily living in MCI patients. The advantage of using a scale of activities of daily living to have an objective assessment appears obvious. However, some subjective judgement appears necessary, because of major differences in styles and requirements of daily living in aged people. We have considered maintained activities of daily living when the patient keeps the professional, social and familial activities by clinical judgement, and has no or only mild impairment in the IADL scale (Lawton and Brody, 1969), i.e., no more than one item from the IADL scale suffered any changes (Ribeiro et al., 2007).

8.1.5. Absence of Dementia

The criteria used to define dementia will certainly determine the population of patients with the diagnosis of MCI. A few years ago, a consecutive series of 897 subjects referred to the Laboratory of Language, Faculty of Medicine of Lisbon, with possible cognitive impairment was studied (de Mendonça et al., 2004). These subjects were evaluated for the presence of dementia by applying two commonly used criteria: one from the DSM-IV (American Psychiatric Association, 1994), and the other from the International Classification of Diseases (ICD-10, World Health Organization, 1992). Interestingly, 82.6% of the subjects were demented according to the DSM-IV criteria, and only 63.8% followed the ICD-10 criteria (Guerreiro, 1998). Concordance between the two criteria was good, as they agreed in classifying 81.3% of the subjects. The discordance between the criteria was essentially attributable to subjects who were considered to be demented with the DSM-IV criteria but not with the ICD-10 criteria. The ICD-10 criteria are thus more stringent in diagnosing dementia, essentially because they require the presence of thinking impairment and the presence of symptoms for at least 6 months (Guerreiro, 1998). In practical terms, we find preferable to use the DSM-IV criteria in order to exclude from MCI those patients who may or may not have dementia according to the dementia criteria used.

8.1.6. Exclusion Criteria

Patients with many neurological or medical conditions may have a cognitive condition that is in the transition between normality and dementia. However the concept of MCI should be better reserved to patients in whom no clear cause for cognitive impairment is known (de Mendonça et al., 2004). Exclusion criteria should involve clinical history, imaging, or laboratory tests indicating other neurological or psychiatric disorder. Specifically, patients with history of stroke or transient ischemic attack, brain images showing cortical or cortico-subcortical large vessel infarcts, brain hemorrhage or extensive age related white matter changes are usually excluded, as well as patients with a diagnosis of major depression (for instance according to the DSM-IV criteria, American Psychiatric Association, 1994), or a high score on a depression scale (the Hamilton Depression scale (Hamilton, 1960) is commonly used, a scale designed for aged people, like the Geriatric Depression Scale (Yesavage et al., 1983), might be preferable). Subjects with any systemic disease with possible impact on cognition as well as chronic alcohol or drug abuse should also be excluded.

8.2. Clinical Trials with Anti-Inflammatory Drugs in Mild Cognitive Impairment

The results of a double-blind, randomized, controlled study to investigate whether the selective COX-2 inhibitor, rofecoxib, could delay the diagnosis of AD in patients with MCI was recently published (Thal et al., 2005). About 1,500 MCI patients ≥65 years old were randomized to rofecoxib 25 mg or placebo daily for up to 4 years. The primary end point was the percentage of patients with a clinical diagnosis of AD. The estimated annual AD diagnosis rate was in fact slightly higher, 6.4%, in the rofecoxib group, as compared to the placebo group, 4.5% (rofecoxib/placebo hazard ratio = 1.46 (1.09–1.94, 95% CI, $P = 0.011$)). Analysis of secondary end points, including measures of cognition and global function, did not demonstrate differences between treatment groups. The results from this MCI study did not support the hypothesis that rofecoxib would delay a diagnosis of AD.

We recently participated in a double-blind, randomized, controlled study to investigate whether the aspirin derivative triflusal could stabilize or delay the progression of MCI. About 560 patients were planned to be randomized to triflusal (900 mg daily) or placebo, and followed for 18 months, but the final number of participants was less. The primary end point was a change in ADAS-cog score of 2.5 points. Several secondary efficacy variables were also considered. The results of this study have been partially presented very recently, and, as compared to the placebo group, there were significant differences in the probability of progression to dementia in the triflusal group (Gómez-Isla et al., 2006). However, there was a significant difference in baseline characteristics of the APOE genotype, with a significant proportion of patients with at least one ε4 allele randomized to the placebo group. Anyway, the complete results of this trial are not yet known.

Another study evaluated the effect of vitamin E, as well as the acetylcholinesterase inhibitor, donepezil, in delaying progression of MCI patients to Alzheimer's disease (Petersen et al., 2005). Vitamin E has known antioxidant properties. A previous study had provided evidence that vitamin E could delay the time to important milestones in patients with AD, but this study had some controversial methodological issues (Sano et al., 1997). Vitamin E did not reduce the probability that patients with MCI will progress to AD.

Certainly, several compounds other than anti-inflammatory compounds were recently subjected to clinical trials, and all failed to show efficacy in preventing or slowing the progression of MCI to dementia (Jelic et al., 2005). Some suggestions to improve the design of clinical trials were advanced, concerning the selection of more homogenous samples at entry, longer duration of the trials, and more relevant outcome measures (Jelic et al., 2005; Visser et al., 2005).

9. NEW ANTI-INFLAMMATORY STRATEGIES

It has been suggested that the neuroprotective effect of anti-inflammatory compounds may not be due to the inhibition of COX, but to the allosteric modulation of γ-secretase and decrease in the production of the Aβ peptide, a mechanism that is shared by some anti-inflammatory drugs, like ibuprofen, indomethacin, and diclofenac, but not by others, like aspirin, naproxen, and celecoxib (Weggen et al., 2001, Imbimbo, 2004). The investigation of new mechanisms of action of these compounds is thus worthwhile pursuing.

In conclusion, the use of anti-inflammatory drugs remains an interesting approach, in the hope that they may attenuate the inflammatory changes associated with Alzheimer's disease pathology (Fig. 3). Should an anti-inflammatory compound, or any other neuroprotective drug, prove effective in delaying progression to AD, the detection and treatment of the very early stages of the disease will certainly become a public health priority.

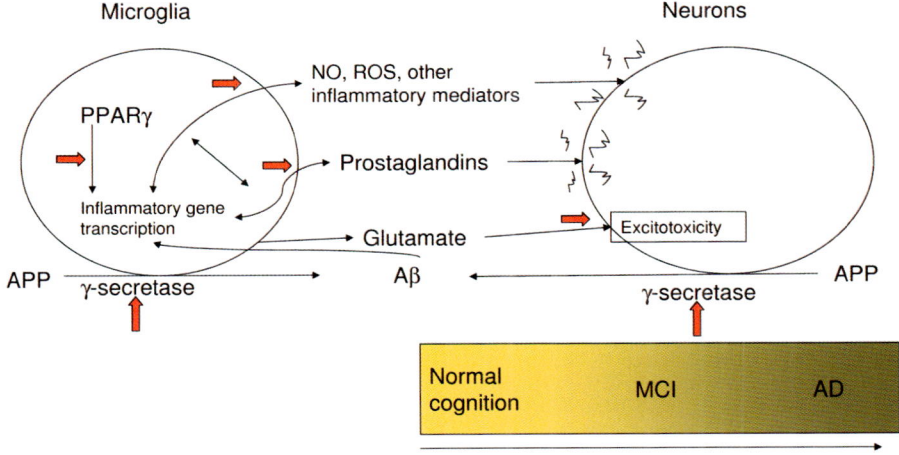

Fig. 3. Microglial cells and astrocytes produce several inflammatory mediators, in response to different stimuli, namely Aβ. These inflammatory mediators would damage neurons and perpetuate the inflammatory stress. Nonsteroidal anti-inflammatory drugs may attenuate the inflammatory stress, by several mechanisms (*red arrows*). However, in Alzheimer's disease (AD), neuronal loss hampers the benefit of anti-inflammatory strategies and this maybe one reason for failure of these strategies in AD. MCI patients thus represent an excellent target for clinical trials

10. ACKNOWLEDGMENTS

The authors thank Manuela Guerreiro, Filipa Ribeiro, and Tiago Mendes for helpful comments.

11. REFERENCES

Aisen, P.S., 2002, The potential of anti-inflammatory drugs for the treatment of Alzheimer's disease. *The Lancet Neurology* **1**: 279.

Aisen, P.S., Davis, K.L., Berg, J.D., Schafer, K., Campbell, K., Thomas, R.G., Weiner, M.F., Farlow, M.R., Sano, M., Grundman, M. and Thal, L.J., 2000, A randomized controlled trial of prednisone in Alzheimer's disease. Alzheimer's Disease Cooperative Study. *Neurology* **54**: 588.

Aisen, P.S., Schmeidler, J., Pasinetti, G.M., 2002, Randomized pilot study of nimesulide treatment in Alzheimer's disease. *Neurology* **58**: 1050.

Aisen, P.S., Schafer, K.A., Grundman, M., Pfeiffer, E., Sano, M., Davis, K.L., Farlow, M.R., Jin, S., Thomas, R.G. and Thal, L.J., 2003, Alzheimer's Disease Cooperative Study. Effects of rofecoxib or naproxen vs. placebo on Alzheimer's disease progression: a randomized controlled trial. *JAMA* **289**: 2819.

Akiyama, H., Barger, S., Barnum, S., Bradt, B., Bauer, J., Cole, G.M., Cooper, N.R., Eikelenboom, P., Emmerling, M., Fiebich, B.L., Finch, C.E., Frautschy, S., Griffin, W.S., Hampel, H., Hull, M., Landreth, G., Lue, L., Mrak, R., Mackenzie, I.R., McGeer, P.L., O'Banion, M.K., Pachter, J., Pasinetti, G., Plata-Salaman, C., Rogers, J., Rydel, R., Shen, Y., Streit, W., Strohmeyer, R., Tooyoma, I., Van Muiswinkel, F.L., Veerhuis, R., Walker, D., Webster, S., Wegrzyniak, B., Wenk, G. and Wyss-Coray, T., 2000, Inflammation and Alzheimer's disease. *Neurobiol. Aging* **21**: 383.

Akwa, Y., Allain, H., Bentue-Ferrer, D., Berr, C., Bordet, R., Geerts, H., Nieoullon, A., Onteniente, B. and Vercelletto, M., 2005, Neuroprotection and neurodegenerative diseases: from biology to clinical practice. *Alzheimer Dis. Assoc. Disord.* **19**: 226.

Alarcon, R., Fuenzalida, C., Santibanez, M. and von Bernhardi, R., 2005, Expression of scavenger receptors in glial cells. Comparing the adhesion of astrocytes and microglia from neonatal rats to surface-bound beta-amyloid. *J. Biol. Chem.* **280**: 30406.

American Psychiatric Association, 1994, *Diagnostic and Statistical Manual of Mental Disorders*, 4th edition. American Psychiatric Association, Washington.

Andersen, K., Launer, L.J., Ott, A., Hoes, A.W., Breteler, M.M. and Hofman, A., 1995, Do nonsteroidal anti-inflammatory drugs decrease the risk for Alzheimer's disease? The Rotterdam Study. *Neurology* **45**: 1441.

Anthony, J.C., Breitner, J.C., Zandi, P.P., Meyer, M.R., Jurasova, I., Norton, M.C. and Stone, S.V., 2000, Reduced prevalence of AD in users of NSAIDs and H2 receptor antagonists: the Cache County study. *Neurology* **54**: 2066.

Bauer, J., Konig, G., Strauss, S., Jonas, U., Ganter, U., Weidemann, A., Monning, U., Masters, C.L., Volk, B., Berger, M., et al., 1991, In vitro matured human macrophages express Alzheimer's beta A4-amyloid precursor protein indicating synthesis in microglial cells. *FEBS Lett.* **282**: 335.

Beard, C.M., Waring, S.C., O'Brien, P.C., Kurland, L.T. and Kokmen, E., 1998, Nonsteroidal anti-inflammatory drug use and Alzheimer's disease: a case-control study in Rochester, Minnesota, 1980 through 1984. *Mayo Clin. Proc.* **73**: 951.

Bennett, D.A., Wilson, R.S., Schneider, J.A., Evans, D.A., Beckett, L.A., Aggarwal, N.T., Barnes, L.L., Fox, J.H. and Bach, J., 2002, Natural history of mild cognitive impairment in older persons. *Neurology* **59**: 198.

Bitting, L., Naidu, A., Cordell, B. and Murphy, G.M. Jr., 1996, Beta-amyloid peptide secretion by a microglial cell line is induced by beta-amyloid-(25–35) and lipopolysaccharide. *J. Biol. Chem.* **271**: 16084.

Bodick, N., Forette, F., Hadler, D., Harvey, R.J., Leber, P., McKeith, I.G., Riekkinen, P.J., Rossor, M.N., Scheltens, P., Shimohama, S., Spiegel, R., Tanaka, S., Thal, L.J., Urata, Y., Whitehouse, P. and Wilcock, G., 1997, Protocols to demonstrate slowing of Alzheimer's disease progression. Position paper from the International Working Group on Harmonization of Dementia Drug Guidelines. The Disease Progression Sub-Group. *Alzheimer Dis. Assoc. Disord.* **11**: 50.

Bour, A.M., Westendorp, R.G., Laterveer, J.C., Bollen, E.L. and Remarque, E.J., 2000, Interaction of indomethacin with cytokine production in whole blood. Potential mechanism for a brain-protective effect. *Exp. Gerontol.* **35**: 1017.

Bradt, B.M., Kolb, W.P. and Cooper, N.R., 1998, Complement dependent proinflammatory properties of the Alzheimer's disease beta-peptide. *J. Exp. Med.* **188**: 431.

Breitner, J.C., Gau, B.A., Welsh, K.A., Plassman, B.L., McDonald, W.M., Helms, M.J. and Anthony, J.C., 1994, Inverse association of anti-inflammatory treatments and Alzheimer's disease: initial results of a co-twin control study. *Neurology* **44**: 227.

Breitner, J.C., Welsh, K.A., Helms, M.J., Gaskell, P.C., Gau, B.A., Roses, A.D, Pericak-Vance, M.A. and Saunders, A.M., 1995, Delayed onset of Alzheimer's disease with nonsteroidal anti-inflammatory and histamine H2 blocking drugs. *Neurobiol. Aging* **16**: 523.

Broe, G.A., Henderson, A.S., Creasey, H., McCusker, E., Korten, A.E., Jorm, A.F., Longley, W. and Anthony, J.C., 1990, A case-control study of Alzheimer's disease in Australia. *Neurology* **40**: 1698.

Butterfield, D.A., Drake, J., Pocernich, C. and Castegna, A., 2001, Evidence of oxidative damage in Alzheimer's disease brain: central role for amyloid beta-peptide. *Trends Mol. Med.* **7**: 548.

Cohen, O., Zylber-Katz, E., Caraco, Y., Granit, L. and Levy, M., 1998, Cerebrospinal fluid and plasma concentrations of dipyrone metabolites after a single oral dose of dipyrone. *Eur. J. Clin. Pharmacol.* **54**: 549.

Colton, C.A., Chernyshev, O.N., Gilbert, D.L. and Vitek, M.P., 2000, Microglial contribution to oxidative stress in Alzheimer's disease. *Ann. NY Acad. Sci.* **899**: 292.

Combs, C.K., Johnson, D.E., Cannady, S.B., Lehman, T.M. and Landreth, G.E., 1999, Identification of microglial signal pathways mediating a neurotoxic response to amyloidogenic fragments of beta-amyloid and prion proteins. *J. Neurosci.* **19**: 928.

Combs, C.K., Johnson, D.E., Karlo, J.C., Cannady, S.B. and Landreth, G.E., 2000, Inflammatory mechanisms in Alzheimer's disease: inhibition of beta-amyloid-stimulated proinflammatory responses and neurotoxicity by PPARgamma agonists. *J. Neurosci.* **20**: 558.

Cornelius, C., Fratiglioni, L., Fastbom, J., Guo, Z., Viitanem, M. and Winblad, B., 1998, No support for a protective role of NSAIDS against Alzheimer's disease – a follow-up population-based study. *Neurobiol. Aging* **19**: S28.

Cornelius, C., Fastbom, J., Winblad, B. and Viitanen, M., 2004, Aspirin, NSAIDs, risk of dementia, and influence of the apolipoprotein E epsilon 4 allele in an elderly population. *Neuroepidemiology* **23**: 135.

Cotman, C.W., Tenner, A.J. and Cummings, B.J., 1996, Beta-amyloid converts an acute phase injury response to chronic injury responses. *Neurobiol. Aging* **17**: 723.

Davis, J.B., McMurray, H.F. and Schubert, D., 1992, The amyloid Beta-protein of Alzheimer's disease is chemotactic for mononuclear phagocytes. *Biochem. Biophys. Res. Commun.* **189**: 1096.

de Castro, M., Mota-Filipe, H., Caneira, M., Rico, J.M., Scott-Burden, T. and Vanhoutte, P.M., 1994, DL-propranolol augments production of NO induced by cytokines in cultured aortic smooth muscle of the rat. *Eur. J. Pharmacol.* **261**: 199.

de Mendonça, A., Sebastião, A.M. and Ribeiro, J.A., 2000, Adenosine: does it have a neuroprotective role after all? *Brain Res. Rev.* **33**: 258.

de Mendonça, A., Guerreiro, M., Ribeiro, F., Mendes, T. and Garcia, C., 2004, Mild cognitive impairment: focus on the diagnosis. *J. Mol. Neurosci.* **23**: 13.

Delis, D.C., Kramer, J.H., Kaplan, E. and Ober, B.A., 1987, *California Verbal Learning Test: Adult Version Manual.* The Psychological Corporation, San Antonio, Texas.

Dembo, G., Park, S.B. and Kharasch, E.D., 2005, Central nervous system concentrations of cyclooxygenase-2 inhibitors in humans. *Anesthesiology* **102**: 409.

DeWitt, D.A., Perry, G., Cohen, M., Doller, C. and Silver, J., 1998, Astrocytes regulate microglial phagocytosis of senile plaque cores of Alzheimer's disease. *Exp. Neurol.* **149**: 329.

Dickson, D.W., 1997, The pathogenesis of senile plaques. *J. Neuropathol. Exp. Neurol.* **56**: 321.

Dickson, D.W., Lee, S.C., Mattiace, L.A., Yen, S.H. and Brosnan, C., 1993, Microglia and cytokines in neurological disease, with special reference to AIDS and Alzheimer's disease. *Glia* **7**: 75.

Eikelenboom, P., Veerhuis, R., 1996, The role of complement and activated microglia in the pathogenesis of Alzheimer's disease. *Neurobiol. Aging* **17**: 673.

El Khoury, J., Hickman, S.E., Thomas, C.A., Cao, L., Silverstein, S.C. and Loike, J.D., 1996, Scavenger receptor-mediated adhesion of microglia to beta-amyloid fibrils. *Nature* **382**: 716.

El Khoury, J., Hickman, S.E., Thomas, C.A., Loike, J.D. and Silverstein, S.C., 1998, Microglia, scavenger receptors, and the pathogenesis of Alzheimer's disease. *Neurobiol. Aging* **19**: S81.

Endoh, M., Kunishita, T. and Tabira, T., 1999, No effect of anti-leprosy drugs in the prevention of Alzheimer's disease and beta-amyloid neurotoxicity. *J. Neurol. Sci.* **165**: 28.

Eriksen, J.L., Sagi, S.A., Smith, T.E., Weggen, S., Das, P., McLendon, D.C., Ozols, V.V., Jessing, K.W., Zavitz, K.H., Koo, E.H. and Golde, T.E., 2003, NSAIDs and enantiomers of flurbiprofen target gamma-secretase and lower Abeta 42 in vivo. *J. Clin. Invest.* **112**: 440.

Etminan, M., Gill, S. and Samii, A., 2003, Effect of nonsteroidal anti-inflammatory drugs on risk of Alzheimer's disease: systematic review and meta-analysis of observational studies. *BMJ* **327**: 128.

Farrell, M.J., Katz, B. and Helme, R.D., 1996, The impact of dementia on the pain experience. *Pain* **67**: 7.

FitzGerald, G.A. and Patrono, C., 2001, The coxibs, selective inhibitors of cyclooxygenase-2. *N. Engl. J. Med.* **345**: 433.

Fourrier, A., Letenneur, L., Begaud, B. and Dartigues, J.F., 1996, Nonsteroidal antiinflammatory drug use and cognitive function in the elderly: inconclusive results from a population-based cohort study. *J. Clin. Epidemiol.* **49**: 1201.
French, L.R., Schuman, L.M., Mortimer, J.A., Hutton, J.T., Boatman, R.A. and Christians, B., 1985, A case-control study of dementia of the Alzheimer type. *Am. J. Epidemiol.* **121**: 414.
Friedman, W.J., 2001, Cytokines regulate expression of the type 1 interleukin-1 receptor in rat hippocampal neurons and glia. *Exp. Neurol.* **68**: 23.
Garcia, C., 1984, *A doença de Alzheimer. Problemas de diagnóstico clínico* (*Alzheimer's disease. Difficulties of clinical diagnosis*), Ph.D. Thesis, Universidade de Lisboa.
Gauthier, S., Touchon, J., 2005, Mild cognitive impairment is not a clinical entity and should not be treated. *Arch. Neurol.* **6**: 1164.
Gewurz, H., Ying, S-C., Jiang, H. and Lint, T.E., 1993, Nonimmune activation of the classical complement pathway. *Behring. Inst. Mitt.* **93**: 138.
Giulian, D., 1999, Microglia and the immune pathology of Alzheimer's disease. *Am. J. Hum. Genet.* **65**: 13.
Giulian, D., Haverkamp, L.J., Yu, J.H., Karshin, W., Tom, D., Li, J., Kirkpatrick, J., Kuo, L.M. and Roher, A.E., 1996, Specific domains of beta-amyloid from Alzheimer plaque elicit neuron killing in human microglia. *J. Neurosci.* **16**: 6021.
Gómez-Isla, T., Price, J.L., McKeel, D.W., Jr., Morris, J.C., Growdon, J.H. and Hyman, B.T., 1996, Profound loss of layer II entorhinal cortex neurons occurs in very mild Alzheimer's disease. *J. Neurosci.* **16**: 4491.
Gómez-Isla, T., Muñoz, G., Ferro, J.M., Lage, J.M.M. and Ramírez, J.C.N., 2006, A randomized, double-blind, placebo controlled-trial of triflusal in mild cognitive impairment. *Alzheimers Dement* **2**: S512.
Goodwin, J.L., Uemura, E. and Cunnick, J.E., 1995, Microglial release of nitric oxide by the synergistic action of beta-amyloid and IFN-gamma. *Brain Res.* **692**: 207.
Gordon, M.L., Mirza, N., Bauer, L., Spoor, E., Overman, G., Dustin, I., Fleischer, T.A., Putman, K., Cohen, R.M. and Sunderland, T., 2006, Intravenous pulse cyclophosphamide in Alzheimer's disease: results of a pilot dose-finding study. *Alzheimers Dement* **2**: S360.
Graves, A.B., White, E., Koepsell, T.D., Reifler, B.V., van Belle, G., Larson, E.B. and Raskind, M., 1990, A case-control study of Alzheimer's disease. *Ann. Neurol.* **28**: 766.
Griffin, W.S., 2006, Inflammation and neurodegenerative diseases. *Am. J. Clin. Nutr.* **83**: 470.
Griffin, W.S., Sheng, J.G., Roberts, G.W. and Mrak, R.E., 1995, Interleukin-1 expression in different plaque types in Alzheimer's disease: significance in plaque evolution. *J. Neuropathol. Exp. Neurol.* **54**: 276.
Guerreiro, M., 1998, *Contributo da Neuropsicologia para o Estudo das Demências* (*Contribution of neuropsychology to the study of dementia*), Ph.D. Thesis, Universidade de Lisboa.
Halliday, G., Robinson, S.R., Shepherd, C. and Kril, J., 2000, Alzheimer's disease and inflammation: a review of cellular and therapeutic mechanisms. *Clin. Exp. Pharmacol. Physiol.* **27**: 1.
Hamilton, M., 1960, A rating scale for depression. *J. Neurol. Neurosurg. Psychiatry* **23**: 56.
Hayden, K.M., Zandi, P.P., Khachaturian, A.S., Pieper, C.F., Sanders, L., Ostbye, T., Tschanz, J.T., Norton, M.C., Munger, R., Lyketsos, C.G., Breitner, J.C.S. and Welsh-Bohmer, K.A., 2006, Modification of cognitive trajectories: NSAID use in the Cache County Study. *Alzheimers Dement* **2**:S174.
Henderson, A.S., Jorm, A.F., Korten, A.E., Creasey, H., McCusker, E., Broe, G.A., Longley, W. and Anthony, J.C., 1992, Environmental risk factors for Alzheimer's disease: their relationship to age of onset and to familial or sporadic types. *Psychol. Med.* **22**: 429.
Henderson, A.S., Jorm, A.F., Christensen, H., Jacomb, P.A. and Korten, A.E., 1997, Aspirin, anti-inflammatory drugs and risk of dementia. *Int. J. Geriatr. Psychiatry* **12**: 926.
Heneka, M.T., Klockgether, T. and Feinstein, D.L., 2000, Peroxisome proliferator-activated receptor-gamma ligands reduce neuronal inducible nitric oxide synthase expression and cell death in vivo. *J. Neurosci.* **20**: 6862.
Hewett, S.J., Uliasz, T.F., Vidwans, A.S. and Hewett, J.A., 2000, Cyclooxygenase-2 contributes to N-methyl-D-aspartate-mediated neuronal cell death in primary cortical cell culture. *J. Pharmacol. Exp. Ther.* **293**: 417.
Heyman, A., Wilkinson, W.E., Stafford, J.A., Helms, M.J., Sigmon, A.H. and Weinberg, T., 1984, Alzheimer's disease: a study of epidemiological aspects. *Ann. Neurol.* **15**: 335.
Ho, L., Pieroni, C., Winger, D., Purohit, D.P., Aisen, P.S. and Pasinetti, G.M.J., 1999, Regional distribution of cyclooxygenase-2 in the hippocampal formation in Alzheimer's disease. *Neurosci. Res.* **57**: 295.
Hüll, M., Lieb, K. and Fiebich, B.L., 2002, Pathways of inflammatory activation in Alzheimer's disease: potential targets for disease modifying drugs. *Curr. Med. Chem.* **9**: 83.
Hurley, S.D., Walter, A.S., Semple-Rowland, S.L. and Streit, W.J., 1999, Cytokine transcripts expressed by microglia in vitro are not expressed by ameboid microglia at the developing rat central nervous system. *Glia* **25**: 304.
Imbimbo, B.P., 2004, The potential role of nonsteroidal anti-inflammatory drugs in treating Alzheimer's disease. *Expert. Opin. Investig. Drugs* **13**: 1469.

International Classification of Disease. Word Health Organization, 1992, *The ICD-10 Classification of Mental and Behavioral Disorders.* World Health Organization, Geneva, Switzerland.

in't Veld, B.A., Launer, L.J., Hoes, A.W., Ott, A., Hofman, A., Breteler, M.M. and Stricker, B.H., 1998, NSAIDs and incident Alzheimer's disease. The Rotterdam Study. *Neurobiol. Aging* **19**: 607.

in't Veld, B.A., Launer, L.J., Breteler, M.M., Hofman, A. and Stricker, B.H., 2002, Pharmacologic agents associated with a preventive effect on Alzheimer's disease: a review of the epidemiologic evidence. *Epidemiol. Rev.* **24**: 248.

Ishizuka, K., Kimura, T., Igata-yi, R., Katsuragi, S., Takamatsu, J. and Miyakawa, T., 1997, Identification of monocyte chemoattractant protein-1 in senile plaques and reactive microglia of Alzheimer's disease. *Psychiatry Clin. Neurosci.* **51**: 135.

Jelic, V., Kivipelto, M. and Winblad, B., 2005, Clinical trials in mild cognitive impairment: lessons for the future. *J Neurol. Neurosurg. Psychiatry*, Published on-line, 23 Nov 2005.

Jenkinson, M.L., Bliss, M.R., Brain, A.T. and Scott, D.L., 1989, Rheumatoid arthritis and senile dementia of the Alzheimer's type. *Br. J. Rheumatol.* **28**: 86.

Johnstone, M., Gearing, A.J. and Miller, K.M., 1999, A central role for astrocytes in the inflammatory response to beta-amyloid; chemokines, cytokines and reactive oxygen species are produced. *J. Neuroimmunol.* **93**: 182.

Kanowski, S. and Hoerr, R., 2003, *G. biloba* extract Egb 761 in dementia intent-to-treat analysis of a 24-week, multi-center, double-blind, placebo controlled, randomized trial. *Pharmacopsychiatry* **63**: 297.

Kelley, K.A., Ho, L., Winger, D., Freire-Moar, J., Borelli, C.B., Aisen, P.S. and Pasinetti, G.M., 1999, Potentiation of excitotoxicity in transgenic mice overexpressing neuronal cyclooxygenase-2. *Am. J. Pathol.* **155**: 995.

Kitamura, Y., Shimohama, S., Koike, H., Kakimura, J., Matsuoka, Y., Nomura, Y., Gebicke-Haerter, P.J. and Taniguchi, T., 1999, Increased expression of cyclooxygenases and peroxisome proliferator-activated receptor-gamma in Alzheimer's disease brains. *Biochem. Biophys. Res. Commun.* **254**: 582.

Kittner, B., Rossner, M. and Rother, M., 1997, Clinical trials in dementia with propentofylline. *Ann. NY Acad. Sci.* **826**: 307.

Klegeris, A. and McGeer, P.L., 1997, Beta-amyloid protein enhances macrophage production of oxygen free radicals and glutamate. *J. Neurosci. Res.* **49**: 229.

Knopman, D., Kahn, J. and Miles, S., 1998, Clinical Research Designs for Emerging Treatments for Alzheimer's disease. *Arch. Neurol.* **55**: 1425.

Kopec, K.K. and Carroll, R.T., 1998, Alzheimer's beta-amyloid peptide 1-42 induces a phagocytic response in murine microglia. *J. Neurochem.* **71**: 2123.

Kunz, T. and Oliw, E.H., 2001, The selective cyclooxygenase-2 inhibitor rofecoxib reduces kainate-induced cell death in the rat hippocampus. *Eur. J. Neurosci.* **13**: 569.

Lawton, M.P. and Brody, E.M., 1969, Assessment of older people: self-maintaining and instrumental activities of daily living. *Gerontologist* **9**: 179.

Lee, Y.B., Nagai, A. and Kim, S.U., 2002, Cytokines, chemokines, and cytokine receptors in human microglia. *J. Neurosci. Res.* **69**: 94.

Leszek, J. and Gasiorowski, K., 1996, Therapeutic efficay of cyclophosphamide in Alzheimer's disease. *Alzheimers Res.* **2**: 43.

Levi, M.S. and Brimble, M.A., 2004, A review of neuroprotective agents. *Curr. Med. Chem.* **11**: 2383.

Lue, L.F., Yan, S.D. and Stern, D.M., 2005, Preventing activation of receptor for advanced glycation endproducts in Alzheimer's disease. *Curr. Drug Targets CNS Neurol. Disord.* **4**: 249.

Lukiw, W.J. and Bazan, N.G., 1997, Cyclooxygenase 2 RNA message abundance, stability, and hypervariability in sporadic Alzheimer neocortex. *J. Neurosci. Res.* **50**: 937.

McDonald, D.R., Brunden, K.R. and Landreth, G.E., 1997, Amyloid fibrils activate tyrosine kinase-dependent signaling and superoxide production in microglia. *J. Neurosci.* **17**: 2284.

McDonald, D.R., Bamberger, M.E., Combs, C.K. and Landreth, G.E., 1998, beta-amyloid fibrils activate parallel mitogen-activated protein kinase pathways in microglia and THP1 monocytes. *J. Neurosci.* **18**: 4451.

McGeer, P.L., Schulzer, M. and McGeer, E.G., 1996, Arthritis and anti-inflammatory agents as possible protective factors for Alzheimer's disease: a review of 17 epidemiologic studies. *Neurology* **47**: 425.

McGeer, E.G., Yasojima, K., Schwab, C. and McGeer, P.L., 2001, The pentraxins: possible role in Alzheimer's disease and other innate inflammatory diseases. *Neurobiol. Aging* **22**: 843.

Meda, L., Cassatella, M.A., Szendrei, G.I., Otvos, L., Jr., Baron, P., Villalba, M., Ferrari, D. and Rossi, F., 1995, Activation of microglial cells by beta-amyloid protein and interferon-gamma. *Nature* **374**: 647.

Meda, L., Baron, P. and Scarlato, G., 2001, Glial activation in Alzheimer's disease: the role of Abeta and its associated proteins. *Neurobiol. Aging* **22**: 885.

Mendes, T., 2003, Memory and Metamemory: from the young to the aged. Master degree thesis, Faculdade de Medicina da Universidade de Lisboa.

Moreira, P.I., Honda, K., Liu, Q., Santos, M.S., Oliveira, C.R., Aliev, G., Nunomura, A., Zhu, X., Smith, M.A. and Perry, G., 2005, Oxidative stress: the old enemy in Alzheimer's disease pathophysiology. *Curr. Alzheimer Res.* **2**: 403.
Mrak, R.E. and Griffin, W.S., 2001, Interleukin-1, neuroinflammation, and Alzheimer's disease. *Neurobiol. Aging* **22**: 903.
Nakayama, M., Uchimura, K., Zhu, R.L., Nagayama, T., Rose, M.E., Stetler, R.A., Isakson, P.C., Chen, J. and Graham, S.H., 1998, Cyclooxygenase-2 inhibition prevents delayed death of CA1 hippocampal neurons following global ischemia. *Proc. Natl. Acad. Sci. USA.* **95**: 10954.
Ogawa, O., Umegaki, H., Sumi, D., Hayashi, T., Nakamura, A., Thakur, N.K., Yoshimura, J., Endo., H. and Iguchi, A., 2000, Inhibition of inducible nitric oxide synthase gene expression by indomethacin or ibuprofen in beta-amyloid protein-stimulated J774 cells. *Eur. J. Pharmacol.* **408**: 137.
Pasinetti, G.M. and Aisen, P.S., 1998, Cyclooxygenase-2 expression is increased in frontal cortex of Alzheimer's disease brain. *Neuroscience* **87**: 319.
Perry, R.T., Collins, J.S., Wiener, H., Acton, R. and Go, R.C., 2001, The role of TNF and its receptors in Alzheimer's disease. *Neurobiol. Aging* **22**: 873.
Petersen, R.C., Smith, G.E., Waring, S.C., Ivnik, R.J., Tangalos, E.G. and Kokmen, E., 1999, Mild cognitive impairment. *Arch. Neurol.* **56**: 303.
Petersen, R.C., Doody, R., Kurz, A., Mohs, R.C., Morris, J.C., Rabins, P.V., Ritchie, K., Rossor, M., Thal, L. and Winblad, B., 2001a, Current concepts in mild cognitive impairment. *Arch. Neurol.* **58**: 1985.
Petersen, R.C., Stevens, J.C., Ganguli, M., Tangalos, E.G., Cummings, J.L. and DeKosky, S.T., 2001b, Practice parameter: early detection of dementia, mild cognitive impairment (an evidence-based review). *Neurology* **56**: 1133.
Petersen, R.C., Thomas, R.G., Grundman, M., Bennett, D., Doody, R., Ferris, S., Galasko, D., Jin, S., Kaye, J., Levey, A., Pfeiffer, E., Sano, M., van Dyck, C.H. and Thal, L.J.; Alzheimer's Disease Cooperative Study Group, 2005, Vitamin E and donepezil for the treatment of mild cognitive impairment. *N. Engl. J. Med.* **352**: 2379.
Raskind, M.A., Peskind, E.R., Truyen, L., Kershaw, P. and Damaraju, C.V., 2004, The cognitive benefits of galantamine are sustained for at least 36 months: a long-term extension trial. *Arch. Neurol.* **61**: 252.
Reines, S.A., Block, G.A., Morris, J.C., Liu, G., Nessly, M.L., Lines, C.R., Norman, B.A. and Baranak, C.C., 2004, Rofecoxib Protocol 091 Study Group. Rofecoxib: no effect on Alzheimer's disease in a 1-year, randomized, blinded, controlled study. *Neurology* **62**: 66.
Ribeiro, F., de Mendonça, A. and Guerreiro, M., 2006, Mild Cognitive Impairment: deficits in cognitive domains other than memory. *Dement. Geriatr. Cogn. Disord.* **21**: 284.
Ribeiro, F., Guerreiro, M. and de Mendonça, A., 2007, Verbal learning and memory deficits in Mild Cognitive Impairment. *J. Clin. Exp. Neuropsychol.* **29**: 187.
Ritchie, K., Artero, S. and Touchon, J., 2001, Classification criteria for mild cognitive impairment – A population-based validation study. *Neurology* **56**: 37.
Robertson, M., 2003, Effect of NSAIDs on risk of Alzheimer's disease: confounding factors were not discussed. *BMJ* **327**: 751.
Rogers, J., Luber-Narod, J., Styren, S.D. and Civin, W.H., 1988, Expression of immune system-associated antigens by cells of the human central nervous system: relationship to the pathology of Alzheimer's disease. *Neurobiol. Aging* **9**: 339.
Rogers, J., Cooper, N.R., Webster, S. et al., 1992, Complement activation by beta-amyloid in Alzheimer disease. *Proc. Natl. Acad. Sci. USA* **89**: 10016.
Rogers, J., Kirby, L.C., Hempelman, S.R., Berry, D.L., McGeer, P.L., Kaszniak, A.W., Zalinski, J., Cofield, M., Mansukhani, L., Willson, P. et al., 1993, Clinical trial of indomethacin in Alzheimer's disease. *Neurology* **43**: 1609.
Rother, M., Erkinjuntti, T., Roessner, M., Kittner, B., Marcusson, J. and Karlsson, I., 1998, Propentofylline in the treatment of Alzheimer's disease and vascular dementia: a review of phase III trials. *Dement. Geriatr. Cogn. Disord.* **9 S1**: 36.
Rudolphi, K.A., Park, C.K. and Rother, M., 1997, Propentofylline (HWA 285), a neuroprotective glial cell modulator: pharmacologic profile. *CNS Drug. Rev.* **3**: 260.
Sano, M., Ernesto, C., Thomas, R.G., Klauber, M.R., Schafer, K., Grundman, M., Woodbury, P., Growdon, J., Cotman, C.W., Pfeiffer, E., Schneider, L.S. and Thal, L.J., 1997, A controlled trial of selegiline, alpha-tocopherol, or both as treatment for Alzheimer's disease. The Alzheimer's Disease Cooperative Study. *N. Engl. J. Med.* **336**: 1216.
Sastre, M., Dewachter, I., Rossner, S., Bogdanovic, N., Rosen, E., Borghgraef, P., Evert, B.O., Dumitrescu-Ozimek, L., Thal, D.R., Landreth, G., Walter, J., Klockgether, T., van Leuven, F. and Heneka, M.T., 2006,

Nonsteroidal anti-inflammatory drugs repress beta-secretase gene promoter activity by the activation of PPARgamma. *Proc. Natl. Acad. Sci. USA* **103**: 443.

Scharf, S., Mander, A., Ugoni, A., Vajda, F. and Christophidis, N., 1999, A double-blind, placebo-controlled trial of diclofenac/misoprostol in Alzheimer's disease. *Neurology* **53**: 197.

Schimmer, B., and Parker, K.L., 2006, Adrenocorticotropic hormone: adrenocortical steroids and their synthetic analogs; inhibitors of the synthesis and actions of adrenocortical hormones. In: *Goodman & Gilman's The Pharmacological Basis of Therapeutics*. L.L. Brunton (ed.). McGraw Hill, NY, pp. 653–670.

Schmand, B., Jonker, C., Hooijer, C. and Lindeboom, J., 1996, Subjective memory and memory complaints may announce dementia. *Neurology* **46**: 121.

Schmidt, L.A., Fox, A., Goldberg, M.C., Smith, C.C. and Schulkin, J., 1999, Effects of acute prednisone administration on memory, attention, and emotion in healthy human adults. *Psychoneuroendocrinology* **24**: 461.

Schneider, L.S., DeKosky, S.T., Farlow, M.R., Tariot, P.N., Hoerr, R. and Kieser, M., 2005, A randomized, double-blind, placebo control trial of two doses of *G. biloba* extract in dementia of Alzheimer's type. *Curr. Alzheimer Res.* **2**: 495.

Shaffer, L.M., Dority, M.D., Gupta-Bansal, R., Frederickson, R.C., Younkin, S.G. and Brunden, K.R., 1995, Amyloid beta protein (A beta) removal by neuroglial cells in culture. *Neurobiol. Aging* **16**: 737.

Shen, Y. and Meri, S., 2003, Yin and Yang: complement activation and regulation in Alzheimer's disease. *Prog. Neurobiol.* **70**: 463.

Shen, Y., Li, R., McGeer, E.G. and McGeer, P.L., 1997, Neuronal expression of mRNAs for complement proteins of the classical pathway in Alzheimer brain. *Brain Res.* **769**: 391.

Shen, Y., Lue, L., Yang, L., Roher, A., Kuo, Y., Strohmeyer, R., et al., 2001, Complement activation by neurofibrillary tangles in Alzheimer's disease. *Neurosci. Lett.* **305**: 165.

Shirahama, T., Miura, K., Ju, S.T., Kisilevsky, R., Gruys, E. and Cohen, A.S., 1990, Amyloid enhancing factor-loaded macrophages in amyloid fibril formation. *Lab. Invest.* **62**: 61.

Smits, H.A., Rijsmus, A., van Loon, J.H., Wat, J.W., Verhoef, J., Boven, L.A. and Nottet, H.S., 2002, Amyloid-beta-induced chemokine production in primary human macrophages and astrocytes. *J. Neuroimmunol.* **127**: 160.

Smyth, E.M., Burke, A. and FitzGerald, G.A., 2006, Lipid-derived autacoids: eisosanoids and platett activating factor. In: *Goodman & Gilman's The Pharmacological Basis of Therapeutics*. L.L. Brunton, (ed.). McGraw Hill, NY, pp. 653–670.

Stern, R.G., Mohs, R.C., Davidson, M., Schmeidler, J., Silverman, J., Kramer-Ginsberg, E., Searcey, T., Bierer, L. and Davis, K.L., 1994, A longitudinal study of Alzheimer's disease: measurement, rate, and predictors of cognitive deterioration. *Am. J. Psychiatry* **151**: 390.

Stewart, W.F., Kawas, C., Corrada, M. and Metter, E.J., 1997, Risk of Alzheimer's disease and duration of NSAID use. *Neurology* **48**: 626.

Streit, W.J., Walter, S.A., Permell, N.A., 1999, Reactive microgliosis. *Prog. Neurobiol.* **57**: 563.

Strohmeyer, R., Shen, Y. and Rogers, J., 2000, Detection of complement alternative pathway mRNA and proteins in the Alzheimer's disease brain. *Brain Res. Mol. Brain Res.* **81**: 7.

Stuchbury, G. and Munch, G., 2005, Alzheimer's associated inflammation, potential drug targets and future therapies. *J. Neural Transm.* **112**: 429.

Styren, D.S., Civin, W.H. and Rodgers, J., 1990, Molecular, cellular, and pathologic characterization of HLA-DR immunoreactivity in normal elderly and Alzheimer's disese brain. *Exp. Neurol.* **110**: 93.

Suzuki, S., Tanaka, K., Nagata, E., Ito, D., Dembo, T. and Fukuuchi, Y., 1999, Cerebral neurons express interleukin-6 after transient forebrain ischemia in gerbils. *Neurosci. Lett.* **262**: 117.

Szekely, C.A., Thorne, J.E., Zandi, P.P., Ek, M., Messias, E., Breitner, J.C. and Goodman, S.N., 2004, Nonsteroidal anti-inflammatory drugs for the prevention of Alzheimer's disease: a systematic review. *Neuroepidemiology* **23**: 159.

Tabet, N. and Feldman, H., 2002, Indomethacin for the treatment of Alzheimer's disease patients. *Cochrane Database Syst. Rev.* **2**: CD003673.

Tabet, N. and Feldman, H., 2003, Ibuprofen for Alzheimer's disease. *Cochrane Database Syst Rev.* **2**:CD004031.

Tarkowski, E., Liljeroth, A.M., Minthon, L., Tarkowski, A., Wallin, A. and Blennow, K., 2003, Cerebral pattern of cytokines in dementias. *Brain Research Bulletin* **61**: 255.

Taylor, D., Paton, C. and Kerwin, R., 2005, Use of psychotropics in special patient groups. *The Maudseley 2005–2006 Prescribing Guidelines*, 8th edition. Taylor and Francis, London and New York, pp. 259–335.

Tchelingerian, J.L., Vignais, L. and Jacque, C., 1994, TNF alpha gene expression is induced in neurones after a hippocampal lesion. *Neuroreport* **5**: 585.

Terai, K., Walker, D.G., McGeer, M.G. and McGeer, P.L., 1997, Neurons express proteins of the classic complement pathway in Alzheimer's disease. *Brain Res.* **769**: 385.

Thal, L.J., Grundman, M., Berg, J., Ernstrom, K., Margolin, R., Pfeiffer, E., Weiner, M.F., Zamrini, E. and Thomas, R.G., 2003, Idebenone fails to slow cognitive decline in Alzheimer's disease. *Neurology* **61**: 1498.

Thal, L.J., Ferris, S.H., Kirby, L., Block, G.A., Lines, C.R., Yuen, E., Assaid, C., Nessly, M.L., Norman, B.A., Baranak, C.C. and Reines, S.A., Rofecoxib Protocol 078 study group, 2005, A randomized, double-blind, study of rofecoxib in patients with mild cognitive impairment. *Neuropsychopharmacology* **30**: 1204.

The Canadian Study of Health and Aging, 1994, Risk factors for Alzheimer's disease in Canada. *Neurology* **44**: 2073.

Tobinick, E.L., Gross, H., Weinberger, A. and Cohen, H., 2006, TNF-alpha modulation for treatment of Alzheimer's disease: a six month pilot study. *Alzheimers Dement* **2**: S364.

Tuppo, E.E. and Arias, H.R., 2005, The role of inflammation in Alzheimer's disease. *Int. J. Biochem. Cell. Biol.* **37**: 289.

Van Gool, W.A., Weinstein, H.C., Scheltens, P., Walstra, G.J., 2001, Effect of hydroxychloroquine on progression of dementia in early Alzheimer's disease: an 18-month randomised, double-blind, placebo-controlled study. *Lancet* **358**: 455.

Vane, J. and Botting, R., 1987, Inflammation and the mechanism of action of anti-inflammatory drugs. *FASEB J.* **1**: 89.

Visser, P.J., Scheltens, P. and Verhey, F.R., 2005, Do MCI criteria in drug trials accurately identify subjects with predementia Alzheimer's disease? *J. Neurol. Neurosurg. Psychiatry* **76**: 1348.

Watson, G.S., Cholerton. B.A., Reger, M.A., Baker, L.D., Plymate, S.R., Asthana, S., Fishel, M.A., Kulstad, J.J., Green, P.S., Cook, D.G., Kahn, S.E., Keeling, M.L. and Craft, S., 2005, Preserved cognition in patients with early Alzheimer's disease and amnestic mild cognitive impairment during treatment with rosiglitazone: a preliminary, study. *Am. J. Geriatr. Psychiatry* **13**: 950.

Webster, S., Lue, L.F., Brachova, L., Tenner, A.J., McGeer, P.L., Terai, K., Walker, D.G., Bradt, B., Cooper, N.R. and Rogers, J., 1997, Molecular and cellular characterization of the membrane attack complex, C5b-9, in Alzheimer's disease. *Neurobiol. Aging* **18**: 415.

Wechsler, D., 1969, Manuel de l'Échelle Clinique de Mémoire. Centre de Psychologie Appliquée, Paris.

Weggen, S., Eriksen, J.L., Das, P., Sagi, S.A., Wang, R., Pietrzik, C.U., Findlay, K.A., Smith, T.E., Murphy, M.P., Bulter, T., Kang, D.E., Marquez-Sterling, N., Golde, T.E. and Koo, E.H., 2001, A subset of NSAIDs lower amyloidogenic Abeta42 independently of cyclooxygenase activity. *Nature* **414**: 212.

Weggen, S., Eriksen, J.L., Sagi, S.A., Pietrzik, C.U., Ozols, V., Fauq, A., Golde, T.E. and Koo, E.H., 2003, Evidence that nonsteroidal anti-inflammatory drugs decrease amyloid beta 42 production by direct modulation of gamma-secretase activity. *J. Biol. Chem.* **278**: 31831.

Whitehouse, P.J., Kittner, B., Roessner, M., Rossor, M., Sano, M., Thal, L. and Winblad, B., 1998, Clinical trial designs for demonstrating disease-course-altering effects in dementia. *Alzheimer Dis. Assoc. Disord.* **12**: 281.

Wilson, C.J., Finch, C.E. and Cohen, H.J., 2002, Cytokines and cognition – the case for a head-to-toe inflammation paradigm. *J. Am. Geriat. Soc.* **50**: 2041.

Wolkowitz, O.M., Reus, V.I., Canink, J., Levin, B., Lupien, S., 1997, Glucocorticoid medication, memory and steriod psychosis in medical illness. *Ann. NY Acad. Sci.* **832**: 37.

Yan, S.D., Stern, D., Kane, M.D., Kuo, Y.M., Lampert, H.C. and Roher, A.E., 1998, RAGE-Abeta interactions in the pathophysiology of Alzheimer's disease. *Restor. Neurol. Neurosci.* **12**: 167.

Yasojima, K., Schwab, C., McGeer, E.G. and McGeer, P.L., 1999a, Up-regulated production and activation of the complement system in Alzheimer's disease brain. *Am. J. Pathol.* **154**: 927.

Yasojima, K., Schwab, C., McGeer, E.G. and McGeer, P.L., 1999b, Distribution of cyclooxygenase-1 and cyclooxygenase-2 mRNAs and proteins in human brain and peripheral organs. *Brain Res.* **830**: 226

Yasojima, K., Schwab, C., McGeer, E.G. and McGeer, P.L., 2000, Human neurons generate C-reactive protein and amyloid P: upregulation in Alzheimer's disease. *Brain Res.* **887**: 80.

Yesavage, J.A., Brink, T.L., Rose, T.L., Lum, O., Huang, V., Adey, M. and Leirer, O., 1983, Development and validation of a geriatric depression scale: a preliminary report. *J. Psych. Res.* **17**: 37.

Zandi, P.P. and Breitner, J.C., 2001, Do NSAIDs prevent Alzheimer's disease? And, if so, why? The epidemiological evidence. *Neurobiol. Aging* **22**: 811.

Zhang, Y., Raud, J., Hedqvist, P. and Fredholm, B.B., 1996, Propentofylline inhibits polymorphonuclear leukocyte recruitment in vivo by a mechanism involving adenosine A2A receptors. *Eur. J. Pharmacol.* **313**: 237.

Section 3

NEURODEGENERATION AND INFLAMMATION IN AGE RELATED DISEASES

Ana Cristina Rego and Catarina R. Oliveira

With the increase in life expectancy, the incidence of dementia, involving the impairment in intellectual abilities, also tends to increase. Therefore, a high prevalence of brain pathologies associated with aging constitutes serious public health problems in the next years. This section is devoted to the analysis of mechanisms of neuronal death involving protein aggregation, inflammatory responses, and changes in neurogenesis in diseases of the peripheral (PNS) or the central nervous system (CNS).

Chapter 12 describes the toxic nature of nonfibrillar aggregates of mutant transthyretin (TTR). This plasma protein is responsible for the transport of thyroid hormones and vitamin A and its mutations have been associated with autosomal dominant amyloidosis in the PNS, leading to Familial Amyloid Polyneuropathy. Nonmature TTR aggregates have been associated with oxidative stress and inflammatory responses, two mechanisms activated in presymptomatic individuals (Chap. 12).

Inflammatory processes in the CNS are also caused by infection by the human immunodeficiency virus-1 (HIV-1), namely due to the migration of HIV-1 infected peripheral cells to the CNS, as reported in Chap. 13. Inflammation in the periphery may also affect the CNS. Thus, the progression of the acquired immunodeficiency syndrome is characterized by several neurological problems that lead to dementia. Although the mechanisms that contribute to the development of dementia are not completely understood, neuroinflammation may act in concert with the overactivation of glutamate receptors (excitotoxicity).

Dementia associated with cerebral amyloid deposits, inflammatory responses mainly coordinated by microglia and oxidative stress (which aggravate neuronal damage) are also characteristic of at least two neurodegenerative diseases that affect the CNS, namely AD and the prion-related encephalopathies (Chap. 14). Importantly, therapeutic strategies for these disorders have been directed against the neuroinflammatory process. Moreover, mitochondrial dysfunction, production of reactive oxygen species and neuronal apoptosis are important features of these two neurodegenerative diseases. The progression of AD is also influenced by increased levels of metal ions, such as iron, copper, or zinc (Chap. 15). These metals are altered during aging and mediate the generation of reactive oxygen species. Thus, metal-modulating agents, such as specific metal chelators, are promising compounds to counteract the imbalance in transition metal homeostasis occurring in AD.

Severe neurodegeneration of the hippocampus (CA1, CA3, and dentate hilus interneurons) is a feature of the most frequent type of epilepsy, the mesial temporal lobe epilepsy. Initial triggers of the disease are related with repeated neuronal stimulation and anoxia-hypoglycemia, leading to excitotoxicity, neuronal death, and gliosis. Importantly, rearrangement of hippocampal circuitry takes place during epilepsy (Chap. 16). Epilepsy is also associated with neurogenesis in the hippocampus, as described in Chap. 17. Moreover, neurogenesis in the subgranular zone of the dentate gyrus is altered during aging. Indeed, in normal, aged, and epileptic-associated situations, glial cells have emerged as important components and modulators of neurogenesis.

Selective neurodegeneration in the CNS is also caused by at least nine polyglutamine expansion diseases so far identified, for which Machado-Joseph disease (MJD) is an example. MJD is caused by a mutation in ataxin-3, a small, polyubiquitin-binding protein that displays ubiquitin hydrolase activity. In the recent years, investigation has been directed toward understanding the pathogenesis of MJD. Chapter 18 describes the structure and function of ataxin-3, the formation of protein aggregates, and the current MJD animal models. On the basis of knowledge of cellular and molecular mechanisms of MJD, several therapeutic strategies are identified, providing ways to specifically prevent the expression of the protein carrying the polyglutamine expansion or to stabilize the protein, preventing its oligomerization, and/or to slow the disease progression by interacting with altered transcription, impaired quality control mechanisms, apoptosis, and inflammatory processes (Chap. 18).

Demyelinating diseases such as Multiple Sclerosis (MS) are also associated with axonal loss. Remyelination of axons in the CNS is a spontaneous regenerative process that prevents chronic disability in MS. Interestingly, remyelination is associated with inflammation and a large macrophage response (Chap. 19).

In several neurodegenerative disorders of the CNS, such as AD, Parkinson's disease, or Huntington's disease both decreased and increased adult neurogenesis have been observed (Chap. 20). In the normal adult brain, neurogenesis occurs in two brain regions: the subventricular zone in the ventricular walls and the subgranular zone of the dentate gyrus in the hippocampus. As reviewed in Chap. 20, generation of new neurons can replace the neurons lost during each one of these diseases and thus promote functional recovery, but these brain pathologies may affect the genesis and differentiation of neurons.

In neuropathic pain, a consequence of nerve injury due to surgery, bone compression, diabetes or infection, activated microglia has been found to have an important role. Indeed, ATP P2X (P2X4 and P2X7) receptors signaling in microglia mediates proinflammatory responses, providing new targets for the development of therapeutic agents (Chap. 21).

Chapter 22 outlines the evidences of early chronic inflammatory processes, which along with hyperglycemia-induced dysfunction of the ubiquitin-proteasome pathway have been observed in diabetic retinopathy, a leading cause of blindness in adults. Several cells in the retina are affected by hyperglycemia, inducing the formation of advanced glycation end-products. Inflammatory mediators and apoptosis also play important roles in diabetes. Moreover, the proteasome may be implicated in the inflammatory response. Thus, several therapeutic targets have been identified for the prevention and treatment of diabetic retinopathy (Chap. 22).

Finally, Chap. 23 proposes an AD therapy based on the removal of intracellular aggregates with bacterial/fungal enzymes, which implicates the delivery of enzymes or its genes into the brain or into the blood circulation.

12

INFLAMMATION AND APOPTOTIC PATHWAYS IN THE PERIPHERAL NERVOUS SYSTEM RELATED TO PROTEIN MISFOLDING

Maria João Saraiva

1. ABSTRACT

Familial amyloidotic polyneuropathy (FAP) is an autosomal dominant neuro-degenerative disorder related to the systemic deposition of mutated transthyretin (TTR) amyloid fibrils, particularly in peripheral nervous system (PNS). Recently, evidence for the presence of toxic non-fibrillar TTR aggregates early in FAP nerves constituted a first step to unravel molecular signaling related to neurodegeneration in FAP. The toxic nature of TTR non-fibrillar aggregates, and not mature TTR fibrils, was evidenced by their ability to induce the expression of oxidative stress and inflammation-related molecules in neuronal cells, driving them into apoptotic pathways. How these TTR aggregates exert their effects is debatable; interaction with cellular receptors, namely the receptor for advanced glycation end-products is a probable candidate mechanism. The pathology and the yet unknown molecular signaling mechanisms responsible for neurodegeneration in FAP will be discussed.

2. INTRODUCTION

Human plasma transthyretin, TTR, has been actively studied in the last 25 years both in health and disease. TTR transports thyroid hormones and vitamin A, in the latter case by the formation of a complex with retinol binding protein, and plays an important role in the nervous system. However, TTR can undergo conformational changes and form amyloid fibrils, in both acquired and hereditary forms of systemic amyloidosis. More than 80 TTR mutations have been associated with autosomal dominant amyloidosis, usually presenting peripheral and autonomic neuropathy (FAP) and/or cardiomyopathy (Familial amyloidotic cardiomyopathy (FAC)). The mechanisms underlying protein aggregation and cell death need to be addressed for the development of future therapies in FAP. These issues are the subject of this review.

Maria João Saraiva – mjsaraiv@ibmc.up.pt
Molecular Neurobiology, IBMC & ICBAS, University of Porto, Portugal

3. FAMILIAL AMYLOIDOTIC POLYNEUROPATHY – FAP

The amyloidoses comprise a spectrum of acquired and hereditary diseases caused by the deposition of characteristic fibrillar material in various organs and tissues throughout the body. These fibrils are 7–10 nm wide, rigid, non-branching and are of variable length with a common twisted β-pleated-sheet structure (Eanes and Glenner, 1968). These deposits are eosinophilic, PAS positive, and mildly stained by silver impregnation. They have unique tinctorial staining properties including apple-green birefringence when viewed under polarised light after staining with Congo red. The accumulation of amyloid fibrils ultimately accompanies the lesion of the structure and function of affected organs. Immunoglobulin light chains related amyloidosis (AL amyloidosis), formerly known as primary amyloidosis, is caused by the accumulation of monoclonal immunoglobulin (Ig) light chains as amyloid fibrils (Glenner et al., 1971). Igs are synthesised by plasma cells and are usually maintained at a balanced level; however, in AL amyloidosis, clonal plasma cells express Ig light chains with amyloidogenic potential, i.e. those which can polymerise as amyloid fibrils in multiple organs of the body, including kidney, heart, liver, intestines, skin, spleen and lungs. As these deposits build up, organ function begins to be affected. This disease affects men and women, with a peak occurrence at 60–65 years, but even quite young adults can be affected.

Systemic AA amyloidosis is associated with chronic inflammatory diseases (e.g. rheumatoid arthritis) and involves amyloid deposition in organs such as the kidneys, liver and spleen (Benditt et al., 1971), in which the fibrils are derived from the circulating acute phase reactant serum amyloid A protein.

Hereditary transthyretin (TTR) amyloidosis is a genetically transmitted disease that results from a mutation in the gene encoding the plasma TTR protein (Saraiva et al., 1984). TTR is a transport protein for thyroid hormones and vitamin A and is predominantly synthesised in the liver. Most known TTR mutations increase the potential for the protein to destabilise and aggregate as amyloid fibrils extracellularly in different organs and tissues, with predominance in the PNS. Brain and liver are spared from deposition. The first symptoms relate to a sensory neuropathy beginning at the extremities of the lower limbs, with paresthesias, disesthesias and loss of thermal and pain sensations. The disease advances proximally and the upper extremities are affected late, when the loss of sensation has reached the level of the knee. Usually 2 or 3 years after the first sensory manifestations, motor disturbances are noticeable, with atrophy and muscular weakness, beginning also by the lower extremities. In the more advanced stages, patients are confined to a wheel-chair (Andrade, 1952). Orthostatic hypotension is common and renal involvement varies with mild to moderate proteinuria occasionally observed. Ocular abnormalities deriving from vitreous opacities may develop, as well as bilateral scalloping of the pupils. The progression of the disease is slow and relentless, leading to cachexia and death in 10–15 years from its onset.

Portugal has the highest focus of the disease with over 500 kindreds. The genetic defect in Portuguese FAP kindreds is heterozygosity for a single point mutation in TTR, giving rise to variant TTR Val30Met (Saraiva et al., 1984). In the region of Portugal where FAP is common, the gene carrier frequency has been estimated to be 1 in 625 (Alves et al., 1997). The *TTR* gene is located on chromosome 18 (Wallace et al., 1985) and is composed of 4 exons, each of approximately 200 base pairs (Sasaki et al., 1985); the Val-Met substitution is due to a single point mutation of guanine for adenine, creating a restriction site for the enzyme *Nsi* I, facilitating the molecular diagnosis of TTR Val30Met FAP by restriction fragment length polymorphism (RFLP) analysis of PCR amplified DNA.

FAP kindreds with TTR Val30Met have been identified throughout the world, with particular foci in Northern Sweden, Japan and Maiorca (Andersson 1970; Araki, 1984; Munar-Qués et al., 1997). In addition, over 80 other *TTR* mutations have been identified throughout the *TTR* gene (http://www.ibmc.up.pt/mjsaraiva/ttrmut.html). Some mutations are associated with FAP that is indistinguishable clinically from the original description of the disease; others give rise to phenotypes that variously include neuropathy, cardiomyopathy, carpal tunnel syndrome (entrapment of the median nerve at the carpal bones of the wrist), vitreous TTR deposition and leptomeningeal involvement.

A few TTR mutations are related predominantly to cardiomyopathy. The most common TTR mutation associated with cardiac amyloidosis is Val122Ile, described in the American black population. After the age of 60 apparently isolated cardiac amyloidosis is four times more common among blacks than whites in the USA, and 3.9% of blacks are heterozygous for Val122Ile (Jacobson et al., 1997).

Susceptibility to amyloidosis is governed by heterozygosity for TTR mutations, underlying autosomal dominant inheritance with variable penetrance. However, some TTR mutations appear to be non-amyloidogenic, and can be responsible for hyperthyroxinemia (Moses et al., 1990). In addition, compound heterozygotes have been identified, in whom two TTR mutations are present (Alves et al., 1997).

Interestingly, in some cases where a pathogenic and a non-pathogenic mutation occur, the non-pathogenic mutation may protect against the development of FAP (Coelho et al., 1996). Normal wild-type TTR is itself weakly amyloidogenic and is deposited as amyloid predominantly in the hearts of up to 25% of elderly people, a condition termed senile systemic amyloidosis (Westermark et al., 1990).

4. WHY DOES TTR AGGREGATE AS AMYLOID AND WHERE?

TTR is a well-characterised molecule that consists of a tetramer of identical subunits of 127 amino acids each; the molecular structure has been determined by X-ray analysis (Blake et al., 1974). It is synthesised mainly by liver and the choroid plexuses of brain (Soprano et al., 1985).

Major questions in the aggregation pathway include:
    (a) What are the building blocks of fibrils?
    (b) What are the structures of intermediates?
    (c) Where does the process take place?

The three-dimensional structure of TTR revealed an extensive β-sheet structure. Each monomer contains two β-sheets, composed of strands DAGH and CBEF, which interact face-to-face through hydrogen bonds between strands HH' and FF' to form a dimer. In the tetramer, hydrogen bonds between main chain atoms belonging to loop AB of one monomer and strand H' from the other monomer as well as hydrophobic contacts are important. The CD loop is a hot spot for amyloidogenic mutations.

The effects introduced by amyloidogenic mutations have been the subject of intensive study mainly by X-ray crystallography, but with the exception of the Leu55Pro mutation, did not reveal drastic changes; so far, the solved structures point to a clear destabilisation of the tetrameric structure of the protein. The structural studies by X-ray diffraction on the particularly aggressive mutant TTR – Leu55Pro – revealed aggregation of TTR having as

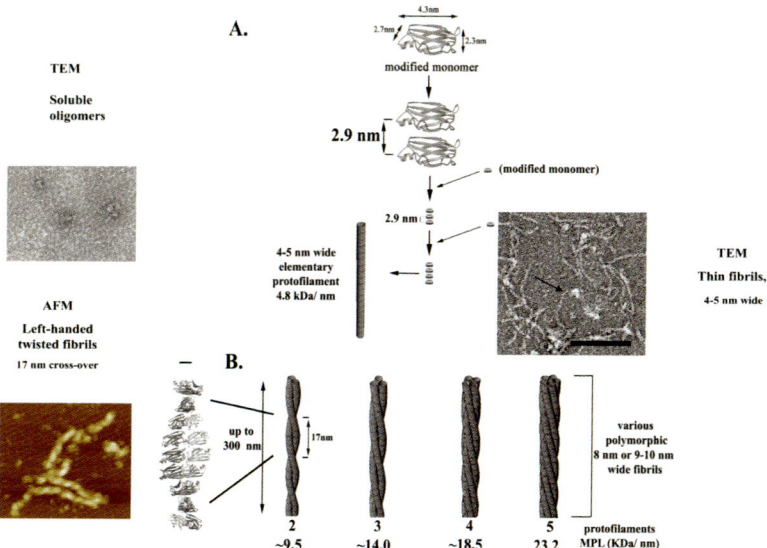

Fig. 1. Structural model describing in vitro TTR fibril assembly. (a) Structurally modified monomers (discs) form initially soluble oligomers and then stack vertically with monomer–monomer spacing to form the smallest filamentous structure, a 4–5 nm wide elementary protofilament which has a MPL of 4.8 kDa nm^{-1}. (b) Fibrils with the distinct MPLs measured by STEM. The indicated intertwining of their elementary protofilaments would give rise to the 17 nm axial repeat observed for some 8 nm fibrils. These fibrils either form by the lateral association of 2–4 or more of elementary protofilaments or nucleate and elongate separately but maintain a common protofilament substructure

building blocks monomers and indicated important changes in secondary structure by the disruption of strand D which becomes part of a long loop that connects strands C and E. Disruption of the D strand affects the hydrogen bonding with the A strand, exposing new surfaces involved in aggregation (Sebastião et al., 1998).

In vitro assembly properties as observed by transmission electron microscopy (TEM), atomic force microscopy (AFM) and quantitative scanning transmission electron microscopy (STEM) for both TTR wild-type fibrils produced by acidification, and TTR Leu55Pro fibrils assembled at physiological pH have been investigated; soluble oligomeric structures were followed by fibrillar forms, with fibrils up to 300 nm long, and a 8 nm-wide fibril being the most prominent species in both cases. Figure 1 shows fibril assembly of Leu55Pro over time. Mass-per-length (MPL) measurements by STEM revealed discrete fibril types with masses of 9.5 and 14.0 ± 1.4 kDa nm^{-1} for TTR wild-type fibrils and 13.7, 18.5 and 23.2 ± 1.5 kDa nm^{-1} for TTR Leu55Pro fibrils. The obtained MPL values are consistent with a model in which fibrillar TTR structures are composed of 2, 3, 4 or 5 elementary protofilaments, with each protofilament being a vertical stack of structurally modified TTR monomers assembled with the 2.9 nm axial monomer–monomer spacing indicated by X-ray fibre diffraction data (Cardoso et al., 2002).

Thus, soluble tetrameric TTR suffers modification leading to monomeric species that are highly unstable and tend to aggregate; elucidation of the structures of the intermediaries and knowledge of factors triggering this process will in the future provide important clues in TTR fibrillogenesis.

Electron microscopy performed in nerve specimens in the early days of FAP research described "bundles of amyloid fibrils that appeared contiguous with collagen and Schwann cells or fused with the Schwann cell basement membrane although did not penetrate beyond" (Coimbra and Andrade, 1971). Thus, the interstitial millieu is likely to be the site of TTR aggregation. In fact, upon diluting TTR "in vitro" to concentrations compatible with the interstitial milieu, tetrameric TTR dissociates into monomeric species which then aggregate as described above (Quintas et al., 2001).

A monoclonal antibody, mab 39-44, reacting with high-molecular-weight aggregates of TTR, but not with tetrameric TTR, was generated and characterised (Goldsteins et al., 1999). This antibody recognizes a cryptic epitope that is expressed in some isolated recombinant amyloidogenic mutants and in ex vivo amyloid; furthermore, it specifically recognizes, in a direct enzyme-linked immunoassay (ELISA), plasma TTR from carriers of different mutations associated with FAP, both in asymptomatic individuals and patients. By contrast, it does not react with plasma TTR from healthy individuals, or carriers of non-pathogenic mutations (Palha et al., 2001). Since neither monomeric species nor aggregates circulate in plasma of FAP patients, mab 39-44 may recognize circulating modified tetrameric TTR in plasma of carriers of amyloidogenic mutations, which subsequently dissociates and aggregates in the tissues.

5. THE AGGREGATION PATHWAY AND PERSPECTIVE FOR TREATMENT

Substantial differences in clinical presentation and severity of symptoms among Portuguese Val30Met kindreds are rare; only a few cases have been reported in which a more benign clinical course is observed. DNA analyses of the individuals showing reduced symptoms has indicated the presence, in a different allele, of a second mutation in the TTR gene on exon 4, resulting in a substitution of threonine for methionine at position 119.

TTR Thr119Met is an apparently non-pathogenic mutation that has been found with high frequency in the Portuguese population in the same area where FAP prevails; its presence in compound heterozygotes ameliorates the pathogenic mutation, reducing symptoms. It will be important in the future to analyse in more detail the clinical outcome of the compound heterozygotes carrying pathogenic and non-pathogenic mutations. To date, the self-assembly properties of TTR Thr119Met have been investigated: Comparative studies of the amyloidogenic TTR Val30Met and the non-amyloidogenic TTR Thr119Met by semi-denaturing isoelectric focusing revealed that TTR Val30Met has a higher tendency for dissociation of the tetramer into monomers than the wild-type TTR; by contrast, Thr119Met showed higher resistance to dissociation into monomers than the wild-type protein; and finally, TTR from compound heterozygotes Val30 Met/Thr119Met behaved like wild-type TTR (Alves et al., 1997). Thus, one possible way by which Thr119Met can exert anti-amyloidogenic effects is by counteracting the weaker subunit interactions of Val30Met tetramers.

Modulators responsible for phenotypic diversity can be addressed by TTR transgenic mice carrying the human full length TTR Val30Met gene, and lacking the endogenous murine TTR, showed that serum levels of the human mutant did not correlate with amyloid deposition. Deposition in this strain occurred at 12 months in three out of six animals investigated (Kohno et al., 1997). This incomplete penetrance seems to depend on the environment, as mice reared under specific-pathogen-free conditions do not develop amyloid, suggesting that inflammatory stimuli might be part of the cascade of events leading to amyloidogenesis.

At the present time, the only treatment for FAP of proven efficacy is orthotopic liver transplantation. This approach was shown to virtually eliminate variant TTR from the plasma of patients with FAP (Holmgren et al., 1991), and serial serum amyloid P component scintigraphy subsequently demonstrated regression of amyloid in large solid organs such as spleen in which TTR amyloid deposits could be imaged (Holmgren et al., 1993).

Clinically, improvement has been reported in autonomic function, such as gastrointestinal symptoms, following liver transplantation (Holmgren et al., 1993). Many centres worldwide now routinely perform liver transplantation for FAP, and the procedure has been developed to include split liver transplants, partial related donor transplants, and domino transplants in which the explanted FAP liver is re-utilised in patients with life threatening liver diseases (Lewis and Skinner, 1994). Early reports from a world registry based in Sweden indicated a transplant related mortality rate of approximately 20%, although this has lately been reduced by experience and earlier timing of surgery. Ideally, surgery is recommended in the first year of clinical disease (Lewis and Skinner, 1994). Liver transplantation does not provide a practical means of treating a large number of patients and other forms of therapy are being pursued.

Procedures that specifically block the synthesis of mutant TTR, (e.g. by gene therapy), might prove effective in treating TTR amyloidoses. Other potential alternatives include stabilisation of the native conformation of TTR, with the aim of providing protective effects similar to those observed with the non-pathogenic Thr119Met mutation described above. Efforts have been made to design drugs aimed at binding TTR in the central hydrophobic channel that runs through the molecule, at the position where the hydrophobic hormone T4 binds, to prevent dissociation into monomers; among tested molecules, for ex-vivo effect on tetramer stabilisation, derivatives of diflunisal seem to be the most efficient (Almeida et al., 2004) but "in vivo" studies are eagerly awaited to confirm their potential application in patient's treatment.

Another class of drugs with potential application to TTR amyloidosis are fibril disrupters. Very recently, doxycycline was shown to act in vitro as a TTR fibril disrupter (Cardoso et al., 2003). When activity of this drug was assessed in vivo in transgenic animals, fibrillar material was only observed in the control group (non-treated) whereas none of the animals treated with doxycycline was CR-positive. In virtue of the safety profile of doxycycline, this drug has immediate potential for clinical trials.

6. HOW DOES AGGREGATION CONTRIBUTE TO CELL DEATH?

Cellular toxicity and neurodegeneration in FAP are poorly understood. Nerves from FAP patients in different stages of disease progression (0–3) were assessed for TTR deposition by immunohistochemistry, and for the presence of amyloid fibrils by Congo red staining. The nature of the deposited material was further studied by immunocytochemistry. At the pre-symptomatic stage, (FAP 0), TTR is already deposited in an aggregated non-fibrillar form, negative by Congo red staining, suggesting that, in vivo, pre-amyloidogenic forms of TTR exist in the nerve, in a stage prior to fibril formation.

Toxicity of synthetic TTR fibrils formed in vitro at physiological pH was studied on a Schwanoma cell line by caspase-3 activation assays and showed that early aggregates but not mature fibrils activate caspases, indicative of cytotoxicity by non-fibrillar deposits occurring in early stages of FAP (Sousa et al., 2001).

Increased expression of RAGE has been observed in TTR amyloid – laden tissues; binding of RAGE by fibrillar TTR triggers activation of the transcription factor NFκB (Sousa et al., 2000). RAGE is a member of the immunoglobulin superfamily with a broad repertoire of ligands in addition to amyloid-associated macromolecules, including products of non-enzymatic glycoxidation (advanced glycation endproducts, AGEs), proinflammatory mediators (S100/calgranulins) and amphoterin (Schmidt et al., 2000). In each case, the receptor recruits signal-transduction mechanisms, often resulting in a sustained and pathogenic inflammatory and stress response. This response might underlie peripheral nerve dysfunction; analyses of nerve biopsy samples from Portuguese FAP patients at different stages of the disease, including the pre-symptomatic stage, compared with age-matched controls by semi-quantitative immunohistology and "in situ" hybridisation, showed upregulation in axons of proinflammatory cytokines (tumor necrosis factor and interleukin 1), and the inducible form of nitric oxide synthase (iNOS), as well as increased tyrosine nitration and activated caspase-3 (Sousa et al., 2001a). Recent work showed that extracellular signal-regulated kinases 1/2 (ERK1/2) displayed increased activation in FAP tissues and in transgenic mice. Cultured rat Schwannoma cell line treatment with TTR aggregates stimulated ERK1/2 activation, which was partially mediated by RAGE and abrogated by specific ERK1/2 inhibitors; these data suggest that abnormally sustained activation of ERK in FAP may represent an early signaling cascade leading to neurodegeneration (Monteiro et al., 2006).

7. CELL AND TISSUE DEATH AND PERSPECTIVES FOR TREATMENT

Although inflammation and oxidative stress are already triggered in pre-symptomatic individuals, clinical disease appears only after amyloid deposition. Gene arrays on clinical material (salivary glands) and control tissues allowed analyses of differential gene expression between pathogenic and healthy tissues. Among the differentially expressed genes, upregulation of genes related to extracellular matrix – ECM – remodeling was evident in patient's tissues. Matrix metalloproteinase 9 – MMP-9 – is overexpressed only in amyloid laden tissues, but not in tissues from pre-symptomatic individuals (Sousa et al., 2005), suggesting that fibrils trigger specific signaling pathways leading to ECM remodeling and that this process causes tissue damage with pathological consequences. Doxycycline treated tissues from transgenic mice, where disaggregation of fibrils occurred, disclosed inhibition of MMP-9 expression as compared to untreated animals presenting fibrillar material (Cardoso and Saraiva, 2006).

Taken together, the data on cell and tissue death presented above give insights into additional treatments besides those based on the aggregation pathway, with potential application in the treatment of FAP (depicted in Fig. 2). Thus, modulation of MMP-9 activity and counteracting RAGE-mediated cascades either by RAGE analogues or ERK1/2 kinase inhibitors are avenues to explore. Finally, a recent immunisation protocol using a Tyr78Phe

TTR mutant that exposes amyloid cryptic epitopes was successful in eliciting an immunological response in transgenic mice for the Val30Met human mutant, leading to total clearance of fibrillar and non-fibrillar deposits (Terazaki et al., 2006) further enlarging therapeutic perspectives.

8. CONCLUDING REMARKS

Protein misfolding occurring in PNS triggers inflammatory, oxidative stress and matrix remodeling pathways that resemble in many aspects, common molecular players and scenarios to those described in CNS for misfolding disorders, as in Alzheimer's disease. Thus, similarities and dissimilarities between the two systems are very useful to pinpoint and guide us to the treatment of age related neurodegenerative disorders.

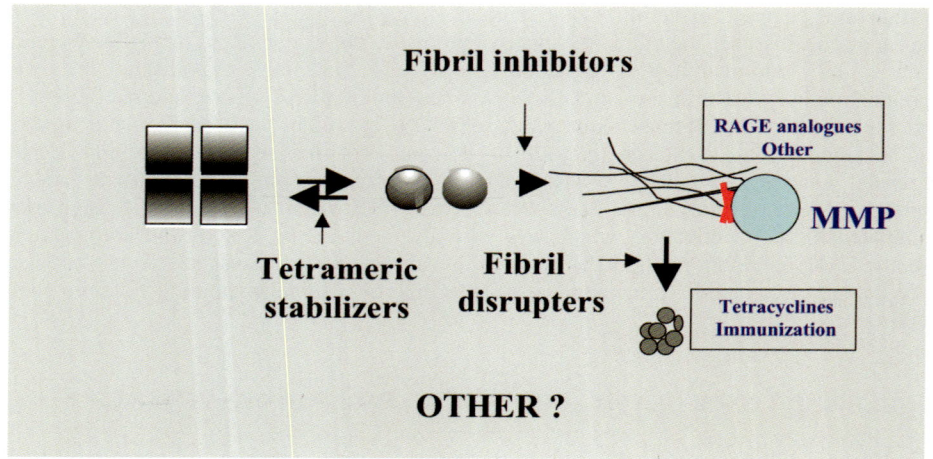

Fig. 2. Insights into new targets for treatment of familial amyloidotic polyneuropathy. Modulation of MMP-9 activity and counteracting RAGE-mediated cascades either by RAGE analogues or ERK1/2 kinase inhibitors are avenues to explore. Moreover, immunization strategy using a Tyr78Phe TTR mutant eliciting immunological response can lead to clearance of fibrillar and non-fibrillar deposits further enlarging therapeutic perspectives

9. REFERENCES

Almeida, M.R., Macedo, B., Cardoso, I., Alves, I., Valencia, G., Arsequell, G., Planas, A. and Saraiva, M.J., 2004, Effective and selective action of a iodinated diflunisal derivative in transthyretin binding and tetramer stabilization in serum from familial amyloidotic polyneuropathy patients. *Biochem. J.* **381**: 351.

Alves, I.L., Altland, K., Almeida, M.R., Winter, P. and Saraiva, M.J.M., 1997, Screening and biochemical characterization of transthyretin variants in the Portuguese population. *Hum. Mutat.* **9**: 226.

Andersson, R.,1970, Hereditary amyloidosis with polyneuropathy. *Acta Med. Scand.* **188**: 85.

Andrade, C.,1952, A peculiar form of peripheral neuropathy. Familial atypical generalized amyloidosis with special involvement of the peripheral nerves. *Brain* **175**: 408.

Araki, S., 1984, Type I familial amyloidotic polyneuropathy (Japanese type). *Brain Develop.* **6**: 128.

Benditt, E.P., Eriksen, N., Hermodson, M.A. and Ericsson, L.H., 1971, The major proteins of human and monkey substance: common properties including unusual N-terminal amino acid sequences. *FEBS Lett.* **19**: 169.

Blake, C.C.F., Geisow, M.J., Swan, I.D.A., Rérat, C. and Rérat, B., 1974, Structure of human plasma prealbumin at 2.5 A resolution. A preliminary report on the polypeptide chain conformation quaternary structure and thyroxine binding. *J. Mol. Biol.* **88**: 1.

Cardoso, I., Goldsbury, C., Muller, S.A., Olivieri, V., Wirtz, S., Damas, A.M., Aebi, U. and Saraiva, M.J., 2002, Transthyretin fibrillogenesis entails the assembly of monomers: A molecular model for in vitro assembled transthyretin amyloid-like fibrils. *J. Mol. Biol.* **317**: 687.

Cardoso, I., Merlini, G. and Saraiva, M.J., 2003, 4'-iodo-4'-Deoxydoxorubicin and tetracyclines disrupt transthyretin amyloid fibrils in vitro producing non-cytotoxic species. Screening for TTR fibril disrupters. *FASEB J.* **17**: 803.

Cardoso, I. and Saraiva, M.J., 2006, Doxycycline disrupts transthyretin amyloid: evidence from studies in a FAP transgenic mice model. *FASEB J.* **20**: 234.

Coelho, T., Chorão, R., Sousa, A., Alves, I.L., Torres, M.F. and Saraiva, M.J.M., 1996, Compound heterozygotes of transthyretin Met 30 and transthyretin Met 119 are protected from the devastating effects of familial amyloid polyneuropathy. *Neuromuscular Disord.* (Suppl) **6**: S20.

Coimbra, A. and Andrade, C., 1971, Familial amyloid polyneuropathy: and electron microscope study of peripheral nerve in five cases. I. Interstitial changes. *Brain* **94**: 199.

Eanes, E.D. and Glenner, G.G., 1968, X-ray diffraction studies on amyloid filaments. *J. Histochem. Cytochem.* **16**: 673.

Glenner, G.G., Terry, W.D., Harada, M., Isersky, C. and Page, D.L., 1971, Amyloid fibril proteins: proof of homology with immunoglobulin light chains by sequence analysis. *Science* **172**: 1150.

Goldsteins, G., Persson, H., Andersson, K., Olofsson, A., Dacklin, I., Edvinsson, Saraiva, M.J. and Lundgren, E., 1999, Exposure of cryptic epitopes on transthyretin only in amyloid and in amyloidogenic mutants. *Proc. Natl Acad. Sci. USA* **96**: 3108.

Holmgren, G., Steen, L., Ekstedt, J., Groth, C.G., Ericzon, B.G., Eriksson, S., Andersen, O., Karlberg, I., Norden, G. and Nakazato, M., 1991, Biochemical effect of liver transplantation in two Swedish patients with familial amyloidotic polyneuropathy (FAP-met30). *Clin. Genet.* **40**: 242.

Holmgren, G., Ericzon, B.G., Groth, C.G., Steen, L., Suhr, O., Andersen, O., Wallin, B.G., Seymour, A., Richardson, S., Hawkins, P.N. and Pepys, M.B., 1993, Clinical improvement and amyloid regression after liver transplantation in hereditary transthyretin amyloidosis. *Lancet* **341**: 1113.

Jacobson, D.R., Pastore, R.D., Yaghoubian, R., Kane, I., Gallo, G., Buck, F.S. and Buxbaum, J., 1997, Variant-sequence transthyretin (isoleucine 122) in late-onset cardiac amyloidosis in black Americans. *N. Engl. J. Med.* **336**: 466.

Kohno, K., Palha, J.A., Miyakawa, K., Saraiva, M.J., Ito, S., Mabuchi, T., Blaner, W.S., Iijima, H., Tsukahara, S., Episkopou, V., Gottesman, M.E., Shimada, K., Takahashi, K., Yamamura, K. and Maeda, S., 1997, Analysis of amyloid deposition in a transgenic mouse model of homozygous familial amyloidotic polyneuropathy. *Am. J. Pathol.* **150**: 1497.

Lewis, W.D. and Skinner, M., 1994, Liver transplantation for familial amyloidotic polyneuropathy: a potentially curative treatment. *Amyloid: Int. J. Exp. Clin. Invest.* **1**: 143.

Monteiro, F., Sousa, M.M., Cardoso, I., Barbas do Amaral, J., Guimarães, A. and Saraiva, M.J., 2006, Activation of ERK1/2 MAP kinases in Familial Amyloidotic Polyneuropathy. *J. Neurochem.* **97**: 151.

Moses, A.C., Rosen, H.N., Moller, D.E., Tsuzaki, S., Haddow, J.E., Lawlor, J., Liepnieks, J.J., Nichols, W.C. and Benson, M.D., 1990, A point mutation in transthyretin increases affinity for thyroxine and produces euthyroid hyperthyroxinemia. *J. Clin. Invest.* **86**: 2025.

Munar-Qués, M., Costa, P.P., Saraiva, M.J.M., Farré, C.V., Bernat, C.M., Luna, C.C. and Alberti, J.F.F., 1997, Familial amyloidotic polyneuropathy. TTR Met 30 in Majorca (Spain). *Amyloid* **4**: 181.

Palha, J.A., Moreira, P., Olofsson, A., Lundgren, E. and Saraiva, M.J., 2001, Antibody recognition of amyloidogenic transthyretin variants in serum of patients with familial amyloidotic polyneuropathy. *J. Mol. Med.* **78**: 703.

Quintas, A., Vaz, D., Cardoso, I., Saraiva, M.J. and Brito, R.M., 2001, Tetramer dissociation and monomer partial unfolding precedes protofibril formation in amyloidogenic transthyretin variants. *J. Biol. Chem.* **276**: 27207.

Saraiva, M.J.M., Birken, S., Costa, P.P. and Goodman, D.S., 1984, Amyloid fibril protein in familial amyloidotic polyneuropathy, Portuguese type. Definition of a molecular abnormality in transthyretin (prealbumin). *J. Clin. Invest.* **74**: 104.

Sasaki, H., Yoshioka, N., Takagi, Y. and Sakaki, Y., 1985, Structure of the chromosomal gene for human serum prealbumin. *Gene* **37**: 191.

Schmidt, A.M., Yan, S.D., Yan, S.F. and Stern, D.M., 2000, The biology of the receptor for advanced glycation end products and its ligands. *Biochim. Biophys. Acta* **1498**: 99.

Sebastião, M.P., Saraiva, M.J. and Damas, A.M., 1998, The crystal structure of amyloidogenic Leu55 --> Pro transthyretin variant reveals a possible pathway for transthyretin polymerization into amyloid fibrils. *J. Biol. Chem.* **273**: 24715.

Soprano, D.R., Herbert, J., Soprano, K.J., Schon, E.A. and Goodman, D.S., 1985, Demonstration of transthyretin mRNA in the brain and other extrahepatic tissues in the rat. *J. Biol. Chem.* **260**: 11793.

Sousa, M.M., Yan, S.D., Stern, D., Saraiva, M.J., 2000, Interaction of the receptor for advanced glycation end products (RAGE) with transthyretin triggers nuclear transcription factor kB (NF-kB) activation. *Lab. Invest.* **80**: 1101.

Sousa, M.M., Cardoso, I., Fernandes, R., Guimarães, A. and Saraiva, M.J., 2001, Deposition of transthyretin in early stages of familial amyloidotic polyneuropathy: evidence for toxicity of non-fibrillar aggregates. *Am. J. Pathol.* **159**: 1993.

Sousa, M.M., Yan, S.D., Fernandes, R., Guimarães, A., Stern, D. and Saraiva, M.J.M., 2001a, Familial amyloid polyneuropathy: RAGE-dependent triggering of neuronal inflammatory and apoptotic pathways. *J. Neurosci.* **21**: 7576.

Sousa, M.M., Barbas do Amaral, J., Guimarães, A. and Saraiva, M.J., 2005, Upregulation of the extracellular matrix remodelling genes, biglycan, neutrophil gelatinase-associated lipocalin and matrix metalloproteinase-9 in familial amyloid polyneuropathy. *FASEB J.* **19**: 124.

Terazaki, H., Ando, Y., Fernandes, R., Yamamura, K., Maeda, S. and Saraiva, M.J., 2006, Immunization in familial amyloidotic polyneuropathy: Counteracting deposition by immunization with a Y78F TTR mutant. *Lab. Invest.* **86**: 23.

Wallace, M.R., Naylor, S.L., Kluve-Beckerman, B., Long, G.L., McDonald, L., Shows, T.B. and Benson, M.D., 1985, Localization of the human prealbumin gene to chromosome 18, *Biochem. Biophys. Res. Commun.* **129**: 753.

Westermark, P., Sletten, K., Johansson, B. and Cornwell, G.G. III, 1990, Fibril in senile systemic amyloidosis is derived from normal transthyretin. *Proc. Natl Acad. Sci. USA* **87**: 2843.

13

NEUROINFLAMMATION AND EXCITOTOXICITY IN NEUROBIOLOGY OF HIV-1 INFECTION AND AIDS: TARGETS FOR NEUROPROTECTION

Marcus Kaul and Stuart A. Lipton

1. ABSTRACT

Infection with the human immunodeficiency virus-1 (HIV-1) and acquired immuno-deficiency syndrome (AIDS) pose a persistent health problem worldwide. Infected peripheral immune-competent cells, in particular macrophages, appear to infiltrate the central nervous system (CNS) and provoke a neuropathological and inflammatory response involving all cell types in the brain. In fact, HIV-1 seems to enter the brain very soon after infection and can subsequently induce severe and debilitating neurological problems that range from mild behavioral abnormalities and motor dysfunction to frank dementia. The course of HIV-1 disease is strongly influenced by viral and host factors, such as the viral strain and the response of the host's immune system. In addition, HIV-1-dependent disease processes in the periphery, such as inflammation, have a substantial effect on the pathological changes in the CNS, despite the fact that the brain seems to harbor a distinctive viral population of its own. In the CNS, HIV-1 incites activation of chemokine receptors, inflammatory mediators, extracellular matrix-degrading enzymes, and glutamate receptor-mediated excitotoxicity, all of which can initiate numerous downstream signaling pathways and perturb neuronal and glial function. Despite many major improvements in the control of viral infection in the periphery, an effective therapy for HIV-1 associated dementia remains to be developed. This chapter will address recently uncovered pathologic and degenerative mechanisms contributing to neuronal damage induced by HIV-1 and discuss experimental and potential future therapeutic approaches.

2. INTRODUCTION

The worldwide development of disease associated with infection by the human immunodeficiency virus-1 (HIV-1) and acquired immunodeficiency syndrome (AIDS) is alarming, with estimated numbers having grown from more than 35 million existing infections in 2001 to 38 millions in 2003, and more than 20 million deaths since 1981

Marcus Kaul – mkaul@burnham.org; Stuart A. Lipton – slipton@burnham.org
Center for Neuroscience and Aging Research, The Burnham Institute for Medical Research, 10901 North Torrey Pines Road, La Jolla, CA 92037, USA

(UNAIDS, 2004). Not only can HIV-1 destroy the immune system of its host and eventually lead to AIDS, the virus can also cause a variety of neurological problems that culminate in frank dementia. AIDS-related opportunistic infections may affect the central nervous system (CNS) more often in the absence of treatment than in the presence of medication, but HIV infection itself can also induce a number of neurological syndromes (Petito et al., 1986). Neuropathological conditions directly triggered by HIV-1 include peripheral neuropathies, vacuolar myelopathy, and a clinical syndrome of cognitive and motor dysfunction that has been designated HIV-associated dementia (HAD) (Glass et al., 1993; Kaul et al., 2001; Power et al., 2002; Gendelman et al., 2005). A mild form of HAD is termed minor cognitive motor disorder (MCMD) (Ellis et al., 1997; Kaul et al., 2001; Gendelman et al., 2005). Interestingly, a high risk of neuropsychological impairment in HIV-1 infection seems to be indicated early on by anemia (McArthur et al., 1993).

The mechanisms contributing to the development of MCMD and HAD remain incompletely understood, but the discovery in the brain of cellular binding sites for HIV-1, the chemokine receptors, and recent progress in understanding neuroinflammation and neural stem cell biology continue to provide new and surprising insights (Lavi et al., 1998; Miller and Meucci, 1999; Kaul et al., 2001, 2005; Gonzalez-Scarano and Martin-Garcia, 2005; Kramer-Hammerle et al., 2005; Minghetti, 2005; Jones and Power, 2006). The present chapter reviews recent developments regarding the understanding of HIV-1's neurotoxic effect in the CNS and potential approaches for therapy and prevention of HAD.

3. EPIDEMIOLOGY OF HAD BEFORE AND IN THE ERA OF HIGHLY ACTIVE ANTIRETROVIRAL THERAPY

HIV-1 productively infects macrophages and lymphocytes, first in the periphery and then in the brain, after binding of the viral envelope protein gp120 to one of several possible chemokine receptors in conjunction with CD4. Depending on the primary sequence of their gp120, different HIV-1 strains may use CCR5 (CD195) and CCR3, or CXCR4 (CD184), or a combination of these chemokine receptors to enter target cells (Dragic et al., 1996; Oberlin et al., 1996; He et al., 1997).

Since most transmitted viruses use CCR5, deficiency in a functional receptor molecule (Δ32-CCR5) can provide substantial protection against HIV-1 infection (Liu et al., 1996). Some individuals who become infected though remain asymptomatic long-term and do not progress to AIDS have been found to express high levels of certain CCR5-binding β-chemokines (Paxton et al., 1996). Again a few people never show seroconversion and seem to mount an unconventional, very effective humoral immune response that includes IgA antibodies against viral glycoprotein 41 (gp41) and IgG recognizing a CD4-gp120 complex (Lopalco et al., 2005).

Initially, the majority of severe neurological symptoms occurred in advanced stages of systemic HIV-1 disease and the prevalence of HAD was estimated to be 20–30% in individuals with low CD4 T cell counts (McArthur et al., 1993). In addition, anemia associated with HIV-1 infection presented itself as an early predictor for a high risk of neuropsychological impairment (McArthur et al., 1993). The introduction of highly active antiretroviral therapy (HAART) has increased the life expectancy of people infected with HIV-1 and resulted in an at least temporary decrease in the incidence of HAD to as low as 10.5% (McArthur et al., 2003). This transient effect attests to the point that the effects of

HIV-1 infection in the brain should always be considered in conjunction with the systemic conditions and it is now widely understood that a peripheral infection and an associated immune response and inflammatory processes can influence all cell types in the CNS (Turrin and Rivest, 2004; Chakravarty and Herkenham, 2005). Indeed, improved control of peripheral viral replication and the treatment of opportunistic infections continue to extend survival times, but HAART fails to provide protection from MCMD or HAD, or to reverse the disease in most cases (Cunningham et al., 2000). Although MCMD may be more prevalent than frank dementia in the HAART era, HAD constitutes a significant independent risk factor for death due to AIDS and it is assumed to be the most common cause of dementia worldwide among people of age 40 or less, (Ellis et al., 1997). Moreover, the proportion of new cases of HAD displaying a CD4 cell count greater then 200 μl^{-1} is growing (McArthur et al., 2003), and another recent study found that in a group of 669 HIV patients who died between 1996 and 2001 more than 90% had been diagnosed with HAD as an AIDS-defining condition within the last 12 months of life (Welch and Morse, 2002). This situation might at least in part be due to poor penetration into the CNS of HIV protease inhibitors and several of the nucleoside analogues, and distinct patterns of viral drug resistance in plasma and cerebrospinal fluid (CSF) compartments have also been observed (Cunningham et al., 2000; Kaul et al., 2005; Kramer-Hammerle et al., 2005). While HIV seems to penetrate into the CNS soon after infection in the periphery, and then resides primarily in perivascular macrophages and microglia (Ho et al., 1985; Gartner, 2000; Gonzalez-Scarano and Martin-Garcia, 2005), current therapeutic guidelines for AIDS suggest to start HAART only when the number of $CD4^+$ T cells begin to decline. Since this might occur up to some years after peripheral infection, HAART is unlikely to prevent the entry of HIV-1 into the CNS (Kramer-Hammerle et al., 2005). Consequently, as people live longer with HIV-1 and AIDS the prevalence of dementia might be rising and in recent years the incidence of HAD as an AIDS-defining illness has actually increased (Lipton, 1997b; Cunningham et al., 2000; Kaul et al., 2001, 2005; McArthur et al., 2003; Kramer-Hammerle et al., 2005; Jones and Power, 2006). Therefore, a better understanding of the pathogenesis of HAD, including viral and host factors, is urgently required in order to identify additional therapeutic targets for the prevention and treatment of this neurodegenerative disease.

4. FROM HIV ENTRY INTO THE BRAIN TO DEVELOPMENT OF MINOR COGNITIVE-MOTOR DISORDER AND HAD

Soon after infection in the periphery, HIV penetrates into the CNS where the virus primarily resides in microglia and macrophages (Koenig et al., 1986; Gartner, 2000). Viral load in brain can be measured by quantitative PCR, and the highest concentrations of virus are detected in those subcortical structures most often affected in patients with HAD (McArthur et al., 1997; Wiley et al., 1998). However, infection of macrophages and microglia alone does not seem to initiate neurodegeneration, and it has therefore been proposed that additional factors associated with advanced HIV infection in the periphery, thus outside the CNS, provide important triggers for events leading to dementia (Gartner, 2000). An elevated number of circulating monocytes that express CD16 and CD69 could constitute one such factor. These activated cells tend to adhere to and transmigrate through the normal endothelium of the brain microvasculature and might then initiate processes deleterious to neurons (Gartner, 2000).

The blood–brain barrier (BBB) also plays a crucial role in HIV infection of the CNS (Nottet et al., 1996; Persidsky et al., 1997; Asensio and Campbell, 1999; Gartner, 2000). Microglia and astrocytes produce chemokines – cell migration/chemotaxis inducing cytokines – such as monocyte chemoattractant protein (MCP)-1, which appear to regulate migration of peripheral blood mononuclear cells through the BBB (Asensio and Campbell, 1999). In fact a mutant MCP-1 allele that causes increased infiltration of mononuclear phagocytes into tissues has recently been implicated in an increased risk of HAD (Gonzalez et al., 2002).

Histological studies in specimens from HIV-1-infected humans and simian immunodeficiency virus (SIV)-infected rhesus macaques found that lymphocytes and monocytes enter the brain (Prospero-Garcia et al., 1996; Kalams and Walker, 1995). The pathophysiological relevance of CNS invading lymphocytes in HAD is not clearly established (Kalams and Walker, 1995; Mennicken et al., 1999). However, infiltrating lymphocytes and activated microglia in brains with HIV-1 encephalitis showed strong immunoreactivity for IL-16, a natural ligand of CD4. Since this cytokine inhibits HIV-1 propagation, lymphocytes might contribute to an innate antiviral immune response in the CNS in addition to microglia (Zhao et al., 2004b).

Cell migration also engages adhesion molecules, and increased expression of vascular cell adhesion molecule-1 (VCAM-1) has been implicated in mononuclear cell migration into the brain during HIV and SIV infections (Sasseville et al., 1994; Nottet et al., 1996; Persidsky et al., 1997). As an alternative to entry via infected macrophages, it has been suggested that the inflammatory cytokine, tumor necrosis factor-alpha (TNF-α), promotes a paracellular route for HIV-1 across the BBB (Fiala et al., 1997). Interestingly, alterations in the BBB occur even in the absence of intact virus in transgenic mice expressing the HIV envelope protein gp120 in a form that circulates in plasma (Marshall et al., 1998). This finding suggests that circulating virus or envelope proteins may provoke BBB dysfunction during the viremic phase of primary infection. On the part of the host, a vicious cycle of immune dysregulation and BBB dysfunction might be required to achieve sufficient entry of infected or activated immune cells into the brain to cause neuronal injury (Bazan et al., 1997; Kaul et al., 2001). On the side of the virus, variations of the envelope protein gp120 might also influence the timing and extent of events allowing viral entry into the CNS and leading to neuronal injury (Power et al., 1998).

5. POTENTIAL LINKS BETWEEN NEUROPATHOLOGY OF HIV INFECTION AND PATHOGENESIS OF HAD

The neuropathological hallmarks of HIV infection in the brain are termed HIV encephalitis and include widespread reactive astrocytosis, myelin pallor, microglial nodules, activated resident microglia, multinucleated giant cells, and infiltration predominantly by monocytoid cells, including blood-derived macrophages (Budka, 1991). Surprisingly, measures of cognitive function do not correlate well with numbers of HIV-infected cells, multinucleated giant cells, or viral antigens in CNS tissue (Glass et al., 1995; Masliah et al., 1997). In contrast, increased numbers of microglia (Glass et al., 1995), elevated TNF-α mRNA in microglia and astrocytes (Wesselingh et al., 1997), evidence of excitotoxins (Heyes et al., 1991; Giulian et al., 1996; Jiang et al., 2001), decreased synaptic and dendritic density (Masliah et al., 1997; Everall et al., 1999), and selective neuronal loss (Masliah et al., 1992; Fox et al., 1997) constitute the pathologic features most closely associated with the clinical signs of HAD. Furthermore, signs of neuronal apoptosis have been linked to HAD (Adle-Biassette et al., 1995; Gelbard et al., 1995; Petito and Roberts, 1995), although this finding is not clearly associated with viral burden (Adle-Biassette et al., 1995) or a history of dementia (Adle-Biassette et al., 1999). The localization of apoptotic neurons is correlated with evidence of structural atrophy and closely associated with signs of microglial activation, especially within subcortical deep gray structures (Adle-Biassette et al., 1999), which may show a predilection for atrophy in HAD.

The neuropathology observed in postmortem specimens from HAD patients in combination with extensive studies using both in vitro and animal models of HIV-induced neurodegeneration have led to a fairly complex model for the pathogenesis of HAD. The available information strongly suggests that the pathogenesis of HAD might be most effectively explained when viewed as similar to the multi-hit model of oncogenesis. Fig. 1 shows a model of potential intercellular interactions and alterations of normal cell functions that can lead to neuronal injury and death in the setting of HIV infection (Kaul et al., 2001). Macrophages and microglia can be infected by HIV-1, but they can also be activated by factors released from infected cells. These factors include cytokines and shed viral proteins such as gp120. Variations of the HIV-1 envelope protein gp120, in particular in its V1, V2, and V3 loop sequences, have been implicated in modulating the activation of macrophages and microglia (Power et al., 1998). Factors released by activated microglia affect all cell types in the CNS, resulting in upregulation of cytokines, chemokines, and endothelial adhesion molecules (Lipton and Gendelman, 1995; Gartner, 2000; Kaul et al., 2001). Some of these factors may directly or indirectly contribute to neuronal damage and apoptosis. Neurotoxic factors directly released from activated microglia include excitatory amino acids (EAAs) and related substances, such as quinolinate, cysteine, and a not completely characterized amine compound named "NTox" (Giulian et al., 1990, 1993; Lipton et al., 1991; Brew et al., 1995; Yeh et al., 2000; Jiang et al., 2001; Zhao et al., 2004a). EAAs induce neuronal apoptosis through a process known as excitotoxicity. This detrimental process engenders excessive Ca^{2+} influx and free radical (nitric oxide and superoxide anion) formation by overstimulation of glutamate receptors (Lipton et al., 1991; Bonfoco et al., 1995). Certain HIV proteins, such as gp120 and Tat, have also been reported to be directly neurotoxic, although high concentrations of viral protein may be needed, or neurons may have to be cultured in isolation to see these direct effects (Meucci et al., 1998; Liu et al., 2000). It is important to note that toxic viral proteins, as well as injurious factors released from microglia and glutamate, free radicals, or other substances released from astrocytes, may act in concert to promote neurodegeneration, even in the absence of extensive viral invasion of the CNS.

6. CHEMOKINE RECEPTORS IN HIV-1 INFECTION AND HAD

Chemokine receptors are seven transmembrane-spanning domain, G-protein-coupled receptors, and as such trigger intracellular signaling events. While chemokines and their receptors were originally shown to mediate leukocyte trafficking and to contribute intimately to the organization of inflammatory responses of the immune system, they are now known to contribute to far more physiological and pathological processes (Oberlin et al., 1996; Bazan et al., 1997; Tran and Miller, 2003). The additional functions include the intricate control of

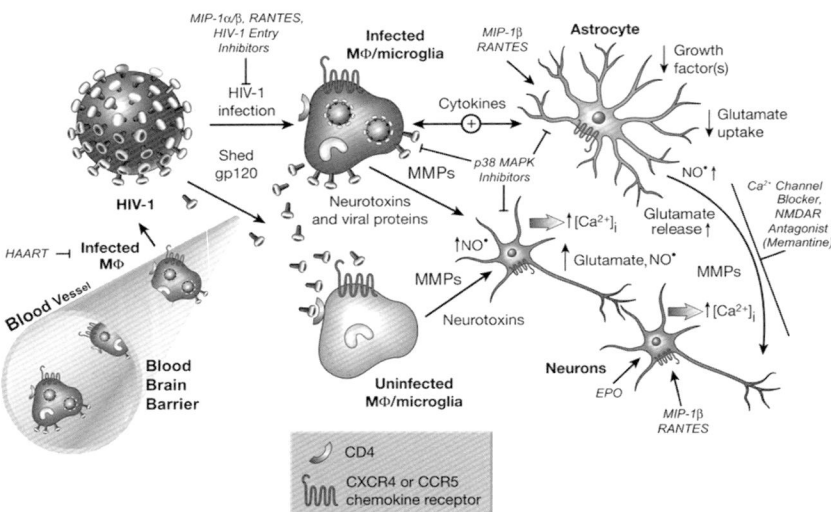

Fig. 1. Current model of HIV-1 neuropathology indicating presumably toxic or protective factors and potential sites for therapeutic intervention (*protective factors are shown in italic*). Neuronal injury and death induced by HIV-1 infection: Immune-activated and HIV-infected, brain-infiltrating macrophages (MΦ) and microglia release potentially neurotoxic substances. These substances include quinolinic acid and other excitatory amino acids such as glutamate and L-cysteine, arachidonic acid, PAF, NTox, free radicals, TNF-α, and probably others. These factors from MΦ/microglia and also possibly from reactive astrocytes contribute to neuronal injury, dendritic and synaptic damage, and apoptosis as well as to astrocytosis. Entry of HIV-1 into MΦ/microglia occurs via gp120 binding, and therefore it is not surprising that gp120 (or a fragment thereof) is capable of activating uninfected MΦ/microglia to release similar factors to those secreted in response to productive HIV infection. MΦ/microglia express CCR5 and CXCR4 chemokine receptors on their surface in addition to CD4 and viral gp120 binds via these receptors. Some populations of neurons and astrocytes have been reported to also possess CXCR4 and CCR5 receptors on their surface, raising the possibility of direct interaction with gp120. MΦ/microglia and astrocytes have mutual feedback loops (bidirectional arrow). Cytokines participate in this multicellular network in several ways. For example, HIV-infection or gp120-stimulation of MΦ/microglia enhances their production of TNF-α and IL-1ß (cytokines-arrow). The TNF-α and IL-1ß produced by MΦ/microglia stimulate astrocytosis. Arachidonate released from MΦ/microglia impairs astrocyte clearing of the neurotransmitter glutamate and thus contributes to excitotoxicity. In conjunction with cytokines, the α-chemokine SDF-1 stimulates reactive astrocytes to release glutamate in addition to the free-radical nitric oxide [NO$^{\bullet}$], which in turn may react with superoxide (O$_2^{\bullet-}$) to form the neurotoxic molecule peroxynitrite (ONOO$^-$). NO might also activate extracellular matrix metalloproteinases (MMPs), which can then proteolytically affect neurons, and also cleave membrane-anchored fractalkine (Kaul et al., 2005). Neuronal injury is primarily mediated by overactivation of NMDARs with resultant excessive influx of Ca^{2+}. This, in turn, leads to overactivation of a variety of potentially harmful signaling systems, the formation of free radicals, and release of additional neurotransmitter glutamate. Glutamate subsequently overstimulates NMDARs on neighboring neurons, resulting in further injury. This final common pathway of neurotoxic action can be blocked by NMDAR antagonists. For certain neurons, depending on their exact repertoire of ionic channels, this form of damage can also be ameliorated to some degree by calcium channel antagonists or non-NMDAR antagonists. Additionally, MIP-1ß and RANTES, agonists of β-chemokine receptors, which are present in the CNS on neurons, astrocytes, and microglia, can confer partial protection against neuronal apoptosis induced by HIV/gp120 or NMDA (modified from Kaul et al. (2005)).

organogenesis, including hematopoiesis, angiogenesis, and development of heart and brain (Ma et al., 1998; Tachibana et al., 1998; Zou et al., 1998; Locati and Murphy, 1999). Furthermore, chemokines and their receptors are essential for maintenance, maturation, and migration of hematopoietic and neural stem cells (Lapidot and Petit, 2002; Tran and Miller, 2003). However, the most prominent pathological function of certain chemokine receptors seems to be the mediation of HIV-1 infection (Alkhatib et al., 1996; Bleul et al., 1996; Locati and Murphy, 1999).

Infection of macrophages and lymphocytes by HIV-1 can occur after binding of the viral envelope protein gp120 to one of several possible chemokine receptors in conjunction with CD4. Generally, T cells are infected via the α-chemokine receptor CXCR4 and/or the β-chemokine receptor CCR5. In contrast, macrophages and microglia are primarily infected via the β-chemokine receptor CCR5 or CCR3, but the α-chemokine receptor CXCR4 may also be involved (He et al., 1997; Michael and Moore, 1999; Ohagen et al., 1999; Chen et al., 2002). The HIV co-receptors CCR5 and CXCR4, among other chemokine receptors, are also present on neurons and astrocytes (Rottman et al., 1997; Zhang et al., 1998), although these cells are not thought to harbor productive infection. Several in vitro studies strongly suggest that CXCR4 is directly involved in HIV-associated neuronal damage while CCR5 may additionally serve a protective role (Hesselgesser et al., 1998; Meucci et al., 1998; Kaul and Lipton, 1999).

In cerebrocortical neurons and neuronal cell lines from humans and rodents, picomolar concentrations of HIV-1/gp120, as well as intact virus, can induce neuronal death via CXCR4 receptors (Hesselgesser et al., 1998; Kaul and Lipton, 1999; Ohagen et al., 1999; Chen et al., 2002; Garden et al., 2004). In mixed neuronal/glial cerebrocortical cultures that mimic the cellular composition of the intact brain, this apoptotic death appears to be mediated predominantly via the release of microglial toxins rather than by direct neuronal damage (Kaul and Lipton, 1999; Chen et al., 2002; Garden et al., 2004). However, nanomolar concentrations of SDF-1α/β interacting with CXCR4 can induce apoptotic death of neurons in the absence of microglia, suggesting a possible direct interaction with neurons while interaction with astrocytes can also occur (Kaul and Lipton, 1999; Zheng et al., 1999; Bezzi et al., 2001). In contrast to these findings, it has been reported that somewhat higher concentrations of SDF-1α provide neuroprotection from X4-preferring gp120-induced damage of isolated hippocampal neurons (Meucci et al., 1998).

Using mixed neuronal/glial cerebrocortical cultures from rat and mouse, we have further investigated the role of chemokine receptors in the neurotoxicity of gp120. We found that gp120 from CXCR4 (X4)-preferring as well as CCR5 (R5)-preferring and dual tropic HIV-1 strains all were able to trigger neuronal death. While gp120 from one out of two X4-preferring HIV-1 strains showed no longer neurotoxicity in CXCR4-deficient cerebrocortical cultures, dual tropic gp120$_{SF2}$ showed surprisingly even greater neurotoxicity in CCR5 knockout cultures compared to wild-type or CXCR4-deficient cultures (Kaul, 2002; Kaul et al., 2006). These findings are consistent with the primarily neurotoxic effect of CXCR4 activation by gp 120. In contrast, activation of CCR5 might at least in part be neuroprotective depending on the HIV-1 strain from which a given gp120 originated. Furthermore, we observed earlier that the CCR5 ligands MIP-1β and "regulated upon activation T cell expressed and secreted" (RANTES) protect neurons against gp120-induced toxicity (Kaul and Lipton, 1999).

Since in vitro inhibition of microglial activation is sufficient to prevent neuronal death after gp120 exposure, it seems likely that stimulation of CXCR4 in macrophages/microglia is a prerequisite for the neurotoxicity of gp120 (Kaul and Lipton, 1999; Ohagen et al., 1999). In contrast, SDF-1 might directly activate CXCR4 in astrocytes and neurons to trigger neuronal death, for example, by reversing glutamate uptake in astrocytes (Hesselgesser et al., 1998; Kaul and Lipton, 1999; Bezzi et al., 2001; Kaul et al., 2001). SDF-1 is produced by astrocytes, macrophages, neurons, and Schwann cells (McGrath et al., 1999; Zheng et al., 1999; Gleichmann et al., 2000; Stumm et al., 2002). An increase in SDF-1 mRNA has been detected in HIV encephalitis (Zhang et al., 1998), and protein expression of SDF-1 also appears to be elevated in the brains of HIV patients (Langford et al., 2002). To what degree the increased expression of SDF-1 aggravates neuronal damage by HIV-1 remains to be shown. We had reported previously that intact SDF-1 can be toxic to mature neurons in a CXCR4-dependent manner, at least in culture (Kaul and Lipton, 1999; Zheng et al., 1999; Kaul, 2002; Kaul et al., 2006). Additionally, it was recently reported that cleavage of SDF-1 by MMPs may contribute to neuronal injury and thus HAD via a non-CXCR4-mediated mechanism (Zhang et al., 2003). Importantly, increased expression and activation of MMPs, including MMP-2 and MMP-9, were detected in HIV-infected macrophages and also in postmortem brain specimens from AIDS patients when compared with uninfected controls (Johnston et al., 2000). As elegantly shown by Power and colleagues, MMP-2 released from HIV-infected macrophages is able to proteolytically remove four amino acids from the N-terminus of SDF-1. This truncated form of SDF-1 no longer binds CXCR4 and is an even more powerful neurotoxin than full length SDF-1 (Zhang et al., 2003).

7. EFFECT OF CHEMOKINES AND HIV/GP120 ON NEURAL STEM AND PROGENITOR CELLS

The CXCR4-SDF-1 receptor–ligand axis plays an important role in the physiological function of hematopoietic and neural stem cells (Asensio and Campbell, 1999; Tran and Miller, 2003). This fact indicates a potential of HIV-1 and its envelope protein to directly interfere with biological functions of neural stem and progenitor cells.

In cultures of primary mouse and human neural progenitor cells (NPCs) obtained from fetal tissue, cells stain positively for the neural stem cell marker nestin and readily undergo cell division. After several rounds of proliferation, the progenitors exit the cell cycle and express neuronal markers such as βIII-tubulin (TuJ1). Our immunocytochemical studies showed that the progenitors expressed CXCR4 and CCR5. Treatment with HIV-1/gp120 reduced the number of progenitors and differentiating neurons. Accounting for these observations, we found that gp120 inhibited proliferation of NPCs without producing apoptosis. The resulting decrease in neural stem cell proliferation engendered by gp120 also meant that there were fewer progenitor cells present to differentiate in neurons, thus impairing neurogenesis (Okamoto, et al., unpublished). These findings were complemented and extended by others using commercially generated human NPCs (Krathwohl and Kaiser, 2004a, b). In those experiments, chemokines promoted the quiescence and survival of human NPCs via stimulation of CXCR4 and CCR3 and a mechanism that involves downregulation of extracellularly regulated kinase-1 and -2 (ERK-1/2) with simultaneous upregulation of the neuronal glycoprotein reelin (Krathwohl and Kaiser, 2004a). Exposure to HIV-1 caused quiescence of neural progenitors, again through engagement of CXCR4 and CCR3. The coat protein HIV-1/gp120 reportedly downregulated ERK-1/2 but had no effect on reelin (Krathwohl and Kaiser, 2004b). Interestingly, the effects of both the chemokines and HIV-1/gp120 were reversible and could be inhibited with recombinant Apolipoprotein E3

(ApoE3), but not ApoE4. Although it is widely accepted that HIV-1 fails to productively infect neurons, it has been reported that NPCs are permissive to the virus (Mattson et al., 2005). The apparent ability of HIV-1/gp120 to interfere with the normal function of NPCs suggested the possibility that HAD might develop as a consequence of not only injury and death of existing neurons but also due to virus-induced disturbance of potential repair mechanisms in the CNS (Fig. 2).

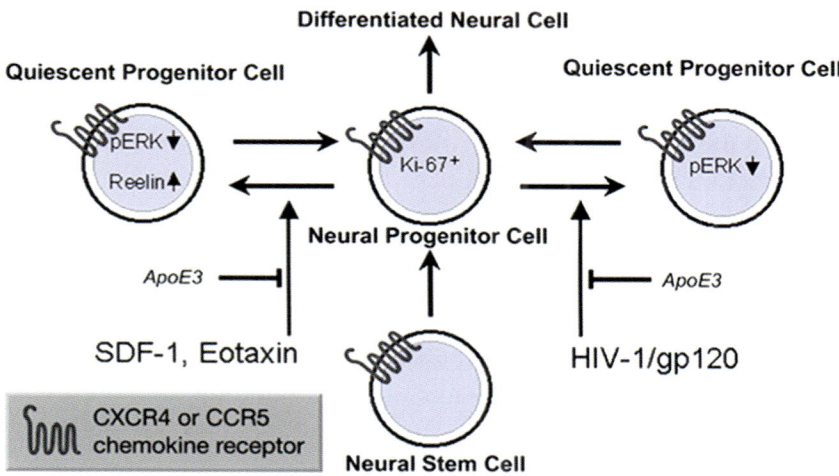

Fig. 2. Current model of HIV-1 interference with the function of neural progenitor cells (NPCs) and potential sites for therapeutic intervention *(protective factors are shown in italic)*: Exposure to chemokines, SDF-1 and Eotaxin, or HIV-1/gp120 of mouse or human NPCs reduces proliferation and promotes quiescence. ApoE3 inhibits these effects on NPCs. NPCs express nestin and show decreased proliferation as judged by decreased BrdU incorporation. However, NPCs do not undergo apoptosis, as evidenced by lack of TUNEL staining and nuclear condensation under the same conditions (Krathwohl and Kaiser, 2004a, b; Okamoto et al., unpublished). Modified from Kaul et al. (2001, 2005)

8. MACROPHAGES, MICROGLIA, AND NEURONAL INJURY IN HAD

Macrophages and microglia play a pivotal role, although somewhat paradoxical, in the pathobiology of HAD (Milligan et al., 1991; Kaul et al., 2001; Luo et al., 2003). Under steady-state conditions, mononuclear phagocytes, macrophages, and microglia act as scavengers and sentinel cells, nonspecifically eliminating foreign material, and secreting trophic factors critical for maintenance of homeostasis within the CNS microenvironment (Elkabes et al., 1996; Lazarov-Spiegler et al., 1996; Rapalino et al., 1998; Zheng et al., 1999; Gras et al., 2003). These protective functions, however, can evolve into destructive ones. A number of neurotrophins are secreted by macrophages (Robinson et al., 1986). These factors include but are not limited to, brain-derived neurotrophic factor (BDNF) (Miwa et al., 1997), insulin-like growth factor-2 (IGF-2) (Nicholas et al., 2002), β-nerve growth factor (βNGF) (Garaci et al., 1999), transforming growth factor beta (TGF-β) (Chao et al., 1995), neurotrophin-3 (NT3) (Kullander et al., 1997), and glial-derived neurotrophic factor (GDNF) (Batchelor et al., 1999). Clearly, a dysregulation of macrophage neurotrophic factors by viral infection and/or immune activation may occur during disease. This dysregulation may be as important as the production of neurotoxins for eliciting neuronal damage. Additionally, some neurotrophic factors are regulated by cytokines. For example, TNF-α (a candidate HIV-1-induced neurotoxin) produced by immune competent microglia can play a neurotrophic role by inducing biologically active TGF-β (Chao et al., 1995). TGF-β is a protective cytokine for mammalian neurons, particularly in protection against glutamate neurotoxicity, hypoxia, and gp120-mediated neural injury (Meucci and Miller, 1996). This cytokine also affects long-term synaptic facilitation (Milligan et al., 1991).

HIV establishes a latent and persistent infection within macrophages (Koenig et al., 1986). The majority of HIV within the CNS appears to be localized within perivascular and blood-derived parenchymal brain macrophages and microglia (Koenig et al., 1986). Astrocytes, oligodendrocytes, and brain endothelial cells are rarely infected, if at all (Tornatore et al., 1994). As a result of viral infection and resultant immune activation macrophages produce and release a variety of neurotoxins within the brain (Gendelman et al., 1997; Nath, 1999; Kaul et al., 2001). These products comprise not only viral proteins, such as gp120 (Brenneman et al., 1988), gp41 (Adamson et al., 1996), and Tat (Nath et al., 1996), but also host cell-encoded products including platelet activating factor (PAF) (Gelbard et al., 1994), glutamate (Jiang et al., 2001), arachidonic acid and its metabolites (Nottet et al., 1995), proinflammatory cytokines, such as interleukin-1 beta (IL-1β), TNF-α, TNF-related apoptosis inducing ligand (TRAIL) (Gelbard et al., 1993; Ryan et al., 2001), quinolinic acid (Heyes et al., 1991; Kerr et al., 1997), NTox (Giulian et al., 1996), and nitric oxide (Adamson et al., 1996) among others. In this manner, macrophages, which were once pillars of the immune system, are now responsible for tissue damage, though it is still unclear how macrophages evolve from producing neurotrophins to producing neurotoxins. Perhaps, HIV-1 infection and immune activation induces a transition between neurotrophic and neurotoxic activities. In any case, it seems that activation of p53 in microglia plays a crucial role for neurotoxicity to occur upon exposure of the cells to HIV-1/gp120 (Garden et al., 2004).

9. MECHANISMS OF NEURONAL INJURY AND DEATH IN HIV-ASSOCIATED DEMENTIA

9.1. Physiological and Pathological Roles of *N*-Methyl-D-Aspartate Receptors (NMDARs)

A recurring question has been whether HIV-1 or its component proteins induce neuronal damage predominantly by an indirect route (e.g., via toxins produced by infected or immune-stimulated macrophages and/or astrocytes), or by a direct route (e.g., via binding to neuronal receptors) (Lipton and Gendelman, 1995; Lipton, 1997a; Kaul and Lipton, 1999; Kaul et al., 2001). Several lines of evidence suggest that HIV-associated neuronal injury involves predominantly an indirect route from macrophage and astrocyte toxins resulting in excessive activation of NMDARs with consequent excitotoxicity (Fig. 1) (Olney, 1969; Giulian et al., 1990; Lipton et al., 1991; Doble, 1999; Kaul and Lipton, 1999; Sardar et al., 1999; Kaul et al., 2001).

Under physiological conditions, activation of ionotropic glutamate receptors in neurons initiates transient depolarization and excitation. AMPARs mediate a fast component of excitatory postsynaptic potentials, and NMDARs underlie a slower component. Presynaptic release of glutamate and consequent depolarization of the postsynaptic neuronal membrane via AMPAR-coupled channels relieve the Mg^{2+} block of the NMDAR-associated ion channel that occurs under resting conditions. This effect allows subsequent controlled Ca^{2+} influx through the NMDAR-coupled ion channel. This voltage-dependent modulation of the NMDAR results in activity-driven synaptic modulation (Bigge, 1999; Doble, 1999). However, extended and/or excessive NMDAR activation and consequent excitotoxicity are triggered by sustained elevation of the intracellular Ca^{2+} concentration, compromised cellular energy metabolism, and resultant free radical formation (Olney, 1969; Lipton and Rosenberg, 1994; Doble, 1999).

A role for excitotoxicity in brain disorders was first suggested by the work of Olney following the pioneering work of Lucas and Newhouse in the retina (Olney, 1969; Olney and Sharpe, 1969). Subsequently, several lines of evidence indicated that excessive stimulation of glutamate receptors contributes to the neuropathological process in a large number of disorders, including stroke, head and spinal cord injury, seizures, Huntington's disease, Parkinson's disease, possibly Alzheimer's disease, amyotrophic lateral sclerosis, multiple sclerosis, glaucoma, and HAD (Lipton and Gendelman, 1995; Doble, 1999; Brauner-Osborne et al., 2000). Indeed, excitotoxicity seems to represent a common final pathway of neuronal injury and death in a wide variety of neurodegenerative disorders (Lipton and Rosenberg, 1994).

The NMDAR has attracted particular interest as a major player in excitotoxicity because this receptor, in contrast to most non-NMDARs (AMPA and kainate receptors), is highly permeable to Ca^{2+}, and excessive Ca^{2+} influx can trigger excitotoxic neuronal injury (Choi, 1988; Weiss and Sensi, 2000). In addition, NMDAR antagonists effectively prevent some forms of glutamate neurotoxicity, both in vitro and in vivo in animal studies (Choi et al., 1988a; Bigge, 1999; Doble, 1999). This potential as a therapeutic agent was recently borne out in human phase III clinical trials with the NMDAR open-channel blocker, memantine, based upon pioneering work by our group (see below). However, AMPA and kainate receptors can also mediate excitotoxicity and contribute to neuronal damage under certain conditions (Bigge, 1999; Doble, 1999). For example, a subpopulation of $Ca^{2+}-$ or $Zn^{2+}-$ permeable AMPA receptor-coupled channels have been implicated in selective neurodegenerative disorders, such as ischemia, epilepsy, Alzheimer's disease, and amyotrophic lateral sclerosis (Weiss and Sensi, 2000). Also transgenic mice overexpressing AMPARs display increased damage subsequent to ischemia when compared to control animals (Le et al., 1997).

9.2. Downstream Signaling Pathways from NMDARs

Excessive stimulation of the NMDAR induces several detrimental intracellular signals that contribute to neuronal cell death by apoptosis or necrosis, depending on the intensity of the initial insult (Nicotera et al., 1997). For example, excessive Ca^{2+} influx through NMDAR-coupled ion channels leads to an elevation of the intracellular free Ca^{2+} concentration to a point that results in Ca^{2+} overload of mitochondria, depolarization of the mitochondrial membrane potential, and a decrease in ATP synthesis. The scaffolding protein PSD-95 (postsynaptic density-95) links the principal subunit of the NMDAR (NR1) with neuronal nitric oxide synthase (nNOS), a Ca^{2+}-activated enzyme, and thus brings nNOS into close proximity to Ca^{2+} via the NMDAR-operated ion channel (Sattler et al., 1999). Excessive intracellular Ca^{2+} overstimulates nNOS and protein kinase cascades with consequent generation of deleterious levels of free radicals, including reactive oxygen species (ROS) and nitric oxide (NO) (Nicotera et al., 1997). NO can react with ROS to form cytotoxic peroxynitrite ($ONOO^-$) (Nicotera et al., 1997). However, in alternative redox states, NO can activate p21ras (Gonzalez-Zulueta et al., 2000) and inhibit caspases (Tenneti et al., 1997) via S-nitrosylation (transfer of the NO group to critical cysteine thiols), thereby attenuating apoptosis in cerebrocortical neurons.

Importantly, excessive Ca^{2+} influx and free radicals also activate stress-related p38 mitogen-activated protein kinase (p38 MAPK) and, via c-Jun N-terminal kinase (JNK), c-Jun in cerebrocortical or hippocampal neurons. In turn, p38 MAPK also phosphorylates/activates transcription factors, including direct activation of myocyte enhancer factor 2 (MEF2). Activation of these pathways has been implicated in neuronal apoptosis, probably in conjunction with caspase activation (Kaul and Lipton, 1999; Mukherjee et al., 1999). As stated above, excessive intracellular Ca^{2+} accumulation after NMDAR stimulation leads to depolarization of the mitochondrial membrane potential ($\Delta\psi_m$) and a drop in the cellular ATP concentration. If the initial excitotoxic insult is fulminant, the neuronal cells do not recover their ATP levels and die at this point because of the loss of ionic homeostasis, resulting in acute swelling and lysis (necrosis). If the insult is more mild, ATP levels recover, and the neuronal cells enter a delayed death pathway requiring energy, known as apoptosis (Nicotera et al., 1997). Interestingly, Zn^{2+} can substitute for Ca^{2+} and lead to neuronal death by these and other pathways (Choi et al., 1988b; Aizenman et al., 2000; Weiss and Sensi, 2000).

It has been reported that NMDAR-mediated excitotoxicity leading to neuronal apoptosis also involves activation of the Ca^{2+}/calmodulin-regulated protein phosphatase calcineurin (Nicotera et al., 1997), mitochondrial permeability transition, release of cytochrome c from mitochondria (Budd et al., 2000), activation of caspase-3 (Tenneti et al., 1998), lipid peroxidation (Tenneti et al., 1998), and cytoskeletal breakdown (Nicotera et al., 1997). Inhibition of calcineurin and caspase-3 by FK506 and caspase inhibitors, respectively, can attenuate this form of excitotoxicity (Nicotera et al., 1997; Tenneti et al., 1998). It has been proposed that the adenine nucleotide translocator (ANT) is a part of the mitochondrial permeability transition pore (PTP) and participates in mitochondrial depolarization. Indeed, our group found that pharmacologic blockade of the ANT with bongkrekic acid prevented collapse of the mitochondrial membrane potential ($\Delta\psi_m$), as well as subsequent caspase-3 activation and NMDA-induced neuronal apoptosis. However, treatment with bongkrekic acid failed to inhibit the transient drop in ATP concentration (although it hastened the recovery of ATP levels) and did not prevent the liberation of cytochrome c into the cytosol. Thus, initiation of caspase-3 activation and resultant neuronal apoptosis after NMDAR activation require a factor(s) in addition to cytochrome c release (Budd et al., 2000).

9.3. HIV-1 in the Brain and Activation of NMDARs

Analysis of specimens from AIDS patients (Sardar et al., 1999) as well as in vivo and in vitro experiments indicate that HIV-1 infection creates excitotoxic conditions, predominantly via an indirect route. HIV-1 infection induces soluble factors in macrophage/microglia and/or astrocytes, such as glutamate and glutamate-like molecules, viral proteins, cytokines, chemokines, and arachidonic acid metabolites (Lipton and Gendelman, 1995; Lipton, 1997a; Lipton, 1998; Bezzi et al., 2001; Kaul et al., 2001).

However, it has also been suggested that HIV-1 or its protein components can directly interact with neurons and modulate NMDAR function, at least under some conditions (Savio and Levi, 1993; Meucci et al., 1998). Picomolar concentrations of soluble HIV/gp120 induce injury and apoptosis in primary rodent and human neurons both in vitro and in vivo (Brenneman et al., 1988; Lannuzel et al., 1995). Additionally, our group and subsequently several others have shown that gp120 contributes to NMDAR-mediated neurotoxicity (Lipton et al., 1991). Both voltage-gated Ca^{2+} channel blockers and NMDAR antagonists can ameliorate gp120-induced neuronal cell death in vitro (Dreyer et al., 1990; Lipton et al., 1991). Transgenic mice expressing gp120 manifest neuropathological features that are similar to the findings in brains of AIDS patients, and in these mice neuronal damage is ameliorated by the NMDAR antagonist memantine (Toggas et al., 1994, 1996) (see below). It is also conceivable that other glutamate receptors in addition to NMDARs influence HIV-associated neuronal damage. Interestingly, stimulation of specific subtypes of the G protein-coupled mGluRs interferes with excitotoxic NMDAR-mediated activation of MAPKs and can attenuate subsequent neuronal cell death (Mukherjee et al., 1999).

In the case of HAD, macrophages and microglia play a crucial role because they are the predominant cells productively infected with HIV-1 in the brain (Lipton and Gendelman, 1995) (Fig. 1), although infection of astrocytes has also been rarely observed in pediatric cases (reviewed in Brack-Werner and Bell, 1999). In accordance with the report that the presence of macrophages/microglia correlates with the severity of HAD (Glass et al., 1995), in our hands, the predominant mode of neurotoxicity of HIV-1 or gp120 requires the presence and activation of macrophages/microglia (Lipton, 1992c, 1994; Kaul and Lipton, 1999). Moreover, HIV-1-infected or gp120-stimulated mononuclear phagocytes have been shown to release neurotoxins that directly stimulate the NMDAR (Giulian et al., 1990; Lipton et al., 1991; Kaul and Lipton, 1999). Those macrophage toxic factors include molecules that directly or indirectly act as NMDAR agonists, such as quinolinic acid, cysteine, PAF, and a low-molecular weight compound designated NTox (Lipton and Gendelman, 1995; Lipton, 1998; Yeh et al., 2000).

Additionally, HIV-infected or immune-activated macrophages/microglia and possibly astrocytes produce inflammatory cytokines, including TNF-α and IL-1β, arachidonic acid metabolites, and free radicals (ROS and NO) that may indirectly contribute to excitotoxic neuronal damage (Fig. 1) (Lipton and Gendelman, 1995; Lipton, 1998; Bezzi et al., 2001). TNF-α and IL-1β may amplify neurotoxin production by stimulating adjacent glial cells and by increasing immunologic NOS activity (Adamson et al., 1996; Lipton, 1998).

In contrast to these indirect neurotoxic pathways, it has been reported that gp120 can directly interact with neurons in the absence of glial cells. Recently, gp120 was found to act at chemokine receptors directly on isolated neurons in culture to induce their death (Meucci et al., 1998). Additionally, higher nanomolar concentrations of gp120 have been reported to interact with the glycine binding site of the NMDAR (Fontana et al., 1997). Furthermore, gp120 may produce a direct excitotoxic influence via NMDAR-mediated Ca^{2+} oscillations in rat hippocampal neurons (Lo et al., 1992), and may bind to noradrenergic axon terminals in neocortex, where it possibly potentiates NMDA-evoked noradrenaline release (Pittaluga et al., 1996). Nonetheless, many if not all of these direct effects on neurons were observed in vitro in the absence of glial cells. Since glial cells are known to modify these death pathways, we feel that under in vivo conditions, the indirect route to neuronal injury is the predominant one, based on studies in mixed neuronal–glial cultures and on work in a gp120-transgenic mouse (see below).

Along these lines, gp120 has been found to aggravate excitotoxic conditions by impairing astrocyte uptake of glutamate via arachidonic acid that is released from activated macrophages/microglia (Dreyer and Lipton, 1995; Lipton, 1997a). The α-chemokine SDF-1, the cytokine TNF-α, and metabolites of arachidonic acid, such as prostaglandins, also stimulate a Ca^{2+}-dependent release of glutamate by astrocytes (Bezzi et al., 1998, 2001). Moreover, HIV-1 can induce astrocytic expression of the β-chemokine known as macrophage chemotactic protein-1 (MCP-1). This β-chemokine in turn attracts additional mononuclear phagocytes and microglia to further enhance the potential for indirect neuronal injury via the release of macrophage toxins (Conant et al., 1998).

In our view, therefore, HIV-1 infection and its associated neurological dysfunction involve both chemokine receptors and NMDAR-mediated excitotoxicity. This dual receptor involvement raises the question of whether G protein-coupled chemokine receptors and ionotropic glutamate receptors might influence each other's activity. Indeed, the β-chemokine RANTES, which binds to chemokine receptors CCR1, CCR3, and CCR5, can abrogate neurotoxicity induced by gp120 (Kaul and Lipton, 1999) or by excessive NMDAR stimulation (Bruno et al., 2000). In turn, excitotoxic stimulation can enhance expression of CCR5 (Galasso et al., 1998). Whether or not these findings reflect a mechanism of feedback or crosstalk in which chemokines indirectly antagonize the stimulation of the NMDAR awaits to be elucidated.

10. PREVIOUS AND POTENTIAL FUTURE STRATEGIES FOR PREVENTION OR THERAPY OF HAD

10.1. Previous Pharmacotherapy for HAD

A truly effective pharmacotherapy for HAD has yet to be developed. Previous approaches to cope with HAD reflect the challenging complexity inherent in the treatment of patients with AIDS (reviewed in Melton et al., 1997; Clifford, 1999). Previous and current therapeutic approaches include various antiretroviral compounds, alone or in combination (1) reverse transcriptase inhibitors, including Zidovudine, Didanosine, Zalcitabine, Stavudine, and Lamivudine; and (2) protease inhibitors, such as Saquinavir, Ritonavir, and Indinavir. Of these only Zidovudine has been shown to cross the BBB to some extent, and Zidovudine has a beneficial effect on HAD but the effect is not long lasting. The other antiretroviral drugs may not penetrate the brain sufficiently to eradicate the virus in the CNS. Thus an adjunctive treatment besides antiretroviral drugs is needed.

10.2. Current and Future Strategies Targeting Receptors for Glutamate, Chemokines, and Erythropoietin

In recent therapeutic attempts, Pentoxyfylline, an inhibitor of production and action of TNF-α, and the neurotrophic peptide T were tested as "investigational agents" (Melton et al., 1997), but clinical studies assessing their therapeutic potential did not prove substantial benefit. Previous, small clinical trials of the voltage-activated (L-type) calcium channel blocker, nimodipine, and a PAF inhibitor suggested some therapeutic benefit but were not conclusive (Navia et al., 1998; Clifford, 1999; Schifitto et al., 1999).

From the pathogenesis of HAD as described above, several potential therapeutic strategies appear viable (Fig. 1). NMDAR antagonists are among the agents under consideration. Others include certain chemokines and cytokines, and antagonists for their receptors, p38 MAPK inhibitors, caspase inhibitors, and antioxidants (free-radical scavengers or other inhibitors of excessive nitric oxide or ROS) (Clifford, 1999; Kaul et al., 2001; Turchan et al., 2003; Digicaylioglu et al., 2004b; Lipton, 2004).

Chemokine receptors allow HIV-1 to enter cells and as such are major potential therapeutic targets in the fight against AIDS in general (Michael and Moore, 1999). Antagonists of CXCR4 and CCR5 inhibit HIV-1 entry and are being assessed in clinical trials (Michael and Moore, 1999). However, the benefit of inhibitors of chemokine receptors for HIV-associated neurological complications awaits study (Gartner, 2000; Kaul et al., 2001, 2005). Interestingly, certain chemokines have been shown to protect neurons from injury, even though the virus does not productively infect neurons. In particular, β-chemokines and fractalkine prevent gp120-induced neuronal apoptosis in vitro (Kaul and Lipton, 1999; Bruno et al., 2000), and similarly, some β-chemokines can ameliorate NMDAR-mediated neurotoxicity (Bruno et al., 2000; Kaul et al., 2006). Additionally, the CCR5 ligands MIP-1α, MIP-1β, and RANTES are able to suppress HIV-1 infection in the periphery and are highly expressed in long-term HIV-1 infected individuals who do not, or only very slowly, progress to AIDS (Cocchi et al., 1995; Scala et al., 1997; Paxton et al., 1998; Zagury et al., 1998). HIV-infected patients with relatively higher CSF concentrations of MIP-1α/β and RANTES performed better on neuropsychological measures than those with low or undetectable levels (Letendre et al., 1999). These findings support the hypothesis that selected β-chemokines may represent a potential treatment modality for AIDS and HAD. One of the efforts underway aims at modification of natural CCR5 ligands in order to avoid adverse inflammatory side effects upon application (Verani and Lusso, 2002; De Clercq, 2004; Pierson et al., 2004).

Previously, we have shown that the cytokine erythropoietin (EPO) may not only be effective in treating anemia but also for protecting neurons, since it prevents NMDAR-mediated and HIV-1/gp120-induced neuronal death in mixed cerebrocortical cultures (Digicaylioglu and Lipton, 2001; Digicaylioglu et al., 2004b). Since EPO is already clinically approved for the treatment of anemia, human trials of EPO as a neuroprotectant from HAD may be expedited (Lipton, 2004). Additionally, EPO plus IGF-1 act synergistically as neuroprotectants by activating the PI3K/Akt pathway (Digicaylioglu et al., 2004a), so the use of these two cytokines in conjunction has been advocated for clinical trials (Lipton, 2004).

NMDAR antagonists have been shown to attenuate neuronal damage due to either HIV-infected macrophages or HIV/gp120, both in vitro and in vivo (Dreyer et al., 1990; Lipton, 1992a; Toggas et al., 1996; Chen et al., 2002). Both voltage-gated Ca^{2+} channel blockers and NMDAR antagonists can ameliorate gp120-induced neuronal cell death in vitro (Dreyer et al., 1990; Lipton et al., 1991). Transgenic mice expressing gp120 in their CNS manifest neuropathological features that are similar to the findings in brains of AIDS patients, and in these mice neuronal damage is ameliorated by the NMDAR antagonist memantine (Toggas et al., 1994, 1996). Memantine-treated gp120 transgenics and nontransgenic control mice retain a density of presynaptic terminals and dendrites that is similar to untreated nontransgenic/wild-type controls but significantly higher than in untreated gp120 transgenic animals (Toggas et al., 1996). This finding supports the hypothesis that the HIV-1 surface glycoprotein is sufficient to initiate downstream of chemokine receptor activation excitotoxic neuronal injury and death. It also shows that an antagonist of NMDAR overstimulation can ameliorate HIV-associated neuronal damage in vivo, an observation that another group recently confirmed (Anderson et al., 2004).

However, the majority of NMDAR antagonists have unacceptable psychotomimetic side effects in humans, and this problem and its solution are discussed below.

NMDAR antagonist drugs with fewer adverse effects were thought to include the glycine site antagonists, but these can cause dizziness and sedation in healthy human volunteers (Lees, 1997). In general, competitive antagonists for the glutamate or glycine co-agonist sites may be doomed to failure because they inhibit normal brain function (which occurs at lower levels of agonist) before they block pathological actions (which occur at higher levels of agonist); hence, normal brain areas are affected in an adverse manner prior to the drugs becoming effective in pathologically injured brain regions. Thus, in our view, uncompetitive open-channel blockers have the best chance of emerging as acceptable agents in clinical practice, as discussed below.

As alluded to above, many NMDAR antagonists are not clinically tolerated, while some others appear to be tolerated by humans at concentrations that are effective neuroprotectants (Lipton, 1993; Lipton and Rosenberg, 1994; Parsons et al., 1999). Several NMDAR antagonists prevent neuronal injury in animal models of a variety of neurological disorders, including HAD, focal stroke, Parkinson's disease, Huntington's disease, Alzheimer's disease, amyotrophic lateral sclerosis, neuropathic pain, glaucoma, and others (Choi, 1988; Lipton and Rosenberg, 1994; Doble, 1999; Parsons et al., 1999). Of these drugs, two of the most promising, because of their long experience in patients with other diseases, are memantine (Bormann, 1989; Parsons et al., 1999) and nitroglycerin (Lipton, 1993; Lipton and Rosenberg, 1994; Lipton and Gendelman, 1995), as well as new combinatorial agents combining features of both of these drugs (Lipton and Kieburtz, 1998).

Our group was the first to show that memantine blocks the NMDAR-associated ion channel only when it is open for pathological periods of time; conversely, we showed that during normal neurotransmission, when there is less NMDAR-operated channel activity, memantine has relatively little effect on this activity (Chen et al., 1992; Lipton, 1992a, b, 1993, 1998; Pellegrini and Lipton, 1993; Lipton and Rosenberg, 1994; Lipton and Gendelman, 1995; Chen and Lipton, 1997; Chen et al., 1998; Lipton and Kieburtz, 1998; Stieg et al., 1999; Kaul et al., 2001; Le and Lipton, 2001). We found that unlike other NMDAR open-channel blockers, such as dizocilpine (MK-801), memantine does not remain in the channel for an excessively long time, and hence we discovered that this short dwell time is the key to memantine's lack of clinical side effects. For example, we found that the relatively short dwell time accounts for the fact that neuroprotective concentrations of memantine manifest little or no effect on the NMDAR component of excitatory postsynaptic potentials (EPSPs), on long-term potentiation (LTP), and on performance in the Morris water maze behavioral task (Chen et al., 1992, 1998; Chen and Lipton, 1997). Interestingly, while Mg^{2+} has an even shorter dwell time in the channel than memantine, the Mg^{2+} effect is so short lived that it does not effectively block the NMDAR-associated channel during insult, and thus does not afford significant neuroprotection under most conditions. In contrast, MK-801 can afford neuroprotection by effectively blocking the NMDAR-associated channel, but is not clinically tolerated because its block is too prolonged, contributing to its very high affinity of action, and thus MK-801 blocks all normal physiological activity (Chen et al., 1992, 1998; Chen and Lipton, 1997). Thus, it has been possible to use memantine safely in humans for over fifteen years in Europe as a treatment for Parkinson's disease and spasticity (Chen et al., 1992, 1998; Chen and Lipton, 1997; Parsons et al., 1999). Our fortuitous breakthrough was the realization that a low-affinity agent such as memantine can afford significant neuroprotection while leaving normal physiological function relatively unaffected. The affinity of a channel-blocking drug is related to the ratio of its on-time to its off-rate (the latter representing the inverse of its dwell time in the channel). Importantly, the on-time is influenced by the concentration of the drug, but the off-rate (or dwell time) is not concentration related and instead is purely an intrinsic property of the antagonist. Hence, we realized that the off-rate (and hence the dwell time) was the key property of an open-channel blocking drug that contributes to its affinity, to its efficacy as a neuroprotectant, and to its safety or tolerability in the brain. Another important realization in our work was that memantine was selective for NMDARs at a neuroprotective concentration despite its relatively low affinity (IC_{50} = ~1 µM), and, in fact, a high-affinity agent, such as MK-801, would be toxic. In other words, one does not need high affinity in a drug in order to have high selectivity for its target. Quite the opposite is desired in the brain: a low-affinity agent such as memantine is preferred because this results in relative sparing of normal neurotransmission; however, the drug also needs to be selective for its target receptor in order to avoid side effects stemming from interactions with unwanted targets.

We realized that there were benefits of such a low-affinity agent. For example, increasing concentrations of glutamate/glycine or other NMDA agonists cause NMDAR channels to remain open on average for a greater fraction of time. Under pathological conditions of increased glutamate (and glycine), we discovered that the open-channel blocking drug, memantine, has a better chance to enter the channel and block it (after all, the drug can only get into the channel when it is open, and statistically, more drug will get into the channel when, on average, the channel is open longer). It is because of this mechanism of action that the destructive effects of greater (pathological) concentrations of glutamate are prevented to a greater extent than the effects of lower (physiological) concentrations, which are relatively spared (Lipton, 1993; Lipton and Rosenberg, 1994; Chen and Lipton, 1997; Chen et al., 1998). This mechanism of inhibition is termed uncompetitive antagonism, defined as the action of the antagonist being contingent upon prior activation of the receptor by the agonist. Moreover, in animal model systems, clinically tolerated concentrations of memantine can ameliorate neuronal injury associated with either focal cerebral ischemia or HIV-1 proteins, both in vitro and in vivo (Seif el Nasr et al., 1990; Erdo and Schafer, 1991; Chen et al., 1992; Keilhoff and Wolf, 1992; Lipton, 1992a; Lipton and Jensen, 1992; Muller et al., 1992; Osborne and Quack, 1992; Pellegrini and Lipton, 1993; Sathi et al., 1993; Toggas et al., 1996). Thus, because of its lack of major adverse effects under pathological conditions and recent evidence of efficacy in human clinical trials, memantine might be the first NMDAR antagonist that will be successful in the market.

Our group has also shown that another potentially clinically useful modulatory agent of the NMDAR is nitroglycerin, which produces nitric oxide-related molecules. Nitric oxide ($NO^{\bullet}$, where the dot represents one unpaired electron in the outer molecular orbital) can contribute to neuronal damage. One of the pathways to neurotoxicity involves the reaction of $NO^{\bullet}$ with $O_2^{\bullet -}$ to form peroxynitrite ($ONOO^-$) (Beckman et al., 1990; Dawson et al., 1991, 1993; Lipton et al., 1993). In contrast, $NO^{\bullet}$ can be converted to a chemical state that has just the opposite effect, i.e., one that protects neurons from injury due to NMDA receptor-mediated overstimulation. The change in chemical state is dependent on the removal or addition of an electron to $NO^{\bullet}$. We and our colleagues have demonstrated that with one less electron, $NO^{\bullet}$ acts like nitrosonium ion (NO^+), which facilitates reaction with critical thiol group(s) (R–SH or, more properly, thiolate anion, R–S–) comprising a redox modulatory site(s) on the NMDA receptor–channel complex, which decreases channel activity (Lei et al.,

1992; Lipton et al., 1993; Kim et al., 1999; Choi et al., 2000, 2001). Our group has further shown that this reaction can afford neuronal protection from overstimulation of NMDA receptors, as well as other reactions, which would otherwise result in excessive Ca^{2+} influx (Lipton and Stamler, 1994). One such drug that can react with NMDA receptors in this redox-related manner is the common vasodilator nitroglycerin (Lei et al., 1992; Lipton, 1993; Lipton et al., 1993; Lipton and Rosenberg, 1994). Chronic use of nitroglycerin induces tolerance to the drug's effects on the cardiovascular system, thus avoiding systemic adverse effects such as hypotension. However, during chronic use, nitroglycerin still appears to work in the brain to attenuate NMDA receptor-mediated neurotoxicity (Lipton, 1993). Nonetheless, the exact dosing regimen has yet to be worked out for the neuroprotective effects of nitroglycerin in the brain; therefore, caution has to be exercised before attempting to implement this form of therapy. In preliminary experiments, including those using animal models of focal ischemia, high concentrations of nitroglycerin were neuroprotective during various NMDA receptor-mediated insults (Lipton and Wang, 1996; Stieg et al., 1999). Our in vivo data suggest that this effect of nitroglycerin may, at least in part, be due to a direct effect on neurons, consistent with an action at the NMDAR redox modulatory site(s) (Lipton and Wang, 1996; Stieg et al., 1999). As the structural basis for redox modulation of the NMDA receptor has recently been further elucidated (Kohr et al., 1994; Sullivan et al., 1994; Das et al., 1998; Kim et al., 1999; Choi et al., 2000, 2001), it has become possible to design even better redox reactive reagents of clinical value, for example, with the NO group in appropriate redox state, targeted specifically to the NMDAR. This targeting strategy avoids hypotensive and other adverse effects of acute systemic administration of NO-related drugs. We and our colleagues have accomplished this goal by synthesizing a series of nitro-memantine compounds, i.e., using the NMDA channel blockade by memantine to target the NO group to the NMDA receptor. However, for this to work in an efficient manner, one should know the correct length of the "arm" or "bridge" that chemically links memantine with an NO group. To determine the length of this bridge, one should ideally know the location of both the ion pore and the redox site(s) on the NMDA receptor/channel complex. In recent years, the channel pore has been localized to the second membrane loop of NMDA receptor subunits. Our laboratory has now characterized the redox modulatory sites of NMDAR at a molecular level using site-directed mutagenesis of recombinant NMDAR subunits (NR1; NR2A-D; NR3A, B) as well as crystallographic modeling techniques (Kohr et al., 1994; Sullivan et al., 1994; Das et al., 1998; Kim et al., 1999; Choi et al., 2000, 2001). These approaches have facilitated the design strategy of NO-group targeting to the NMDA receptor. Unlike many, if not all, of the other drugs currently under investigation, memantine, nitroglycerin, and combinatorial nitro-memantine compounds have a high degree of clinical tolerability at neuroprotective doses. These facts should expedite clinical studies for the use of these drugs in patients with a variety of neurological disorders mediated, at least in part, by excessive NMDA receptor activity (Lipton and Rosenberg, 1994).

Because of the apparent clinical safety of memantine, nitroglycerin, and combinatorial nitro-memantine compounds, they have the potential for expeditious trials in humans. In fact, memantine was proven effective in a phase III multicenter clinical trial in patients with severe Alzheimer's disease in the USA and was also shown to hold promise for HAD in a recent phase II clinical trial in the USA (Jain, 2000; Susman, 2001). Memantine revealed a trend toward improvement on neuropsychological test scores above the control group, and significant improvement in a last-observation-carried-forward (LOCF) analysis.

Concomitantly, magnetic resonance spectroscopy (MRS) values for the *N*-acetylaspartate (NAA) to choline ratio were also significantly improved, suggesting neuronal protection. Finally, p38 MAPK inhibitors have been shown to reduce or abrogate neuronal apoptosis due to exposure to HIV/gp120 or SDF-1, or excitotoxicity (Kaul and Lipton, 1999; Kikuchi et al., 2000). The pharmaceutical industry is currently developing p38 inhibitors for a variety of inflammatory- and stress-related conditions, such as arthritis, and this may expedite trials for CNS indications such as HAD.

The most recent experimental evidence regarding HAD indicates that synergy between excitatory and inflammatory pathways to neuronal injury and death may, at least in part, be common to other CNS disorders including stroke, spinal cord injury, and Alzheimer's disease. It seems likely therefore that the development of new therapeutic strategies for HAD will impact several other neurodegenerative diseases and possibly vice-versa.

11. ACKNOWLEDGMENTS

M.K. and S.A.L. are supported by the National Institutes of Health, R01 NS050621 (to M.K.), P01 HD029587, R01 EY09024, R01 NS046994, R01 EY05477 and R01 NS41207 (to S.A.L.). S.A.L. is or has been a consultant to Allergan, Alcon, Merck, Johnson & Johnson, Forest Laboratories, NeuroMolecular Pharmaceuticals, Inc., and Neurobiological Technologies, Inc. in the field of neuroprotective agents.

12. REFERENCES

Adamson, D.C., Wildemann, B., Sasaki, M., Glass, J.D., McArthur, J.C., Christov, V.I., Dawson, T.M. and Dawson, V.L., 1996, Immunologic NO synthase: elevation in severe AIDS dementia and induction by HIV-1 gp41. *Science* **274**: 1917.

Adle-Biassette, H., Levy, Y., Colombel, M., Poron, F., Natchev, S., Keohane, C. and Gray, F., 1995, Neuronal apoptosis in HIV infection in adults. *Neuropathol. Appl. Neurobiol.* **21**: 218.

Adle-Biassette, H., Chretien, F., Wingertsmann, L., Hery, C., Ereau, T., Scaravilli, F., Tardieu, M. and Gray, F., 1999, Neuronal apoptosis does not correlate with dementia in HIV infection but is related to microglial activation and axonal damage. *Neuropathol. Appl. Neurobiol.* **25**: 123.

Aizenman, E., Stout, A.K., Hartnett, K.A., Dineley, K.E., McLaughlin, B. and Reynolds, I.J., 2000, Induction of neuronal apoptosis by thiol oxidation: putative role of intracellular Zinc release. *J. Neurochem.* **75**: 1878.

Alkhatib, G., Combadiere, C., Broder, C.C., Feng, Y., Kennedy, P.E., Murphy, P.M. and Berger, E.A., 1996, CC CKR5: a RANTES, MIP-1alpha, MIP-1beta receptor as a fusion cofactor for macrophage-tropic HIV-1. *Science* **272**: 1955.

Anderson, E.R., Gendelman, H.E. and Xiong, H., 2004, Memantine protects hippocampal neuronal function in murine human immunodeficiency virus type 1 encephalitis. *J. Neurosci.* **24**: 7194.

Asensio, V.C. and Campbell, I.L., 1999, Chemokines in the CNS: plurifunctional mediators in diverse states. *Trends Neurosci.* **22**: 504.

Batchelor, P.E., Liberatore, G.T., Wong, J.Y., Porritt, M.J., Frerichs, F., Donnan, G.A. and Howells, D.W., 1999, Activated macrophages and microglia induce dopaminergic sprouting in the injured striatum and express brain-derived neurotrophic factor and glial cell line-derived neurotrophic factor. *J. Neurosci.* **19**: 1708.

Bazan, J.F., Bacon, K.B., Hardiman, G., Wang, W., Soo, K., Rossi, D., Greaves, D.R., Zlotnik, A. and Schall, T.J., 1997, A new class of membrane-bound chemokine with a CX3C motif. *Nature* **385**: 640.

Beckman, J.S., Beckman, T.W., Chen, J., Marshall, P.A. and Freeman, B.A., 1990, Apparent hydroxyl radical production by peroxynitrite: implications for endothelial injury from nitric oxide and superoxide. *Proc. Natl. Acad. Sci. USA* **87**: 1620.

Bezzi, P., Carmignoto, G., Pasti, L., Vesce, S., Rossi, D., Rizzini, B.L., Pozzan, T. and Volterra, A., 1998, Prostaglandins stimulate calcium-dependent glutamate release in astrocytes. *Nature* **391**: 281.

Bezzi, P., Domercq, M., Brambilla, L., Galli, R., Schols, D., De Clercq, E., Vescovi, A., Bagetta, G., Kollias, G., Meldolesi, J. and Volterra, A., 2001, CXCR4-activated astrocyte glutamate release via TNFalpha: amplification by microglia triggers neurotoxicity. *Nat. Neurosci.* **4**: 702.

Bigge, C.F., 1999, Ionotropic glutamate receptors. *Curr. Opin. Chem. Biol.* **3**: 441.

Bleul, C.C., Farzan, M., Choe, H., Parolin, C., Clark-Lewis, I., Sodroski, J. and Springer, T.A., 1996, The lymphocyte chemoattractant SDF-1 is a ligand for LESTR/fusin and blocks HIV-1 entry. *Nature* **382**: 829.

Bonfoco, E., Krainc, D., Ankarcrona, M., Nicotera, P. and Lipton, S.A., 1995, Apoptosis and necrosis: two distinct events induced, respectively, by mild and intense insults with N-methyl-D-aspartate or nitric oxide/superoxide in cortical cell cultures. *Proc. Natl. Acad. Sci. USA* **92**: 7162.

Bormann, J., 1989, Memantine is a potent blocker of N-methyl-D-aspartate (NMDA) receptor channels. *Eur. J. Pharmacol.* **166**: 591.

Brack-Werner, R. and Bell, J.E., 1999, Replication of HIV-1 in human astrocytes. *Science Online: NeuroAids (www. sciencemag. org/NAIDS)* **2**: 1.

Brauner-Osborne, H., Egebjerg, J., Nielsen, E.O., Madsen, U. and Krogsgaard-Larsen, P., 2000, Ligands for glutamate receptors: design and therapeutic prospects. *J. Med. Chem.* **43**: 2609.

Brenneman, D.E., Westbrook, G.L., Fitzgerald, S.P., Ennist, D.L., Elkins, K.L., Ruff, M.R. and Pert, C.B., 1988, Neuronal cell killing by the envelope protein of HIV and its prevention by vasoactive intestinal peptide. *Nature* **335**: 639.

Brew, B.J., Corbeil, J., Pemberton, L., Evans, L., Saito, K., Penny, R., Cooper, D.A. and Heyes, M.P., 1995, Quinolinic acid production is related to macrophage tropic isolates of HIV-1. *J. Neurovirol.* **1**: 369.

Bruno, V., Copani, A., Besong, G., Scoto, G. and Nicoletti, F., 2000, Neuroprotective activity of chemokines against N-methyl-D-aspartate or beta-amyloid-induced toxicity in culture. *Eur. J. Pharmacol.* **399**: 117.

Budd, S.L., Tenneti, L., Lishnak, T. and Lipton, S.A., 2000, Mitochondrial and extramitochondrial apoptotic signaling pathways in cerebrocortical neurons. *Proc. Natl. Acad. Sci. USA* **97**: 6161.

Budka, H., 1991, Neuropathology of human immunodeficiency virus infection. *Brain Pathol.* **1**: 163.

Chakravarty, S. and Herkenham, M., 2005, Toll-like receptor 4 on nonhematopoietic cells sustains CNS inflammation during endotoxemia, independent of systemic cytokines. *J. Neurosci.* **25**: 1788.

Chao, C.C., Hu, S., Sheng, W.S. and Peterson, P.K., 1995, Tumor necrosis factor-alpha production by human fetal microglial cells: regulation by other cytokines. *Dev. Neurosci.* **17**: 97.

Chen, H.S. and Lipton, S.A., 1997, Mechanism of memantine block of NMDA-activated channels in rat retinal ganglion cells: uncompetitive antagonism. *J. Physiol.* **499**: 27.

Chen, H.S., Pellegrini, J.W., Aggarwal, S.K., Lei, S.Z., Warach, S., Jensen, F.E. and Lipton, S.A., 1992, Open-channel block of N-methyl-D-aspartate (NMDA) responses by memantine: therapeutic advantage against NMDA receptor-mediated neurotoxicity. *J. Neurosci.* **12**: 4427.

Chen, H.S., Wang, Y.F., Rayudu, P.V., Edgecomb, P., Neill, J.C., Segal, M.M., Lipton, S.A. and Jensen, F.E., 1998, Neuroprotective concentrations of the N-methyl-D-aspartate open- channel blocker memantine are effective without cytoplasmic vacuolation following post-ischemic administration and do not block maze learning or long-term potentiation. *Neuroscience* **86**: 1121.

Chen, W., Sulcove, J., Frank, I., Jaffer, S., Ozdener, H. and Kolson, D.L., 2002, Development of a human neuronal cell model for human immunodeficiency virus (HIV)-infected macrophage-induced neurotoxicity: apoptosis induced by HIV type 1 primary isolates and evidence for involvement of the Bcl-2/Bcl-xL-sensitive intrinsic apoptosis pathway. *J. Virol.* **76**: 9407.

Choi, D.W., 1988, Glutamate neurotoxicity and diseases of the nervous system. *Neuron* **1**: 623.

Choi, D.W., Koh, J.Y. and Peters, S., 1988a, Pharmacology of glutamate neurotoxicity in cortical cell culture: attenuation by NMDA antagonists. *J. Neurosci.* **8**: 185.

Choi, D.W., Yokoyama, M. and Koh, J., 1988b, Zinc neurotoxicity in cortical cell culture. *Neuroscience* **24**: 67.

Choi, Y.B., Tenneti, L., Le, D.A., Ortiz, J., Bai, G., Chen, H.S. and Lipton, S.A., 2000, Molecular basis of NMDA receptor-coupled ion channel modulation by S-nitrosylation. *Nat. Neurosci.* **3**: 15.

Choi, Y.B., Chen, H.S. and Lipton, S.A., 2001, Three pairs of cysteine residues mediate both redox and Zn^{2+} modulation of the NMDA receptor. *J. Neurosci.* **21**: 392.

Clifford, D.B., 1999, Central neurologic complications of HIV infection. *Curr. Infect. Dis. Rep.* **1**: 187.

Cocchi, F., Devico, A.L., Garzino-Demo, A., Arya, S.K., Gallo, R.C. and Lusso, P., 1995, Identification of RANTES, MIP-1 alpha, and MIP-1 beta as the major HIV- suppressive factors produced by $CD8^+$ T cells. *Science* **270**: 1811.

Conant, K., Garzino-Demo, A., Nath, A., McArthur, J.C., Halliday, W., Power, C., Gallo, R.C. and Major, E.O., 1998, Induction of monocyte chemoattractant protein-1 in HIV-1 Tat- stimulated astrocytes and elevation in AIDS dementia. *Proc. Natl. Acad. Sci. USA* **95**: 3117.

Cunningham, P.H., Smith, D.G., Satchell, C., Cooper, D.A. and Brew, B., 2000, Evidence for independent development of resistance to HIV-1 reverse transcriptase inhibitors in the cerebrospinal fluid. *AIDS* **14**: 1949.

Das, S., Sasaki, Y.F., Rothe, T., Premkumar, L.S., Takasu, M., Crandall, J.E., Dikkes, P., Conner, D.A., Rayudu, P.V., Cheung, W., Chen, H.S., Lipton, S.A. and Nakanishi, N., 1998, Increased NMDA current and spine density in mice lacking the NMDA receptor subunit NR3a. *Nature* **393**: 377.

Dawson, V.L., Dawson, T.M., London, E.D., Bredt, D.S. and Snyder, S.H., 1991, Nitric oxide mediates glutamate neurotoxicity in primary cortical cultures. *Proc. Natl. Acad. Sci. USA* **88**: 6368.

Dawson, V.L., Dawson, T.M., Bartley, D.A., Uhl, G.R. and Snyder, S.H., 1993, Mechanisms of nitric oxide-mediated neurotoxicity in primary brain cultures. *J. Neurosci.* **13**: 2651.

De Clercq, E., 2004, HIV-chemotherapy and -prophylaxis: new drugs, leads and approaches. *Int. J. Biochem. Cell Biol.* **36**: 1800.

Digicaylioglu, M. and Lipton, S.A., 2001, Erythropoietin-mediated neuroprotection involves cross-talk between Jak2 and NF-kappaB signaling cascades. *Nature* **412**: 641.

Digicaylioglu, M., Garden, G., Timberlake, S., Fletcher, L. and Lipton, S.A., 2004a, Acute neuroprotective synergy of erythropoietin and insulin-like growth factor I. *Proc. Natl. Acad. Sci. USA* **101**: 9855.

Digicaylioglu, M., Kaul, M., Fletcher, L., Dowen, R. and Lipton, S.A., 2004b, Erythropoietin protects cerebrocortical neurons from HIV-1/gp120-induced damage. *Neuroreport* **15**: 761.

Doble, A., 1999, The role of excitotoxicity in neurodegenerative disease: implications for therapy. *Pharmacol. Ther.* **81**: 163.

Dragic, T., Litwin, V., Allaway, G.P., Martin, S.R., Huang, Y., Nagashima, K.A., Cayanan, C., Maddon, P.J., Koup, R.A., Moore, J.P. and Paxton, W.A., 1996, HIV-1 entry into CD4$^+$ cells is mediated by the chemokine receptor CC-CKR-5. *Nature* **381**: 667.

Dreyer, E.B. and Lipton, S.A., 1995, The coat protein gp120 of HIV-1 inhibits astrocyte uptake of excitatory amino acids via macrophage arachidonic acid. *Eur. J. Neurosci.* **7**: 2502.

Dreyer, E.B., Kaiser, P.K., Offermann, J.T. and Lipton, S.A., 1990, HIV-1 coat protein neurotoxicity prevented by calcium channel antagonists. *Science* **248**: 364.

Elkabes, S., DiCicco-Bloom, E.M. and Black, I.B., 1996, Brain microglia/macrophages express neurotrophins that selectively regulate microglial proliferation and function. *J. Neurosci.* **16**: 2508.

Ellis, R.J., Deutsch, R., Heaton, R.K., Marcotte, T.D., McCutchan, J.A., Nelson, J.A., Abramson, I., Thal, L.J., Atkinson, J.H., Wallace, M.R. and Grant, I., 1997, Neurocognitive impairment is an independent risk factor for death in HIV infection. San Diego HIV Neurobehavioral Research Center Group. *Arch. Neurol.* **54**: 416.

Erdo, S.L. and Schafer, M., 1991, Memantine is highly potent in protecting cortical cultures against excitotoxic cell death evoked by glutamate and N-methyl-D-aspartate. *Eur. J. Pharmacol.* **198**: 215.

Everall, I.P., Heaton, R.K., Marcotte, T.D., Ellis, R.J., McCutchan, J.A., Atkinson, J.H., Grant, I., Mallory, M. and Masliah, E., 1999, Cortical synaptic density is reduced in mild to moderate human immunodeficiency virus neurocognitive disorder. HNRC group. HIV Neurobehavioral Research Center. *Brain Pathol.* **9**: 209.

Fiala, M., Looney, D.J., Stins, M., Way, D.D., Zhang, L., Gan, X., Chiappelli, F., Schweitzer, E.S., Shapshak, P., Weinand, M., Graves, M.C., Witte, M. and Kim, K.S., 1997, TNF-alpha opens a paracellular route for HIV-1 invasion across the blood-brain barrier. *Mol. Med.* **3**: 553.

Fontana, G., Valenti, L. and Raiteri, M., 1997, gp120 can revert antagonism at the glycine site of NMDA receptors mediating GABA release from cultured hippocampal neurons. *J. Neurosci. Res.* **49**: 732.

Fox, L., Alford, M., Achim, C., Mallory, M. and Masliah, E., 1997, Neurodegeneration of somatostatin-immunoreactive neurons in HIV encephalitis. *J. Neuropathol. Exp. Neurol.* **56**: 360.

Galasso, J.M., Harrison, J.K. and Silverstein, F.S., 1998, Excitotoxic brain injury stimulates expression of the chemokine receptor CCR5 in neonatal rats. *Am. J. Pathol.* **153**: 1631.

Garaci, E., Caroleo, M.C., Aloe, L., Aquaro, S., Piacentini, M., Costa, N., Amendola, A., Micera, A., Calio, R., Perno, C.F. and Levi-Montalcini, R., 1999, Nerve growth factor is an autocrine factor essential for the survival of macrophages infected with HIV. *Proc. Natl. Acad. Sci. USA* **96**: 14013.

Garden, G.A., Guo, W., Jayadev, S., Tun, C., Balcaitis, S., Choi, J., Montine, T.J., Moller, T. and Morrison, R.S., 2004, HIV associated neurodegeneration requires p53 in neurons and microglia. *FASEB J.* **18**: 1141.

Gartner, S., 2000, HIV infection and dementia. *Science* **287**: 602.

Gelbard, H.A., Dzenko, K.A., DiLoreto, D., del Cerro, C., del Cerro, M. and Epstein, L.G., 1993, Neurotoxic effects of tumor necrosis factor alpha in primary human neuronal cultures are mediated by activation of the glutamate AMPA receptor subtype: implications for AIDS neuropathogenesis. *Dev. Neurosci.* **15**: 417.

Gelbard, H.A., Nottet, H.S., Swindells, S., Jett, M., Dzenko, K.A., Genis, P., White, R., Wang, L., Choi, Y.B., Zhang, D. et al., 1994, Platelet-activating factor: A candidate human immunodeficiency virus type 1-induced neurotoxin. *J. Virol.* **68**: 4628.

Gelbard, H.A., James, H.J., Sharer, L.R., Perry, S.W., Saito, Y., Kazee, A.M., Blumberg, B.M., and Epstein, L.G., 1995, Apoptotic neurons in brains from paediatric patients with HIV-1 encephalitis and progressive encephalopathy. *Neuropathol. Appl. Neurobiol.* **21**: 208.

Gendelman, H.E., Persidsky, Y., Ghorpade, A., Limoges, J., Stins, M., Fiala, M. and Morrisett, R., 1997, The neuropathogenesis of the AIDS dementia complex. *AIDS* **11** suppl A: S35.

Gendelman, H.E., Grant, I., Lipton, S.A., Everall, I. and Swindells, S., 2005, *The Neurology of AIDS*. Oxford University Press, London.

Giulian, D., Vaca, K. and Noonan, C.A., 1990, Secretion of neurotoxins by mononuclear phagocytes infected with HIV-1. *Science* **250**: 1593.

Giulian, D., Wendt, E., Vaca, K. and Noonan, C.A., 1993, The envelope glycoprotein of human immunodeficiency virus type 1 stimulates release of neurotoxins from monocytes. *Proc. Natl. Acad. Sci. USA* **90**: 2769.

Giulian, D., Yu, J., Li, X., Tom, D., Li, J., Wendt, E., Lin, S.N., Schwarcz, R. and Noonan, C., 1996, Study of receptor-mediated neurotoxins released by HIV-1 infected mononuclear phagocytes found in human brain. *J. Neurosci.* **16**: 3139.

Glass, J.D., Wesselingh, S.L., Selnes, O.A. and McArthur, J.C., 1993, Clinical-neuropathologic correlation in HIV-associated dementia. *Neurology* **43**: 2230.

Glass, J.D., Fedor, H., Wesselingh, S.L. and McArthur, J.C., 1995, Immunocytochemical quantitation of human immunodeficiency virus in the brain: correlations with dementia. *Ann. Neurol.* **38**: 755.

Gleichmann, M., Gillen, C., Czardybon, M., Bosse, F., Greiner-Petter, R., Auer, J. and Muller, H.W., 2000, Cloning and characterization of SDF-1gamma, a novel SDF-1 chemokine transcript with developmentally regulated expression in the nervous system. *Eur. J. Neurosci.* **12**: 1857.

Gonzalez, E., Rovin, B.H., Sen, L., Cooke, G., Dhanda, R., Mummidi, S., Kulkarni, H., Bamshad, M.J., Telles, V., Anderson, S.A., Walter, E.A., Stephan, K.T., Deucher, M., Mangano, A., Bologna, R., Ahuja, S.S., Dolan, M.J. and Ahuja, S.K., 2002, HIV-1 infection and AIDS dementia are influenced by a mutant MCP-1 allele linked to increased monocyte infiltration of tissues and MCP-1 levels. *Proc. Natl. Acad. Sci. USA* **99**: 13795.

Gonzalez-Scarano, F. and Martin-Garcia, J., 2005, The neuropathogenesis of AIDS. *Nat. Rev. Immunol.* **5**: 69.

Gonzalez-Zulueta, M., Feldman, A.B., Klesse, L.J., Kalb, R.G., Dillman, J.F., Parada, L.F., Dawson, T.M. and Dawson, V.L., 2000, Requirement for nitric oxide activation of p21(ras)/extracellular regulated kinase in neuronal ischemic preconditioning. *Proc. Natl. Acad. Sci. USA* **97**: 436.

Gras, G., Chretien, F., Vallat-Decouvelaere, A.V., Le Pavec, J., Porcheray, F., Bossuet, C., Leone, C., Mialocq, P., Dereuddre-Bosquet, N., Clayette, P., Le Grand, R., Creminon, C., Dormont, D., Rimaniol, A.C. and Gray, F., 2003, Regulated expression of sodium-dependent glutamate transporters and synthetase: a neuroprotective role for activated microglia and macrophages in HIV infection? *Brain Pathol.* **13**: 211.

He, J., Chen, Y., Farzan, M., Choe, H., Ohagen, A., Gartner, S., Busciglio, J., Yang, X., Hofmann, W., Newman, W., Mackay, C.R., Sodroski, J. and Gabuzda, D., 1997, CCR3 and CCR5 are co-receptors for HIV-1 infection of microglia. *Nature* **385**: 645.

Hesselgesser, J., Taub, D., Baskar, P., Greenberg, M., Hoxie, J., Kolson, D.L. and Horuk, R., 1998, Neuronal apoptosis induced by HIV-1 gp120 and the chemokine SDF-1 alpha is mediated by the chemokine receptor CXCR4. *Curr. Biol.* **8**: 595.

Heyes, M.P., Brew, B.J., Martin, A., Price, R.W., Salazar, A.M., Sidtis, J.J., Yergey, J.A., Mouradian, M.M., Sadler, A.E., Keilp, J., Rubinow, D. and Markey, S.P., 1991, Quinolinic acid in cerebrospinal fluid and serum in HIV-1 infection: relationship to clinical and neurological status. *Ann. Neurol.* **29**: 202.

Ho, D.D., Rota, T.R., Schooley, R.T., Kaplan, J.C., Allan, J.D., Groopman, J.E., Resnick, L., Felsenstein, D., Andrews, C.A. and Hirsch, M.S., 1985, Isolation of HTLV-III from cerebrospinal fluid and neural tissues of patients with neurologic syndromes related to the acquired immunodeficiency syndrome. *N. Engl. J. Med.* **313**: 1493.

Jain, K.K., 2000, Evaluation of memantine for neuroprotection in dementia. *Expert. Opin. Investig. Drugs* **9**: 1397.

Jiang, Z.G., Piggee, C., Heyes, M.P., Murphy, C., Quearry, B., Bauer, M., Zheng, J., Gendelman, H.E. and Markey, S.P., 2001, Glutamate is a mediator of neurotoxicity in secretions of activated HIV-1-infected macrophages. *J. Neuroimmunol.* **117**: 97.

Johnston, J.B., Jiang, Y., van Marle, G., Mayne, M.B., Ni, W., Holden, J., McArthur, J.C. and Power, C., 2000, Lentivirus infection in the brain induces matrix metalloproteinase expression: role of envelope diversity. *J. Virol.* **74**: 7211.

Jones, G. and Power, C., 2006, Regulation of neural cell survival by HIV-1 infection. *Neurobiol. Dis.* **21**: 1.

Kalams, S.A. and Walker, B.D., 1995, Cytotoxic T lymphocytes and HIV-1 related neurologic disorders. *Curr. Top. Microbiol. Immunol.* **202**: 79.

Kaul, M., 2002, Chemokines and their receptors in HIV-associated dementia. *J. Neurovirol.* **8** suppl. 1: 41.

Kaul, M. and Lipton, S.A., 1999, Chemokines and activated macrophages in gp120-induced neuronal apoptosis. *Proc. Natl. Acad. Sci. USA* **96**: 8212.

Kaul, M., Garden, G.A. and Lipton, S.A., 2001, Pathways to neuronal injury and apoptosis in HIV-associated dementia. *Nature* **410**: 988.

Kaul, M., Zheng, J., Okamoto, S., Gendelman, H.E. and Lipton, S.A., 2005, HIV-1 infection and AIDS: consequences for the central nervous system. *Cell Death Differ.* **12** suppl. 1: 878.

Kaul, M., Ma, Q., Medders, K.E., Desai, M.K. and Lipton, S.A., 2006, HIV-1 coreceptors CCR5 and CXCR4 both mediate neuronal cell death but CCR5 paradoxically can also contribute to protection. *Cell Death Differ.* July 14; [Epub ahead of print].

Keilhoff, G. and Wolf, G., 1992, Memantine prevents quinolinic acid-induced hippocampal damage. *Eur. J. Pharmacol.* **219**: 451.

Kerr, S.J., Armati, P.J., Pemberton, L.A., Smythe, G., Tattam, B. and Brew, B.J., 1997, Kynurenine pathway inhibition reduces neurotoxicity of HIV-1- infected macrophages. *Neurology* **49**: 1671.

Kikuchi, M., Tenneti, L. and Lipton, S.A., 2000, Role of p38 mitogen-activated protein kinase in axotomy-induced apoptosis of rat retinal ganglion cells. *J. Neurosci.* **20**: 5037.

Kim, W.K., Choi, Y.B., Rayudu, P.V., Das, P., Asaad, W., Arnelle, D.R., Stamler, J.S. and Lipton, S.A., 1999, Attenuation of NMDA receptor activity and neurotoxicity by nitroxyl anion, NO-. *Neuron* **24**: 461.

Koenig, S., Gendelman, H.E., Orenstein, J.M., Dal Canto, M.C., Pezeshkpour, G.H., Yungbluth, M., Janotta, F., Aksamit, A., Martin, M.A. and Fauci, A.S., 1986, Detection of AIDS virus in macrophages in brain tissue from AIDS patients with encephalopathy. *Science* **233**: 1089.

Kohr, G., Eckardt, S., Luddens, H., Monyer, H. and Seeburg, P.H., 1994, NMDA receptor channels: subunit-specific potentiation by reducing agents. *Neuron* **12**: 1031.

Kramer-Hammerle, S., Rothenaigner, I., Wolff, H., Bell, J.E. and Brack-Werner, R., 2005, Cells of the central nervous system as targets and reservoirs of the human immunodeficiency virus. *Virus Res.* **111**: 194.

Krathwohl, M.D. and Kaiser, J.L., 2004a, Chemokines promote quiescence and survival of human neural progenitor cells. *Stem Cells* **22**: 109.

Krathwohl, M.D. and Kaiser, J.L., 2004b, HIV-1 promotes quiescence in human neural progenitor cells. *J. Infect. Dis.* **190**: 216.

Kullander, K., Kylberg, A. and Ebendal, T., 1997, Specificity of neurotrophin-3 determined by loss-of-function mutagenesis. *J. Neurosci. Res.* **50**: 496.

Langford, D., Sanders, V.J., Mallory, M., Kaul, M. and Masliah, E., 2002, Expression of stromal cell-derived factor 1alpha protein in HIV encephalitis. *J. Neuroimmunol.* **127**: 115.

Lannuzel, A., Lledo, P.M., Lamghitnia, H.O., Vincent, J.D. and Tardieu, M., 1995, HIV-1 envelope proteins gp120 and gp160 potentiate NMDA [Ca2+]i increase, alter [Ca2+]i homeostasis and induce neurotoxicity in human embryonic neurons. *Eur. J. Neurosci.* **7**: 2285.

Lapidot, T. and Petit, I., 2002, Current understanding of stem cell mobilization: the roles of chemokines, proteolytic enzymes, adhesion molecules, cytokines, and stromal cells. *Exp. Hematol.* **30**: 973.

Lavi, E., Kolson, D.L., Ulrich, A.M., Fu, L. and Gonzalez-Scarano, F., 1998, Chemokine receptors in the human brain and their relationship to HIV infection. *J. Neurovirol.* **4**: 301.

Lazarov-Spiegler, O., Solomon, A.S., Zeev-Brann, A.B., Hirschberg, D.L., Lavie, V. and Schwartz, M., 1996, Transplantation of activated macrophages overcomes central nervous system regrowth failure. *FASEB J.* **10**: 1296.

Le, D.A. and Lipton, S.A., 2001, Potential and current use of N-methyl-D-aspartate (NMDA) receptor antagonists in diseases of aging. *Drugs Aging* **18**: 717.

Le, D., Das, S., Wang, Y.F., Yoshizawa, T., Sasaki, Y.F., Takasu, M., Nemes, A., Mendelsohn, M., Dikkes, P., Lipton, S.A. and Nakanishi, N., 1997, Enhanced neuronal death from focal ischemia in AMPA-receptor transgenic mice. *Brain Res. Mol. Brain Res.* **52**: 235.

Lees, K.R., 1997, Cerestat and other NMDA antagonists in ischemic stroke. *Neurology* **49**: S66.

Lei, S.Z., Pan, Z.H., Aggarwal, S.K., Chen, H.S., Hartman, J., Sucher, N.J. and Lipton, S.A., 1992, Effect of nitric oxide production on the redox modulatory site of the NMDA receptor–channel complex. *Neuron* **8**: 1087.

Letendre, S.L., Lanier, E.R. and McCutchan, J.A., 1999, Cerebrospinal fluid beta chemokine concentrations in neurocognitively impaired individuals infected with human immunodeficiency virus type 1. *J. Infect. Dis.* **180**: 310.

Lipton, S.A., 1992a, Memantine prevents HIV coat protein-induced neuronal injury in vitro. *Neurology* **42**: 1403.

Lipton, S.A., 1992b, Models of neuronal injury in AIDS: another role for the NMDA receptor? *Trends Neurosci.* **15**: 75.

Lipton, S.A., 1992c, Requirement for macrophages in neuronal injury induced by HIV envelope protein gp120. *Neuroreport* **3**: 913.

Lipton, S.A., 1993, Prospects for clinically tolerated NMDA antagonists: open- channel blockers and alternative redox states of nitric oxide. *Trends Neurosci.* **16**: 527.

Lipton, S.A., 1994, HIV coat protein gp120 induces soluble neurotoxins in culture medium. *Neurosci. Res. Commun.* **15**: 31.

Lipton, S.A., 1997a, Neuropathogenesis of acquired immunodeficiency syndrome dementia. *Curr. Opin. Neurol.* **10**: 247.

Lipton, S.A., 1997b, Treating AIDS dementia [letter; comment]. *Science* **276**: 1629.

Lipton, S.A., 1998, Neuronal injury associated with HIV-1: approaches to treatment. *Annu. Rev. Pharmacol. Toxicol.* **38**: 159.

Lipton, S.A., 2004, Erythropoietin for neurologic protection and diabetic neuropathy. *N. Engl. J. Med.* **350**: 2516.

Lipton, S.A. and Gendelman, H.E., 1995, Seminars in Medicine of the Beth Israel Hospital, Boston. Dementia associated with the acquired immunodeficiency syndrome. *N. Engl. J. Med.* **332**: 934.

Lipton, S.A. and Jensen, F.E., 1992, Memantine, a clinically-tolerated NMDA open-channel blocker, prevents HIV coat protein-induced neuronal injury in vitro and in vivo. *Soc. Neurosci. Abstr.* **18**: 757.

Lipton, S.A. and Kieburtz, K., 1998, Development of adjunctive therapies for the neurologic manifestations of AIDS: dementia and painful neuropathy. In: *The Neurology of AIDS*. H.E. Gendelman, S.A. Lipton, L.G. Epstein and S. Swindells, eds, Chapman and Hall, New York, pp. 377.

Lipton, S.A. and Rosenberg, P.A., 1994, Excitatory amino acids as a final common pathway for neurologic disorders [see comments]. *N. Engl. J. Med.* **330**: 613.

Lipton, S.A. and Stamler, J.S., 1994, Actions of redox-related congeners of nitric oxide at the NMDA receptor. *Neuropharmacology* **33**: 1229.

Lipton, S.A. and Wang, Y.F., 1996, NO-related species can protect from focal cerebral ischemia/reperfusion. In: *Pharmacology of cerebral ischemia*. J. Krieglstein, ed., Medpharm Scientific Publisher, Stuttgart, pp. 183.

Lipton, S.A., Sucher, N.J., Kaiser, P.K. and Dreyer, E.B., 1991, Synergistic effects of HIV coat protein and NMDA receptor-mediated neurotoxicity. *Neuron* **7**: 111.

Lipton, S.A., Choi, Y.B., Pan, Z.H., Lei, S.Z., Chen, H.S., Sucher, N.J., Loscalzo, J., Singel, D.J. and Stamler, J.S., 1993, A redox-based mechanism for the neuroprotective and neurodestructive effects of nitric oxide and related nitroso-compounds. *Nature* **364**: 626.

Liu, R., Paxton, W.A., Choe, S., Ceradini, D., Martin, S.R., Horuk, R., MacDonald, M.E., Stuhlmann, H., Koup, R.A. and Landau, N.R., 1996, Homozygous defect in HIV-1 coreceptor accounts for resistance of some multiply-exposed individuals to HIV-1 infection. *Cell* **86**: 367.

Liu, Y., Jones, M., Hingtgen, C.M., Bu, G., Laribee, N., Tanzi, R.E., Moir, R.D., Nath, A. and He, J.J., 2000, Uptake of HIV-1 Tat protein mediated by low-density lipoprotein receptor-related protein disrupts the neuronal metabolic balance of the receptor ligands. *Nat. Med.* **6**: 1380.

Lo, T.M., Fallert, C.J., Piser, T.M. and Thayer, S.A., 1992, HIV-1 envelope protein evokes intracellular calcium oscillations in rat hippocampal neurons. *Brain Res.* **594**: 189.

Locati, M. and Murphy, P.M., 1999, Chemokines and chemokine receptors: biology and clinical relevance in inflammation and AIDS. *Annu. Rev. Med.* **50**: 425.

Lopalco, L., Barassi, C., Paolucci, C., Breda, D., Brunelli, D., Nguyen, M., Nouhin, J., Luong, T.T., Truong, L.X., Clerici, M., Calori, G., Lazzarin, A., Pancino, G. and Burastero, S.E., 2005, Predictive value of anti-cell and anti-human immunodeficiency virus (HIV) humoral responses in HIV-1-exposed seronegative cohorts of European and Asian origin. *J. Gen. Virol.* **86**: 339.

Luo, X., Carlson, K.A., Wojna, V., Mayo, R., Biskup, T.M., Stoner, J., Anderson, J., Gendelman, H.E. and Melendez, L.M., 2003, Macrophage proteomic fingerprinting predicts HIV-1-associated cognitive impairment. *Neurology* **60**: 1931.

Ma, Q., Jones, D., Borghesani, P.R., Segal, R.A., Nagasawa, T., Kishimoto, T., Bronson, R.T. and Springer, T.A., 1998, Impaired B-lymphopoiesis, myelopoiesis, and derailed cerebellar neuron migration in CXCR4- and SDF-1-deficient mice. *Proc. Natl. Acad. Sci. USA* **95**: 9448.

Marshall, D.C., Wyss-Coray, T. and Abraham, C.R., 1998, Induction of matrix metalloproteinase-2 in human immunodeficiency virus-1 glycoprotein 120 transgenic mouse brains. *Neurosci. Lett.* **254**: 97.

Masliah, E., Ge, N., Achim, C.L., Hansen, L.A. and Wiley, C.A., 1992, Selective neuronal vulnerability in HIV encephalitis. *J. Neuropathol. Exp. Neurol.* **51**: 585.

Masliah, E., Heaton, R.K., Marcotte, T.D., Ellis, R.J., Wiley, C.A., Mallory, M., Achim, C.L., McCutchan, J.A., Nelson, J.A., Atkinson, J.H. and Grant, I., 1997, Dendritic injury is a pathological substrate for human immunodeficiency virus-related cognitive disorders. HNRC group. The HIV Neurobehavioral Research Center. *Ann. Neurol.* **42**: 963.

Mattson, M.P., Haughey, N.J. and Nath, A., 2005, Cell death in HIV dementia. *Cell Death Differ.* **12**: 893.

McArthur, J.C., Hoover, D.R., Bacellar, H., Miller, E.N., Cohen, B.A., Becker, J.T., Graham, N.M., McArthur, J.H., Selnes, O.A., Jacobson, L.P. et al., 1993, Dementia in AIDS patients: incidence and risk factors. Multicenter AIDS Cohort Study. *Neurology* **43**: 2245.

McArthur, J.C., McClernon, D.R., Cronin, M.F., Nance-Sproson, T.E., Saah, A.J., St Clair, M. and Lanier, E.R., 1997, Relationship between human immunodeficiency virus-associated dementia and viral load in cerebrospinal fluid and brain. *Ann. Neurol.* **42**: 689.

McArthur, J.C., Haughey, N., Gartner, S., Conant, K., Pardo, C., Nath, A. and Sacktor, N., 2003, Human immunodeficiency virus-associated dementia: an evolving disease. *J. Neurovirol.* **9**: 205.

McGrath, K.E., Koniski, A.D., Maltby, K.M., McGann, J.K. and Palis, J., 1999, Embryonic expression and function of the chemokine SDF-1 and its receptor, CXCR4. *Dev. Biol.* **213**: 442.

Melton, S.T., Kirkwood, C.K. and Ghaemi, S.N., 1997, Pharmacotherapy of HIV dementia. *Ann. Pharmacother.* **31**: 457.

Mennicken, F., Maki, R., de Souza, E.B. and Quirion, R., 1999, Chemokines and chemokine receptors in the CNS: a possible role in neuroinflammation and patterning. *Trends Pharmacol. Sci.* **20**: 73.

Meucci, O. and Miller, R.J., 1996, Gp120-induced neurotoxicity in hippocampal pyramidal neuron cultures: protective action of TGF-beta1. *J. Neurosci.* **16**: 4080.

Meucci, O., Fatatis, A., Simen, A.A., Bushell, T.J., Gray, P.W. and Miller, R.J., 1998, Chemokines regulate hippocampal neuronal signaling and gp120 neurotoxicity. *Proc. Natl. Acad. Sci. USA* **95**: 14500.

Michael, N.L. and Moore, J.P., 1999, HIV-1 entry inhibitors: evading the issue [news] [see comments]. *Nat. Med.* **5**: 740.

Miller, R.J. and Meucci, O., 1999, AIDS and the brain: Is there a chemokine connection? *Trends Neurosci.* **22**: 471.

Milligan, C.E., Cunningham, T.J. and Levitt, P., 1991, Differential immunochemical markers reveal the normal distribution of brain macrophages and microglia in the developing rat brain. *J. Comp. Neurol.* **314**: 125.

Minghetti, L., 2005, Role of inflammation in neurodegenerative diseases. *Curr. Opin. Neurol.* **18**: 315.

Miwa, T., Furukawa, S., Nakajima, K., Furukawa, Y. and Kohsaka, S., 1997, Lipopolysaccharide enhances synthesis of brain-derived neurotrophic factor in cultured rat microglia. *J. Neurosci. Res.* **50**: 1023.

Mukherjee, P.K., DeCoster, M.A., Campbell, F.Z., Davis, R.J. and Bazan, N.G., 1999, Glutamate receptor signaling interplay modulates stress- sensitive mitogen-activated protein kinases and neuronal cell death. *J. Biol. Chem.* **274**: 6493.

Muller, W.E., Schroder, H.C., Ushijima, H., Dapper, J. and Bormann, J., 1992, Gp120 of HIV-1 induces apoptosis in rat cortical cell cultures: prevention by memantine. *Eur. J. Pharmacol.* **226**: 209.

Nath, A., 1999, Pathobiology of human immunodeficiency virus dementia. *Semin. Neurol.* **19**: 113.

Nath, A., Psooy, K., Martin, C., Knudsen, B., Magnuson, D.S., Haughey, N. and Geiger, J.D., 1996, Identification of a human immunodeficiency virus type 1 Tat epitope that is neuroexcitatory and neurotoxic. *J. Virol.* **70**: 1475.

Navia, B.A., Dafni, U., Simpson, D., Tucker, T., Singer, E., McArthur, J.C., Yiannoutsos, C., Zaborski, L. and Lipton, S.A., 1998, A phase I/II trial of nimodipine for HIV-related neurologic complications. *Neurology* **51**: 221.

Nicholas, R.S., Stevens, S., Wing, M.G. and Compston, D.A., 2002, Microglia-derived IGF-2 prevents TNFalpha induced death of mature oligodendrocytes in vitro. *J. Neuroimmunol.* **124**: 36.

Nicotera, P., Ankarcrona, M., Bonfoco, E., Orrenius, S. and Lipton, S.A., 1997, Neuronal necrosis and apoptosis: two distinct events induced by exposure to glutamate or oxidative stress. *Adv. Neurol.* **72**: 95.

Nottet, H.S., Jett, M., Flanagan, C.R., Zhai, Q.H., Persidsky, Y., Rizzino, A., Bernton, E.W., Genis, P., Baldwin, T., Schwartz, J., LaBenz, C.J. and Gendelman, H.E., 1995, A regulatory role for astrocytes in HIV-1 encephalitis. An overexpression of eicosanoids, platelet-activating factor, and tumor necrosis factor-alpha by activated HIV-1-infected monocytes is attenuated by primary human astrocytes. *J. Immunol.* **154**: 3567.

Nottet, H.S., Persidsky, Y., Sasseville, V.G., Nukuna, A.N., Bock, P., Zhai, Q.H., Sharer, L.R., McComb, R.D., Swindells, S., Soderland, C. and Gendelman, H.E., 1996, Mechanisms for the transendothelial migration of HIV-1-infected monocytes into brain. *J. Immunol.* **156**: 1284.

Oberlin, E., Amara, A., Bachelerie, F., Bessia, C., Virelizier, J.L., Arenzana-Seisdedos, F., Schwartz, O., Heard, J.M., Clark-Lewis, I., Legler, D.F., Loetscher, M., Baggiolini, M. and Moser, B., 1996, The CXC chemokine SDF-1 is the ligand for LESTR/fusin and prevents infection by T-cell-line-adapted HIV-1. *Nature* **382**: 833.

Ohagen, A., Ghosh, S., He, J., Huang, K., Chen, Y., Yuan, M., Osathanondh, R., Gartner, S., Shi, B., Shaw, G. and Gabuzda, D., 1999, Apoptosis induced by infection of primary brain cultures with diverse human immunodeficiency virus type 1 isolates: evidence for a role of the envelope. *J. Virol.* **73**: 897.

Olney, J.W., 1969, Brain lesions, obesity, and other disturbances in mice treated with monosodium glutamate. *Science* **164**: 719.

Olney, J.W. and Sharpe, L.G., 1969, Brain lesions in an infant rhesus monkey treated with monosodium glutamate. *Science* **166**: 386.

Osborne, N.N. and Quack, G., 1992, Memantine stimulates inositol phosphates production in neurones and nullifies N-methyl-D-aspartate-induced destruction of retinal neurons. *Neurochem. Int.* **21**: 329.

Parsons, C.G., Danysz, W. and Quack, G., 1999, Memantine is a clinically well tolerated N-methyl-D-aspartate (NMDA) receptor antagonist-a review of preclinical data. *Neuropharmacology* **38**: 735.

Paxton, W.A., Martin, S.R., Tse, D., O'Brien, T.R., Skurnick, J., VanDevanter, N.L., Padian, N., Braun, J.F., Kotler, D.P., Wolinsky, S.M. and Koup, R.A., 1996, Relative resistance to HIV-1 infection of CD4 lymphocytes from persons who remain uninfected despite multiple high-risk sexual exposure. *Nat. Med.* **2**: 412.

Paxton, W.A., Liu, R., Kang, S., Wu, L., Gingeras, T.R., Landau, N.R., Mackay, C.R. and Koup, R.A., 1998, Reduced HIV-1 infectability of $CD4^+$ lymphocytes from exposed- uninfected individuals: Association with low expression of CCR5 and high production of beta-chemokines. *Virology* **244**: 66.

Pellegrini, J.W. and Lipton, S.A., 1993, Delayed administration of memantine prevents N-methyl-D- aspartate receptor-mediated neurotoxicity. *Ann. Neurol.* **33**: 403.

Persidsky, Y., Stins, M., Way, D., Witte, M.H., Weinand, M., Kim, K.S., Bock, P., Gendelman, H.E. and Fiala, M., 1997, A model for monocyte migration through the blood-brain barrier during HIV-1 encephalitis. *J. Immunol.* **158**: 3499.

Petito, C.K. and Roberts, B., 1995, Evidence of apoptotic cell death in HIV encephalitis. *Am. J. Pathol.* **146**: 1121.

Petito, C.K., Cho, E.S., Lemann, W., Navia, B.A. and Price, R.W., 1986, Neuropathology of acquired immunodeficiency syndrome (AIDS): an autopsy review. *J Neuropathol. Exp. Neurol.* **45**: 635.

Pierson, T.C., Doms, R.W. and Pohlmann, S., 2004, Prospects of HIV-1 entry inhibitors as novel therapeutics. *Rev. Med. Virol.* **14**: 255.

Pittaluga, A., Pattarini, R., Severi, P. and Raiteri, M., 1996, Human brain N-methyl-D-aspartate receptors regulating noradrenaline release are positively modulated by HIV-1 coat protein gp120. *AIDS* **10**: 463.

Power, C., McArthur, J.C., Nath, A., Wehrly, K., Mayne, M., Nishio, J., Langelier, T., Johnson, R.T. and Chesebro, B., 1998, Neuronal death induced by brain-derived human immunodeficiency virus type 1 envelope genes differs between demented and nondemented AIDS patients. *J. Virol.* **72**: 9045.

Power, C., Gill, M.J. and Johnson, R.T., 2002, Progress in clinical neurosciences: the neuropathogenesis of HIV infection: host-virus interaction and the impact of therapy. *Can. J. Neurol. Sci.* **29**: 19.

Prospero-Garcia, O., Gold, L.H., Fox, H.S., Polis, I., Koob, G.F., Bloom, F.E. and Henriksen, S.J., 1996, Microglia-passaged simian immunodeficiency virus induces neurophysiological abnormalities in monkeys. *Proc. Natl. Acad. Sci. USA* **93**: 14158.

Rapalino, O., Lazarov-Spiegler, O., Agranov, E., Velan, G.J., Yoles, E., Fraidakis, M., Solomon, A., Gepstein, R., Katz, A., Belkin, M., Hadani, M. and Schwartz, M., 1998, Implantation of stimulated homologous macrophages results in partial recovery of paraplegic rats. *Nat. Med.* **4**: 814.

Robinson, A.P., White, T.M. and Mason, D.W., 1986, Macrophage heterogeneity in the rat as delineated by two monoclonal antibodies MRC OX-41 and MRC OX-42, the latter recognizing complement receptor type 3. *Immunol.* **57**: 239.

Rottman, J.B., Ganley, K.P., Williams, K., Wu, L., Mackay, C.R. and Ringler, D.J., 1997, Cellular localization of the chemokine receptor CCR5. Correlation to cellular targets of HIV-1 infection. *Am. J. Pathol.* **151**: 1341.

Ryan, L.A., Zheng, J., Brester, M., Bohac, D., Hahn, F., Anderson, J., Ratanasuwan, W., Gendelman, H.E. and Swindells, S., 2001, Plasma levels of soluble CD14 and tumor necrosis factor-alpha type II receptor correlate with cognitive dysfunction during human immunodeficiency virus type 1 infection. *J. Infect. Dis.* **184**: 699.

Sardar, A.M., Hutson, P.H. and Reynolds, G.P., 1999, Deficits of NMDA receptors and glutamate uptake sites in the frontal cortex in AIDS. *Neuroreport* **10**: 3513.

Sasseville, V.G., Newman, W., Brodie, S.J., Hesterberg, P., Pauley, D. and Ringler, D.J., 1994, Monocyte adhesion to endothelium in simian immunodeficiency virus-induced AIDS encephalitis is mediated by vascular cell adhesion molecule-1/alpha 4 beta 1 integrin interactions. *Am. J. Pathol.* **144**: 27.

Sathi, S., Edgecomb, P., Warach, S., Manchester, K., Donaghey, T., Stieg, P.E., Jensen, F.E. and Lipton, S.A., 1993, Chronic transdermal nitroglycerin (NTG) is neuroprotective in experimental rodent stroke models. *Soc. Neurosci. Abstr.* **19**: 849.

Sattler, R., Xiong, Z., Lu, W.Y., Hafner, M., MacDonald, J.F. and Tymianski, M., 1999, Specific coupling of NMDA receptor activation to nitric oxide neurotoxicity by PSD-95 protein. *Science* **284**: 1845.

Savio, T. and Levi, G., 1993, Neurotoxicity of HIV coat protein gp120, NMDA receptors, and protein kinase C: a study with rat cerebellar granule cell cultures. *J. Neurosci. Res.* **34**: 265.

Scala, E., D'Offizi, G., Rosso, R., Turriziani, O., Ferrara, R., Mazzone, A.M., Antonelli, G., Aiuti, F. and Paganelli, R., 1997, C-C chemokines, IL-16, and soluble antiviral factor activity are increased in cloned T cells from subjects with long-term nonprogressive HIV infection. *J. Immunol.* **158**: 4485.

Schifitto, G., Sacktor, N., Marder, K., McDermott, M.P., McArthur, J.C., Kieburtz, K., Small, S. and Epstein, L.G., 1999, Randomized trial of the platelet-activating factor antagonist lexipafant in HIV-associated cognitive impairment. Neurological AIDS Research Consortium. *Neurology* **53**: 391.

Seif el Nasr, M., Peruche, B., Rossberg, C., Mennel, H.D. and Krieglstein, J., 1990, Neuroprotective effect of memantine demonstrated in vivo and in vitro. *Eur. J. Pharmacol.* **185**: 19.

Stieg, P.E., Sathi, S., Warach, S., Le, D.A. and Lipton, S.A., 1999, Neuroprotection by the NMDA receptor-associated open-channel blocker memantine in a photothrombotic model of cerebral focal ischemia in neonatal rat. *Eur. J. Pharmacol.* **375**: 115.

Stumm, R.K., Rummel, J., Junker, V., Culmsee, C., Pfeiffer, M., Krieglstein, J., Hollt, V. and Schulz, S., 2002, A dual role for the SDF-1/CXCR4 chemokine receptor system in adult brain: isoform-selective regulation of SDF-1 expression modulates CXCR4- dependent neuronal plasticity and cerebral leukocyte recruitment after focal ischemia. *J. Neurosci.* **22**: 5865.

Sullivan, J.M., Traynelis, S.F., Chen, H.S., Escobar, W., Heinemann, S.F. and Lipton, S.A., 1994, Identification of two cysteine residues that are required for redox modulation of the NMDA subtype of glutamate receptor. *Neuron* **13**: 929.

Susman, E., 2001, Memantine improves function and cognition in advanced Alzheimer's. *Inpharma Weekly* **1292**: 5.

Tachibana, K., Hirota, S., Iizasa, H., Yoshida, H., Kawabata, K., Kataoka, Y., Kitamura, Y., Matsushima, K., Yoshida, N., Nishikawa, S., Kishimoto, T. and Nagasawa, T., 1998, The chemokine receptor CXCR4 is essential for vascularization of the gastrointestinal tract. *Nature* **393**: 591.

Tenneti, L., D'Emilia, D.M. and Lipton, S.A., 1997, Suppression of neuronal apoptosis by S-nitrosylation of caspases. *Neurosci. Lett.* **236**: 139.

Tenneti, L., D'Emilia, D.M., Troy, C.M. and Lipton, S.A., 1998, Role of caspases in N-methyl-D-aspartate-induced apoptosis in cerebrocortical neurons. *J. Neurochem.* **71**: 946.

Toggas, S.M., Masliah, E., Rockenstein, E.M., Rall, G.F., Abraham, C.R. and Mucke, L., 1994, Central nervous system damage produced by expression of the HIV-1 coat protein gp120 in transgenic mice. *Nature* **367**: 188.

Toggas, S.M., Masliah, E. and Mucke, L., 1996, Prevention of HIV-1 gp120-induced neuronal damage in the central nervous system of transgenic mice by the NMDA receptor antagonist memantine. *Brain Res.* **706**: 303.

Tornatore, C., Chandra, R., Berger, J.R. and Major, E.O., 1994, HIV-1 infection of subcortical astrocytes in the pediatric central nervous system. *Neurology* **44**: 481.

Tran, P.B. and Miller, R.J., 2003, Chemokine receptors: signposts to brain development and disease. *Nat. Rev. Neurosci.* **4**: 444.

Turchan, J., Sacktor, N., Wojna, V., Conant, K. and Nath, A., 2003, Neuroprotective therapy for HIV dementia. *Curr. HIV. Res.* **1**: 373.

Turrin, N.P. and Rivest, S., 2004, Unraveling the molecular details involved in the intimate link between the immune and neuroendocrine systems. *Exp. Biol. Med. (Maywood.)* **229**: 996.

UNAIDS., 2004, Report on the global AIDS epidemic; executive summary.

Verani, A. and Lusso, P., 2002, Chemokines as natural HIV antagonists. *Curr. Mol. Med.* **2**: 691.

Weiss, J.H. and Sensi, S.L., 2000, Ca^{2+}-Zn^{2+} permeable AMPA or kainate receptors: possible key factors in selective neurodegeneration. *Trends Neurosci.* **23**: 365.

Welch, K. and Morse, A., 2002, The clinical profile of end-stage AIDS in the era of highly active antiretroviral therapy. *AIDS Patient. Care STDS.* **16**: 75.

Wesselingh, S.L., Takahashi, K., Glass, J.D., McArthur, J.C., Griffin, J.W. and Griffin, D.E., 1997, Cellular localization of tumor necrosis factor mRNA in neurological tissue from HIV-infected patients by combined reverse transcriptase/polymerase chain reaction in situ hybridization and immunohistochemistry. *J. Neuroimmunol.* **74**: 1.

Wiley, C.A., Soontornniyomkij, V., Radhakrishnan, L., Masliah, E., Mellors, J., Hermann, P. and Achim, C.L., 1998, Distribution of brain HIV load in AIDS. *Brain Pathol.* **8**: 277.

Yeh, M.W., Kaul, M., Zheng, J., Nottet, H.S., Thylin, M., Gendelman, H.E. and Lipton, S.A., 2000, Cytokine-stimulated, but not HIV-infected, human monocyte-derived macrophages produce neurotoxic levels of L – cysteine. *J. Immunol.* **164**: 4265.

Zagury, D., Lachgar, A., Chams, V., Fall, L.S., Bernard, J., Zagury, J.F., Bizzini, B., Gringeri, A., Santagostino, E., Rappaport, J., Feldman, M., O'Brien, S.J., Burny, A. and Gallo, R.C., 1998, C-C chemokines, pivotal in protection against HIV type 1 infection. *Proc. Natl. Acad. Sci. USA* **95**: 3857.

Zhang, L., He, T., Talal, A., Wang, G., Frankel, S.S. and Ho, D.D., 1998, In vivo distribution of the human immunodeficiency virus/simian immunodeficiency virus coreceptors: CXCR4, CCR3, and CCR5. *J. Virol.* **72**: 5035.

Zhang, K., McQuibban, G.A., Silva, C., Butler, G.S., Johnston, J.B., Holden, J., Clark-Lewis, I., Overall, C.M. and Power, C., 2003, HIV-induced metalloproteinase processing of the chemokine stromal cell-derived factor-1 causes neurodegeneration. *Nat. Neurosci.* **6**: 1064.

Zhao, J., Lopez, A.L., Erichsen, D., Herek, S. Cotter, R.L., Curthoys, N.P. and Zheng, J., 2004a, Mitochondrial glutaminase enhances extracellular glutamate production in HIV-1-infected macrophages: Linkage to HIV-1 associated dementia. *J. Neurochem.* **88**: 169.

Zhao, M.L., Si, Q. and Lee, S.C., 2004b, IL-16 expression in lymphocytes and microglia in HIV-1 encephalitis. *Neuropathol. Appl. Neurobiol.* **30**: 233.

Zheng, J., Thylin, M.R., Ghorpade, A., Xiong, H., Persidsky, Y., Cotter, R., Niemann, D., Che, M., Zeng, Y.C., Gelbard, H.A., Shepard, R.B., Swartz, J.M. and Gendelman, H.E., 1999, Intracellular CXCR4 signaling, neuronal apoptosis and neuropathogenic mechanisms of HIV-1-associated dementia. *J. Neuroimmunol.* **98**: 185.

Zou, Y.R., Kottmann, A.H., Kuroda, M., Taniuchi, I. and Littman, D.R., 1998, Function of the chemokine receptor CXCR4 in haematopoiesis and in cerebellar development. *Nature* **393**: 595.

14

NEUROINFLAMMATION AND MITOCHONDRIAL DYSFUNCTION IN ALZHEIMER'S AND PRION'S DISEASES

Paula Agostinho and Catarina R. Oliveira

1. ABSTRACT

Alzheimer's disease (AD) and prion-related encephalopathies (PRE) are neurodegenerative disorders linked to the aberrant extracellular deposition of amyloidogenic proteins, amyloid-beta (Aβ), and pathogenic scrapie prion (PrP^{Sc}), respectively. In both disorders, cerebral amyloid deposits are associated with a local inflammatory response, which is initiated by the activation of microglia and recruitment of astrocytes. Activated microglia, particularly those in the vicinity of amyloid deposits can produce and release proinflammatory cytokines, chemokines, complement proteins, acute-phase proteins, and reactive oxygen and nitrogen species that can damage the neighboring neurons. Thus, activated microglia may be the link between Aβ and PrP^{Sc} deposition and neurodegeneration. The neuronal damage caused by inflammation, that is, neuroinflammation, can be both a cause and a consequence of chronic oxidative stress and mitochondrial dysfunction. In this review, we describe the similarities and differences between the neuroinflammatory pathways that are thought to be present in AD and PRE, based on findings from patients, animal models, and cell cultures. The role of mitochondrial dysfunction and chronic oxidative conditions in these diseases is also outlined. Furthermore, we focus on the current knowledge about the therapeutic strategies targeting neuroinflammation, such as nonsteroidal anti-inflammatory drugs, antioxidants, and passive and active immunotherapies, which may be a hope to treat or prevent Alzheimer's and Prion's diseases.

2. HALLMARKS OF ALZHEIMER'S DISEASE AND PRION-RELATED ENCEPHALOPATHIES

Alzheimer's disease (AD) and prion-related encephalopathies (PRE) are fatal progressive neurological disorders characterized by the extracellular deposition of pathological proteins in the form of amyloid plaques. The PRE, also called transmissible spongiform encephalopathies, include Creutzfeldt-Jakob disease (CJD), Gerstmann-Sträussler-Scheinker

---

Paula Agostinho – pagost@cnc.cj.uc.pt; Catarina R. Oliveira – Catarina@cnc.cj.uc.pt
Center for Neuroscience and Cell Biology, Institute of Biochemistry, Faculty of Medicine, University of Coimbra, Portugal

syndrome, fatal familial insomnia, and Kuru in humans, as well as scrapie in sheep and bovine spongiform encephalopathy (BSE). The symptoms of AD and human prion disorders, in which the most known is the CJD, are usually global cognitive dysfunctions, especially memory loss, leading to dementia and finally to death (Prusiner, 1996; Wisniewski et al., 1997). Animals with PRE can have variable symptoms, which often involve progressive ataxia and dementia. Both diseases can be either sporadic or of genetic origin, but PRE, unlike AD, can also be of infectious etiology (Selkoe and Lansbury, 1999). Most of AD cases are sporadic and result probably from the synergistic action of environmental and genetic factors. Advanced age and inheritance of ε4 allele of the polymorphic apolipoprotein E gene (APOE) are the major risk factors (Selkoe and Lansbury, 1999; Pereira et al., 2005). In contrast to AD, sporadic PRE is very rare, and most of the cases are hereditary or caused by prion infection due to ingestion of contaminated food or to iatrogenic actions. This may explain the characteristically different frequencies with which these diseases occur. While AD is the most common cause of dementia in the elderly, affecting 7–10% of individuals over the age of 65 and about 40% of persons over 80 years of age, the cases of prion diseases are extremely rare (one individual in a million each year) (Aguzzi et al., 2001; Sleegers and van Duijn, 2001).

Although PRE differs from AD in that they are horizontally transmissible and of very low incidence, the key pathogenic factors in both diseases are the membrane proteins (1) cellular prion protein (PrP^C) and (2) amyloid precursor protein (APP), respectively. The abnormal proteolytic processing of APP by β- and γ-secretases gives rise to β-sheet-containing peptides with ~4 kDa, made up of either 40 or 42 amino acids, known as amyloid-beta (Aβ) peptides, which can form fibrils and aggregates. The accumulation of Aβ fibrils, resulting from an imbalance between Aβ production and Aβ clearance, is the initiating molecular event that triggers neurodegeneration in sporadic and familiar AD. Currently, it is also thought that some nonfibrillar forms of Aβ, such as small oligomers and proto-fibrils, might be responsible for neuropathological alterations in AD (Pereira et al., 2005). The crucial pathogenic event in PRE is believed to be the conformational transition of the cellular prion protein (PrP^C) into the β-sheet rich isoform PrP^{Sc} (scrapie prion protein), which is particularly resistant to proteases digestion and accumulates in the brain as extracellular amorphous aggregates or as amyloid fibrils. The PrP^{Sc}, which could serve as a "template" for further PrP^C transformation, is thought to be the agent responsible for disease transmission (Prusiner, 1996; Aguzzi and Haass, 2003). The "protein-only theory" of prion infection, formulated by Stanley Prusiner in 1982, postulated that the infectious agent of PRE is a 27–30 kDa protein, which was denominated as prion or as PrP (from proteinaceous infectious particles). According to this theory, the conversion of PrP^C to PrP^{Sc} is induced by infectious PrP^{Sc}. This theory was strengthened by experimental studies that showed the transmission of this atypical infectious agent to laboratory rodents. In rare cases of sporadic and inherited human PRE, the conversion of PrP^C to PrP^{Sc} is thought to be caused by mutations in gene that encodes the cellular prion protein (Prusiner, 1996; Selkoe and Lansbury, 1999; Aguzzi and Haass, 2003). Although the mechanism of PrP^{Sc} neurotoxicity remains to be clarified, PrP^C was shown to be crucial for PrP^{Sc} replication and toxicity. In fact, tissue devoid of PrP^C, which subsequently is infected by PrP^{Sc}, remains healthy and free of pathology (Brown et al., 1994). As is the case in AD, the precise mechanism whereby accumulation of misfolded PrP^{Sc} leads to neuronal death is unclear. Nevertheless, the role of Aβ and PrP^{Sc} in the pathogenesis of these diseases was validated by the finding that mutations on genes encoding APP (localized on chromosome 21 in humans) and PrP^C (PRNP, localized on chromosome 20 in humans) result in autosomal-dominant AD and PRE (Aguzzi and Haass, 2003; Pereira et al., 2005).

The amyloid plaques formed in the brain of AD and PRE patients are composed mainly by Aβ and PrP^{Sc}. AD plaques have some histochemical properties similar to those formed in PRE: birefringent staining with Congo red and immunoreactivity for ubiquitin and heparin sulfate proteoglycans. Moreover, the amyloid deposits formed in these disorders are frequently co-localized with clusters of activated glia cells (astrocytes and microglia) and inflammation-related factors, as well as with dystrophic neurites (Selkoe and Lansbury, 1999; Eikelenboom et al., 2002; Veerhuis et al., 2005a). Besides the presence of PrP^{Sc} amyloid plaques, PRE diseases are characterized by neuronal vacuolation and loss, astrogliosis, and accumulation of activated microglia cells in affected brain areas. Cerebral accumulation of PrP^{Sc} may, through microglia activation, lead to vacuolization of neurons, which correlates with disease progression. The neuropathological hallmarks of AD include (1) profound neuronal loss in the hippocampus, entorhinal, and temporoparietal cortex, (2) the presence of intraneuronal neurofibrillary tangles (NFTs) composed mainly by hyperphosphorylated tau protein, (3) gliosis, and (4) extracellular deposits of Aβ peptides, which exist in the vessel wall or organized as amyloid plaques in the brain parenchyma (Wisniewski et al., 1997; Pereira et al., 2005).

Different types of Aβ plaques can be distinguished morphologically and based on composition as well as on mechanisms of formation. This can explain the lack of an established and comprehensible amyloid plaque nomenclature in AD research literature, where descriptive plaque terms as senile, compact, star, diffuse mature, and dense-core are common. Usually, the amyloid plaque terms most used are the "diffuse plaques" and the "neuritic plaques." These latter plaques are composed by highly fibrillar Aβ, often have a dense amyloid core, and are associated with clusters of activated microglia and astrocytes, as well as dystrophic neuronal processes. The diffuse plaques, composed mainly by non- or low-fibrillar Aβ, are not associated with a strong glial activation or with neuronal changes (Veerhuis et al., 2003, 2005a; D'Andrea et al., 2004). Contrary to the wide thought that all amyloid plaques arise from extracellular deposition, several evidences support that plaques may originate from vessels, neurons, Purkinje cell dendrites, and astrocytes (D'Andrea et al., 2004). Therefore, recent attention has turned toward the intracellular pathological mechanisms that can lead to the lysis of Aβ-filled cells and to the deposit of their content as fibrillar plaques in the brain (Wegiel et al., 2003; D'Andrea et al., 2004).

3. NEUROINFLAMATION

In the past, the brain was considered to be an immune privileged organ because of its compartmentalization and separation from peripheral blood system, as provided by blood–brain barrier. Currently, the CNS is known to have an endogenous immune system that is coordinated mainly by the immunocompetent cells, microglia. The inflammation of CNS, neuroinflammation, differs from that found in the periphery. The brain, due to lack of pain fibers, has difficulty to recognize the occurrence of inflammation and, thus, the classic signs of inflammation (rubor, tumor, calor, and dolor) are usually not seen in the CNS. Neuroinflammation is related to the neuronal death or deterioration caused by inflammation. It is unlikely that neuroinflammation initiates a neurological disease, because something else must first trigger the inflammatory process (Block and Hong, 2005; Tuppo and Arias, 2005).

However, the inflammatory process, if excessive and long-lasting, may contribute to cell death and consequently to pathology, turning a relatively benign pathology into a dangerous one. This does not mean that all neuroinflammation is harmful. The extent to which inflammation is positive or negative in a particular pathology may vary at different stages of diseases.

The major players of the inflammatory process in neurodegenerative disorders are the microglia, the astrocytes, and in a less extent the neurons, all of which having crucial roles in brain homeostasis and function. Microglia are the most important cells of the innate immune system in the CNS, representing on average 12% of the adult brain cell population. These cells are morphologically different from astrocytes and neurons, and derived mainly from precursor cells of monocytes–macrophage lineage (Kim and de Vellis, 2005; Tuppo and Arias, 2005). In mature brains and under nonpathological conditions, microglia normally exist in a "resting state," exhibiting small soma and ramified processes. However, upon activation in response to brain homeostasis disruption or immunological stimulus, microglia cells undergo notable morphological alterations, changing from resting ramified morphology into activated ameboid phenotype (small processes and large soma). Activated microglia also upregulate several cell surface antigens/receptors with significant functional proprieties, such as the major histocompatibility complex (MHC) class I and II (HLA in humans), the complement receptors and the immunoglobulin Fc receptors (Eikelenboom et al., 1994; Kim and de Vellis 2005). These cells are considered as antigen presenting cells due to expression of MHC class II antigens and pronounced phagocytic capacity. In addition, reactive microglia produce and secrete a variety of soluble proinflammatory factors potentially cytotoxic, such as cytokines, chemokines, and complement proteins, as well as reactive oxygen species (ROS) and nitric oxide (Kim and de Vellis, 2005). There is considerable debate as whether activated microglia are beneficial or harmful, given that they can also release throphic and anti-inflammatory factors. It is likely that microglia can exercise both functions depending on several factors, such as the type and duration of stimulus (Block and Hong, 2005).

Astrocytes are crucial to provide glia–neuron contact, preserving the functional integrity of neuronal synapses and modulating the activity of neurons. They are responsible for buffering the excess of neurotransmitters, maintaining ionic homeostasis and by the secretion of neurotrophic factors. Although the role of astrocytes in the inflammatory process is not as prominent as that of microglia, these cells become activated in response to immunologic challenges or brain injuries. Reactive astrocytes become hyperthrophic, increase the expression of glial fibrillary acidic protein (GFAP), secrete several proinflammatory factors, and form glial scars that disturb axonal regeneration. There is a clear interaction between astrocytes and microglia; however, the detailed mechanisms of this complex interaction remain to be completely established (Tuppo and Arias, 2005).

Neurons may also play a role in the inflammatory processes. They can produce several cytokines, acute-phase proteins, and complement proteins, mainly under pathological conditions (Tuppo and Arias, 2005). The production of these proinflammatory factors by neurons may trigger further inflammatory processes that exacerbate neuronal damage. The presence of phosphatidylserine on the surface of apoptotic neurons has been shown to be involved in the recognition and phagocytosis of these cells by microglia. A rapid and effective phagocytosis of apoptotic neurons by microglia is crucial to protect the adjacent cells from an undesirable inflammatory reaction that could aggravate neurodegeneration (Minghetti et al., 2005).

3.1. Neuroinflammation Associated to AD and PRE

The activation of glial cells, mainly microglia, is a general hallmark of neurodegenerative amyloidoses. Increasing evidences suggest that Aβ or PrP^{Sc} deposition, as amyloid plaques, is concomitant with microglia activation and precedes neuronal death (Marella and Chabry, 2004; Veerhuis et al., 2005a). In AD and PRE, these amyloid deposits are co-localized with a variety of inflammation-related factors (cytokines, chemokines, complement factors, acute-phase proteins, and reactive oxygen and nitrogen species) that are released by activated glial cells or dying neurons. These amyloid-associated factors are able to further activate and recruit glial cells that can contribute to neuronal damage, triggering a chronic neuroinflammatory process (Fig. 1).

3.1.1. Microglia Activation and Migration

The amyloid plaques found in AD and PRE brains are surrounded by clusters of reactive microglia (Block and Hong, 2005; Kim and de Vellis, 2005), suggesting that microglia activation induced by Aβ or PrP^{Sc} is associated with chemotactic responses to these amyloidogenic proteins. Most of the studies concerning microglia activation by Aβ and PrP^{Sc} were performed using synthetic peptides derived from these proteins, such as Aβ$_{1-40}$, Aβ$_{1-42}$, and PrP$_{106-126}$. Our group showed that Aβ$_{1-40}$ and PrP$_{106-126}$ are able to trigger microglia activation, by themselves, in vitro. Aβ- and PrP-activated microglia have a rounded ameboid morphology, phase-bright cell soma, exhibit small branches, and are immunopositive for ED-1, a marker of microglia activation (Garção et al., 2006; Fig. 2).

Reactive microglia associated with Aβ deposits in AD brain were shown to be immunopositive for members of the β2 integrin family (CD11a, CD11b, CD11c), the leukocyte common antigen (CD45), and MHC class II glycoprotein (HLA-DR9), a classic marker for the activation of scavenger cells. It was also reported an increased expression of immunoglobulin Fcγ receptors (CD64) and complement receptors (CR3 and CR4) in microglia, which may direct these cells to phagocyte complement- and immunoglobulin-opsonized cell debris (Akiyama et al., 2000; Kim and de Vellis, 2005).

Rogers et al. (2002) showed that microglia obtained from AD patients, as well as from rat brain, exhibit a great chemotaxis to fibrillar Aβ synthetic peptides. Multiple mediators of chemoattraction to Aβ were found in microglia that include the formyl peptide receptor-like 1 (FPRL1, Le et al., 2001a), the scavenger receptor A and B1 (Husemann et al., 2001, 2002), and the receptor for advanced glycation end-products (RAGE; Yan et al., 1996). Recently, a cell surface complex receptor, composed of scavenger receptor class B (CD36), α6β1 integrin, and CD47 (integrin-associated protein), was pointed out to bind Aβ in vitro. This complex receptor is also responsible for Aβ internalization through a β1 integrin-linked process, which is mechanistically different from the classical phagocytic process mediated by Fc and complement receptors (Koenigsknecht and Landreth, 2004). In addition, it was also shown that astrocytes can interact directly with Aβ fibrils, probably via a scavenger receptor (Wyss-Coray et al., 2003). The astrocytes surrounding amyloid plaques, as well as activated microglia or dying neurons, can trigger microglial chemotaxis to Aβ deposits through the upregulation and/or secretion of chemokines, such as transforming growth factor beta (TGF-β) and monocyte chemoattractant protein 1 (MCP-1), and regulate on activated normal T cell expressed and secreted (RANTES; reviewed in Rogers et al., 2002).

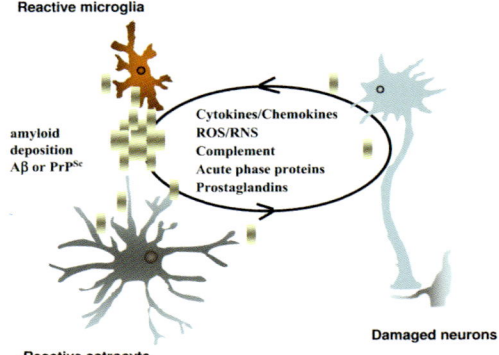

Fig. 1. Schematic representation of neuroinflammatory process in AD and PRE. The deposition of Aβ or PrP^{Sc}, as amyloid plaques, can trigger microglia and astrocytes activation, initiating an inflammatory process. Reactive glial cells release potential neurotoxic factors, such as cytokines, chemokines, complement factors, acute-phase proteins, prostaglandins, and reactive oxygen and nitrogen species. The damaged neurons can also release some pro-inflammatory factors, which can further cause glial activation and proliferation. Thus, a vicious circle is initiated, triggering a chronic neuroinflammatory process

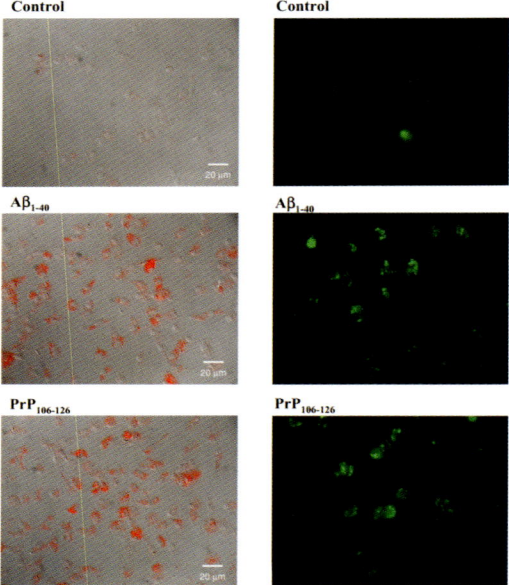

Fig. 2. Effect of Aβ and PrP peptides on microglia activation. Cultured rat brain microglia exposed to fibrillar Aβ$_{1-40}$ and PrP$_{106-126}$ peptides have rounded ameboid morphology, phase-bright cell soma, and exhibit small and thick branches. Control (*untreated*) cells have elongated shape and ramified cell bodies. An intense immunoreactivity against ED-1 (*red fluorescence*) was also observed in microglia treated with peptides, as compared with control cells. The cells were observed in contrast phase and in fluorescent view on an Axiovert 200 fluorescence microscope (*on right*). The iNOS expression in microglia cells, treated with Aβ and PrP peptides, assessed using an anti-iNOS antibody (*green fluorescence*), is also shown. Aβ and PrP peptides augmented iNOS expression (*on left*). These data indicate that these peptides are able to activate microglia cells. Scale bars 20 μm

Microglia migration to PrP^{Sc} amyloid plaques is at least partly controlled by chemokines. PrP^{Sc} itself triggers the recruitment of microglia by interacting with neurons, leading to the up-regulation of the expression level of chemokines, mainly RANTES (Marella et al., 2005). Although many studies have reported microglia activation by PrP$_{106-126}$, the actual cell surface receptors involved in the recognition of the peptide remain largely unknown (Burwinkel et al., 2004). The FPRL1, which also binds Aβ, was identified as a likely binding receptor for PrP$_{106-126}$ (Le et al., 2001b). Combs et al. (1999) have compared the cytokines production and activation of signal transduction pathways by Aβ and PrP synthetic peptides in microglia cells, and observed that both peptides trigger virtually identical activation cascades, which suggest that these peptides are recognized by the same receptors.

The glial activation, which is associated with the induction of cytokines, chemokines, and pattern-recognition receptors expression, may also mediate phagocytosis of bound ligands, contributing thus for the clearance and subsequent degradation of amyloid deposits in amyloidosis diseases (Wyss-Coray et al., 2003).

3.1.2. Microglial Phagocytic Role

The presence of activated microglia surrounding Aβ plaques suggests a phagocytic effort by these cells in attempt to remove Aβ deposits. Indeed, there are several evidences supporting that these cells can phagocyte fibrillar aggregates of Aβ (1) microglia in the AD cortex contain intracytoplasmic Aβ fibrils and C-terminal fragments of Aβ (Lewandowska et al., 1999); (2) rat brain microglia in culture migrate to synthetic fibrillar Aβ$_{1-42}$ and remove it in a dose- and time-dependent manner (Kopec and Carroll, 1998; Rogers et al., 2002); (3) fibrillar Aβ, as well as soluble Aβ, injected into rat cerebral cortex is rapidly phagocytosed by microglia (Weldon et al., 1998); and (4) Aβ immunization, in AD transgenic animals, leads to a significant reduction in the number of Aβ deposits (Bard et al., 2000). Several agents have been shown to stimulate microglial phagocytosis of Aβ, such as estrogens, Aβ peptides themselves, complement proteins that opsonize Aβ and anti-Aβ antibodies generated by active or passive immunization (reviewed in Rogers et al., 2002; D'Andrea et al., 2004). Although some studies postulated that microglial Aβ phagocytosis is neuroprotective, it was recently demonstrated that microglial chemotaxis and phagocytic activities can be associated with neurotoxic consequences (Schenk and Yednock, 2002). Since one of the neuroprotective functions of microglia is the amyloid deposits clearance, the age-related dysfunction of these cells may contribute to the formation of amyloid plaques in AD (Streit et al., 2005).

There are also evidences suggesting that microglia can phagocyte PrP^{Sc}. Misfolded PrP can be found in lysosomes of microglia and astrocytes surrounding the PrP^{Sc} plaque (Jeffrey et al., 1994), suggesting that these cells are involved in the phagocytosis of this pathogenic protein. In vitro studies showed that the synthetic PrP$_{106-126}$ peptide and isolated PrP^{Sc} compromise microglial phagocytic activity, probably by affecting the later steps of the internalization process (Ciesielski-Treska et al., 2004). The chronic activation of microglia may compromise PrP^{Sc} clearance by phagocytosis, thus promoting the formation of PrP^{Sc} plaques.

Participation of Glia Cells in Aβ Degradation

Besides the phagocytic role of microglia and astrocytes, these cells can also degrade Aβ fibrils (reviewed in Nagele et al., 2004). Most of the Aβ peptides present in amyloid plaques are truncated, probably because they were degraded by reactive microglia and astrocytes (D' Andrea et al., 2004; Tuppo and Arias, 2005). Neprilysin and insulin degrading enzyme (IDE) are candidate proteases responsible for fibrillar Aβ degradation in astrocytes and microglia, respectively (Guénette, 2003). Neprilysin is expressed as extracellular membrane-bound and as soluble forms and, thus, can directly affect brain interstitial Aβ levels. IDE is abundant in cytosol and peroxisomes and can also be secreted by microglia, acting on internal Aβ peptides, as well as on extracellular Aβ deposits. Purified IDE from brain homogenates was shown to degrade different forms of Aβ peptides: Aβ40, Aβ42, and an Aβ mutant in one type of familial AD (Dutch variant 1–40 Q) (Perez et al., 2000; Morelli et al., 2002). IDE knockout mice display high Aβ levels in the brain, suggesting that IDE activity is critical in determining the amount of this protein in vivo (Farris et al., 2003). It has also been reported that Aβ, after being phagocytosed by microglia, can accumulate in endosomal/lysosomal systems where they can be degraded by cathepsin D (Nakanishi, 2003). The apolipoprotein E was shown to promote Aβ clearance by facilitating the capacity of astrocytes to migrate, internalize, and degrade amyloid deposits (Koistinaho et al., 2004). In AD brain, the levels of neprilysin and IDE are reduced, as compared with age-matched controls, and the levels of IDE-related Aβ degradation products are also decreased (LeVine, 2004).

3.1.3. Molecular Mediators of Neuroinflammation

Cytokines and Chemokines

Several cytokines (including growth factors) and their receptors are present in the CNS. This ever growing family of proteins includes the tumor necrosis factor-α (TNF-α), the interferons (INF), the interleukin-1 (IL-1), IL-2, -3, -4, -6, -10, -12, -15, and -18. The transforming growing factor-β (TGF-β), the colony-stimulating factors, and the neurotrophic factors, such as nerve growth factor, neurotrophins (NT-3 and NT-4), or brain-derived neurotrophic factor (BDNF), are also cytokines. In general, these proteins regulate the intensity and duration of the inflammatory response, and are produced mainly by microglia and astrocytes (Hanisch, 2002).

A number of cytokines have been shown to be elevated in the brain of AD patients. The IL-1, which can exist in a soluble (IL-1β) or cell-associated (IL-1α) form, is overexpressed by microglia and astrocytes associated with Aβ deposits (Johnstone et al., 1999; Tuppo and Arias, 2005). This proinflammatory cytokine (possibly synergistically with other factors) can regulate APP synthesis and Aβ40/Aβ42 peptides production (Blasko et al., 1999; Rogers et al., 1999) and set up a vicious circle whereby the Aβ deposits stimulate further cytokine production by activated microglia (Eikelenboom et al., 2002). In addition, IL-1 can activate astrocytes, inducing the cytokine S100B expression by these cells. This interleukin also modulates APP expression in neurons, and has been shown to be associated with NFTs (Mrak and Griffin, 2001). TNF-α, a cytokine produced by microglia known to promote neuronal survival and death, has also been shown to be increased in AD brain. Overexpression of TNF-α in the brain alters the levels of nerve growth factors and the choline acetyltransferase activity (Aloe et al., 1999; Wenk et al., 2003), which suggests that this cytokine may have a role in cognitive aspects of AD and PRE. This cytokine seems also to regulate Aβ production (Tuppo and Arias, 2005), although the mechanisms by which cytokines stimulate Aβ production are not completely clarified; it is likely that APP maturation (e.g. glycosylation) can be affected (Blasko et al., 2004). The IL-6, like IL-1 and TNF-α, acts on the initiation and coordination of inflammatory responses. This cytokine can have both pro and anti-inflammatory outcomes, depending on the presence of other factors (e.g. cytokines). IL-6 is secreted by microglia in an early phase of brain injury, and seems to mediate subsequent glial activation and to engage astrocytes in the tissue repair (Hanish, 2002). Recently, it was suggested that IL-6 induces Alzheimer-type tau phosphorylation (Quintanilla et al., 2004). Genetic studies have shown that polymorphisms of some plaque-associated cytokines (e.g. IL-1, IL-6, IL-10, and TNF-α) are genetic risk factors for AD (Eikelenboom et al., 2002; Tuppo and Arias, 2005). Similarly to AD, activated microglia and astrocytes associated with PrP^{Sc} deposits show a high immunoreactivity for IL-1 (α and β), IL-6, and TNF-α (Eikelenboom et al., 2002). Moreover, several in vitro studies have shown that cultured microglia cells exposed to Aβ or PrP synthetic peptides increase the secretion of several cytokines, such as IL-1β, IL-6, and TNF-α (Veerhuis et al., 2002; Kim and de Vellis, 2005). Schultz et al. (2004) reported that IL-1 contributes to astrocytes activation in a PRE animal model in vivo, during early stages of the disease. A recent study has shown that the fibrillar state of Aβ peptides (monomers, oligomers, and fibrils) determines the type of cytokines released by cultured microglia cells (Lindberg et al., 2005). The existence of fibrillar PrP deposits was also found to be a prerequisite for in vitro induction of cytokine production by microglia (Peyrin et al., 1999).

Chemokines are a family of proinflammatory cytokines that participate in inflammatory cell recruitment. The chemokine gene family is represented by more than 40 members. Like their ligands, chemokine receptors are also numerous and exert their biological effects via interaction with G-protein-coupled receptors (Cartier et al., 2005). The chemokines are secreted by the different cells of CNS in response to injury and attract microglia and astrocytes to sites of inflammation, contributing to exacerbate neurodegeneration (Tuppo and Arias, 2005). In AD brain, it was observed that chemokines and chemokine receptors are increased and associated with amyloid plaques. Additionally, in vitro studies show that Aβ peptides increase the levels of the monocyte chemotactic protein (MCP-1) in cultured microglia and astrocytes, the production of this chemokine being regulated by cytokines IL-10, IL-4, and IL-13. Other chemokines have also been reported to be overexpressed in AD brain or in response to Aβ stimulation in vitro (Streit et al., 2001). Although few data exist about the expression of chemokine receptors in AD brain, it was shown an overexpression of CXCR3, CXCR5, and β-chemokine receptors in activated microglia (Streit et al., 2001; Cartier et al., 2005). In neurons of neuritic plaques, the chemokine receptor CXCR2 is upregulated (Xia et al., 1997). The ligand of this receptor, GROα-KC, can activate the ERK1/2 and PI3 kinase pathways, promoting tau hyperphosphorylation and, consequently, the formation of NFTs in cultured cortical neurons (Xia and Hyman, 2002). Similarly, the expression of RANTES and its receptors CXCR1, CXCR3, and CXCR5 is increased in reactive astrocytes of PrP^{Sc}-infected mice (Lee et al., 2005). The genes encoding the chemokines, CXCL9, CXCL10, and CXCL13 are already upregulated in asymptomatic stages of the PRE disease (Burwinkel et al., 2004). In the brains of animals infected with PrP^{Sc}, the expression of MCP-1 increases throughout disease progression and is positively correlated with microglial activation (Felton et al., 2005). These data suggest that chemokines and its receptors may participate in amplifying proinflammatory responses and, thereby, exacerbate the neurodegeneration in AD and PRE.

The Complement System

The activation of complement system, which compromise at least 30 soluble and cellular proteins, facilitates the opsonization of pathogens, promotes the formation of ROS, and induces proinflammatory responses. Glial cells and neurons can express complement proteins and regulatory factors of the complement cascade (e.g. CD46, CD59, C1 inhibitor, and C4 binding protein), particularly in response to an infectious challenge.

Overexpression and activation of complement system has been shown to occur in AD and PRE (Emmerling et al., 2000; Mabbot, 2004). The complement components C1q, C3B, C3C, and C4, as well as complement regulators have been detected associated to amyloid plaques in the brain of individuals affected with sporadic and familial form of PRE and AD. The ultimate goal of the complement system is the formation of MAC (membrane attach complex), composed of complement proteins C5-9. This complex forms a pore in the membrane of target cells, leading to its lysis and death. High levels of MAC have been detected in neurons of AD and PRE patients (Mabbot, 2004; Tuppo and Arias, 2005). Complement activation products C3d and C1q were found associated to reactive microglia in PrP^{Sc} deposits of PRE brains (Veerhuis et al., 2005b). The PrP^{Sc} brain infection (neuro-invasion) is accompanied by a significant upregulation of complement regulatory molecules or inhibitors, and it is likely that these complement regulators protect neurons from complement-mediated lysis (reviewed in Mabbot, 2004).

Acute-Phase Proteins

Some acute-phase proteins, such as α1-antichymotrypsin (ACT), α1-antitrypsin, α2-macroglobulin, and pentraxins, can be found in amyloid plaques of AD brain. These inflammation-related proteins are probably produced by brain cells, because there is no evidence of blood–brain barrier dysfunction in AD (Eikelenboom and van Gool, 2004). The classical pentraxins, C-reactive protein (CRP) and the serum amyloid P (SAP), are involved in the innate immune defense and can activate the classical complement system by an antibody-independent pathway. CRP and SAP are overexpressed in pyramidal neurons of AD brain. SAP is a component of all amyloid deposits and has a widespread association with NFTs (reviewed in Tuppo and Arias, 2005). In the brain of CJD patients and of PrP^{Sc}-infected animals, an evident SAP accumulation in the PrP^{Sc} deposits was also observed. This protein was shown to influence the fibrillogenic properties of PrP^{Sc} peptides in vitro, and concomitantly their microglial stimulatory effects (Veerhuis et al., 2005b). The overexpression of acute-phase proteins in these neurodegenerative diseases may reflect a brain "acute-phase" response comparable to the acute-phase response elicited by inflammation in the peripheral immune system.

Cyclooxygenases and Prostaglandins

Prostaglandins (PGs) are inflammatory mediators that are produced during the metabolism of arachidonic acid by cyclooxygenase (COX) and prostaglandin synthases. COX can exist as constitutive (COX-1) and inducible (COX-2) isoforms (Hurley et al., 2002; Minghetti, 2004). COX-1 is widely distributed by virtually all cell types and mediates several physiological functions, whereas COX-2 is rapidly expressed in cells in response to cytokines, growth factors, and proinflammatory factors. This inducible COX isoform has emerged as primarily responsible for PGs and ROS production in acute and chronic inflammatory conditions. The inflammatory effects of PGs are mediated by prostanoid binding G-protein-coupled receptors. Although in the mammalian brain, COX-2 expression has been associated with proinflammatory processes, this isoform can also be constitutively expressed in specific neuronal populations under "normal" conditions, contributing to vital brain functions, such as synaptic activity, memory, and functional hyperemia (reviewed in Minghetti, 2004).

During the last decade, several histological analyses of COX-1 and COX-2 expression have been performed in brain tissues of AD patients and AD animal models. These investigations have provided a substantial, but still controversial, body of evidences suggesting the involvement of COX-2 in the cascade of events leading to neuroinflamation in AD. Several studies reported an increased neuronal COX-2 immunoreactivity in AD brains, compared to control brain tissues, whereas other studies described a reduction in COX-2 positive neurons. Recently, it was suggested that COX-2 expression varies along with AD progression, which may explain the controversial data reported in the literature. The increase in COX-2 expression and activity at an early stage of AD could explain the primary protective effect of NSAIDs (outlined below). In accordance, it was shown that the PGE_2 levels in the CSF decline with the increasing of dementia severity. The expression of COX-1, unlike to COX-2, seems not to be affected in AD (reviewed in Minghetti, 2004; Hoozemans and O'Banion, 2005).

In CJD brain, the increase of PGE_2 levels in hippocampus is associated with a strong induction of COX-2 expression, mainly in microglia, that augments with the disease progression (Walsh et al., 2000). In the brains of mice infected with human PrP^{Sc}, a microglial COX-2 upregulation was also observed (Minghetti, 2004). In the cerebral cortex of CJD patients, both COX-1 and COX-2 are upregulated. In this study, COX-2 expression was predominantly observed in neurons, whereas COX-1 immunoreactivity was mainly present in microglia (Deininger et al., 2003). High levels of PGE_2 in CSF were associated with a short survival of sporadic CJD patients. However, it remains to be established whether PGE_2, a product of COX-2 activity, contributes to neuronal death or is a consequence of neuronal apoptosis (Minghetti, 2004).

Reactive Oxygen and Nitrogen Species

Other mechanisms by which activated microglia cause neuronal damage have been suggested to include the release of nitric oxide (NO), ROS, and mitochondrial damage (outlined below). These distinct events may be part of a single casual chain resulting in cell death (reviewed in Colton et al., 2000; Mhatre et al., 2004). Van Everbroeck et al. (2004) showed a correlation between amyloid deposition, glial activation, and oxidative stress in the brains of AD and CJD patients.

In brain, NO can act as a cellular signaling molecule or as a cytotoxic factor, killing both pathogenic cells and "healthy" host cells. NO is produced, during the conversion of arginine in citrulline, by nitric oxide synthases (NOS). Three isoforms of this enzyme are known (1) neuronal NOS (nNOS) that is constitutively expressed in some neurons, (2) endothelial NOS (eNOS) that can exist in brain endothelial cells and some astrocytes, and (3) inducible NOS (iNOS), which is not normally expressed in the healthy brain cells, but is induced in glial cells by proinflammatory cytokines (Brown and Bal-Price, 2003). The expression of iNOS is one of the characteristic changes of activated microglia and astrocytes in AD and PRE brains (Guentchev et al., 2000; Brown and Bal-Price, 2003). The stimulation of iNOS expression causes a long-lasting generation of high amounts of NO, which may be converted into a number of more reactive derivates, collectively known as reactive nitrogen species (RNS). Peroxynitrite ($ONOO^-$), which results from the reaction of NO with the radical anion superoxide ($O_2^{\cdot-}$), is highly reactive and cytotoxic. RNS has been implicated in AD, because the levels of nitrotyrosine, a product of the reaction of $ONOO^-$ with tyrosine, are significantly increased in AD (Smith et al., 1997; Luth et al., 2002). Recent studies have also reported increased levels of nitrosylated proteins in the cerebral cortex of CJD patients and PRE animal models (Guentchev et al., 2000; Freixes et al., 2006).

The radical derived from molecular oxygen, $O_2^{\cdot-}$, is generated as a by-product of the mitochondrial respiratory chain or as product of enzymes, such as xanthine oxidase and NADPH oxidase. This free radical can rapidly form hydrogen peroxide (H_2O_2) by a reaction catalyzed by superoxide dismutase (SOD). In the presence of redox-active iron, highly reactive toxic hydroxyl radicals ($HO^{\cdot}$) are generated from H_2O_2 and $O_2^{\cdot-}$ in the Fenton reaction and the Haber-Weiss cycle. Inevitably, if the amount of these ROS overhelms the antioxidant capacity of cells, oxidative stress occurs (reviewed in Dringen, 2005). Activated microglia and astrocytes can produce $O_2^{\cdot-}$ via a reaction catalyzed by NADPH oxidase (Mhatre et al., 2004; Abramov et al., 2005). This enzyme utilizes electrons from NADPH to reduce O_2 to $O_2^{\cdot-}$. In vitro studies reported that fibrillar Aβ peptides activate NADPH oxidase, promoting the production of several ROS. The activity of this enzyme can be modulated by interferon-γ and TNF-α (Bianca et al., 1999; Andersen et al., 2003). Aβ peptides per se can directly induce ROS production, via a metal ion reductive reaction, which may be essential to trigger glial activation (Mhatre et al., 2004). Several oxidative stress markers, such as 4-hydroxynonenal, 8-hydroxy-deoxyguanosine, and nitrotyrosine have been found in AD brain (reviewed in Mhatre et al., 2004).

An increasing number of studies support a correlation between the PrP^{Sc} deposition, microglia activation, and oxidative stress in PRE (Mhatre et al., 2004; Van Everbroeck et al., 2004; Freixes et al., 2006). Since PrP^C can have a SOD-like activity, the PrP^C deficiency, due to its conversion in PrP^{Sc}, may result in a reduced antioxidant capacity; thus, cells become more susceptible to oxidant injury (Brown et al., 1997; Klamt et al., 2001). Indeed, it was reported that the mice devoid of PrP^C have high levels of oxidative stress (Brown et al., 1997).

The advanced glycation end-products (AGE), formed by the nonenzymatic glycation of proteins with reducing sugars, can also contribute to a state of oxidative stress. These products through the activation of their specific receptor (RAGE) induce iNOS and proinflammatory cytokines expression, via nuclear translocation of NFκB and subsequent transcription of target genes (Munch et al., 2003). Given that they accumulate on long-lived proteins, high concentrations of AGE can be found in amyloid plaques. The AGE present in amyloid plaques were shown to be potent activators of microglia (Munch et al., 2003). In mature amyloid plaques of AD brain, astrocytes, and microglia co-localized with AGE also express iNOS (Wong et al., 2001). AGE and their receptors were also found in astrocytes associated to PrP^{Sc} deposits of CJD patients (Sasaki et al., 2002).

Microglia are equipped with efficient antioxidant defense systems for self-protection against ROS and RNS. These cells contain glutathione in high concentrations and several antioxidant enzymes, such as SOD, catalase, glutathione peroxidase, and glutathione reductase, as well as NADP-regenerating enzymes (reviewed in Dringen, 2005).

4. MITOCHONDRIAL DYSFUNCTION IN AD AND PRE

Mitochondria are essential organelles for production of ATP through oxidative phosphorylation. They also regulate intracellular Ca^{2+} homeostasis and are the principal sources of intracellular ROS. When the electron transport chain is inhibited, electrons accumulate in complex I and coenzyme Q where they can be donated to molecular oxygen to form $O_2^{\cdot-}$. This radical can be detoxified by the mitochondrial manganese superoxide dismutase (MnSOD) to form H_2O_2, which can then be reduced to H_2O by either glutathione peroxidase (in mitochondria and cytosol) or catalase (in peroxisomes). Alternatively, H_2O_2 in the presence of reduced transition metals, such as Fe^{2+} and Cu^+, can be converted into a highly potent radical, $HO^{\cdot}$ (reviewed in Eckert et al., 2003; Moreira et al., 2005). ROS formed by mitochondria can attack several cellular targets, including mitochondrial components themselves (lipids, proteins, and DNA). Mitochondria are particularly susceptible to oxidative stress, due to the lack of histones in mitochondrial DNA (mtDNA) and to reduced capacity for DNA repair (Swerdlow and Khan, 2004; Moreira et al., 2005).

Reactive oxygen and nitrogen species can induce the formation of mitochondria permeability transition pore (MPTP), due to oxidation and/or nitration of proteins that regulate pore opening. MPTP opening may be a physiological process and usually does not cause cell death, while longer sustained MPTP opening may cause either apoptosis or necrosis. The MPTP can dissipate proton motive force, inducing uncoupling of oxidative phosphorylation and ATP synthase reversion and, consequently, cellular ATP hydrolysis and depletion that potentially trigger necrosis. The formation of this pore may also allow the release of cytochrome c, and other apoptogenic intermembrane proteins, such as apoptosis-inducing factor (AIF), and SMAC/Diablo that potentially trigger apoptosis. The mode of cell death after MPTP opening is likely to depend on additional factors, such as activation of Bid/Bax/Bad pathway or availability of ATP (ATP depletion favors necrosis). The execution of apoptosis usually involves the activation of the cysteine proteases, called caspases (reviewed in Eckert et al., 2003; Moreira et al., 2005; Pereira et al., 2005).

Impairment of mitochondrial energy metabolism and increased production of ROS have been proposed as potential mechanisms in the pathogenesis of AD. Brain imaging studies have clearly showed deficits in glucose consumption in living AD patients, and this metabolic impairment seems to occur before the onset of clinical symptoms (Blass, 2002). However, the mechanisms by which Aβ peptides impair mitochondrial function remain to be completely clarified. Increased ROS production and lipid peroxidation products, induced by Aβ peptides, have been shown to affect several enzyme complexes of the respiratory chain, as well as pyruvate dehydrogenase and α-ketoglutarate dehydrogenase (Pereira et al., 1999; Pocernich and Butterfield, 2003). Aβ peptides have also multiple direct effects on isolated mitochondria, causing alterations in enzyme activity and MPTP opening (Moreira et al., 2002). Further evidence for the relationship between Aβ peptides, oxidative stress, and mitochondrial dysfunction comes from recent studies demonstrating that Aβ interacts with a binding protein in mitochondria, called Aβ-binding alcohol dehydrogenase (ABAD), and directly causes oxidative stress and impairs mitochondrial function (Lustbader et al., 2004; Takuma et al., 2005). Aβ peptides cause a loss of mitochondrial membrane potential via NADPH oxidase activation in astrocytes (Abramov et al., 2004). Several evidences suggest that altered proteolytic processing of APP is synergistically related with impaired energetic metabolism. Moreover, recent studies showed that (1) APP can be targeted to mitochondria affecting directly mitochondrial function (Anandatheerthavarada et al., 2003), and (2) active γ-secretase complexes are located in mitochondria and can cleave APP (Hansson et al., 2004). These data suggest that Aβ peptides can be formed in mitochondria; however, whether Aβ is produced in mitochondria or reaches the mitochondria from other subcellular locations is still a controversial issue.

Mitochondrial dysfunction is one of the earliest signs of AD, occurring before NFTs are evident (Hirai et al., 2001). Several in vitro studies showed that Aβ peptides activate neuronal apoptosis through the activation of a mitochondrial pathway, involving intracellular Ca^{2+} homeostasis disruption, oxidative stress, cytochrome c release, and caspase activation (reviewed in Pereira et al., 2005). High levels of mitochondrial DNA (mtDNA) mutations, linked to cytochrome c oxidase deficiency, were also observed in hippocampal neurons of AD brain (Cottrell et al., 2002). Human cell lines expressing mtDNA from AD patients contain reduced cytochrome c oxidase activity, elevated ROS and reduced ATP levels, compared to age-matched control subjects. These data suggest that alterations in mtDNA, probably due to ROS attack, may play a role in mitochondrial dysfunction in AD (Cardoso et al., 2001).

The role of mitochondrial dysfunction in PRE has not been as intensively studied as for AD. Abnormal morphology of mitochondria and reduction of their number have been shown in hippocampal neurons of PrP^C-knockout mice. In these animals, high levels of antioxidants in mitochondria were found, suggesting that these organelles were subjected to oxidative stress. These findings support the hypothesis that PrP^C has an antioxidant function and that oxidative stress occurs in its absence, particularly affecting mitochondria (Miele et al., 2002). In the brain of PrP^{Sc}-infected animals, electron microscopy studies showed the loss of mitochondrial cristae and matrix, particularly in hippocampus and cortex. Moreover, biochemical analysis of these brains have shown a reduced activity of MnSOD concomitant with a decreased activity of cytochrome c oxidase and ATP synthase, as well as increased levels of oxidative stress markers (Choi et al., 1998; Lee et al., 1999). Several in vitro studies have reported that $PrP_{106-126}$, like Aβ synthetic peptides, triggers apoptotic neuronal death via a mitochondrial pathway, involving impairment of intracellular Ca^{2+} homeostasis, cytochrome c release and activation of caspases and calpains (O'Donovan et al., 2001; Agostinho and Oliveira, 2003).

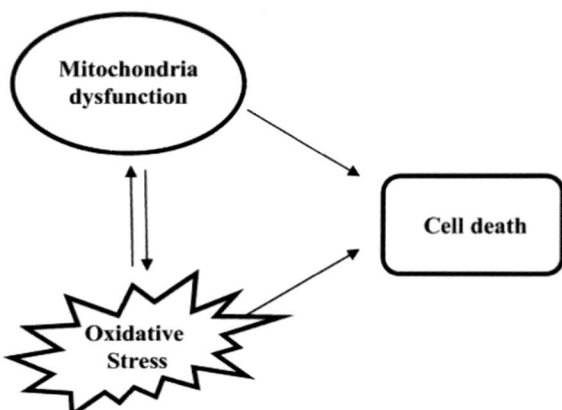

Fig. 3. Mitochondrial dysfunction as a major player in neurodegeneration. Mitochondria can be a source and a target of ROS. The excessive production of ROS, by mitochondria and/or by reactive glial cells trigger mitochondria dysfunction and (apoptotic or necrotic) cell death. The impairment of mitochondrial function can activate intracellular signaling pathways, leading to cellular stress and demise. Mitochondria dysfunction and the oxidative stress associated with an active immune response are prominent events in AD and PRE pathology

In summary, mitochondrial dysfunction and the resulting ROS production can trigger cycle events that cause the onset of neuronal death (apoptotic or necrotic) in AD and PRE (Fig. 3).

5. POTENTIAL THERAPEUTIC STRATEGIES TARGETING THE NEUROINFLAMMATORY PROCESS

5.1. Nonsteroidal Anti-Inflammatory Drugs

There is compelling epidemiological evidence that long-term nonsteroidal anti-inflammatory drugs (NSAIDs) therapies, which are widely used for a variety of inflammatory pathologies, reduce the incidence and risk of AD. In AD brain, the use of NSAIDs results in a substantial reduction in the number of microglia associated to senile plaques, supporting the idea that these cells are the targets of these drugs (Aisen, 2002). The common feature of most NSAIDs is their capacity to inhibit COX and, consequently, the decrease PGs production, in particular PGE2. Prolonged use of traditional NSAIDs that inhibit COX-1 usually leads to gastrointestinal perturbations, such as ulcers and bleeding. Since several evidences support the involvement of COX-2 with inflammation, selective COX-2 inhibitors were developed. These NSAIDS do not cause gastrointestinal toxicity and are able to prevent not only the inflammatory process, but also the cholinergic dysfunction associated to AD memory loss (Giovannini et al., 2003). During the past decade, several randomized, double-blind, placebo-controlled trials have been carried out to investigate the efficacy of some NSAIDs in AD treatment. Clinical trials of the COX-2 inhibitor, rofecoxib, and of the COX-1 and COX-2 inhibitor, naproxen, have not found any significant efficacy in cognitive decline of patients with mild to moderate AD. However, it was demonstrated that some NSAIDs, such as ibuprofen, flurbiprofen, indometacine, and sulindac sulfide, provide benefits for AD patients (reviewed in Aisen, 2002; Pereira et al., 2005; Stuchbury and Münch, 2005). The contradictory results between NSAIDs trials seems to indicate that the action of COX in AD may not be as significant as previously predicted, and that neuroprotection may be due to other mechanisms.

Currently, it is known that ibuprofen, flurbiprofen, indometacine, and sulindac sulfide (but not naproxen, celecoxib, or aspirin) decreased $A\beta_{1-42}$ production and led to an increase in $A\beta_{1-38}$ levels. These NSAIDs act by changing γ-secretase activity, independently of COX inhibition (Weggen et al., 2003). Some NSAIDs can also lower the levels of $A\beta$ (1) by inhibiting the protein Rho, a protein that belongs to a family of small GTP proteins that are involved in the regulation of multiple cellular functions (Zhou et al., 2003) or (2) by activating the peroxisomal proliferators-activated receptors (PPARs), a group of nuclear hormone receptors that act to negatively inhibit the transcription of proinflammatory genes. PPARα agonists have been shown to inhibit iNOS and COX-2 expression, as well as IL-6, IL-1β and TNF-α production in vitro. In addition, PPARγ activation is able to regulate β-secretase expression and activity (reviewed in Tuppo and Arias, 2005). Although the clinical trials of NSAIDs effects in AD have not confirmed their benefit, these drugs are likely to be more useful in preventing or reverting early AD neuropathology than in treating advanced stages of this disease.

Preventing the conversion of PrP^C into PrP^{Sc} is the most obvious therapeutic approach to halt PRE and to reverse early degenerative changes. Although therapeutic targeting of inflammatory responses would not directly address PrP^{Sc} replication, an anti-inflammatory therapy might prolong survival times by limiting secondary inflammatory reactions. It is known that interfering with IL-1 system may delay PRE onset. Moreover, Klein et al. (2001) reported that mice devoid of C1qa, a constituent of complement system, and mice deficient for FcRγ were partially or fully protected against PrP^{Sc} infection, supporting the potential therapeutic value of anti-inflammatory approaches. However, to achieve a better therapeutic effect, the anti-inflammatory therapies should be combined with other strategies, such as inhibition of PrP^{Sc} accumulation, enhancement of PrP^{Sc} clearance or neuroprotective drugs (reviewed in Mallucci and Collinge, 2005).

5.2. Antioxidants

Antioxidants are able to scavenge or neutralize both extracellular and, depending on their membrane-permeability, intracellular ROS, before they damage cellular constituents or act as second messengers on inflammation. The AD and PRE brains are under significant oxidative stress, mainly due to activated glial cells and mitochondrial dysfunction (outlined above). Thus, it is likely that antioxidants use might attenuate the neuronal damage associated to these neurodegenerative disorders.

Vitamin E (a chain-breaking lipid soluble antioxidant) is one of the most extensively antioxidant studied. An in vitro study showed that vitamin E may exert beneficial effects, independently of its antioxidant capacity, through the downregulation of specific cell signaling pathways in microglia that control the production of cytokines and NO (Li et al., 2001). Transgenic mice of AD (Tg2576) maintained on a vitamin E-supplemented diet showed a decrease in $A\beta_{1-40}$ and $A\beta_{1-42}$ levels, suggesting that vitamin E may have a direct effect on the attenuation of symptoms in early AD (Sung et al., 2004). However, epidemiological studies have questioned the efficacy of antioxidants on AD treatment and prevention. Clinical trials showed that vitamin E and selegiline (a selective-monoamine type B inhibitor with antioxidant properties) are beneficial in patients with moderate AD severity slowing the disease progression (Sano et al., 1997). Recently, a cross-sectional and large prospective study indicated that the use of vitamin E and C supplements in combination was associated with a decreased incidence and prevalence of AD (Zandi et al., 2004). The efficacy of other antioxidants against AD, such as 17β-estradiol, α-lipoic acid, aromatic amine/imines, and Cu/Zn chelators, has been studied, but the results are still inconclusive (Kitazawa et al., 2004; Pereira et al., 2005). More clinical and research studies are needed to assess their true potential in AD. However, given the low toxicity of antioxidants and the wide variety available, mainly in vegetables and fruits, they are promising candidates until more specific therapeutic strategies are developed.

5.3. Immunotherapy

Immunization against Aβ is a possible therapeutic strategy against AD that has gained considerable attention in the last decade. This approach to reduce the presence of Aβ in brain has been carried out through active immunization or passive administration of antibodies against Aβ. Schenk et al. (1999) were the first to show that immunization of AD animal models (PDAPP) with human fibrillar $A\beta_{1-42}$ reduced the extent of cortical Aβ deposits and plaques development. This study reported an almost complete prevention of Aβ deposition in vaccinated six-week-old mice and a slowing of AD pathology progression in older mice.

These findings were associated with the appearance of serum antibody against $A\beta_{1-42}$. In addition, it was also found a decrease in astrocytosis and microglial activation, as well as a reduced number of dystrophic neurites, suggesting a benefit beyond the reduction of $A\beta$ deposits. The generation of anti-$A\beta$ antibodies combined with the presence of $A\beta$-immunoreactive microglia in regions with amyloid plaques led to propose that clearance of $A\beta$ may be through Fc receptor-mediated microglial phagocytosis of $A\beta$ (Schenk et al., 1999). Other study, using the same AD mice model, showed that peripherally administered antibodies (passive immunization) enter the CNS and bind $A\beta$ fibrils, with subsequent recruitment of microglia and the phagocytosis of the complex antibodies/$A\beta$ through Fc receptor ligation (Bard et al., 2000). The involvement of microglia in $A\beta$ clearance by active and passive immunization has been further supported by other studies (reviewed in Gelinas et al., 2004; Kitazawa et al., 2004). In contrast, it was also suggested that $A\beta$ clearance may occur independently of microglia activation, and depend on the antibody capacity to inhibit $A\beta$ aggregation (Solomon et al., 1996) or to bind and withdraw soluble $A\beta$ from brain to plasma (DeMattos et al., 2002). Although the mechanisms involved in $A\beta$ clearance by immunization remain to be completely clarified, several studies confirmed that immunotherapy contributes to clear $A\beta$ plaques and to reduce the cognitive deficit in a variety of AD animal models.

After the promising results obtained in several species (mice, rabbits, guinea pigs, and monkeys) clinical trials using a synthetic $A\beta_{1-42}$ peptide (AN-1792) in conjunction with the T helper 1 adjuvant, QS-21, were initiated in mild to moderate AD patients. Although the results of phase I trials (in a small group) did not show significant side effects, phase II trials involving 375 patients were halted due to the development of signs and symptoms of meningoencephalitis in about 6% of patients. Despite the cessation of clinical trials, several further findings provide hope for AD patients. Postmortem histological examinations of one AD patient immunized with AN-1792, who developed meningoencephalitis, revealed (1) a decrease of $A\beta$ plaques from neocortex and a reduced number of reactive astrocytes and dystrophic neuritis, and (2) in some regions devoid of plaques, $A\beta$-immunoreactivity was associated with microglia. These data suggest that an effective immune response was generated that results in the clearance of $A\beta$ plaques by microglia from immunized AD brain (Nicoll et al., 2003). Moreover, Hock et al. (2003) showed that patients who generate antibodies to $A\beta$ exhibit slower rates of cognitive decline compared to the group who did not develop such antibodies. This beneficial effect was even observed in patients who experienced transient episodes of meningoencephalitis. Although it is unclear whether these lesions have a direct link to immunization, studies in transgenic mice models of AD reported that these animals developed microhemorrhages after passive $A\beta$ immunotherapy (Pfeifer et al., 2002). In general, the researchers believe that immunotherapy may still hold potential promise to treat or prevent AD. However, the immunization protocols must be refined before further clinical approaches are conducted.

Recent studies showed that antibodies against certain PrP^C epitopes are able to inhibit PrP^{Sc} formation in cell culture. These antibodies can bind PrP^C, decreasing its availability for endogenous PrP^{Sc} synthesis (Enari et al., 2001; Peretz et al., 2001). In mice peripherally infected with PrP^{Sc}, it was shown that passive immunization with anti-PrP^C decreases PrP^{Sc} levels in the spleen and prolongs their survival more than 300 days. However, this protective effect was not observed in intracerebrally infected mice because antibodies do not readily cross the blood–brain barrier (White et al., 2003). Solforosi et al. (2004) showed that the intracerebroventricular injection of anti-PrP^C antibodies caused a rapid and extensive neuronal apoptosis, suggesting a vital role for PrP^C in the control of neuronal survival. Nevertheless, the understanding of the molecular and cellular basis of PrP^{Sc} replication and neurotoxicity is advancing rapidly, and the development of effective and safe therapeutics to treat PRE seems to be feasible (reviewed in Mallucii and Collinge, 2005).

6. REFERENCES

Abramov, A.Y., Canevari, L. and Duchen, M.R., 2004, Beta-amyloid peptides induce mitochondrial dysfunction and oxidative stress in astrocytes and death of neurons through activation of NADPH oxidase. *J. Neurosci.* **24**: 565.

Abramov, A.Y., Jacobson, J., Wientjes, F., Hothersall, J., Canevari, L. and Duchen, M.R., 2005, Expression and modulation of an NADPH oxidase in mammalian astrocytes. *J. Neurosci.* **25**: 9176.

Agostinho, P. and Oliveira, C.R., 2003, Involvement of calcineurin in the neurotoxic effects induced by amyloid-beta and prion peptides. *Eur. J. Neurosci.* **17**: 1189.

Aguzzi, A. and Haass, C., 2003, Games played by rogue proteins in prion disorders and Alzheimer's disease. *Science* **302**: 814.

Aguzzi, A., Glatzel, M., Montrasio, F., Prinz, M. and Heppner, F.L., 2001, Interventional strategies against prion diseases. *Nat. Rev. Neurosci.* **2**: 745.

Aisen, P.S., 2002, The potential of anti-inflammatory drugs for the treatment of Alzheimer's disease. *Lancet Neurol.* **1**: 279.

Akiyama, H., Barger, S., Barnum, S., Bradt, B., Bauer, J., Cole, G.M., Cooper, N.R., Eikelenboom, P., Emmerling, M., Fiebich, B.L., Finch, C.E., Frautschy, S., Griffin, W.S., Hampel, H., Hull, M., Landreth, G., Lue, L., Mrak, R., Mackenzie, I.R., McGeer, P.L., O'Banion, M.K., Pachter, J., Pasinetti, G., Plata-Salaman, C., Rogers, J., Rydel, R., Shen, Y., Streit, W., Strohmeyer, R., Tooyoma, I., Van Muiswinkel, F.L., Veerhuis, R., Walker, D., Webster, S., Wegrzyniak, B., Wenk, G. and Wyss-Coray, T., 2000, Inflammation and Alzheimer's disease. *Neurobiol Aging* **21**: 383.

Aloe, L., Fiore, M., Probert, L., Turrini, P. and Tirassa, P., 1999, Overexpression of tumour necrosis factor-alpha in the brain of transgenic mice differentially alters nerve growth factor levels and choline acetyltransferase activity. *Cytokine* **11**: 45.

Anandatheerthavarada, H.K., Biswas, G., Robin, M.A. and Avadhani, N.G., 2003, Mitochondrial targeting and a novel transmembrane arrest of Alzheimer's amyloid precursor protein impairs mitochondrial function in neuronal cells. *J. Cell Biol.* **161**: 41.

Andersen, J.M., Myhre, O., Aarnes, H., Vestad, T.A. and Fonnum, F., 2003, Identification of the hydroxyl radical and other reactive oxygen species in human neutrophil granulocytes exposed to a fragment of the amyloid beta peptide. *Free Radic. Res.* **37**: 269.

Bard, F., Cannon, C., Barbour, R., Burke, R.L., Games, D., Grajeda, H., Guido, T., Hu, K., Huang, J., Johnson-Wood, K., Khan, K., Kholodenko, D., Lee, M., Lieberburg, I., Motter, R., Nguyen, M., Soriano, F., Vasquez, N., Weiss, K., Welch, B., Seubert, P., Schenk, D. and Yednock, T., 2000, Peripherally administered antibodies against amyloid beta-peptide enter the central nervous system and reduce pathology in a mouse model of Alzheimer's disease. *Nat. Med.* **6**: 916.

Bianca, V.D., Dusi, S., Bianchini, E., Dal Pra, I. and Rossi, F., 1999, Beta-amyloid activates the O-2 forming NADPH oxidase in microglia, monocytes, and neutrophils. A possible inflammatory mechanism of neuronal damage in Alzheimer's disease. *J. Biol. Chem.* **274**: 15493.

Blasko, I., Marx, F., Steiner, E., Hartmann, T. and Grubeck-Loebenstein, B., 1999, TNFalpha plus IFNgamma induce the production of Alzheimer beta-amyloid peptides and decrease the secretion of APPs. *FASEB J.* **13**: 63.

Blasko, I., Stampfer-Kountchev, M., Robatscher, P., Veerhuis, R., Eikelenboom, P. and Grubeck-Loebenstein, B., 2004, How chronic inflammation can affect the brain and support the development of Alzheimer's disease in old age: the role of microglia and astrocytes. *Aging Cell.* **3**: 169.

Blass, J.P., 2002, Glucose/mitochondria in neurological conditions. *Int. Rev. Neurobiol.* **51**: 325.

Block, M.L. and Hong, J.-S., 2005, Microglia and inflammation-mediated neurodegeneration: multiple triggers with a common mechanism. *Prog. Neurobiol.* **76**: 77.

Brown, D.R., Herms, J. and Kretzschmar, H.A., 1994, Mouse cortical cells lacking cellular PrP survive in culture with a neurotoxic PrP fragment. *Neuroreport* **5**: 2057.

Brown, D.R., Qin, K., Herms, J.W., Madlung, A., Manson, J., Strome, R., Fraser, P.E., Kruck, T., von Bohlen, A., Schulz-Schaeffer, W., Giese, A., Westaway, D. and Kretzschmar, H., 1997, The cellular prion protein binds copper in vivo. *Nature* **390**: 684.
Brown, G.C. and Bal-Price, A., 2003, Inflammatory neurodegeneration mediated by nitric oxide, glutamate, and mitochondria. *Mol. Neurobiol.* **27**: 325.
Burwinkel, M., Riemer, C., Schwarz, A., Schultz, J., Neidhold, S., Bamme, T. and Baier, M., 2004, Role of cytokines and chemokines in prion infections of the central nervous system. *Int. J. Dev. Neurosci.* **22**: 497.
Cardoso, S.M., Santos, S., Swerdlow, R.H. and Oliveira, C.R., 2001, Functional mitochondria are required for amyloid beta-mediated neurotoxicity. *FASEB J.* **15**: 1439.
Cartier, L., Hartley, O., Dubois-Dauphin, M. and Krause, K.H., 2005, Chemokine receptors in the central nervous system: role in brain inflammation and neurodegenerative diseases. *Brain Res. Brain Res. Rev.* **48**: 16.
Choi, S.I., Ju, W.K., Choi, E.K., Kim, J., Lea, H.Z., Carp, R.I., Wisniewski, H.M. and Kim, Y.S., 1998, Mitochondrial dysfunction induced by oxidative stress in the brains of hamsters infected with the 263 K scrapie agent *Acta Neuropathol. (Berl).* **96**: 279.
Ciesielski-Treska, J., Grant, N.J., Ulrich, G., Corrotte, M., Bailly, Y., Haeberle, A.M., Chasserot-Golaz, S. and Bader, M.F., 2004, Fibrillar prion peptide (106-126) and scrapie prion protein hamper phagocytosis in microglia. *Glia* **46**: 101.
Colton, C.A., Chernyshev, O.N., Gilbert, D.L. and Vitek, M.P., 2000, Microglial contribution to oxidative stress in Alzheimer's disease. *Ann. NY Acad. Sci.* **899**: 292.
Combs, C.K., Johnson, D.E., Cannady, S.B., Lehman, T.M. and Landreth, G.E., 1999, Identification of microglial signal transduction pathways mediating a neurotoxic response to amyloidogenic fragments of beta-amyloid and prion proteins. *J. Neurosci.* **19**: 928.
Cottrell, D.A., Borthwick, G.M., Johnson, M.A., Ince, P.G., and Turnbull, D.M., 2002, The role of cytochrome c oxidase deficient hippocampal neurones in Alzheimer's disease. *Neuropathol. Appl. Neurobiol.* **28**: 390.
D'Andrea, M.R., Cole, G.M. and Ard, M.D., 2004, The microglial phagocytic role with specific plaque types in the Alzheimer's disease brain. *Neurobiol. Aging* **25**: 675.
Deininger, M.H., Bekure-Nemariam, K., Trautmann, K., Morgalla, M., Meyermann, R. and Schluesener, H.J., 2003, Cyclooxygenase-1 and -2 in brains of patients who died with sporadic Creutzfeldt-Jakob disease. *J. Mol. Neurosci.* **20**: 25.
DeMattos, R.B., Bales, K.R., Cummins, D.J., Paul, S.M. and Holtzman, D.M., 2002, Brain to plasma amyloid-beta efflux: a measure of brain amyloid burden in a mouse model of Alzheimer's disease. *Science* **295**: 2264.
Dringen, R., 2005, Oxidative and antioxidative potential of brain microglial cells. *Antioxid. Redox Signal* **7**: 1223.
Eckert, A., Keil, U., Marques, C.A., Bonert, A., Frey, C., Schussel, K. and Muller, W.E., 2003, Mitochondrial dysfunction, apoptotic cell death, and Alzheimer's disease. *Biochem. Pharmacol.* **66**: 1627.
Eikelenboom, P. and van Gool, W.A., 2004, Neuroinflammatory perspectives on the two faces of Alzheimer's disease. *J. Neural Transm.* **111**: 281.
Eikelenboom, P., Zhan, S.S., van Gool, W.A. and Allsop, D., 1994, Inflammatory mechanisms in Alzheimer's disease. *Trends Pharmacol. Sci.* **15**: 447.
Eikelenboom, P., Bate, C., van Gool, W.A., Hoozemans, J.J., Rozemuller, J.M., Veerhuis, R. and Williams, A., 2002, Neuroinflammation in Alzheimer's disease and prion disease. *Glia* **40**: 232.
Emmerling, M.R., Watson, M.D., Raby, C.A. and Spiegel, K., 2000, The role of complement in Alzheimer's disease pathology. *Biochim. Biophys. Acta* **1502**: 158.
Enari, M., Flechsig, E. and Weissmann, C., 2001, Scrapie prion protein accumulation by scrapie-infected neuroblastoma cells abrogated by exposure to a prion protein antibody. *Proc. Natl Acad. Sci. USA* **98**: 9295.
Farris, W., Mansourian, S., Chang, Y., Lindsley, L., Eckman, E.A., Frosch, M.P., Eckman, C.B., Tanzi, R.E., Selkoe, D.J. and Guenette, S., 2003, Insulin-degrading enzyme regulates the levels of insulin, amyloid beta-protein, and the beta-amyloid precursor protein intracellular domain in vivo. *Proc. Natl Acad. Sci. USA* **100**: 4162.
Felton, L.M., Cunningham, C., Rankine, E.L., Waters, S., Boche, D. and Perry, V.H., 2005, MCP-1 and murine prion disease: separation of early behavioural dysfunction from overt clinical disease. *Neurobiol. Dis.* **20**: 283.
Freixes, M., Rodriguez, A., Dalfo, E. and Ferrer, I., 2006, Oxidation, glycoxidation, lipoxidation, nitration, and responses to oxidative stress in the cerebral cortex in Creutzfeldt-Jakob disease. *Neurobiol. Aging* **27**: 1807.
Garção, P., Oliveira, C.R. and Agostinho P., 2006, A comparative study of microglia activation induced by amyloid-beta and prion peptides. The role in neurodegeneration. *J. Neurosci. Res.* **84**: 182.
Gelinas, D.S., DaSilva, K., Fenili, D., St. George-Hyslop, P. and McLaurin, J., 2004, Immunotherapy for Alzheimer's disease. *Proc. Natl Acad. Sci. USA* **101** (suppl. 2): 14657.
Giovannini, M.G., Scali, C., Prosperi, C., Bellucci, A., Pepeu, G. and Casamenti, F., 2003, Experimental brain inflammation and neurodegeneration as model of Alzheimer's disease: protective effects of selective COX-2 inhibitors. *Int. J. Immunopathol. Pharmacol.* **16** (suppl. 2): 31.
Guénette, S.Y., 2003, Mechanisms of Abeta clearance and catabolism. *Neuromolecular Med.* **4**: 147.

Guentchev, M., Voigtlander, T., Haberler, C., Groschup, M.H. and Budka, H., 2000, Evidence for oxidative stress in experimental prion disease. *Neurobiol. Dis.* **7**: 270.
Hanisch, U.-K., 2002, Microglia as a source and target of cytokines. *Glia* **40**: 140.
Hansson, C.A., Frykman, S., Farmery, M.R., Tjernberg, L.O., Nilsberth, C., Pursglove, S.E., Ito, A., Winblad, B., Cowburn, R.F., Thyberg, J. and Ankarcrona, M., 2004, Nicastrin, presenilin, APH-1, and PEN-2 form active gamma-secretase complexes in mitochondria. *J. Biol. Chem.* **279**: 51654.
Hirai, K., Aliev, G., Nunomura, A., Fujioka, H., Russell, R.L., Atwood, C.S., Johnson, A.B., Kress, Y., Vinters, H.V., Tabaton, M., Shimohama, S., Cash, A.D., Siedlak, S.L., Harris, P.L., Jones, P.K., Petersen, R.B., Perry, G. and Smith, M.A., 2001, Mitochondrial abnormalities in Alzheimer's disease. *J. Neurosci.* **21**: 3017.
Hock, C., Konietzko, U., Streffer, J.R., Tracy, J., Signorell, A., Muller-Tillmanns, B., Lemke, U., Henke, K., Moritz, E., Garcia, E., Wollmer, M.A., Umbricht, D., de Quervain, D.J., Hofmann, M., Maddalena, A., Papassotiropoulos, A. and Nitsch, R.M., 2003, Antibodies against beta-amyloid slow cognitive decline in Alzheimer's disease. *Neuron* **38**: 547.
Hoozemans, J.J. and O'Banion, M.K., 2005, The role of COX-1 and COX-2 in Alzheimer's disease pathology and the therapeutic potentials of nonsteroidal anti-inflammatory drugs. *Curr. Drug Targets CNS Neurol. Disord.* **4**: 307.
Hurley, S.D., Olschowka, J.A. and O'Banion, M.K., 2002, Cyclooxygenase inhibition as a strategy to ameliorate brain injury. *J. Neurotrauma.* **19**: 1.
Husemann, J., Loike, J.D., Kodama, T. and Silverstein, S.C., 2001, Scavenger receptor class B type I (SR-BI) mediates adhesion of neonatal murine microglia to fibrillar beta-amyloid. *J. Neuroimmunol.* **114**: 142.
Husemann, J., Loike, J.D., Anankov, R., Febbraio, M. and Silverstein, S.C., 2002, Scavenger receptors in neurobiology and neuropathology: their role on microglia and other cells of the nervous system. *Glia* **40**: 195.
Jeffrey, M., Goodsir, C.M., Bruce, M.E., McBride, P.A. and Farquhar, C., 1994, Morphogenesis of amyloid plaques in 87V murine scrapie. *Neuropathol. Appl. Neurobiol.* **20**: 535.
Johnstone, M., Gearing, A.J. and Miller, K.M., 1999, A central role for astrocytes in the inflammatory response to beta-amyloid; chemokines, cytokines and reactive oxygen species are produced. *J. Neuroimmunol.* **93**: 182.
Kim, S.U. and de Vellis, J., 2005, Microglia in health and disease. *J. Neurosci. Res.* **81**: 302.
Kitazawa, M., Yamasaki, T.R. and LaFerla, F.M., 2004, Microglia as a potential bridge between the amyloid {beta}-peptide and tau. *Ann. NY Acad. Sci.* **1035**: 85.
Klamt, F., Dal-Pizzol, F., Conte da Frota, M.L., Walz, R., Andrades, M.E., da Silva, E.G., Brentani, R.R., Izquierdo, I. and Fonseca Moreira, J.C., 2001, Imbalance of antioxidant defense in mice lacking cellular prion protein. *Free Radic. Biol. Med.* **30**: 1137.
Klein, M.A., Kaeser, P.S., Schwarz, P., Weyd, H., Xenarios, I., Zinkernagel, R.M., Carroll, M.C., Verbeek, J.S., Botto, M., Walport, M.J., Molina, H., Kalinke, U., Acha-Orbea, H. and Aguzzi, A., 2001, Complement facilitates early prion pathogenesis. *Nat. Med.* **7**: 488.
Koenigsknecht, J. and Landreth, G., 2004, Microglial phagocytosis of fibrillar β-amyloid through a β1-integrin-dependent mechanism. *J. Neurosci.* **24**: 9838.
Koistinaho, M., Lin, S., Wu, X., Esterman, M., Koger, D., Hanson, J., Higgs, R., Liu, F., Malkani, S., Bales, K.R. and Paul, S.M., 2004, Apolipoprotein E promotes astrocyte colocalization and degradation of deposited amyloid-beta peptides. *Nat. Med.* **10**: 719.
Kopec, K.K. and Carroll, R.T., 1998, Alzheimer's beta-amyloid peptide 1–42 induces a phagocytic response in murine microglia. *J. Neurochem.* **71**: 2123.
Le, Y., Gong, W., Tiffany, H.L., Tumanov, A., Nedospasov, S., Shen, W., Dunlop, N.M., Gao, J.L., Murphy, P.M., Oppenheim, J.J. and Wang, J.M., 2001a, Amyloid (beta)42 activates a G-protein-coupled chemoattractant receptor, FPR-like-1. *Neurosci.* **21**: RC123.
Le, Y., Yazawa, H., Gong, W., Yu, Z., Ferrans, V.J., Murphy, P.M. and Wang, J.M., 2001b, The neurotoxic prion peptide fragment PrP(106-126) is a chemotactic agonist for the G-protein-coupled receptor formyl peptide receptor-like 1. *J. Immunol.* **166**: 1448.
Lee, D.W., Sohn, H.O., Lim, H.B., Lee, Y.G, Kim, Y.S., Carp, R.I. and Wisniewski, H.M., 1999, Alteration of free radical metabolism in the brain of mice infected with scrapie agent. *Free Radic. Res.* **30**: 499.
Lee, H.P., Jun, Y.C., Choi, J.K., Kim, J.I., Carp, R.I. and Kim, Y.S., 2005, The expression of RANTES and chemokine receptors in the brains of scrapie-infected mice. *J. Neuroimmunol.* **158**: 26.
LeVine, H. 3rd., 2004, The Amyloid Hypothesis and the clearance and degradation of Alzheimer's beta-peptide. *J. Alzheimers Dis.* **6**: 303.
Lewandowska, E., Bertrand, E., Kulczycki, J., Lipczynska-Lojkowska, W., Lechowicz, W. and Stankiewicz, J., 1999, Microglia and neuritic plaques in familial Alzheimer's disease induced by a new mutation of presenilin-1 gene. An ultrastructural study. *Folia Neuropathol.* **37**: 243.
Li, Y., Liu, L., Barger, S.W., Mrak, R.E. and Griffin, W.S., 2001, Vitamin E suppression of microglial activation is neuroprotective. *J. Neurosci. Res.* **66**: 163.

Lindberg, C., Selenica, M.L., Westlind-Danielsson, A. and Schultzberg, M., 2005, Beta-amyloid protein structure determines the nature of cytokine release from rat microglia. *J. Mol. Neurosci.* **27**: 1.
Lustbader, J.W., Cirilli, M., Lin, C., Xu, H.W., Takuma, K., Wang, N., Caspersen, C., Chen, X., Pollak, S., Chaney, M., Trinchese, F., Liu, S., Gunn-Moore, F., Lue, L.F., Walker, D.G., Kuppusamy, P., Zewier, Z.L., Arancio, O., Stern, D., Yan, S.S. and Wu, H., 2004, ABAD directly links Abeta to mitochondrial toxicity in Alzheimer's disease. *Science* **304**: 448.
Luth, H.J., Munch, G. and Arendt, T., 2002, Aberrant expression of NOS isoforms in Alzheimer's disease is structurally related to nitrotyrosine formation. *Brain Res.* **953**: 135.
Mabbot, N.A., 2004, The complement system in prion disease. *Curr. Opinion Immunol.* **16**: 587.
Mallucci, G. and Collinge, J., 2005, Rational targeting for prion therapeutics. *Nat. Rev. Neurosci.* **6**: 23.
Marella, M. and Chabry, J., 2004, Neurons and astrocytes respond to prion infection by inducing microglia recruitment. *J. Neurosci.* **24**: 620.
Marella, M., Gaggioli, C., Batoz, M., Deckert, M., Tartare-Deckert, S. and Chabry, J., 2005, Pathological prion protein exposure switches on neuronal mitogen-activated protein kinase pathway resulting in microglia recruitment. *J. Biol. Chem.* **280**: 1529.
Mhatre, M., Floyd, R.A. and Hensley, K., 2004, Oxidative stress and neuroinflammation in Alzheimer's disease and amyotrophic lateral sclerosis: common links and potential therapeutic targets. *J. Alzheimers Dis.* **6**: 147.
Miele, G., Jeffrey, M., Turnbull, D., Manson, J. and Clinton, M., 2002, Ablation of cellular prion protein expression affects mitochondrial numbers and morphology. *Biochem. Biophys. Res. Commun.* **291**: 372.
Minghetti, L., 2004, Cyclo-oxygenase-2 (COX-2) in inflammatory and degenerative brain diseases. *J. Neuropathol. Exp. Neurol.* **63**: 901.
Minghetti, L., Ajmone-Cat, M.A., De Berardinis, M.A. and De Simone, R., 2005, Microglial activation in chronic neurodegenerative diseases: roles of apoptotic neurons and chronic stimulation. *Brain Res. Brain Res. Rev.* **48**: 251.
Moreira, P.I., Honda, K., Liu, Q., Santos, M.S., Oliveira, C.R., Aliev, G., Nunomura, A., Zhu, X., Smith, M.A. and Perry, G., 2005, Oxidative stress: the old enemy in Alzheimer's disease pathophysiology. *Curr. Alzheimer Res.* **2**: 403.
Moreira, P.I., Santos, M.S., Moreno, A., Rego, A.C. and Oliveira, C., 2002, Effect of amyloid beta-peptide on permeability transition pore: a comparative study. *J. Neurosci. Res.* **69**: 257.
Morelli, L., Llovera, R., Ibendahl, S. and Castano, E.M., 2002, The degradation of amyloid beta as a therapeutic strategy in Alzheimer's disease and cerebrovascular amyloidoses. *Neurochem. Res.* **27**: 1387.
Mrak, R.E. and Griffin, W.S., 2001, Interleukin-1, neuroinflammation, and Alzheimer's disease. *Neurobiol. Aging* **22**: 903.
Munch, G., Gasic-Milenkovic, J. and Arendt, T., 2003, Effect of advanced glycation end-products on cell cycle and their relevance for Alzheimer's disease. *J. Neural Transm. Suppl.* **65**: 63.
Nagele, R.G., Wegiel, J., Venkataraman, V., Imaki, H., Wang, K.C. and Wegiel, J., 2004, Contribution of glial cells to the development of amyloid plaques in Alzheimer's disease. *Neurobiol. Aging* **25**: 663.
Nakanishi, H., 2003, Microglial functions and proteases. *Mol. Neurobiol.* **27**: 163.
Nicoll, J.A., Wilkinson, D., Holmes, C., Steart, P., Markham, H. and Weller, R.O., 2003, Neuropathology of human Alzheimer's disease after immunization with amyloid-beta peptide: a case report. *Nat. Med.* **9**: 448.
O'Donovan, C.N., Tobin, D. and Cotter, T.G., 2001, Prion protein fragment PrP-(106-126) induces apoptosis via mitochondrial disruption in human neuronal SH-SY5Y cells. *J. Biol. Chem.* **276**: 43516.
Pereira, C., Santos, M.S. and Oliveira, C., 1999, Involvement of oxidative stress on the impairment of energy metabolism induced by Abeta peptides on PC12 cells: protection by antioxidants. *Neurobiol. Dis.* **6**: 209.
Pereira, C., Agostinho, P., Moreira, P.I., Cardoso, S.M. and Oliveira C.R., 2005, Alzheimer's disease-associated neurotoxic mechanisms and neuroprotective strategies. *Curr. Drug Targets CNS Neurol. Disord.* **4**: 383.
Peretz, D., Williamson, R.A., Kaneko, K., Vergara, J., Leclerc, E., Schmitt-Ulms, G., Mehlhorn, I.R., Legname, G., Wormald, M.R., Rudd, P.M., Dwek, R.A., Burton, D.R. and Prusiner, S.B., 2001, Antibodies inhibit prion propagation and clear cell cultures of prion infectivity. *Nature* **412**: 739.
Perez, A., Morelli, L., Cresto, J.C. and Castano, E.M., 2000, Degradation of soluble amyloid beta-peptides 1-40, 1-42, and the Dutch variant 1–40Q by insulin degrading enzyme from Alzheimer's disease and control brains. *Neurochem. Res.* **25**: 247.
Peyrin, J.M., Lasmezas, C.I., Haik, S., Tagliavini, F., Salmona, M., Williams, A., Richie, D., Deslys, J.P. and Dormont, D., 1999, Microglial cells respond to amyloidogenic PrP peptide by the production of inflammatory cytokines. *Neuroreport* **10**: 723.
Pfeifer, M., Boncristiano, S., Bondolfi, L., Stalder, A., Deller, T., Staufenbiel, M., Mathews, P.M. and Jucker, M., 2002, Cerebral hemorrhage after passive anti-Abeta immunotherapy. *Science* **298**: 1379.
Pocernich, C.B. and Butterfield, D.A., 2003, Acrolein inhibits NADH-linked mitochondrial enzyme activity: implications for Alzheimer's disease. *Neurotox. Res.* **5**: 515.
Prusiner, S.B., 1996, Molecular biology and pathogenesis of prion diseases. *Trends Biochem. Sci.* **21**: 482.

Quintanilla, R.A., Orellana, D.I., Gonzalez-Billault, C. and Maccioni, R.B., 2004, Interleukin-6 induces Alzheimer-type phosphorylation of tau protein by deregulating the cdk5/p35 pathway. *Exp. Cell Res.* **295**: 245.

Rogers, J.T., Leiter, L.M., McPhee, J., Cahill, C.M., Zhan, S.S., Potter, H. and Nilsson, L.N., 1999, Translation of the alzheimer amyloid precursor protein mRNA is up-regulated by interleukin-1 through 5'-untranslated region sequences. *J. Biol. Chem.* **274**: 6421.

Rogers, J., Strohmeyer, R., Kovelowski, C.J. and Li, R., 2002, Microglia and inflammatory mechanisms in the clearance of amyloid beta peptide. *Glia* **40**: 260.

Sano, M., Ernesto, C., Thomas, R.G., Klauber, M.R., Schafer, K., Grundman, M., Woodbury, P., Growdon, J., Cotman, C.W., Pfeiffer, E., Schneider, L.S. and Thal, L.J., 1997, A controlled trial of selegiline, alpha-tocopherol, or both as treatment for Alzheimer's disease. The Alzheimer's Disease Cooperative Study. *N. Engl. J. Med.* **336**: 1216.

Sasaki, N., Takeuchi, M., Chowei, H., Kikuchi, S., Hayashi, Y., Nakano, N., Ikeda, H., Yamagishi, S., Kitamoto, T., Saito, T. and Makita, Z., 2002, Advanced glycation end-products (AGE) and their receptor (RAGE) in the brain of patients with Creutzfeldt-Jakob disease with prion plaques. *Neurosci. Lett.* **326**: 117.

Schenk, D.B. and Yednock, T., 2002, The role of microglia in Alzheimer's disease: friend or foe? *Neurobiol. Aging* **23**: 677.

Schenk, D., Barbour, R., Dunn, W., Gordon, G., Grajeda, H., Guido, T., Hu, K., Huang, J., Johnson-Wood, K., Khan, K., Kholodenko, D., Lee, M., Liao, Z., Lieberburg, I., Motter, R., Mutter, L., Soriano, F., Shopp, G., Vasquez, N., Vandevert, C., Walker, S., Wogulis, M., Yednock, T., Games, D. and Seubert, P., 1999, Immunization with amyloid-beta attenuates Alzheimer-disease-like pathology in the PDAPP mouse. *Nature* **400**: 173.

Schultz, J., Schwarz, A., Neidhold, S., Burwinkel, M., Riemer, C., Simon, D., Kopf, M., Otto, M. and Baier, M., 2004, Role of interleukin-1 in prion disease-associated astrocyte activation. *Am. J. Pathol.* **165**: 671.

Selkoe, D.J. and Lansbury Jr., P.J., 1999, Biochemistry of Alzheimer's and prion diseases. In: *Basic Neurochemistry: Molecular, Cellular and Medical Aspects* 6th edition, G.J. Siegel, (ed.), Lippincott-Raven, Philadelphia, pp. 949–968.

Sleegers, K. and van Duijn, C.M., 2001, Alzheimer's disease: genes, pathogenesis and risk prediction. *Community Genet.* **4**: 197.

Smith, M.A., Harris, P.L., Sayre, L.M. and Perry, G., 1997, Iron accumulation in Alzheimer's disease is a source of redox-generated free radicals. *Proc. Natl Acad. Sci. USA* **94**: 9866.

Solforosi, L., Criado, J.R., McGavern, D.B., Wirz, S., Sanchez-Alavez, M., Sugama, S., DeGiorgio, L.A., Volpe, B.T., Wiseman, E., Abalos, G., Masliah, E., Gilden, D., Oldstone, M.B., Conti, B. and Williamson. R.A., 2004, Cross-linking cellular prion protein triggers neuronal apoptosis in vivo. *Science* **303**: 1514.

Solomon, B., Koppel, R., Hanan, E. and Katzav, T., 1996, Monoclonal antibodies inhibit in vitro fibrillar aggregation of the Alzheimer beta-amyloid peptide. *Proc. Natl Acad. Sci. USA* **93**: 452.

Streit, W.J., Conde, J.R. and Harrison, J.K., 2001, Chemokines and Alzheimer's disease. *Neurobiol. Aging* **22**: 909.

Streit, W.J., Conde, J.R., Fendrick, S.E., Flanary, B.E. and Mariani, C.L., 2005, Role of microglia in the central nervous system's immune response. *Neurol. Res.* **27**: 685.

Stuchbury, G. and Münch, G., 2005, Alzheimer's associated inflammation, potential drug targets and future therapies. *J. Neural Transm.* **112**: 429.

Sung, S., Yao, Y., Uryu, K., Yang, H., Lee, V.M., Trojanowski, J.Q. and Pratico, D., 2004, Early vitamin E supplementation in young but not aged mice reduces Abeta levels and amyloid deposition in a transgenic model of Alzheimer's disease. *FASEB J.* **18**: 323.

Swerdlow, R.H. and Khan, S.M., 2004, A "mitochondrial cascade hypothesis" for sporadic Alzheimer's disease. *Med. Hypotheses* **63**: 8.

Takuma, K., Yao, J., Huang, J., Xu, H., Chen, X., Luddy, J., Trillat, A.C., Stern, D.M., Arancio, O. and Yan, S.S., 2005, ABAD enhances Abeta-induced cell stress via mitochondrial dysfunction. *FASEB J.* **19**: 597.

Tuppo, E.E. and Arias, H.R., 2005, The role of inflammation in Alzheimer's disease. *Int. J. Biochem. Cell Biol.* **37**: 289.

Van Everbroeck, B., Dobbeleir, I., De Waele, M., De Leenheir, E., Lubke, U., Martin, J.J. and Cras, P., 2004, Extracellular protein deposition correlates with glial activation and oxidative stress in Creutzfeldt-Jakob and Alzheimer's disease. *Acta Neuropathol. (Berl).* **108**: 194.

Veerhuis, R., Hoozemans, J.J., Janssen, I., Boshuizen, R.S., Langeveld, J.P. and Eikelenboom, P., 2002, Adult human microglia secrete cytokines when exposed to neurotoxic prion protein peptide: no intermediary role for prostaglandin E2. *Brain Res.* **925**: 195.

Veerhuis, R., Van Breemen, M.J., Hoozemans, J.M., Morbin, M., Ouladhadj, J., Tagliavini, F. and Eikelenboom, P., 2003, Amyloid beta plaque-associated proteins C1q and SAP enhance the Abeta1-42 peptide-induced cytokine secretion by adult human microglia in vitro. *Acta Neuropathol. (Berl)* **105**: 135.

Veerhuis, R., Boshuizen, R.S. and Familian, A., 2005a, Amyloid associated proteins in Alzheimer's and prion disease. *Curr. Drug Targets CNS Neurol. Disord.* **4**: 235.

Veerhuis, R., Boshuizen, R.S., Morbin, M., Mazzoleni, G., Hoozemans, J.J., Langedijk, J.P., Tagliavini, F., Langeveld, J.P. and Eikelenboom, P., 2005b, Activation of human microglia by fibrillar prion protein-related peptides is enhanced by amyloid-associated factors SAP and C1q. *Neurobiol. Dis.* **19**: 273.

Walsh, D.T., Perry, V.H. and Minghetti, L., 2000, Cyclooxygenase-2 is highly expressed in microglial-like cells in a murine model of prion disease. *Glia* **29**: 392.

Weggen, S., Eriksen, J.L., Sagi, S.A., Pietrzik, C.U., Ozols, V., Fauq, A., Golde, T.E. and Koo, E.H., 2003, Evidence that nonsteroidal anti-inflammatory drugs decrease amyloid beta 42 production by direct modulation of gamma-secretase activity. *J. Biol. Chem.* **278**: 31831.

Wegiel, J., Imaki, H., Wang, K.C., Wegiel, J., Wronska, A., Osuchowski, M. and Rubenstein, R., 2003, Origin and turnover of microglial cells in fibrillar plaques of APPsw transgenic mice. *Acta Neuropathol. (Berl.)* **105**: 393.

Weldon, D.T., Rogers, S.D., Ghilardi, J.R., Finke, M.P., Cleary, J.P., O'Hare, E., Esler, W.P., Maggio, J.E. and Mantyh, P.W., 1998, Fibrillar beta-amyloid induces microglial phagocytosis, expression of inducible nitric oxide synthase, and loss of a select population of neurons in the rat CNS in vivo. *J. Neurosci.* **18**: 2161.

Wenk, G.L., McGann, K., Hauss-Wegrzyniak, B. and Rosi, S., 2003, The toxicity of tumor necrosis factor-alpha upon cholinergic neurons within the nucleus basalis and the role of norepinephrine in the regulation of inflammation: implications for Alzheimer's disease. *Neuroscience* **121**: 719.

White, A.R., Enever, P., Tayebi, M., Mushens, R., Linehan, J., Brandner, S., Anstee, D., Collinge, J. and Hawke, S., 2003, Monoclonal antibodies inhibit prion replication and delay the development of prion disease. *Nature* **422**: 80.

Wisniewski, T., Ghiso, J. and Frangione, B., 1997, Biology of A-beta-amyloid in Alzheimer's disease. *Neurobiol. Dis.* **4**: 313.

Wong, A., Luth, H.J., Deuther-Conrad, W., Dukic-Stefanovic, S., Gasic-Milenkovic, J., Arendt, T. and Munch G., 2001, Advanced glycation end-products co-localize with inducible nitric oxide synthase in Alzheimer's disease. *Brain Res.* **920**: 32.

Wyss-Coray, T., Loike, J.D., Brionne, T.C., Lu, E., Anankov, R., Yan, F., Silverstein, S.C. and Husemann, J., 2003, Adult mouse astrocytes degrade amyloid-beta in vitro and in situ. *Nat. Med.* **9**: 453.

Xia, M. and Hyman, B.T., 2002, GROalpha/KC, a chemokine receptor CXCR2 ligand, can be a potent trigger for neuronal ERK1/2 and PI-3 kinase pathways and for tau hyperphosphorylation-a role in Alzheimer's disease? *J. Neuroimmunol.* **122**: 55.

Xia, M., Qin, S., McNamara, M., Mackay, C. and Hyman, B.T., 1997, Interleukin-8 receptor B immunoreactivity in brain and neuritic plaques of Alzheimer's disease. *Am. J. Pathol.* **150**: 1267.

Yan, S.D., Chen, X., Fu, J., Chen, M., Zhu, H., Roher, A., Slattery, T., Zhao, L., Nagashima, M., Morser, J., Migheli, A., Nawroth, P., Stern, D. and Schmidt, A.M., 1996, RAGE and amyloid-beta peptide neurotoxicity in Alzheimer's disease. *Nature* **382**: 685.

Zandi, P.P., Anthony, J.C., Khachaturian, A.S., Stone, S.V., Gustafson, D., Tschanz, J.T., Norton, M.C., Welsh-Bohmer, K.A. and Breitner, J.C., 2004, Reduced risk of Alzheimer's disease in users of antioxidant vitamin supplements: the Cache County Study. *Arch. Neurol.* **61**: 82.

Zhou, Y., Su, Y., Li, B., Liu, F., Ryder, J.W., Wu, X., Gonzalez-DeWhitt, P.A., Gelfanova, V., Hale, J.E., May, P.C., Paul, S.M. and Ni, B., 2003, Nonsteroidal anti-inflammatory drugs can lower amyloidogenic Abeta42 by inhibiting Rho. *Science* **302**: 1215.

15

METAL IONS AND ALZHEIMER'S DISEASE

Paul A. Adlard[1] and Ashley I. Bush[1,2]

1. ABSTRACT

The role of metal ions in the evolution and progression of Alzheimer's disease (AD) is becoming increasingly apparent. Indeed, the interactions of age-associated increases in metals with the amyloid precursor protein (APP) and its proteolytic enzymes and subsequent proteolytic fragments (such as β-amyloid (Aβ)) are well characterized. Likewise, the metal-associated and age-related formation of free radicals, the subsequent generation of oxidative stress and its interactions on process whose dysfunction may contribute to the development of AD have also been highly characterized. Metal dyshomeostasis may thus initiate, and propagate, the development of AD. As science continues to gain greater resolution into the molecular underpinnings of AD, the potential for the use of metal-modulation in the treatment of this disorder is gaining greater acceptance.

2. INTRODUCTION

Alzheimer's disease is a progressive neurodegenerative disorder that is reported to affect almost 2% of the population of industrialized countries (Mattson, 2004) and approximately 12 million individuals worldwide (Citron, 2004). The risk for the development of AD increases significantly with advancing age, with the prevalence increasing from ~5% in people aged 65–74 years to 50% in people aged 85 years and older (Desai and Grossberg, 2005). As the world's geriatric population increases, the incidence of AD will continue to increase, such that within the next 50 years the number of AD cases is predicted to quadruple (Brookmeyer et al., 1998).

The onset of AD is insidious and typically involves subtle symptoms of memory loss. While the clinical presentations of AD do vary, they can include psychosis, depression, and behavioral disturbances. Ultimately, there is a gradual decline in higher order functions resulting in significant cognitive impairment, dysfunction, and eventually death. The clinical course can extend up to 20 years, but the median survival time is between 5 and 10 years (Walsh et al., 1990).

The onset of dementia in AD is often preceded by a preclinical phase that may last for several years and which is characterized by deficits in cognitive function (Frisoni et al., 2004). In addition,

---

Ashley I. Bush – BUSH@helix.mgh.harvard.edu
1. The Oxidation Disorders Laboratory, The Mental Health Research Institute of Victoria, Parkville, Victoria, 3052, Australia. 2. Genetics and Aging Research Unit, Massachusetts General Hospital, Charlestown, MA, USA

approximately 12% of individuals with mild cognitive impairment (MCI), a condition in which individuals have no demonstrable symptoms of dementia but do have decrements in particular cognitive domains (Chong and Sahadevan, 2005), will convert to AD in any given year (Petersen et al., 1999) and this increases to 80% at 6 years follow up (Petersen, 2004).

3. ALZHEIMER'S DISEASE NEUROPATHOLOGY

The AD brain is characterized by two pathological structures whose spatial and temporal evolution and distribution have been well described (Braak et al., 1998; Schonheit et al., 2004). These structures are the extracellular plaque and the intracellular neurofibrillary tangle (NFT). These protein aggregates are present primarily in brain regions such as the entorhinal cortex, hippocampus, basal forebrain, and amygdala and are broadly associated with the loss of specific neuronal cell populations and the degeneration of synapses and synaptic connections. As such, these areas of the brain show a loss of function which then drives the clinical presentation of AD. The postmortem neuropathological diagnosis of AD is based upon the presence of both plaques and tangles, although the presence of these structures is not unique to AD (Adlard and Cummings, 2004).

3.1. Aβ Plaques

The extracellular plaques found in AD are primarily comprised of aggregations of an insoluble fibrillar metallopeptide, Aβ. These small structures have been variously classified, but primarily exist in three different morphological subtypes – the diffuse, fibrillar, and dense-cored plaque. These structures are also associated with varying numbers of abnormal dystrophic neurites (Dickson and Vickers, 2001).

The Aβ protein is derived from the proteolytic cleavage of a larger type 1 transmembrane protein, APP, by competing protease pathways. While the precise functional role of APP is yet to be established, the ablation of APP results in impairments in spatial learning and locomotor/exploratory behavior (Muller et al., 1994; Zheng et al., 1995) in addition to reactive gliosis (Zheng et al., 1995). The combined knockout of multiple members of the APP gene family (which also consists of the amyloid-precursor-like protein (APLP)–1 and –2), however, results in a more significant phenotype characterized by early postnatal lethality (Heber et al., 2000).

The processing of APP is well documented and involves a number of activities termed α-, β- and γ-secretase (for review, Suh and Checler, 2002; Ling et al., 2003; Gandy, 2005). Candidate α-secretases, tumor necrosis factor-α-converting enzyme (TACE or ADAM-17) and ADAM-10 (Buxbaum et al., 1998), are members of the cell surface metalloproteinase family of disintegrin and metalloprotease (ADAM) proteins. β-secretase has been identified as a membrane anchored aspartyl protease termed β-APP-site cleaving enzyme or BACE (Sinha et al., 1999; Vassar et al., 1999; Yan et al., 1999; Cai et al., 2001; Luo et al., 2001) while the high molecular weight active γ-secretase complex is comprised of at least four components, presenilin, APH-1, PEN-2, and nicastrin (De Strooper, 2003).

In AD, the α-secretase and β-secretase cleavage of APP results in the generation of large ectodomain fragments, sAPP$_\alpha$ and sAPP$_\beta$ and membrane embedded fragments, C83 (CTF$_\alpha$) and C99 (CTF$_\beta$), respectively, (β-secretase cleavage can also be displaced by ten amino acids, to result in sAPP$_\beta$, and C89). The C-terminal fragments of both pathways are substrates for γ-secretase and are likely to undergo sequential ε-, ζ-, and finally γ-cleavage in the transmembrane domain. ε-cleavage generates the APP intracellular domain (AICD) (both α and β pathways), ε P3 (α) and Aβ49 (β) fragments. ζ-cleavage then results in ζ P3 (α) and Aβ46 (β) fragments. Finally, cleavage in the middle of the transmembrane domain at the γ site results in the production of a secreted peptide, P3 (α) and a varying length (between 37 and 43 residues) Aβ peptide (β). Thus, the α-secretase pathway is "nonamyloidogenic," while the β-secretase pathway is "amyloidogenic." The Aβ peptides can form a variety of structures, such as oligomeric Aβ (Aβ dimers, trimers, and others), various fibrillar intermediates (such as protofibrils and Aβ-derived diffusible ligands (ADDLs)) and finally, fibrils which ultimately aggregate into the classical amyloid plaque structure (Gandy, 2005; Reinhard et al., 2005; Zhao et al., 2005).

While a number of studies have examined the relationship between the extracellular Aβ plaques and the severity of AD (for review, Adlard and Cummings, 2004), the emerging view is that the various soluble Aβ intermediates, such as oligomers and ADDLs, represent the toxic moiety in the disease and that the Aβ plaques show a poor correlation with the clinical severity. The best pathological correlate of cognitive decline is, in fact, the degree of synaptic loss (Coleman et al., 2004).

3.2. Neurofibrillary Tangles

The intracellular NFT is comprised primarily of paired-helical filaments (PHFs) formed from abnormally phosphorylated forms of the microtubule-associated protein, tau (for review, Friedhoff et al., 2000). Recent studies, however, have demonstrated that there are more than 72 proteins found within the NFT (Wang et al., 2005). Tau subserves many cellular processes and is involved in, stabilization of axonal microtubules, signal transduction, neurite outgrowth, vesicle transport and other functions (Friedhoff et al., 2000). Tau is a soluble protein encoded by a single gene on chromosome 17 that contains 16 exons, of which 11 are found in the human brain (Hyman et al., 2005). Exons 2, 3, and 10 are alternatively spliced, such that the amino-terminal projection domain contains either 0, 1, or 2 inserts (splicing of exon 2 and 3) while the carboxy-terminal contains either three or four repeats (3R and 4R, respectively) in the microtubule binding domain (splicing of exon 10). Thus, there are "short" (no insert), "medium" (1 insert), and "long" (2 inserts) isoforms of both 3R and 4R tau. The repeat regions are believed to contribute to the formation of PHFs (Wille et al., 1992; Friedhoff et al., 1998) and an alteration in their ratio can lead to neurodegenerative disorders (Goode et al., 2000). In AD, both 3R and 4R isoforms contribute to the formation of NFT, although the majority of NFTs contain either 3R or a combination of 3R and 4R tau (Liu et al., 2001; de Silva et al., 2003). While there are currently no known mutations in the tau gene that are associated with AD, there are more than 20 that are associated with the clinical phenotype of frontotemporal dementia and which give rise to neuropathological features (Hyman et al., 2005).

The most striking feature associated with tau in AD is its phosphorylation, which affects the affinity of tau for microtubules (Mandelkow et al., 1995). The phosphorylation of tau typically dissociates it from the microtubule to then destabilize the cytoskeleton and result in a disruption of normal cellular transport and trafficking. There are multiple kinases, such as GSK3β, that are known to phosphorylate tau (for review, Geschwind, 2003) at more than 40 different sites on the protein (Hyman et al., 2005). Once dissociated from the microtubule, tau can then aggregate and form the classical PHFs that contribute to the generation of NFTs (for review, Friedhoff et al., 2000). Ultimately the aggregations of hyperphosphorylated tau will fill the neuron, which eventually dies and leaves the insoluble NFT in the neuropil. The progressive accumulation of NFTs is associated with frank neuronal degeneration and correlates with a loss of neuronal integrity (Augustinack et al., 2002).

4. ALZHEIMER'S DISEASE RISK FACTORS

The primary risk factor for the development of AD is age, although advanced age is no guarantee that an individual will eventually develop this disease. In contrast, a small percentage (<5%) of AD cases result from various missense mutations in three different genes – APP (chromosome 21; 21q21), Presenilin-1 (PSEN1) (chromosome 14; 14q24), and Presenilin-2 (PSEN2) (chromosome 1; 1q42) – that gives rise to an early-onset form of the disease known as familial AD (FAD) (Tanzi and Bertram, 2005). The majority of studies have demonstrated that these mutations result in an altered ratio of Aβ fragments that favors the formation of the extracellular plaques that characterize AD (although more recent studies have highlighted a putative pathogenic mutation in PSEN1 which is unrelated to Aβ and that may give rise to non-AD dementia (Dermaut et al., 2004)). In addition, the APP mutation increases the generation and aggregation of Aβ. Such mutations are fully penetrant, guaranteeing the onset of disease and also resulting in a Mendelian pattern of inheritance. Other risk factors include carriage of the E4 variant of the apolipoprotein E gene (chromosome 19; 19q13), the presence of Down syndrome and traumatic brain injury.

While the precise etiology of AD remains unknown, there is an emerging view that one of the key cellular processes that may become dysregulated with age and participate both directly and indirectly in the evolution of AD, is metal homeostasis and the neurochemistry of metalloproteins.

5. METALS – TOXICOLOGICAL VS. FUNCTIONAL

Metals, which are distinguished by their ionization and binding properties, are widely distributed in nature. Within biological systems they can be broadly classified as either "biometals" or "toxicological metals" based upon whether they have a functional role, or are detrimental to the organism. Biometals such as copper, zinc, and iron for example, are found in living organisms and have defined physiological functions that serve to maintain normal cellular processes. Toxicological metals such as lead and aluminum, however, have no known normal biological function. As with many proteins, however, the mismetabolism of any metal ion to result in levels outside the normal physiological range can result in biological damage (Frausto da Silva and Williams, 2001).

Lead is an example of a neurotoxic heavy metal that is associated with intellectual impairment. A recent population-based study demonstrated an inverse relationship between blood lead concentrations and full-scale IQ scores (Lanphear et al., 2005). Likewise, higher bone lead levels are associated with accelerated declines on neuropsychological tests (Weisskopf et al., 2004), and occupational inorganic lead exposure is associated with impairments in delayed recall, poorer performance on verbal learning tests and deficits in logical memory (Bleecker et al., 2005). While the effect of such heavy metals on cognitive impairment is well established, less well studied is the effect of such heavy metals in AD. Studies have only recently highlighted a possible interaction between early life exposure to lead and the subsequent elevation of APP and Aβ in old age (Basha et al., 2005a) as well as in the aggregation of synthetic Aβ1-40 (Basha et al., 2005b). Theories concerning the involvement of another "toxicological" metal, aluminum, in the pathogenesis of AD arose approximately 40 years ago following the observation of NFT-like structures in the brains of rabbits that had received injections of aluminum salts (Klatzo et al., 1965). There have subsequently been a variety of roles ascribed to aluminum in the pathogenesis of AD, including involvement in the formation of PHF and the aggregation and toxicity of Aβ (for review, Gupta et al., 2005). This non-redox active metal can also promote oxidative stress through reducing the rate of oxidation of the ferrous ion, which can subsequently generate oxidative species (see Gupta et al., 2005). However, while aluminum, which has no known biological function, does have toxic properties, its relevance to AD remains controversial (Forbes and Hill, 1998; Munoz, 1998).

In this chapter, we will focus on a subset of those essential metals that comprise the endogenous pools of transition metals within the body. Transition metals are integral to hundreds of normal cellular and metabolic processes within the body (for review, Valko et al., 2005) and are defined as those elements which form at least one ion with a partially filled d shell of electrons. Transition metals form ions with a wide variety of oxidation states, are good catalysts and, broadly speaking, include metals such as copper, iron, and zinc. These metals are present in the brain at concentrations ranging from 100–1,000 μM, and therefore should not be termed "trace metals." The dysfunction in homeostasis of a number of these endogenous transition metals can result in significant neurological abnormalities (Keen et al., 2003). Zinc, copper, and iron, for example, have a well documented interaction with the biochemical substrates of AD and their modulation with age may initiate AD-related cascades/pathology. Conversely, the formation of many of the biochemical substrates of AD may relate to the regulation of metal ion homeostasis which may be corrupted in the disease.

6. METALS – MODULATION WITH AGE

The modulation of various metal ions species, in addition to a number of different metal transport and storage proteins, as a function of age have all been extensively studied in both animals and man. In rat species, for example, inductively coupled plasma mass spectrometry (ICP-MS) has been utilized to examine the levels of Mn, Fe, Cu, and Zn in 18 different brain regions, and shown that all four elements show a region-specific increase from postnatal day 1 to day 147 (Tarohda et al., 2004). Other studies in normal mice (such as BL6/SJL) have also demonstrated significant age-related increases in copper (46% increase from 2.8 to 18 months), iron (51% increase from 2.8 to 18 months) and cobalt (66% increase from 2.8 to 18 months) levels in whole brain (without olfactory bulb, cerebellum, or brain stem). Other metals such as zinc and manganese, however, did not change (Maynard et al., 2002). Increases in iron over the first 28 weeks of postnatal life are also associated with region-specific alterations in a newly discovered proton-coupled metal-ion transporter, DMT1 that contains an IRE in the 3′-UTR (Ke et al., 2005).

Reports of plasma level changes in copper and zinc suggest that copper levels are low at birth and increase sharply after postnatal day 10, concomitantly with the copper storage protein ceruloplasmin. In contrast, plasma zinc levels were highest at birth and decreased slowly to adult values (Martinez Lista et al., 1993). Older rats (20–22 months) have also been shown to have lower total plasma iron content than middle aged rats (8–10 months), in addition to lower levels of the iron-containing protein, hemoglobin (Ahluwalia et al., 2000).

In the human brain, studies of healthy men (8–89 years, $n = 408$) have demonstrated that plasma copper concentrations increase steadily, whereas zinc levels tend to remain constant throughout life (up to the age of 75). In subjects greater than 75 years of age, there are increases in serum copper and decreases in zinc (Madaric et al., 1994). This is consistent with the bulk of the literature that reports that aging is characterized by elevated plasma copper levels (McMaster et al., 1992; Iskra et al., 1993; Menditto et al., 1993; Milne and Johnson, 1993; Prasad et al., 1993; Ekmekcioglu, 2001) and decreased plasma zinc concentrations (Lindeman et al., 1971; Bunker et al., 1987; Munro et al., 1987; Monget et al., 1996; Ravaglia et al., 2000; Ekmekcioglu, 2001). These changes may be associated with concomitant alterations in the levels of metal ions within the brain, possibly reflected in an age-related increase in ceruloplasmin in the superior temporal gyrus of normal individuals (Connor et al., 1993) (and also in the serum, Milne and Johnson, 1993). Furthermore, electron paramagnetic imaging has demonstrated an age-related increase in clusters of copper and iron ions within the brain (Wender et al., 1992). It is important to note that these gross changes in trace metal levels may be region specific, as the examination of 12 different brain regions by ICP-MS has demonstrated that the 15 trace metals studied show significant between-area variability in their metal content (Rajan et al., 1997). Since age is the major risk factor for AD, alterations in the distribution or levels of metal ions with age might be important in the underlying disease pathogenesis.

7. METALS – MODULATION IN ALZHEIMER'S DISEASE

There is considerable evidence for there being an imbalance in transition metal homeostasis in AD, which is reflected by very specific alterations in metal levels in both the brain (in areas such as the hippocampus, amygdala, neocortex, and olfactory bulb) and the peripheral compartments of the CSF and serum.

Microparticle-induced X-ray emission (micro-PIXE) analysis demonstrated that there are elevations (3–5 fold increase in the cortical and accessory basal nuclei of the amygdala) in zinc, copper, and iron in the neuropil of AD patients, as compared to age-matched controls (Lovell et al., 1998). Furthermore, metals are significantly increased specifically within Aβ plaques, with copper (390 μM), zinc (1055 μM), and iron (940 μM) all elevated (although iron is found primarily complexed with ferritin in the plaque-associated neuritic processes (Grundke-Iqbal et al., 1990) and within neurons and NFTs (Morris et al., 1994; Bouras et al., 1997; LeVine, 1997)) as compared to the normal age-matched neuropil (copper (70 μM), zinc (350 μM), and iron (340 μM)) (Lovell et al., 1998). Histochemically reactive zinc deposits are also found specifically localized to cerebral amyloid angiopathy deposits and NFT-bearing neurons (Suh et al., 2000).

While there have been a number of studies examining the presence of metal ions in both the serum and CSF of AD patients, there is little consistency among studies, although most show decreases in metal ions (Basun et al., 1991; Molina et al., 1998). Interestingly, a recent study examining a pair of elderly monozygotic female twins that were discordant for AD demonstrated that the twin who met the criteria for the diagnosis of AD and who performed significantly worse on all cognitive testing had significantly higher serum copper and total peroxide levels (44% increase) (Squitti et al., 2004). This is consistent with this groups' previous data demonstrating significantly higher levels of serum copper in a cohort of AD subjects ($n = 79$) as compared to controls ($n = 76$) (Squitti et al., 2002).

A number of the alterations in metal ions may be underscored by alterations in the different metal storage and transport proteins that are suggested to be dysregulated in the AD brain. Ceruloplasmin (CP), for example, that transports and oxidizes iron, is variably reported to be altered in AD, with reports suggesting that it is increased in the brain (but unaltered in neurons) and CSF of AD patients (for review, Moreira et al., 2005), although other studies suggest that it may be decreased in brain tissue (Connor et al., 1993). Zinc transporter protein-1 (ZnT-1), which transports zinc into the extracellular space, also shows brain region-specific alterations in mild cognitive impairment and both early and late AD (Lovell et al., 2005). Other proteins, such as the metal-binding (e.g., zinc) and reactive oxygen species (ROS)-scavenging proteins metallothioneins (MT), show isoform-specific regulation in the AD brain, with MT-I/II increased (Adlard et al., 1998) and MT-III decreased in AD (Uchida et al., 1991; Yu et al., 2001). There is also an extensive literature on the dysregulation of iron binding proteins in AD (Bishop et al., 2002), with the iron regulatory protein-2, for example, localized specifically in NFTs, plaques, and neuropil threads of AD cases (Smith et al., 1998). These are just some examples of metal binding protein dysregulation in AD which further supports the notion that there is a significant disruption in the normal metal ion homeostasis in the AD brain.

8. HOW ARE METALS INVOLVED IN THE AD CASCADE?

The evolving field of AD-associated metallochemistry is rapidly highlighting the role that different metals may play in multiple aspects of the disease process, such as the regulation of APP gene expression, mRNA translation and proteolytic processing to generate Aβ, and the aggregation of Aβ into plaques. In consideration of the two hallmark neuropathologies that characterize AD – the NFT and the Aβ plaque – metals have been reported to affect both.

8.1. Neurofibrillary Tangles and the Cytoskeleton

Recent data, utilizing circular dichroism, NMR, MALDI-TOF, and UV-visible absorption spectra, has demonstrated that two peptides corresponding to the second and third repeat region of the microtubule binding domain of tau bind copper (Cu^{2+}) in a pH- and stoichiometric-dependent fashion. This results in a conformational change in tau that may be important in the formation of PHF (Ma et al., 2005, 2006). Other studies have also demonstrated that tau is capable of adventitious binding of copper and iron in a redox-competent manner (Sayre et al., 2000), thereby adding to oxidative stress within the neuron.

It is also known that many of the kinases that phosphorylate tau, such as the extracellular signal-regulated kinase (Erk1/2), can be induced by metal ions such as zinc (Harris et al., 2004; An et al., 2005). Indeed, tau hyperphosphorylation has been induced in both SH-SY5Y and N2a cells by zinc (Bjorkdahl et al., 2005). This is important since zinc levels rise dramatically in AD-affected neocortex (Danscher et al., 1997; Lovell et al., 1998; Opazo et al., 2006). In contrast, the treatment of hippocampal neuron cultures with iron citrate results in a decrease in tau phosphorylation at AD-related epitopes (as assessed using the antibody AT8). It is suggested that this results from a decrease in the activity of the Cdk5/p25 complex, where p25 may be a regulator of the activity of protein kinase Cdk5 (Egana et al., 2003). Furthermore, iron (III), but not iron (II), induces an aggregation of hyperphosphorylated tau that can be reversed by the reduction of iron (III) to iron (II) (Yamamoto et al., 2002). Similarly, the treatment of tau aggregates from the AD brain with reducing agents results in the resolubilization of tau and a release of iron (II), further demonstrating the potential role of iron in NFT formation (Yamamoto et al., 2002).

While tau is the principal component of the NFT, there are other cytoskeletal components found within NFTs that are also known to interact with metals. Neurofilaments (NF), for example, show phosphorylation-dependent alterations very early in the AD cascade and are found within the NFT and in association with the dystrophic neurites surrounding Aβ plaques (Vickers et al., 2000). Purification of this multisubunit protein from bovine spinal cord has demonstrated that it stoichiometrically binds at least one mole of copper and four moles of zinc (Pierson and Evenson, 1988). The assembly of NFs may be mediated, in part, by metal ions such as copper which foster the assembly of the light NF subunit (NF-L), which is a prerequisite for formation of the mature NF protein. Furthermore, copper ions have been shown to be involved in the Cu, Zn-SOD/H_2O_2 mediated aggregation of NF-L, and this can be inhibited by copper chelators such as DTPA and penicillamine (Kim and Kang, 2003; Kim et al., 2004). Finally, NFs can also be phosphorylated by a number of the tau-kinases which, as discussed above, can be modulated by metal ions and indeed, zinc can induce the phosphorylation of NFs at a number of different epitopes (Bjorkdahl et al., 2005).

In addition to NFs, it has long been known that functional tubulin polymerization and subsequent assembly into microtubules involve metal ions (Gaskin, 1981), such as zinc (Gaskin and Shelanski, 1976; Gaskin and Kress, 1977) and magnesium (Weisenberg, 1972). This has been functionally demonstrated *in vivo*, where zinc deficiency results in impairments in the polymerization of tubulin and microtubule formation (Oteiza et al., 1990a, b) that can be restored by low level zinc replacement (Hesketh, 1982). There are multiple zinc binding sites on tubulin (Eagle et al., 1983), including one localized to the first 150 amino acids of both α- and β-tubulin (Serrano et al., 1988). Copper can also disrupt tubulin polymerization and microtubule assembly (Liliom et al., 1999). It has recently been reported that microtubules may also play a significant role in regulating the intracellular concentration and release of the iron storage protein, ferritin (Hasan et al., 2005). Interestingly, studies of progressive supranuclear palsy affected brain specimens have also demonstrated that tau-containing filaments isolated from the brain were associated with ferritin (Perez et al., 1998).

The role played by microtubules and microtubule-associated proteins, such as tau, in the spectrum of changes associated with AD has been well characterized and it is likely, therefore, that a mismetabolism of metals may interact with these pathways to contribute to the course of AD. Perhaps a more self-evident contribution to the AD cascade, however, comes from the interaction of metals with APP and Aβ.

8.2. Amyloid Precursor Protein

Prior to discussing the role that metals may play in the regulation of APP, we will first discuss the notion that the physiological function of APP may have evolved as a metal chaperone and, specifically, that it is a candidate copper chaperone.

Transition metals are a key component in the generation of the deleterious ROS and as such, cells have evolved a complex array of processes to maintain tight homeostatic control over such potentially reactive redox metals as copper (see Lutsenko and Petris, 2003; Prohaska and Gybina, 2004). In this way, the levels of unbound copper in the cytoplasm remain minimal, thereby limiting ROS chemistry (Rae et al., 1999). APP is a ubiquitously expressed protein that may function to maintain metal ion homeostasis (for reviews, Cerpa et al., 2005; Inestrosa et al., 2005). This notion has arisen, in part, based upon the discovery of metal binding sites for copper (Hesse et al., 1994; Atwood et al., 2000a; Simons et al., 2002; Barnham et al., 2003a; Valensin et al., 2004) and zinc (Bush et al., 1993, 1994a–c; Multhaup et al., 1994) in the APP sequence (at the amino-terminal ectodomain of APP and APLP1&2 and also within the Aβ sequence of APP) (see Fig. 1). The structure of the copper binding domain has also been solved and shown to consist of four ligands, provided by the amino acids His-147, His-151, Tyr-168, and Met-170 (Barnham et al., 2003a), which show structural homology to other copper chaperones.

These data support the candidacy of APP as a metallotransporter. In addition, there are a number of in vivo studies that highlight the sensitivity of APP to copper levels. Modulation of copper levels by either chronic copper overload or by copper deficiency for example, can both up and down regulate APP mRNA expression (Armendariz et al., 2004; Bellingham et al., 2004a). This is also supported, in part, by a study of healthy postmenopausal women ($n = 25$) that consumed either a low (1 mg day^{-1}) or a high (3 mg day^{-1}) copper content diet that was also supplemented with zinc. The low copper diet was associated with a significant decrease in APP expression in platelets (Davis et al., 2000).

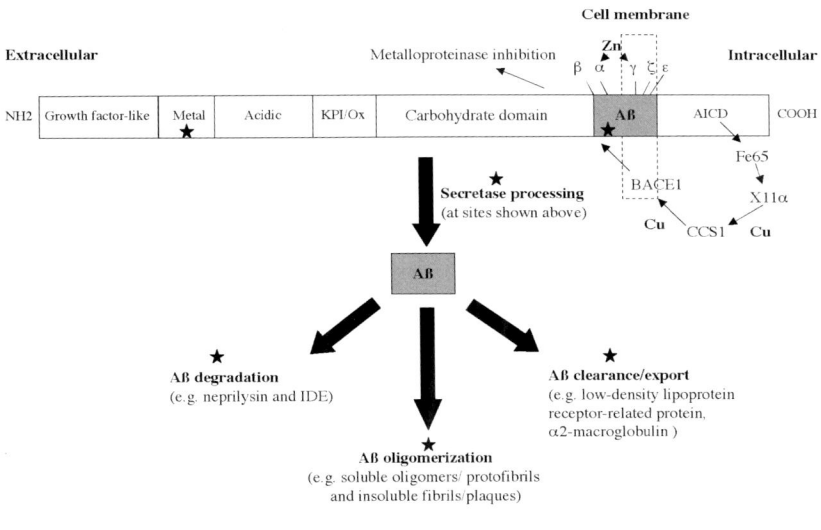

Fig. 1. Domain organization of APP and schematic of Aβ processing. Also shown are the metal binding domains of APP (one at the N-terminus and one in the Aβ sequence itself). As outlined in this chapter, metals are known to affect multiple steps in the Aβ pathway. KPI, Kunitz protease inhibitor domain; AICD, APP intracellular domain; CCS1, copper chaperone for superoxide dismutase

Conversely, both in vivo and in vitro studies in APP and APLP2 knockout mice have demonstrated that there are significant increases in copper levels in both the brain, liver and primary cortical neurons, and embryonic fibroblasts derived from these animals (White et al., 1999; Bellingham et al., 2004b). Furthermore, the overexpression of APP in various transgenic mouse lines decreased copper levels both *in vivo* and *in vitro* (Maynard et al., 2002; Bayer et al., 2003; Phinney et al., 2003). While copper is the transition metal most affected in these experimental models, other metals also bind to, and regulate, APP.

The binding of zinc to APP, and its homologues APLP1 and APLP2 for example, is well characterized and crucially involves a conserved region of amino acids between position 170 and 188 of APP (Bush et al., 1993, 1994a). This domain consists of two key Cys ligands (Cys-186 and Cys-187) that are proposed to be crucial for zinc binding, in addition to other potential ligands (e.g., Cys-174, Met-170, Asp-177, and Glu-184) (Ciuculescu et al., 2005). The binding of zinc may result in the dimerization of APP (Ciuculescu et al., 2005) which may occur *in vivo* and play an important role in its function (Scheuermann et al., 2001). Furthermore, the secreted form of APP aggregates in the presence of zinc (Brown et al., 1997).

In addition to regulation by copper and zinc, APP translation is also mediated by the iron-regulatory element (IRE type II) present in its 5′ untranslated region (Venti et al., 2004). APP 5′ UTR conferred translation is selectively downregulated in response to intracellular metal chelation and is upregulated by exposure to iron (Rogers et al., 2002).

APP expression may, therefore be regulated by metals and also participate in their homeostatic regulation. As such, the direct effect of metals on APP may also act downstream to affect Aβ levels. Metals may also indirectly affect APP and Aβ generation by affecting the processing of APP.

8.3. Metals and Secretase Processing of APP

All three secretases are known to have an interaction with different metal species. The α-secretase, TACE, is a multidomain protein that includes a zinc ion in its catalytic domain (Cross et al., 2002). Following TACE synthesis, it is maintained in an inactive state mediated by an intramolecular bond between a cysteine in the prodomain and a zinc atom in the catalytic site (referred to as the "cysteine-switch" motif) of the protein. It has subsequently been shown, however, that TACE enzymatic activity can be mediated independently of this classic cysteine-switch motif, such that other regions of the prodomain can interact with the catalytic domain (and the zinc ion) to inhibit enzymatic activity (Gonzales et al., 2004; Buckley et al., 2005). Similarly, the activity of ADAM-10 can be inhibited by a dominant negative form of this candidate α-secretase that has a point mutation in its zinc binding site (Lammich et al., 1999).

BACE1 has been shown to bind a copper atom in its C-terminal domain (via cysteine linkages) and to interact and coimmunoprecipitate with CCS in rat brain homogenates (Angeletti et al., 2005).

Presenilin, which comprises part of the γ-secretase complex, is sensitive to metal levels with exogenous zinc administration (but not copper or iron) increasing the C-terminal fragments of Presenilin 1 (PS1-CTF) by enhancing de novo synthesis of PS1 (Park et al., 2001). More recent data suggest that zinc chloride causes the oligomerization of an APP γ-secretase substrate and inhibits its cleavage *in vitro*, supporting a role for zinc dysregulation in abnormal Aβ processing (Hoke et al., 2005).

These data demonstrate that there may be a direct interaction of metals on the secretases and, therefore, also on APP processing. In addition, the zinc binding domain in the Aβ region of the APP sequence spans the α-secretase cleavage site and may, therefore, modulate the cleavage of Aβ from APP and also protect Aβ from proteolytic degradation (Bush et al., 1994b).

8.4. Metal-Mediated Regulation of Aβ

In support of the notion that the metal-mediated modulation of APP processing may result in a physiological change in Aβ generation, recent in vivo studies have examined the effect of copper augmentation in two APP transgenic models. Importantly, these manipulations restored the homeostatic copper deficit present in both of the background APP mouse lines utilized (deficits which were consistent with that shown in other APP transgenic models (Maynard et al., 2002)). Specifically, toxic-milk mice harbor an autosomal recessive mutation in the gene encoding the copper transport protein, ATPase7b. Consequently, there is an inability to load copper into secretory vesicles and a subsequent elevation in copper within the cytoplasm. These mice were crossed with the TgCRND8 mouse line (overexpressing human APP with both the Swedish and Indiana mutations) to assess the role of copper metabolism in the evolution of Aβ pathology (Phinney et al., 2003). The TgCRND8 animals that were homozygous for the ATPase7b mutation, and which had elevated copper levels, demonstrated significantly reduced numbers of dense-cored amyloid plaques, a decrease in the surface area occupied by plaques and a trend to decreased levels of both soluble and formic-acid extractable Aβ40 and Aβ42. There were also reduced levels of Aβ40 and Aβ42 in the plasma and a reduction in the levels of endogenous murine Aβ40 and Aβ42 in young mice also harboring the copper transporter mutation.

In addition to this study, the effect of 3 months of dietary copper, supplementation in APP23 mice (that overexpress the Swedish mutation of human APP) has also been examined (Bayer et al., 2003). A similar magnitude increase in copper levels was observed in this study as compared to the previous one and resulted in a normalization of the activity of the copper-dependent enzyme, SOD1. Furthermore, the increase in bioavailable copper significantly lowered PBS-soluble Aβ40 and Aβ42 in male mice and there was also a trend to decreased formic acid extractable Aβ42. One possibility to explain these findings is that copper may promote the nonamyloidogenic processing of APP by mediating α-secretase cleavage (see also Borchardt et al., 1999).

Since copper promotes the aggregation, redox activity, and toxicity of Aβ, it at first seems counter-intuitive that increasing brain copper levels would inhibit Aβ pathology. However, while copper binds to Aβ in the AD brain (Lovell et al., 1998; Opazo et al., 2002; Dong et al., 2003) most evidence indicates that there is a decrease in bulk tissue levels in AD-affected neocortex (Plantin et al., 1987; Deibel et al., 1996; Loeffler et al., 1996; Rao et al., 1999). At variance with a picture of copper deficiency in AD cortex, Lovell et al. (1998) reported increased copper levels. However, their analysis was confined to the amygdala, and used micro-PIXE to probe heavily amyloid-burdened tissue. The remaining reports utilized bulk tissue analysis. Taken together, a complex picture emerges where copper abnormally redistributes in AD and collects outside of the cell, leaving the tissue relatively deficient. This may explain the deficiency of copper-dependent enzymatic activities in AD such as cytochrome c oxidase (Maurer et al., 2000; Cottrell et al., 2001; Cardoso et al., 2004; Sullivan and Brown, 2005) and SOD1 (De Deyn et al., 1998; Omar et al., 1999) (SOD activity is also decreased in APP transgenic animals (Schuessel et al., 2005)), as well as the reported elevation of serum copper (Squitti et al., 2002, 2004).

In contrast to the above-mentioned studies in transgenic models, however, it has also been reported that rabbits maintained on an elevated cholesterol diet, that also received elevated copper levels in their drinking water, demonstrated an almost 50% increase in Aβ-positive neurons and the appearance of plaque-like structures in the hippocampus and temporal lobe concomitant with impairments in trace conditioning (Sparks and Schreurs, 2003). This study perhaps demonstrates the potentially pleiotropic role that metals may have in AD, as under conditions of high cholesterol, copper supplementation may have an inverse effect to that seen under conditions of APP overexpression and brain copper deficiency.

The pool of chemically exchangeable zinc that is released during synaptic transmission of neocortical glutamatergic fibers (Quinta-Ferreira and Matias, 2005) has been shown to play an essential role in amyloid formation in APP transgenic models for AD. The genetic ablation of the transport protein required for zinc passage into synaptic vesicles, zinc transporter-3 (ZnT3), in the Tg2576 mouse causes a profound reduction in cerebral plaque load (Lee et al., 2002). Furthermore, there is a depletion of an exchangeable zinc pool in the cerebrovascular wall which is also associated with a significant reduction in cerebral amyloid angiopathy (Friedlich et al., 2004). These data suggest that synaptically released zinc underlies amyloid pathology.

These studies provide proof of principle demonstrations that a mismetabolism of metal ion homeostasis is sufficient to significantly alter the normal course of APP processing and Aβ generation and deposition in biologically relevant animal models.

8.5. Metal Binding to Aβ

The high affinity binding of transition metals, particularly zinc, copper and iron, to Aβ (Bush et al., 1994b, c) has been extensively studied and shown to result in a pH-dependent induction of protease resistant aggregation and precipitation of Aβ (Miura et al., 2000). There are up to 3.5 metal binding sites on Aβ (probably as an oligomer) (Atwood et al., 2000a). At least two of these sites are selective for zinc and copper, but some substitution can occur at each site (Atwood et al., 2000a; Yang et al., 2000). Copper binds to soluble Aβ and reversibly precipitates it under mildly acidic conditions (pH 6.6), whereas zinc-induced aggregation of Aβ occurs over a broad pH range (pH 5.5–7.5) (Bush et al., 1994c; Multhaup et al., 1994; Atwood et al., 1998; Yoshiike et al., 2001) including physiological pH (pH 7.4) (Cherny et al., 1999). The Aβ affinity for copper is well below physiological concentrations achieved within the brain, and is reported to be as high as attomolar for Aβ1-42 (Atwood et al., 2000a), although some reports have lower apparent affinities in the submicromolar range (Garzon-Rodriguez et al., 1999; Syme et al., 2004). The differences between reported affinities may be due to the sensitivity the peptide demonstrates to the ionic environment, particularly the concentration of NaCl (Huang et al., 1997). The affinity for zinc is in the low micromolar range (Bush et al., 1994b; Huang et al., 1997). Iron will also precipitate Aβ under acidic conditions, although only copper and zinc have been shown to be specifically associated with Aβ plaques and to copurify with Aβ extracted from the human brain (Opazo et al., 2002; Dong et al., 2003).

Histidine residues are essential for the binding of these metal ions to Aβ. Rat Aβ1-40, which differs from human Aβ by three substitutions including one at the histidine at position 13 (substituted for Arg13) (Shivers et al., 1988), and histidine-modified human Aβ1-40 both have attenuated aggregation in the presence of zinc, copper, or iron (Bush et al., 1994b, c; Atwood et al., 1998). The coordination of zinc to histidine at position 13 is crucial for aggregation of Aβ to occur (Liu et al., 1999). The affinity of rat/mouse Aβ for metals is reported to be reduced as compared to the human peptide (Bush et al., 1994b, c).

Raman spectroscopic analysis of Aβ plaques has demonstrated that copper and zinc ions coordinate to histidine residues located at the N-terminal end of the human Aβ sequence (Dong et al., 2003), while both NMR and EPR-spectroscopy have demonstrated that the binding of copper, zinc, and iron to synthetic Aβ in either aqueous solution or in membrane mimetic environments is mediated by the three histidine residues at positions 6, 13, and 14, along with an oxygen ligand (Curtain et al., 2001, 2003). More recent studies have examined the copper binding domain in the Aβ sequence and shown that it is sensitive to small changes at the N-terminus, as peptides lacking one to three amino acids do not bind copper in the same fashion as the native peptide (Karr et al., 2005). This report also excludes the oxygen atom donor ligand as being the tyrosine at position 10 of the Aβ sequence (Karr et al., 2005). Other studies, however, suggest that the aspartate at position 1 may act as a fourth ligand for zinc under more basic conditions, in addition to the histidines at positions 6, 13, and 14 (Mekmouche et al., 2005). (For a review of the metal binding domain of Aβ (see Barnham et al., 2003a)).

Furthermore, the typical in vitro aggregation of Aβ is inhibited when there is a strict exclusion of metals from buffers, suggesting that common metal contaminants of laboratory buffers are sufficient to accelerate the aggregation of Aβ (Huang et al., 2004).

The metal concentrations required to precipitate Aβ in vitro have been reported to be in the low micromolar range. This is important because, while metal ions are tightly coordinated within the cell, micromolar levels of free ionic zinc and copper are released during neurotransmission and may account for the synaptic localization of Aβ precipitation (Terry, 1996; Lee et al., 2002; Schlief et al., 2005).

The metal-induced aggregation of both synthetic Aβ and Aβ derived from human postmortem brain specimens can be reversed not only by alterations in pH, but also by the use of copper- and zinc-specific chelators (Cherny et al., 1999). Importantly, this has also been translated to *in vivo* studies using APP overexpressing transgenic animals. The use of the hydrophobic Cu/Zn chelator clioquinol was shown to reduce brain Aβ burden in Tg2576 mice by 49% after just 9 weeks of treatment (Cherny et al., 2001). Likewise, studies using the hydrophobic metal chelator DP109 have demonstrated marked reductions in brain Aβ burden after 3 months of treatment in the same transgenic mouse line (Lee et al., 2004). Due to the potential for such metal-protein-attenuating compounds to affect Aβ, a phase 2 clinical trial was conducted in which 36 patients with moderately severe AD were randomized into treatment (clioquinol) and placebo groups. The outcome measurements showed a significant stabilization in the more severely affected group that received clioquinol, as compared to those taking the placebo that showed a substantial worsening on the neuropsychological evaluations (Ritchie et al., 2003). The precise mechanism underlying these effects has not been established. One likely scenario is that clioquinol can facilitate the disaggregation of Aβ plaques via metal chelation and also subsequently suppress Aβ fibril growth (Raman et al., 2005). Another possibility is that clioquinol is mediating neuronal cell death via interactions on calcium influx to ultimately prevent calcium-dependent depletion of glutathione (Abramov et al., 2003). As suggested earlier, however, it is likely that there is an abnormal distribution of metal ions in the AD brain that may then interact on AD-related pathways and which may be augmented by drugs such as clioquinol. In support of this, studies in a yeast model system have suggested that clioquinol may act to alter the distribution of copper, or to facilitate copper uptake, rather than simply decreasing copper levels by chelation (Treiber et al., 2004).

8.6. Metal-Mediated Degradation of Aβ

Two of the primary Aβ degrading enzymes identified are insulin-degrading enzyme (IDE) and neprilysin (NEP), a neutral endopeptidase (for review on Aβ catabolism, see Carson and Turner, 2002; Ling et al., 2003). The effect of IDE on Aβ catabolism has been examined in IDE-/- mice, which show a significant decrease (>50%) in Aβ degradation as well as an increase in endogenous Aβ accumulation (Farris et al., 2003). NEP has also been identified as a principal mediator in the regulation of brain Aβ (Iwata et al., 2000) and is capable of degrading both monomeric and pathological oligomeric forms of Aβ40 and Aβ42 (Kanemitsu et al., 2003). The levels of endogenous Aβ40 and Aβ42 in NEP-deficient mice are significantly elevated in a gene dose-dependent manner (Iwata et al., 2001). The transgenic overexpression of either IDE or NEP is sufficient to reduce Aβ levels, slow or completely prevent plaque formation and extend lifespan (Leissring et al., 2003).

Both these candidate proteases are members of the zinc metallopeptidase family of proteins that all share a common primary structure in their sequence that is involved in the binding of zinc. The majority of members of this family possess the short consensus sequence, HEXXH, in which the two histidines act as the ligand for the catalytic zinc. Other members may also have an elongated C-terminal motif, HEXXHXXGXXH/D, in which there is an additional zinc-binding histidine (or aspartate) (Gomis-Ruth, 2003). This suggests, therefore, that IDE and NEP activities have the potential to be influenced by abnormal zinc metabolism. In addition, it has recently been demonstrated that both IDE and NEP are oxidatively modified in the AD brain (Wang et al., 2003; Caccamo et al., 2005) and that the generation of ROS may serve to inactivate these two enzymes (Shinall et al., 2005).

Furthermore, another potential mediator of Aβ catabolism is plasmin, which can be activated by an Aβ-induced upregulation of the precursor of plasmin, tissue-plasminogen activator (tPA) (see Ling et al., 2003). While plasmin is not a zinc-dependent enzyme in the same fashion as NEP and IDE, it has been demonstrated that it may also be regulated by site-specific oxidation of its active site, perhaps via a modification of the histidine residue in its active site (Lind et al., 1993). Furthermore, the ability of tPA to cleave substrates was also prevented by coincubation with copper and ascorbate (Lind et al., 1993). The tPA/plasmin system may also be involved in the activation of matrix metalloproteinases (MMPs), which are a family of zinc-dependent enzymes involved in the degradation of particular extracellular substrates such as Aβ (for review, see Yong et al., 1998).

Taken together, therefore, metal dysregulation may directly interact on a number of AD-related pathways to affect the course of the disease. There are other cellular processes and downstream events, however, that are modulated by metals and which may interact with and mediate the aging process and subsequent disorders such as AD. The primary candidate is oxidative stress and free radical generation.

9. AD AND THE METAL-MEDIATED PRODUCTION OF ROS

ROS are derived from the univalent reduction of oxygen, and include superoxide anions, hydrogen peroxide, and hydroxyl radicals (Halliwell and Gutteridge, 1984). ROS can be generated from molecular oxygen by a variety of mechanisms. However, as a general principle, the reaction of molecular oxygen with the redox-active metals, copper and iron (in a protein-bound, or nonbound state), is the principle cellular source of ROS (Halliwell and Gutteridge, 1984).

Small quantities of ROS may serve a vital cellular role by acting as second messengers, gene regulators, and mediators for cell activation. A net increase in ROS, however, can generate a wide spectrum of cellular changes including lipid peroxidation, inactivation of enzymes, tertiary structural alterations to proteins, and DNA damage (Halliwell and Gutteridge, 1984; Squier, 2001).

In order to minimize the effect of ROS there is a complex array of defense mechanisms designed to protect the cell, including several metal containing enzymes, copper/zinc-superoxide dismutase (Cu/Zn-SOD or SOD-1), mitochondrial manganese-SOD (or SOD-2), catalase, glutathione peroxidase, and substrates: glutathione and vitamin E/C (Squier, 2001). AD, however, is associated with a general overall decrease in antioxidant status, although there are specific elevations in the levels of Mn-SOD mRNA and Cu/Zn-SOD protein in addition to increases in catalase, glutathione peroxidase, and glutathione reductase in the hippocampus and amygdala (Zemlan et al., 1989; Pappolla et al., 1992). Other enzymes, such as methionine sulfoxide reductase, which are responsible for the repair of oxidized methionine residues, are also decreased in the AD brain (Moskovitz et al., 2001).

As such, the AD brain demonstrates several features that highlight the existence of uncontrolled oxidative processes, including the presence of advanced glycation end-products (AGE), nitration, lipid peroxidation adduction products (such as 4-hydroxy-2,3-nonenal (HNE) and malondialdehyde), carbonyl-modified proteins, oxysterols, and oxidatively damaged RNA (Gabbita et al., 1998; Campbell et al., 2001; Arlt et al., 2002; Butterfield et al., 2002; Perry et al., 2002; Selley et al., 2002; Ischiropoulos and Beckman, 2003; Puglielli et al., 2005). While these processes may occur through a variety of mechanisms, it is known that Aβ is a source of ROS generation.

9.1. ROS Generation by Metal: Aβ Mediated Toxicity

The coordination of both copper and iron to Aβ also results in the chemical reduction of both metals (from Cu^{2+} to Cu^+ and Fe^{3+} to Fe^{2+}) and the subsequent generation of hydrogen peroxide from molecular oxygen together with other available biological reducing agents such as cholesterol, in a catalytic manner (Huang et al., 1999a, b; Cuajungco et al., 2000; Opazo et al., 2002). In the case of Aβ1-40, the reduction of copper is independent of the aggregation state of the peptide, as both soluble and fibrillar forms show copper-reducing ability (Opazo et al., 2000).

APP also displays copper reducing ability (Multhaup et al., 1996) and this is likely to involve key amino acid residues, His, Cys, and Trp in APP (135–156). Utilizing analogues of APP (135–156) that contain specific amino acid substitutions, it has been shown that mutations of His147 (to Ala) and also the mutation of both His147 (to Ala) and His149 (to Ala) can decrease the level of copper reduction. However, mutation of Cys144-Ser abolishes the ability of the peptide to reduce copper (II) to copper (I), suggesting that this is the key amino acid involved in copper reduction (Ruiz et al., 1999).

The generation of hydrogen peroxide in the presence of reduced metals, and in the absence of sufficient detoxifying enzymes such as catalase and glutathione peroxidase, gives rise to the toxic hydroxyl radical via Fenton chemistry (Huang et al., 1999a). The generation of hydrogen peroxide contributes to Aβ toxicity. In support of this, cellular toxicity can be rescued by the addition of catalase (Behl et al., 1994; Opazo et al., 2002; Ciccotosto et al., 2004), resistance to Aβ toxicity is associated with an enhanced ability to degrade hydrogen peroxide (Sagara et al., 1996), and catalase inhibitors can enhance Aβ toxicity. The potentiation of Aβ toxicity by copper is greatest for Aβ1-42 > Aβ1-40 > rodent Aβ1-40, which corresponds to the peptides relative activities in reducing copper (II) to copper (I) (Huang et al., 1999a).

Iron also facilitates Aβ toxicity on cultured cells (Schubert and Chevion, 1995), and the pretreatment of Aβ with an iron chelator, deferoxamine, results in a significant attenuation of cellular toxicity that can be restored by incubation with excess iron (Rottkamp et al., 2001). This toxicity may be linked to the ability of Aβ to reduce iron (III) in generating hydrogen peroxide (Huang et al., 1999a).

The oxidative modification of the methionine at position 35 in the Aβ sequence by copper or hydrogen peroxide generates modified forms of Aβ that have a lower affinity for, and may be released from membranes but retain their toxic properties (Barnham et al., 2003b). Soluble forms of Aβ have been shown to correlate with the severity of the clinical symptoms of AD and to dementia (Lue et al., 1999; McLean et al., 1999; Wang et al., 1999). Soluble dimers and oligomers of Aβ have enhanced toxicity (Dahlgren et al., 2002; Walsh et al., 2002a, b) and form rapidly upon Aβ reduction with copper.

Dityrosine crosslinking and formation of soluble oligomers can be mediated by direct interactions between copper and Aβ (Atwood et al., 2004; Barnham et al., 2004). Hydrogen peroxide promotes copper-induced dityrosine cross-linking of Aβ1-28, Aβ1-40, and Aβ1-42 to result in soluble oligomers. The dityrosine bond cannot be biologically degraded, making these oligomers stable to proteolysis (Atwood et al., 2004). In addition, other oxidative reactions may also potentiate the cross linking of Aβ (Atwood et al., 2000b; Loske et al., 2000). Hence, the oxidatively modified forms of Aβ that comprise the primary species of Aβ populating the AD brain (Head et al., 2001), can be generated by copper binding or metal-mediated oxidative stress.

The catalytic production of hydrogen peroxide by Aβ copper and the production of dityrosine has been further delineated and shown to crucially involve the tyrosine10 (Y10) residue in the Aβ peptide (Barnham et al., 2004). When tyrosine10 is substituted for alanine (Y10A) there is a significant reduction in the ability of Aβ to reduce Cu^{2+}.

These data support a role for Aβ in the generation of hydrogen peroxide via metal ion reduction and for oxidative processes in the augmentation of Aβ to potentiate the AD cascade. A summary of the proposed role of metal ions in AD is shown in Fig. 2.

10. THE INTERACTION BETWEEN AD, METAL IONS, AND GLIAL CELLS

In this chapter we have outlined a number of AD-related cascades that may be sensitive to metal ions. Another prominent cellular target that may participate in the onset and progression of AD, and perhaps intersect with metal ion dyshomeostasis, is glial cells.

Broadly speaking, glia within the CNS are non neuronal cells that include both microglia and macroglia (such as astrocytes and oligodendrocytes). They are crucial in the normal growth and maintenance of the nervous system, providing chemical and structural support to neuronal cells, and have an extensive range of functions in both health and disease (Streit, 2004; Kim and de Vellis, 2005; Mrak and Griffin, 2005; Wojtera et al., 2005).

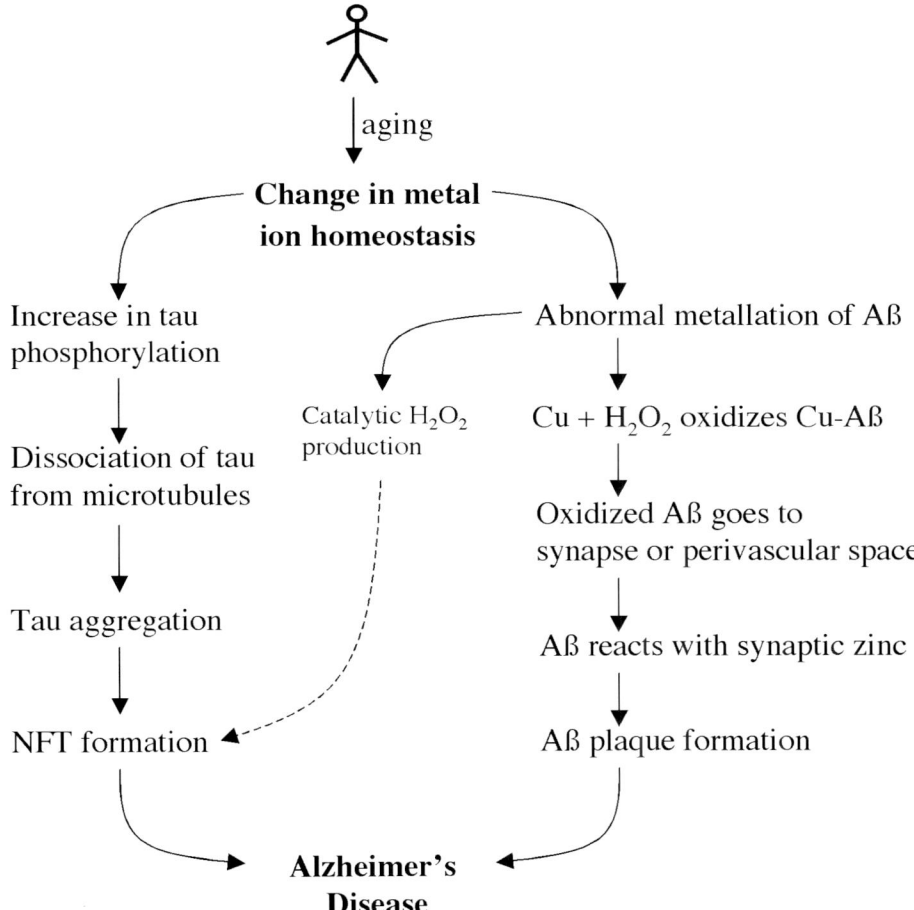

Fig. 2. Proposed Alzheimer's disease pathway involving metal ions. Metals, particularly zinc, copper, and iron, have been demonstrated to affect the two major hallmark neuropathologies of AD – the NFT and the Aβ plaque. The literature supports the notion that metal dyshomeostasis may be a common mediator of the pathways involved in the formation of these abnormal pathological structures (the role that hydrogen peroxide may play in the tau cascade is less clear, as hydrogen peroxide is known to transiently dephosphorylate tau)

In AD, microglial cells are found clustered around Aβ plaques (Terry and Wisniewski, 1975) and upon activation by Aβ, will produce a host of pro-inflammatory cytokines (such as IL-1, IL-6, TNFα) and chemokines (such as IL-8). Likewise, astrocytes can be activated by Aβ peptides, release cytokines, and growth factors and are found in association with Aβ plaques (Blasko et al., 2004). The activation of astrocytes is suggested to occur after that of microglia. However, the phagocytic activity of the microglia may be regulated by astrocytes. In addition, while microglia are considered to be the functional equivalent of macrophages within the brain, both microglia and astrocytes are able to phagocytose and degrade Aβ (Blasko et al., 2004). As such, the presence of both of these activated glial cell types in association with plaques within the CNS can be considered to be beneficial in limiting Aβ accumulation/pathology. However, glia may have a number of deleterious effects within the CNS that may actually contribute to the progression of AD. It has been suggested, for example, that astrocytes may accumulate Aβ species over time (and perhaps even contribute to the production of Aβ due to the presence of low levels of endogenous APP in astrocytes, but not in microglia) and then undergo lysis to result in astrocyte-derived Aβ plaques (Nagele et al., 2003, 2004). Microglia may also contribute to the evolution of AD by driving Aβ peptides toward a more fibrillar morphology that may precipitate the plaque formation characteristic of this disease (Nagele et al., 2004).

A more evident scenario, however, is that glial cells may negatively impact AD as a function of their chronic activation and subsequent release of reactive oxygen intermediates, nitric oxide, and other toxic products. The brain is sensitive to such potent molecules, which may kill neurons and also directly affect both APP and Aβ cascades. For example, activated microglia may produce membrane permeable hydrogen peroxide which, on reaction with reduced metal ions, can lead to the formation of hydroxyl radicals which can subsequently induce oxidative modifications of macromolecules and potentiate AD (as discussed earlier in this chapter).

Glia themselves are also sensitive not only to Aβ (Aβ for example can cause calcium dysregulation in neuronal and glial cells) but also to metal ions. Copper for example, at a concentration approximately equivalent to that found in the CSF of AD patients, can induce oxidative damage in human astrocytes (Ferretti et al., 2003). Thus, metal- or Aβ-mediated damage to glia may impair their normal function. As it is now recognized that there is a significant amount of neuron-glia cross-talk (Haydon, 2001; Chung et al., 2004), and given the known roles of glia in supporting neuronal cells (such as by providing antioxidant support by the provision of metabolic substrates and precursors of glutathione (Abramov et al., 2003)), the glial cell abnormalities may subsequently impair normal neuronal function.

Thus, metal ion dyshomeostasis may be a precipitating event in disruptions to the neuron-glia axis which may drive a degenerative phenotype within the CNS and also contribute to the pathogenesis of AD. The onset/progression of AD may then directly affect glial cells, which in turn can feed back onto various Aβ/AD cascades and also affect surrounding neurons both through the release of neuromodulators and a diminished level of chemical support (potentially involving metals). Glial inflammatory processes may mediate neurotoxicity and functional decline in AD, and may in fact precede neuronal damage in AD (Rosenberg, 2005), but the precise role and interactions of metal ions in the function of glial cells, and their subsequent interaction with neurons in health and disease, is one that remains to be fully explored.

11. ACKNOWLEDGMENTS

We are grateful for the support funds from the Australian Research Council Federation Fellowship (A.B), the Howard Florey Centenary Research Fellowship (P.A) and the National Health and Medical Research Council, the Wood Family Foundation, the Percy Baxter Charitable Trust, the National Institute on Aging (RO1AG12686), the American Health Assistance Foundation and the Alzheimer's Association.

12. REFERENCES

Abramov, A.Y., Canevari, L. and Duchen, M.R., 2003, Changes in intracellular calcium and glutathione in astrocytes as the primary mechanism of amyloid neurotoxicity. *J. Neurosci.* **23**: 5088.

Adlard, P.A. and Cummings, B.J., 2004, Alzheimer's disease – a sum greater than its parts? *Neurobiol. Aging* **25**: 725.

Adlard, P.A., West, A.K. and Vickers, J.C., 1998, Increased density of metallothionein I/II-immunopositive cortical glial cells in the early stages of Alzheimer's disease. *Neurobiol. Dis.* **5**: 349.

Ahluwalia, N., Gordon, A., Handte, G., Mahlon, M., Li, N.Q., Beard, J.L., Weinstock, D. and Ross, A.C., 2000, Iron status and stores decline with age in Lewis rats. *J. Nutr.* **130**: 2378.

An, W.L., Bjorkdahl, C., Liu, R., Cowburn, R.F., Winblad, B. and Pei, J.J., 2005, Mechanism of zinc-induced phosphorylation of p70 S6 kinase and glycogen synthase kinase 3beta in SH-SY5Y neuroblastoma cells. *J. Neurochem.* **92**: 1104.

Angeletti, B., Waldron, K.J., Freeman, K.B., Bawagan, H., Hussain, I., Miller, C.C., Lau, K.F., Tennant, M.E., Dennison, C., Robinson, N.J. and Dingwall, C., 2005, BACE1 cytoplasmic domain interacts with the copper chaperone for superoxide dismutase-1 and binds copper. *J. Biol. Chem.* **280**: 17930.

Arlt, S., Beisiegel, U. and Kontush, A., 2002, Lipid peroxidation in neurodegeneration: new insights into Alzheimer's disease. *Curr. Opin. Lipidol.* **13**: 289.

Armendariz, A.D., Gonzalez, M., Loguinov, A.V. and Vulpe, C.D., 2004, Gene expression profiling in chronic copper overload reveals upregulation of Prnp and App. *Physiol. Genomics.* **20**: 45.

Atwood, C.S., Moir, R.D., Huang, X., Scarpa, R.C., Bacarra, N.M., Romano, D.M., Hartshorn, M.A., Tanzi, R.E. and Bush, A.I., 1998, Dramatic aggregation of Alzheimer abeta by Cu(II) is induced by conditions representing physiological acidosis. *J. Biol. Chem.* **273**: 12817.

Atwood, C.S., Huang, X., Khatri, A., Scarpa, R.C., Kim, Y.S., Moir, R.D., Tanzi, R.E., Roher, A.E. and Bush, A.I., 2000a, Copper catalyzed oxidation of Alzheimer Abeta. *Cell Mol. Biol. (Noisy-le-grand)* **46**: 777.

Atwood, C.S., Scarpa, R.C., Huang, X., Moir, R.D., Jones, W.D., Fairlie, D.P., Tanzi, R.E. and Bush, A.I., 2000b, Characterization of copper interactions with alzheimer amyloid beta peptides: identification of an attomolar-affinity copper binding site on amyloid beta1–42. *J. Neurochem.* **75**: 1219.

Atwood, C.S., Perry, G., Zeng, H., Kato, Y., Jones, W.D., Ling, K.Q., Huang, X., Moir, R.D., Wang, D., Sayre, L.M., Smith, M.A., Chen, S.G. and Bush, A.I., 2004, Copper mediates dityrosine cross-linking of Alzheimer's amyloid-beta. *Biochemistry* **43**: 560.

Augustinack, J.C., Schneider, A., Mandelkow, E.M. and Hyman, B.T., 2002, Specific tau phosphorylation sites correlate with severity of neuronal cytopathology in Alzheimer's disease. *Acta Neuropathol. (Berl).* **103**: 26.

Barnham, K.J., McKinstry, W.J., Multhaup, G., Galatis, D., Morton, C.J., Curtain, C.C., Williamson, N.A., White, A.R., Hinds, M.G., Norton, R.S., Beyreuther, K., Masters, C.L., Parker, M.W. and Cappai, R., 2003a, Structure of the Alzheimer's disease amyloid precursor protein copper binding domain. A regulator of neuronal copper homeostasis. *J. Biol. Chem.* **278**: 17401.

Barnham, K.J., Ciccotosto, G.D., Tickler, A.K., Ali, F.E., Smith, D.G., Williamson, N.A., Lam, Y.H., Carrington, D., Tew, D., Kocak, G., Volitakis, I., Separovic, F., Barrow, C.J., Wade, J.D., Masters, C.L., Cherny, R.A., Curtain, C.C., Bush, A.I. and Cappai, R., 2003b, Neurotoxic, redox-competent Alzheimer's beta-amyloid is released from lipid membrane by methionine oxidation. *J. Biol. Chem.* **278**: 42959.

Barnham, K.J., Haeffner, F., Ciccotosto, G.D., Curtain, C.C., Tew, D., Mavros, C., Beyreuther, K., Carrington, D., Masters, C.L., Cherny, R.A., Cappai, R. and Bush, A.I., 2004, Tyrosine gated electron transfer is key to the toxic mechanism of Alzheimer's disease beta-amyloid. *FASEB J.* **18**: 1427.

Basha, M.R., Wei, W., Bakheet, S.A., Benitez, N., Siddiqi, H.K., Ge, Y.W., Lahiri, D.K. and Zawia, N.H., 2005a, The fetal basis of amyloidogenesis: exposure to lead and latent overexpression of amyloid precursor protein and beta-amyloid in the aging brain. *J. Neurosci.* **25**: 823.

Basha, M.R., Murali, M., Siddiqi, H.K., Ghosal, K., Siddiqi, O.K., Lashuel, H.A., Ge, Y.W., Lahiri, D.K. and Zawia, N.H., 2005b, Lead (Pb) exposure and its effect on APP proteolysis and Abeta aggregation. *FASEB J.* **19**: 2083.

Basun, H.L., Forssell, G., Wetterberg, L. and Winblad, B., 1991, Metals and trace elements in plasma and cerebrospinal fluid in normal aging and Alzheimer's disease. *J. Neural Transm. Park. Dis. Dement. Sect.* **3**: 231.

Bayer, T.A., Schafer, S., Simons, A., Kemmling, A., Kamer, T., Tepest, R., Eckert, A., Schussel, K., Eikenberg, O., Sturchler-Pierrat, C., Abramowski, D., Staufenbiel, M. and Multhaup, G., 2003, Dietary Cu stabilizes brain superoxide dismutase 1 activity and reduces amyloid Abeta production in APP23 transgenic mice. *Proc. Natl. Acad. Sci. USA* **100**: 14187.

Behl, C., Davis, J.B., Lesley, R. and Schubert, D., 1994, Hydrogen peroxide mediates amyloid beta protein toxicity. *Cell* **77**: 817.

Bellingham, S.A., Lahiri, D.K., Maloney, B., La Fontaine, S., Multhaup, G. and Camakaris, J., 2004a, Copper depletion down-regulates expression of the Alzheimer's disease amyloid-beta precursor protein gene. *J. Biol. Chem.* **279**: 20378.

Bellingham, S.A., Ciccotosto, G.D., Needham, B.E., Fodero, L.R., White, A.R., Masters, C.L., Cappai, R. and Camakaris, J., 2004b, Gene knockout of amyloid precursor protein and amyloid precursor-like protein-2 increases cellular copper levels in primary mouse cortical neurons and embryonic fibroblasts. *J. Neurochem.* **91**: 423.

Bishop, G.M., Robinson, S.R., Liu, Q., Perry, G., Atwood, C.S. and Smith, M.A., 2002, Iron: a pathological mediator of Alzheimer disease? *Dev. Neurosci.* **24**: 184.

Bjorkdahl, C., Sjogren, M.J., Winblad, B. and Pei, J.J., 2005, Zinc induces neurofilament phosphorylation independent of p70 S6 kinase in N2a cells. *Neuroreport* **16**: 591.

Blasko, I., Stampfer-Kountchev, M., Robatscher, P., Veerhuis, R., Eikelenboom, P. and Grubeck-Loebenstein, B., 2004, How chronic inflammation can affect the brain and support the development of Alzheimer's disease in old age: the role of microglia and astrocytes. *Aging Cell* **3**: 169.

Bleecker, M.L., Ford, D.P., Lindgren, K.N., Hoese, V.M., Walsh, K.S. and Vaughan, C.G., 2005, Differential effects of lead exposure on components of verbal memory. *Occup. Environ. Med.* **62**: 181.

Borchardt, T., Camakaris, J., Cappai, R.C., Masters, L., Beyreuther, K. and Multhaup, G., 1999, Copper inhibits beta-amyloid production and stimulates the nonamyloidogenic pathway of amyloid-precursor-protein secretion. *Biochem. J.* **344**: 461.

Bouras, C., Giannakopoulos, P., Good, P.F., Hsu, A., Hof, P.R. and Perl, D.P., 1997, A laser microprobe mass analysis of brain aluminum and iron in dementia pugilistica: comparison with Alzheimer's disease. *Eur. Neurol.* **38**: 53.

Braak, H., Braak, E., Bohl, J. and Bratzke, H., 1998, Evolution of Alzheimer's disease related cortical lesions. *J. Neural Transm. Suppl.* **54**: 106.

Brookmeyer, R., Gray, S. and Kawas, C., 1998, Projections of Alzheimer's disease in the United States and the public health impact of delaying disease onset. *Am. J. Public Health* **88**: 1337.

Brown, A.M., Tummolo, D.M., Rhodes, K.J., Hofmann, J.H., Jacobsen, J.S. and Sonnenberg-Reines, J., 1997, Selective aggregation of endogenous beta-amyloid peptide and soluble amyloid precursor protein in cerebrospinal fluid by zinc. *J. Neurochem.* **69**: 1204.

Buckley, C.A., Rouhani, F.N., Kaler, M., Adamik, B., Hawari, F.I. and Levine, S.J., 2005, Amino-terminal TACE prodomain attenuates TNFR2 cleavage independently of the cysteine switch. *Am. J. Physiol. Lung Cell Mol. Physiol.* **288**: L1132.

Bunker, V.W., Hinks, L.J., Stansfield, M.F., Lawson, M.S. and Clayton, B.E., 1987, Metabolic balance studies for zinc and copper in housebound elderly people and the relationship between zinc balance and leukocyte zinc concentrations. *Am. J. Clin. Nutr.* **46**: 353.

Bush, A.I., Multhaup, G., Moir, R.D., Williamson, T.G., Small, D.H., Rumble, B., Pollwein, P., Beyreuther, K. and Masters, C.L., 1993, A novel zinc(II) binding site modulates the function of the beta A4 amyloid protein precursor of Alzheimer's disease. *J. Biol. Chem.* **268**: 16109.

Bush, A.I., Pettingell, W.H., Jr., de Paradis, M., Tanzi, R.E. and Wasco, W., 1994a, The amyloid beta-protein precursor and its mammalian homologues. Evidence for a zinc-modulated heparin-binding superfamily. *J. Biol. Chem.* **269**: 26618.

Bush, A.I., Pettingell, W.H., Multhaup, G., Paradis, M., Vonsattel, J.P., Gusella, J.F., Beyreuther, K., Masters, C.L. and Tanzi, R.E., 1994b, Rapid induction of Alzheimer A beta amyloid formation by zinc. *Science* **265**: 1464.

Bush, A.I., Pettingell, W.H., Jr., Paradis, M.D. and Tanzi, R.E., 1994c, Modulation of A beta adhesiveness and secretase site cleavage by zinc. *J. Biol. Chem.* **269**: 12152.

Butterfield, D.A., Castegna, A., Lauderback, C.M. and Drake, J., 2002, Evidence that amyloid beta-peptide-induced lipid peroxidation and its sequelae in Alzheimer's disease brain contribute to neuronal death. *Neurobiol. Aging* **23**: 655.

Buxbaum, J.D., Liu, K.N., Luo, Y., Slack, J.L., Stocking, K.L., Peschon, J.J., Johnson, R.S., Castner, B.J., Cerretti, D.P. and Black, R.A., 1998, Evidence that tumor necrosis factor alpha converting enzyme is involved in regulated alpha-secretase cleavage of the Alzheimer amyloid protein precursor. *J. Biol. Chem.* **273**: 27765.

Caccamo, A., Oddo, S., Sugarman, M.C., Akbari, Y. and LaFerla, F.M., 2005, Age- and region-dependent alterations in Abeta-degrading enzymes: implications for Abeta-induced disorders. *Neurobiol. Aging* **26**: 645.

Cai, H., Wang, Y., McCarthy, D., Wen, H., Borchelt, D.R., Price, D.L. and Wong, P.C., 2001, BACE1 is the major beta-secretase for generation of Abeta peptides by neurons. *Nat. Neurosci.* **4**: 233.

Campbell, A.M., Smith, A., Sayre, L.M., Bondy, S.C. and Perry, G., 2001, Mechanisms by which metals promote events connected to neurodegenerative diseases. *Brain Res. Bull.* **55**: 125.

Cardoso, S.M., Proenca, M.T., Santos, S., Santana, I. and Oliveira, C.R., 2004, Cytochrome c oxidase is decreased in Alzheimer's disease platelets. *Neurobiol. Aging* **25**: 105.

Carson, J.A. and Turner, A.J., 2002, Beta-amyloid catabolism: roles for neprilysin (NEP) and other metallopeptidases? *J. Neurochem.* **81**: 1.

Cerpa, W., Varela-Nallar, L., Reyes, A.E., Minniti, A.N. and Inestrosa, N.C., 2005, Is there a role for copper in neurodegenerative diseases? *Mol. Aspects Med.* **26**: 405.

Cherny, R.A., Legg, J.T., McLean, C.A., Fairlie, D.P., Huang, X., Atwood, C.S., Beyreuther, K., Tanzi, R.E., Masters, C.L. and Bush, A.I., 1999, Aqueous dissolution of Alzheimer's disease Abeta amyloid deposits by biometal depletion. *J. Biol. Chem.* **274**: 23223.

Cherny, R.A., Atwood, C.S., Xilinas, M.E., Gray, D.N., Jones, W.D., McLean, C.A., Barnham, K.J., Volitakis, I., Fraser, F.W., Kim, Y., Huang, X., Goldstein, L.E., Moir, R.D., Lim, J., Beyreuther, T.K., Zheng, H., Tanzi, R.E., Masters, C.L. and Bush, A.I., 2001, Treatment with a copper–zinc chelator markedly and rapidly inhibits beta-amyloid accumulation in Alzheimer's disease transgenic mice. *Neuron* **30**: 665.

Chong, M.S. and Sahadevan, S., 2005, Preclinical Alzheimer's disease: diagnosis and prediction of progression. *Lancet Neurol.* **4**: 576.

Chung, R.S., Adlard, P.A., Dittmann, J., Vickers, J.C., Chuah, M. and West, A., 2004, Neuron-glia communication: metallothionein expression is specifically upregulated by astrocytes in response to neuronal injury. *J. Neurochem.* **88**: 454.

Ciccotosto, G.D., Tew, D., Curtain, C.C., Smith, D., Carrington, D., Masters, C.L., Bush, A.I., Cherny, R.A., Cappai, R. and Barnham, K.J., 2004, Enhanced toxicity and cellular binding of a modified amyloid beta peptide with a methionine to valine substitution. *J. Biol. Chem.* **279**: 42528.

Citron, M., 2004, Strategies for disease modification in Alzheimer's disease. *Nat. Rev. Neurosci.* **5**: 677.

Ciuculescu, E.D., Mekmouche, Y. and Faller, P., 2005, Metal-binding properties of the peptide APP(170–188): a model of the Zn(II)-binding site of amyloid precursor protein (APP). *Chemistry* **11**: 903.

Coleman, P., Federoff, H. and Kurlan, R., 2004, A focus on the synapse for neuroprotection in Alzheimer's disease and other dementias. *Neurology* **63**: 1155.

Connor, J.R., Tucker, P., Johnson, M. and Snyder, B., 1993, Ceruloplasmin levels in the human superior temporal gyrus in aging and Alzheimer's disease. *Neurosci. Lett.* **159**: 88.

Cottrell, D.A., Blakely, E.L., Johnson, M.A., Ince, P.G. and Turnbull, D.M., 2001, Mitochondrial enzyme-deficient hippocampal neurons and choroidal cells in AD. *Neurology* **57**: 260.

Cross, J.B., Duca, J.S., Kaminski, J.J. and Madison, V.S., 2002, The active site of a zinc-dependent metalloproteinase influences the computed pK(a) of ligands coordinated to the catalytic zinc ion. *J. Am. Chem. Soc.* **124**: 11004.

Cuajungco, M.P., Goldstein, L.E., Nunomura, A., Smith, M.A., Lim, J.T., Atwood, C.S., Huang, X., Farrag, Y.W., Perry, G. and Bush, A.I., 2000, Evidence that the beta-amyloid plaques of Alzheimer's disease represent the redox-silencing and entombment of abeta by zinc. *J. Biol. Chem.* **275**: 19439.

Curtain, C.C., Ali, F., Volitakis, I., Cherny, R.A., Norton, R.S., Beyreuther, K., Barrow, C.J., Masters, C.L., Bush, A.I. and Barnham, K.J., 2001, Alzheimer's disease amyloid-beta binds copper and zinc to generate an allosterically ordered membrane-penetrating structure containing superoxide dismutase-like subunits. *J. Biol. Chem.* **276**: 20466.

Curtain, C.C., Ali, F.E., Smith, D.G., Bush, A.I., Masters, C.L. and Barnham, K.J., 2003, Metal ions, pH, and cholesterol regulate the interactions of Alzheimer's disease amyloid-beta peptide with membrane lipid. *J. Biol. Chem.* **278**: 2977.

Dahlgren, K.N., Manelli, A.M., Stine, W.B., Jr., Baker, L.K., Krafft, G.A. and LaDu, M.J., 2002, Oligomeric and fibrillar species of amyloid-beta peptides differentially affect neuronal viability. *J. Biol. Chem.* **277**: 32046.

Danscher, G., Jensen, K.B., Frederickson, C.J., Kemp, K. Andreasen, A., Juhl, S., Stoltenberg, M. and Ravid, R., 1997, Increased amount of zinc in the hippocampus and amygdala of Alzheimer's diseased brains: a proton-induced X-ray emission spectroscopic analysis of cryostat sections from autopsy material. *J. Neurosci. Methods* **76**: 53.

Davis, C.D., Milne, D.B. and Nielsen, F.H., 2000, Changes in dietary zinc and copper affect zinc-status indicators of postmenopausal women, notably, extracellular superoxide dismutase and amyloid precursor proteins. *Am. J. Clin. Nutr.* **71**: 781.

De Deyn, P.P., Hiramatsu, M., Borggreve, F., Goeman, J., D'Hooge, R., Saerens, J. and Mori, A., 1998, Superoxide dismutase activity in cerebrospinal fluid of patients with dementia and some other neurological disorders. *Alzheimer Dis. Assoc. Disord.* **12**: 26.

de Silva, R., Lashley, T., Gibb, G., Hanger, D., Hope, A., Reid, A., Bandopadhyay, R., Utton, M., Strand, C., Jowett, T., Khan, N., Anderton, B., Wood, N., Holton, J., Revesz, T. and Lees, A., 2003, Pathological inclusion bodies in tauopathies contain distinct complements of tau with three or four microtubule-binding repeat domains as demonstrated by new specific monoclonal antibodies. *Neuropathol. Appl. Neurobiol.* **29**: 288.

De Strooper, B., 2003, Aph-1, Pen-2, and Nicastrin with Presenilin generate an active gamma-Secretase complex. *Neuron* **38**: 9.

Deibel, M.A., Ehmann, W.D. and Markesbery, W.R., 1996, Copper, iron, and zinc imbalances in severely degenerated brain regions in Alzheimer's disease: possible relation to oxidative stress. *J. Neurol. Sci.* **143**: 137.

Dermaut, B., Kumar-Singh, S., Engelborghs, S., Theuns, J., Rademakers, R., Saerens, J., Pickut, B.A., Peeters, K., van den Broeck, M., Vennekens, K., Claes, S., Cruts, M., Cras, P., Martin, J.J., Van Broeckhoven, C. and De Deyn, P.P., 2004, A novel presenilin 1 mutation associated with Pick's disease but not beta-amyloid plaques. *Ann. Neurol.* **55**: 617.

Desai, A.K. and Grossberg, G.T., 2005, Diagnosis and treatment of Alzheimer's disease. *Neurology* **64**: S34.

Dickson, T.C. and Vickers, J.C., 2001, The morphological phenotype of beta-amyloid plaques and associated neuritic changes in Alzheimer's disease. *Neuroscience* **105**: 99.

Dong, J., Atwood, C.S., Anderson, V.E., Siedlak, S.L., Smith, M.A., Perry, G. and Carey, P.R., 2003, Metal binding and oxidation of amyloid-beta within isolated senile plaque cores: Raman microscopic evidence. *Biochemistry* **42**: 2768.

Eagle, G.R., Zombola, R.R. and Himes, R.H., 1983, Tubulin-zinc interactions: binding and polymerization studies. *Biochemistry* **22**: 221.

Egana, J.T., Zambrano, C., Nunez, M.T., Gonzalez-Billault, C. and Maccioni, R.B., 2003, Iron-induced oxidative stress modify tau phosphorylation patterns in hippocampal cell cultures. *Biometals* **16**: 215.

Ekmekcioglu, C., 2001, The role of trace elements for the health of elderly individuals. *Nahrung* **45**: 309.

Farris, W., Mansourian, S., Chang, Y., Lindsley, L., Eckman, E.A., Frosch, M.P., Eckman, C.B., Tanzi, R.E., Selkoe, D.J. and Guenette, S., 2003, Insulin-degrading enzyme regulates the levels of insulin, amyloid beta-protein, and the beta-amyloid precursor protein intracellular domain in vivo. *Proc. Natl. Acad. Sci. USA* **100**: 4162.

Ferretti, G., Bacchetti, T., Moroni, C., Vignini, A. and Curatola, G., 2003, Copper-induced oxidative damage on astrocytes: protective effect exerted by human high density lipoproteins. *Biochim. Biophys. Acta* **1635**: 48.

Forbes, W.F. and Hill, G.B., 1998, Is exposure to aluminum a risk factor for the development of Alzheimer's disease? – Yes. *Arch. Neurol.* **55**: 740.

Frausto da Silva, J.J.R. and Williams, R.J.P., 2001, *The Biological Chemistry of the Elements*, Oxford University Press, Oxford.

Friedhoff, P., von Bergen, M., Mandelkow, E.M., Davies, P. and Mandelkow, E., 1998, A nucleated assembly mechanism of Alzheimer paired helical filaments. *Proc. Natl. Acad. Sci. USA* **95**: 15712.

Friedhoff, P., von Bergen, M., Mandelkow, E.M. and Mandelkow, E., 2000, Structure of tau protein and assembly into paired helical filaments. *Biochim. Biophys. Acta* **1502**: 122.

Friedlich, A.L., Lee, J.Y., van Groen, T., Cherny, R.A., Volitakis, I., Cole, T.B., Palmiter, R.D., Koh, J.Y. and Bush, A.I., 2004, Neuronal zinc exchange with the blood vessel wall promotes cerebral amyloid angiopathy in an animal model of Alzheimer's disease. *J. Neurosci.* **24**: 3453.

Frisoni, G.B., Padovani, A. and Wahlund, L.O., 2004, The predementia diagnosis of Alzheimer's disease. *Alzheimer Dis. Assoc. Disord.* **18**: 51.

Gabbita, S.P., Lovell, M.A. and Markesbery, W.R., 1998, Increased nuclear DNA oxidation in the brain in Alzheimer's disease. *J. Neurochem.* **71**: 2034.

Gandy, S., 2005, The role of cerebral amyloid beta accumulation in common forms of Alzheimer's disease. *J. Clin. Invest.* **115**: 1121.

Garzon-Rodriguez, W., Yatsimirsky, A.K. and Glabe, C.G., 1999, Binding of Zn(II), Cu(II), and Fe(II) ions to Alzheimer's A beta peptide studied by fluorescence. *Bioorg. Med. Chem. Lett.* **9**: 2243.

Gaskin, F., 1981, In vitro microtubule assembly regulation by divalent cations and nucleotides. *Biochemistry* **20**: 1318.

Gaskin, F. and Kress, Y., 1977, Zinc ion-induced assembly of tubulin. *J. Biol. Chem.* **252**: 6918.

Gaskin, F. and Shelanski, M.L., 1976, Microtubules and intermediate filaments. *Essays Biochem.* **12**: 115.

Geschwind, D.H., 2003, Tau phosphorylation, tangles, and neurodegeneration: the chicken or the egg? *Neuron* **40**: 457.

Gomis-Ruth, F.X., 2003, Structural aspects of the metzincin clan of metalloendopeptidases. *Mol. Biotechnol.* **24**: 157.

Gonzales, P.E., Solomon, A., Miller, A.B., Leesnitzer, M.A., Sagi, I. and Milla, M.E., 2004, Inhibition of the tumor necrosis factor-alpha-converting enzyme by its prodomain. *J. Biol. Chem.* **279**: 31638.

Goode, B.L., Chau, M., Denis, P.E. and Feinstein, S.C., 2000, Structural and functional differences between 3-repeat and 4-repeat tau isoforms. Implications for normal tau function and the onset of neurodegenetative disease. *J. Biol. Chem.* **275**: 38182.

Grundke-Iqbal, I., Fleming, J., Tung, Y.C., Lassmann, H., Iqbal, K. and Joshi, J.G., 1990, Ferritin is a component of the neuritic (senile) plaque in Alzheimer dementia. *Acta Neuropathol.* (*Berl*) **81**: 105.

Gupta, V.B., Anitha, S., Hegde, M.L., Zecca, L.R., Garruto, M., Ravid, R., Shankar, S.K., Stein, R., Shanmugavelu, P. and Jagannatha Rao, K.S., 2005, Aluminium in Alzheimer's disease: are we still at a crossroad? *Cell Mol. Life Sci.* **62**: 143.

Halliwell, B. and Gutteridge, J.M., 1984, Oxygen toxicity, oxygen radicals, transition metals and disease. *Biochem. J.* **219**: 1.

Harris, F.M., Brecht, W.J., Xu, Q., Mahley, R.W. and Huang, Y., 2004, Increased tau phosphorylation in apolipoprotein E4 transgenic mice is associated with activation of extracellular signal-regulated kinase: modulation by zinc. *J. Biol. Chem.* **279**: 44795.

Hasan, M.R., Morishima, D., Tomita, K., Katsuki, M. and Kotani, S., 2005, Identification of a 250 kDa putative microtubule-associated protein as bovine ferritin. Evidence for a ferritin-microtubule interaction. *FEBS J.* **272**: 822.

Haydon, P.G., 2001, GLIA: listening and talking to the synapse. *Nat. Rev. Neurosci.* **2**: 185.

Head, E., Garzon-Rodriguez, W., Johnson, J.K., Lott, I.T., Cotman, C.W. and Glabe, C., 2001, Oxidation of Abeta and plaque biogenesis in Alzheimer's disease and Down syndrome. *Neurobiol. Dis.* **8**: 792.

Heber, S., Herms, J., Gajic, V., Hainfellner, J., Aguzzi, A., Rulicke, T., von Kretzschmar, H., von Koch, C., Sisodia, S., Tremml, P., Lipp, H.P., Wolfer, D.P. and Muller, U., 2000, Mice with combined gene knock-outs reveal essential and partially redundant functions of amyloid precursor protein family members. *J. Neurosci.* **20**: 7951.

Hesketh, J.E., 1982, Zinc-stimulated microtubule assembly and evidence for zinc binding to tubulin. *Int. J. Biochem.* **14**: 983.

Hesse, L., Beher, D., Masters, C.L. and Multhaup, G., 1994, The beta A4 amyloid precursor protein binding to copper. *FEBS Lett.* **349**: 109.

Hoke, D.E., Tan, J.L., Ilaya, N.T., Culvenor, J.G., Smith, S.J., White, A.R., Masters, C.L. and Evin, G.M., 2005, In vitro gamma-secretase cleavage of the Alzheimer's amyloid precursor protein correlates to a subset of presenilin complexes and is inhibited by zinc. *FEBS J.* **272**: 5544.

Huang, X., Atwood, C.S., Moir, R.D., Hartshorn, M.A., Vonsattel, J.P., Tanzi, R.E. and Bush, A.I., 1997, Zinc-induced Alzheimer's Abeta1–40 aggregation is mediated by conformational factors. *J. Biol. Chem.* **272**: 26464.

Huang, X., Atwood, C.S., Hartshorn, M.A., Multhaup, G., Goldstein, L.E., Scarpa, R.C., Cuajungco, M.P., Gray, D.N., Lim, J., Moir, R.D., Tanzi, R.E. and Bush, A.I., 1999a, The A beta peptide of Alzheimer's disease directly produces hydrogen peroxide through metal ion reduction. *Biochemistry* **38**: 7609.

Huang, X., Cuajungco, M.P., Atwood, C.S., Hartshorn, M.A., Tyndall, J.D., Hanson, G.R., Stokes, K.C., Leopold, M., Multhaup, G., Goldstein, L.E., Scarpa, R.C., Saunders, A.J., Lim, J., Moir, R.D., Glabe, C., Bowden, E.F., Masters, C.L., Fairlie, D.P., Tanzi, R.E. and Bush, A.I., 1999b, Cu(II) potentiation of alzheimer abeta neurotoxicity. Correlation with cell-free hydrogen peroxide production and metal reduction. *J. Biol. Chem.* **274**: 37111.

Huang, X., Atwood, C.S., Moir, R.D., Hartshorn, M.A., Tanzi, R.E. and Bush, A.I., 2004, Trace metal contamination initiates the apparent auto-aggregation, amyloidosis, and oligomerization of Alzheimer's Abeta peptides. *J. Biol. Inorg. Chem.* **9**: 954.

Hyman, B.T., Augustinack, J.C. and Ingelsson, M., 2005, Transcriptional and conformational changes of the tau molecule in Alzheimer's disease. *Biochim. Biophys. Acta* **1739**: 150.

Inestrosa, N.C., Cerpa, W. and Varela-Nallar, L., 2005, Copper brain homeostasis: role of amyloid precursor protein and prion protein. *IUBMB Life* **57**: 645.

Ischiropoulos, H. and Beckman, J.S., 2003, Oxidative stress and nitration in neurodegeneration: cause, effect, or association? *J. Clin. Invest.* **111**: 163.

Iskra, M., Patelski, J. and Majewski, W., 1993, Concentrations of calcium, magnesium, zinc and copper in relation to free fatty acids and cholesterol in serum of atherosclerotic men. *J. Trace Elem. Electrolytes Health Dis.* **7**: 185.

Iwata, N., Tsubuki, S., Takaki, Y., Watanabe, K., Sekiguchi, M., Hosoki, E., Kawashima-Morishima, M., Lee, H.J., Hama, E., Sekine-Aizawa, Y. and Saido, T.C., 2000, Identification of the major Abeta1–42-degrading catabolic pathway in brain parenchyma: suppression leads to biochemical and pathological deposition. *Nat. Med.* **6**: 143.

Iwata, N., Tsubuki, S., Takaki, Y., Shirotani, K., Lu, B., Gerard, N.P., Gerard, C., Hama, E., Lee, H.J. and Saido, T.C., 2001, Metabolic regulation of brain Abeta by neprilysin. *Science* **292**: 1550.

Kanemitsu, H., Tomiyama, T. and Mori, H., 2003, Human neprilysin is capable of degrading amyloid beta peptide not only in the monomeric form but also the pathological oligomeric form. *Neurosci. Lett.* **350**: 113.

Karr, J.W., Akintoye, H., Kaupp, L.J. and Szalai, V.A., 2005, N-Terminal deletions modify the Cu^{2+} binding site in amyloid-beta. *Biochemistry* **44**: 5478.

Ke, Y., Chang, Z., Duan, X.L., Du, J.R., Zhu, L., Wang, K., Yang, X.D., Ho, K.P. and Qian, Z.M., 2005, Age-dependent and iron-independent expression of two mRNA isoforms of divalent metal transporter 1 in rat brain. *Neurobiol. Aging* **26**: 739.

Keen, C.L., Hanna, L.A., Lanoue, L., Uriu-Adams, J.Y., Rucker, R.B. and Clegg, M.S., 2003, Developmental consequences of trace mineral deficiencies in rodents: acute and long-term effects. *J. Nutr.* **133**: 1477S.

Kim, S.U. and de Vellis, J., 2005, Microglia in health and disease. *J. Neurosci. Res.* **81**: 302.

Kim, N.H. and Kang, J.H., 2003, Oxidative modification of neurofilament-L by copper-catalyzed reaction. *J. Biochem. Mol. Biol.* **36**: 488.

Kim, N.H., Jeong, M.S., Choi, S.Y. and Hoon Kang, J., 2004, Oxidative modification of neurofilament-L by the Cu,Zn-superoxide dismutase and hydrogen peroxide system. *Biochimie* **86**: 553.

Klatzo, I., Wisniewski, H. and Streicher, E., 1965, Experimental production of neurofibrillary degeneration. I. Light microscopic observations. *J. Neuropathol. Exp. Neurol.* **24**: 187.

Lammich, S., Kojro, E., Postina, R., Gilbert, S., Pfeiffer, R., Jasionowski, M., Haass, C. and Fahrenholz, F., 1999, Constitutive and regulated alpha-secretase cleavage of Alzheimer's amyloid precursor protein by a disintegrin metalloprotease. *Proc. Natl. Acad. Sci. USA* **96**: 3922.

Lanphear, B.P., Hornung, R., Khoury, J., Yolton, K., Baghurst, P., Bellinger, D.C., Canfield, R.L., Dietrich, K.N., Bornschein, R., Greene, T., Rothenberg, S.J., Needleman, H.L., Schnaas, L., Wasserman, G., Graziano, J. and Roberts, R., 2005, Low-level environmental lead exposure and children's intellectual function: an international pooled analysis. *Environ. Health Perspect.* **113**: 894.

Lee, J.Y., Cole, T.B., Palmiter, R.D., Suh, S.W. and Koh, J.Y., 2002, Contribution by synaptic zinc to the gender-disparate plaque formation in human Swedish mutant APP transgenic mice. *Proc. Natl. Acad. Sci. USA* **99**: 7705.

Lee, J.Y., Friedman, J.E., Angel, I., Kozak, A. and Koh, J.Y., 2004, The lipophilic metal chelator DP-109 reduces amyloid pathology in brains of human beta-amyloid precursor protein transgenic mice. *Neurobiol. Aging* **25**: 1315.

Leissring, M.A., Farris, W., Chang, A.Y., Walsh, D.M., Wu, X., Sun, X., Frosch, M.P. and Selkoe, D.J., 2003, Enhanced proteolysis of beta-amyloid in APP transgenic mice prevents plaque formation, secondary pathology, and premature death. *Neuron* **40**: 1087.

LeVine, S.M., 1997, Iron deposits in multiple sclerosis and Alzheimer's disease brains. *Brain Res.* **760**: 298.

Liliom, K., Wagner, G., Kovacs, J., Comin, B., Cascante, M., Orosz, F. and Ovadi, J., 1999, Combined enhancement of microtubule assembly and glucose metabolism in neuronal systems in vitro: decreased sensitivity to copper toxicity. *Biochem. Biophys. Res. Commun.* **264**: 605.

Lind, S.E., McDonagh, J.R. and Smith, C.J., 1993, Oxidative inactivation of plasmin and other serine proteases by copper and ascorbate. *Blood* **82**: 1522.

Lindeman, R.D., Clark, M.L. and Colmore, J.P., 1971, Influence of age and sex on plasma and red-cell zinc concentrations. *J. Gerontol.* **26**: 358.

Ling, Y., Morgan, K. and Kalsheker, N., 2003, Amyloid precursor protein (APP) and the biology of proteolytic processing: relevance to Alzheimer's disease. *Int. J. Biochem. Cell Biol.* **35**: 1505.

Liu, S.T., Howlett, G. and Barrow, C.J., 1999, Histidine-13 is a crucial residue in the zinc ion-induced aggregation of the A beta peptide of Alzheimer's disease. *Biochemistry* **38**: 9373.

Liu, W.K., Le, T.V., Adamson, J., Baker, M., Cookson, N., Hardy, J., Hutton, M., Yen, S.H. and Dickson, D.W., 2001, Relationship of the extended tau haplotype to tau biochemistry and neuropathology in progressive supranuclear palsy. *Ann. Neurol.* **50**: 494.

Loeffler, D.A., LeWitt, P.A., Juneau, P.L., Sima, A.A., Nguyen, H.U., DeMaggio, A.J., Brickman, C.M., Brewer, G.J., Dick, R.D., Troyer, M.D. and Kanaley, J., 1996, Increased regional brain concentrations of ceruloplasmin in neurodegenerative disorders. *Brain Res.* **738**: 265.

Loske, C., Gerdemann, A., Schepl, W., Wycislo, M., Schinzel, R., Palm, D., Riederer, P. and Munch, G., 2000, Transition metal-mediated glycoxidation accelerates cross-linking of beta-amyloid peptide. *Eur. J. Biochem.* **267**: 4171.

Lovell, M.A., Robertson, J.D., Teesdale, W.J., Campbell, J.L. and Markesbery, W.R., 1998, Copper, iron and zinc in Alzheimer's disease senile plaques. *J. Neurol. Sci.* **158**: 47.

Lovell, M.A., Smith, J.L., Xiong, S. and Markesbery, W.R., 2005, Alterations in zinc transporter protein-1 (ZnT-1) in the brain of subjects with mild cognitive impairment, early, and late-stage Alzheimer's disease. *Neurotox. Res.* **7**: 265.

Lue, L.F., Kuo, Y.M., Roher, A.E., Brachova, L., Shen, Y., Sue, L., Beach, T., Kurth, J.H., Rydel, R.E. and Rogers, J., 1999, Soluble amyloid beta peptide concentration as a predictor of synaptic change in Alzheimer's disease. *Am. J. Pathol.* **155**: 853.

Luo, Y., Bolon, B., Kahn, S., Bennett, B.D., Babu-Khan, S., Denis, P., Fan, W., Kha, H., Zhang, J., Gong, Y., Martin, L.J., Louis, C., Yan, Q., Richards, W.G., Citron, M. and Vassar, R., 2001, Mice deficient in BACE1, the Alzheimer's beta-secretase, have normal phenotype and abolished beta-amyloid generation. *Nat. Neurosci.* **4**: 231.

Lutsenko, S. and Petris, M.J., 2003, Function and regulation of the mammalian copper-transporting ATPases: insights from biochemical and cell biological approaches. *J. Membr. Biol.* **191**: 1.

Ma, Q.F., Li, Y.M., Du, J.T., Kanazawa, K., Nemoto, T., Nakanishi, H. and Zhao, Y.F., 2005, Binding of copper (II) ion to an Alzheimer's tau peptide as revealed by MALDI-TOF MS, CD, and NMR. *Biopolymers* **79**: 74.

Ma, Q., Li, Y., Du, J., Liu, H., Kanazawa, K., Nemoto, T., Nakanishi, H. and Zhao, Y., 2006, Copper binding properties of a tau peptide associated with Alzheimer's disease studied by CD, NMR, and MALDI-TOF MS. *Peptides* **27**: 841.

Madaric, A., Ginter, E. and Kadrabova, J., 1994, Serum copper, zinc, and copper/zinc ratio in males: influence of aging. *Physiol. Res.* **43**: 107

Mandelkow, E.M., Biernat, J., Drewes, G., Gustke, N., Trinczek, B. and Mandelkow, E., 1995, Tau domains, phosphorylation, and interactions with microtubules. *Neurobiol. Aging* **16**: 355.

Martinez Lista, E., Sole, J., Arola, L. and Mas, A., 1993, Changes in plasma copper and zinc during rat development. *Biol. Neonate*. **64**: 47.

Mattson, M.P., 2004, Pathways toward and away from Alzheimer's disease. *Nature* **430**: 631.

Maurer, I., Zierz, S. and Moller, H.J., 2000, A selective defect of cytochrome c oxidase is present in brain of Alzheimer's disease patients. *Neurobiol. Aging* **21**: 455.

Maynard, C.J., Cappai, R., Volitakis, I., Cherny, R.A., White, A.R., Beyreuther, K., Masters, C.L., Bush, A.I. and Li, Q.X., 2002, Overexpression of Alzheimer's disease amyloid-beta opposes the age-dependent elevations of brain copper and iron. *J. Biol. Chem.* **277**: 44670.

McLean, C.A., Cherny, R.A., Fraser, F.W., Fuller, S.J., Smith, M.J., Beyreuther, K., Bush, A.I. and Masters, C.L., 1999, Soluble pool of Abeta amyloid as a determinant of severity of neurodegeneration in Alzheimer's disease. *Ann. Neurol.* **46**: 860.

McMaster, D., McCrum, E., Patterson, C.C., Kerr, M.M., O'Reilly, D., Evans, A.E. and Love, A.H., 1992, Serum copper and zinc in random samples of the population of Northern Ireland. *Am. J. Clin. Nutr.* **56**: 440.

Mekmouche, Y., Coppel, Y., Hochgrafe, K., Guilloreau, L., Talmard, C., Mazarguil, H. and Faller, P., 2005, Characterization of the ZnII binding to the peptide amyloid-beta1–16 linked to Alzheimer's disease. *Chembiochem* **6**: 1663.

Menditto, A., Morisi, G., Alimonti, A., Caroli, S., Petrucci F., Spagnolo, A. and Menotti, A., 1993, Association of serum copper and zinc with serum electrolytes and with selected risk factors for cardiovascular disease in men aged 55–75 years. NFR Study Group. *J. Trace Elem. Electrolytes Health Dis*. **7**: 251.

Milne, D.B. and Johnson, P.E., 1993, Assessment of copper status: effect of age and gender on reference ranges in healthy adults. *Clin. Chem.* **39**: 883.

Miura, T., Suzuki, K., Kohata, N. and Takeuchi, H., 2000, Metal binding modes of Alzheimer's amyloid beta-peptide in insoluble aggregates and soluble complexes. *Biochemistry* **39**: 7024.

Molina, J.A., Jimenez-Jimenez, F.J., Aguilar, M.V., Meseguer, I., Mateos-Vega, C.J., Gonzalez-Munoz, M.J., de Bustos, F., Porta, J., Orti-Pareja, M., Zurdo, M., Barrios, E. and Martinez-Para, M.C., 1998, Cerebrospinal fluid levels of transition metals in patients with Alzheimer's disease. *J. Neural Transm.* **105**: 479.

Monget, A.L., Galan, P., Preziosi, P., Keller, H., Bourgeois, C., Arnaud, J., Favier, A. and Hercberg, S., 1996, Micronutrient status in elderly people. Geriatrie/Min. Vit. Aux Network. *Int. J. Vitam. Nutr. Res*. **66**: 71.

Moreira, P.I., Siedlak, S.L., Aliev, G., Zhu, X., Cash, A.D., Smith, M.A. and Perry, G., 2005, Oxidative stress mechanisms and potential therapeutics in Alzheimer's disease. *J. Neural Transm.* **112**: 921.

Morris, C.M., Kerwin, J.M. and Edwardson, J.A., 1994, Nonhaem iron histochemistry of the normal and Alzheimer's disease hippocampus. *Neurodegeneration* **3**: 267.

Moskovitz, J., Bar-Noy, S., Williams, W.M., Requena, J., Berlett, B.S. and Stadtman, E.R., 2001, Methionine sulfoxide reductase (MsrA) is a regulator of antioxidant defense and lifespan in mammals. *Proc. Natl. Acad. Sci. USA* **98**: 12920.

Mrak, R.E. and Griffin, S.W., 2005, Glia and their cytokines in progression of neurodegeneration. *Neurobiol. Aging* **26**: 349.

Muller, U., Cristina, N., Li, Z.W., Wolfer, D.P., Lipp, H.P., Rulicke, T., Brandner, S., Aguzzi, A. and Weissmann, C., 1994, Behavioral and anatomical deficits in mice homozygous for a modified beta-amyloid precursor protein gene. *Cell* **79**: 755.

Multhaup, G., Bush, A.I., Pollwein, P. and Masters, C.L., 1994, Interaction between the zinc (II) and the heparin binding site of the Alzheimer's disease beta A4 amyloid precursor protein (APP). *FEBS Lett.* **355**: 151.

Multhaup, G., Schlicksupp, A., Hesse, L., Beher, D., Ruppert, T., Masters, C.L. and Beyreuther, K., 1996, The amyloid precursor protein of Alzheimer's disease in the reduction of copper(II) to copper(I). *Science* **271**: 1406.

Munoz, D.G., 1998, Is exposure to aluminum a risk factor for the development of Alzheimer disease? – No. *Arch. Neurol.* **55**: 737.

Munro, H.N., Suter, P.M. and Russell, R.M., 1987, Nutritional requirements of the elderly. *Annu. Rev. Nutr.* **7**: 23.

Nagele, R.G., D'Andrea, M.R., Lee, H., Venkataraman, V. and Wang, H.Y., 2003, Astrocytes accumulate A beta 42 and give rise to astrocytic amyloid plaques in Alzheimer's disease brains. *Brain Res.* **971**: 197.

Nagele, R.G., Wegiel, J., Venkataraman, V., Imaki, H. and Wang, K.C., 2004, Contribution of glial cells to the development of amyloid plaques in Alzheimer's disease. *Neurobiol. Aging* **25**: 663.

Omar, R.A., Chyan, Y.J., Andorn, A.C., Poeggeler, B., Robakis, N.K. and Pappolla, M.A., 1999, Increased expression but reduced activity of antioxidant enzymes in Alzheimer's disease. *J. Alzheimers Dis.* **1**: 139.

Opazo, C., Ruiz, F.H. and Inestrosa, N.C., 2000, Amyloid-beta-peptide reduces copper(II) to copper(I) independent of its aggregation state. *Biol. Res.* **33**: 125.

Opazo, C., Huang, X., Cherny, R.A., Moir, R.D., Roher, A.E., White, A.R., Cappai, R., Masters, C.L., Tanzi, R.E., Inestrosa, N.C. and Bush, A.I., 2002, Metalloenzyme-like activity of Alzheimer's disease beta-amyloid. Cu-dependent catalytic conversion of dopamine, cholesterol, and biological reducing agents to neurotoxic H_2O_2. *J. Biol. Chem.* **277**: 40302.

Opazo, C., Luza, S., Villemagne, V.L., Volitakis, I., Rowe, C., Barnham, K.J., Strozyk, D., Masters, C.L., Cherny, R.A. and Bush, A.I., 2006, Radioiodinated clioquinol as a biomarker for ß-amyloid:Zn^{2+} complexes in Alzheimer's disease. *Aging Cell* **5**: 69.

Oteiza, P.I., Cuellar, S., Lonnerdal, B., Hurley, L.S. and Keen, C.L., 1990a, Influence of maternal dietary zinc intake on in vitro tubulin polymerization in fetal rat brain. *Teratology* **41**: 97.

Oteiza, P.I., Hurley, L.S., Lonnerdal, B. and Keen, C.L., 1990b, Effects of marginal zinc deficiency on microtubule polymerization in the developing rat brain. *Biol. Trace Elem. Res.* **24**: 13.

Pappolla, M.A., Omar, R.A., Kim, K.S. and Robakis, N.K., 1992, Immunohistochemical evidence of oxidative (corrected) stress in Alzheimer's disease. *Am. J. Pathol.* **140**: 621.

Park, I.H., Jung, M.W., Mori, H. and Mook-Jung, I., 2001, Zinc enhances synthesis of presenilin 1 in mouse primary cortical culture. *Biochem. Biophys. Res. Commun.* **285**: 680.

Perez, M., Valpuesta, J.M., de Garcini, E.M., Quintana, C., Arrasate, M., Lopez Carrascosa, J.L., Rabano, A., Garcia de Yebenes, J. and Avila, J., 1998, Ferritin is associated with the aberrant tau filaments present in progressive supranuclear palsy. *Am. J. Pathol.* **152**: 1531.

Perry, G., Cash, A.D. and Smith, M.A., 2002, Alzheimer's disease and oxidative stress. *J. Biomed. Biotechnol.* **2**: 120.

Petersen, R.C., 2004, Mild cognitive impairment as a diagnostic entity. *J. Intern. Med.* **256**: 183.

Petersen, R.C., Smith, G.E., Waring, S.C., Ivnik, R.J., Tangalos, E.G. and Kokmen, E., 1999, Mild cognitive impairment: clinical characterization and outcome. *Arch. Neurol.* **56**: 303.

Phinney, A.L., Drisaldi, B., Schmidt, S.D., Lugowski, S., Coronado, V., Liang, Y., Horne, P., Yang, J., Sekoulidis, J., Coomaraswamy, J., Chishti, M.A., Cox, D.W., Mathews, P.M., Nixon, R.A., Carlson, G.A., St. George-Hyslop, P. and Westaway, D., 2003, In vivo reduction of amyloid-beta by a mutant copper transporter. *Proc. Natl. Acad. Sci. USA* **100**: 14193.

Pierson, K.B. and Evenson, M.A., 1988, 200 Kd neurofilament protein binds Al, Cu, and Zn. *Biochem. Biophys. Res. Commun.* **152**: 598.

Plantin, L.-O., Lysing-Tunnell, U. and Kristensson, K., 1987, Trace elements in the human central nervous system studied with neutron activation analysis. *Biol. Trace Elem. Res.* **13**: 69.

Prasad, A.S., Fitzgerald, J.T., Hess, J.W., Kaplan, J., Pelen, F. and Dardenne, M., 1993, Zinc deficiency in elderly patients. *Nutrition* **9**: 218.

Prohaska, J.R. and Gybina, A.A., 2004, Intracellular copper transport in mammals. *J. Nutr.* **134**: 1003.

Puglielli, L., Friedlich, A.L., Setchell, K.D., Nagano, S., Opazo, C., Cherny, R.A., Barnham, K.J., Wade, J.D., Melov, S., Kovacs, D.M. and Bush, A.I., 2005, Alzheimer disease beta-amyloid activity mimics cholesterol oxidase. *J. Clin. Invest.* **115**: 2556.

Quinta-Ferreira, M.E. and Matias, C.M., 2005, Tetanically released zinc inhibits hippocampal mossy fiber calcium, zinc, and synaptic responses. *Brain Res.* **1047**: 1.

Rae, T.D., Schmidt, P.J., Pufahl, R.A., Culotta, V.C. and O'Halloran, T.V., 1999, Undetectable intracellular free copper: the requirement of a copper chaperone for superoxide dismutase. *Science* **284**: 805.

Rajan, M.T., Jagannatha Rao, K.S., Mamatha, B.M., Rao, R.V., Shanmugavelu, P., Menon, R.B. and Pavithran, M.V., 1997, Quantification of trace elements in normal human brain by inductively coupled plasma atomic emission spectrometry. *J. Neurol. Sci.* **146**: 153.

Raman, B., Ban, T., Yamaguchi, K., Sakai, M., Kawai, T., Naiki, H. and Goto, Y., 2005, Metal ion-dependent effects of clioquinol on the fibril growth of an amyloid-beta peptide. *J. Biol. Chem.* **280**: 16157.

Rao, K.S.J., Rao, R.V., Shanmugavelu, P. and Menon, R.B., 1999, Trace elements in Alzheimer's disease brain: A new hypothesis. *Alzheimers Rep.* 241.

Ravaglia, G., Forti, P., Maioli, F., Nesi, B., Pratelli, L., Savarino, L., Cucinotta, D. and Cavalli, G., 2000, Blood micronutrient and thyroid hormone concentrations in the oldest-old. *J. Clin. Endocrinol. Metab.* **85**: 2260.

Reinhard, C., Hebert, S.S. and De Strooper, B., 2005, The amyloid-beta precursor protein: integrating structure with biological function. *EMBO J.* **24**: 3996.

Ritchie, C.W., Bush, A.I., Mackinnon, A., Macfarlane, S., Mastwyk, M., MacGregor, L., Kiers, L., Cherny, R., Li, Q.X., Tammer, A., Carrington, D., Mavros, C., Volitakis, I., Xilinas, I.M., Ames, D., Davis, S., Beyreuther, K., Tanzi, R.E. and Masters, C.L., 2003, Metal-protein attenuation with iodochlorhydroxyquin (clioquinol) targeting Abeta amyloid deposition and toxicity in Alzheimer's disease: a pilot phase 2 clinical trial. *Arch. Neurol* **60**: 1685.

Rogers, J.T., Randall, J.D., Cahill, C.M., Eder, P.S., Huang, X., Gunshin, H., Leiter, L., McPhee, J., Sarang, S.S., Utsuki, T., Greig, N.H., Lahiri, D.K., Tanzi, R.E., Bush, A.I., Giordano, T. and Gullans, S.R., 2002, An iron-responsive element type II in the 5'-untranslated region of the Alzheimer's amyloid precursor protein transcript. *J. Biol. Chem.* **277**: 45518.

Rosenberg, P.B., 2005, Clinical aspects of inflammation in Alzheimer's disease. *Int. Rev. Psychiatry* **17**: 503.

Rottkamp, C.A., Raina, A.K., Zhu, X., Gaier, E., Bush, A.I., Atwood, C.S., Chevion, M., Perry, G. and Smith, M.A., 2001, Redox-active iron mediates amyloid-beta toxicity. *Free. Radic. Biol. Med.* **30**: 447.

Ruiz, F.H., Gonzalez, M., Bodini, M., Opazo, C. and Inestrosa, N.C., 1999, Cysteine 144 is a key residue in the copper reduction by the beta-amyloid precursor protein. *J. Neurochem.* **73**: 1288.

Sagara, Y., Dargusch, R., Klier, F.G., Schubert, D. and Behl, C., 1996, Increased antioxidant enzyme activity in amyloid beta protein-resistant cells. *J. Neurosci.* **16**: 497.

Sayre, L.M., Perry, G., Harris, P.L., Liu, Y., Schubert, K.A. and Smith, M.A., 2000, In situ oxidative catalysis by neurofibrillary tangles and senile plaques in Alzheimer's disease: a central role for bound transition metals. *J. Neurochem.* **74**: 270.

Scheuermann, S., Hambsch, B., Hesse, L., Stumm, J., Schmidt, C., Beher, D., Bayer, T.A., Beyreuther, K. and Multhaup, G., 2001, Homodimerization of amyloid precursor protein and its implication in the amyloidogenic pathway of Alzheimer's disease. *J. Biol. Chem.* **276**: 33923.

Schlief, M.L., Craig, A.M. and Gitlin, J.D., 2005, NMDA receptor activation mediates copper homeostasis in hippocampal neurons. *J. Neurosci.* **25**: 239.

Schonheit, B., Zarski, R. and Ohm, T.G., 2004, Spatial and temporal relationships between plaques and tangles in Alzheimer-pathology. *Neurobiol. Aging* **25**: 697.

Schubert, D. and Chevion, M., 1995, The role of iron in beta amyloid toxicity. *Biochem. Biophys. Res. Commun.* **216**: 702.

Schuessel, K., Schafer, S., Bayer, T.A., Czech, C., Pradier, L., Muller-Spahn, F., Muller, W.E. and Eckert, A., 2005, Impaired Cu/Zn–SOD activity contributes to increased oxidative damage in APP transgenic mice. *Neurobiol. Dis.* **18**: 89.

Selley, M.L., Close, D.R. and Stern, S.E., 2002, The effect of increased concentrations of homocysteine on the concentration of (E)-4-hydroxy-2-nonenal in the plasma and cerebrospinal fluid of patients with Alzheimer's disease. *Neurobiol. Aging* **23**: 383.

Serrano, L., Dominguez, J.E. and Avila, J., 1988, Identification of zinc-binding sites of proteins: zinc binds to the amino-terminal region of tubulin. *Anal. Biochem.* **172**: 210.

Shinall, H., Song, E.S. and Hersh, L.B., 2005, Susceptibility of amyloid ß peptide degrading enzymes to oxidative damage: a potential Alzheimer's disease spiral. *Biochemistry* **44**: 15345.

Shivers, B.D., Hilbich, C., Multhaup, G., Salbaum, M., Beyreuther, K. and Seeburg, P.H., 1988, Alzheimer's disease amyloidogenic glycoprotein: expression pattern in rat brain suggests a role in cell contact. *EMBO J.* **7**: 1365.

Simons, A., Ruppert, T., Schmidt, C., Schlicksupp, A., Pipkorn, R., Reed, J., Masters, C.L., White, A.R., Cappai, R., Beyreuther, K., Bayer, T.A. and Multhaup, G., 2002, Evidence for a copper-binding superfamily of the amyloid precursor protein. *Biochemistry* **41**: 9310.

Sinha, S., Anderson, J.P., Barbour, R., Basi, G.S., Caccavello, R., Davis, D., Doan, M., Dovey, H.F., Frigon, N., Hong, J., Jacobson-Croak, K., Jewett, N., Keim, P., Knops, J., Lieberburg, I., Power, M., Tan, H., Tatsuno, G., Tung, J., Schenk, D., Seubert, P., Suomensaari, S.M., Wang, S., Walker, D., Zhao, J., McConlogue, L. and John, V., 1999, Purification and cloning of amyloid precursor protein beta-secretase from human brain. *Nature* **402**: 537.

Smith, M.A., Wehr, K., Harris, P.L., Siedlak, S.L., Connor, J.R. and Perry, G., 1998, Abnormal localization of iron regulatory protein in Alzheimer's disease. *Brain Res.* **788**: 232.

Sparks, D.L. and Schreurs, B.G., 2003, Trace amounts of copper in water induce beta-amyloid plaques and learning deficits in a rabbit model of Alzheimer's disease. *Proc. Natl. Acad. Sci. USA* **100**: 11065.

Squier, T.C., 2001, Oxidative stress and protein aggregation during biological aging. *Exp. Gerontol.* **36**: 1539.

Squitti, R., Lupoi, D., Pasqualetti, P., Dal Forno, G., Vernieri, F., Chiovenda, P., Rossi, L., Cortesi, M., Cassetta, E. and Rossini, P.M., 2002, Elevation of serum copper levels in Alzheimer's disease. *Neurology* **59**: 1153.

Squitti, R., Cassetta, E., Dal Forno, G., Lupoi, D., Lippolis, G., Pauri, F., Vernieri, F., Cappa, A. and Rossini, P.M., 2004, Copper perturbation in 2 monozygotic twins discordant for degree of cognitive impairment. *Arch. Neurol.* **61**: 738.

Streit, W.J., 2004, Microglia and Alzheimer's disease pathogenesis. *J. Neurosci. Res.* **77**: 1.

Suh, S.W., Jensen, K.B., Jensen, M.S., Silva, D.S., Kesslak, P.J., Danscher, G. and Frederickson, C.J., 2000, Histochemically-reactive zinc in amyloid plaques, angiopathy, and degenerating neurons of Alzheimer's diseased brains. *Brain Res.* **852**: 274.

Suh, Y.H. and Checler, F., 2002, Amyloid precursor protein, presenilins, and alpha-synuclein: molecular pathogenesis and pharmacological applications in Alzheimer's disease. *Pharmacol. Rev.* **54**: 469.

Sullivan, P.G. and Brown, M.R., 2005, Mitochondrial aging and dysfunction in Alzheimer's disease. *Prog. Neuropsychopharmacol. Biol. Psychiatry* **29**: 407.

Syme, C.D., Nadal, R.C., Rigby, S.E. and Viles, J.H., 2004, Copper binding to the amyloid-beta (Abeta) peptide associated with Alzheimer's disease: folding, coordination geometry, pH dependence, stoichiometry, and affinity of Abeta-(1–28): insights from a range of complementary spectroscopic techniques. *J. Biol. Chem.* **279**: 18169.

Tanzi, R.E. and Bertram, L., 2005, Twenty years of the Alzheimer's disease amyloid hypothesis: a genetic perspective. *Cell* **120**: 545.

Tarohda, T., Yamamoto, M. and Amamo, R., 2004, Regional distribution of manganese, iron, copper, and zinc in the rat brain during development. *Anal. Bioanal. Chem.* **380**: 240.

Terry, R.D., 1996, The pathogenesis of Alzheimer's disease: an alternative to the amyloid hypothesis. *J. Neuropathol. Exp. Neurol.* **55**: 1023.

Terry, R.D. and Wisniewski, H.M., 1975, Structural and chemical changes of the aged human brain. *Psychopharmacol. Bull.* **11**: 46.

Treiber, C., Simons, A., Strauss, M., Hafner, M., Cappai, R., Bayer, T.A. and Multhaup, G., 2004, Clioquinol mediates copper uptake and counteracts copper efflux activities of the amyloid precursor protein of Alzheimer's disease. *J. Biol. Chem.* **279**: 51958.

Uchida, Y., Takio, K., Titani, K., Ihara, Y. and Tomonaga, M., 1991, The growth inhibitory factor that is deficient in the Alzheimer's disease brain is a 68 amino acid metallothionein-like protein. *Neuron* **7**: 337.

Valensin, D.F., Mancini, M., Luczkowski, M., Janicka, A., Wisniewska, K., Gaggelli, E., Valensin, G., Lankiewicz, L. and Kozlowski, H., 2004, Identification of a novel high affinity copper binding site in the APP(145–155) fragment of amyloid precursor protein. *Dalton Trans.* **1**: 16.

Valko, M., Morris, H. and Cronin, M.T., 2005, Metals, toxicity, and oxidative stress. *Curr. Med. Chem.* **12**: 1161.

Vassar, R., Bennett, B.D., Babu-Khan, S., Kahn, S., Mendiaz, E.A., Denis, P., Teplow, D.B., Ross, S., Amarante, P., Loeloff, R., Luo, Y., Fisher, S., Fuller, J., Edenson, S., Lile, J., Jarosinski, M.A., Biere, A.L., Curran, E., Burgess, T., Louis, J.C., Collins, F., Treanor, J., Rogers, G. and Citron, M., 1999, Beta-secretase cleavage of Alzheimer's amyloid precursor protein by the transmembrane aspartic protease BACE. *Science* **286**: 735.

Venti, A., Giordano, T., Eder, P., Bush, A.I., Lahiri, D.K., Greig, N.H. and Rogers, J.T., 2004, The integrated role of desferrioxamine and phenserine targeted to an iron-responsive element in the APP-mRNA 5′-untranslated region. *Ann. NY Acad. Sci.* **1035**: 34.

Vickers, J.C., Dickson, T.C., Adlard, P.A., Saunders, H.L., King, C.E. and McCormack, G., 2000, The cause of neuronal degeneration in Alzheimer's disease. *Prog. Neurobiol.* **60**: 139.

Walsh, J.S., Welch, H.G. and Larson, E.B., 1990, Survival of outpatients with Alzheimer-type dementia. *Ann. Intern. Med.* **113**: 429.

Walsh, D.M., Klyubin, I., Fadeeva, J.V., Cullen, W.K., Anwyl, R., Wolfe, M.S., Rowan, M.J. and Selkoe, D.J., 2002a, Naturally secreted oligomers of amyloid beta protein potently inhibit hippocampal long-term potentiation in vivo. *Nature* **416**: 535.

Walsh, D.M., Klyubin, I., Fadeeva, J.V., Rowan, M.J. and Selkoe, D.J., 2002b, Amyloid-beta oligomers: their production, toxicity, and therapeutic inhibition. *Biochem. Soc. Trans.* **30**: 552.

Wang, J., Dickson, D.W., Trojanowski, J.Q. and Lee, V.M., 1999, The levels of soluble versus insoluble brain Abeta distinguish Alzheimer's disease from normal and pathologic aging. *Exp. Neurol.* **158**: 328.

Wang, D.S., Iwata, N., Hama, E., Saido, T.C. and Dickson, D.W., 2003, Oxidized neprilysin in aging and Alzheimer's disease brains. *Biochem. Biophys. Res. Commun.* **310**: 236.

Wang, Q., Woltjer, R.L., Cimino, P.J., Pan, C., Montine, K.S., Zhang, J. and Montine, T.J., 2005, Proteomic analysis of neurofibrillary tangles in Alzheimer's disease identifies GAPDH as a detergent-insoluble paired helical filament tau binding protein. *FASEB J.* **19**: 869.

Weisenberg, R.C., 1972, Microtubule formation in vitro in solutions containing low calcium concentrations. *Science* **177**: 1104.

Weisskopf, M.G., Wright, R.O., Schwartz, J., Spiro, A. 3rd, Sparrow, D., Aro, A. and Hu, H., 2004, Cumulative lead exposure and prospective change in cognition among elderly men: the VA normative aging study. *Am. J. Epidemiol.* **160**: 1184.

Wender, M., Szczech, J., Hoffmann, S. and Hilczer, W., 1992, Electron paramagnetic resonance analysis of heavy metals in the aging human brain. *Neuropatol. Pol.* **30**: 65.

White, A.R., Reyes, R., Mercer, J.F., Camakaris, J., Zheng, H., Bush, A.I., Multhaup, G., Beyreuther, K., Masters, C.L. and Cappai, R., 1999, Copper levels are increased in the cerebral cortex and liver of APP and APLP2 knockout mice. *Brain Res.* **842**: 439.

Wille, H., Drewes, G., Biernat, J., Mandelkow, E.M. and Mandelkow, E., 1992, Alzheimer-like paired helical filaments and antiparallel dimers formed from microtubule-associated protein tau in vitro. *J. Cell Biol.* **118**: 573.

Wojtera, M., Sikorska, B., Sobow, T. and Liberski, P.P., 2005, Microglial cells in neurodegenerative disorders. *Folia Neuropathol.* **43**: 311.

Yamamoto, A., Shin, R.W., Hasegawa, K., Naiki, H., Sato, H., Yoshimasu, F. and Kitamoto, T., 2002, Iron (III) induces aggregation of hyperphosphorylated tau and its reduction to iron (II) reverses the aggregation: implications in the formation of neurofibrillary tangles of Alzheimer's disease. *J. Neurochem.* **82**: 1137.

Yan, R., Bienkowski, M.J., Shuck, M.E., Miao, H., Tory, M.C., Pauley, A.M., Brashier, J.R., Stratman, N.C., Mathews, W.R., Buhl, A.E., Carter, D.B., Tomasselli, A.G., Parodi, L.A., Heinrikson, R.L. and Gurney, M.E., 1999, Membrane-anchored aspartyl protease with Alzheimer's disease beta-secretase activity. *Nature* **402**: 533.

Yang, D.S., McLaurin, J., Qin, K., Westaway, D. and Fraser, P.E., 2000, Examining the zinc binding site of the amyloid-beta peptide. *Eur. J. Biochem.* **267**: 6692.

Yong, V.W., Krekoski, C.A., Forsyth, P.A., Bell, R. and Edwards, D.R., 1998, Matrix metalloproteinases and diseases of the CNS. *Trends Neurosci.* **21**: 75.

Yoshiike, Y., Tanemura, K., Murayama, O., Akagi, T., Murayama, M., Sato, S., Sun, X., Tanaka, N. and Takashima, A., 2001, New insights on how metals disrupt amyloid beta-aggregation and their effects on amyloid-beta cytotoxicity. *J. Biol. Chem.* **276**: 32293.

Yu, W.H., Lukiw, W.J., Bergeron, C., Niznik, H.B. and Fraser, P.E., 2001, Metallothionein III is reduced in Alzheimer's disease. *Brain Res.* **894**: 37.

Zemlan, F.P., Thienhaus, O.J. and Bosmann, H.B., 1989, Superoxide dismutase activity in Alzheimer's disease: possible mechanism for paired helical filament formation. *Brain Res.* **476**: 160.

Zhao, G., Cui, M.Z., Mao, G., Dong, Y., Tan, J., Sun, L. and Xu, X., 2005, Gamma-cleavage is dependent on zeta-cleavage during the proteolytic processing of amyloid precursor protein within its transmembrane domain. *J. Biol. Chem.* **280**: 37689.

Zheng, H., Jiang, M., Trumbauer, M.E., Sirinathsinghji, D.J., Hopkins, R., Smith, D.W., Heavens, R.P., Dawson, G.R., Boyce, S., Conner, M.W., Stevens, K.A., Slunt, H.H., Sisoda, S.S., Chen, H.Y. and Van der Ploeg, L.H., 1995, beta-Amyloid precursor protein-deficient mice show reactive gliosis and decreased locomotor activity. *Cell* **81**: 525.

16

EPILEPSY, BRAIN INJURY, AND CELL DEATH

Günther Sperk, Meinrad Drexel, Ramon Tasan, and Anna Wieselthaler

1. ABSTRACT

Mesial temporal lobe epilepsy (TLE) represents the most frequent type of focal epilepsies. The most prominent neuropathological characteristics of TLE are severe neurodegenerations in the hippocampus (termed Ammon's horn sclerosis) and in related brain areas such as the amygdala and entorhinal cortex. Within the hippocampus, pyramidal cells of the areas CA1 and CA3 and interneurons of the dentate hilus are most vulnerable, whereas granule cells of the dentate gyrus, CA2 pyramidal cells and the subiculum are comparatively preserved. TLE can be initiated by prolonged febrile convulsions in early life or by a *status epilepticus*. It often takes years until the disease becomes clinically manifest (latent period). In animal models an initial *status epilepticus* is induced by application of a convulsant drug (e.g., kainic acid, pilocarpine) or by sustained electrical stimulation. Spontaneous seizures start thereafter within 1–3 weeks. The primary molecular mechanism underlying neuronal cell losses after *status epilepticus* are excitotoxic mechanisms caused by excessive release of glutamate and is presumably mediated by calcium influx through NMDA, kainate, or AMPA receptors. In addition, clear signs of apoptosis (DNA laddering, condensation of chromatin, and expression of caspases) can be seen. The initial neuropathological triggers for neurodegeneration are repeated neuronal stimulation during the epileptic seizures, but presumably also local ischaemia, hypoxia, or hypoglycemia induced by the strong neuronal activity. As a consequence of neurodegeneration, proliferation of astrocytes and of microglia (termed reactive gliosis) are typical signs of Ammon's horn sclerosis and contribute to scar formation in the lesioned areas. In animal models, opening of the blood–brain barrier, formation of local edema, and hemorrhages are seen upon the *status epilepticus* and significantly contribute to the necrosis in TLE brains. Secondarily to neurodegeneration, impressive neurochemical and morphological plasticity takes place in the epileptic hippocampus leading to a rearrangement of hippocampal circuitry.

2. EPILEPSIES AND TEMPORAL LOBE EPILEPSY

Epilepsies are characterized by recurrent spontaneous seizures due to uncontrolled discharges of cerebral neurons that are the basis of a variety of clinical symptoms like loss of consciousness, psychological changes, motor convulsions and altered perception, or a

Günther Sperk – Guenther.Sperk@i-med.ac.at
Department of Pharmacology, Innsbruck Medical University, Innsbruck, Austria

combination of these symptoms (Browne and Holmes, 2001). Epileptic seizures are classified as focal (partial) or generalized seizures. In focal seizures, epileptic discharges start in a specific area of one hemisphere. In generalized seizures, epileptic discharges are generated in both hemispheres. Focal seizures may become secondarily generalized. Temporal lobe epilepsy (TLE) is the most common form of focal epilepsies. In TLE the seizure focus (and also epileptogenic focus) is localized in the mesial temporal lobe, an area including the hippocampal formation and the entorhinal cortex, and in functionally related brain areas like the amygdala. Clinical seizures typically start with an aura, often described as a rising epigastric sensation, or hallucinations of memory. Auras are followed by disturbed consciousness, and oral or manual automatisms. Vegetative symptoms like ictal urinary urgency, ictal water drinking, changes of heart frequency, sweating, paleness, and others are common, in the same way as emotional and cognitive symptoms. (Margerison and Corsellis, 1966; Liu et al., 1995; Engel, 1996).

The neuropathology of mesial TLE was already described in the nineteenth century (Bouchet and Cazauvieilh, 1825; Sommer, 1880; Bratz, 1898). Sommer termed the typical degeneration of hippocampal structures as Ammon's horn sclerosis. Hallmarks of Ammon's horn sclerosis are the loss of pyramidal cells in hippocampal sectors CA1 and CA3 and of interneurons of the dentate gyrus associated with marked proliferation of astrocytes. Granule cells of the dentate gyrus, CA2 pyramidal cells, and neurons of the subiculum are less severely affected. In addition to the degenerations in the hippocampus varying damage is also seen in other brain areas, especially in the amygdala and entorhinal cortex (pyramidal cell layer III), but also in the thalamus, cerebellum, and other brain areas. Apart from mesial TLE there are also temporal lobe epilepsies without hippocampal sclerosis, called lesional TLE. In these patients, temporal lobe seizures can result, for e.g., from tumors, vascular malformations or scars after trauma.

Common etiologies of TLE are prolonged febrile convulsions in early childhood or a *status epilepticus*. A *status epilepticus* is defined as intermittent seizure activity lasting for 30 min or longer. In this case endogenous regulatory mechanisms may fail and secondarily excitotoxic events occur (Lowenstein and Alldredge, 1998; Treiman et al., 1998). These initial events are followed by a so-called latent period without seizure activity during which hippocampal sclerosis evolves. In TLE patients with a history of febrile convulsions, unprovoked seizures usually start not before the second decade of life (Liu et al., 1995; Engel, 1996). The process of the epilepsy developing during this period is called "epileptogenesis." Genetic factors may have a role in the susceptibility for febrile convulsions and in the process of epileptogenesis.

3. ANIMAL MODELS OF MESIAL TEMPORAL LOBE EPILEPSY

Animal models of TLE are based on the initial induction of prolonged seizures or a *status epilepticus* and subsequent development of spontaneously recurrent seizures. These models closely resemble the pathological and clinical picture of TLE in humans. The initial *status epilepticus* is generally induced by injection of a potent convulsant drug such as the glutamate receptor agonist kainic acid (Sperk, 1994), the cholinomimetic drug pilocarpine (often in combination with lithium salts) (Turski et al., 1989), or of tetanus toxin (Mellanby et al., 1984), or by sustained electrical stimulation of the amygdala, the hippocampus, or of the perforant path projecting from the entorhinal cortex to the hippocampus (Sloviter, 1983; Lothman et al., 1989; Nissinen et al., 2000). In these models, the toxins only induce seizures which then propagate to a *status epilepticus* (Sperk, 1994). During the acute *status epilepticus*, seizures include generalized clonic convulsions with rearing, foam at the mouth, and loss of balance. Depth electrode EEG or ex vivo autoradiography using [^{14}C]2-desoxyglucose shows involvement of hippocampus, amygdala, and limbic cortex (piriform and entorhinal cortex). The term "limbic seizures" is often used for this type of seizures. Following the *status epilepticus*, rats are exhausted but recover after several days. After 5–20 days they develop spontaneously recurrent seizures, persisting lifelong. The extent of hippocampal sclerosis likely depends on the duration and severity of the *status epilepticus*. Like in humans, hippocampal sclerosis involves pyramidal cells of CA1 and CA3 and interneurons of the dentate hilus also in animal models (Sperk, 1994). Granule cells and CA2 pyramidal cells are considerably less affected. In addition, parts of the amygdala, piriform, and entorhinal cortex, thalamus, septum, and olfactory nuclei are damaged. Interestingly, stimulation of kainic acid receptors can also lead to hippocampal sclerosis and TLE in humans. After consumption of domoic acid, a kainic acid receptor agonist contaminating sea food, patients developed acute seizures and, 1 year later, TLE with hippocampal sclerosis (Teitelbaum et al., 1990; Cendes et al., 1995).

At least in animal models, the brain damage is primarily associated with the initial *status epilepticus* but may progress during and after development of chronic epilepsy. There are striking similarities of brain damage seen in anoxia and ischemia and those after a *status epilepticus*. It has been therefore proposed that major aspects of "epilepsy-induced" brain damage may be caused by local or general ischemia, anoxia, hypoxia, or hypoglycemia induced by the severely increased neuronal activity (Spielmeyer, 1927; Blennow et al., 1978; Tanaka et al., 1992). On the other hand, in animal models of TLE, the pattern of brain damage is somewhat different from that in models of transient ischemia. Whereas in ischemia the sector CA1 is the most vulnerable area in the hippocampus, in epilepsy-induced brain damage pyramidal neurons of the sector CA3 appear to be more vulnerable than those in CA1 (Sperk, 1994). The initial mechanisms ultimately leading to neuronal cell death may be rather similar in brain trauma, ischemia, and in epilepsy (Lowenstein et al., 1992). Under all these conditions, extensive release of glutamate is thought to induce excitotoxicity (Choi, 1990; Mody and MacDonald, 1995). This mechanism includes activation of NMDA receptors by glutamate, extensive influx of Ca^{2+} ions into neurons, activation of the Na^+/K^+-ATPases for keeping up the ion gradients and reuptake of glutamate, and finally a breakdown of energy supply within the cell. The initial mechanisms induce cascades of other events leading ultimately to neuronal cell death (Siesjo and Wieloch, 1986; Simon et al., 1986; Meldrum, 1993). The neuropathology seen in animal TLE models can also be produced by local injection of glutamate (Sloviter and Dempster, 1985).

4. WHY ARE SOME NEURONS MORE VULNERABLE THAN OTHERS?

It has been proposed that in animal models of TLE the sector CA3 may have a pacemaker role in the generation of seizures in the hippocampus which may depend on kainic acid receptors on CA3 neurons innervated by mossy fiber terminals (Ben-Ari and Cossart, 2000). This would be one explanation for the high vulnerability of CA3 pyramidal neurons in animal models of *status epilepticus* and in TLE. Kainic acid receptors are also ligand-operated ion channels and their activation may induce excitotoxicity in the same way as activation of NMDA receptors. Protection of CA1 neurons in epilepsy may be attributed to the loss of

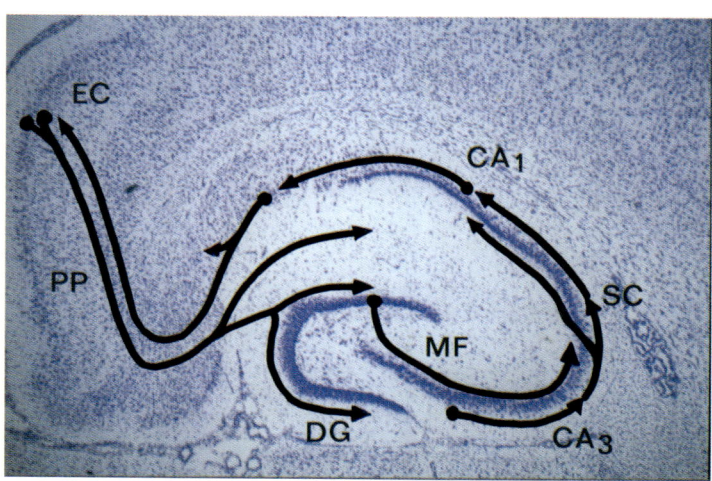

Fig. 1. Hippocampal pathways. A photomicrograph of the ventral hippocampus of the rat depicting the major excitatory pathway, the so-called tri-synaptic circuitry. It shows the perforant path (PP) projections from the entorhinal cortex (EC) to the molecular layer of the dentate gyrus (DG) with its branch to the stratum lacunosum moleculare in sector CA1. The molecular layer of the dentate gyrus contains the dendrites of granule cells that project with their mossy fibers (MF) to the sector CA3. Pyramidal cells of the sector CA3 send their projections, the so-called Schaffer collaterals (SC) to CA1 pyramidal neurons. These target to the subiculum, from where projections lead back to the entorhinal cortex (EC). All these pathways are glutamatergic (excitatory). The respective neurons (granule cells in the dentate gyrus and pyramidal cells in the other areas) carry different types of glutamate receptors, through which excitotoxic mechanisms can be induced

excitatory input from CA3 (Schaffer collaterals). Interrupting a kainate-induced *status epilepticus* in rats by early treatment with an anticonvulsant dose of diazepam results in a reduced damage in sector CA3 but increased pyramidal cell loss in CA1 (see Figs. 1 and 2). Similarly, increasing the dose of kainic acid in rats (partially protected from seizures by keeping them hypothermic) paradoxically increases damage in the sector CA1 and not in CA3 (Balchen et al., 1993). Both experiments indicate that preventing seizure activity and consequently CA3 neuronal death leads to increased neurodegeneration in CA1.

Similar mechanisms may be considered for the lack of damage in the subiculum seen in TLE patients. Again this may be a consequence of the extensive loss of CA1 pyramidal cells and therefore of the excitatory subicular input arising from CA1 pyramids. In contrast, in kainate-treated rats with generally less damage of CA1, subicular neurons are more affected. In the same line, lesioning the excitatory projections from the pre and parasubiculum to the entorhinal cortex protects layer 3 of the entorhinal cortex from seizure-induced cell loss (Eid et al., 2001).

Not clear is the high resistance of granule cells and of CA2 pyramidal neurons in epilepsy. In rat models granule cells and CA2 pyramids remain entirely intact, whereas in human TLE they are relatively protected (Pirker et al., 2001). The stability of CA2 pyramidal neurons may be due to the lack of a major excitatory input (the mossy fiber pathway terminates at the border to CA2). Granule cells are under strong control of inhibitory interneurons (e.g., basket cells) that may control their over excitation (Scharfman, 1991). On the other hand, target neurons of granule cells, mossy cells, and CA3 pyramidal neurons are vulnerable in epilepsy.

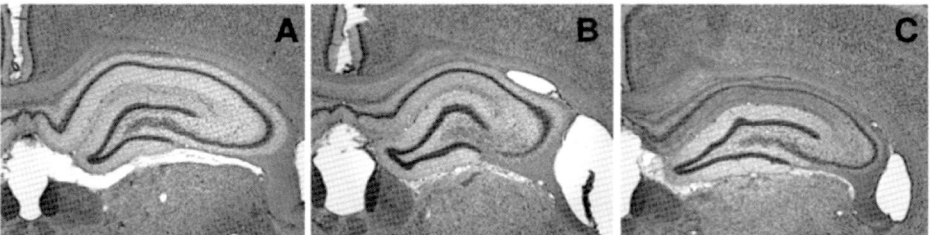

Fig. 2. Epilepsy-induced cell death. The photomicrographs depict Nissl stained sections of the dorsal hippocampus from a control rat (**a**) and from rats injected with kainate 8 days before (**b, c**) that had experienced either full *status epilepticus* (b) or were injected with an anticonvulsive dose of diazepam 1 h after the first stage 3 seizure (c). In the rat with full status (b), seizure-induced damage is seen in CA3. In the other rat (c) the CA3 sector was protected by the diazepam injection; damage was most pronounced in CA1

Among the most sensitive neurons are glutamatergic mossy cells giving rise to the so-called assocional/commissural fibers projecting to the inner molecular layers of the ipsi- and contralateral hippocampus. The high vulnerability of mossy cells leads to an often cited hypothesis for the generation of TLE, the so-called dormant basket cell hypothesis (Sloviter, 1991). It postulates that inhibitory basket cells loose their excitatory input from mossy cells and therefore become "dormant" in the epileptic hippocampus. An argument against this hypothesis is the observation that basket cells are rather active in animal TLE models (Sperk et al., 1992; Sperk, 1994; Bernard et al., 1998; Sloviter et al., 2003).

5. MECHANISMS LEADING TO EPILEPSY-INDUCED BRAIN DAMAGE

5.1. Excitotoxic Mechanisms

The primary molecular and cellular mechanism underlying neurodegeneration after a *status epilepticus* is excitotoxicity initiated by the excessive release of glutamate during acute seizures, as demonstrated in animal models and humans (Lehmann, 1987; During and Spencer, 1993; Mody and MacDonald, 1995). As discussed above, this extensive release of glutamate can be triggered by sustained epileptic stimulation or by hypoxic–hypoglycemic mechanisms. Glutamate induces an extensive Ca^{2+}-influx through strongly stimulated NMDA or kainate receptors. In chronic epilepsy, it has been suggested that the AMPA receptor subunit GluR2 becomes partially substituted by GluR1 in the AMPA receptor ion channel, making the channel also permeable for Ca^{2+}-ions (in addition to Na^+) (Pellegrini-Giampietro et al., 1997). The strong Ca^{2+}-influx is the first trigger for excitotoxicity. It induces sustained depolarization of the neuron. Physiological mechanisms like reuptake of glutamate and other neurotransmitters, restoration of the ion gradients within the neurons require energy and thus glucose and oxygen. Extensive activation of these mechanisms finally leads to a breakdown of the cell's energy supply, formation of oxygen radicals and of NO, lipid peroxidation and a sequelae of other events (Choi, 1990; Meldrum, 1993; Iadecola, 1997). Finally proteolytic enzymes, like the calcium-activated calpains, contribute to the protein degradation in the cell.

5.2. Apoptosis

Either concomitantly or after initiation of excitotoxicity, mechanisms of programmed cell death (apoptosis) become activated. Markers for apoptosis in the epileptic brain are condensation of chromatin, fragmentation of the DNA, and expression of caspases (Pollard et al., 1994; Faherty et al., 1999). Apoptosis is mediated by complex cascades of events. There are two principal pathways that may be activated, an extrinsic and an intrinsic pathway. The extrinsic pathway is triggered by activation of so-called death receptors in the cell membrane (members of the TNF-α receptor family, e.g., Fas). The intrinsic apoptosis pathway is for example activated by intracellular stressors like increased Ca^{2+} concentrations or reactive oxygen species (Orrenius et al., 2003; Polster and Fiskum, 2004) also present in excitotoxic brain damage. Key mechanisms are then activation of members of the Bcl-2 family activating (Bax) or inhibiting (Bcl-2) programmed cell death (Gillardon et al., 1995), release of cytochrome c from mitochondria (among various other mechanisms), and activation of members of the caspase family (Nagata, 1997; Skulachev, 1998; Henshall and Simon, 2005). Caspases belong to a family of aspartate-specific cysteine proteases taking part in the cascade of events of both the extrinsic and the intrinsic apoptotic pathway. Notably caspases 3 and 6 are expressed in models of epilepsy and thus may be involved in seizure-related cell death (Faherty et al., 1999; Henshall and Simon, 2005). Whether apoptosis is an independent or even primary mechanism in epilepsy-induced neuronal cell death, or whether it follows excitotoxic cell death for removing cells that had lost their functions within the neuronal circuitries (delayed neuronal cell death) is not entirely settled yet.

6. OTHER PATHOLOGICAL EVENTS INDUCING BRAIN DAMAGE

During the initial *status epilepticus*, a variety of pathological signs can be seen in animal models of TLE. Early histopathological changes include dendritic swelling, shrinkage, and pycnosis of neuronal perikarya, and signs of general edema with swelling of astrocytes. These changes are seen in most brain areas. During or immediately after acute seizures they are mostly reversible and may be signs of transient hypoglycemia due to the increased glucose consumption. About one day after the *status epilepticus*, severe pathological changes are seen that are, however, largely restricted to brain areas directly involved in the propagation of the acute *status epilepticus*, such as the hippocampus, amygdala, entorhinal and piriform cortices, thalamus, and septal nuclei. In the kainate model, this pathology consists of edema, opening of the blood–brain barrier, hemorrhages, and necrosis of neurons and nonneuronal elements (Sperk et al., 1983; Nitsch and Hubauer, 1986; Jensen et al., 1997). In addition to neuronal loss and reactive gliosis, loss of oligodendrocytes, demyelination, capillary sprouting, and perivascular hemorrhages are seen (Blennow et al., 1978; Sperk et al., 1983; Lassmann et al., 1984; Siesjo and Wieloch, 1986). Thus, epilepsy-related cell death is not only mediated by excitotoxicity and apoptosis, but may also comprise brain damage induced by edema and hemorrhages.

EPILEPSY, BRAIN INJURY, AND CELL DEATH

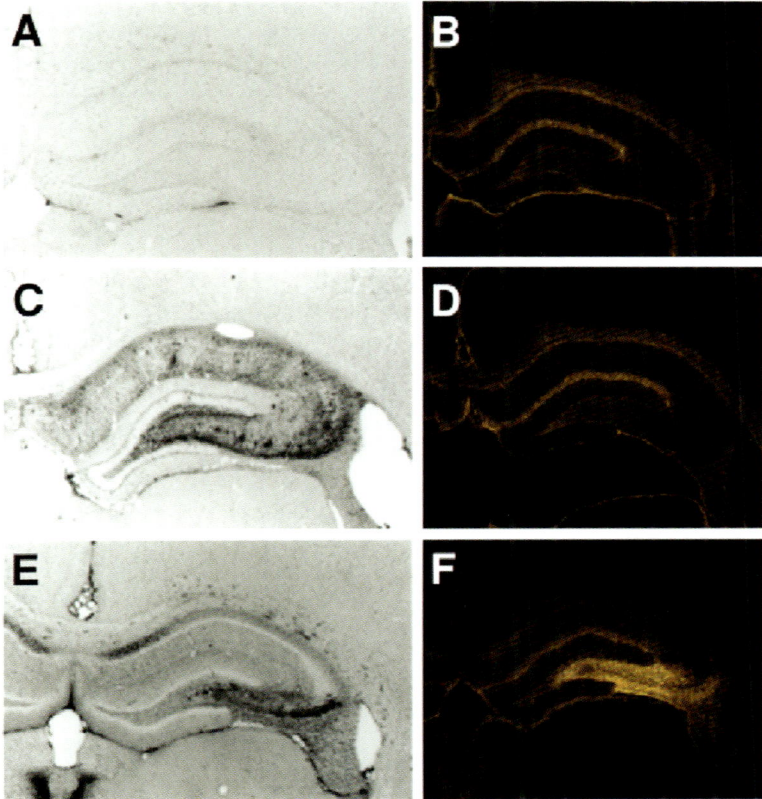

Fig. 3. Activation of astrocytes and microglia. Photomicrographs of sections labeled for reactive microglia, using immunohistochemistry for ox-42 (a marker for class II MHC antigens; **a, c, e**) and for reactive astrocytes, using GFAP- immunohistochemistry (**b, d, f**) are depicted. In panels (a) and (b) sections of a control rat are shown. Panels (c)/(d) and (e)/(f) show sections from two rats after kainic acid-induced *status epilepticus*, but with different pathology. Note the broad microglial reaction, shown by overexpression of ox-42 in (c), with minor astrocytic reaction (d) in a rat injected with kainic acid 8 days before. In contrast ox-42 expression was restricted to the sector CA3 and the subiculum of another rat (28 days after kainic acid injection) as depicted in (e), with rather strong activation of GFAP in the corresponding parts of the hippocampus (f)

7. PROLIFERATION OF ASTROCYTES AND MICROGLIA

Proliferation of astrocytes is one of the most prominent neuropathological reactions in neurodegeneration. It is long known as a hallmark symptom of Ammon's horn sclerosis contributing to hippocampal scar formation (Fig. 3). In animal models of TLE as in models of lesioning certain fiber tracts, increased expression of glial fibrillary acid protein (GFAP) or of vimentin are commonly used as markers for reactive astrocytes (Finsen et al., 1993; Mitchell et al., 1993; Represa et al., 1993). Reactive astrocytes may serve multiple pathophysiological functions in the degenerating neuronal tissue. They may be involved in the guidance of axon terminals during sprouting by releasing growth-inducing factors or guidance factors such as GAP-43 or cell adhesion molecules (Nakic et al., 1996; Guthrie et al., 1997; Seki and Rutishauser, 1998). Some of the functions of astrocytes seem to be considerably altered in epilepsy. Thus subtypes of "epileptic astrocytes" contain a reduced number of glutamate transporters or altered glutamate receptors (Seifert et al., 2004) or enzymes of the kynurenic acid pathway (Wu et al., 1991).

In addition, prominent and widespread proliferation of microglia expressing class I and class II MHC antigens has been demonstrated in TLE models and ischemia (Finsen et al., 1993; Jorgensen et al., 1993; Mitchell et al., 1993; Jensen et al., 1997). Also the microglia may serve multiple roles in acute neurodegeneration and in chronic situation of the recurrently stimulated epileptic hippocampus. Thus, microglia may have crucial phagocytic functions in acute injury but may also release immune mediators such as interleukins (IL-1β) or of TGF-β. Interleukins may even be involved in seizure propagation (Dube et al., 2005).

8. GRANULE CELL DISPERSION IN AMMON'S HORN SCLEROSIS

A striking characteristic of some human TLE specimens is a dispersion of granule cells (Houser, 1990). The granule cell layer is widened and the number of granule cells is reduced resulting in disarrangement of the cell layer. In some cases the granule cell layer appears to separate even into two distinct layers. The role of granule cell dispersion for epileptogenesis is not known. Local unilateral injections of kainic acid into the hippocampus reproduces granule cell dispersion in mice and induces recurrent spontaneous seizures (Bouilleret et al., 1999). Both in the mouse and in human TLE, impairment of reelin secretion may be a molecular trigger for granule cell dispersion (Haas et al., 2002; Frotscher et al., 2003).

9. NEURONAL PLASTICITY AS A RESPONSE TO NEURODEGENERATION

Severe disarrangement of the Ammon's horn may also lead to considerable functional changes of hippocampal pathways in TLE. Interestingly the input region of the hippocampus, the dentate granule cell layer (receiving excitation from entorhinal cortex), and its output region, the subiculum, remain relatively intact. In contrast, pathways between these two regions (interneurons and mossy cells of the dentate gyrus and CA1 and CA3 pyramidal cells) sustain extensive damage. As a response to neurodegeneration, and presumably also due to the recurrent seizures, pronounced neurochemical and extended morphological plasticity of the hippocampus is observed indicating morphological and functional rearrangement of the hippocampal circuitries. Plastic changes are indicated by changes in the expression of immediate early genes (Morgan and Curran, 1991; Herdegen et al., 1993), in the expression of growth factors (Gall et al., 1991), neuropeptides (Sperk, 1994; Vezzani et al., 1999), neurotransmitter receptors (Pellegrini-Giampietro et al., 1997; Sperk et al., 2004), and many other proteins.

The best-demonstrated and most consistent example for morphological plasticity in TLE is sprouting of mossy fibers. Collaterals of mossy fibers (the axons of granule cells) sprout to the inner molecular layer, where the dendrites of granule cells are located. This is typically demonstrated by Timm's staining (labeling of Zn^{2+} in axons) or by immunohistochemistry for dynorphin, chromogranin B, or neuronal adhesion molecules (NCAM) (Sutula et al., 1988,

1989; Houser, 1990; Sperk et al., 1992; Mikkonen et al., 1998; Pirker et al., 2001). Glutamatergic granule cells seem to take over the terminal region of the degenerating (glutamatergic) mossy cells in the inner molecular layer. There is little evidence that epileptic activity *per se* leads to mossy fiber sprouting. In contrast, lesioning of assocional–commissural fibers lead to a similar picture of mossy fiber rearrangement (Laurberg and Zimmer, 1981).

Using, e.g., anterogradely transported fluorescence dyes, formation of new axon collaterals was shown also in other parts of the hippocampus in animal models of TLE. CA1 pyramidal cells form local axon collaterals in the *stratum oriens* and in the alveus, but also in areas CA1 and CA3 and in the subiculum (Perez et al., 1996; Lehmann et al., 2001). In human TLE specimens intense sprouting of axons from GABAergic neurons is found in all parts of the hippocampus (Davenport et al., 1990; Mathern et al., 1995; Furtinger et al., 2001; Pirker et al., 2001).

10. CONCLUSION

Neuronal cell death in TLE may be initiated by a variety of, often interrelated mechanisms. Severe seizures (*status epilepticus*) induce overexcitation due to augmented release of glutamate leading to anoxia–hypoglycemia and excitotoxicity. Accumulation of intermediary metabolites and breakdown of energy supply are associated with severe edema and breakdown of the blood–brain barrier in brain areas most severely affected by the acute limbic seizures. Possibly as a secondary mechanism, apoptosis is activated for clearing away damaged neurons. Neurodegeneration is also accompanied by prominent reactive activation, and/or proliferation of astrocytes and of microglia (reactive gliosis), serving roles not only in the acute pathology but also in tissue repair and guidance of sprouting neurons.

11. REFERENCES

Balchen, T., Berg, M. and Diemer, N.H., 1993, A paradox after systemic kainate injection in rats: lesser damage of hippocampal CA1 neurons after higher doses. *Neurosci. Lett.* **163**: 151.

Ben-Ari, Y. and Cossart, R., 2000, Kainate, a double agent that generates seizures: two decades of progress. *Trends Neurosci.* **23**: 580.

Bernard, C., Esclapez, M., Hirsch, J.C. and Ben-Ari, Y., 1998, Interneurones are not so dormant in temporal lobe epilepsy: a critical reappraisal of the dormant basket cell hypothesis. *Epilepsy Res.* **32**: 93.

Blennow, G., Brierley, J.B., Meldrum, B.S., and Siesjo, B.K., 1978, Epileptic brain damage: the role of systemic factors that modify cerebral energy metabolism. *Brain* **101**: 687.

Bouchet, C. and Cazauvieilh, A., 1825, De l'épilepsie considéré dans ses rappoert avec l'alenienation mentale. *Archives Générales de Medicine (Paris)* **9**: 519.

Bouilleret, V., Ridoux, V., Depaulis, A., Marescaux, C., Nehlig, A. and Le Gal La Salle, G., 1999, Recurrent seizures and hippocampal sclerosis following intrahippocampal kainate injection in adult mice: electroencephalography, histopathology and synaptic reorganization similar to mesial temporal lobe epilepsy. *Neuroscience* **89**: 717.

Bratz, E.,1898, Ammonshornbefunde bei Epileptikern. *Arch. Psychiatr. Nervenkrankh* **32**: 820.

Browne, T.R. and Holmes, G.L., 2001, Epilepsy. *N. Engl. J. Med.* **344**: 1145

Cendes, F., Andermann, F., Carpenter, S., Zatorre, R.J. and Cashman, N.R., 1995, Temporal lobe epilepsy caused by domoic acid intoxication: evidence for glutamate receptor-mediated excitotoxicity in humans. *Ann. Neurol.* **37**: 123.

Choi, D.W., 1990, Cerebral hypoxia: some new approaches and unanswered questions. *J. Neurosci.* **10**: 2493.

Davenport, C.J., Brown, W.J. and Babb, T.L., 1990, Sprouting of GABAergic and mossy fiber axons in dentate gyrus following intrahippocampal kainate in the rat. *Exp. Neurol.* **109**: 180.

Dube, C., Vezzani, A., Behrens, M., Bartfai, T. and Baram, T.Z., 2005, Interleukin-1beta contributes to the generation of experimental febrile seizures. *Ann. Neurol.* **57**: 152.

During, M.J. and Spencer, D.D., 1993, Extracellular hippocampal glutamate and spontaneous seizure in the conscious human brain. *Lancet* **341**: 1607.

Eid, T., Du, F. and Schwarcz, R., 2001, Ibotenate injections into the pre- and parasubiculum provide partial protection against kainate-induced epileptic damage in layer III of rat entorhinal cortex. *Epilepsia* **42**: 817.

Engel, J., Jr., 1996, Introduction to temporal lobe epilepsy. *Epilepsy Res.* **26**: 141.

Faherty, C.J., Xanthoudakis, S. and Smeyne, R.J., 1999, Caspase-3-dependent neuronal death in the hippocampus following kainic acid treatment. *Brain Res. Mol. Brain Res.* **70**: 159.

Finsen, B.R., Jorgensen, M.B., Diemer, N.H. and Zimmer, J., 1993, Microglial MHC antigen expression after ischemic and kainic acid lesions of the adult rat hippocampus. *Glia* **7**: 41.

Frotscher, M., Haas, C.A. and Forster, E., 2003, Reelin controls granule cell migration in the dentate gyrus by acting on the radial glial scaffold. *Cereb. Cortex* **13**: 634.

Furtinger, S., Pirker, S., Czech, T., Baumgartner, C., Ransmayr, G. and Sperk, G., 2001, Plasticity of Y1 and Y2 receptors and neuropeptide Y fibers in patients with temporal lobe epilepsy. *J. Neurosci.* **21**: 5804.

Gall, C., Lauterborn, J., Bundman, M., Murray, K. and Isackson, P., 1991, Seizures and the regulation of neurotrophic factor and neuropeptide gene expression in brain. *Epilepsy Res.* **4** suppl: 225.

Gillardon, F., Wickert, H. and Zimmermann, M., 1995, Up-regulation of bax and down-regulation of bcl-2 is associated with kainate-induced apoptosis in mouse brain. *Neurosci. Lett.* **192**: 85.

Guthrie, K.M., Woods, A.G., Nguyen, T. and Gall, G.M., 1997, Astroglial ciliary neurotrophic factor mRNA expression is increased in fields of axonal sprouting in deafferented hippocampus. *J. Comp. Neurol.* **386**: 137.

Haas, C.A., Dudeck, O., Kirsch, M., Huszka, C., Kann, G., Pollak, S., Zentner, J. and Frotscher, M., 2002, Role for reelin in the development of granule cell dispersion in temporal lobe epilepsy. *J. Neurosci.* **22**: 5797.

Henshall, D.C. and Simon, R.P., 2005, Epilepsy and apoptosis pathways. *J. Cereb. Blood Flow Metab.* **25**: 1557.

Herdegen, T., Sandkuhler, J., Gass, P., Kiessling, M., Bravo, R. and Zimmermann, M., 1993, JUN, FOS, KROX, and CREB transcription factor proteins in the rat cortex: basal expression and induction by spreading depression and epileptic seizures. *J. Comp. Neurol.* **333**: 271.

Houser, C.R., 1990, Granule cell dispersion in the dentate gyrus of humans with temporal lobe epilepsy. *Brain Res.* **535**: 195.

Iadecola, C., 1997, Bright and dark sides of nitric oxide in ischemic brain injury. *Trends Neurosci.* **20**: 132.

Jensen, M.B., Finsen, B. and Zimmer, J., 1997, Morphological and immunophenotypic microglial changes in the denervated fascia dentata of adult rats: correlation with blood–brain barrier damage and astroglial reactions. *Exp. Neurol.* **143**: 103.

Jorgensen, M.B., Finsen, B.R., Jensen, M.B., Castellano, B., Diemer, N.H. and Zimmer, J., 1993, Microglial and astroglial reactions to ischemic and kainic acid-induced lesions of the adult rat hippocampus. *Exp. Neurol.* **120**: 70.

Lassmann, H., Petsche, U., Kitz, K., Baran, H., Sperk, G., Seitelberger, F. and Hornykiewicz, O., 1984, The role of brain edema in epileptic brain damage induced by systemic kainic acid injection. *Neuroscience* **13**: 691.

Laurberg, S. and Zimmer, J., 1981, Lesion-induced sprouting of hippocampal mossy fiber collaterals to the fascia dentata in developing and adult rats. *J. Comp. Neurol.* **200**: 433.

Lehmann, A., 1987, Alterations in hippocampal extracellular amino acids and purine catabolites during limbic seizures induced by folate injections into the rabbit amygdala. *Neuroscience* **22**: 573.

Lehmann, T.N., Gabriel, S., Eilers, A., Njunting, M., Kovacs, R., Schulze, K., Lanksch, W.R. and Heinemann, U., 2001, Fluorescent tracer in pilocarpine-treated rats shows widespread aberrant hippocampal neuronal connectivity. *Eur. J. Neurosci.* **14**: 83.

Liu, Z., Mikati, M. and Holmes, G.L., 1995, Mesial temporal sclerosis: pathogenesis and significance. *Pediatr. Neurol.* **12**: 5

Lothman, E.W., Bertram, E.H., Bekenstein, J.W. and Perlin, J.B., 1989, Self-sustaining limbic status epilepticus induced by 'continuous' hippocampal stimulation: electrographic and behavioral characteristics. *Epilepsy Res.* **3**:107.

Lowenstein, D.H. and Alldredge, B.K., 1998, Status epilepticus. *N. Engl. J. Med.* **338**: 970.

Lowenstein, D.H., Thomas, M.J., Smith, D.H. and McIntosh, T.K., 1992, Selective vulnerability of dentate hilar neurons following traumatic brain injury: a potential mechanistic link between head trauma and disorders of the hippocampus. *J. Neurosci.* **12**: 4846.

Margerison, J.H. and Corsellis, J.A., 1966, Epilepsy and the temporal lobes. A clinical, electroencephalographic and neuropathological study of the brain in epilepsy, with particular reference to the temporal lobes. *Brain* **89**: 499.

Mathern, G.W., Babb, T.L., Pretorius, J.K. and Leite, J.P., 1995, Reactive synaptogenesis and neuron densities for neuropeptide Y, somatostatin, and glutamate decarboxylase immunoreactivity in the epileptogenic human fascia dentata. *J. Neurosci.* **15**: 3990.

Meldrum, B.S., 1993, Excitotoxicity and selective neuronal loss in epilepsy. *Brain Pathol.* **3**: 405.

Mellanby, J., Hawkins, C., Mellanby, H., Rawlins, J.N. and Impey, M.E., 1984, Tetanus toxin as a tool for studying epilepsy. *J. Physiol. (Paris)* **79**: 207.
Mikkonen, M., Soininen, H., Kalvianen, R., Tapiola, T., Ylinen, A., Vapalahti, M., Paljarvi, L. and Pitkanen, A., 1998, Remodeling of neuronal circuitries in human temporal lobe epilepsy: increased expression of highly polysialylated neural cell adhesion molecule in the hippocampus and the entorhinal cortex. *Ann. Neurol.* **44**: 923.
Mitchell, J., Sundstrom, L.E. and Wheal, H.V., 1993, Microglial and astrocytic cell responses in the rat hippocampus after an intracerebroventricular kainic acid injection. *Exp. Neurol.* **121**: 224.
Mody, I. and MacDonald, J.F., 1995, NMDA receptor-dependent excitotoxicity: the role of intracellular Ca2+ release. *Trends Pharmacol. Sci.* **16**: 356.
Morgan, J.I. and Curran, T., 1991, Proto-oncogene transcription factors and epilepsy. *Trends Pharmacol. Sci.* **12**: 343.
Nagata, S., 1997, Apoptosis by death factor. *Cell* **88**: 355.
Nakic, M., Mitrovic, N., Sperk, G. and Schachner, M., 1996, Kainic acid activates transient expression of tenascin-C in the adult rat hippocampus. *J. Neurosci. Res.* **44**: 355.
Nissinen, J., Halonen, T., Koivisto, E. and Pitkanen, A., 2000, A new model of chronic temporal lobe epilepsy induced by electrical stimulation of the amygdala in rat. *Epilepsy Res.* **38**: 177.
Nitsch, C. and Hubauer, H., 1986, Distant blood–brain barrier opening in subfields of the rat hippocampus after intrastriatal injections of kainic acid but not ibotenic acid. *Neurosci. Lett.* **64**: 53.
Orrenius, S., Zhivotovsky, B. and Nicotera, P., 2003, Regulation of cell death: the calcium-apoptosis link. *Nat. Rev. Mol. Cell. Biol.* **4**: 552.
Pellegrini-Giampietro, D.E., Gorter, J.A., Bennett, M.V. and Zukin, R.S., 1997, The GluR2 (GluR-B) hypothesis: Ca(2+)-permeable AMPA receptors in neurological disorders. *Trends Neurosci.* **20**: 464.
Perez, Y., Morin, F., Beaulieu, C. and Lacaille, J.C., 1996, Axonal sprouting of CA1 pyramidal cells in hyperexcitable hippocampal slices of kainate-treated rats. *Eur. J. Neurosci.* **8**: 736.
Pirker, S., Czech, T., Baumgartner, C., Maier, H., Novak, K., Furtinger, S., Fischer-Colbrie, R. and Sperk, G., 2001, Chromogranins as markers of altered hippocampal circuitry in temporal lobe epilepsy. *Ann. Neurol.* **50**: 216.
Pollard, H., Charriaut-Marlangue, C., Cantagrel, S., Represa, A., Robain, O., Moreau, J. and Ben-Ari, Y., 1994, Kainate-induced apoptotic cell death in hippocampal neurons. *Neuroscience* **63**: 7.
Polster, B.M. and Fiskum, G., 2004, Mitochondrial mechanisms of neural cell apoptosis. *J. Neurochem.* **90**: 1281.
Represa, A., Niquet, J., Charriaut-Marlangue, C. and Ben-Ari, Y., 1993, Reactive astrocytes in the kainic acid-damage hippocampus have the phenotypic features of type-2 astrocytes. *J. Neurocytol.* **22**: 299.
Scharfman, H.E., 1991, Dentate hilar cells with dendrites in the molecular layer have lower thresholds for synaptic activation by perforant path than granule cells. *J. Neurosci.* **11**: 1660.
Seifert, G., Huttmann, K., Schramm, J. and Steinhauser, C., 2004, Enhanced relative expression of glutamate receptor 1 flip AMPA receptor subunits in hippocampal astrocytes of epilepsy patients with Ammon's horn sclerosis. *J. Neurosci.* **24**: 1996.
Seki, T. and Rutishauser, U., 1998, Removal of polysialic acid-neural cell adhesion molecule induces aberrant mossy fiber innervation and ectopic synaptogenesis in the hippocampus. *J. Neurosci.* **18**: 3757.
Siesjo, B.K. and Wieloch, T., 1986, Epileptic brain damage: pathophysiology and neurochemical pathology. *Adv. Neurol.* **44**: 813.
Simon, R.P., Schmidley, J.W., Meldrum, B.S., Swan, J.H. and Chapman, A.G., 1986, Excitotoxic mechanisms in hypoglycaemic hippocampal injury. *Neuropathol. Appl. Neurobiol.* **12**: 567.
Skulachev, V.P., 1998, Cytochrome c in the apoptotic and antioxidant cascades. *FEBS Lett.* **423**: 275.
Sloviter, R.S., 1983, "Epileptic" brain damage in rats induced by sustained electrical stimulation of the perforant path. I. Acute electrophysiological and light microscopic studies. *Brain Res. Bull.* **10**: 675.
Sloviter, R.S., 1991, Permanently altered hippocampal structure, excitability, and inhibition after experimental status epilepticus in the rat: the "dormant basket cell" hypothesis and its possible relevance to temporal lobe epilepsy. *Hippocampus* **1**: 41.
Sloviter, R.S. and Dempster, D.W., 1985, "Epileptic" brain damage is replicated qualitatively in the rat hippocampus by central injection of glutamate or aspartate but not by GABA or acetylcholine. *Brain Res. Bull.* **15**: 39.
Sloviter, R.S., Zappone, C.A., Harvey, B.D., Bumanglag, A.V., Bender, R.A. and Frotscher, M., 2003, "Dormant basket cell" hypothesis revisited: relative vulnerabilities of dentate gyrus mossy cells and inhibitory interneurons after hippocampal status epilepticus in the rat. *J. Comp. Neurol.* **459**: 44.
Sommer, W., 1880, Erkrankung des Ammonshornes als ätiologisches Moment der Epilepsie. *Arch. Psychiatr. Nervenkrankh.* **10**: 631.
Sperk, G., 1994, Kainic acid seizures in the rat. *Prog. Neurobiol.* **42**: 1.
Sperk, G., Lassmann, H., Baran, H., Kish, S.J., Seitelberger, F. and Hornykiewicz, O., 1983, Kainic acid induced seizures: Neurochemical and histopathological changes. *Neuroscience* **10**: 1301.

Sperk, G., Marksteiner, J., Gruber, B., Bellmann, R., Mahata, M. and Ortler, M. 1992, Functional changes in neuropeptide Y- and somatostatin-containing neurons induced by limbic seizures in the rat. *Neuroscience* **50**: 831.
Sperk, G., Furtinger, S., Schwarzer, C. and Pirker, S., 2004, GABA and its receptors in epilepsy. *Adv. Exp. Med. Biol.* **548**: 92.
Spielmeyer, W., 1927, Die Pathogenese des epileptischen Krampfes. *Z. ges Neurol. Psychiatr.* **109**: 501.
Sutula, T., Xiao-Xian, H., Cavazos, J. and Scott, G., 1988, Synaptic reorganization in the hippocampus induced by abnormal functional activity. *Science* **239**: 1147.
Sutula, T., Cascino, G., Cavazos, J. Parada, I. and Ramirez, L., 1989, Mossy fiber synaptic reorganization in the epileptic human temporal lobe. *Ann. Neurol.* **26**: 321.
Tanaka, T., Tanaka, S., Fujita, T., Takano, K., Fukuda, H., Sako, K. and Yonemasu, Y., 1992, Experimental complex partial seizures induced by a microinjection of kainic acid into limbic structures. *Prog. Neurobiol.* **38**: 317.
Teitelbaum, J.S., Zatorre, R.J., Carpenter, S., Gendron, D., Evans, A.C., Gjedde, A., and Cashman, N.R., 1990, Neurologic sequelae of domoic acid intoxication due to the ingestion of contaminated mussels. *N. Engl. J. Med.* **322**: 1781.
Treiman, D.M., Meyers, P.D., Walton, N.Y., Collins, J.F., Colling, C., Rowan, A.J., Handforth, A., Faught, E., Calabrese, V.P., Uthman, B.M., Ramsay, R.E. and Mamdani, M.B., 1998, A comparison of four treatments for generalized convulsive status epilepticus. Veterans Affairs Status Epilepticus Cooperative Study Group. *N. Engl. J. Med.* **339**: 792.
Turski, L., Ikonomidou, C., Turski, W.A., Bortolotto, Z.A. and Cavalheiro, E.A., 1989, Review: Cholinergic mechanisms and epileptogenesis. The seizures induced by pilocarpine: a novel experimental model of intractable epilepsy. *Synapse* **3**: 154.
Vezzani, A., Sperk, G. and Colmers, W.F., 1999, Neuropeptide Y: emerging evidence for a functional role in seizure modulation. *Trends Neurosci.* **22**: 25.
Wu, H.Q., Turski, W.A., Ungerstedt, U. and Schwarcz, R., 1991, Systemic kainic acid administration in rats: effects on kynurenic acid production in vitro and in vivo. *Exp. Neurol.* **113**: 47.

17

GLIA AND HIPPOCAMPAL NEUROGENESIS IN THE NORMAL, AGED AND EPILEPTIC BRAIN

William P. Gray and Alexandra Laskowski

1. ABSTRACT

In this chapter, we review the role of glial cells in hippocampal neurogenesis under normal conditions, in aging and in the epileptic brain. Astrocytes or astrocyte-like cells are emerging as key components of the neurogenic niche in health and disease, with roles ranging from being stem cells to regulating almost all aspects of neurogenesis and synaptic integration of the newly generated neurons. It is likely that astrocytes and microglial cells are key sensors of local environmental changes, modulating neurogenesis appropriately. This is likely to be a fruitful area of research for extending our understanding of the role of stem cells in the normal and diseased brain.

2. INTRODUCTION

Over the last 15 years, there has been an explosion of interest in stem-cell biology, particularly in the adult brain, since the realization of the importance of Joseph Altman's work demonstrating neurogenesis in the post-natal dentate gyrus of the hippocampus (Altman and Das, 1965). The hippocampus is involved in the learning and memory of explicit information (Eichenbaum, 1997) and, with the amygdalar formation, is important in the generation of behavioural responses to stress. Given the emerging roles of post-natal neurogenesis in hippocampal-dependent learning and behaviour, much research has concentrated on elucidating the function of hippocampal (or more properly granule cell layer) neurogenesis as well as determining factors that influence it (Gage, 2002). More recently, there is an emerging awareness of the effect of pathological processes on dentate neurogenesis and on possible roles of abnormal neurogenesis in the pathophysiology of disease.

Disease may alter the production and/or integration of newly born neurons within the dentate gyrus and hippocampus and therefore may be viewed as perturbating a physiological function such as mood (Jacobs et al., 2000) or hippocampal-dependent learning (Shors, 2004), or creating abnormal circuits that might subserve altered network activity, as in epilepsy (Parent, 2002; Scharfman and Gray, 2006). Central to these questions is the study of the

---

William P. Gray – w.p.gray@soton.ac.uk
Division of Clinical Neurosciences, University of Southampton, Room 6207, Level 6, Biomedical Sciences Building, Bassett Crescent East, Southampton SO16 7PX, UK

constituent cells that generate new neurons in the dentate gyrus. Chief among these cells is the radial glia-like astrocyte in the dentate gyrus and the subependymal astrocyte in the caudal subventricular zone. This chapter first explains the architecture of neurogenesis in the post-natal hippocampus and summarises the role of glial fibrillary acidic protein (GFAP)-expressing cells in this process. It then describes the putative roles of adult neurogenesis in the normal brain and then details the role of glia in neurogenesis under conditions of aging and pathology, mainly focusing on temporal lobe epilepsy.

3. PROGENITOR NICHES IN THE POST-NATAL HIPPOCAMPUS

3.1. Granule Cell Layer of the Dentate Gyrus

Unlike the rest of the hippocampus, which is generated from the hippocampal anlage at embryonic day 16 (E16), the granule cell layer of the dentate is largely generated post-natally. The initial granule cell layer is formed from the cells of the first dentate migration at E18, followed shortly after birth by the second migration to form the post-natal granule cell layer. Cells of the second dentate migration form the tertiary matrix in the hilus of the developing dentate by post-natal day 3 (P3), which generates ~50% of the granule cell layer after P5, and from where precursor cells migrate to the inner border of the granule cell layer to form the adult neurogenic subgranular zone (SGZ) by P20 (Fig. 1). Recently, a separate contribution to the formation of the SGZ and granule cell layer has been identified from the transient hippocampal subventricular zone, a secondary germinal matrix also derived from the dentate neuroepithelium (Navarro-Quiroga et al., 2006). The granule cell layer therefore develops as an inside-out structure with the newer cells being added successively along the inner margin initially from the tertiary germinal matrix and later from the SGZ (Schlessinger et al., 1975; Altman and Bayer, 1990) (see Fig. 1). This neurogenic niche is unique in that it is the site of ongoing neurogenesis throughout adult life in all mammals studied to date, including humans, (Eriksson et al., 1998) and it has been estimated to generate 6% of the total granule cell population every month (Cameron and McKay, 2001).

3.2. Caudal Subventricular Zone

Proliferating cells continue to be seen in the subventricular zone (SVZ), a remnant of the embryonic ventricular zones, which persists throughout life. In vivo, in rodents progenitors in the anterior SVZ give rise to chains of rostrally migrating cells called the rostral migratory stream (RMS) that eventually reach the olfactory bulb where they differentiate into GABAergic and dopaminergic interneurons (Lois and Alvarez-Buylla, 1993). The caudal SVZ of the lateral ventricle lines the superficial layer of the hippocampus over areas CA1 and CA3 and contains neural progenitors that develop into glia, giving rise to white matter astrocytes and oligodendrocytes in neonatal and juvenile rodents (Levison and Goldman, 1993; Levison et al., 1999).

3.3. GFAP-Expressing Cells are Progenitors in the DG and cSVZ

In the late nineties, controversy initially surrounded the issue of whether SVZ stem cells were ciliated ependymal cells (Johansson et al., 1999) or subependymal astrocytes

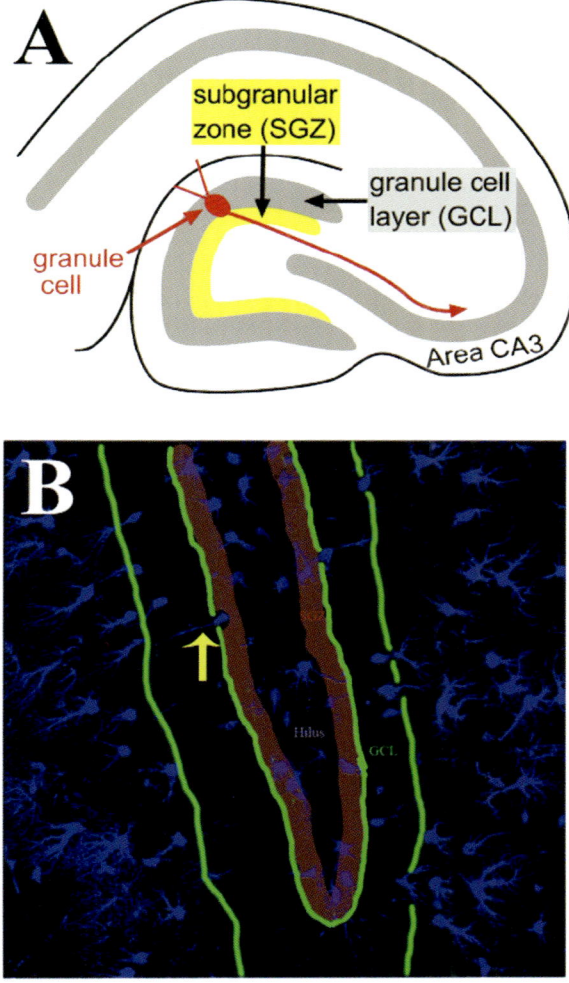

Fig. 1. (a) A cross section through the dorsal hippocampus showing the principal layers of the hippocampus and the neurogenic subgranular zone (SGZ). The cell body of a mature granule cell neuron is detailed within the granule cell layer with its dendritic tree arborising into the molecular layer and its axon extending out to area CA3 of the hippocampus. (b) A 1μM optical section taken with a confocal microscope through the dentate granule cell layer (green outline) and immunostained for the astrocytic marker S100β showing the orientation of S100β-positive cells in the subgranular zone (SGZ – RED). Note that while most S100β+ cells are orientated horizontally in the SGZ, some have a radial glia-like morphology and are oriented radially in the SGZ extending a single process into the granule cell layer (yellow arrow). GCL, granule cell layer

(GFAP-expressing cells). Using transient mitotic inhibition, Doetch et al. (1999) elegantly demonstrated a reconstitution of the SVZ neurogenic niche by the initial division of GFAP+ subependymal astrocytes (Type B cells), followed by GFAP-negative nestin-positive population of rapidly cycling cells (Type C cells) and lastly by proliferating neuroblasts (Type A cells). These A cells then continued proliferating whilst migrating towards the olfactory bulb surrounded by Type B and Type A cells in the RMS. Dopaminergic periglomerular neurons originate predominantly from precursors in the RMS, while granule interneurons arise from progenitors in the SVZ demonstrating specific regionalization of olfactory bulb interneuron precursors (Hack et al., 2005).

Within the neurogenic SGZ of the dentate gyrus, astrocytes with a radial glia-like morphology extending a single main process into the granule cell layer are primary progenitor cells (Type B cells) (Seri et al., 2004). Although initial reports suggested that in the adult SGZ Type B cells appear to give rise to D cells (committed neuroblasts) without the intervening transiently amplifying Type C cells, a recent study by Encinas and colleagues (Encinas et al., 2006) have confirmed a weakly nestin-positive/GFAP-negative, rapidly cycling precursor, which has also been demonstrated in adrenalectomy-induced neurogenesis (Battista et al., 2006). Transiently amplifying cells may exist in the early post-natal period when the granule cell layer is largely being formed (see below) (Namba et al., 2005). Seri et al. classified D cells into three subtypes based on morphology, D1 cells are mitotic and divide symmetrically to give rise to two post-mitotic D2 cells, only one of which usually survives (Dayer et al., 2003; Seri et al., 2004), to mature via a D3 stage into a granule cell neuron. B and D cells appear to form a functional unit with D1 cells being held in the SGL within a basket of processes arising from the cell body of the radial B cell, while D2 cells lie without the SGL but with an intimate relationship between its apical dendrite and the radial process of the B cell, which is thought to guide its migration into the molecular layer (Seki and Arai, 1999; Seri et al., 2004; Shapiro et al., 2005). This arrangement is consistent with the hypothesis that primary progenitors and their progeny interact bidirectionally to influence both progeny differentiation and maturation (Song et al., 2002a) as well as primary progenitor proliferation (Liu et al., 2005).

A proportion of radial glia-like astrocytes express the intermediate filament nestin, which is expressed in progenitor/stem cells (Lendahl et al., 1990), and experiments using transgenic mice expressing fluorescent protein on the neural nestin promoter show that this subpopulation of astrocytes is neurogenic (Filippov et al., 2003; Fukuda et al., 2003). On the basis of nestin-driven fluorescent protein expression and doublecortin immunostaining (a microtubule associated protein transiently expressed by neuroblasts and immature neurons (Francis et al., 1999)), Kempermann and colleagues have classified stem/progenitor cells in the SGZ as bipotent GFAP-expressing Type 1 cells that have astrocytic properties and GFAP-negative Type 2 and Type 3 lineage determined progenitor cells (for review see (Kempermann et al., 2004)). Type 1 cells show electrophysiological characteristics of astrocytes (Filippov et al., 2003; Fukuda et al., 2003), whereas Type 2 cells expressed voltage-dependent sodium currents and the early neuronal marker polysialylated NCAM and appeared to be proliferative progeny of Type 1 cells (Fukuda et al., 2003) consistent with the findings of Seri et al. (2004).

A subpopulation of proliferative astrocytes in the SGZ also express the calcium binding protein S100β (Cocchia, 1981; Hinterkeuser et al., 1999). Studies in CD1 and female C57/BL6 mice have shown that all astrocytes with a horizontal process running along the SGZ, rather than radial processes extending into the granule cell layer, express both GFAP and S100β but not nestin (Seri et al., 2004). Short survival BrdU experiments from these studies have shown that GFAP+/S100β+ cells do not appear to be in the S-phase of cell division and may be either mature astrocytes or quiescent stem cells, under normal conditions (Steiner et al., 2004). However, studies in transgenic FVB/N mice expressing eGFP on the human GFAP promoter have shown both radially and horizontally orientated S100β-positive cells in the SGZ (Huttmann et al., 2003) (Fig. 2), suggesting strain variation in S100β expression. Interestingly, in the early post-natal developing dentate gyrus in Wistar rats, the majority of proliferating cells express S100β, and lineage tracing shows neuronal as well as GFAP+ radial glial-like and star-shaped astrocyte phenotypes in the SGZ (Namba et al., 2005) in keeping with a stem-cell function of S100β+ cells. SGZ S100β expression under conditions of brain injury is detailed below. Clearly, more investigation of this interesting cell phenotype is warranted.

Elegant in vivo experiments using transgenic fate mapping and ablation of proliferating GFAP-expressing cells (Garcia et al., 2004b) have confirmed that uni- or bipolar GFAP-expressing progenitors are the principal source of constitutive neurogenesis in the adult mouse forebrain. However, these studies also demonstrated that developmental neurogenesis in the forebrain and hippocampus does not arise from GFAP-expressing cells. This is supported by the finding that the predominant neural progenitor isolated from the mouse forebrain does not express GFAP in early development, but does so in late development and adulthood (Imura et al., 2003). This is consistent with the late expression of GFAP by ventricular zone radial glia (Schmechel and Rakic, 1979; Levitt and Rakic, 1980), a subpopulation of which do not differentiate into astrocytes, but retain multipotent potential in the favourable environments of the SVZ and possibly after migration to the SGZ in late development (Rickmann et al., 1987).

The designation of all GFAP-expressing cells as astrocytes must be treated with caution, as GFAP-positive cells are found in the liver, gut, kidney and lung (Neubauer et al., 1996; Buniatian et al., 1998; Bush et al., 1998). Indeed, GFAP-expressing adult progenitor cells also express the intermediate filaments *nestin* and *vimentin*, which are not found in mature protoplasmic astrocytes, and differentiated astrocytes harvested from adult tissue do not generate neurons in vitro (Laywell et al., 2000).

3.4. Microglia and the Neurogenic Niche

Recently, CNS inflammation has been shown to decrease neurogenesis in the SGZ (Ekdahl et al., 2003; Monje et al., 2003) as has the pro-inflammatory cytokine IL-6 (Vallieres et al., 2002). Immune cytokines are synthesized under physiological conditions and TGF-β has a key neurogenic effect in vivo and in vitro (Vallieres et al., 2002), the levels of which correlate significantly with the number of activated microglia. Autoimmune T cells program microglia to make their phenotype supportive of neuronal survival (Butovsky et al., 2005) and renewal (Butovsky et al., 2006) and contribute to the maintenance of neurogenesis and spatial learning in adulthood (Ziv et al., 2006). The emerging role of the immune system, beyond the role of microglia, in modulating CNS maintenance and repair, is outside the scope of this review but the interested reader is referred to recent reviews by Schwartz and colleagues (Kipnis and Schwartz, 2005; Schwartz et al., 2006).

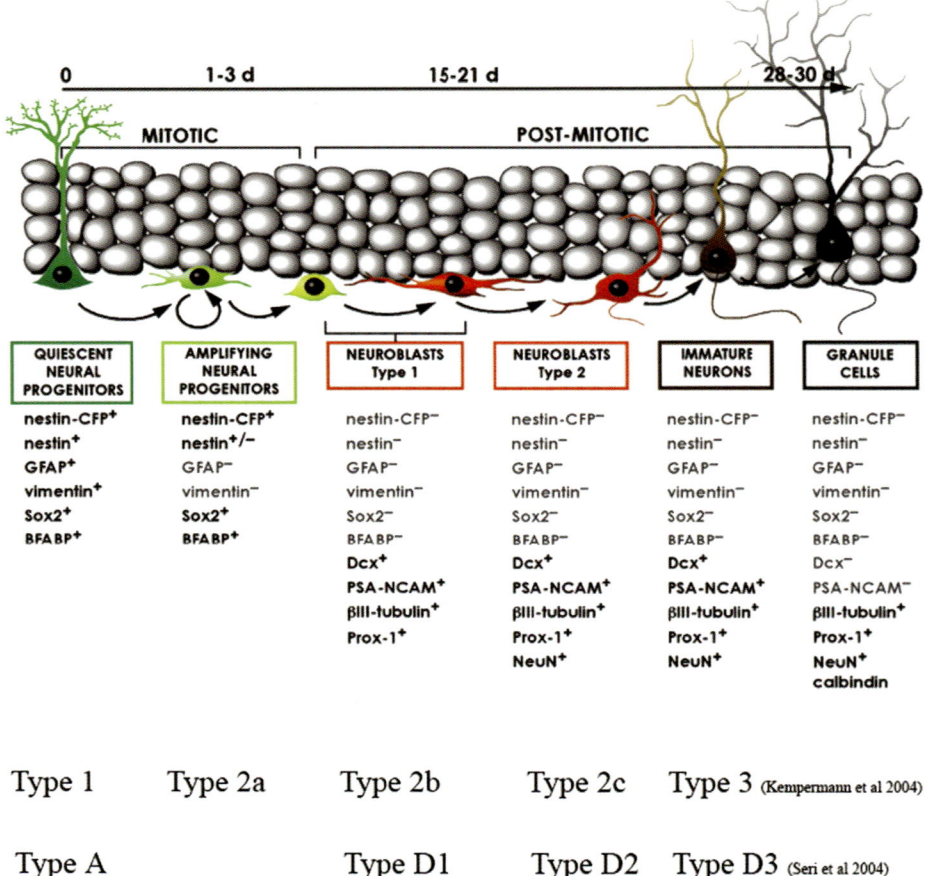

Fig. 2. Different progenitor cells in the SGZ and the phases of maturation in the subgranular zone (SGZ) together with the markers expressed and the timescale of their differentiation. An attempted translation between this and other classifications is also shown. Modified and reproduced with permission from Encinas et al. (2006). Copyright 2006 National Academy of Sciences, USA

4. HIPPOCAMPAL NEUROGENESIS

Neurogenesis in the hippocampus is found predominantly in the dentate gyrus (Altman and Das, 1965), although there is also evidence of ongoing hippocampal neurogenesis in the cornu ammonis, albeit at a much slower rate (Rietze et al., 2000). The production of new neurons in the SGZ exceeds the number that survive and integrate into hippocampal circuitry

(Dayer et al., 2003), net production depending on the number and proliferation rate of SGZ precursors as well as the survival of the newly generated neurons (Prickaerts et al., 2004). Doublecortin immunoreactivity reliably estimates immature neurons in the first 12 days after neuronal birth, while NeuN staining identifies a more mature subset that survives and integrates into hippocampal circuitry (Rao and Shetty, 2004). Dentate neurogenesis is highest in the post-natal period and decreases exponentially throughout life in rodents (Seki and Arai, 1995; Kuhn et al., 1996) and also probably in humans (Eriksson et al., 1998; Blumcke et al., 2001). The reasons for this age-related decline are unclear, but appear to be related to suppression of precursor proliferation by increasing levels of circulating adrenal corticosteroids with age (Cameron and McKay, 1999) vide infra.

The total number of granule cell neurons (Bayer et al., 1982) and the distance from the hilus of cells labelled by injection of 3H-thymidine at P10 increase throughout life (Crespo et al., 1986), in keeping with the successive addition of new granule cells from the SGZ, and therefore implying a functional role for these adult-generated neurons. The hippocampus has been repeatedly implicated in the modulation of cognition in association with learning and memory, and the demonstration of functional integration of these newly born dentate granule cells into hippocampal circuitry (Scharfman et al., 2000; van Praag et al., 2002) has lead to the hypothesis that neurogenesis may be important in learning and memory formation (Gould et al., 1999; Shors et al., 2001; Snyder et al., 2001). Hippocampal-dependent associative learning tasks increase dentate neurogenesis (Gould et al., 1999) and pharmacological reduction of newly born neurons in the dentate impairs certain types of hippocampal-dependent learning such as trace conditioning (Shors et al., 2001). Moreover, prenatal stress produces learning deficits associated with inhibition of dentate neurogenesis (Lemaire et al., 2000) and the level of cell proliferation in the SGZ predicts the spatial memory performance of rats in the water maze (Drapeau et al., 2003). Exposure to a novel or enriched environment increases neurogenesis largely by increasing the proportion of cells that survive and integrate and also by increasing the population of proliferating SGZ cells, but in a strain-dependent manner (Kempermann et al., 1997) (for a critical review see Prickaerts et al. (2004)).

In addition, several avenues of research have suggested that decreased hippocampal neurogenesis may be important in the pathogenesis of depression (D'Sa and Duman, 2002). Stress, which is often a requisite for depression, reduces hippocampal neurogenesis (Gould et al., 1997), tri-cyclic antidepressants increase hippocampal neurogenesis (Malberg et al., 2000) and the antidepressant-induced increase in dentate neurogenesis appear to be necessary for their behavioural effects (Santarelli et al., 2003). As newly born neurons functionally integrate into hippocampal circuitry 2–3 weeks after birth (Hastings and Gould, 1999; Markakis and Gage, 1999), this may explain the puzzling 2–3 week delay in mood elevation after commencing antidepressants.

Our knowledge of the control mechanisms underlying hippocampal neurogenesis remains far from complete despite the substantial literature generated by current research, which is identifying an ever increasing number of factors that influence it. For reasons of space, a comprehensive review cannot be given here and instead some salient factors pertinent to GFAP-positive primitive precursors and disease are highlighted, while the interested reader is directed to a recent comprehensive review (Abrous et al., 2005).

Glucocorticoids depress cell proliferation and neurogenesis in the SGZ, whereas adrenalectomy has the opposite effect (Cameron and Gould, 1994). In contrast, neurosteroids such as DHEA increase SGZ cell proliferation (Karishma and Herbert, 2002). Glutamatergic afferent input via the perforant path negatively regulates dentate neurogenesis (Cameron et al., 1995) and acts downstream of steroid control (Cameron et al., 1998), while both serotonin and NPY increase SGZ cell proliferation (Brezun and Daszuta, 2000; Howell et al., 2003, 2005). Circulating growth factors such as IGF-1 also increase hippocampal neurogenesis (O'Kusky et al., 2000) and may partly mediate the neurogenic effect of prolonged exercise, such as running, on dentate neurogenesis. Cytokines can either promote or reduce neurogenesis (vide supra), as can a wide variety of hormones, growth factors and other neurotransmitters.

Stem-cell maintenance and self-renewal are coordinated by Notch and mitogen signaling (Gaiano and Fishell, 2002) and once maintained, quiescent stem cells may require mitogen signaling to engage in self-renewal. Granule cell layer astrocytes show FGF (Shetty et al., 2005) ligands, and cystatin C, a fibroblastic growth factor 2 (FGF-2) cofactor expressed by adult stem cells and astrocytes, modulates neurogenesis (Muotri et al., 2005). Astroglial derived Wnt signaling is a key pathway in the promotion of SGZ neurogenesis (Lie et al., 2005). In contrast, signaling from the bone morphogenetic protein (BMP) family instructs progenitor cells towards a glial fate (Lim et al., 2000). GFAP-positive Type 1 cells also secrete Neurogenesin-1 (Ng-1), a factor that shares similarity with chordin, and which antagonizes BMP-4, altering the fate commitment of progenitor cells from gliogenesis to neurogenesis (Ueki et al., 2003). Ng-1 is also secreted, albeit at lower levels, from granule cell neurons around Type 1 cells, indicating an interaction between progenitor cells and their differentiated progeny. Astrocytes also play an important role in the regulation of neuronal maturation and synaptic integration and function (Ullian et al., 2001). Co-culture experiments have shown that hippocampal astrocytes promote functional maturation and synaptic integration of neural progeny derived from adult hippocampal neural stem cells (Song et al., 2002a, b) and activity-dependent neurotrophic factor secreted by VIP-stimulated astrocytes promotes functional maturation and synaptogenesis of embryonic hippocampal neurons in culture (Blondel et al., 2000). For a more comprehensive review of glial influences on neural stem cells, the reader is directed to a review by Song (Ma et al., 2005).

5. GLIA, NEUROGENESIS AND AGING

As introduced above, dentate neurogenesis exhibits an exponential age-related decline throughout life in rodents (Seki and Arai, 1995; Kuhn et al., 1996) and probably in humans. Although attractive as a hypothesis, a direct relationship between reduced neurogenesis and impairments in learning and memory in old age is contentious (Bizon and Gallagher, 2003; Merrill et al., 2003). However, aged rats with preserved spatial memory show greater levels of neurogenesis than spatial memory impaired rats (Drapeau et al., 2003), supporting a mechanistic relationship. Twelve-month-old (aged) rats show a 94% reduction in neurogenesis compared to six-week-old animals due almost entirely to decreased cell production (McDonald and Wojtowicz, 2005). Older animals are significantly more vulnerable to the adverse effect of stress on SGZ cell proliferation (Simon et al., 2005) and newly born neurons in the middle-aged dentate and onwards also show slower maturation (Rao et al., 2005).

The mechanism of this age-related decrease in precursor proliferation and neurogenesis is unknown. Interestingly, there is a constitutive decline in the levels of intrinsic mitogenic factors in the hippocampus with increasing age (Shetty et al., 2005) and many of these factors are released by hippocampal astrocytes. Intriguingly, the number of GFAP-positive cells in the SGZ does not decrease with increasing age, tentatively suggesting that reduced proliferation may not be due to loss of primary progenitors. However, the proportion of GFAP cells in the SGZ expressing FGF-2, a potent factor for dentate neurogenesis that is largely produced by hippocampal astrocytes, is significantly reduced with increasing age (Shetty et al., 2005). Significantly for a causal hypothesis, FGF-2 also reverses age-declined neurogenesis (Jin et al., 2003).

Steroid hormones are potent regulators of adult neurogenesis and appear to play a prominent role in age-related neurogenic decline, as circulating steroid levels increase with age and complete reversal of the age-related reduction in SGZ proliferation is achieved by adrenalectomy (Cameron and McKay, 1999). Fifty percent of radial glia-like Type 1 and Type 2a cells express immunohistochemical evidence of glucocorticoid receptors in young mice, while post-mitotic precursor cells in the SGZ do not (Garcia et al., 2004a). Significantly, age increases the expression of glucocorticoid receptors on Type 1 and 2a cells, which may contribute to their age-related reduction in proliferation. The causes of age-related reduction in FGF-2 production and increase in glucocorticoid receptor expression are unknown but are likely to yield to further experimental study in this important area.

6. GLIA AND NEUROGENESIS IN THE DISEASED BRAIN

Putative roles of glia in disease with respect to hippocampal neurogenesis may be considered from two perspectives, that of effects of pathology on GFAP-positive primary progenitors both in the dentate SGZ and caudal SVZ and the effects of glial cells on the neurogenic niche.

6.1. GFAP-Positive Primary Progenitors in Disease

Because the study of adult neurogenesis is a recent endeavour, which has initially concentrated on describing the phenomenon and delineating factors that influence it, determining the contribution of its alteration to disease has only recently begun to be investigated after the realization of its importance in hippocampal-dependent learning and the behavioural response to antidepressants.

6.2. Seizures and Mesial Temporal Lobe Epilepsy

The majority of studies support the concept that increased neuronal activity increases neurogenesis in the dentate gyrus. Therefore, it is not surprising that seizures also increase neurogenesis. One of the first studies of this kind examined effects of a single after-discharge, which was sufficient to increase SGZ cell proliferation (Bengzon et al., 1997). Subsequent studies illustrated that *status epilepticus* (SE) produced by many different methods could increase SGZ cell proliferation and neurogenesis, e.g. after administration of the chemoconvulsant pilocarpine i.p. (Parent et al., 1997) or bilaterally increased proliferation and neurogenesis after unilateral intracerebroventricular kainic acid (Gray and Sundstrom, 1998). The time course of proliferation in most studies shows an increase above baseline 2–3 days after SE (Parent et al., 1997; Nakagawa et al., 2000), although one study has shown an earlier increase (Radley and Jacobs, 2003). The mechanism of the increased cell proliferation is largely unknown, although 5HT1A (Radley and Jacobs, 2003; Zucchini et al., 2005) and Galanin Type 2 receptors (Mazarati et al., 2004) appear important, and Neuropeptide Y may also have a role (Scharfman and Gray, 2006) via its Y1 receptor (Howell et al., 2005). FGF-2 partially mediates progenitor proliferation after seizures (Yoshimura et al., 2001) and seizure-induced injury may precede the effect (Sadgrove et al., 2005) and modulate it via neurotrophins

(Hagihara et al., 2005). However, cell death may not be necessary for a proliferative progenitor response (Smith et al., 2005). Sonic hedgehog (Shh) signaling is important for seizure-induced progenitor proliferation (Banerjee et al., 2005). Interestingly, the cysteine protease cystatin C, which is expressed on astrocytes and microglia after SE may have a role in baseline and seizure-induced neurogenesis (Pirttila et al., 2005).

As Type 1 and Type 2a (or amplifying neural progenitors *vide supra* and Fig. 2) are proliferative, then seizure-induced neurogenesis may be achieved by increased proliferation rate or recruitment of one or both progenitor cell types. Pertinent to the role of glial progenitor cells in seizure-induced proliferation is the observation that NMDA receptor antagonism induces a long lasting increase in the number of proliferating cells, and newly born neurons in the dentate gyrus that is associated with a persistent increase in the number of radial glia-like GFAP-positive Type 1 cells (Nacher et al., 2001). To examine the mechanisms underlying seizure-induced proliferation, which begins on day three after SE, Huttman et al. examined the proliferation of Type 1 cells 72 h after kainate-induced SE, using transgenic FVB/n mice expressing eGFP on the human GFAP promoter (Huttman et al., 2003). They found an increased number of SGZ Type 1 cells, a greater proportion of which were proliferating, consistent with an early proliferative response of these cells after seizures. Interestingly, they also found that the greatest proportional change was in Type 1 cells expressing both GFAP and S100β, reminiscent of the developing dentate when most proliferating precursor cells in the hilar matrix also express S100β (Namba et al., 2005). An increase in the number of proliferating astrocytes with a radial process extending into the granule cell layer has also been reported 7 days after kainate-induced SE using the antibody expression pattern for ribonucleotide reductase, an endogenously expressed cytoplasmic marker of cell proliferation (Zhu et al., 2005). Interestingly, they also found that the number of proliferating clusters in the SGZ increased after seizures but that the ratio of GFAP+ cells to GFAP− cells in each cluster was unchanged, suggesting that more Type 1 cells were recruited into the cell cycle after seizures.

Jessberger et al. (2005) examined doublecortin-positive cells after kainate-induced SE and found that SE stimulated the division of late Type 3 progenitor cells. They did not find evidence of increased numbers of Type 1 cells using a nestin transgenic mouse but their time frame for examining cell phenotype was later, at 9 days post SE by which time many of the increased number of Type 1 cells may have become Type 3 cells after terminal symmetric division to yield neurons. There is therefore evidence that seizures can influence the proliferation of both Type 1 and Type 2 cells. Interestingly, there is also evidence that seizures accelerate the functional integration of newly born neurons (Overstreet-Wadiche et al., 2006). There is also evidence that specific subclasses of progenitors can respond selectively to exogenous stimuli with the recent finding that amplifying precursor or Type 2a cells uniquely respond to the antidepressant fluoxetine while Type 1 cells do not (Encinas et al., 2006).

Jessberger et al. (2005) also found that doublecortin-positive cells were not taking up their normal positions in the granule cell layer but were "dispersed" in an appearance consistent with "granule cell dispersion", a morphological change in the dentate gyrus of patients who underwent surgery for intractable epilepsy, originally reported by Houser (1990). Ectopic newly born granule cell neurons in the hilus have been observed after SE (Scharfman et al., 2000) and may contribute to epileptogenesis, a topic outside our remit here, but for which the reader is referred to a review by Helen Scharfman (Scharfman and Gray, 2006). While dispersion of the granule cell layer has been hypothesized to be due to an initial excessive production of new neurons with later pruning (Thom et al., 2002), there is no evidence to support this. Indeed, reelin deficiency and displacement of mature neurons, rather than neurogenesis, underlies granule cell dispersion in the epileptic hippocampus, as elegantly demonstrated by Heinrich et al. (2006). Interestingly, neurogenesis in the chronically dispersed dentate gyrus has been reported to be significantly reduced (Kralic et al., 2005), even after the onset of recurrent spontaneous seizures, and this reduction appears to occur early on with a shift towards a non-neuronal phenotype (Heinrich et al., 2006). The cause of this is unknown, but may reflect disruption of the neurogenic niche, and may partly explain the reduction in neurogenesis seen in chronic models of temporal lobe epilepsy (Hattiangady et al., 2004) and in patients with long standing epilepsy (Mathern et al., 2002; Crespel et al., 2005). Given the role of neurogenesis in hippocampal-dependent learning and memory, it is tempting to associate chronically reduced neurogenesis with cognitive dysfunction, but at the moment this hypothesis remains entirely speculative.

Stem cells in the caudal SVZ also respond to SE but here there is specific recruitment of glial progenitors, which migrate into the damaged hippocampus to generate astrocytes rather than neurons (Parent et al., 2006). This is in contrast to neuroblasts in the anterior SVZ where SE increases their proliferation (Parent et al., 2002). Whether this difference is a result of stimulation of different lineage committed progenitors or whether environmental cues direct fate down different lineages remains to be determined.

6.3. Effect of Glial Cells on the Neurogenic Niche in Epilepsy

While inflammation is detrimental to neurogenesis in the adult SGZ (Ekdahl et al., 2003) and blockade of inflammation restores neurogenesis (Monje et al., 2003), the role of neuroinflammation in seizure-induced neurogenesis is only beginning to be investigated (Bernardino et al., 2005). As alluded to above, the role of microglia in SGZ neurogenesis is complex, depending on the different pro- and anti-inflammatory cytokines released (Battista et al., 2006). The pro-inflammatory cytokine Leukaemia-inhibitory factor (LIF) appears to be an important modulator of seizure-induced injury (Holmberg and Patterson, 2006). Cystatin C modulates seizure-induced neurogenesis and migration (Pirttila et al., 2005) and is localized to microglia and astrocytes after SE. Interestingly, although the survival of seizure-induced neurons is initially compromised by the acute inflammatory response induced by electrically evoked SE, a substantial proportion survive despite chronic inflammation (Bonde et al., 2006) and it remains to be seen if chronic inflammation is detrimental to long-term neurogenesis as has been suggested from studies using the kainate model (Hattiangady et al., 2004).

7. ACKNOWLEDGEMENT

This work was supported by the Medical Research Council, Grant G0300356.

8. REFERENCES

Abrous, D.N., Koehl, M. and Le Moal, M., 2005, Adult neurogenesis: from precursors to network and physiology. *Physiol. Rev.* **85**: 523.

Altman, J. and Bayer, S.A., 1990, Migration and distribution of two populations of hippocampal granule cell precursors during the perinatal and postnatal periods. *J. Comp. Neurol.* **301**: 365.

Altman, J. and Das, G.D., 1965, Autoradiographic and histological evidence of postnatal hippocampal neurogenesis in rats. *J. Comp. Neurol.* **124**: 319.

Banerjee, S.B., Rajendran, R., Dias, B.G., Ladiwala, U., Tole, S. and Vaidya, V.A., 2005, Recruitment of the Sonic hedgehog signaling cascade in electroconvulsive seizure-mediated regulation of adult rat hippocampal neurogenesis. *Eur. J. Neurosci.* **22**: 1570.

Battista, D., Ferrari, C.C., Gage, F.H. and Pitossi, F.J., 2006, Neurogenic niche modulation by activated microglia: transforming growth factor beta increases neurogenesis in the adult dentate gyrus. *Eur. J. Neurosci.* **23**: 83.

Bayer, S.A., Yackel, J.W. and Puri, P.S., 1982, Neurons in the rat dentate gyrus granular layer substantially increase during juvenile and adult life. *Science* **216**: 890.

Bengzon, J., Kokaia, Z., Elmer, E., Nanobashvili, A., Kokaia, M. and Lindvall, O., 1997, Apoptosis and proliferation of dentate gyrus neurons after single and intermittent limbic seizures. *Proc. Natl Acad. Sci. USA* **94**: 10432.

Bernardino, L., Ferreira, R., Cristóvão, A.J., Sales, F. and Malva, J.O., 2005, Inflammation and neurogenesis in temporal lobe epilepsy. *Curr. Drug Targets CNS Neurol. Disord.* **4**: 349.

Bizon, J.L. and Gallagher, M., 2003, Production of new cells in the rat dentate gyrus over the lifespan: relation to cognitive decline. *Eur. J. Neurosci.* **18**: 215.

Blondel, O., Collin, C., McCarran, W.J., Zhu, S., Zamostiano, R., Gozes, I., Brenneman, D.E. and McKay, R.D., 2000, A glia-derived signal regulating neuronal differentiation. *J. Neurosci.* **20**: 8012.

Blumcke, I., Schewe, J.C., Normann, S., Brustle, O., Schramm, J., Elger, C.E. and Wiestler, O.D., 2001, Increase of nestin-immunoreactive neural precursor cells in the dentate gyrus of pediatric patients with early-onset temporal lobe epilepsy. *Hippocampus* **11**: 311.

Bonde, S., Ekdahl, C.T. and Lindvall, O., 2006, Long-term neuronal replacement in adult rat hippocampus after status epilepticus despite chronic inflammation. *Eur. J. Neurosci.* **23**: 965.

Brezun, J.M. and Daszuta, A., 2000, Serotonin may stimulate granule cell proliferation in the adult hippocampus, as observed in rats grafted with foetal raphe neurons. *Eur. J. Neurosci.* **12**: 391.

Buniatian, G., Traub, P., Albinus, M., Beckers, G., Buchmann, A., Gebhardt, R. and Osswald, H., 1998, The immunoreactivity of glial fibrillary acidic protein in mesangial cells and podocytes of the glomeruli of rat kidney in vivo and in culture. *Biol. Cell* **90**: 53.

Bush, T.G., Savidge, T.C., Freeman, T.C., Cox, H.J., Campbell, E.A., Mucke, L., Johnson, M.H. and Sofroniew, M.V., 1998, Fulminant jejuno-ileitis following ablation of enteric glia in adult transgenic mice. *Cell* **93**: 189.

Butovsky, O., Talpalar, A.E., Ben-Yaakov, K. and Schwartz, M., 2005, Activation of microglia by aggregated beta-amyloid or lipopolysaccharide impairs MHC-II expression and renders them cytotoxic whereas IFN-gamma and IL-4 render them protective. *Mol. Cell Neurosci.* **29**: 381.

Butovsky, O., Ziv, Y., Schwartz, A., Landa, G., Talpalar, A.E., Pluchino, S., Martino, G. and Schwartz, M., 2006, Microglia activated by IL-4 or IFN-gamma differentially induce neurogenesis and oligodendrogenesis from adult stem/progenitor cells. *Mol. Cell Neurosci.* **31**: 149.

Cameron, H.A. and Gould, E., 1994, Adult neurogenesis is regulated by adrenal steroids in the dentate gyrus. *Neuroscience* **61**: 203.

Cameron, H.A. and McKay, R.D., 1999, Restoring production of hippocampal neurons in old age. *Nat. Neurosci.* **2**: 894.

Cameron, H.A. and McKay, R.D., 2001, Adult neurogenesis produces a large pool of new granule cells in the dentate gyrus. *J. Comp. Neurol.* **435**: 406.

Cameron, H.A., McEwen, B.S. and Gould, E., 1995, Regulation of adult neurogenesis by excitatory input and NMDA receptor activation in the dentate gyrus. *J. Neurosci.* **15**: 4687.

Cameron, H.A., Tanapat, P. and Gould, E., 1998, Adrenal steroids and N-methyl-D-aspartate receptor activation regulate neurogenesis in the dentate gyrus of adult rats through a common pathway. *Neuroscience* **82**: 349.

Cocchia, D., 1981, Immunocytochemical localization of S-100 protein in the brain of adult rat. An ultrastructural study. *Cell Tissue Res.* **214**: 529.

Crespel, A., Rigau, V., Coubes, P., Rousset, M.C., de Bock, F., Okano, H., Baldy-Moulinier, M., Bockaert, J. and Lerner-Natoli, M., 2005, Increased number of neural progenitors in human temporal lobe epilepsy. *Neurobiol. Dis.* **19**: 436.

Crespo, D., Stanfield, B.B. and Cowan, W.M., 1986, Evidence that late-generated granule cells do not simply replace earlier formed neurons in the rat dentate gyrus. *Exp. Brain Res.* **62**: 541.

D'Sa, C. and Duman, R.S., 2002, Antidepressants and neuroplasticity. *Bipolar Disord.* **4**: 183.

Dayer, A.G., Ford, A.A., Cleaver, K.M., Yassaee, M. and Cameron, H.A., 2003, Short-term and long-term survival of new neurons in the rat dentate gyrus. *J. Comp. Neurol.* **460**: 563.

Doetsch, F., Caille, I., Lim, D.A., Garcia-Verdugo, J.M. and Alvarez-Buylla, A., 1999, Subventricular zone astrocytes are neural stem cells in the adult mammalian brain. *Cell* **97**: 703.

Drapeau, E., Mayo, W., Aurousseau, C., Le Moal, M., Piazza, P.V. and Abrous, D.N., 2003, Spatial memory performances of aged rats in the water maze predict levels of hippocampal neurogenesis. *Proc. Natl Acad. Sci. USA* **100**: 14385.

Eichenbaum, H., 1997, Declarative memory: insights from cognitive neurobiology. *Annu. Rev. Psychol.* **48**: 547.
Ekdahl, C.T., Claasen, J.H., Bonde, S., Kokaia, Z. and Lindvall, O., 2003, Inflammation is detrimental for neurogenesis in adult brain. *Proc. Natl Acad. Sci. USA* **100**: 13632.
Encinas, J.M., Vaahtokari, A. and Enikolopov, G. , 2006, Fluoxetine targets early progenitor cells in the adult brain. *Proc. Natl Acad. Sci. USA* **103**: 8233.
Eriksson, P.S., Perfilieva, E., Bjork-Eriksson, T., Alborn, A.M., Nordborg, C., Peterson, D.A. and Gage, F.H., 1998, Neurogenesis in the adult human hippocampus. *Nat. Med.* **4**: 1313.
Filippov, V., Kronenberg, G., Pivneva, T., Reuter, K., Steiner, B., Wang, L.P., Yamaguchi, M., Kettenmann, H. and Kempermann, G., 2003, Subpopulation of nestin-expressing progenitor cells in the adult murine hippocampus shows electrophysiological and morphological characteristics of astrocytes. *Mol. Cell. Neurosci.* **23**: 373.
Francis, F., Koulakoff, A., Boucher, D., Chafey, P., Schaar, B., Vinet, M.C., Friocourt, G., McDonnell, N., Reiner, O., Kahn, A., McConnell, S.K., Berwald-Netter, Y., Denoulet, P. and Chelly, J., 1999, Doublecortin is a developmentally regulated, microtubule-associated protein expressed in migrating and differentiating neurons. *Neuron* **23**: 247.
Fukuda, S., Kato, F., Tozuka, Y., Yamaguchi, M., Miyamoto, Y. and Hisatsune, T., 2003, Two distinct subpopulations of nestin-positive cells in adult mouse dentate gyrus. *J. Neurosci.* **23**: 9357.
Gage, F.H., 2002, Neurogenesis in the adult brain. *J. Neurosci.* **22**: 612.
Gaiano, N. and Fishell, G., 2002, The role of notch in promoting glial and neural stem cell fates. *Annu. Rev. Neurosci.* **25**: 471.
Garcia, A., Steiner, B., Kronenberg, G., Bick-Sander, A. and Kempermann, G., 2004a, Age-dependent expression of glucocorticoid- and mineralocorticoid receptors on neural precursor cell populations in the adult murine hippocampus. *Aging Cell* **3**: 363.
Garcia, A.D., Doan, N.B., Imura, T., Bush, T.G. and Sofroniew, M.V., 2004b, GFAP-expressing progenitors are the principal source of constitutive neurogenesis in adult mouse forebrain. *Nat. Neurosci.* **7**: 1233.
Gould, E., McEwen, B.S., Tanapat, P., Galea, L.A. and Fuchs, E., 1997, Neurogenesis in the dentate gyrus of the adult tree shrew is regulated by psychosocial stress and NMDA receptor activation. *J. Neurosci.* **17**: 2492.
Gould, E., Beylin, A., Tanapat, P., Reeves, A. and Shors, T.J., 1999, Learning enhances adult neurogenesis in the hippocampal formation. *Nat. Neurosci.* **2**: 260.
Gray, W.P. and Sundstrom, L.E., 1998, Kainic acid increases the proliferation of granule cell progenitors in the dentate gyrus of the adult rat. *Brain Res.* **790**: 52.
Hack, M.A., Saghatelyan, A., de Chevigny, A., Pfeifer, A., Ashery-Padan, R., Lledo, P.M. and Gotz, M., 2005, Neuronal fate determinants of adult olfactory bulb neurogenesis. *Nat. Neurosci.* **8**: 865.
Hagihara, H., Hara, M., Tsunekawa, K., Nakagawa, Y., Sawada, M. and Nakano, K., 2005, Tonic-clonic seizures induce division of neuronal progenitor cells with concomitant changes in expression of neurotrophic factors in the brain of pilocarpine-treated mice. *Brain Res. Mol. Brain Res.* **139**: 258.
Hastings, N.B. and Gould, E., 1999, Rapid extension of axons into the CA3 region by adult-generated granule cells. *J. Comp. Neurol.* **413**: 146.
Hattiangady, B., Rao, M.S. and Shetty, A.K., 2004, Chronic temporal lobe epilepsy is associated with severely declined dentate neurogenesis in the adult hippocampus. *Neurobiol. Dis.* **17**: 473.
Heinrich, C., Nitta, N., Flubacher, A., Muller, M., Fahrner, A., Kirsch, M., Freiman, T., Suzuki, F., Depaulis, A., Frotscher, M. and Haas, C.A., 2006, Reelin deficiency and displacement of mature neurons, but not neurogenesis, underlie the formation of granule cell dispersion in the epileptic hippocampus. *J. Neurosci.* **26**: 4701.
Hinterkeuser, S., Gray, W., Hager, G., Sundstrom, L. and Steinhauser, C., 1999, Characterization of astrocytic proliferation in rat dentate gyrus. *J. Neurochem.* **73**: S70.
Holmberg, K.H. and Patterson, P.H., 2006, Leukemia inhibitory factor is a key regulator of astrocytic, microglial and neuronal responses in a low-dose pilocarpine injury model. *Brain Res.* **1075**: 26.
Houser, C.R., 1990, Granule cell dispersion in the dentate gyrus of humans with temporal lobe epilepsy. *Brain Res.* **535**: 195.
Howell, O.W., Scharfman, H.E., Herzog, H., Sundstrom, L.E., Beck-Sickinger, A. and Gray, W.P., 2003, Neuropeptide Y is neuroproliferative for post-natal hippocampal precursor cells. *J. Neurochem.* **86**: 646.
Howell, O.W., Doyle, K., Goodman, J.H., Scharfman, H.E., Herzog, H., Pringle, A., Beck-Sickinger, A.G. and Gray, W.P., 2005, Neuropeptide Y stimulates neuronal precursor proliferation in the post-natal and adult dentate gyrus. *J. Neurochem.* **93**: 560.
Huttmann, K., Sadgrove, M., Wallraff, A., Hinterkeuser, S., Kirchhoff, F., Steinhauser, C. and Gray, W.P., 2003, Seizures preferentially stimulate proliferation of radial glia-like astrocytes in the adult dentate gyrus: functional and immunocytochemical analysis. *Eur. J. Neurosci.* **18**: 2769.
Imura, T., Kornblum, H.I. and Sofroniew, M.V., 2003, The predominant neural stem cell isolated from postnatal and adult forebrain but not early embryonic forebrain expresses GFAP. *J. Neurosci.* **23**: 2824.

Jacobs, B.L., Praag, H. and Gage, F.H., 2000, Adult brain neurogenesis and psychiatry: a novel theory of depression. *Mol. Psychiatry* **5**: 262.
Jessberger, S., Romer, B., Babu, H. and Kempermann, G., 2005, Seizures induce proliferation and dispersion of doublecortin-positive hippocampal progenitor cells. *Exp. Neurol.* **196**: 342.
Jin, K., Sun, Y., Xie, L., Batteur, S., Mao, X.O., Smelick, C., Logvinova, A. and Greenberg, D.A., 2003, Neurogenesis and aging: FGF-2 and HB-EGF restore neurogenesis in hippocampus and subventricular zone of aged mice. *Aging Cell* **2**: 175.
Johansson, C.B., Momma, S., Clarke, D.L., Risling, M., Lendahl, U. and Frisen, J., 1999, Identification of a neural stem cell in the adult mammalian central nervous system. *Cell* **96**: 25.
Karishma, K.K. and Herbert, J., 2002, Dehydroepiandrosterone (DHEA) stimulates neurogenesis in the hippocampus of the rat, promotes survival of newly formed neurons and prevents corticosterone-induced suppression. *Eur. J. Neurosci.* **16**: 445.
Kempermann, G., Kuhn, H.G. and Gage, F.H., 1997, More hippocampal neurons in adult mice living in an enriched environment. *Nature* **386**: 493.
Kempermann, G., Jessberger, S., Steiner, B. and Kronenberg, G., 2004, Milestones of neuronal development in the adult hippocampus. *Trends Neurosci.* **27**: 447.
Kipnis, J. and Schwartz, M., 2005, Controlled autoimmunity in CNS maintenance and repair: naturally occurring CD4+CD25+ regulatory T-Cells at the crossroads of health and disease. *Neuromol. Med.* **7**: 197.
Kralic, J.E., Ledergerber, D.A. and Fritschy, J.M., 2005, Disruption of the neurogenic potential of the dentate gyrus in a mouse model of temporal lobe epilepsy with focal seizures. *Eur. J. Neurosci.* **22**: 1916.
Kuhn, H.G., Dickinson-Anson, H. and Gage, F.H., 1996, Neurogenesis in the dentate gyrus of the adult rat: age-related decrease of neuronal progenitor proliferation. *J. Neurosci.* **16**: 2027.
Laywell, E.D., Rakic, P., Kukekov, V.G., Holland, E.C. and Steindler, D.A., 2000, Identification of a multipotent astrocytic stem cell in the immature and adult mouse brain. *Proc. Natl Acad. Sci. USA* **97**: 13883.
Lemaire, V., Koehl, M., Le Moal, M. and Abrous, D.N., 2000, Prenatal stress produces learning deficits associated with an inhibition of neurogenesis in the hippocampus. *Proc. Natl Acad. Sci. USA* **97**: 11032.
Lendahl, U., Zimmerman, L.B. and McKay, R.D., 1990, CNS stem cells express a new class of intermediate filament protein. *Cell* **60**: 585.
Levison, S.W. and Goldman, J.E., 1993, Both oligodendrocytes and astrocytes develop from progenitors in the subventricular zone of postnatal rat forebrain. *Neuron* **10**: 201.
Levison, S.W., Young, G.M. and Goldman, J.E., 1999, Cycling cells in the adult rat neocortex preferentially generate oligodendroglia. *J. Neurosci. Res.* **57**: 435.
Levitt, P. and Rakic, P., 1980, Immunoperoxidase localization of glial fibrillary acidic protein in radial glial cells and astrocytes of the developing rhesus monkey brain. *J. Comp. Neurol.* **193**: 815.
Lie, D.C., Colamarino, S.A., Song, H.J., Desire, L., Mira, H., Consiglio, A., Lein, E.S., Jessberger, S., Lansford, H., Dearie, A.R. and Gage, F.H., 2005, Wnt signaling regulates adult hippocampal neurogenesis. *Nature* **437**: 1370.
Lim, D.A., Tramontin, A.D., Trevejo, J.M., Herrera, D.G., Garcia-Verdugo, J.M. and Alvarez-Buylla, A., 2000, Noggin antagonizes BMP signaling to create a niche for adult neurogenesis. *Neuron* **28**: 713.
Liu, X., Wang, Q., Haydar, T.F. and Bordey, A., 2005, Nonsynaptic GABA signaling in postnatal subventricular zone controls proliferation of GFAP-expressing progenitors. *Nat. Neurosci.* **8**: 1179.
Lois, C. and Alvarez-Buylla, A., 1993, Proliferating subventricular zone cells in the adult mammalian forebrain can differentiate into neurons and glia. *Proc. Natl Acad. Sci. USA* **90**: 2074.
Ma, D.K., Ming, G.L. and Song, H., 2005, Glial influences on neural stem cell development: cellular niches for adult neurogenesis. *Curr. Opin. Neurobiol.* **15**: 514.
Malberg, J.E., Eisch, A.J., Nestler, E.J. and Duman, R.S., 2000, Chronic antidepressant treatment increases neurogenesis in adult rat hippocampus. *J. Neurosci.* **20**: 9104.
Markakis, E.A. and Gage, F.H., 1999, Adult-generated neurons in the dentate gyrus send axonal projections to field CA3 and are surrounded by synaptic vesicles. *J. Comp. Neurol.* **406**: 449.
Mathern, G.W., Leiphart, J.L., De Vera, A., Adelson, P.D., Seki, T., Neder, L. and Leite, J.P., 2002, Seizures decrease postnatal neurogenesis and granule cell development in the human fascia dentata. *Epilepsia* **43** (suppl. 5): 68.
Mazarati, A., Lu, X., Kilk, K., Langel, U., Wasterlain, C. and Bartfai, T., 2004, Galanin type 2 receptors regulate neuronal survival, susceptibility to seizures and seizure-induced neurogenesis in the dentate gyrus. *Eur. J. Neurosci.* **19**: 3235.
McDonald, H.Y. and Wojtowicz, J.M., 2005, Dynamics of neurogenesis in the dentate gyrus of adult rats. *Neurosci. Lett.* **385**: 70.
Merrill, D.A., Karim, R., Darraq, M., Chiba, A.A. and Tuszynski, M.H., 2003, Hippocampal cell genesis does not correlate with spatial learning ability in aged rats. *J. Comp. Neurol.* **459**: 201.

Monje, M.L., Toda, H. and Palmer, T.D., 2003, Inflammatory blockade restores adult hippocampal neurogenesis. *Science* **302**: 1760.
Muotri, A.R., Chu, V.T., Marchetto, M.C., Deng, W., Moran, J.V. and Gage, F.H., 2005, Somatic mosaicism in neuronal precursor cells mediated by L1 retrotransposition. *Nature* **435**: 903.
Nacher, J., Rosell, D.R., Alonso-Llosa, G. and McEwen, B.S., 2001, NMDA receptor antagonist treatment induces a long-lasting increase in the number of proliferating cells, PSA-NCAM-immunoreactive granule neurons and radial glia in the adult rat dentate gyrus. *Eur. J. Neurosci.* **13**: 512.
Nakagawa, E., Aimi, Y., Yasuhara, O., Tooyama, I., Shimada, M., McGeer, P.L. and Kimura, H., 2000, Enhancement of progenitor cell division in the dentate gyrus triggered by initial limbic seizures in rat models of epilepsy. *Epilepsia* **41**: 10.
Namba, T., Mochizuki, H., Onodera, M., Mizuno, Y., Namiki, H. and Seki, T., 2005, The fate of neural progenitor cells expressing astrocytic and radial glial markers in the postnatal rat dentate gyrus. *Eur. J. Neurosci.* **22**: 1928.
Navarro-Quiroga, I., Hernandez-Valdes, M., Lin, S.L. and Naegele, J.R., 2006, Postnatal cellular contributions of the hippocampus subventricular zone to the dentate gyrus, corpus callosum, fimbria, and cerebral cortex. *J. Comp. Neurol.* **497**: 833.
Neubauer, K., Knittel, T., Aurisch, S., Fellmer, P. and Ramadori, G., 1996, Glial fibrillary acidic protein-a cell type specific marker for Ito cells in vivo and in vitro. *J. Hepatol.* **24**: 719.
O'Kusky, J.R., Ye, P. and D'Ercole, A.J., 2000, Insulin-like growth factor-I promotes neurogenesis and synaptogenesis in the hippocampal dentate gyrus during postnatal development. *J. Neurosci.* **20**: 8435.
Overstreet-Wadiche, L.S., Bromberg, D.A., Bensen, A.L. and Westbrook, G.L., 2006, Seizures accelerate functional integration of adult-generated granule cells. *J. Neurosci.* **26**: 4095.
Parent, J.M., 2002, The role of seizure-induced neurogenesis in epileptogenesis and brain repair. *Epilepsy Res.* **50**: 179.
Parent, J.M., Yu, T.W., Leibowitz, R.T., Geschwind, D.H., Sloviter, R.S. and Lowenstein, D.H., 1997, Dentate granule cell neurogenesis is increased by seizures and contributes to aberrant network reorganization in the adult rat hippocampus. *J. Neurosci.* **17**: 3727.
Parent, J.M., Valentin, V.V. and Lowenstein, D.H., 2002, Prolonged seizures increase proliferating neuroblasts in the adult rat subventricular zone-olfactory bulb pathway. *J. Neurosci.* **22**: 3174.
Parent, J.M., von dem Bussche, N. and Lowenstein, D.H., 2006, Prolonged seizures recruit caudal subventricular zone glial progenitors into the injured hippocampus. *Hippocampus* **16**: 321.
Pirttila, T.J., Lukasiuk, K., Hakansson, K., Grubb, A., Abrahamson, M. and Pitkanen, A., 2005, Cystatin C modulates neurodegeneration and neurogenesis following status epilepticus in mouse. *Neurobiol. Dis.* **20**: 241.
Prickaerts, J., Koopmans, G., Blokland, A. and Scheepens, A., 2004, Learning and adult neurogenesis: survival with or without proliferation? *Neurobiol. Learn. Mem.* **81**: 1.
Radley, J.J. and Jacobs, B.L., 2003, Pilocarpine-induced status epilepticus increases cell proliferation in the dentate gyrus of adult rats via a 5-HT1A receptor-dependent mechanism. *Brain Res.* **966**: 1.
Rao, M. and Shetty, A.K., 2004, Efficacy of doublecortin as a marker to analyse the absolute number and dendritic growth of newly generated neurons in the adult dentate gyrus. *Eur. J. Neurosci.* **19**: 234.
Rao, M.S., Hattiangady, B., Abdel-Rahman, A., Stanley, D.P. and Shetty, A.K., 2005, Newly born cells in the ageing dentate gyrus display normal migration, survival and neuronal fate choice but endure retarded early maturation. *Eur. J. Neurosci.* **21**: 464.
Rickmann, M., Amaral, D.G. and Cowan, W.M., 1987, Organization of radial glial cells during the development of the rat dentate gyrus. *J. Comp. Neurol.* **264**: 449.
Rietze, R., Poulin, P. and Weiss, S., 2000, Mitotically active cells that generate neurons and astrocytes are present in multiple regions of the adult mouse hippocampus. *J. Comp. Neurol.* **424**: 397.
Sadgrove, M.P., Chad, J.E. and Gray, W.P., 2005, Kainic acid induces rapid cell death followed by transiently reduced cell proliferation in the immature granule cell layer of rat hippocampal slice cultures. *Brain Res.* **1035**: 111.
Santarelli, L., Saxe, M., Gross, C., Surget, A., Battaglia, F., Dulawa, S., Weisstaub, N., Lee, J., Duman, R., Arancio, O., Belzung, C. and Hen, R., 2003, Requirement of hippocampal neurogenesis for the behavioral effects of antidepressants. *Science* **301**: 805.
Scharfman, H.E. and Gray, W.P., 2006, Plasticity of neuropeptide Y in the dentate gyrus after seizures, and its relevance to seizure-induced neurogenesis. *EXS* 193.
Scharfman, H.E., Goodman, J.H. and Sollas, A.L., 2000, Granule-like neurons at the hilar/CA3 border after status epilepticus and their synchrony with area CA3 pyramidal cells: functional implications of seizure-induced neurogenesis. *J. Neurosci.* **20**: 6144.
Schlessinger, A.R., Cowan, W.M. and Gottlieb, D.I., 1975, An autoradiographic study of the time of origin and the pattern of granule cell migration in the dentate gyrus of the rat. *J. Comp. Neurol.* **159**: 149.

Schmechel, D.E. and Rakic, P., 1979, A Golgi study of radial glial cells in developing monkey telencephalon: morphogenesis and transformation into astrocytes. *Anat. Embryol. (Berl.)* **156**: 115.

Schwartz, M., Butovsky, O., Bruck, W. and Hanisch, U.K., 2006, Microglial phenotype: is the commitment reversible? *Trends Neurosci.* **29**: 68.

Seki, T. and Arai, Y., 1995, Age-related production of new granule cells in the adult dentate gyrus. *Neuroreport* **6**: 2479.

Seki, T. and Arai, Y., 1999, Temporal and spacial relationships between PSA-NCAM-expressing, newly generated granule cells, and radial glia-like cells in the adult dentate gyrus. *J. Comp. Neurol.* **410**: 503.

Seri, B., Garcia-Verdugo, J.M., Collado-Morente, L., McEwen, B.S. and Alvarez-Buylla, A., 2004, Cell types, lineage, and architecture of the germinal zone in the adult dentate gyrus. *J. Comp. Neurol.* **478**: 359.

Shapiro, L.A., Korn, M.J., Shan, Z. and Ribak, C.E., 2005, GFAP-expressing radial glia-like cell bodies are involved in a one-to-one relationship with doublecortin-immunolabeled newborn neurons in the adult dentate gyrus. *Brain Res.* **1040**: 81.

Shetty, A.K., Hattiangady, B. and Shetty, G.A., 2005, Stem/progenitor cell proliferation factors FGF-2, IGF-1, and VEGF exhibit early decline during the course of aging in the hippocampus: role of astrocytes. *Glia* **51**: 173.

Shors, T.J., 2004, Memory traces of trace memories: neurogenesis, synaptogenesis and awareness. *Trends Neurosci.* **27**: 250.

Shors, T.J., Miesegaes, G., Beylin, A., Zhao, M., Rydel, T. and Gould, E., 2001, Neurogenesis in the adult is involved in the formation of trace memories. *Nature* **410**: 372.

Simon, M., Czeh, B. and Fuchs, E., 2005, Age-dependent susceptibility of adult hippocampal cell proliferation to chronic psychosocial stress. *Brain Res.* **1049**: 244.

Smith, P.D., McLean, K.J., Murphy, M.A., Turnley, A.M. and Cook, M.J., 2005, Seizures, not hippocampal neuronal death, provoke neurogenesis in a mouse rapid electrical amygdala kindling model of seizures. *Neuroscience* **136**: 405.

Snyder, J.S., Kee, N. and Wojtowicz, J.M., 2001, Effects of adult neurogenesis on synaptic plasticity in the rat dentate gyrus. *J. Neurophysiol.* **85**: 2423.

Song, H., Stevens, C.F. and Gage, F.H., 2002a, Astroglia induce neurogenesis from adult neural stem cells. *Nature* **417**: 39.

Song, H.J., Stevens, C.F. and Gage, F.H., 2002b, Neural stem cells from adult hippocampus develop essential properties of functional CNS neurons. *Nat. Neurosci.* **5**: 438.

Steiner, B., Kronenberg, G., Jessberger, S., Brandt, M.D., Reuter, K. and Kempermann, G., 2004, Differential regulation of gliogenesis in the context of adult hippocampal neurogenesis in mice. *Glia* **46**: 41.

Thom, M., Sisodiya, S.M., Beckett, A., Martinian, L., Lin, W.R., Harkness, W., Mitchell, T.N., Craig, J., Duncan, J. and Scaravilli, F., 2002, Cytoarchitectural abnormalities in hippocampal sclerosis. *J. Neuropathol. Exp. Neurol.* **61**: 510.

Ueki, T., Tanaka, M., Yamashita, K., Mikawa, S., Qiu, Z., Maragakis, N.J., Hevner, R.F., Miura, N., Sugimura, H. and Sato, K., 2003, A novel secretory factor, Neurogenesin-1, provides neurogenic environmental cues for neural stem cells in the adult hippocampus. *J. Neurosci.* **23**: 11732.

Ullian, E.M., Sapperstein, S.K., Christopherson, K.S. and Barres, B.A., 2001, Control of synapse number by glia. *Science* **291**: 657.

Vallieres, L., Campbell, I.L., Gage, F.H. and Sawchenko, P.E., 2002, Reduced hippocampal neurogenesis in adult transgenic mice with chronic astrocytic production of interleukin-6. *J. Neurosci.* **22**: 486.

van Praag, H., Schinder, A.F., Christie, B.R., Toni, N., Palmer, T.D. and Gage, F.H., 2002, Functional neurogenesis in the adult hippocampus. *Nature* **415**: 1030.

Yoshimura, S., Takagi, Y., Harada, J., Teramoto, T., Thomas, S.S., Waeber, C., Bakowska, J.C., Breakefield, X.O. and Moskowitz, M.A., 2001, FGF-2 regulation of neurogenesis in adult hippocampus after brain injury. *Proc. Natl Acad. Sci. USA* **98**: 5874.

Zhu, H., Dahlstrom, A. and Hansson, H.A., 2005, Characterization of cell proliferation in the adult dentate under normal conditions and after kainate induced seizures using ribonucleotide reductase and BrdU. *Brain Res.* **1036**: 7.

Ziv, Y., Ron, N., Butovsky, O., Landa, G., Sudai, E., Greenberg, N., Cohen, H., Kipnis, J. and Schwartz, M., 2006, Immune cells contribute to the maintenance of neurogenesis and spatial learning abilities in adulthood. *Nat. Neurosci.* **9**: 268.

Zucchini, S., Barbieri, M. and Simonato, M., 2005, Alterations in seizure susceptibility and in seizure-induced plasticity after pharmacologic and genetic manipulation of the fibroblast growth factor-2 system. *Epilepsia* **46** (Suppl. 5): 52.

18

POLYGLUTAMINE EXPANSION DISEASES – THE CASE OF MACHADO-JOSEPH DISEASE

Sandra Macedo-Ribeiro, Luís Pereira de Almeida, Ana Luísa Carvalho and A. Cristina Rego

1. ABSTRACT

Polyglutamine expansion diseases are inherited neurodegenerative disorders caused by the expansion of CAG repeat mutations in the coding region of genes encoding for specific proteins, mostly of unknown function. One example is Machado-Joseph disease (MJD) or spinocerebellar ataxia 3, which was described in people of Portuguese descendents and is caused by expanded ataxin-3, a polyubiquitin-binding protein. Like other neurodegenerative diseases, MJD exhibits gradual progression of symptoms that finally result in the death of the patients. Despite the identification of the genetic defects, the molecular mechanisms by which the mutant protein initiates the pathogenic process remain to be elucidated. This chapter resumes some of the most important features of polyglutamine expansion diseases with a special emphasis on MJD pathogenesis. Particular relevance is given to ataxin-3 structure and function, the formation of aggregates of mutant ataxin-3, the characteristics of current disease animal models and the most recent therapeutic strategies proposed for the treatment of MJD.

2. POLYGLUTAMINE EXPANSION DISEASES

Proteins with expanded polyglutamine tracts are a common feature in at least nine human inherited neurodegenerative disorders. These include dentatorubral-pallidoluysian atrophy (DRPLA), Huntington's disease (HD), spinal bulbar muscular atrophy (SBMA), and spinocerebellar ataxias (SCA) 1, 2, 3, 6, 7, and 17. The prevalence of these pathologies is generally low and varies depending on the disease and the geographic region. With the exception of SBMA or Kennedy's disease, an X-linked recessive disease, all other polyglutamine disorders are autosomal dominant (Zoghbi and Orr, 2000). Polyglutamine expansion diseases are caused by unstable CAG-repeat expansions within the coding region of a specific gene that encodes polyglutamine expansions in the disease protein (Zoghbi and Orr, 2000). The proteins affected by polyglutamine expansion are unrelated in terms of protein sequence, structure, and function. The different protein context in which the

---

Ana Cristina Rego – acrego@cnc.cj.uc.pt
Center for Neuroscience and Cell Biology, Institute of Biochemistry, Faculty of Medicine
University of Coimbra, 3004-504 Coimbra, Portugal

polyglutamine tract is inserted determines the brain regions and neurons that are affected (Gusella and MacDonald, 2000). Although the mutant proteins carrying the polyglutamine expansions are ubiquitously expressed, the neurodegeneration is selective to specific brain areas. Patterns of affected brain regions vary among the diseases, but common features include progressive neuronal cell loss and decline in motor and cognitive functions. Despite presenting a wide variety of clinical features, polyglutamine expansion diseases onset occurs in midlife, invariably leading to the death of the patients 10–30 years after the appearance of the first symptoms.

In general, the age of onset of polyglutamine expansion diseases inversely correlates with the size of expanded CAG repeats, and the normal repeat appears to interact with the expanded CAG repeat influencing the age of onset (Gusella and MacDonald, 2000; Djousse et al., 2003). Nevertheless, for a given CAG repeat, there is a great variation in the age of onset, suggesting that other factors may determine the onset and progression of polyglutamine expansion diseases. CAG expansions exceeding the 100 repeats, although rare, are always associated with very early onset (infantile) and more severe forms of the diseases.

Cases of homozygosity and heterozygosity may exhibit similar age of onset, clinical features, and disease progression. However, in some cases, such as in SCA6, homozygosity was shown to induce an earlier age of onset (Kato et al., 2000) as well as a more severe clinical manifestation than a heterozygote carrying an expanded allele with the same repeat length (Matsumura et al., 1997).

Early onset in successive generations resulting from paternal transmission has been observed, a trend named "anticipation" (Myers et al., 1982). New mutations (CAG expansions) may occur at the high end of the normal range, in the intermediate allele, typically resulting from paternal transmission. This instability in transmission from one generation to the other has been attributed to gametic instability. The SCA7 mutation is highly unstable. In this disease repeat expansions, paternal transmissions are greater than those from maternal transmissions, but maternal transmission is more common (Gouw et al., 1998). Somatic mosaicism has been also found in different brain regions, resulting from instability at the cellular level.

The normal and pathogenic length of polyglutamine stretches almost overlap in HD, SBMA, SCA1, SCA2, SCA7, and SCA17, as depicted in Table 1. The pathogenic range of SCA6 differs most from the other polyglutamine expansion diseases, showing a mutant CAG repeat ranging from 20 to 33 (normal range 4–19), which is within normal range for the other 8 diseases. Excluding SBMA, SCA6, and SCA17, which affect the androgen receptor (AR), the α_{1A} subunit of the P/Q type voltage-dependent Ca^{2+}-channel (Kato et al., 2000) and TATA-binding protein (TBP, a transcription factor that regulates the expression of most eukaryotic genes), respectively, all other diseases affect proteins of ubiquitous expression (in the brain and peripheral tissues) but still with unknown function.

Polyglutamine aggregates or inclusion bodies are a common feature of these diseases, present in the nucleus and/or in the cytoplasm (Table 1; See Sect. 5). Aggregates of misfolded mutant proteins that are not degraded by the proteasome (despite the presence of ubiquitin) tend to accumulate into neuronal inclusions, which are most frequent in the nucleus (Table 1; Davies et al., 1997). Although mainly affecting neurons, nuclear aggregates have also been found in glial cells in DRPLA, a very rare disorder that occurs most frequently in Japan

Table 1. Characteristics of proteins affected in polyglutamine expansion diseases

Polyglutamine expansion disease	Protein affected	Gene location	MW (kDa)	Normal CAG repeat	Pathogenic CAG repeat	Position of polyglutamine repeat/total amino acids	Intracellular protein localization	Intracellular localization of polyglutamine aggregates
SBMA	AR	Xq11-12	104	9–36	38–65	58/918	Cytoplasm/nucleus	Nuclear
HD	Huntingtin	4p16.3	348	6–35	36–121	18/3144	Cytoplasm/nucleus	Nuclear/cytoplasmic
DRPLA	Atrophin-1	12p13	190	3–35	49–88	474/1184	Cytoplasm/nucleus	Nuclear
SCA1	Ataxin-1	6p23	87	6–38	39–83	197/816	Cytoplasm/nucleus	Nuclear
SCA2	Ataxin-2	12q24.1	145	14–31	32–77	166/1313	Cytoplasm	Nuclear/cytoplasmic
SCA3	Ataxin-3	14q32.1	42	10–51	55–87	292/359	Cytoplasm/nucleus	Nuclear
SCA6	CACNA1A	19p13.1	280	4–19	20–33	2328/2410	Cytoplasm	Nuclear/cytoplasmic
SCA7	Ataxin-7	3p12-13.5	95	4–35	37–306	30/893	Cytoplasm/nucleus	Nuclear
SCA17	TBP	6q27	38	29–42	47–55	58/339	Nucleus	Nuclear

Data were collated from a series of literature reporting the affected proteins, the sizes of human CAG repeats for each polyglutamine expansion disease and the localization of the proteins affected. *Abbreviations*: AR, androgen receptor; CACNA1A, α_{1A} subunit of the P/Q type voltage-dependent Ca^{2+}-channel; DRPLA, Dentatorubral-pallidoluysian atrophy; HD, Huntington's disease; SBMA, spinal bulbar muscular atrophy; SCA, spinocerebellar ataxia, TBP, TATA-binding protein.

(Hayashi et al., 1998). In SCA1, cytoplasmic aggregates were also detected in glia (Gilman et al., 1996). Several mechanisms of pathogenesis have been proposed to lead to neuronal dysfunction, later progressing to neurodegeneration in polyglutamine expansion diseases, including protein aggregation, transcription dysregulation, decreased axonal transport, inhibition of ubiquitin proteasome system, synaptic and mitochondrial dysfunction, and apoptotic cell death. Factors that modulate aggregate formation and toxicity are therefore potential targets for therapeutic intervention in polyglutamine expansion diseases (see Sect. 7).

In Sects. 2.1.–2.4., we resume the main features of SBMA, the first polyglutamine expansion neurological disease to be identified, HD, the most common polyglutamine expansion disease, DRPLA, a very rare disorder, and SCAs, which include Machado-Joseph disease (MJD) or spinocerebellar ataxia 3 (SCA3), which is further developed in this chapter.

2.1. Spinal Bulbar Muscular Atrophy

The origin of SBMA was first identified by the group of Kenneth Fischbeck (La Spada et al., 1991). SBMA is an adult-onset motoneuron disease caused by a CAG-repeat expansion in the first exon of the AR gene in the X chromosome (Xq11-12). The AR is a nuclear receptor that acts as a ligand-inducible transcriptional factor. The carriers of SBMA exhibit muscle weakness and atrophy; dysarthria; dysphagia; fasciculations of the tongue, lips, or face; and sensory neuropathy (Lieberman and Fischbeck, 2000). Juvenile onset of SBMA showing rapid neurologic progression was reported to occur at 8–15 years in seven patients from the same family with 50–54 CAG repeats (Echaniz-Laguna et al., 2005).

Polyglutamine pathogenesis has been attributed to a gain of function due to the toxicity of the proteins carrying a polyglutamine expansion, rather than to a loss of function. Accordingly, polyglutamine expansion of AR in SBMA causes weakness and motor neuron degeneration, whereas loss of function of the AR causes feminisation. Nevertheless, SBMA patients show androgen insensitivity, testicular atrophy, and reduced fertility (Lieberman and Fischbeck, 2000). Mutant AR accumulates in the nucleus (Table 1) causing a loss of motor neurons in the anterior horn of spinal cord as well as in brainstem motor nuclei and the degeneration of sensory neurons in the dorsal root ganglia. The disease affects only males (female carriers are usually asymptomatic or may exhibit mild symptoms), mainly adult, and is characterized by slowly progressive muscle weakness and atrophy of bulbar, facial, and limb muscles (Katsuno et al., 2006).

Presently SBMA is the only polyglutamine expansion disease for which effective therapeutic approaches were developed in mice models. These involve elimination of androgens by surgical or chemical castration of severely affected male mice, partially restoring motor function (Chevalier-Larsen et al., 2004). Leuprorelin, a luteinizing hormone-releasing hormone (LHRH) agonist that reduces testosterone release from the testis, seems to be a promising candidate for SBMA treatment, since it rescued motor dysfunction and nuclear accumulation of mutant AR in SBMA male transgenic mice (Katsuno et al., 2003).

2.2. Huntington's Disease

HD is the most common and most studied autosomal dominant polyglutamine disorder, affecting 5–10 per 100,000 individuals in Western Europe and North America (Vonsattel and DiFiglia, 1998). HD was first described by George Huntington in 1872 who, based on the observation of patients from his father's practice, described the main clinical features of disease and the familial transmission. The disorder was classically described as Huntington's chorea due to presence of dance-like movements, the major motor abnormality occurring in HD. Disease symptomatology also includes cognitive impairment and psychiatric problems. In the late stages of the disease, chorea tends to be replaced by bradykinesia and rigidity, and death may occur due to heart failure or pneumonia. HD patients also suffer from weight loss and muscle wasting. Moreover, endocrine abnormalities accompany the disease, such as an increase in corticosteroids (Heuser et al., 1991; Leblhuber et al., 1995) and a reduction in testosterone levels (Markianos et al., 2005), and 10–15% of HD patients exhibit diabetes mellitus (Farrer, 1985). Similar to other polyglutamine expansion diseases, no cure or effective treatment is yet available to prevent or delay HD progression.

HD is caused by abnormal expansion of the CAG triplet in the coding region of the HD (*IT15*, Interesting Transcript 15) gene on the short arm of chromosome 4 (4p16.3), which encodes for a polyglutamine tract at the N-terminal of huntingtin (HDCRG, 1993). HD is characterized by a selective and progressive loss of GABAergic medium spiny neurons (MSNs) with atrophy of caudate-putamen and also with cerebral cortex dysfunction (Vonsattel and DiFiglia, 1998). Nevertheless, the mechanisms by which mutant huntingtin exerts its deleterious effects on MSNs are poorly understood.

Polyglutamine expanded huntingtin is cleaved in the cytoplasm and then translocates to the nucleus, forming ubiquitin-immunopositive nuclear aggregates, which have been related with transcription dysregulation (Cha, 2000; Bates, 2001; Sugars and Rubinsztein, 2003). Mutant huntingtin aggregates were also found in neurites, which could account for the disturbance of the microtubule system and the effect on the axonal transport (Gunawardena et al., 2003; Lee et al., 2004). In fact, mutant huntingtin disturbs the transport of brain-derived-neurotrophic factor (BDNF)-containing vesicles, affecting neuronal survival (Gauthier et al., 2004). Furthermore, dysregulation of BDNF transcription has been linked to the loss of huntingtin normal function (Zuccato et al., 2001). Evidence also suggests that mutant huntingtin induces abnormalities in neuronal calcium homeostasis (Bezprozvanny and Hayden, 2004). Mitochondrial dysfunction, loss of calcium homeostasis, and cell death features of HD are described in Chap. 9.

2.3. Dentatorubral and Pallidoluysian Atrophy (DRPLA)

DRPLA is an autosomal dominant cerebellar ataxia clinically characterized by myoclonus, epilepsy, cerebellar ataxia, choreoathetosis, and dementia with personality changes. The disease is very rare, extremely uncommon in Caucasians but fairly common in Japan (Hayashi et al., 1998). Histopathologically, DRPLA is characterized by the degeneration of the dentatofugal and the pallidofugal systems. Furthermore, a variability in the clinical features depending on the age of onset has been observed (Kanazawa, 1998). DRPLA presents prominent "anticipation" shown by a marked instability of the expanded CAG repeat length during spermatogenesis. Moreover, the instability of the CAG repeat length also seems to occur in the somatic cells, resulting in "somatic mosaicism" (Kanazawa, 1999).

The *drpla* gene (located on chromosome 12) product atrophin-1 is a widely expressed protein, with no known function that is found in both the nucleus and cytoplasm of neurons. Like in many other polyglutamine-expanded proteins, truncated fragments of atrophin-1 containing the polyglutamine expansion accumulate in the nucleus. Hence, caspase cleavage of atrophin-1 at Asp109 modulates aggregate formation and cytotoxicity (Ellerby et al., 1999).

2.4. Spinocerebellar Ataxias

Ataxia is defined by the inability to coordinate voluntary muscle movements, showing unsteady movements and staggering gait. Although at least 17 SCA mutations have been found, only SCA1, SCA2, SCA3, SCA6, SCA7, and SCA17 are caused by CAG repeat expansion in the coding regions of ataxin-1 (6p23), ataxin-2 (12q24.1), ataxin-3 (14q32.1), the CACNA1A calcium channel (19p13.1), ataxin-7 (3p12-13.5), and TBP (6q27), respectively (Table 1).

SCAs are clinically very heterogeneous diseases. SCA 1, 2, 3, 6, and 7 patients show symptoms of progressive ataxia associated with cerebellar syndrome, dysarthria, spasticity, nystagmus, and dysphagia, among others. SCA7 also affects the retina (Stevanin et al., 2000). SCA17 exhibits epilepsy with cerebellar ataxia, dementia, psychosis, hyperreflexia and bradykinesia (Koide et al., 1999; Nakamura et al., 2001; Zuhlke et al., 2001). Huntington's disease-like phenotype or chorea (Stevanin et al., 2003) and dystonia (Hagenah et al., 2004) were also found in groups of patients carrying the TBP gene mutation.

The CAG repeat expansion is generally not interrupted, with the exception of SCA1, discontinued by 1-3 CAT triplets (that code for histidines residues) (Chong et al., 1995), and the CAG repeat expansion in SCA2 (Imbert et al., 1996; Pulst et al., 1996) and SCA17 (Koide et al., 1999) containing the CAA triplet, which also codes for glutamine, thus not altering the polyglutamine stretch in the respective translated proteins. These interruptions in the CAG expansion have been reported to stabilize the long (CAG)n stretches (Chong et al., 1995; Koide et al., 1999). The histidine interruptions within the polyglutamine tract of ataxin-1 have been reported to increase the solubility of model polyglutamine peptides in vitro (Sharma et al., 1999) and to mitigate the pathological effects of long glutamine stretches, decreasing disease severity and resulting in later onset in SCA1 patients (Matsuyama et al., 1999).

3. CLINICAL AND PATHOLOGICAL ASPECTS OF MJD

MJD or SCA3 is the most common dominantly inherited ataxia worldwide (Coutinho and Andrade, 1978; Rosenberg, 1984; Ranum et al., 1995). This neurodegenerative disorder of adult onset was named after Antone Joseph and William Machado, of Portuguese Azorean origin, who migrated to the USA. MJD was subsequently identified in Brazil, Japan, China, and Australia. In the islands of the Azores, namely São Miguel and Flores, MJD reaches the highest prevalence (1:140 in the small island of Flores) reported worldwide (Sudarsky and Coutinho, 1995).

MJD is associated with an unstable expansion of a (CAG)n tract in the coding region of the *MJD1* gene localized on chromosome 14q32.1 (Kawaguchi et al., 1994). *MJD1* encodes ataxin-3, a polyubiquitin-binding protein whose physiological function has been recently linked to ubiquitin-mediated proteolysis (Burnett et al., 2003; Donaldson et al., 2003; Doss-Pepe et al., 2003; Scheel et al., 2003; Chai et al., 2004). The mutation results in an expanded polyglutamine tract at the C-terminus of ataxin-3 (Durr et al., 1996). The CAG repeats range from 10 to 51 in the normal population and from 55 to 87 in MJD patients (Cummings and Zoghbi, 2000; Maciel et al., 2001). This high threshold of pathogenicity is a special characteristic of this disorder, since in most other polyglutamine disorders trinucleotide repeats over 36 to 40 become pathogenic. As in the other polyglutamine disorders, there is a correlation between the age of onset and CAG repeat numbers.

The clinical hallmark of MJD is progressive ataxia, a dysfunction of motor coordination that can affect gaze, speech, gait, and balance (Taroni and DiDonato, 2004). Other clinical features include extrapyramidal features, severe spasticity, dysarthria and difficulty with swallowing, postural instability, nystagmus, eyelid retraction, vision problems, facial fasciculations, axonal polyneuropathy and dystonia particularly prominent in young patients (Sudarsky and Coutinho, 1995). Levodopa-responsive parkinsonism symptoms resembling Parkinson's disease were also reported (Gwinn-Hardy et al., 2001).

There are three types of MJD, distinguished by the age of onset and the symptomatology (1) type I has an onset at 10–30 years, shows fast progression and severe dystonia and rigidity; (2) type II begins at 20–50 years, shows an intermediate progression and causes spasticity, spastic gait, and increased reflex responses; (3) type III has an onset at 40–70 years, shows slow progression, muscle twitching, muscle atrophy, and unpleasant sensations such as numbness, tingling, cramps, and pain in the hands, feet, and limbs.

Neuropathology involves cerebellar systems (particularly dentate nucleus and pontine neurons), *substantia nigra*, and cranial nerve motor nuclei, with relative preservation of cerebellar cortex, particularly Purkinje cells and inferior olive (Sudarsky and Coutinho, 1995; Durr et al., 1996). A marked degeneration of Clarke's column nuclei and vestibular and pontine nuclei is observed (Durr et al., 1996). Marked neuronal loss is also observed in the anterior horn of the spinal cord, and motor nuclei of the brainstem. Variable degrees of dopamine terminal damage and metabolic impairment have been reported for MJD (Shinotoh et al., 1997; Soong et al., 1997; Yen et al., 2000, 2002). Yen and collaborators investigated the presence of early alterations in asymptomatic MJD gene carriers. Dopamine transporter binding was assessed with single-photon emission computed tomography coupled to a radio-tracer that binds specifically to the dopamine transporter on the nigrostriatal terminals. Interestingly, asymptomatic MJD gene carriers showed reduction of dopamine uptake in putamen, suggesting that impairment of presynaptic dopamine function occurs at an early stage. This imaging technique seems promising for early detection of the neuronal dysfunction in MJD, monitoring of disease progression, and therapeutic approaches (Yen et al., 2002).

Positron emission tomography has also been useful in showing neuronal dysfunction, particularly decreased regional cerebral glucose metabolism, not only in the regions with apparent pathological involvement such as cerebellum, brainstem, thalamus, and nigro-striatal dopaminergic system, but also in the cerebral cortex and the striatum where no pathology could be observed using conventional morphological techniques (Taniwaki et al., 1997; Wullner et al., 2005).

The presence of neuronal intranuclear inclusions (NIIs) is a hallmark of neurodegeneration in the brains of patients with MJD (as in most other polyglutamine disorders). Since the association between polyglutamine expansions and neurodegeneration (La Spada et al., 1994), a number of mechanisms have been put forward (Aronin et al., 1999), such as altered gene transcription (reviewed in Okazawa, 2003), impairment of the cell's quality control machinery: proteasome and chaperones (reviewed in Ferrigno and Silver, 2000), and perturbation of intracellular trafficking and apoptotic cell death (Rego and de Almeida, 2005).

In MJD, mutant ataxin-3 and CREB-binding protein (CBP) have been shown to colocalize in nuclear inclusions, suggesting a role for transcription impairment in MJD pathogenesis (McCampbell et al., 2000). Mutant ataxin-3 and a cleavage fragment of the polyglutamine-expanded protein have also been implicated in perturbation of the ubiquitin–proteasome system (Ikeda et al., 1996; Goti et al., 2004) and in impaired fast axonal transport (Gunawardena et al., 2003). Furthermore, expression of expanded full-length mutant ataxin-3 caused a decrease in the expression of heat shock protein 27 (Hsp27) (Wen et al., 2003), previously reported to increase neuronal survival (Lewis et al., 1999). In addition, the cytotoxicity of mutant ataxin-3 is enhanced by cell cycle arrest in the G(0)/G(1) phase (Yoshizawa et al., 2000).

Despite the report of nonapoptotic cell death evidenced by ultrastructural analysis in a rat mesencephalic dopaminergic cell line expressing expanded full-length ataxin-3 (Evert et al., 1999), early evidences of apoptosis in cultured cells expressing a portion of mutant ataxin-3 and the participation of caspases in ataxin-3 cleavage highly support the involvement of apoptotic cell death (Ikeda et al., 1996; Berke et al., 2004). Mitochondrial-dependent apoptosis was further ascribed to a decrease in Bcl-2 (no changes in Bax were observed), associated with increased release of cytochrome c in cells expressing mutant (Q78) ataxin-3, or in MJD fetus fibroblasts (Tsai et al., 2004). Similar to other polyglutamine-expanded disease proteins, ataxin-3 can be cleaved by caspase-1, producing a 28 kDa fragment containing the glutamine repeat. Caspase-cleavage of ataxin-3 promotes its aggregation. A major caspase cleavage site was ascribed to aspartate residues within an ubiquitin interacting motif (UIM) (Berke et al., 2004). Recently, overexpression of ataxin-3, with 69 glutamines, mediated by adenoviral vectors in cultured cerebellar, striatal, and *substantia nigra* neurons upregulated Bax and downregulated Bcl-xl expression leading to mitochondrial release of cytochrome c and Smac, caspase-3 activation, and apoptotic neuronal death (Chou et al., 2006). However, no evidences were found of proteolytic fragments in human brain tissue of two MJD patients (Berke et al., 2004). Analysis of cerebellar degeneration in brains of MJD patients revealed TUNEL-positive granular cells in the cerebellar cortex in one case of MJD and GFAP-immunopositive glial cells in the granular layer (Kumada et al., 2000).

4. ATAXIN-3 STRUCTURE AND FUNCTION

4.1. Ataxin-3 Primary Structure

Ataxin-3 is a modular protein with an overall molecular weight of 42 kDa, containing a conserved N-terminal Josephin domain (Masino et al., 2003; Scheel et al., 2003; Albrecht et al., 2004), followed by two UIM domains and the polyglutamine repeat region (Fig. 1). Alternative splicing of the *MJD* gene results in the production of different isoforms of ataxin-3 varying at the C-terminal portion of the protein (Goto et al., 1997), one of them containing a third UIM domain after the polyglutamine region. The relevance of the different ataxin-3 isoforms in protein function and pathology is still unknown. The protein is expressed in various tissues, suggesting that it plays an important role in eukaryotic cells, and it has been detected both in the nucleus and in the cytoplasm (Paulson et al., 1997a; Trottier et al., 1998; Ichikawa et al., 2001). A putative nuclear localization signal (NLS) has been identified upstream the polyglutamine repeat region (Tait et al., 1998).

Protein sequences with high homology to human ataxin-3 were found in the genomes of mouse (do Carmo Costa et al., 2004), rat (Schmitt et al., 1997), chicken (Linhartova et al., 1999), frog, *Fugu, D. melanogaster*, *C. Elegans*, and *P. falciparum,* as well as in the genomes of rice and *A. thaliana*. The N-terminal portion of the protein, the Josephin domain, is the most highly conserved region of ataxin-3, and short proteins of unknown function, Josephins, containing almost exclusively this domain have been identified in different species (Scheel et al., 2003; Albrecht et al., 2004). The Josephin domain was predicted to have cysteine protease fold and to be distantly related to a class of ubiquitin-specific proteases (Scheel et al., 2003). The C-terminal portion of ataxin-3 homologues is not so well conserved; however, except for the ataxin-3-like protein found in *P. falciparum*, all ataxin-3 homologs contain at least two UIM domains. UIMs are short conserved motifs of around 11 aminoacids that occur in many proteins involved in protein ubiquitination and ubiquitin recognition pathways (Andersen et al., 2005). All these data highlighted a close link between ataxin-3 cellular role and the ubiquitin–proteasomal pathways.

4.2. Ataxin-3 Molecular Partners: Clues into the Function of Ataxin-3

Although the precise cellular role of ataxin-3 and how it is altered upon polyglutamine expansion is presently unknown, ataxin-3 was shown to be a polyubiquitin-binding protein (Donaldson et al., 2003; Doss-Pepe et al., 2003), interacting via the first two UIM domains with K48-linked tetraubiquitin chains (Burnett et al., 2003; Chai et al., 2004), independently of the polyQ size, but failing to pull-down both mono- or diubiquitin. Additionally, as predicted by bioinformatics analysis of the protein sequence (Scheel et al., 2003), both normal and expanded ataxin-3 display ubiquitin hydrolase activity that can be inhibited by the irreversible inhibitor, ubiquitin aldehyde (Burnett et al., 2003). The N-terminal Josephin domain also interacts with the N-terminal ubiquitin-like (UBL) domain of the DNA-repair proteins HHR23A and HHR23B and this interaction is independent of the UIMs (Wang et al., 2000). Recently, it has been shown that both the proteolytic activity of ataxin-3 and its UIMs are required for aggresome formation by the cystic fibrosis transmembrane regulator mutant CFTRΔF508 (Burnett and Pittman, 2005). Furthermore, ataxin-3 interacts with p97/valosin-containing protein (VCP) (Doss-Pepe et al., 2003; Matsumoto et al., 2004), and this interaction has been mapped to the arginine/lysine rich motif predicted to be a NLS in ataxin-3 (Boeddrich et al., 2006). The major histone acetyltransferases cAMP-response-element binding protein (CREB)-binding protein (CBP), p300, and p300/CBP-associated factor have also been reported to interact with ataxin-3 through the polyglutamine-containing C-terminus (Li et al., 2002). Novel insights on the physiological role of ataxin-3 came from studies in *Drosophila* showing that ataxin-3 ubiquitin hydrolase activity modulates polyglutamine-induced neurodegeneration in vivo, and that this suppressor activity is dependent on proteasome function (Warrick et al., 2005).

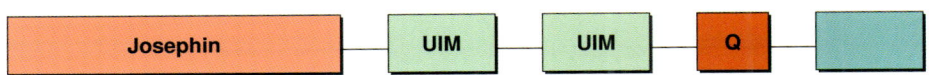

Fig. 1. Schematic diagram of ataxin-3 domain structure. The N-terminal catalytic domain containing the active site cysteine 14 is termed Josephin domain. Two ubiquitin interacting motifs (UIM) are found between the Josephin domain and the polyglutamine tract (Q). An alternatively spliced form has a third UIM C-terminal to the polyglutamine repeat region

4.3. Structure of the Josephin Domain

Detailed structural information on polyglutamine proteins is crucial to understand the mechanisms underlying the normal functions of these proteins as well as the pathogenesis induced by polyglutamine expansion. Ataxin-3 is the smallest of the proteins associated with polyglutamine expansion diseases and has been a target of numerous structural studies (Bevivino and Loll, 2001; Masino et al., 2003). However, determination of the detailed three-dimensional structure of ataxin-3 has been hampered by the tendency of the full-length protein to aggregate (see Sect. 5.4). Nevertheless, it was possible to show that ataxin-3 is composed of a structured N-terminal domain, followed by a flexible tail (Masino et al., 2003). Recently, the structure of the isolated N-terminal Josephin domain has been determined (Nicastro et al., 2005), providing further clues about the catalytic functions of this novel ubiquitous domain. The solution structure of the Josephin domain of ataxin-3 (Fig. 2) showed that it belongs to the family of papain-like cysteine proteases, with the structurally conserved Cys14, Asn134, His119 catalytic triad forming the enzyme's active site (Mao et al., 2005; Nicastro et al., 2005). The Josephin domain has a mixed α/β fold with high structural homology to staphopain from *Staphylococcus aureus* (Filipek et al., 2003), to the *Pseudomonas* avirulence protease AVRPPH3 (Zhu et al., 2004), and to the deubiquitinating enzymes, human UCH-L3 (Misaghi et al., 2005), and yeast YUH1 (Johnston et al., 1999). Several groups have shown that the Josephin domain alone displays proteolytic activity both towards linear substrates containing a ubiquitin molecule and towards K48-linked tetraubiquitin (Chow et al., 2004b; Mao et al., 2005; Nicastro et al., 2005). The described proteolytic activity, although low, is comparable to the activity shown by the full-length protein, indicating that either it requires additional cofactors for optimal activity or that the substrates used in vitro are still far from ideal (Chow et al., 2004b). The binding site of the substrate ubiquitin has been proposed to be located just above the active site of the Josephin domain, in a cleft bordered by the central catalytic subdomain and by a flexible helical hairpin that is likely to move to better accommodate the substrate (Mao et al., 2005; Nicastro et al., 2005). A possible model for ataxin-3 action is that the C-terminal UIM domains of ataxin-3 recruits polyubiquitinated substrates, which are then presented to the proteolytic Josephin domain (Berke et al., 2005; Mao et al., 2005; Nicastro et al., 2005). The low efficiency of proteolytic cleavage towards monomeric ubiquitin would increase thanks to the additional anchoring. Moreover, the structural studies also proved that despite the sequence and structural similarities between ubiquitin and the UBL domain of HHR23A/B, their binding sites on the surface of the Josephin domain are not overlapping. Indeed, the binding site for UBL domain is located on a large hydrophobic patch just opposite to the catalytic site of the Josephin domain (Nicastro et al., 2005), highlighting the multifaceted nature of this protein and the inherent complexity of determining its physiological role(s).

4.4. Ataxin-3 Interaction with Polyubiquitinated Proteins

A common feature of the diseases caused by polyglutamine expansion is the formation of ubiquitin-positive protein aggregates in the affected neurons, suggesting that the ubiquitin-proteasome system may be involved in mechanisms common to these otherwise unrelated diseases. In what concerns ataxin-3, several evidences indicate that it can bind polyubiquitin and polyubiquitinated proteins. Ataxin-3 can bind polyubiquitin chains in vitro and in vivo, through the UIM domains at its C-terminal region (Donaldson et al., 2003; Doss-Pepe et al., 2003; Chai et al., 2004). Moreover, ataxin-3 recruitment to polyglutamine aggregates was found to depend on functional UIM domains (Donaldson et al., 2003). Surface plasmon resonance binding analysis of the ubiquitin/ataxin-3 interaction confirmed that ataxin-3 binds

K48-linked tetra-ubiquitin chains, and showed that normal and expanded ataxin-3 display similar submilimolar dissociation constants for tetra-ubiquitin (Chai et al., 2004).

A possible model for ataxin-3 action is that the C-terminal UIM domains of ataxin-3 recruit ubiquitinated substrates, which are cleaved by the N-terminal protease domain of the protein. In fact, cells transfected with catalytically inactive constructs for normal and expanded ataxin-3 showed accumulation of ubiquitinated proteins, to levels similar to those observed when the proteasome is inhibited pharmacologically (Berke et al., 2005). Interestingly, this buildup of ubiquitinated proteins localized primarily to the cell nucleus, and no longer occurred when ataxin-3 mutants for the catalytic site were also mutated at the UIMs, strongly suggesting that substrates are presented to the protease domain of ataxin-3 through binding to the UIMs (Berke et al., 2005).

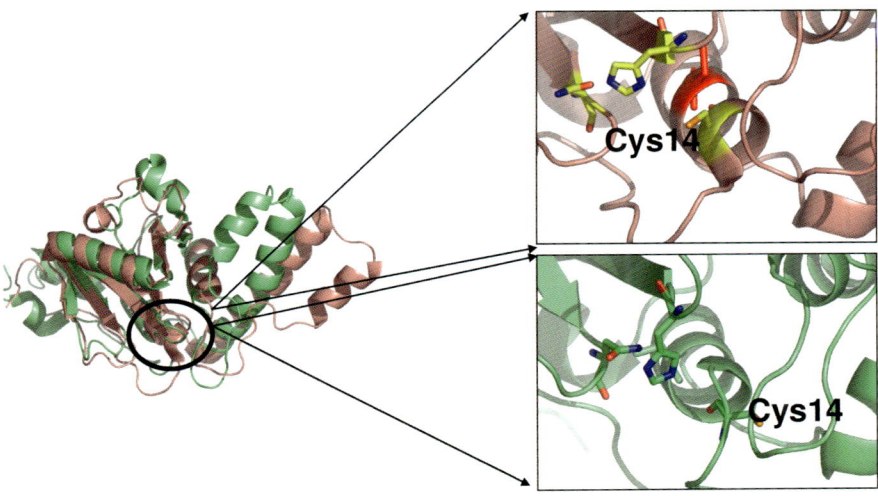

Fig. 2. Cartoon representation of the solution structures of the Josephin domain of ataxin-3. Superposition of the Josephin structures determined by NMR by Nicastro and colleagues (2005), colored in salmon, and Mao and colleagues (2005), colored in green. The major difference between both structures is the conformation of the flexible helical hairpin, located towards the right of the catalytic site, and proposed to be responsible for substrate specificity. The insets depict the catalytic site highlighting the catalytic triad represented as sticks (Cys14, His119, Asn134). Atoms are color-coded: oxygen in red, nitrogen in blue, sulfur in orange, carbon in yellow (top inset, Nicastro et al., 2005) or in green (bottom inset, Mao et al., 2005). Figures prepared with PyMol (http://www.pymol.org)

4.5. Ataxin-3 Ubiquitination and Clearance

Besides binding polyubiquitinated proteins, ataxin-3 is itself a substrate for ubiquitination. E4B, an ubiquitin chain assembly factor (E4), is required for the polyubiquitination and degradation of ataxin-3 (Matsumoto et al., 2004), and ataxin-3 is degraded by the ubiquitin–proteasome pathway (Matsumoto et al., 2004; Berke et al., 2005). Moreover, the expanded form of ataxin-3 is more stable than the normal form, but overexpression of E4B markedly increases the rates of degradation of normal and expanded ataxin-3, by similar extents, and both proteins are efficiently ubiquitinated (Matsumoto et al., 2004). In a *Drosophila* model, expanded ataxin-3-induced neurotoxicity was suppressed by E4B, eventually by reduction of the abundance of the expanded protein through ubiquitin-mediated degradation (Matsumoto et al., 2004).

A recent study by Berke and colleagues showed that the formation of large ubiquitinated forms of ataxin-3 is largely dependent on functional UIM domains, whereas the smallest ubiquitinated species of ataxin-3, which may represent monoubiquitinated protein, form even in the absence of functional UIMs (Berke et al., 2005). This study also suggests that ataxin-3 can act to remove ubiquitin in trans from other ataxin-3 molecules, via its ubiquitin protease activity, and can therefore be a target for its own deubiquitinating activity (Berke et al., 2005). The mono-/polyubiquitination of ataxin-3 is an intriguing modification, which function may go beyond the targeting of ataxin-3 for degradation.

A role for a cochaperone and ubiquitin ligase, the C-terminal Heat-shock protein 70-interacting protein (CHIP), in ataxin-3 degradation has also been reported (Jana et al., 2005). CHIP was found to interact with a truncated fragment of expanded ataxin-3 and to associate with cellular aggregates formed by this fragment. Moreover, one study found that CHIP enhances the ubiquitination of truncated ataxin-3 with an expanded polyQ tract, and promotes its degradation (Jana et al., 2005), whereas others found no effect of CHIP on the degradation of ataxin-3 (Matsumoto et al., 2004) or of other polyQ-containing proteins (Miller et al., 2005a). Interestingly, CHIP rescues polyQ aggregation and toxicity in primary cortical neurons, and in animal models (Miller et al., 2005a). These evidences point to an involvement of the two major components of cellular quality control mechanisms, molecular chaperones, and the ubiquitin proteasome pathway, in the cellular response to polyQ toxicity.

4.6. Ataxin-3 and the Proteasome

Redistribution of the 26S proteasome complex to nuclear inclusions of expanded ataxin-3 has been shown both in cellular models and upon analysis of human tissues, and proteasome inhibition has been directly associated with increased polyglutamine aggregation (Chai et al., 1999b). Wild-type ataxin-3 and its expanded form were found to associate with the proteasome, through the N-terminal part of the protein (Doss-Pepe et al., 2003). Moreover, the expanded form of ataxin-3 was shown to affect proteasome function, and to interfere with the degradation of proteolytic substrates (Doss-Pepe et al., 2003).

Ataxin-3 interacts with the human homologs of the yeast DNA repair protein RAD23, HHR23A, and HHR23B (Wang et al., 2000). RAD23 contains an N-terminal UBL domain that interacts with catalytically active proteasomes (Schauber et al. 1998), and two ubiquitin-associated domains (UBA) that bind ubiquitin and multiubiquitinated substrates (Chen et al., 2001a). Binding of RAD23 to the proteasome is required for efficient nucleotide excision repair, the primary mechanism for repairing UV-induced DNA lesions (Watkins et al., 1993). Moreover, RAD23 is thought to play a role as a "shuttle factor," for translocating proteolytic ubiquitinated substrates to the proteasome (Chen et al., 2002a). Normal and expanded ataxin-3 both bind to the N-terminal UbL domains of HHR23A/B, and HHR23A is recruited to intranuclear inclusions formed by mutant ataxin-3 (Wang et al., 2000). The structural bases of the interaction between the Josephin domain of ataxin-3 and HHR23B have been investigated, and Josephin was found to form a stable complex with HHR23B. The binding site at the surface of Josephin lies on the opposite face to the active site, strongly suggesting that this interaction is distinct from the protease activity of Josephin (Nicastro et al., 2005).

VCP (or p97) is an AAA-ATPase, and another protein with a role in the degradation of proteins by the ubiquitin–proteasome system (Ye et al., 2003). Several studies revealed an involvement of VCP in the pathogenesis of protein misfolding diseases (Higashiyama et al., 2002). VCP colocalizes with insoluble protein aggregates in several neurodegenerative diseases, including aggregated forms of ataxin-3 (Hirabayashi et al., 2001). VCP was recently found to be an interaction partner of ataxin-3 (Doss-Pepe et al., 2003; Matsumoto et al., 2004; Boeddrich et al., 2006), and to bind to an arginine/lysine-rich motif in ataxin-3 (Boeddrich et al., 2006). Moreover, VCP forms a ternary complex with ataxin-3 and E4B, an enzyme required for the ubiquitination of ataxin-3, thereby mediating the interaction between ataxin-3 and E4B (Matsumoto et al., 2004). Interestingly, the interaction between VCP and ataxin-3 modulates fibrillogenesis of pathogenic forms of ataxin-3. In vitro assays showed that equimolar concentrations of VCP enhance the formation of SDS-insoluble aggregates of expanded ataxin-3, whereas a fourfold molar excess of VCP completely suppresses the in vitro aggregation of expanded ataxin-3 (Boeddrich et al., 2006). In a *Drosophila* model, coexpression of VCP and full-length ataxin-3 with an expanded polyglutamine tract suppressed the degeneration of photoreceptors coupled with the formation of protein aggregates that is observed in transgenic flies that express expanded ataxin-3 (Boeddrich et al., 2006), in a way reminiscent of the suppression exerted by expression of E4B (Matsumoto et al., 2004). The functional relevance of the VCP/ataxin-3 interaction remains to be clarified, but an interesting hypothesis is that the association of ataxin-3 to VCP may help VCP present ubiquitinated proteins to the proteasome for degradation (Doss-Pepe et al., 2003; Boeddrich et al., 2006), taken that ataxin-3 binds polyubiquitinated chains and has ubiquitin protease activity (see earlier sections). The deubiquitinating activity of ataxin-3 may be important in removing the polyubiquitin chains from substrates before digestion in the proteasome. In fact, the capacity of ataxin-3 to suppress neurotoxicity caused by polyglutamines in a *Drosophila* model depends on the catalytic activity of ataxin-3, on functional UIM domains, and requires normal proteasome activity (Warrick et al., 2005).

It is at present still unclear how the action of VCP and HHR23A/B, the two binding partners of ataxin-3 with a functional role in the ubiquitin/proteasome pathway, is concerted to impinge upon ataxin-3. In the model proposed by Doss-Pepe and collaborators, multiubiquitinated proteolytic substrates bind to the UBA domains of HHR23, while VCP associates with ataxin-3 at the proteasome. HHR23 then associates with the proteasome through its UbL domain, and VCP mediates the transfer of multiubiquitinated substrates from Rad23 to ataxin-3 (Doss-Pepe et al., 2003). Ataxin-3 with its deubiquitinating activity might contribute to proteasomal degradation of ubiquitinated proteins by removing the polyubiquitin chains from substrates prior to digestion (Boeddrich et al., 2006).

5. ATAXIN-3 AGGREGATION IN MJD

5.1. Toxic Polyglutamine Tracts

A common feature of polyglutamine diseases is the deposition of insoluble intracellular ubiquitinated inclusions containing the misfolded disease protein (Paulson, 1999). Although their correlation with pathology is controversial (Bates, 2003; Michalik and Van Broeckhoven, 2003), nuclear and sometimes cytoplasmic inclusions are considered the identifying fingerprint of polyglutamine diseases and may provide clues to pathogenesis. These ubiquitin-rich aggregates recruit chaperone proteins such as Hsp40 and Hsp70, transcription factors, and proteasome subunits (Chai et al., 1999a, b). This implies underlying misfolding events that are critical to pathogenesis (Paulson, 1999; Taylor et al., 2002). Antibodies recognizing the polyglutamine expansion, independently of the disease protein, reflect the existence of conformational changes induced upon polyglutamine expansion (Trottier et al., 1998; Lunkes et al., 1999).

Early experiments on the elucidation of the pathogenicity associated with polyglutamine expansions have been focused on the structural, biochemical, and cytotoxic features of the polyglutamine repeats alone, independently of the protein context. The hypothesis that long polyglutamine tracts induce a common dominant "toxic gain of function" to the carrier proteins was based on the observation that (a) there is a lack of sequence similarity among proteins affected by polyglutamine expansion; (b) genetically engineered mice expressing proteins with long polyglutamine tracts recapitulate most of the disease features of polyglutamine diseases, regardless of whether the mice express full-length proteins (Clark et al., 1997) or only those portions of the proteins containing the polyglutamine tracts (Ikeda et al., 1996); (c) long polyglutamine tracts exert a dominant toxic effect without the need to be part of a functional protein to cause its damage (Ordway et al., 1999; Yang et al., 2002); and (d) the disease age-of-onset correlates inversely with the length of the polyglutamine tract. The fact that expanded polyglutamine containing fragments are sometimes more potent in inducing neurodegeneration than the full-length human disease proteins sparked a lot of research associating proteolytic events with polyglutamine-induced pathogenesis (Tarlac and Storey, 2003). Nonetheless, the type of aggregated proteins and the regional and cellular distribution of the protein deposits vary from disease to disease, indicating that the individual proteins carrying the polyglutamine expansion modulate the associated pathologies (Paulson, 2003). In this context, differential expression of interacting proteins in neurons could explain the selective neurotoxicity.

5.2. Amyloid-like Properties of Expanded Polyglutamine Aggregates

The structural and functional effects of polyglutamine expansions in the affected proteins and the mechanisms of polyglutamine toxicity have been a subject of intense research (Masino and Pastore, 2002; Ross et al., 2003; Temussi et al., 2003), since the establishment of the link between expansion of glutamine repeats and neurological disease. In the early 1990s, Perutz proposed that polyglutamine sequences could adopt a β structure consisting of antiparallel β-sheets held together by networks of hydrogen bonds between main-chain and side-chain amides. These structures were referred to as "polar zippers" (Perutz, 1994). Considering the extreme insolubility of glutamine peptides and the presence of polyglutamine repeats in transcription factors, Max Perutz also proposed that the expansion of polyglutamine repeats could result in (a) abnormal interactions with polyglutamine-containing transcription factors resulting in transcriptional disorders or (b) self-assembly leading to formation of aggregates in neurons (Stott et al., 1995). A couple of years later, intranuclear inclusions containing the disease proteins were unambiguously linked with polyglutamine expansion diseases (Davies et al., 1997; DiFiglia et al., 1997; Paulson et al., 1997b). Association of expanded polyglutamine proteins with transcription factors has also been shown for most of these polyglutamine disorders (Okazawa, 2003; Schaffar et al., 2004).

Two main strategies have been used in attempts to elucidate these repeat structures, either by trying to determine the structure of polyglutamine-containing peptides of different lengths (Perutz et al., 1994; Altschuler et al., 1997; Chen et al., 2001a, b; Sikorski and Atkins, 2005) or by introducing polyglutamine repeats into proteins with well known structures (Chen et al., 1999; Tanaka et al., 2001, 2003; Masino et al., 2002; Sambashivan et al., 2005). All these experimental data show that, independently of the repeat length, monomeric polyglutamine peptides exist predominantly in a random coil conformation. A conformational transition occurs upon aggregation: in close connection with the pathological threshold in polyglutaminopathies, the tendency to aggregate into β-rich fibrillar structures increases with polyglutamine-length (Chen et al., 2002b). The amyloid nature of these polyglutamine insoluble structures has been demonstrated by numerous in vitro (Scherzinger et al., 1997; Bevivino and Loll, 2001; Poirier et al., 2002) and in vivo (Scherzinger et al., 1997; Huang et al., 1998; McGowan et al., 2000) experiments. Thus, polyglutamine disorders are a part of a larger group of protein misfolding disorders, suggesting parallels with other amyloid-associated diseases such as Alzheimer's and prion diseases. Furthermore, polyglutamine aggregates are recognized by "pan-amyloid" antibodies able to associate with generic amyloid epitopes in fibrils formed from a diverse number of proteins (O'Nuallain and Wetzel, 2002). Whether the cellular toxicity is triggered by the pathogenic disease protein alone, by the putative intermediates of the fibrillization pathway, or by the soluble mature fibrils, is still a matter of debate (Michalik and Van Broeckhoven, 2003). Considering that polyglutamine aggregates exhibit many of the defining features of amyloid fibrils, and the striking similarities among amyloids in their formation and toxicity (Caughey and Lansbury, 2003), we should expect that large microscopically visible polyglutamine aggregates may be less influential than earlier aggregates for disease progression. In this line, protofibrillar intermediates in polyglutamine aggregation pathways have already been identified (Poirier et al., 2002; Tanaka et al., 2003; Gales et al., 2005). The possibility of detecting the existence of soluble toxic precursors of the final insoluble amyloid fibrils encountered in nuclear inclusions, will be essential for identifying the major culprits in polyglutamine-induced pathogenesis. Conformational antibodies recognizing soluble toxic oligomeric intermediates (Glabe and Kayed, 2006) and soluble polyglutamine intermediates (Brooks et al., 2004; Peters-Libeu et al., 2005) constitute an invaluable tool for elucidating the pathway of polyglutamine-mediated fibrillogenesis.

5.3. Polyglutamine Aggregation: Structural Models

Long polyglutamine tracts are unstructured as revealed by circular dichroism and NMR spectroscopy (Chen et al., 1999, 2001b; Gordon-Smith et al., 2001; Masino et al., 2002, 2004). Although they do not fold into a well-defined structure, they aggregate into β-rich structures (Chen et al., 2002a) and induce aggregation of their carrier proteins. A detailed structural analysis of polyglutamine aggregates is hampered by their heterogeneous nature and large insolubility. On the basis of the analysis of the aggregation behavior of a Q_{15} peptide flanked by basic residues, Max Perutz proposed the previously mentioned "polar zipper" model (Perutz, 1999). Since then, numerous models have been proposed to describe the structure of expanded polyglutamine aggregates namely parallel and antiparallel β-sheet, π-helix, and μ-helix (for a review, see Masino and Pastore, 2002; Ross et al., 2003). On the basis of mutational analysis, Thakur and Wetzel (2002) also proposed a model for poly-glutamine aggregates comprising alternating β-strand and β-turn elements, with an optimum of seven consecutive glutamines within each β-strand. Fibrillar assemblies from peptides with varying polyglutamine lengths (Q8, Q15, Q28, and Q45) showed that they all assemble into slablike antiparallel β-structures, suggesting that the Q45 fibril is composed of multiple reverse turns (Sharma et al., 2005). In particular, two of the models proposed for polyglutamine aggregation have sparked intense research on the mechanisms of pathogenesis in this group of disorders: the "channel hypothesis" (Kagan et al., 2001). Those were the μ-helix, proposed to form membrane-depolarizing ion channels in cells (Monoi, 1995; Monoi et al., 2000), and β-helix model proposed to form water-filled nanotubes (Perutz et al., 2002; Singer and Dewji, 2006). The plasma membrane is clearly emerging as a likely target for amyloid toxicity, reinforcing the many commonalities between polyglutamine disorders and other types of amyloids (Kagan, 2005). Recent data indicate that toxicity is probably not due to pore or channel formation, but to amyloid induced alterations in lipid packing, resulting in dysregulation of ion homeostasis (Demuro et al., 2005).

Further data on the structure of polyglutamine aggregates has been obtained by introducing polyglutamine stretches into structurally well-characterized proteins. Introduction of a 4–10 residue polyglutamine stretch in the inhibitory loop of the chymotrypsin inhibitor showed that it induces formation of dimers, trimers, and higher oligomers of this otherwise monomeric protein (Stott et al., 1995). As shown in the crystal structure of the isolated dimer, oligomerization occurred by a domain swapping mechanism (Chen et al., 1999) and, although induced by the introduction of a polyglutamine repeat, it was not mediated by the direct polyglutamine association (Gordon-Smith et al., 2001). Three-dimensional domain swapping has since been implicated in the formation of several protein aggregates (Janowski et al., 2005; Bennett et al., 2006; Guo and Eisenberg, 2006). In fact, Sambashivan and collaborators (2005) introduced as few as 10 glutamine residues into ribonuclease A and showed that domain swapping induced the formation of amyloid fibrils with native-like structure.

A better structural understanding of polyglutamine aggregation and the influence of the context of the expanded repeat is critical in order to elucidate the mechanisms of polyglutamine-induced cellular toxicity, still one of the most challenging tasks in the field of polyglutamine disorders. Clues into the structure of expanded polyglutamines will emerge from the elucidation of the crystal structures of polyglutamine protein complexed with novel conformational antibodies (Peters-Libeu et al., 2005).

5.4. The Emerging Role of the Protein Context: Ataxin-3 Aggregation Pathway

In MJD, expanded ataxin-3 accumulates in ubiquitinated intranuclear neuronal inclusions within selectively affected brain regions (Paulson et al., 1997b). Ataxin-3 was the first full-length polyglutamine protein to be structurally characterized. Bevivino and Loll (2001) showed that expansion of the glutamine repeat from 27 to 78 glutamines in ataxin-3 destabilizes the native protein structure. Fibrillar aggregates of the expanded protein formed readily under physiological conditions and showed increased β-structure as well as the ability to bind the amyloid dye, Congo red.

A puzzling issue in MJD, and in other trinucleotide repeat disorders as well, is that both normal and expanded proteins are ubiquitously expressed throughout the brain and other tissues, but only a particular subset of cells is selectively affected. It is becoming clear that although polyglutamine tracts themselves are toxic, the sequence and structure of the proteins carrying the polyglutamine tracts have important roles in defining the region-specific neuronal death (Orr, 2001). Those sequences determine features such as subcellular localization and post-translational modifications, also specifying interactions with other macromolecules within the cell and thus modulating the disease course. A matter of dispute has been whether pathogenesis is primarily activated in the cytoplasm or in the cell nucleus. Specific nuclear localization sequences have been identified in ataxin-3 (Tait et al., 1998, see earlier) and in other proteins carrying expanded polyglutamine tracts and nuclear-associated mechanisms are being implicated in neuropathogenesis (Ross et al., 1999). Curiously, it has been hypothesized that the nuclear environment drives nonexpanded ataxin-3 into a conformation with higher tendency to aggregate, which is recognizable by the expanded polyglutamine-specific antibody 1C2 (Perez et al., 1998, 1999), predicting that the nuclear environment favors pathology and aggregation. Moreover, ataxin-3 is unique among proteins affected by polyglutamine expansions because wild-type ataxin-3 is recruited into nuclear inclusions formed by other polyglutamine expanded proteins (Uchihara et al., 2001). The association of nonexpanded ataxin-3 with aggregation processes is underlined by its presence in intranuclear aggregates within Marinesco bodies found in human and non-human primate *substantia nigra* (Fujigasaki et al., 2000, 2001; Kettner et al., 2002). Altogether, these data reinforce the role of protein domains outside the expanded polyglutamine-tract in ataxin-3 aggregation and disease pathogenesis. This suggested that, in the adequate environment, nonexpanded ataxin-3 could aggregate and that the nucleus possibly provided the adequate environment for ataxin-3 misfolding resulting in protein aggregation and toxicity.

Remarkably, in vitro studies have shown that both nonpathological ataxin-3 and the isolated Josephin domain can form insoluble amyloid fibrils with increased β-structure upon partial unfolding induced by increased temperature and pressure (Marchal et al., 2003; Shehi et al., 2003; Masino et al., 2004) or by chemical denaturation (Chow et al., 2004c). The intrinsic ability of ataxin-3 to form amyloid fibrils, independently of the polyglutamine tract, has been emphasized by the finding that the full-length human protein can oligomerize and form amyloid-like fibrils under physiological conditions (Gales et al., 2005). Moreover, the propensity towards self-assembly seems to be a property of the N-terminal Josephin domain, the domain that shows the highest degree of conservation across species. Indeed, the human ataxin-3 and its *C. elegans* homolog form dimers and high-molecular mass oligomers, and the Josephin domain alone is also able to dimerize (Gales et al., 2005). Strikingly, nonexpanded ataxin-3 oligomers and, most remarkably, the Josephin dimers share a conformational epitope, absent in the monomeric forms of the protein, common to soluble cytotoxic species found in the fibrillization pathway of expanded polyglutamine proteins and other amyloid forming proteins (Kayed et al., 2003). This indicates that dimerization of the Josephin domain is accompanied by a structural rearrangement that might constitute the first misfolding event of the ataxin-3 aggregation pathway. These data suggest that those dimeric structures might represent early toxic intermediates with relevance for the pathogenesis of MJD as shown for other amyloid-related degenerative diseases. Further oligomerization/aggregation events seem to be mediated and/or induced by the other protein domains, namely by the polyglutamine domain. In fact, a recent study has shown that expanded ataxin-3 fibrillogenesis undergoes a two stage pathway and that each step is mediated by different protein domains (Ellisdon et al., 2006). The first stage, involving the formation of sodium dodecylsulfate (SDS)-soluble structures, is common to expanded and nonexpanded variants of ataxin-3 and is accompanied by a conformational change within the Josephin domain. Expansion of the polyglutamine tract into a pathological length introduces a second step in the aggregation pathway of ataxin-3 leading to the formation of SDS-insoluble aggregates (Ellisdon et al., 2006). In agreement, polyglutamine-expanded ataxin-3 fragments have been shown to form SDS-insoluble aggregates, with the ability to recruit full-length nonexpanded ataxin-3 (Haacke et al., 2006). Recruitment of nonexpanded ataxin-3 into insoluble aggregates formed by truncated polyglutamine fragments was shown to be accompanied by a conformational change involving the Josephin domain (Haacke et al., 2006). The presence of self-association regions with relevance for protein aggregation outside the expanded amino acid tract have been previously described for ataxin-1 (Burright et al., 1997; de Chiara et al., 2005) and also for PABPN1, a protein associated with oculopharyngeal muscular dystrophy upon expansion of a polyalanine region (Fan et al., 2001).

A number of reports stress the importance of oligomeric intermediates as the major cytotoxic species in various forms of amyloidogenesis (Bucciantini et al., 2002; Schaffar et al., 2004; Walsh and Selkoe, 2004; Kazlauskaite et al., 2005), suggesting that formation of insoluble inclusions might constitute a protective mechanism (Arrasate et al., 2004). Intermediate oligomeric structures displaying a epitope common to toxic amyloid precursors have been identified both for nonexpanded ataxin-3 and for the isolated Josephin domain (Gales et al., 2005). The relevance of those oligomeric structures as models for understanding the pathogenic mechanisms in MJD has been emphasized by the finding that the first step in ataxin-3 fibrillogenesis is shared by expanded and nonexpanded ataxin-3, and it is thus independent of the polyglutamine tract (Ellisdon et al., 2006). In this context, it was shown that expansion of the polyglutamine tract of ataxin-3 does not affect the overall stability of the protein (Chow et al., 2004a). An attractive hypothesis could be that, within the normal cell environment, expansion of the polyglutamine tract triggers extremely small conformational changes persistently interfering with macromolecular interactions, and continuously inducing the exposure of protein regions with high propensity to misfold and aggregate. In vivo immediate formation of toxic intermediates of nonexpanded protein and expanded ataxin-3 is possibly controlled by the normal protein homeostasis machinery, and additionally by post-translational modifications, intracellular localization, and/or interaction with specific macromolecular partners (Masino et al., 2004; Gales et al., 2005). With aging and consequent impairment of the mechanisms for cellular quality control, the insults inferred by the buildup of partially misfolded ataxin-3 would become increasingly large leading to neuronal cell death, in accordance with the late onset nature of these disorders. Given the relatively lower rates of nonexpanded ataxin-3 fibrillogenesis in the "test tube," it should be expected that in vivo it would have a lag time for aggregation far beyond the normal human lifespan.

All these data point to the emerging role of the Josephin domain in early misfolding events rendering it an extremely attractive target for the design of stabilizing mutations and compounds with the ability to interfere with early oligomerization events in MJD.

Table 2. Transgenic mouse models of Machado-Joseph disease

Promoter	Transgene	Cellular population expressing the transgene	Age of Onset	Pathology	Reference
L7 promoter	Fragment MJD1 cDNA 79 CAGs	Purkinje cells	4 Weeks	Ataxic behavior and cerebellar atrophy	Ikeda et al. (1996)
L7 promoter (Purkinje cells-specific)	Full-length MJD1 cDNA - 79 CAGs	Purkinje cells	-	Normal behavior, no pathology	Ikeda et al. (1996)
MJD1	Full-length MJD1 gene (YAC); 15, 64, 67, 72, 76 and 84 CAGs	Widespread	4 Weeks	Cerebellar ataxia, degeneration and nuclear aggregates	Cemal et al. (2002)
Prion promoter	Full-length mjd1 cDNA 20 and 71 CAGs	Widespread	2–4 Months	Cerebellar ataxia and nuclear aggregates	Goti et al. (2004)

6. ANIMAL MODELS OF MJD

The first transgenic mouse model of MJD was developed by (Ikeda et al., 1996), as depicted in Table 2. Surprisingly, expression of the full-length *MJD1* cDNA with 79 CAGs driven by the L7 Purkinje cell-specific promoter resulted in a nonphenotypic mouse even at 23 weeks of age. Expression of the transgene in a cell population that is not the main target of the disorder may have been critical for the lack of neurological deficit. Nevertheless, replacement of the full-length *MJD1* gene by a fragment of the gene (C-terminal fragment (HA-Q79C), or the same fragment truncated at the C-terminus (HA-Q79)), resulted in mice with ataxic posture and gait disturbance as early as 4 weeks after birth even in heterozygous animals. Frequent falling-down when moving and failure to rear were reported. The mice had an atrophic cerebellum (about one-eighth of its normal volume), with reduction of granular, Purkinje, and molecular cell layers. Toxicity of truncated ataxin-3 coupled to the absence of pathology of full-length ataxin-3 suggested that cleavage of ataxin into a toxic fragment was involved in the pathology.

The group of Nancy Bonini generated one of the first animal models of MJD, the *Drosophila* model. Initially a segment of the ataxin-3 protein was used. The targeted expression of truncated ataxin-3 with an expanded polyglutamine repeat (78 glutamines) in the eye caused loss of pigmentation and collapse of internal retinal structure, while the wild-type ataxin-3 had no effect. Moreover, expression of mutant ataxin-3 fragment led to nuclear inclusion formation and neuronal degeneration demonstrating conservation of polyglutamine toxicity in invertebrates (Warrick et al., 1998). Behavioral changes, particularly locomotor dysfunction and olfactory dysfunction were also reported in this model (Kim et al., 2004). More recently, both normal and several pathogenic forms of truncated and full-length human ataxin-3 were expressed in *Drosophila*, allowing the study of the function of the N-terminal portion of the protein, which is responsible for its ubiquitin protease activity.

The full-length mutant ataxin-3 caused late-onset progressive neurodegeneration but was less toxic than the fragment due to a detoxifying capability of the functional domains of the normal protein. Moreover, wild-type ataxin-3 supressed neurotoxicity of mutant ataxin-3 by a ubiquitin-associated mechanism (Warrick et al., 2005).

Cemal et al. (2002) generated a transgenic mouse model using a human yeast artificial chromosome containing the full-length mutant *MJD1* gene under the control of its own regulatory elements with CAG triplet expansions of 64, 67, 72, 76, and 84 repeats and a wild-type construct with 15 repeats. The presence of all the natural regulatory elements of the gene is expected to allow cell-specific expression at physiological levels and a better reproduction of the disease. The mice with expanded alleles exhibit a mild and slowly progressive cerebellar deficit, beginning at 4 weeks of age. As the disease progress, the pelvic elevation becomes markedly flattened, accompanied by hypotonia, and motor and sensory loss. NII formation and cell loss is prominent in the pontine and dentate nuclei, with variable cell loss in other regions of the cerebellum from 4 weeks of age. Peripheral nerve demyelination and axonal loss are detected in symptomatic mice from 26 weeks of age. Moreover, a number of metabolic perturbations were identified by NMR-based techniques particularly increases in glutamine, decreases in GABA, choline, phosphocholine, and lactate in cerebrum and cerebellum (Griffin et al., 2004).

A transgenic mouse model carrying a truncated ataxin-3 cDNA (amino acid 286 to the C terminus of elongated ataxin-3) with an expanded polyglutamine stretch fused to the orexin gene has been generated. In these mice orexin-containing neurons are ablated by specific expression of mutant ataxin-3 in orexinergic-neurons in mice, which results in narcolepsy, hypophagia, and obesity. These mice were developed as model of human narcolepsy. Nevertheless, they express mutant ataxin-3 in the hypothalamus and have recently been used to evaluate the effects of the natural osmoprotectant ectoine as a protective agent from polyglutamine-induced toxicity (Hara et al., 2001).

Goti et al. (2004) generated transgenic mice expressing human mutant (Q71) or normal (Q20) ataxin-3 under the control of the mouse prion promoter. Heterozygous animals did not display a phenotype. On the contrary, homozygous transgenic mice expressing mutant ataxin-3 (Q71) developed a pathological phenotype (onset age of 2–4 months) with progressive postural instability, tremor, gait and limb ataxia, weight loss, premature death, NIIs, and decreased tyrosine hydroxylase-positive neurons in the *substantia nigra*. A cleavage fragment of ataxin-3 was abundant in brains of animals displaying a phenotype, and scarce in mutant animals not displaying a phenotype (Q71). Moreover, the ataxin-3 fragment was also detected in postmortem human brain. The fragment was attributed to residues C terminal to amino acid 221 including the polyglutamine expansion. These studies suggest that proteolytic cleavage of mutant ataxin-3 may produce a cleavage fragment, which above a critical concentration becomes toxic (Goti et al., 2004). This theory is also supported by the above-mentioned results of Ikeda et al. (1996) and the report of Haacke et al. (2006) who recently demonstrated that the removal of the N-terminus of polyglutamine-expanded ataxin-3 is required for aggregation and that proteolytic cleavage of full-length pathogenic ataxin-3 into C-terminal fragments initiates the formation of aggregates in neuroblastoma cells.

Recently, the development of an inducible transgenic mouse model using the Tet-Off-System has been reported (Riess et al., 2005). For the ataxin-3 responder mouse lines, full-length constructs containing 15 repeats (control lines) and 77 glutamine repeats (disease model) were used. The model allows turning ataxin-3 expression OFF or ON at different developmental stages and will allow study of the pathology reversibility (Riess et al., 2005). This group also developed a transgenic mice with mutant ataxin-3 coupled to either a NLS or a nuclear export signal. The NLS markedly accentuated the neuropathology induced by mutant ataxin-3 as compared to the phenotype of animals carrying the nuclear export signal or no additional signaling.

An alternative model using lentiviral vectors has also been reported. Wild-type and mutant human ataxin-3 were cloned into the lentiviral vectors backbone in order to develop an in vivo model of the disease. Lentiviral vectors coding for wild-type (27 CAG) or mutant ataxin-3 (72 CAG) driven by the phosphoglycerate kinase 1 (PGK) promoter were injected in the rat *substantia nigra*. Overexpression of mutant but not wild-type human ataxin-3 in the *substantia nigra* of adult rats was associated with the appearance of ubiquitinated ataxin-3 aggregates, loss of tyrosine hydroxylase, and vesicular monoamine transporter (VMAT) expression and behavioral deficit. Lentiviral-mediated expression of mutant ataxin-3 in the rat brain may provide a flexible complementary setting to dissect the function of wild-type ataxin-3 and evaluate potential therapies (Alves et al., 2005).

As different models of MJD become available and with more tools to study the pathogenesis of the disease and possible therapeutic approaches, the absence of another important tool – a knock out animal of MJD – becomes striking.

7. THERAPEUTIC STRATEGIES IN MJD

It is not the object of this chapter to provide a review of symptomatic therapies (reviewed in (Ogawa, 2004) for MJD but rather to review the possible approaches based on the available knowledge over the mechanism of the disorder, which could provide ways to slow or block the disease progression. See Fig. 3 for a schematic representation of the pathogenesis and potential therapeutic strategies for MJD.

7.1. Avoiding Cleavage of Ataxin-3

Blocking the cleavage of the amyloid precursor protein is a strategy presently being pursued for therapy of Alzheimer's disease, and a similar strategy could be envisioned for therapy of MJD once knowledge over this process becomes available. The "toxic cleavage fragment hypothesis" suggests that blocking proteolysis of ataxin-3 could be of therapeutic value (Ikeda et al., 1996; Goti et al., 2004; Colomer Gould, 2005; Haacke et al., 2006). Proteolysis of ataxin-3 by caspases, particularly caspase-1, during apoptosis has been suggested. Moreover, the in vitro caspase-mediated proteolysis and production of polyglutamine-containing fragments of ataxin-3 could be blocked by caspase inhibitors (Wellington et al., 1998; Berke et al., 2004). Nevertheless, in other experimental settings involving cerebellar neurons, mouse brains or brain tissues of MJD patients it has not been possible to identify proteolytic fragments of ataxin-3 (Cemal et al., 2002; Berke et al., 2004; Chou et al., 2006). Further studies of this process are needed.

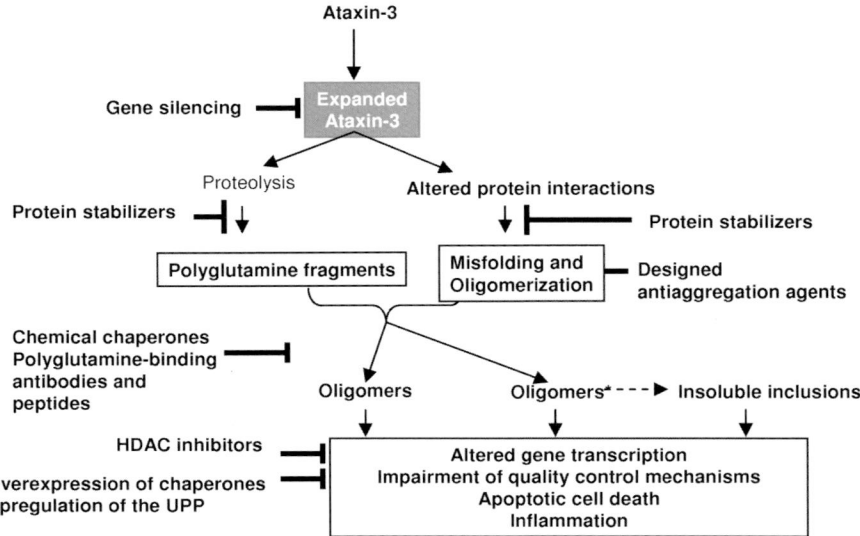

Fig. 3. Flowchart of the pathogenesis and potential therapeutic strategies for MJD. Expansion of the polyglutamine tract of ataxin-3 initiates a cascade of events that culminates with the accumulation of insoluble inclusions in selected neurons. The triggering event in expanded ataxin-3 aggregation can result from (a) its increased sensitivity to proteolysis, leading to generation of a polyglutamine-rich proteolytic fragment, or (b) interference of the polyglutamine tract with stabilizing molecular partners, leading to an increased propensity for misfolding and oligomerization. Either the full-length protein or polyglutamine-rich fragments can generate different oligomeric assemblies, some of which (oligomers*) can later be converted to insoluble inclusions. Although the relative toxicity for each of these oligomeric assemblies is not fully characterized, they are probably structurally and pathologically common to most polyglutamine diseases. Similarly, the downstream events that lead to neurodegeneration, such as altered gene transcription, impairment of the cells quality control mechanisms, apoptotic cell death, and inflammation are common to most other polyglutamine disorders. Early therapeutic approaches to MJD should (1) specifically target the gene (gene silencing) preventing expression of the protein carrying the expanded polyglutamine tract, or (2) precisely stabilize the protein, preventing its proteolyis and/or misfolding and oligomerization. Detailed knowledge of the target gene and protein structure is essential for those approaches. In contrast, therapies based on the use of general protein stabilizers, polyglutamine binding peptides, expanded-polyglutamine specific antibodies, HDAC inhibitors, and upregulation of the cell's quality control machinery, should have a more general application to other polyglutamine disorders, and do not require a detailed knowledge of the gene/protein

7.2. Inhibition of Apoptosis

Despite the evidences of a proteolytic cleavage fragment of ataxin-3 remaining the matter of study, there is evidence of apoptosis and caspase activation mediated by expression of truncated and full-length mutant ataxin-3 containing an expanded polyglutamine tract (Ikeda et al., 1996; Yoshizawa et al., 2000; Chou et al., 2006) as described in Sect. 3. This suggests approaches targeting apoptosis blockade could be envisioned for therapy of MJD.

7.3. Protein Misfolding, Oligomerization, and Aggregation: Chaperones and Proteolysis

The implication of protein misfolding in the pathogenesis of MJD suggested that chaperones could have neuroprotective effects in this disorder.

The groups of Henry Paulson and of Nancy Bonini were the first to suggest that targeted expression of Hsp40 or similar chaperones could become a feasible approach to slow the progression of MJD. Hsp40 and Hsp70 chaperones were shown in vitro to colocalize with ataxin-3 intranuclear aggregates. Overexpression of Hsp40 aided the handling of misfolded or aggregated polyglutamine-expanded ataxin-3 and suppressed polyglutamine aggregation with a parallel decrease in toxicity (Chai et al., 1999a). Furthermore, cell pathology and phenotype were rescued by the overexpression of Hsp70 (Warrick et al., 1999). A mechanism of chaperone detoxification by modification of the solubility properties of mutant ataxin-3 has been proposed on the basis of observations in *Drosophila* (Chan et al., 2000).

Recently, the process of proteasomal degradation of ataxin-3 has been found to be mediated by ubiquitin chain assembly factor (E4), and overexpression of this factor promoted degradation of mutant ataxin-3 both in vitro and in a *Drosophila* model of MJD. In this model expression of E4B specifically suppressed neurodegeneration induced by mutant ataxin-3, suggesting that E4 is a rate-limiting factor in the degradation of polyglutamine-expanded ataxin-3 (see Sect. 4.5). Therefore, a gene therapy-mediated overexpression of E4B could have a potential therapeutic interest in MJD (Matsumoto et al., 2004). Interestingly, CHIP, previously described in Sect. 4.5, has been shown to increase the ubiquitination and the rate of degradation of polyglutamine-expanded huntingtin or ataxin-3 and to suppress the aggregation and cell death mediated by expanded polyglutamine proteins (Jana et al., 2005; Miller et al., 2005a). Further studies are needed to confirm the value of this protein as a molecular target for therapy of MJD.

Another protein that has surprisingly been shown to act as an ubiquitin protease with selective activity against specific disease proteins is ataxin-3 itself. Normal human ataxin-3 was shown to be a striking suppressor of polyglutamine neurodegeneration in *Drosophila*. This suggests that augmenting ataxin-3 activity or enhancing cellular pathways in which it participates could provide therapeutic approaches to MJD and perhaps other neurodegenerative disease states (Warrick et al., 2005).

More studies with these factors are needed to confirm their therapeutic value and better define the proteolytic mechanisms that could promote mutant ataxin-3 degradation. In any case, therapy with the aforementioned factors would probably need a gene transfer approach as large proteic molecules may prove difficult to deliver by other means. An alternative approach to the use of gene transfer could be the use of small molecules that act as chemical chaperones.

7.4. Protein Misfolding, Oligomerization, and Aggregation: Small Molecules and Antibodies

Protein misfolding, oligomerization, and formation of insoluble inclusions represent a common physiological response to pathogenic proteins. Thus, many laboratories have developed high-throughput screening assays aimed at the discovery of molecules with selective binding affinities for polyglutamine expanded proteins, with the ability to modulate their pathogenic properties (such as enrolment in anomalous protein–protein interactions, oligomerization, and aggregation) and with potential therapeutic applications (Desai et al., 2006). Depending on the specific design of the assay, the major targets can be mutant protein misfolding/stability, formation of smaller oligomers or microaggregates, or fibril nucleation and extension. In vitro assays have been developed involving filter retardation of SDS-insoluble polyglutamine aggregates (Heiser et al., 2000, 2002). Compounds identified in those assays such as Congo Red, Thioflavin S, chrysamine G, and benzothiazoles show affinity for amyloid β-sheet structures and may prevent aggregate nucleation. The group of Ronald Wetzel developed an alternative in vitro assay that is based on the recruitment of fluorescently labelled monomeric polyglutamine by preformed polyglutamine aggregates immobilized on a solid surface (Berthelier et al., 2001). This assay monitors aggregate elongation and might yield a different class of inhibitors with the ability to prevent fibril elongation. A FRET (fluorescence resonance energy transfer)-based cellular assay has also been developed with the aim of detecting early intramolecular structural changes in the polyglutamine pathological proteins (Pollitt et al., 2003). Compounds identified in those assays might stabilize the native fold of the pathogenic protein, preventing misfolding, and aggregation into ordered fibrils. Most of the compounds identified in these screens are tested not only in cell culture models of polyglutamine diseases (Apostol et al., 2003) but also in *Drosophila* and mouse models. The azo-dye Congo red is a compound that has shown promising results in the prevention of protein oligomerization in several in vitro and in vivo models (Sanchez et al., 2003). Trehalose, a compound known to affect the hydratation properties of proteins and to stabilize their native fold has been shown to reduce polyglutamine aggregates and to extend the lifespan of a transgenic mouse model of HD (Tanaka et al., 2005).

Another therapeutic approach involves the use of antibodies with specific affinities for the mutant polyglutamine proteins. In fact, the monoclonal antibody 1C2 that recognizes the expanded polyglutamine stretch also inhibits protein aggregation in vitro (Heiser et al., 2000). However, antibodies show poor blood–brain barrier permeability and are highly sensitive to proteolysis, features that might hinder their usage in therapeutical applications. The development of intracellularly expressed single chain Fv antibodies intrabodies (intrabodies), with the ability to specifically bind to mutant proteins and block their toxic effects might be an attractive alternative, although their delivery to the affected brain regions will require gene therapy. In this line, an intrabody binding to a sequence flanking the polyglutamine tract of huntingtin inhibited its aggregation both in a cell culture and in a *Drosophila* model of HD (Lecerf et al., 2001; Miller et al., 2005b; Wolfgang et al., 2005).

Small peptides as well as chemical chaperones with the ability to modulate protein folding and stabilize proteins in their native conformation have been tested for their therapeutical potential in MJD. A tandem repeat of the polyglutamine binding peptide QBP1, that preferencially binds to polyglutamine stretches, has been shown to decrease aggregate formation and rescue survival of a MJD fly model (Nagai et al., 2003). Working with BHK-21 and Neuro2a cells transiently transfected with N-terminal truncated ataxin-3 with an expanded polyglutamine stretch, Yoshida and collaborators observed reductions in aggregate formation and cytotoxicity mediated by chemical chaperones. The organic solvent dimethyl sulfoxide (DMSO), the cellular osmolyte glycerol, and trimethylamine N-oxide (TMAO) all effectively suppressed polyglutamine-induced cytotoxicity (Yoshida et al., 2002). More recently, ectoine, a bacteria osmoprotectant, has been shown to decrease the total amount of aggregates, particularly large cytoplasmic inclusions, but to increase the frequency of nuclear inclusions in cultured neuro2a cells. Despite the presence of nuclear inclusions, apoptotic features were less frequently observed after ectoine application, suggesting it may protect cells from polyglutamine-induced toxicity (Furusho et al., 2005b). This study has recently been transposed in vivo to the orexin mice (Hara et al., 2001; see Sect. 6), which were submitted to daily intraperitoneal injections of ectoine. Although the protective effect over orexin-positive neurons was mild, it suggested that ectoine may reduce polyglutamine-induced neurotoxicity in vivo and opens the way for testing this osmolite in other in vivo models that more closely reproduce the disorder (Furusho et al., 2005a).

In analogy with the structural studies, most of the screenings are based on monitoring the aggregation of polyglutamine peptides or on protein fragments containing the expanded polyglutamine stretch, rather than on full-length proteins that are difficult to obtain in high yields and are very unstable. As the role of the protein context becomes more relevant in these diseases (Sect. 5.4.), therapeutic approaches that target specific structural features of the individual proteins gain relevance to prevent early structural alterations in mutant ataxin-3 and delay pathology in MJD.

7.5. Targeting Transcriptional Dysfunction

Polyglutamine-expanded ataxin-3, as other polyglutamine expanded proteins, has been shown to repress transcription. Ataxin-3 acts through distinct mechanisms involving both the polyglutamine-containing C-terminus and the N-terminus of ataxin-3 (Li et al., 2002). Therefore, transcription dysfunction is a potential target for therapeutic intervention using approaches such as the inhibitors of histone deacetylases already tested with some therapeutic effects in other polyglutamine disorders (Hockly et al., 2003; Gardian et al., 2005).

7.6. Prevention of Inflammation

Inflammatory mediators (pro-inflammatory cytokine interleukin-1 and the chemokine SDF1) are upregulated in ataxin-3-expressing cell lines and disease brains and may contribute to loss of neurons in MJD pathogenesis. Therefore, anti-inflammatory drugs may be envisioned for future treatment of MJD (Evert et al., 2001).

Other therapeutic strategies targeting general mechanisms of neurodegeneration such as excitotoxicity, oxidative stress, and trophic factor deprivation have been proposed and tested in experimental models of other polyglutamine disorders with variable degrees of success and therefore these mechanisms should also be investigated in MJD.

7.7. Gene Silencing

The toxicity of proteins with expanded polyglutamine tracts suggests that blocking the mutant protein production in the cell can be of therapeutic value. Therefore, gene-silencing techniques, aiming at reducing intracellular concentration of polyglutamine carrying proteins, are a promising strategy for therapy of polyglutamine diseases. Significant improvements of the technique of gene silencing by double-stranded RNAs (dsRNAs) have been achieved during the last years (Fire et al., 1998; Elbashir et al., 2001). This strategy, also designated as RNA interference (RNAi), induces the sustained down-regulation of the target gene by promoting mRNA degradation. The technology has been tested in mammalian and *Drosophila* cell culture models of polyglutamine diseases, particularly for SBMA (Caplen et al., 2002) and MJD (Miller et al., 2003). Different dsRNAs were able to inhibit expression of nonrepetitive sequences transcripts of the truncated human AR and of the ataxin-3 protein and to rescue mammalian cells from polyglutamine toxicity. Moreover, Miller et al. (2003) specifically silenced mutant ataxin-3 by taking advantage of a single nucleotide polymorphism in the *MJD1* gene, a G-to-C transition immediately 3' to the CAG repeat (G987C), which is in linkage disequilibrium with the disease-causing expansion, in most families segregating perfectly with the disease allele. Worldwide, 70% of disease chromosomes carry the C variant. In transfected cells, the short interfering RNAs caused allele-specific suppression of the mutant protein and nearly eliminated the accumulation of aggregated mutant ataxin-3, with only modest effects on wild-type ataxin-3. This seminal paper by the groups of Henry Paulson and Beverly Davidson demonstrated the feasibility of specific silencing of the mutant alleles while allowing continued expression of the wild type protein.

More recently, motor and neuropathological abnormalities in mouse models of SCA1 and HD were improved by reducing expression of a pathogenic polyglutamine expanded protein with adenoassociated viral vectors (AAV)-delivered shRNA (Xia et al., 2004; Harper et al., 2005). In both cases, long-term insertion of a RNA template encoding double stranded RNAs within the target cells was achieved with AAV vectors.

A similar strategy could be envisioned for MJD using viral vectors such as AAV vectors or lentiviral vectors. Lentiviral vectors derived from HIV-1 stably infect nondividing cells and have proven to be effective in expressing transgenes over prolonged periods of time (de Almeida et al., 2001). Recently, lentiviral vectors-mediated silencing of polyglutamine-expanded human and rat ataxin-3 in 293T and PC12 cell models has been reported (Alves et al., 2005). The vectors encode for a double stranded RNA-expression construct, targeted at the MJD1 transcripts, and are being used as a gene delivery strategy for in vitro studies with cell lines and investigation of its effects in an animal model of MJD disease will follow. Besides silencing mutant human ataxin-3, the authors also developed vectors targeting rat ataxin-3 to evaluate the role of ataxin-3 in the adult rat brain. Moreover, as no knock-out animal of MJD is available, silencing endogenous rat ataxin-3 will answer whether silencing should be targeted exclusively at the mutant allele or could simply target both mutant and wild-type alleles.

The most recent years were characterized by a proliferation of efforts directed at understanding the pathogenesis of MJD. A considerable assembly of data has been produced. This suggests that, provided these studies are pursued, the next years will hopefully lead to the knowledge and the tools to delay or stop this terrible disease.

8. ACKNOWLEDGMENTS

S.M.-R. acknowledges the financial support from *Fundação para a Ciência e a Tecnologia*, FCT, Portugal (POCTI/MGI/47550/2002 and POCI/SAU-MMO/60156/2004) and from *Crioestaminal/Associação Viver a Ciência*. A.L.C. acknowledges financial support from FCT, Portugal (POCI/SAU-MMO/60156/2004). L.P.A. acknowledges financial support from FCT, Portugal (POCI/SAU-MMO/56055/2004) and from the National Ataxia Foundation (USA). A.C.R. acknowledges financial support from FCT, Portugal (POCI/SAU-NEU/57310/2004) and from the *Instituto de Investigação Interdisciplinar*, University of Coimbra, Portugal (III/BIO/49/2005).

9. REFERENCES

Albrecht, M., Golatta, M., Wullner, U. and Lengauer, T., 2004, Structural and functional analysis of ataxin-2 and ataxin-3. *Eur. J. Biochem.* **271**: 3155.
Altschuler, E.L., Hud, N.V., Mazrimas, J.A. and Rupp, B., 1997, Random coil conformation for extended polyglutamine stretches in aqueous soluble monomeric peptides. *J. Pept. Res.* **50**: 73.
Alves, S., Regulier, E., Deglon, N. and de Almeida, L.P., 2005, Lentiviral-based overexpression and silencing of the ataxin-3 gene. *Abstract Viewer/Itinerary Planner. Washington, DC: Society for Neuroscience, 2005. Online.* Program No. **427.9**.
Andersen, K.M., Hofmann, K. and Hartmann-Petersen, R., 2005, Ubiquitin-binding proteins: similar, but different. *Essays Biochem.* **41**: 49.
Apostol, B.L., Kazantsev, A., Raffioni, S., Illes, K., Pallos, J., Bodai, L., Slepko, N., Bear, J.E., Gertler, F.B., Hersch, S., Housman, D.E., Marsh, J.L. and Thompson, L.M., 2003, A cell-based assay for aggregation inhibitors as therapeutics of polyglutamine-repeat disease and validation in Drosophila. *Proc. Natl .Acad. Sci. USA* **100**: 5950.
Aronin, N., Kim, M., Laforet, G. and DiFiglia, M., 1999, Are there multiple pathways in the pathogenesis of Huntington's disease? *Philos. Trans. R. Soc. Lond. B Biol. Sci.* **354**: 995.
Arrasate, M., Mitra, S., Schweitzer, E.S., Segal, M.R. and Finkbeiner, S., 2004, Inclusion body formation reduces levels of mutant huntingtin and the risk of neuronal death. *Nature* **431**: 805.
Bates, G.P., 2001, Huntington's disease. Exploiting expression. *Nature* **413**: 691.
Bates, G., 2003, Huntingtin aggregation and toxicity in Huntington's disease. *Lancet* **361**: 1642.
Bennett, M.J., Sawaya, M.R. and Eisenberg, D., 2006, Deposition diseases and 3D domain swapping. *Structure* **14**: 811.
Berke, S.J., Schmied, F.A., Brunt, E.R., Ellerby, L.M. and Paulson, H.L., 2004, Caspase-mediated proteolysis of the polyglutamine disease protein ataxin-3. *J. Neurochem.* **89**: 908.
Berke, S.J., Chai, Y., Marrs, G.L., Wen, H. and Paulson, H.L., 2005, Defining the role of ubiquitin-interacting motifs in the polyglutamine disease protein, ataxin-3. *J. Biol. Chem.* **280**: 32026.
Berthelier, V., Hamilton, J.B., Chen, S. and Wetzel, R., 2001, A microtiter plate assay for polyglutamine aggregate extension. *Anal. Biochem.* **295**: 227.
Bevivino, A.E. and Loll, P.J., 2001, An expanded glutamine repeat destabilizes native ataxin-3 structure and mediates formation of parallel beta-fibrils. *Proc. Natl. Acad. Sci. USA* **98**: 11955.
Bezprozvanny, I. and Hayden, M.R., 2004, Deranged neuronal calcium signaling and Huntington disease. *Biochem. Biophys. Res. Commun.* **322**: 1310.
Boeddrich, A., Gaumer, S., Haacke, A., Tzvetkov, N., Albrecht, M., Evert, B.O., Muller, E.C., Lurz, R., Breuer, P., Schugardt, N., et al., 2006, An arginine/lysine-rich motif is crucial for VCP/p97-mediated modulation of ataxin-3 fibrillogenesis. *EMBO J.* **25**: 1547.
Brooks, E., Arrasate, M., Cheung, K. and Finkbeiner, S.M., 2004, Using antibodies to analyze polyglutamine stretches. *Methods. Mol. Biol.* **277**: 103.
Bucciantini, M., Giannoni, E., Chiti, F., Baroni, F., Formigli, L., Zurdo, J., Taddei, N., Ramponi, G., Dobson, C.M. and Stefani, M., 2002, Inherent toxicity of aggregates implies a common mechanism for protein misfolding diseases. *Nature* **416**: 507.
Burnett, B.G. and Pittman, R.N., 2005, The polyglutamine neurodegenerative protein ataxin 3 regulates aggresome formation. *Proc. Natl. Acad. Sci. USA* **102**: 4330.
Burnett, B., Li, F. and Pittman, R.N., 2003, The polyglutamine neurodegenerative protein ataxin-3 binds polyubiquitylated proteins and has ubiquitin protease activity. *Hum. Mol. Genet.* **12**: 3195.
Burright, E.N., Davidson, J.D., Duvick, L.A., Koshy, B., Zoghbi, H.Y. and Orr, H.T., 1997, Identification of a self-association region within the SCA1 gene product, ataxin-1. *Hum. Mol. Genet.* **6**: 513.
Caplen, N.J., Taylor, J.P., Statham, V.S., Tanaka, F., Fire, A. and Morgan, R.A., 2002, Rescue of polyglutamine-mediated cytotoxicity by double-stranded RNA-mediated RNA interference. *Hum. Mol. Genet.* **11**: 175.
Caughey, B. and Lansbury, P.T., 2003, Protofibrils, pores, fibrils, and neurodegeneration: separating the responsible protein aggregates from the innocent bystanders. *Annu. Rev. Neurosci.* **26**: 267.
Cemal, C.K., Carroll, C.J., Lawrence, L., Lowrie, M.B., Ruddle, P., Al-Mahdawi, S., King, R.H., Pook, M.A., Huxley, C. and Chamberlain, S., 2002, YAC transgenic mice carrying pathological alleles of the MJD1 locus exhibit a mild and slowly progressive cerebellar deficit. *Hum. Mol. Genet.* **11**: 1075.
Cha, J.H., 2000, Transcriptional dysregulation in Huntington's disease. *Trends Neurosci.* **23**: 387.
Chai, Y., Koppenhafer, S.L., Bonini, N.M. and Paulson, H.L., 1999a, Analysis of the role of heat shock protein (Hsp) molecular chaperones in polyglutamine disease. *J. Neurosci.* **19**: 10338.
Chai, Y., Koppenhafer, S.L., Shoesmith, S.J., Perez, M.K. and Paulson, H.L., 1999b, Evidence for proteasome involvement in polyglutamine disease: localization to nuclear inclusions in SCA3/MJD and suppression of polyglutamine aggregation in vitro. *Hum. Mol. Genet.* **8**: 673.

Chai, Y., Berke, S.S., Cohen, R.E. and Paulson, H.L., 2004, Poly-ubiquitin binding by the polyglutamine disease protein ataxin-3 links its normal function to protein surveillance pathways. *J. Biol. Chem.* **279**: 3605.

Chan, H.Y., Warrick, J.M., Gray-Board, G.L., Paulson, H.L. and Bonini, N.M., 2000, Mechanisms of chaperone suppression of polyglutamine disease: selectivity, synergy and modulation of protein solubility in Drosophila. *Hum. Mol. Genet.* **9**: 2811.

Chen, Y.W., Stott, K. and Perutz, M.F., 1999, Crystal structure of a dimeric chymotrypsin inhibitor 2 mutant containing an inserted glutamine repeat. *Proc. Natl. Acad. Sci. USA* **96**: 1257.

Chen, L., Shinde, U., Ortolan, T.G. and Madura, K., 2001a, Ubiquitin-associated (UBA) domains in Rad23 bind ubiquitin and promote inhibition of multi-ubiquitin chain assembly. *EMBO Rep.* **2**: 933.

Chen, S., Berthelier, V., Yang, W. and Wetzel, R., 2001b, Polyglutamine aggregation behavior in vitro supports a recruitment mechanism of cytotoxicity. *J. Mol. Biol.* **311**: 173.

Chen, S., Berthelier, V., Hamilton, J.B., O'Nuallain, B. and Wetzel, R., 2002a, Amyloid-like features of polyglutamine aggregates and their assembly kinetics. *Biochemistry* **41**: 7391.

Chen, S., Ferrone, F.A. and Wetzel, R., 2002b, Huntington's disease age-of-onset linked to polyglutamine aggregation nucleation. *Proc. Natl. Acad. Sci. USA* **99**: 11884.

Chevalier-Larsen, E.S., O'Brien, C.J., Wang, H., Jenkins, S.C., Holder, L., Lieberman, A.P. and Merry, D.E., 2004, Castration restores function and neurofilament alterations of aged symptomatic males in a transgenic mouse model of spinal and bulbar muscular atrophy. *J. Neurosci.* **24**: 4778.

Chong, S.S., McCall, A.E., Cota, J., Subramony, S.H., Orr, H.T., Hughes, M.R. and Zoghbi, H.Y., 1995, Gametic and somatic tissue-specific heterogeneity of the expanded SCA1 CAG repeat in spinocerebellar ataxia type 1. *Nat. Genet.* **10**: 344.

Chou, A.H., Yeh, T.H., Kuo, Y.L., Kao, Y.C., Jou, M.J., Hsu, C.Y., Tsai, S.R., Kakizuka, A. and Wang, H.L., 2006, Polyglutamine-expanded ataxin-3 activates mitochondrial apoptotic pathway by upregulating Bax and downregulating Bcl-xL. *Neurobiol. Dis.* **21**: 333.

Chow, M.K., Ellisdon, A.M., Cabrita, L.D. and Bottomley, S.P., 2004a, Polyglutamine expansion in ataxin-3 does not affect protein stability: implications for misfolding and disease. *J. Biol. Chem.* **279**: 47643.

Chow, M.K., Mackay, J.P., Whisstock, J.C., Scanlon, M.J. and Bottomley, S.P., 2004b, Structural and functional analysis of the Josephin domain of the polyglutamine protein ataxin-3. *Biochem. Biophys. Res. Commun.* **322**: 387.

Chow, M.K., Paulson, H.L. and Bottomley, S.P., 2004c, Destabilization of a non-pathological variant of ataxin-3 results in fibrillogenesis via a partially folded intermediate: a model for misfolding in polyglutamine disease. *J. Mol. Biol.* **335**: 333.

Clark, H.B., Burright, E.N., Yunis, W.S., Larson, S., Wilcox, C., Hartman, B., Matilla, A., Zoghbi, H.Y. and Orr, H.T., 1997, Purkinje cell expression of a mutant allele of SCA1 in transgenic mice leads to disparate effects on motor behaviors, followed by a progressive cerebellar dysfunction and histological alterations. *J. Neurosci.* **17**: 7385.

Colomer Gould, V.F., 2005, Mouse models of Machado-Joseph disease and other polyglutamine spinocerebellar ataxias. *NeuroRx* **2**: 480.

Coutinho, P. and Andrade, C., 1978, Autosomal dominant system degeneration in Portuguese families of the Azores Islands. A new genetic disorder involving cerebellar, pyramidal, extrapyramidal and spinal cord motor functions. *Neurology* **28**: 703.

Cummings, C.J. and Zoghbi, H.Y., 2000, Fourteen and counting: unraveling trinucleotide repeat diseases. *Hum. Mol. Genet.* **9**: 909.

Davies, S.W., Turmaine, M., Cozens, B.A., DiFiglia, M., Sharp, A.H., Ross, C.A., Scherzinger, E., Wanker, E.E., Mangiarini, L. and Bates, G.P., 1997, Formation of neuronal intranuclear inclusions underlies the neurological dysfunction in mice transgenic for the HD mutation. *Cell* **90**: 537.

de Almeida, L.P., Zala, D., Aebischer, P. and Deglon, N., 2001, Neuroprotective effect of a CNTF-expressing lentiviral vector in the quinolinic acid rat model of Huntington's disease. *Neurobiol. Dis.* **8**: 433.

de Chiara, C., Menon, R.P., Adinolfi, S., de Boer, J., Ktistaki, E., Kelly, G., Calder, L., Kioussis, D. and Pastore, A., 2005, The AXH domain adopts alternative folds the solution structure of HBP1 AXH. *Structure (Camb.)* **13**: 743.

Demuro, A., Mina, E., Kayed, R., Milton, S.C., Parker, I. and Glabe, C.G., 2005, Calcium dysregulation and membrane disruption as a ubiquitous neurotoxic mechanism of soluble amyloid oligomers. *J. Biol. Chem.* **280**: 17294.

Desai, U.A., Pallos, J., Ma, A.A., Stockwell, B.R., Thompson, L.M., Marsh, J.L. and Diamond, M.I., 2006, Biologically active molecules that reduce polyglutamine aggregation and toxicity. *Hum. Mol. Genet.* **15**: 2114.

DiFiglia, M., Sapp, E., Chase, K.O., Davies, S.W., Bates, G.P., Vonsattel, J.P. and Aronin, N., 1997, Aggregation of huntingtin in neuronal intranuclear inclusions and dystrophic neurites in brain. *Science* **277**: 1990.

Djousse, L., Knowlton, B., Hayden, M., Almqvist, E.W., Brinkman, R., Ross, C., Margolis, R., Rosenblatt, A., Durr, A., Dode, C., Morrison, P.J., Novelletto, A., Frontali, M., Trent, R.J., McCusker, E., Gomez-Tortosa, E., Mayo, D., Jones, R., Zanko, A., Nance, M., Abramson, R., Suchowersky, O., Paulsen, J., Harrison, M., Yang, Q., Cupples, L.A., Gusella, J.F., MacDonald, M.E. and Myers, R.H., 2003, Interaction of normal and expanded CAG repeat sizes influences age at onset of Huntington disease. *Am. J. Med. Genet. A* **119**: 279.

do Carmo Costa, M., Gomes-da-Silva, J., Miranda, C.J., Sequeiros, J., Santos, M.M. and Maciel, P., 2004, Genomic structure, promoter activity and developmental expression of the mouse homologue of the Machado-Joseph disease (MJD) gene. *Genomics* **84**: 361.

Donaldson, K.M., Li, W., Ching, K.A., Batalov, S., Tsai, C.C. and Joazeiro, C.A., 2003, Ubiquitin-mediated sequestration of normal cellular proteins into polyglutamine aggregates. *Proc. Natl. Acad. Sci. USA* **100**: 8892.

Doss-Pepe, E.W., Stenroos, E.S., Johnson, W.G. and Madura, K., 2003, Ataxin-3 interactions with rad23 and valosin-containing protein and its associations with ubiquitin chains and the proteasome are consistent with a role in ubiquitin-mediated proteolysis. *Mol. Cell. Biol.* **23**: 6469.

Durr, A., Stevanin, G., Cancel, G., Duyckaerts, C., Abbas, N., Didierjean, O., Chneiweiss, H., Benomar, A., Lyon-Caen, O., Julien, J., Serdaru, M., Penet, C., Agid, Y. and Brice, A., 1996, Spinocerebellar ataxia 3 and Machado-Joseph disease: clinical, molecular and neuropathological features. *Ann. Neurol.* **39**: 490.

Echaniz-Laguna, A., Rousso, E., Anheim, M., Cossee, M. and Tranchant, C., 2005, A family with early-onset and rapidly progressive X-linked spinal and bulbar muscular atrophy. *Neurology* **64**: 1458.

Elbashir, S.M., Harborth, J., Lendeckel, W., Yalcin, A., Weber, K. and Tuschl, T., 2001, Duplexes of 21-nucleotide RNAs mediate RNA interference in cultured mammalian cells. *Nature* **411**: 494.

Ellerby, L.M., Andrusiak, R.L., Wellington, C.L., Hackam, A.S., Propp, S.S., Wood, J.D., Sharp, A.H., Margolis, R.L., Ross, C.A., Salvesen, G.S., Hayden, M.R. and Bredesen, D.E., 1999, Cleavage of atrophin-1 at caspase site aspartic acid 109 modulates cytotoxicity. *J. Biol. Chem.* **274**: 8730.

Ellisdon, A.M., Thomas, B. and Bottomley, S.P., 2006, The two-stage pathway of ataxin-3 fibrillogenesis involves a polyglutamine-independent step. *J. Biol. Chem.* **281**: 16888.

Evert, B.O., Wullner, U., Schulz, J.B., Weller, M., Groscurth, P., Trottier, Y., Brice, A. and Klockgether, T., 1999, High level expression of expanded full-length ataxin-3 in vitro causes cell death and formation of intranuclear inclusions in neuronal cells. *Hum. Mol. Genet.* **8**: 1169.

Evert, B.O., Vogt, I.R., Kindermann, C., Ozimek, L., de Vos, R.A., Brunt, E.R., Schmitt, I., Klockgether, T. and Wullner, U., 2001, Inflammatory genes are upregulated in expanded ataxin-3-expressing cell lines and spinocerebellar ataxia type 3 brains. *J. Neurosci.* **21**: 5389.

Fan, X., Dion, P., Laganiere, J., Brais, B. and Rouleau, G.A., 2001, Oligomerization of polyalanine expanded PABPN1 facilitates nuclear protein aggregation that is associated with cell death. *Hum. Mol. Genet.* **10**: 2341.

Farrer, L. A., 1985, Diabetes mellitus in Huntington disease. *Clin. Genet.* **27**: 62.

Ferrigno, P. and Silver, P.A., 2000, Polyglutamine expansions: proteolysis, chaperones, and the dangers of promiscuity. *Neuron* **26**: 9.

Filipek, R., Rzychon, M., Oleksy, A., Gruca, M., Dubin, A., Potempa, J. and Bochtler, M., 2003, The Staphostatin-staphopain complex: a forward binding inhibitor in complex with its target cysteine protease. *J. Biol. Chem.* **278**: 40959.

Fire, A., Xu, S., Montgomery, M.K., Kostas, S.A., Driver, S.E. and Mello, C.C., 1998, Potent and specific genetic interference by double-stranded RNA in *Caenorhabditis elegans*. *Nature* **391**: 806.

Fujigasaki, H., Uchihara, T., Koyano, S., Iwabuchi, K., Yagishita, S., Makifuchi, T., Nakamura, A., Ishida, K., Toru, S., Hirai, S., et al., 2000, Ataxin-3 is translocated into the nucleus for the formation of intranuclear inclusions in normal and Machado-Joseph disease brains. *Exp. Neurol.* **165**: 248.

Fujigasaki, H., Uchihara, T., Takahashi, J., Matsushita, M., Nakamura, A., Koyano, S., Iwabuchi, K., Hirai, S. and Mizusawa, H., 2001, Preferential recruitment of ataxin-3 independent of expanded polyglutamine: an immunohistochemical study on Marinesco bodies. *J. Neurol. Neurosurg. Psychiatry* **71**: 518.

Furusho, K., Yoshizawa, T., Hara, J., Yamanaka, A., Sakurai, T., Goto, K. and Shoji, S., 2005a, Effects of intraperitoneal administration of ectoine on the cell death produced by the truncated Machado-Joseph disease gene product with an expanded polyglutamine stretch in the ataxin-3/orexin transgenic mice. *Abstract Viewer/Itinerary Planner. Washington, DC: Society for Neuroscience, 2005. Online* Program No. **427.12**.

Furusho, K., Yoshizawa, T. and Shoji, S., 2005b, Ectoine alters subcellular localization of inclusions and reduces apoptotic cell death induced by the truncated Machado-Joseph disease gene product with an expanded polyglutamine stretch. *Neurobiol. Dis.* **20**: 170.

Gales, L., Cortes, L., Almeida, C., Melo, C.V., do Carmo Costa, M., Maciel, P., Clarke, D.T., Damas, A.M. and Macedo-Ribeiro, S., 2005, Towards a structural understanding of the fibrillization pathway in Machado-Joseph's disease: trapping early oligomers of non-expanded ataxin-3. *J. Mol. Biol.* **353**: 642.

Gardian, G., Browne, S.E., Choi, D.K., Klivenyi, P., Gregorio, J., Kubilus, J.K., Ryu, H., Langley, B., Ratan, R.R., Ferrante, R.J., et al., 2005, Neuroprotective effects of phenylbutyrate in the N171-82Q transgenic mouse model of Huntington's disease. *J. Biol. Chem.* **280**: 556.

Gauthier, L.R., Charrin, B.C., Borrell-Pages, M., Dompierre, J.P., Rangone, H., Cordelieres, F.P., De, M.J., MacDonald, M.E., Lessmann, V., Humbert, S. and Saudou, F., 2004, Huntingtin controls neurotrophic support and survival of neurons by enhancing BDNF vesicular transport along microtubules. *Cell* **118**: 127.

Gilman, S., Sima, A.A., Junck, L., Kluin, K.J., Koeppe, R.A., Lohman, M.E. and Little, R., 1996, Spinocerebellar ataxia type 1 with multiple system degeneration and glial cytoplasmic inclusions. *Ann. Neurol.* **39**: 241.

Glabe, C.G. and Kayed, R., 2006, Common structure and toxic function of amyloid oligomers implies a common mechanism of pathogenesis. *Neurology* **66**: S74.

Gordon-Smith, D.J., Carbajo, R.J., Stott, K. and Neuhaus, D., 2001, Solution studies of chymotrypsin inhibitor-2 glutamine insertion mutants show no interglutamine interactions. *Biochem. Biophys. Res. Commun.* **280**: 855.

Goti, D., Katzen, S.M., Mez, J., Kurtis, N., Kiluk, J., Ben-Haiem, L., Jenkins, N.A., Copeland, N.G., Kakizuka, A., Sharp, A.H., et al., 2004, A mutant ataxin-3 putative-cleavage fragment in brains of Machado-Joseph disease patients and transgenic mice is cytotoxic above a critical concentration. *J. Neurosci.* **24**: 10266.

Goto, J., Watanabe, M., Ichikawa, Y., Yee, S.B., Ihara, N., Endo, K., Igarashi, S., Takiyama, Y., Gaspar, C., Maciel, P., et al., 1997, Machado-Joseph disease gene products carrying different carboxyl termini. *Neurosci. Res.* **28**: 373.

Gouw, L.G., Castaneda, M.A., McKenna, C.K., Digre, K.B., Pulst, S.M., Perlman, S., Lee, M.S., Gomez, C., Fischbeck, K., Gagnon, D., Storey, E., Bird, T., Jeri, F.R. and Ptacek, L.J., 1998, Analysis of the dynamic mutation in the SCA7 gene shows marked parental effects on CAG repeat transmission. *Hum. Mol. Genet.* **7**: 525.

Griffin, J.L., Cemal, C.K. and Pook, M.A., 2004, Defining a metabolic phenotype in the brain of a transgenic mouse model of spinocerebellar ataxia 3. *Physiol. Genomics* **16**: 334.

Gunawardena, S., Her, L.S., Brusch, R.G., Laymon, R.A., Niesman, I.R., Gordesky-Gold, B., Sintasath, L., Bonini, N.M. and Goldstein, L.S., 2003, Disruption of axonal transport by loss of huntingtin or expression of pathogenic polyQ proteins in Drosophila. *Neuron* **40**: 25.

Guo, Z. and Eisenberg, D., 2006, Runaway domain swapping in amyloid-like fibrils of T7 endonuclease I. *Proc. Natl. Acad. Sci. USA* **103**: 8042.

Gusella, J.F. and MacDonald, M.E., 2000, Molecular genetics: unmasking polyglutamine triggers in neurodegenerative disease. *Nat. Rev. Neurosci.* **1**: 109.

Gwinn-Hardy, K., Singleton, A., O'Suilleabhain, P., Boss, M., Nicholl, D., Adam, A., Hussey, J., Critchley, P., Hardy, J. and Farrer, M., 2001, Spinocerebellar ataxia type 3 phenotypically resembling Parkinson's disease in a black family. *Arch. Neurol.* **58**: 296.

Haacke, A., Broadley, S.A., Boteva, R., Tzvetkov, N., Hartl, F.U. and Breuer, P., 2006, Proteolytic cleavage of polyglutamine-expanded ataxin-3 is critical for aggregation and sequestration of non-expanded ataxin-3. *Hum. Mol. Genet.* **15**: 555.

Hagenah, J.M., Zuhlke, C., Hellenbroich, Y., Heide, W. and Klein, C., 2004, Focal dystonia as a presenting sign of spinocerebellar ataxia 17. *Mov. Disord.* **19**: 217.

Hara, J., Beuckmann, C.T., Nambu, T., Willie, J.T., Chemelli, R.M., Sinton, C.M., Sugiyama, F., Yagami, K., Goto, K., Yanagisawa, M. and Sakurai, T., 2001, Genetic ablation of orexin neurons in mice results in narcolepsy, hypophagia, and obesity. *Neuron* **30**: 345.

Harper, S.Q., Staber, P.D., He, X., Eliason, S.L., Martins, I.H., Mao, Q., Yang, L., Kotin, R.M., Paulson, H.L. and Davidson, B.L., 2005, RNA interference improves motor and neuropathological abnormalities in a Huntington's disease mouse model. *Proc. Natl. Acad. Sci. USA* **102**: 5820.

Hayashi, Y., Kakita, A., Yamada, M., Koide, R., Igarashi, S., Takano, H., Ikeuchi, T., Wakabayashi, K., Egawa, S., Tsuji, S. and Takahashi, H., 1998, Hereditary dentatorubral-pallidoluysian atrophy: detection of widespread ubiquitinated neuronal and glial intranuclear inclusions in the brain. *Acta Neuropathol. (Berl.)* **96**: 547.

HDCRG, 1993, A novel gene containing a trinucleotide repeat that is expanded and unstable on Huntington's disease chromosomes. The Huntington's Disease Collaborative Research Group. *Cell* **72**: 971.

Heiser, V., Scherzinger, E., Boeddrich, A., Nordhoff, E., Lurz, R., Schugardt, N., Lehrach, H. and Wanker, E.E., 2000, Inhibition of huntingtin fibrillogenesis by specific antibodies and small molecules: implications for Huntington's disease therapy. *Proc. Natl. Acad. Sci. USA* **97**: 6739.

Heiser, V., Engemann, S., Brocker, W., Dunkel, I., Boeddrich, A., Waelter, S., Nordhoff, E., Lurz, R., Schugardt, N., Rautenberg, S., et al., 2002, Identification of benzothiazoles as potential polyglutamine aggregation inhibitors of Huntington's disease by using an automated filter retardation assay. *Proc. Natl. Acad. Sci. USA* **99** (suppl. 4): 16400.

Heuser, I.J., Chase, T.N. and Mouradian, M.M., 1991, The limbic-hypothalamic-pituitary-adrenal axis in Huntington's disease. *Biol. Psychiatry* **30**: 943.

Higashiyama, H., Hirose, F., Yamaguchi, M., Inoue, Y.H., Fujikake, N., Matsukage, A. and Kakizuka, A., 2002, Identification of ter94, Drosophila VCP, as a modulator of polyglutamine-induced neurodegeneration. *Cell Death Differ.* **9**: 264.

Hirabayashi, M., Inoue, K., Tanaka, K., Nakadate, K., Ohsawa, Y., Kamei, Y., Popiel, A.H., Sinohara, A., Iwamatsu, A., Kimura, Y., et al., 2001, VCP/p97 in abnormal protein aggregates, cytoplasmic vacuoles, and cell death, phenotypes relevant to neurodegeneration. *Cell Death Differ.* **8**: 977.

Hockly, E., Richon, V.M., Woodman, B., Smith, D.L., Zhou, X., Rosa, E., Sathasivam, K., Ghazi-Noori, S., Mahal, A., Lowden, P.A., et al., 2003, Suberoylanilide hydroxamic acid, a histone deacetylase inhibitor, ameliorates motor deficits in a mouse model of Huntington's disease. *Proc. Natl. Acad. Sci. USA* **100**: 2041.

Huang, C.C., Faber, P.W., Persichetti, F., Mittal, V., Vonsattel, J.P., MacDonald, M.E. and Gusella, J.F., 1998, Amyloid formation by mutant huntingtin: threshold, progressivity and recruitment of normal polyglutamine proteins. *Somat. Cell Mol. Genet.* **24**: 217.

Ichikawa, Y., Goto, J., Hattori, M., Toyoda, A., Ishii, K., Jeong, S.Y., Hashida, H., Masuda, N., Ogata, K., Kasai, F., Hirai, M., Maciel, P., Rouleau, G.A., Sakaki, Y. and Kanazawa, I., 2001, The genomic structure and expression of MJD, the Machado-Joseph disease gene. *J. Hum. Genet.* **46**: 413.

Ikeda, H., Yamaguchi, M., Sugai, S., Aze, Y., Narumiya, S. and Kakizuka, A., 1996, Expanded polyglutamine in the Machado-Joseph disease protein induces cell death in vitro and in vivo. *Nat. Genet.* **13**: 196.

Imbert, G., Saudou, F., Yvert, G., Devys, D., Trottier, Y., Garnier, J.M., Weber, C., Mandel, J.L., Cancel, G., Abbas, N., Durr, A., Didierjean, O., Stevanin, G., Agid, Y. and Brice, A., 1996, Cloning of the gene for spinocerebellar ataxia 2 reveals a locus with high sensitivity to expanded CAG/glutamine repeats. *Nat. Genet.* **14**: 285.

Jana, N.R., Dikshit, P., Goswami, A., Kotliarova, S., Murata, S., Tanaka, K. and Nukina, N., 2005, Co-chaperone CHIP associates with expanded polyglutamine protein and promotes their degradation by proteasomes. *J. Biol. Chem.* **280**: 11635.

Janowski, R., Kozak, M., Abrahamson, M., Grubb, A. and Jaskolski, M., 2005, 3D domain-swapped human cystatin C with amyloid like intermolecular beta-sheets. *Proteins* **61**: 570.

Johnston, S.C., Riddle, S.M., Cohen, R.E. and Hill, C.P., 1999, Structural basis for the specificity of ubiquitin C-terminal hydrolases. *EMBO J.* **18**: 3877.

Kagan, B.L., 2005, Amyloidosis and protein folding. *Science* **307**: 42.

Kagan, B.L., Hirakura, Y., Azimov, R. and Azimova, R., 2001, The channel hypothesis of Huntington's disease. *Brain. Res. Bull.* **56**: 281.

Kanazawa, I., 1998, Dentatorubral-pallidoluysian atrophy or Naito-Oyanagi disease. *Neurogenetics* **2**: 1.

Kanazawa, I., 1999, Molecular pathology of dentatorubral-pallidoluysian atrophy. *Philos. Trans. R. Soc. Lond. B Biol. Sci.* **354**: 1069.

Kato, T., Tanaka, F., Yamamoto, M., Yosida, E., Indo, T., Watanabe, H., Yoshiwara, T., Doyu, M. and Sobue, G., 2000, Sisters homozygous for the spinocerebellar ataxia type 6 (SCA6)/CACNA1A gene associated with different clinical phenotypes. *Clin. Genet.* **58**: 69.

Katsuno, M., Adachi, H., Doyu, M., Minamiyama, M., Sang, C., Kobayashi, Y., Inukai, A. and Sobue, G., 2003, Leuprorelin rescues polyglutamine-dependent phenotypes in a transgenic mouse model of spinal and bulbar muscular atrophy. *Nat. Med.* **9**: 768.

Katsuno, M., Adachi, H., Waza, M., Banno, H., Suzuki, K., Tanaka, F., Doyu, M. and Sobue, G., 2006, Pathogenesis, animal models and therapeutics in Spinal and bulbar muscular atrophy (SBMA). *Exp. Neurol.* **200**: 8.

Kawaguchi, Y., Okamoto, T., Taniwaki, M., Aizawa, M., Inoue, R., Katayama, S., Kawakami, H., Nakamura, S., Nishimura, M., Akiguchi, I., et al., 1994, CAG expansions in a novel gene for Machado-Joseph disease at chromosome 14q32.1. *Nat. Genet.* **8**: 221.

Kayed, R., Head, E., Thompson, J.L., McIntire, T.M., Milton, S.C., Cotman, C.W. and Glabe, C.G., 2003, Common structure of soluble amyloid oligomers implies common mechanism of pathogenesis. *Science* **300**: 486.

Kazlauskaite, J., Young, A., Gardner, C.E., Macpherson, J.V., Venien-Bryan, C. and Pinheiro, T.J., 2005, An unusual soluble beta-turn-rich conformation of prion is involved in fibril formation and toxic to neuronal cells. *Biochem. Biophys. Res. Commun.* **328**: 292.

Kettner, M., Willwohl, D., Hubbard, G.B., Rub, U., Dick, E.J., Jr., Cox, A.B., Trottier, Y., Auburger, G., Braak, H. and Schultz, C., 2002, Intranuclear aggregation of nonexpanded ataxin-3 in marinesco bodies of the nonhuman primate substantia nigra. *Exp. Neurol.* **176**: 117.

Kim, Y.T., Shin, S.M., Lee, W.Y., Kim, G.M. and Jin, D.K., 2004, Expression of expanded polyglutamine protein induces behavioral changes in Drosophila (polyglutamine-induced changes in Drosophila). *Cell Mol. Neurobiol.* **24**: 109.

Koide, R., Kobayashi, S., Shimohata, T., Ikeuchi, T., Maruyama, M., Saito, M., Yamada, M., Takahashi, H. and Tsuji, S., 1999, A neurological disease caused by an expanded CAG trinucleotide repeat in the TATA-binding protein gene: a new polyglutamine disease? *Hum. Mol. Genet.* **8**: 2047.

Kumada, S., Hayashi, M., Mizuguchi, M., Nakano, I., Morimatsu, Y. and Oda, M., 2000, Cerebellar degeneration in hereditary dentatorubral-pallidoluysian atrophy and Machado-Joseph disease. *Acta Neuropathol. (Berl.)* **99**: 48.

La Spada, A.R., Wilson, E.M., Lubahn, D.B., Harding, A.E. and Fischbeck, K.H., 1991, Androgen receptor gene mutations in X-linked spinal and bulbar muscular atrophy. *Nature* **352**: 77.

La Spada, A.R., Paulson, H.L. and Fischbeck, K.H., 1994, Trinucleotide repeat expansion in neurological disease. *Ann. Neurol.* **36**: 814.

Leblhuber, F., Peichl, M., Neubauer, C., Reisecker, F., Steinparz, F.X., Windhager, E. and Maschek, W., 1995, Serum dehydroepiandrosterone and cortisol measurements in Huntington's chorea. *J. Neurol. Sci.* **132**: 76.

Lecerf, J.M., Shirley, T.L., Zhu, Q., Kazantsev, A., Amersdorfer, P., Housman, D.E., Messer, A. and Huston, J.S., 2001, Human single-chain Fv intrabodies counteract in situ huntingtin aggregation in cellular models of Huntington's disease. *Proc. Natl. Acad. Sci.USA* **98**: 4764.

Lee, W.C., Yoshihara, M. and Littleton, J.T., 2004, Cytoplasmic aggregates trap polyglutamine-containing proteins and block axonal transport in a Drosophila model of Huntington's disease. *Proc. Natl. Acad. Sci. USA* **101**: 3224.

Lewis, S.E., Mannion, R.J., White, F.A., Coggeshall, R.E., Beggs, S., Costigan, M., Martin, J.L., Dillmann, W.H. and Woolf, C.J., 1999, A role for HSP27 in sensory neuron survival. *J. Neurosci.* **19**: 8945.

Li, F., Macfarlan, T., Pittman, R.N. and Chakravarti, D., 2002, Ataxin-3 is a histone-binding protein with two independent transcriptional corepressor activities. *J. Biol. Chem.* **277**: 45004.

Lieberman, A.P. and Fischbeck, K.H., 2000, Triplet repeat expansion in neuromuscular disease. *Muscle Nerve* **23**: 843.

Linhartova, I., Repitz, M., Draber, P., Nemec, M., Wiche, G. and Propst, F., 1999, Conserved domains and lack of evidence for polyglutamine length polymorphism in the chicken homolog of the Machado-Joseph disease gene product ataxin-3. *Biochim. Biophys. Acta* **1444**: 299.

Lunkes, A., Trottier, Y., Fagart, J., Schultz, P., Zeder-Lutz, G., Moras, D. and Mandel, J.L., 1999, Properties of polyglutamine expansion in vitro and in a cellular model for Huntington's disease. *Philos. Trans. R. Soc. Lond. B Biol. Sci.* **354**: 1013.

Maciel, P., Costa, M.C., Ferro, A., Rousseau, M., Santos, C.S., Gaspar, C., Barros, J., Rouleau, G.A., Coutinho, P. and Sequeiros, J., 2001, Improvement in the molecular diagnosis of Machado-Joseph disease. *Arch. Neurol.* **58**: 1821.

Mao, Y., Senic-Matuglia, F., Di Fiore, P.P., Polo, S., Hodsdon, M.E. and De Camilli, P., 2005, Deubiquitinating function of ataxin-3: insights from the solution structure of the Josephin domain. *Proc. Natl. Acad. Sci. USA* **102**: 12700.

Marchal, S., Shehi, E., Harricane, M.C., Fusi, P., Heitz, F., Tortora, P. and Lange, R., 2003, Structural instability and fibrillar aggregation of non-expanded human ataxin-3 revealed under high pressure and temperature. *J. Biol. Chem.* **278**: 31554.

Markianos, M., Panas, M., Kalfakis, N. and Vassilopoulos, D., 2005, Plasma testosterone in male patients with Huntington's disease: relations to severity of illness and dementia. *Ann. Neurol.* **57**: 520.

Masino, L. and Pastore, A., 2002, Glutamine repeats: structural hypotheses and neurodegeneration. *Biochem. Soc. Trans.* **30**: 548.

Masino, L., Kelly, G., Leonard, K., Trottier, Y. and Pastore, A., 2002, Solution structure of polyglutamine tracts in GST-polyglutamine fusion proteins. *FEBS Lett.* **513**: 267.

Masino, L., Musi, V., Menon, R.P., Fusi, P., Kelly, G., Frenkiel, T.A., Trottier, Y. and Pastore, A., 2003, Domain architecture of the polyglutamine protein ataxin-3: a globular domain followed by a flexible tail. *FEBS Lett.* **549**: 21.

Masino, L., Nicastro, G., Menon, R.P., Dal Piaz, F., Calder, L. and Pastore, A., 2004, Characterization of the structure and the amyloidogenic properties of the Josephin domain and the polyglutamine-containing protein ataxin-3. *J. Mol. Biol.* **344**: 1021.

Matsumoto, M., Yada, M., Hatakeyama, S., Ishimoto, H., Tanimura, T., Tsuji, S., Kakizuka, A., Kitagawa, M. and Nakayama, K.I., 2004, Molecular clearance of ataxin-3 is regulated by a mammalian E4. *EMBO J.* **23**: 659.

Matsumura, R., Futamura, N., Fujimoto, Y., Yanagimoto, S., Horikawa, H., Suzumura, A. and Takayanagi, T., 1997, Spinocerebellar ataxia type 6. Molecular and clinical features of 35 Japanese patients including one homozygous for the CAG repeat expansion. *Neurology* **49**: 1238.

Matsuyama, Z., Izumi, Y., Kameyama, M., Kawakami, H. and Nakamura, S., 1999, The effect of CAT trinucleotide interruptions on the age at onset of spinocerebellar ataxia type 1 (SCA1). *J. Med. Genet.* **36**: 546.

McCampbell, A., Taylor, J.P., Taye, A.A., Robitschek, J., Li, M., Walcott, J., Merry, D., Chai, Y., Paulson, H., Sobue, G. and Fischbeck, K.H., 2000, CREB-binding protein sequestration by expanded polyglutamine. *Hum. Mol. Genet.* **9**: 2197.

McGowan, D.P., van Roon-Mom, W., Holloway, H., Bates, G.P., Mangiarini, L., Cooper, G.J., Faull, R.L. and Snell, R.G., 2000, Amyloid-like inclusions in Huntington's disease. *Neuroscience* **100**: 677.

Michalik, A. and Van Broeckhoven, C., 2003, Pathogenesis of polyglutamine disorders: aggregation revisited. *Hum. Mol. Genet.* **12** (Suppl. 2): R173.

Miller, V.M., Xia, H., Marrs, G.L., Gouvion, C.M., Lee, G., Davidson, B.L. and Paulson, H.L., 2003, Allele-specific silencing of dominant disease genes. *Proc. Natl. Acad. Sci. USA* **100**: 7195.

Miller, V.M., Nelson, R.F., Gouvion, C.M., Williams, A., Rodriguez-Lebron, E., Harper, S.Q., Davidson, B.L., Rebagliati, M.R. and Paulson, H.L., 2005a, CHIP suppresses polyglutamine aggregation and toxicity in vitro and in vivo. *J. Neurosci.* **25**: 9152.

Miller, T.W., Zhou, C., Gines, S., MacDonald, M.E., Mazarakis, N.D., Bates, G.P., Huston, J.S. and Messer, A., 2005b, A human single-chain Fv intrabody preferentially targets amino-terminal Huntingtin's fragments in striatal models of Huntington's disease. *Neurobiol. Dis.* **19**: 47.

Misaghi, S., Galardy, P.J., Meester, W.J.N., Ovaa, H., Ploegh, H.L. and Gaudet, R., 2005, Structure of the ubiquitin hydrolase UCH-L3 complexed with a suicide substrate. *J. Biol. Chem.* **280**: 1512.

Monoi, H., 1995, New tubular single-stranded helix of poly-L-amino acids suggested by molecular mechanics calculations: I. Homopolypeptides in isolated environments. *Biophys. J.* **69**: 1130.

Monoi, H., Futaki, S., Kugimiya, S., Minakata, H. and Yoshihara, K., 2000, Poly-L-glutamine forms cation channels: relevance to the pathogenesis of the polyglutamine diseases. *Biophys. J.* **78**: 2892.

Myers, R.H., Madden, P.J., Teague, J.L. and Falek, A., 1982, Factors related to onset age of Huntington disease. *Am. J. Hum. Genet.* **34**: 481.

Nagai, Y., Fujikake, N., Ohno, K., Higashiyama, H., Popiel, H., Rahadian, J., Yamaguchi, M., Strittmatter, W., Burke, J. and Toda, T., 2003, Prevention of polyglutamine oligomerization and neurodegeneration by the peptide inhibitor QBP1 in Drosophila. *Hum. Mol. Genet.* **12**: 1253.

Nakamura, K., Jeong, S.Y., Uchihara, T., Anno, M., Nagashima, K., Nagashima, T., Ikeda, S., Tsuji, S. and Kanazawa, I., 2001, SCA17, a novel autosomal dominant cerebellar ataxia caused by an expanded polyglutamine in TATA-binding protein. *Hum. Mol. Genet.* **10**: 1441.

Nicastro, G., Menon, R.P., Masino, L., Knowles, P.P., McDonald, N.Q. and Pastore, A., 2005, The solution structure of the Josephin domain of ataxin-3: structural determinants for molecular recognition. *Proc. Natl. Acad. Sci. USA* **102**: 10493.

Ogawa, M., 2004, Pharmacological treatments of cerebellar ataxia. *Cerebellum* **3**: 107.

Okazawa, H., 2003, Polyglutamine diseases: a transcription disorder? *Cell. Mol. Life. Sci.* **60**: 1427.

O'Nuallain, B. and Wetzel, R., 2002, Conformational Abs recognizing a generic amyloid fibril epitope. *Proc. Natl. Acad. Sci. USA* **99**: 1485.

Ordway, J.M., Cearley, J.A. and Detloff, P.J., 1999, CAG-polyglutamine-repeat mutations: independence from gene context. *Philos. Trans. R. Soc. Lond. B Biol. Sci.* **354**: 1083.

Orr, H.T., 2001, Beyond the Qs in the polyglutamine diseases. *Genes Dev.* **15**: 925.

Paulson, H.L., 1999, Protein fate in neurodegenerative proteinopathies: polyglutamine diseases join the (mis)fold. *Am. J. Hum. Genet.* **64**: 339.

Paulson, H., 2003, Polyglutamine neurodegeneration: minding your Ps and Qs. *Nat. Med.* **9**: 825.

Paulson, H.L., Das, S.S., Crino, P.B., Perez, M.K., Patel, S.C., Gotsdiner, D., Fischbeck, K.H. and Pittman, R.N., 1997a, Machado-Joseph disease gene product is a cytoplasmic protein widely expressed in brain. *Ann. Neurol.* **41**: 453.

Paulson, H.L., Perez, M.K., Trottier, Y., Trojanowski, J.Q., Subramony, S.H., Das, S.S., Vig, P., Mandel, J.L., Fischbeck, K.H. and Pittman, R.N., 1997b, Intranuclear inclusions of expanded polyglutamine protein in spinocerebellar ataxia type 3. *Neuron* **19**: 333.

Perez, M.K., Paulson, H.L., Pendse, S.J., Saionz, S.J., Bonini, N.M. and Pittman, R.N., 1998, Recruitment and the role of nuclear localization in polyglutamine-mediated aggregation. *J. Cell. Biol.* **143**: 1457.

Perez, M.K., Paulson, H.L. and Pittman, R.N., 1999, Ataxin-3 with an altered conformation that exposes the polyglutamine domain is associated with the nuclear matrix. *Hum. Mol. Genet.* **8**: 2377.

Perutz, M., 1994, Polar zippers: their role in human disease. *Protein Sci.* **3**: 1629.

Perutz, M.F., 1999, Glutamine repeats and neurodegenerative diseases. *Brain Res. Bull.* **50**: 467.

Perutz, M.F., Johnson, T., Suzuki, M. and Finch, J.T., 1994, Glutamine repeats as polar zippers: their possible role in inherited neurodegenerative diseases. *Proc. Natl. Acad. Sci. USA* **91**: 5355.

Perutz, M.F., Finch, J.T., Berriman, J. and Lesk, A., 2002, Amyloid fibers are water-filled nanotubes. *Proc. Natl. Acad. Sci. USA* **99**: 5591.

Peters-Libeu, C., Newhouse, Y., Krishnan, P., Cheung, K., Brooks, E., Weisgraber, K. and Finkbeiner, S., 2005, Crystallization and diffraction properties of the Fab fragment of 3B5H10, an antibody specific for disease-causing polyglutamine stretches. *Acta Crystallograph. Sect. F. Struct. Biol. Cryst. Commun.* **61**: 1065.

Poirier, M.A., Li, H., Macosko, J., Cai, S., Amzel, M. and Ross, C.A., 2002, Huntingtin spheroids and protofibrils as precursors in polyglutamine fibrilization. *J. Biol. Chem.* **277**: 41032.

Pollitt, S.K., Pallos, J., Shao, J., Desai, U.A., Ma, A.A., Thompson, L.M., Marsh, J.L. and Diamond, M.I., 2003, A rapid cellular FRET assay of polyglutamine aggregation identifies a novel inhibitor. *Neuron* **40**: 685.

Pulst, S.M., Nechiporuk, A., Nechiporuk, T., Gispert, S., Chen, X.N., Lopes-Cendes, I., Pearlman, S., Starkman, S., Orozco-Diaz, G., Lunkes, A., DeJong, P., Rouleau, G.A., Auburger, G., Korenberg, J.R., Figueroa, C. and Sahba, S., 1996, Moderate expansion of a normally biallelic trinucleotide repeat in spinocerebellar ataxia type 2. *Nat. Genet.* **14**: 269.

Ranum, L.P., Lundgren, J.K., Schut, L.J., Ahrens, M.J., Perlman, S., Aita, J., Bird, T.D., Gomez, C. and Orr, H.T., 1995, Spinocerebellar ataxia type 1 and Machado-Joseph disease: incidence of CAG expansions among adult-onset ataxia patients from 311 families with dominant, recessive, or sporadic ataxia. *Am. J. Hum. Genet.* **57**: 603.

Rego, A.C. and de Almeida, L.P., 2005, Molecular targets and therapeutic strategies in Huntington's disease. *Curr. Drug Targets CNS Neurol. Disord.* **4**: 361.

Riess, O., Bichelmeier, U., Boy, J., Schmidt, T., Hbner, J., Holzmann, C., Ibrahim, S., Schmidt, I., Zimmermann, F. and Wilbertz, J., 2005, Transgenic mouse models of SCA3 implicate the nucleus as subcellular site of pathogenesis. *Abstract Viewer/Itinerary Planner. Washington, DC: Society for Neuroscience, 2005. Online.* Program No. **427.11**.

Rosenberg, R.N., 1984, Joseph disease: an autosomal dominant motor system degeneration. *Adv. Neurol.* **41**: 179.

Ross, C.A., Wood, J.D., Schilling, G., Peters, M.F., Nucifora, F.C., Jr., Cooper, J.K., Sharp, A.H., Margolis, R.L. and Borchelt, D.R., 1999, Polyglutamine pathogenesis. *Philos. Trans. R. Soc. Lond. B Biol. Sci.* **354**: 1005.

Ross, C.A., Poirier, M.A., Wanker, E.E. and Amzel, M., 2003, Polyglutamine fibrillogenesis: the pathway unfolds. *Proc. Natl. Acad. Sci. USA* **100**: 1.

Sambashivan, S., Liu, Y., Sawaya, M.R., Gingery, M. and Eisenberg, D., 2005, Amyloid-like fibrils of ribonuclease A with three-dimensional domain-swapped and native-like structure. *Nature* **437**: 266.

Sanchez, I., Mahlke, C. and Yuan, J., 2003, Pivotal role of oligomerization in expanded polyglutamine neurodegenerative disorders. *Nature* **421**: 373.

Schaffar, G., Breuer, P., Boteva, R., Behrends, C., Tzvetkov, N., Strippel, N., Sakahira, H., Siegers, K., Hayer-Hartl, M. and Hartl, F.U., 2004, Cellular toxicity of polyglutamine expansion proteins: mechanism of transcription factor deactivation. *Mol. Cell.* **15**: 95.

Schauber, C., Chen, L., Tongaonkar, P., Vega, I., Lambertson, D., Potts, W. and Madura, K., 1998, Rad23 links DNA repair to the ubiquitin/proteasome pathway. *Nature* **391**: 715.

Scheel, H., Tomiuk, S. and Hofmann, K., 2003, Elucidation of ataxin-3 and ataxin-7 function by integrative bioinformatics. *Hum. Mol. Genet.* **12**: 2845.

Scherzinger, E., Lurz, R., Turmaine, M., Mangiarini, L., Hollenbach, B., Hasenbank, R., Bates, G.P., Davies, S.W., Lehrach, H. and Wanker, E.E., 1997, Huntingtin-encoded polyglutamine expansions form amyloid-like protein aggregates in vitro and in vivo. *Cell* **90**: 549.

Schmitt, I., Brattig, T., Gossen, M. and Riess, O., 1997, Characterization of the rat spinocerebellar ataxia type 3 gene. *Neurogenetics* **1**: 103.

Sharma, D., Sharma, S., Pasha, S. and Brahmachari, S.K., 1999, Peptide models for inherited neurodegenerative disorders: conformation and aggregation properties of long polyglutamine peptides with and without interruptions. *FEBS Lett.* **456**: 181.

Sharma, D., Shinchuk, L.M., Inouye, H., Wetzel, R. and Kirschner, D.A., 2005, Polyglutamine homopolymers having 8–45 residues form slablike beta-crystallite assemblies. *Proteins* **61**: 398.

Shehi, E., Fusi, P., Secundo, F., Pozzuolo, S., Bairati, A. and Tortora, P., 2003, Temperature-dependent, irreversible formation of amyloid fibrils by a soluble human ataxin-3 carrying a moderately expanded polyglutamine stretch (Q36). *Biochemistry* **42**: 14626.

Shinotoh, H., Thiessen, B., Snow, B.J., Hashimoto, S., MacLeod, P., Silveira, I., Rouleau, G.A., Schulzer, M. and Calne, D.B., 1997, Fluorodopa and raclopride PET analysis of patients with Machado-Joseph disease. *Neurology* **49**: 1133.

Sikorski, P. and Atkins, E., 2005, New model for crystalline polyglutamine assemblies and their connection with amyloid fibrils. *Biomacromolecules* **6**: 425.

Singer, S.J. and Dewji, N.N., 2006, Evidence that Perutz's double-beta-stranded subunit structure for beta-amyloids also applies to their channel-forming structures in membranes. *Proc. Natl. Acad. Sci. USA* **103**: 1546.

Soong, B., Cheng, C., Liu, R. and Shan, D., 1997, Machado-Joseph disease: clinical, molecular and metabolic characterization in Chinese kindreds. *Ann. Neurol.* **41**: 446.

Stevanin, G., Durr, A. and Brice, A., 2000, Clinical and molecular advances in autosomal dominant cerebellar ataxias: from genotype to phenotype and physiopathology. *Eur. J. Hum. Genet.* **8**: 4.

Stevanin, G., Fujigasaki, H., Lebre, A.S., Camuzat, A., Jeannequin, C., Dode, C., Takahashi, J., San, C., Bellance, R., Brice, A. and Durr, A., 2003, Huntington's disease-like phenotype due to trinucleotide repeat expansions in the TBP and JPH3 genes. *Brain* **126**: 1599.

Stott, K., Blackburn, J.M., Butler, P.J. and Perutz, M., 1995, Incorporation of glutamine repeats makes protein oligomerize: implications for neurodegenerative diseases. *Proc. Natl. Acad. Sci. USA* **92**: 6509.

Sudarsky, L. and Coutinho, P., 1995, Machado-Joseph disease. *Clin. Neurosci.* **3**: 17.

Sugars, K.L. and Rubinsztein, D.C., 2003, Transcriptional abnormalities in Huntington disease. *Trends Genet.* **19**: 233.

Tait, D., Riccio, M., Sittler, A., Scherzinger, E., Santi, S., Ognibene, A., Maraldi, N.M., Lehrach, H. and Wanker, E.E., 1998, Ataxin-3 is transported into the nucleus and associates with the nuclear matrix. *Hum. Mol. Genet.* **7**: 991.

Tanaka, M., Morishima, I., Akagi, T., Hashikawa, T. and Nukina, N., 2001, Intra- and intermolecular beta-pleated sheet formation in glutamine-repeat inserted myoglobin as a model for polyglutamine diseases. *J. Biol. Chem.* **276**: 45470.

Tanaka, M., Machida, Y., Nishikawa, Y., Akagi, T., Hashikawa, T., Fujisawa, T. and Nukina, N., 2003, Expansion of polyglutamine induces the formation of quasi-aggregate in the early stage of protein fibrillization. *J. Biol. Chem.* **278**: 34717.

Tanaka, M., Machida, Y. and Nukina, N., 2005, A novel therapeutic strategy for polyglutamine diseases by stabilizing aggregation-prone proteins with small molecules. *J. Mol. Med.* **83**: 343.

Taniwaki, T., Sakai, T., Kobayashi, T., Kuwabara, Y., Otsuka, M., Ichiya, Y., Masuda, K. and Goto, I., 1997, Positron emission tomography (PET) in Machado-Joseph disease. *J. Neurol. Sci.* **145**: 63.

Tarlac, V. and Storey, E., 2003, Role of proteolysis in polyglutamine disorders. *J. Neurosci. Res.* **74**: 406.

Taroni, F. and DiDonato, S., 2004, Pathways to motor incoordination: the inherited ataxias, *Nat. Rev. Neurosci.* **5**: 641.

Taylor, J.P., Hardy, J. and Fischbeck, K.H., 2002, Toxic proteins in neurodegenerative disease. *Science* **296**: 1991.

Temussi, P.A., Masino, L. and Pastore, A., 2003, From Alzheimer to Huntington: why is a structural understanding so difficult? *EMBO J.* **22**: 355.

Thakur, A.K. and Wetzel, R., 2002, Mutational analysis of the structural organization of polyglutamine aggregates. *Proc. Natl. Acad. Sci. USA* **99**: 17014.

Trottier, Y., Cancel, G., An-Gourfinkel, I., Lutz, Y., Weber, C., Brice, A., Hirsch, E. and Mandel, J.L., 1998, Heterogeneous intracellular localization and expression of ataxin-3. *Neurobiol. Dis.* **5**: 335.

Tsai, H.F., Tsai, H.J. and Hsieh, M., 2004, Full-length expanded ataxin-3 enhances mitochondrial-mediated cell death and decreases Bcl-2 expression in human neuroblastoma cells. *Biochem. Biophys. Res. Commun.* **324**: 1274.

Uchihara, T., Fujigasaki, H., Koyano, S., Nakamura, A., Yagishita, S. and Iwabuchi, K., 2001, Non-expanded polyglutamine proteins in intranuclear inclusions of hereditary ataxias-triple-labeling immunofluorescence study. *Acta. Neuropathol. (Berl.)* **102**: 149.

Vonsattel, J.P. and DiFiglia, M., 1998, Huntington disease. *J. Neuropathol. Exp. Neurol.* **57**: 369.

Walsh, D.M. and Selkoe, D.J., 2004, Oligomers on the brain: the emerging role of soluble protein aggregates in neurodegeneration. *Protein Pept. Lett.* **11**: 213.

Wang, G., Sawai, N., Kotliarova, S., Kanazawa, I. and Nukina, N., 2000, Ataxin-3, the MJD1 gene product, interacts with the two human homologs of yeast DNA repair protein RAD23, HHR23A and HHR23B. *Hum. Mol. Genet.* **9**: 1795.

Warrick, J.M., Paulson, H.L., Gray-Board, G.L., Bui, Q.T., Fischbeck, K.H., Pittman, R.N. and Bonini, N.M., 1998, Expanded polyglutamine protein forms nuclear inclusions and causes neural degeneration in Drosophila. *Cell* **93**: 939.

Warrick, J.M., Chan, H.Y., Gray-Board, G.L., Chai, Y., Paulson, H.L. and Bonini, N.M., 1999, Suppression of polyglutamine-mediated neurodegeneration in Drosophila by the molecular chaperone HSP70. *Nat. Genet.* **23**: 425.

Warrick, J.M., Morabito, L.M., Bilen, J., Gordesky-Gold, B., Faust, L.Z., Paulson, H.L. and Bonini, N.M., 2005, Ataxin-3 suppresses polyglutamine neurodegeneration in Drosophila by a ubiquitin-associated mechanism. *Mol. Cell.* **18**: 37.

Watkins, J.F., Sung, P., Prakash, L. and Prakash, S., 1993, The Saccharomyces cerevisiae DNA repair gene RAD23 encodes a nuclear protein containing a ubiquitin-like domain required for biological function. *Mol. Cell. Biol.* **13**: 7757.

Wellington, C.L., Ellerby, L.M., Hackam, A.S., Margolis, R.L., Trifiro, M.A., Singaraja, R., McCutcheon, K., Salvesen, G.S., Propp, S.S., Bromm, M., et al., 1998, Caspase cleavage of gene products associated with triplet expansion disorders generates truncated fragments containing the polyglutamine tract. *J. Biol. Chem.* **273**: 9158.

Wen, F.C., Li, Y.H., Tsai, H.F., Lin, C.H., Li, C., Liu, C.S., Lii, C.K., Nukina, N. and Hsieh, M., 2003, Down-regulation of heat shock protein 27 in neuronal cells and non-neuronal cells expressing mutant ataxin-3. *FEBS Lett.* **546**: 307.

Wolfgang, W.J., Miller, T.W., Webster, J.M., Huston, J.S., Thompson, L.M., Marsh, J.L. and Messer, A., 2005, Suppression of Huntington's disease pathology in Drosophila by human single-chain Fv antibodies. *Proc. Natl. Acad. Sci. USA* **102**: 11563.

Wullner, U., Reimold, M., Abele, M., Burk, K., Minnerop, M., Dohmen, B.M., Machulla, H.J., Bares, R. and Klockgether, T., 2005, Dopamine transporter positron emission tomography in spinocerebellar ataxias type 1, 2, 3, and 6. *Arch. Neurol.* **62**: 1280.

Xia, H., Mao, Q., Eliason, S.L., Harper, S.Q., Martins, I.H., Orr, H.T., Paulson, H.L., Yang, L., Kotin, R.M. and Davidson, B.L., 2004, RNAi suppresses polyglutamine-induced neurodegeneration in a model of spinocerebellar ataxia. *Nat. Med.* **10**: 816.

Yang, W., Dunlap, J.R., Andrews, R.B. and Wetzel, R., 2002, Aggregated polyglutamine peptides delivered to nuclei are toxic to mammalian cells. *Hum. Mol. Genet.* **11**: 2905.

Ye, Y., Meyer, H.H. and Rapoport, T.A., 2003, Function of the p97-Ufd1-Npl4 complex in retrotranslocation from the ER to the cytosol: dual recognition of nonubiquitinated polypeptide segments and polyubiquitin chains. *J. Cell. Biol.* **162**: 71.

Yen, T.C., Lu, C.S., Tzen, K.Y., Wey, S.P., Chou, Y.H., Weng, Y.H., Kao, P.F. and Ting, G., 2000, Decreased dopamine transporter binding in Machado-Joseph disease. *J. Nucl. Med.* **41**: 994.

Yen, T.C., Tzen, K.Y., Chen, M.C., Chou, Y.H., Chen, R.S., Chen, C.J., Wey, S.P., Ting, G. and Lu, C.S., 2002, Dopamine transporter concentration is reduced in asymptomatic Machado-Joseph disease gene carriers. *J. Nucl. Med.* **43**: 153.

Yoshida, H., Yoshizawa, T., Shibasaki, F., Shoji, S. and Kanazawa, I., 2002, Chemical chaperones reduce aggregate formation and cell death caused by the truncated Machado-Joseph disease gene product with an expanded polyglutamine stretch. *Neurobiol. Dis.* **10**: 88.

Yoshizawa, T., Yamagishi, Y., Koseki, N., Goto, J., Yoshida, H., Shibasaki, F., Shoji, S. and Kanazawa, I., 2000, Cell cycle arrest enhances the in vitro cellular toxicity of the truncated Machado-Joseph disease gene product with an expanded polyglutamine stretch. *Hum. Mol. Genet.* **9**: 69.

Zhu, M., Shao, F., Innes, R.W., Dixon, J.E. and Xu, Z., 2004, The crystal structure of Pseudomonas avirulence protein AvrPphB: a papain-like fold with a distinct substrate-binding site. *Proc. Natl. Acad. Sci. USA* **101**: 302.

Zoghbi, H.Y. and Orr, H.T., 2000, Glutamine repeats and neurodegeneration. *Annu. Rev. Neurosci.* **23**: 217.

Zuccato, C., Ciammola, A., Rigamonti, D., Leavitt, B.R., Goffredo, D., Conti, L., MacDonald, M.E., Friedlander, R.M., Silani, V., Hayden, M.R., Timmusk, T., Sipione, S. and Cattaneo, E., 2001, Loss of huntingtin-mediated BDNF gene transcription in Huntington's disease. *Science* **293**: 493.

Zuhlke, C., Hellenbroich, Y., Dalski, A., Kononowa, N., Hagenah, J., Vieregge, P., Riess, O., Klein, C. and Schwinger, E., 2001, Different types of repeat expansion in the TATA-binding protein gene are associated with a new form of inherited ataxia. *Eur. J. Hum. Genet.* **9**: 160.

19

REMYELINATION OF THE CENTRAL NERVOUS SYSTEM

Charlotte C. Bruce[1], Robin J. M. Franklin[1] and João B. Relvas[2]

1. ABSTRACT

Myelination in the central nervous system (CNS) is carried out by oligodendrocytes. These cells produce myelin, a lipid-rich biological membrane, which forms multilamellar, spirally wrapped sheets around axons. Myelination allows rapid saltatory conduction of action potentials, and contributes to the maintenance of axonal integrity. The devastating neurological effects caused by demyelinating CNS diseases illustrate the importance of the process.

Demyelination, the process or state resulting from the loss or destruction of myelin, is a hallmark of multiple sclerosis and a characteristic of numerous pathologies such as contusion-type spinal injury and stroke. Remyelination is the process by which myelin sheaths are restored to axons, protecting them from degeneration, and regaining lost function by reviving the ability to carry action potentials by saltatory conduction. In the CNS, which is more generally recognised for its limited repair capacity, this beneficial process stands out as a rare regenerative phenomenon.

2. MULTIPLE SCLEROSIS – A COMMON AND COMPLEX DEMYELINATING DISEASE

The disease we know as multiple sclerosis (MS) was first given its name by Edmé Felix Alfred Vulpian, who in 1866 used the term *sclerose en plaque disseminée* to describe an immune-mediated chronic demyelinating disease of the central nervous system (CNS), characterised by disseminated patches of inflammatory demyelination in the brain and spinal cord. It affects approximately 2.5 million people worldwide, and is a common cause of neurological disturbance in young adults (Compston and Coles, 2002). It has been considered the prototype of a chronic inflammatory disease of the CNS since Charcot's leçons in 1868 prompted its demarcation as a primarily demyelinating disorder. Axonal injury has since been demonstrated in acute and chronic lesions, inciting its contemporary description as both an inflammatory and a neurodegenerative disease (Trapp et al., 1998; Waxman, 1998).

João B. Relvas – joao.relvas@cell.biol.ethz.ch
1- Cambridge Centre for Brain Repair, and Neuroregeneration Laboratory, Department of Veterinary Medicine, University of Cambridge, Madingley Road, Cambridge CB3 0ES, United Kingdom 2- Institute of Cell Biology, Department of Biology, Swiss Federal Institute of Technology, ETH Hönggerberg, CH-8093 Zürich, Switzerland

The aetiology and pathogenesis of MS are still not well understood. Generally, MS follows an initial order of relapses and remissions (RR-MS) for a variable duration, before translating into a progressive course characterised by accumulation of irreversible neurological deficit (SP-MS) (Compston, 2004; Confavreux and Vukusic, 2006). The main theory on the pathogenesis of MS is that inflammatory events cause demyelination and acute injury of axons and myelin. The stages of symptom onset, recovery, persistence and progression of MS can be summarised as functional adversity with intact structure due to direct effects of inflammatory mediators, demyelination and axonal damage with recovery through remyelination and restoration of trophic support, and axonal loss from chronic demyelination in part due to the inadequacy of the healing response of remyelination. The lesions that define MS are now considered to arise from multiple mechanisms, most immunological, but not always cell mediated. Accordingly MS may never be entirely preventable, and new therapeutic approaches must focus on the repair process of remyelination.

3. REMYELINATION – A SPONTANEOUS REGENERATIVE PROCESS

Remyelination of axons restores nearly normal nerve conduction which is sufficiently secure, that few conduction deficits can be detected (Black et al., 2006). Remyelination can occur as a spontaneous regenerative response in the adult human brain. The discriminative features are best recognised ultrastructurally, and are representative of a shortened internode distance and decreased myelin thickness to axonal diameter ratio (Prineas and Connell, 1979; Raine and Wu, 1993; Franklin, 2002). In MS, remyelination of demyelinated plaques may not always be complete; if incomplete, it chiefly occurs at the plaque edge, creating a transition zone between normal appearing white matter and the demyelinated plaque core. Where remyelination proceeds to completion, shadow plaques appear as sharply circumscribed areas of myelin pallor in the normal appearing white matter (pale staining for luxol fast blue or immunocytochemistry for myelin proteins).

The historic argument for promoting remyelination was based on improving conduction efficiency by switching from nonsaltatory to saltatory conduction along axons. The current appreciation of axon loss as a major pathological correlate of progressive functional decline has created an even more compelling case. Promoting myelin repair is potentially a highly effective means of long-term axon protection (Kornek et al., 2000; Rodriguez, 2003). The key therapeutic goal is therefore to remyelinate naked axons rapidly, and prevent slow chronic progressive disability associated with MS (Rodriguez, 2003).

There are two principal strategies to promote remyelination. The first involves the transplantation of exogenous cells with a repair-enhancing or myelinogenic capacity (Pluchino et al., 2003) – a technique that has made considerable progress in experimental models, but is yet to make the transition to clinical therapy. A second avenue is to design methods by which the endogenous process of remyelination can be enhanced or reactivated. The intent of this clinically appealing approach is to identify key factors amenable to pharmacological intervention whose activation or inhibition will lead to the process proceeding more efficiently. Devising proremyelination therapies requires knowledge of why remyelination fails. However, we cannot fully understand why remyelination fails until we disentangle the complex matrix of factors that orchestrate this important repair process (Zhao et al., 2005a).

4. OLIGODENDROCYTE PRECURSOR CELLS – WHICH CELLS MEDIATE REMYELINATION?

Experimental evidence strongly implies that remyelination is not conducted by oligodendrocytes surviving demyelination. Notably, remyelinated white matter contains a greater number of oligodendrocytes than prior to demyelination, implying that, irrespective of any oligodendrocyte survival, a mechanism of generating new oligodendrocytes is necessary (Franklin, 2002). Oligodendrocytes expressing myelin oligodendrocyte glycoprotein (MOG) are present within demyelinated plaques and could possibly form new myelin sheaths (Wolswijk, 2000), but there is no convincing evidence that the differentiated oligodendrocyte can revert to a proliferating state. Furthermore, it has been demonstrated that cells expressing Gal-C (galactocerebrosidase) and other later stage markers of the oligodendrocyte lineage make sheets of compacted myelin, but fail in arranging this membrane around demyelinated axons to complete myelination (Targett et al., 1996; Keirstead and Blakemore, 1997). If such cells were previously myelinating oligodendrocytes, the results would indicate that an oligodendrocyte that has myelinated once cannot repeatedly do so.

The nature of the cells that, in most cases, respond to demyelination and generate new oligodendrocytes are a distinctive phenotype called adult oligodendrocyte precursor cells (OPCs). Early evidence came from cell labelling studies that demonstrated that candidate proliferating cells gave rise to labelled remyelinating oligodendrocytes after the induction of demyelination (Carroll and Jennings, 1994; Gensert and Goldman, 1997). The argument was supported by the demonstration that these cells could remyelinate areas of demyelination after transplantation (Zhang et al., 1999). The cells in question were the adult descendants of a comprehensively researched developmental progenitor, originally named the O2-A progenitor based on its ability in vitro to generate both type 2 astrocytes and oligodendrocytes (Raff et al., 1983; Wren et al., 1992). The type 2 astrocyte seldom occurs, if at all, during development and the cells are therefore regarded simply as OPCs (Levison and Goldman, 1993).

In adult tissue, OPCs have a distinctive multipolar morphology and can be identified by a range of markers that include the platelet-derived growth factor receptor-α (PDGRαR) (Redwine and Armstrong, 1998; Sim et al., 2002a), the proteoglycan NG2 (Dawson et al., 2000), the zinc-finger protein MyT1 (myelin transcription factor 1), transcription factor Nkx2.2 and nuclear marker Olig2, in patterns of expression consistent with being the source of new oligodendrocytes (Dawson et al., 2000). The suitability of the proteoglycan NG2 as a marker for OPCs, and the heterogeneity within this population is hotly debated and is beyond the scope of this review, but certainly whether OPCs in the adult CNS express these markers ubiquitously in all circumstances is uncertain (Hampton et al., 2004); indeed, the degree to which the population is homogeneous throughout the adult neuraxis is currently a popular topic of inquiry. The debate was prompted by tissue-culture studies that revealed that OPC population might be more heterogeneous than previously thought (Gensert and Goldman, 1997). If the results are synonymous in vivo, then different progenitors may react to environmental cues or contribute to remyelination in an altered manner (Mason and Goldman, 2002). Evidence shows that from a developmental outlook progenitor types are diverse (Mallon et al., 2002; Cai et al., 2005; Vallstedt et al., 2005). For instance, two discrete populations can be distinguished on the basis of expression of PDGFαR or DM20, an alternatively spliced isoform of the proteolipid protein gene (Spassky et al., 1998, 2000). The degree to which adult OPCs keep an imprint of their developmental origin remains to be unequivocally determined (Zhao et al., 2005b). OPCs despite their varied ontogeny might be a homogeneous population of cells in the adult CNS. Alternatively, separate classes of OPC may exist, either coexisting or being anatomically specific. Evidence suggests the latter may be the case; in tissue culture the markers O4 and A2B5 appear to identify distinct populations of adult forebrain OPCs that respond differently to a range and combination of growth factors (Mason and Goldman, 2002). This is certainly an important issue to resolve if growth factor based strategies are to be used therapeutically to enhance endogenous remyelination in the clinic (Zhao et al., 2005b).

There is limited evidence to suggest that cells other than OPCs contribute to remyelination, and the contribution of stem cells or whether OPCs themselves are stem cells has also been under review. When applying the strict criteria of a stem cell (a multipotent cell, generally attached to the basal lamina, that is both self-renewing and able to give rise to rapidly proliferating progenitors by asymmetric division) (Zhao et al., 2005b), it is clear that OPCs are not stem cells. Despite multipotency in vitro, their proliferation rate, symmetrical division and the absence of an anatomical relationship with a basal lamina are more consistent with their being a transit amplifying population of progenitors. True stem cells within the adult mammalian CNS are rare, comprising the GFAP-expressing B cells of the subventricular zone (SVZ). Evidence suggests that the component of the total endogenous remyelination attributable to SVZ-derived cells is likely to be small and anatomically restricted given the responsiveness of locally derived OPCs (Zhao et al., 2005b).

5. THE STAGES OF REMYELINATION

In response to demyelination, during the recruitment phase of remyelination OPCs proliferate (Levine and Reynolds, 1999; Sim et al., 2002b) and migrate to rapidly and sufficiently fill the demyelinated area. The differentiation phase completes remyelination, and recruited OPCs engage the axon restoring new myelin sheaths as they differentiate into mature oligodendrocytes (Fig. 1). The key to comprehending the mechanisms of remyelination is through identifying the factors regulating these two phases. In development OPC proliferation and differentiation are mutually exclusive events, and it is likely that separate sets of signals will be involved in the orchestration of the two events during remyelination, a pro-recruitment environment shifting with time to a pro-differentiation environment (Zhao et al., 2005a).

6. WHY DOES REMYELINATION FAIL?

Given a framework by which remyelination involves distinct, well-defined phases of recruitment and differentiation, it is possible to identify critical junctures at which arrest would lead to remyelination failure. This may occur due to an inadequate provision of OPCs (recruitment failure), or via the failure of recruited OPCs to differentiate into remyelinating oligodendrocytes (differentiation failure) (Franklin, 2002). Recruitment may fail if the OPC population is depleted by disease process itself. For example, it has been documented that some patients possess antibodies against OPC proteins. Antibodies against NG2 proteoglycan could also cause pathology by interfering with OPC generation, migration and glial-neuronal

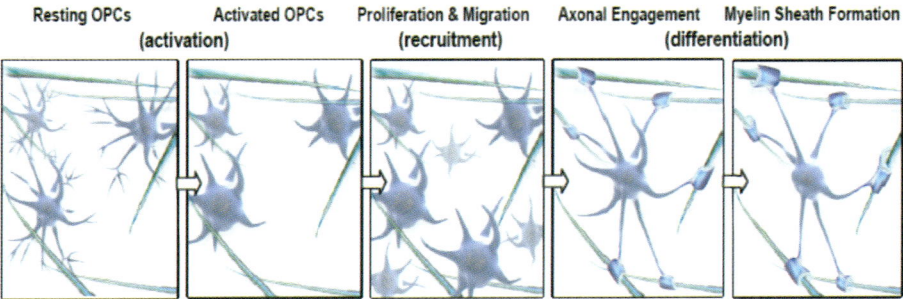

Fig. 1. The stages of remyelination. In response to demyelination OPCs become activated. Distinct morphological changes, such as hypertrophy of the processes are accompanied by shifts in gene expression, and enable the cell to rapidly react to injury (activation phase). OPCs then proliferate and migrate to sufficiently populate the demyelinated region (recruitment phase), before engaging demyelinated axons and differentiating into mature oligodendrocytes (differentiation phase). Given this framework of distinct, well-defined phases it is possible to identify critical points at which arrest would lead to remyelination failure. The key therefore is to understand the mechanism of remyelination by identifying the factors regulating these phases in order to discover why it fails. It is likely that separate sets of signals will be involved; a pro-recruitment environment shifting with time to one that supports differentiation

signaling (Karram et al., 2005). During plaque development, remyelination may occur very rapidly, and occasionally ongoing myelin breakdown may coexist with signs of remyelination (Prineas et al., 1993; Raine and Wu, 1993; Lassmann et al., 1997). This information has evoked review of whether the capacity of transit-amplifying OPCs to proliferate in response to injury may become exhausted if repeatedly tested. The recruitment of OPCs to areas where they are deficient appears to be very robust (Chari and Blakemore, 2002). Exposure to repeated episodes of demyelination/remyelination appears not to affect OPC number or remyelination efficiency (Penderis et al., 2003a), indicating that a failure of remyelination cannot be attributed to the absence of available OPCsN – a hypothesis supported by observations that OPCs (Scolding et al., 1998; Chang et al., 2000; Maeda et al., 2001) and premyelinating oligodendrocytes are present in chronic MS lesions in the absence of remyelination (Wolswijk, 2000; Chang et al., 2002). Additionally, these observations show a failure in producing myelinating oligodendrocytes at chronic stages of lesion development (Wolswijk, 2000). The reasons for the differentiation failure in this quiescent cell population need to be explained. Studies of aging, in which both the recruitment and differentiation phases become protracted (Sim et al., 2002a), fortify these experimental conclusions, and attribute failure to changes in the signaling environment (Hinks and Franklin, 2000) or alterations in the responsiveness of the aged OPCs themselves. Therefore, if the availability of OPCs is not the limiting step to remyelination failure, it must arise from a lack of appropriate exposure to the growth factors required for the conversion to rapidly proliferating, migrating and differentiating OPCs, or because they become less responsive to these factors.

Autopsy and biopsy studies by Lucchinetti and colleagues have described a large, unique, international series of acute MS cases (Lucchinetti et al., 2000; Kornek and Lassmann, 2003). The various patterns identified in the immunopathogenic mechanisms, and extent and topography of oligodendrocyte destruction or preservation have suggested varying aetiologies and pathogenesis between patients, and have reaffirmed the opinion that MS is a heterogeneous disease. Pathological descriptions of MS tissue broadly demarcate two types of non-remyelinating lesion – those in which there are insufficient OPCs, implying recruitment failure, and those in which OPCs are present, implying differentiation failure. The identification of different immunological subtypes is prerequisite for the establishment of therapies based on the underlying mechanisms of myelin destruction, and the factors governing the recruitment and differentiation phases of remyelination must be identified in order to determine whether inappropriate regulation of any of these elements can account for the failure of the regenerative response.

7. FACTORS INVOLVED IN OPC RECRUITMENT

Developmental studies have provided insight into the factors that regulate precursor recruitment and differentiation, but the extent to which these studies apply to remyelination is rarely transparent. Nevertheless, these studies provide useful clues to the regenerative process, unveiling the prominent theme that effects of individual factors interact, and highlighting the likely complexity of the signaling environment mediating remyelination.

Most attention has focused on growth factors, which can have proliferative and pro-migratory effects or be purely mitogenic. PDGF and FGF-2, for example, can act individually as mitogens for OPCs in vitro. The expression of PDGF and FGF-2 ligands has also been shown to increase in demyelinated lesions (Hinks and Franklin, 1999), and OPCs express receptors for PDGF and FGF-2 during the proliferative phase. Furthermore, OPCs that express PDGFαR can be abundant in MS lesions (Maeda et al., 2001), indicating the potential to respond to changes in PDGF expression. The function of endogenous PDGF and FGF-2 activity has been assessed in vivo using gene deletions to impair each signaling pathway (Murtie et al., 2005), and the findings agree with other studies in showing that PDGF is a potent regulator of OPC number following demyelination (Woodruff et al., 2004). The story is less clear for FGF-2, but PDGF and FGF-2 are likely to regulate OPC responses in vivo in the context of multiple signals in the lesion environment, and may therefore be effective therapeutically in circumstances where inadequate provision of OPCs is the primary reason for remyelination failure.

Further to the proliferative response, and to complete the recruitment phase, a proportion of OPCs must migrate to the site of demyelination. The precise fraction is likely to be determined by the extent to which OPCs survive demyelination – an issue that itself is not fully resolved (Franklin, 2002). Neurotransmitters are an integral part of the chemical environment surrounding OPCs, and they could potentially regulate integrin-mediated functions such as proliferation and survival (Barres and Raff, 1999). For example, the $\alpha_6\beta_1$ integrin receptor interacts with PDGFαR to increase OPC proliferation (Blaschuk et al., 2000; Colognato et al., 2002; Baron et al., 2003). It has since been demonstrated in culture that the neurotransmitter glutamate, acting through AMPA receptors, regulates integrin-mediated OPC migration via an α_v integrin/PLP/neurotransmitter receptor complex that reduces binding to the extra-cellular matrix (Gudz et al., 2006). Further to illustrating the synergistic network of factors involved in remyelination, this highlights novel signaling determinants of OPC migration that are potentially amenable to therapeutic intervention.

8. FACTORS INVOLVED IN OPC DIFFERENTIATION

The observation made by Guus Wolswijk in 1998 that relatively quiescent OPCs (recognised as O4-positive, Gal-C-negative cells) were present in chronic MS lesions (Wolswijk, 1998) led to the realisation that the failure of differentiation might be a significant contributor to remyelination failure. Growth factors have been shown to affect proliferation, survival and differentiation of OPCs, and were therefore among the first regulators of OPC differentiation to be studied. Insulin-like growth factor 1 (IGF-1) has been implicated in OPC differentiation principally on the authority of its effects in development (Carson et al., 1993; Ye et al., 1995; Goddard et al., 1999). Its expression profile relative to oligodendrocyte markers in experimental models (Komoly et al., 1992; Hinks and Franklin, 1999; Mason et al., 2000), and MS lesions (Gveric et al., 1999) is agreeable with this role. For example, the peak expression is delayed in aged animals in which there is an impairment of differentiation (Hinks and Franklin, 2000; Sim et al., 2002a). However, the observation that advancing the onset of IGF-1 expression in old animals is not sufficient to bring forward the onset of differentiation (O'Leary et al., 2002) and the knowledge that IGF-1 is also involved in OPC proliferation (Jiang et al., 2001) hint towards a finer degree of intricacy; inhibitors of differentiation might contribute to the recruitment phase by retaining OPCs in a pro-recruitment state until conditions are satisfied to enter the differentiation phase. The cell cycle inhibitor p27Kip-1 for instance has been shown to set the threshold between growth arrest and cell division of OPCs in response to extra-cellular signals (Crockett et al., 2005) and may therefore play a role in remyelination. Furthermore, FGF-2 may play a salient role in governing the timing of terminal differentiation (Ludwin, 2006) – a process also associated with declining levels of basic helix–loop–helix (bHLH) proteins Id2, Id4 (Komoly et al., 1992; Kondo and Raff, 2000; Wang et al., 2001) and the zinc-finger protein MyT1 (Armstrong et al., 1995; Sim et al., 2002a), and increasing levels of AP-1 and the homeodomain transcription factor Gtx (Awatramani et al., 1997; Sim et al., 2000). Retroviral lineage tracing and *FGF-2* knockout mice (Armstrong et al., 2002; Murtie et al., 2005) demonstrate that FGF-2 deletion enhances OPC differentiation, and in vitro, the presence of FGF-2 can inhibit differentiation of OPCs derived from neonatal rat brain (Bansal and Pfeiffer, 1997) or isolated spinal cords of mice with demyelinated lesions (Armstrong et al., 2002). Similarly, axonally expressed glial growth factor 2 (GGF-2), a neuronally derived isoform of neuregulin, prevents differentiation of OPCs (Canoll et al., 1996) and has been shown in animal models (Canoll et al., 1996; Fernandez et al., 2000; Flores et al., 2000) and humans (Wilson et al., 2003) to promote their proliferation. Accordingly, the absence of neuregulin expression in MS lesions has been suggested as a reason for eventual failure of remyelination (Viehover et al., 2001), and its presence has been shown to attenuate autoimmune demyelination and augment remyelination in a chronic relapsing model of MS (Cannella et al., 1998). It has since been discovered, however, that increasing local levels of GGF-2 by direct infusion into areas of demyelination does not alter remyelination in the rat CNS (Penderis et al., 2003b), suggesting that the effects of growth factors are complex, and it is unlikely that the presence or absence of a single factor provides a complete explanation.

The processes of myelination and remyelination both require the axon to be invested with a myelin sheath; one might therefore hypothesise that genes and transcription factors controlling differentiation during development may have similar roles in activated adult OPCs. Consistent with this interpretation, expression of transcription factor Olig2 and homeodomain transcription factor Nkx2.2 in OPCs is likely to be a critical genetic switch required to allow OPCs to differentiate in both processes. In agreement, the expression of these factors converges in OPCs during development (Kohama et al., 2001; Totoiu et al., 2004), they are upregulated in response to demyelination, and their expression is delayed in old animals where the differentiation phase of remyelination is slower (Fancy et al., 2004). The presence of sonic hedgehog (Shh), however, which is responsible for its induction during development, does not appear to be required for CNS remyelination (Fancy et al., 2004), but other factors known to regulate *Olig* expression such as FGF are expressed at high levels following toxin-induced demyelination (Hinks and Franklin, 1999), highlighting that in contrast to development, there is likely to be considerable redundancy during repair.

Studies of mammalian CNS development also show that contact-mediated activation of Notch1 receptors on OPCs by the ligand Jagged1 induces Hes5, which inhibits their differentiation (Wang et al., 1998; Givogri et al., 2002; Dubois-Dalcq et al., 2005). The discovery that both Notch and Jagged are expressed in MS lesions, therefore, vindicated the idea that this signaling pathway may explain remyelination failure (John et al., 2002). Certainly, Jagged1 expression is negligible within remyelinated lesions and TGFβ1, a cytokine upregulated in MS, is found to specifically re-induce Jagged1 in primary cultures of human astrocytes (John et al., 2002). Discordantly, despite expression of Notch1 and Jagged1 in a rodent model of remyelination, lesions of demyelination undergo complete remyelination. Furthermore, OPC-targeted Notch1 ablation in *Plp-creERT2* Notch1$^{lox/lox}$ transgenic mice, and studies in aged rats where remyelination occurs more slowly fail to yield a significant correlation between gene expression and remyelination rate (Stidworthy et al., 2004). These findings clearly advocate that Notch1 and Jagged1 neither prevent nor play a rate-determining role in remyelination, and that re-expression of developmentally expressed genes following injury does not necessarily imply a similar function.

This premise has been well illustrated by the recent demonstration that the *Olig1* gene encoding the basic helix–loop–helix transcription factor Olig1 can be compensated for in development, but has a non-redundant role in repair (Arnett et al., 2004). Its intracellular sublocalisation during the different phases of myelin sheath formation is dynamic. In resting adult OPCs, it resides in the cytoplasm, transfers to the nucleus during OPC activation and relocates to the cytoplasm as cells begin to differentiate. This dynamism is likely to be necessary for successful remyelination, and the subcellular localisation of Olig1 in chronic MS lesions should therefore be investigated. Alleviating differentiation failure by inducing the relocalisation of Olig1 to the nucleus could provide a strategy of repair that bypasses extrinsic signaling events and targets non-redundant genes crucial for repair in MS.

Reports that integrin ligands such as vitronectin, tenascin-R and tenascin-C are expressed in chronic MS lesions (Sobel et al., 1995; Gutowski et al., 1999) suggest that the extracellular matrix (ECM), signaling through integrins, might also regulate the differentiation of OPCs by promoting myelin-sheath formation (Franklin, 2002). The balance of the equilibrium between active and inactive integrins regulates oligodendrocyte morphology, which itself is regulated by extrinsic and intrinsic cues providing a mechanism of signal integration. Integrins, therefore, may also possess non-redundant properties that make them attractive therapeutic targets for repair (Olsen and Ffrench-Constant, 2005), where inadequate process formation is the primary reason for remyelination failure.

It is feasible that both myelination and remyelination per se are prevented by changes in the axonal membrane (Charles et al., 2002) or the ECM. Adhesion molecules for example, have the ability to bring axons and oligodendrocytes into close apposition, and transduce signals between them. Polysialic acid (PSA) moieties on the neural cell adhesion molecule (NCAM) act as inhibitors of myelination by preventing OPCs from attaching the axon. The removal of axonal

PSA-NCAM is therefore a necessary prerequisite for remyelination. Studies show that it is locally re-expressed on chronically demyelinated axons in MS, which could impede attempts to remyelinate chronically demyelinated axons (Coman et al., 2005). Experiments in animal models, however, have demonstrated that chronically demyelinated regions of axons such as those seen in MS may remain competent to be remyelinated (Setzu et al., 2004; Foote and Blakemore, 2005).

Morphological analysis illustrates that remyelinated axons have inappropriately short internodal lengths, suggesting that new nodes are formed. During demyelination, existing axon–glia interactions are destroyed, and remyelination requires the formation of new interactions by differentiating OPCs. Recent research has shown that the paranodin/Caspr-contactin axonal complex interacts with the 155 kDa isoform of neurofascin, NF155, expressed on the oligodendroglial membrane, and that this interaction is likely to be biologically relevant at the axo–glial junction (Tait et al., 2000; Charles et al., 2002). Experiments have also described alterations in the molecular organisation of nodal and perinodal regions during remyelination (Wolswijk and Balesar, 2003; Craner et al., 2004); a default of re-aggregation of these proteins in their specific domains, or the disruption of axo–glial interactions at the paranode could therefore contribute to differentiation failure.

There is also growing appreciation that activity dependent interactions between axons and glia may be important to myelin formation. Neuronal activity influences myelination and is likely to play a part in remyelination. Myelination is decreased in the optic nerve of blind cape mole rats (Omlin, 1997) and by tetrodotoxin (TTX), a toxin that blocks nerve conduction (Coman et al., 2005). It is increased by premature opening of eyelids in rabbits (Tauber et al., 1980) and by application of α-scorpion toxin that stimulates electrical activity in neurons (Coman et al., 2005). Studies in co-cultures of neurons and oligodendrocytes provide a potential mechanism whereby the pro-remyelinating efficacy of electrical activity is related to the extra-synaptic release of the mediator adenosine acting on purinergic receptors expressed on OPCs (Talbott et al., 2005; Fields and Burnstock, 2006). It is therefore emerging that activity-dependent control of CAM expression, such as PSA-NCAM in neurons, as well as activity-dependent release of signaling molecules, such as adenosine or ATP (Fields and Stevens, 2000; Stevens et al., 2002), which act on receptors on OPCs and other cells such as astrocytes (Ishibashi et al., 2006) may influence remyelination.

To further this argument, the neurotransmitter glutamate mediates damage to oligodendrocytes and OPCs and contributes to MS. Contemporary findings have shown that OPCs possess NMDA receptors that have a high sensitivity to small but prolonged increases in extra-cellular glutamate concentration as can occur in MS (Karadottir et al., 2005). Remyelination begins with the extension of multiple processes that make contact with axons where they either retract or differentiate. NMDA receptors on OPCs might therefore influence this decision and dictate the ultimate success of differentiation (Jiang et al., 2001; Salter and Fern, 2005). Therapeutic interventions targeting the unusual subunit composition of these receptors (mainly NR2C and NR3A) (Salter and Fern, 2005) may hence be useful in promoting remyelination where differentiation has been unsuccessful.

9. THE SIGNALING ENVIRONMENT FOR REMYELINATION

A plethora of environmental factors are committed to orchestrating successful remyelination, which, unlike in development, occurs in the face of pathogenesis and inflammation (Fig. 2). Critical junctures are regulated by a detailed fingerprint of signaling events, in which the correct level and temporal profile of expression are probably pivotal. A pro-recruitment environment must be maintained long enough to allow a lesion to be repopulated to an extent adequate for complete remyelination, and only when repopulation is complete should the environment shift to one that supports differentiation (Franklin, 2002).

A further premise that has been guiding recent research is that the initial events following a demyelinating episode may direct the future success of the regenerative response. Myelin debris generated by demyelination, for example, is a prominent component of MS lesions and has been shown in vitro (Miller, 1999) and in vivo (Kotter et al., 2006) to impair remyelination through arresting OPC differentiation. It may also induce apoptosis of differentiating OPCs (Kotter et al., 2006). When investigating the mechanisms by which myelin is acting, two co-receptors associated with myelin inhibition of axonal outgrowth, neurotrophin receptor $p75^{NTR}$ (Wang et al., 2002) as well as LINGO (Wong et al., 2002), are present on OPCs (Mi et al., 2004). LINGO signaling has further been shown to act as a differentiation block to OPCs. These experiments have led to the contemporary hypothesis that myelin plays a central inhibitory role during CNS regenerative processes of both axon re-growth and remyelination (Kotter et al., 2006).

Furthermore, the inflammatory response to demyelination is critical to efficient remyelination. Post-mortem evidence from MS tissue shows that remyelination is usually associated with inflammation and a large macrophage response (Prineas and Connell, 1979; Raine, 1997), and the use of toxin models has made it possible to study its contribution. The age-associated delay in remyelination rate has been linked to changes in the inflammatory response (Zhao et al., 2006), and studies using the *taiep* rat combined with X-irradiation and cell transplantation have shown that a non-remyelinating situation can be transformed to a remyelinating one by the induction of acute inflammation (Foote and Blakemore, 2005). Remyelination is also impaired in transgenic mice lacking inflammatory cytokines such as TNF-α, IL-1β and MCHII (Mason et al., 2001; Arnett et al., 2003), clearly demonstrating its importance. Therefore, although macrophages are mediators of demyelination, they play an important role in repair – a finding that raises questions about immunomodulatory therapies currently used for MS (Li et al., 2005). The activation of microglia and astrocytes set off by demyelination and subsequently by each other elicits a cascade of events that create a pro-remyelinating environment, where numerous mechanisms including cytokines, growth factors, ECM and CAMs are expressed in a mode suitable for both the recruitment and differentiation phases of remyelination to proceed to completion (Franklin, 2002). From this picture, it is clear that the process of remyelination is not simply a recapitulation of development.

10. THE DYSREGULATION HYPOTHESIS OF REMYELINATION FAILURE

The emerging picture illustrates that a manifold of elements contribute to the inadequacy of remyelination in diseases such as MS, uniting axonal damage, OPC default and inappropriate regulation of the cellular environment together into a hypothesis where this elaborate and finely tuned mechanism looses coordination or becomes 'dysregulated' (Franklin, 2002). Stemming from the aftermath of demyelination, inflammation spurs the initiation of a pro-recruitment environment for a sufficient duration to repopulate areas of demyelination, whereupon it switches to inhibit recruitment and enhance differentiation of recruited cells to complete remyelination. Perturbations from this abstruse schema are likely to cause remyelination failure.

11. FUTURE PROSPECTS

The dysregulation hypothesis implies that remyelination is potentially amenable to enhancement by manipulation of the signaling environment. However, there is likely to be substantial redundancy in the system, and therapeutic advances must be made through identifying non-redundant factors or pathways. Given the above, conventional strategies that focus on single molecules or pathways may prove unsuitable, and there is a need for new analytical methods that will provide new opportunities to understand the mechanisms underlying demyelinating disease.

Recent advances in functional genomics methods allowed the concurrent analysis of the expression of multiple mRNA species (genomics) and the expression of multiple protein species (proteomics) to take place (Morris and Wilson, 2004). A distinct advantage of these approaches is that thousands of different biological components involved in the multiple pathways and networks that regulate biological processes such as remyelination can be studied in one go.

Large-scale analysis of mRNA and comparison of mRNA expression levels through the use of gene arrays or micro arrays have been instrumental to elucidate the role played by different genes in neurological disease (reviewed by Morris and Wilson, 2004). However, biological complexity lies primarily at protein level and the analysis of a proteome represents an important and almost obligatory supplementation to the genome analysis (Marcus et al., 2004). One gene may produce several protein products as the result of pre-mRNA and post-translational modifications. For example, in the human genome, which contains approximately 38,000 genes, each gene produces an average of 10 different protein products. Furthermore, protein expression is extremely dynamic in relation to space and time, and there is only a poor correlation between mRNA and the corresponding gene product levels in a given cell (reviewed by Marcus et al., 2004). Although the quantification and identification of all or many differentially expressed proteins between two different samples, e.g., normal and remyelinating CNS tissue is a practical reality, devising remyelination-enhancing strategies from this information is far from easy. High-throughput genomic and proteomic approaches are likely to generate enormous quantities of data from which it will not always be easy to extract relevant information. It is therefore necessary to be aware of the strength and limitations of the different methods and being by using a sequential analytical process in a simple and well-categorised model for remyelination, such as those provided by direct or systemic delivery of demyelinating toxins. From this formulation, significant strides can be taken towards resolving the enigma of remyelination failure in diseases such as MS.

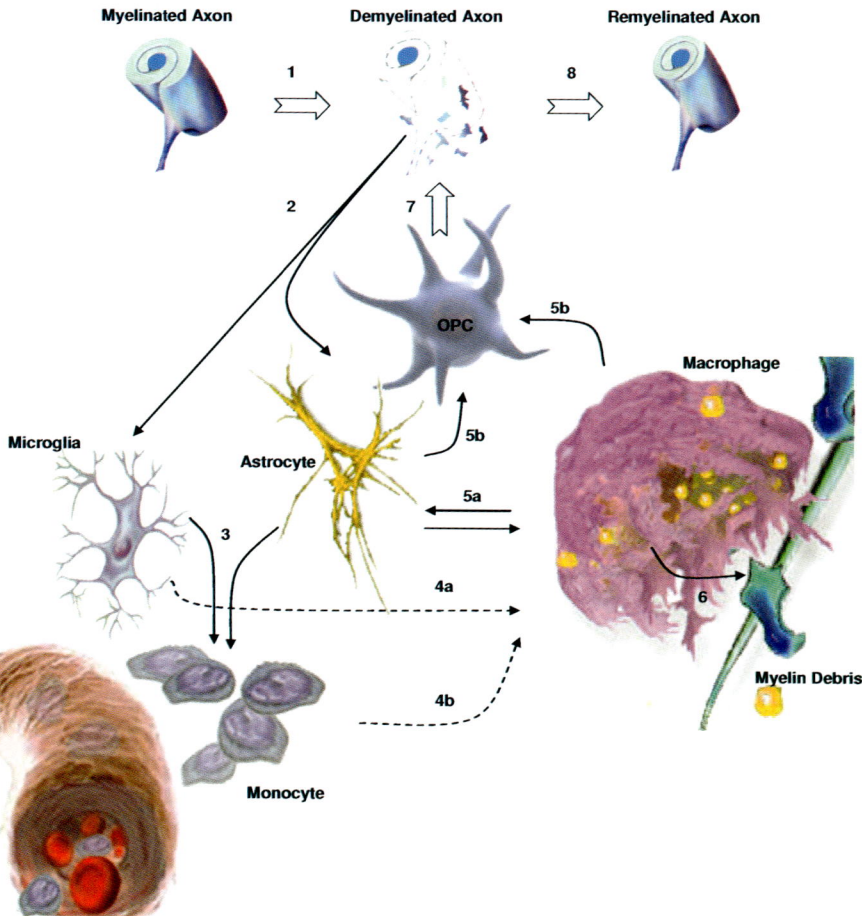

Fig. 2. The environment of remyelination is complex. Unlike developmental myelination, remyelination occurs as a consequence of a demyelinating pathology (1). The activation of microglia and astrocytes set off by demyelination (2) promotes the recruitment of monocytes from the blood (3). Microglia (4a) and recruited monocytes (4b) differentiate into macrophages. Together, this elicits a cascade that aids in the generation of an appropriately regulated pro-remyelination environment. The initial events following a demyelinating episode may direct the future success of remyelination, and the inflammatory response is critical to efficient remyelination. Mutually activated by astrocytes (5a), macrophages are responsible for the clearance of inhibitory myelin (6), and help to spur the initiation of a pro-recruitment environment (5b) for a sufficient duration to repopulate areas of demyelination, before it switches to inhibit recruitment and allow differentiation (7) and the completion of the regenerative response (8). A dysregulation of this precisely orchestrated schema will result in remyelination failure

12. REFERENCES

Armstrong, R.C., Kim, J.G. and Hudson, L.D., 1995, Expression of myelin transcription factor I (MyTI), a "zinc-finger" DNA-binding protein, in developing oligodendrocytes. *Glia* **14**: 303.
Armstrong, R.C., Le, T.Q., Frost, E.E., Borke, R.C. and Vana, A.C., 2002, Absence of fibroblast growth factor 2 promotes oligodendroglial repopulation of demyelinated white matter. *J. Neurosci.* **22**: 8574.
Arnett, H.A., Wang, Y., Matsushima, G.K., Suzuki, K. and Ting, J.P., 2003, Functional genomic analysis of remyelination reveals importance of inflammation in oligodendrocyte regeneration. *J. Neurosci.* **23**: 9824.
Arnett, H.A., Fancy, S.P., Alberta, J.A., Zhao, C., Plant, S.R., Kaing, S., Raine, C.S., Rowitch, D.H., Franklin, R.J.M. and Stiles, C.D., 2004, bHLH transcription factor Olig1 is required to repair demyelinated lesions in the CNS. *Science* **306**: 2111.
Awatramani, R., Scherer, S., Grinspan, J., Collarini, E., Skoff, R., O'Hagan, D., Garbern, J. and Kamholz, J., 1997, Evidence that the homeodomain protein Gtx is involved in the regulation of oligodendrocyte myelination. *J. Neurosci.* **17**: 6657.
Bansal, R. and Pfeiffer, S.E., 1997, Regulation of oligodendrocyte differentiation by fibroblast growth factors. *Adv. Exp. Med. Biol.* **429**: 69.
Baron, W., Decker, L., Colognato, H. and Ffrench-Constant, C., 2003, Regulation of integrin growth factor interactions in oligodendrocytes by lipid raft microdomains. *Curr. Biol.* **13**: 151.
Barres, B.A. and Raff, M.C., 1999, Axonal control of oligodendrocyte development. *J. Cell. Biol.* **147**: 1123.
Black, J.A., Waxman, S.G. and Smith, K.J., 2006, Remyelination of dorsal column axons by endogenous Schwann cells restores the normal pattern of Nav1.6 and Kv1.2 at nodes of Ranvier. *Brain* **129**: 1319.
Blaschuk, K.L., Frost, E.E. and Ffrench-Constant, C., 2000, The regulation of proliferation and differentiation in oligodendrocyte progenitor cells by alphaV integrins. *Development* **127**: 1961.
Cai, J., Qi, Y., Hu, X., Tan, M., Liu, Z., Zhang, J., Li, Q., Sander, M. and Qiu, M., 2005, Generation of oligodendrocyte precursor cells from mouse dorsal spinal cord independent of Nkx6 regulation and Shh signaling. *Neuron* **45**: 41.
Cannella, B., Hoban, C.J., Gao, Y.L., Garcia-Arenas, R., Lawson, D., Marchionni, M., Gwynne, D. and Raine, C.S., 1998, The neuregulin, glial growth factor 2, diminishes autoimmune demyelination and enhances remyelination in a chronic relapsing model for multiple sclerosis. *Proc. Natl. Acad. Sci. USA* **95**: 10100.
Canoll, P.D., Musacchio, J.M., Hardy, R., Reynolds, R., Marchionni, M.A. and Salzer, J.L., 1996, GGF/neuregulin is a neuronal signal that promotes the proliferation and survival and inhibits the differentiation of oligodendrocyte progenitors. *Neuron* **17**: 229.
Carroll, W.M. and Jennings, A.R., 1994, Early recruitment of oligodendrocyte precursors in CNS demyelination. *Brain* **117**: 563.
Carson, M.J., Behringer, R.R., Brinster, R.L. and McMorris, F.A., 1993, Insulin-like growth factor I increases brain growth and central nervous system myelination in transgenic mice. *Neuron* **10**: 729.
Chang, A., Nishiyama, A., Peterson, J., Prineas, J. and Trapp, B.D., 2000, NG2-positive oligodendrocyte progenitor cells in adult human brain and multiple sclerosis lesions. *J. Neurosci.* **20**: 6404.
Chang, A., Tourtellotte, W.W., Rudick, R. and Trapp, B.D., 2002, Premyelinating oligodendrocytes in chronic lesions of multiple sclerosis. *N. Engl. J. Med.* **346**: 165.
Chari, D.M. and Blakemore, W.F., 2002, Efficient recolonisation of progenitor-depleted areas of the CNS by adult oligodendrocyte progenitor cells. *Glia* **37**: 307.
Charles, P., Reynolds, R., Seilhean, D., Rougon, G., Aigrot, M.S., Niezgoda, A., Zalc, B. and Lubetzki, C., 2002, Re-expression of PSA-NCAM by demyelinated axons: an inhibitor of remyelination in multiple sclerosis? *Brain* **125**: 1972.
Colognato, H., Baron, W., Avellana-Adalid, V., Relvas, J.B., Baron-Van Evercooren, A., Georges-Labouesse, E. and Ffrench-Constant, C., 2002, CNS integrins switch growth factor signaling to promote target-dependent survival. *Nat. Cell Biol.* **4**: 833.
Coman, I., Barbin, G., Charles, P., Zalc, B. and Lubetzki, C., 2005, Axonal signals in central nervous system myelination, demyelination and remyelination. *J. Neurol. Sci.* **233**: 67.
Compston, A., 2004, Mechanisms of axon-glial injury of the optic nerve. *Eye* **18**: 1182.
Compston, A. and Coles, A., 2002, Multiple sclerosis. *Lancet* **359**: 1221.
Confavreux, C. and Vukusic, S., 2006, Natural history of multiple sclerosis: a unifying concept. *Brain* **129**: 606.
Craner, M.J., Newcombe, J., Black, J.A., Hartle, C., Cuzner, M.L. and Waxman, S.G., 2004, Molecular changes in neurons in multiple sclerosis: altered axonal expression of Nav1.2 and Nav1.6 sodium channels and Na^+/Ca^{2+} exchanger. *Proc. Natl. Acad. Sci. USA* **101**: 8168.
Crockett, D.P., Burshteyn, M., Garcia, C., Muggironi, M. and Casaccia-Bonnefil, P., 2005, Number of oligodendrocyte progenitors recruited to the lesioned spinal cord is modulated by the levels of the cell cycle regulatory protein p27Kip-1. *Glia* **49**: 301.

Dawson, M.R., Levine, J.M. and Reynolds, R., 2000, NG2-expressing cells in the central nervous system: are they oligodendroglial progenitors? *J. Neurosci. Res.* **61**: 471.

Dubois-Dalcq, M., Ffrench-Constant, C. and Franklin, R.J.M., 2005, Enhancing central nervous system remyelination in multiple sclerosis. *Neuron* **48**: 9.

Fancy, S.P., Zhao, C. and Franklin, R.J.M., 2004, Increased expression of Nkx2.2 and Olig2 identifies reactive oligodendrocyte progenitor cells responding to demyelination in the adult CNS. *Mol. Cell. Neurosci.* **27**: 247.

Fernandez, P.A., Tang, D.G., Cheng, L., Prochiantz, A., Mudge, A.W. and Raff, M.C., 2000, Evidence that axon-derived neuregulin promotes oligodendrocyte survival in the developing rat optic nerve. *Neuron* **28**: 81.

Fields, R.D. and Burnstock, G., 2006, Purinergic signaling in neuron–glia interactions. *Nat. Rev. Neurosci.* **7**: 423.

Fields, R.D. and Stevens, B., 2000, ATP: an extracellular signaling molecule between neurons and glia. *Trends Neurosci.* **23**: 625.

Flores, A.I., Mallon, B.S., Matsui, T., Ogawa, W., Rosenzweig, A., Okamoto, T. and Macklin, W.B., 2000, Akt-mediated survival of oligodendrocytes induced by neuregulins. *J. Neurosci.* **20**: 7622.

Foote, A.K. and Blakemore, W.F., 2005, Inflammation stimulates remyelination in areas of chronic demyelination. *Brain* **128**: 528.

Franklin, R.J.M., 2002, Why does remyelination fail in multiple sclerosis? *Nat. Rev. Neurosci.* **3**: 705.

Gensert, J.M. and Goldman, J.E., 1997, Endogenous progenitors remyelinate demyelinated axons in the adult CNS. *Neuron* **19**: 197.

Givogri, M.I., Costa, R.M., Schonmann, V., Silva, A.J., Campagnoni, A.T. and Bongarzone, E.R., 2002, Central nervous system myelination in mice with deficient expression of Notch1 receptor. *J. Neurosci. Res.* **67**: 309.

Goddard, D.R., Berry, M. and Butt, A.M., 1999, In vivo actions of fibroblast growth factor-2 and insulin-like growth factor-I on oligodendrocyte development and myelination in the central nervous system. *J. Neurosci. Res.* **57**: 74.

Gudz, T.I., Komuro, H. and Macklin, W.B., 2006, Glutamate stimulates oligodendrocyte progenitor migration mediated via an alphav integrin/myelin proteolipid protein complex. *J. Neurosci.* **26**: 2458.

Gutowski, N.J., Newcombe, J. and Cuzner, M.L., 1999, Tenascin-R and C in multiple sclerosis lesions: relevance to extracellular matrix remodelling. *Neuropathol. Appl. Neurobiol.* **25**: 207.

Gveric, D., Cuzner, M.L. and Newcombe, J., 1999, Insulin-like growth factors and binding proteins in multiple sclerosis plaques. *Neuropathol. Appl. Neurobiol.* **25**: 215.

Hampton, D.W., Rhodes, K.E., Zhao, C., Franklin, R.J. and Fawcett, J.W., 2004, The responses of oligodendrocyte precursor cells, astrocytes and microglia to a cortical stab injury, in the brain. *Neuroscience* **127**: 813.

Hinks, G.L. and Franklin, R.J.M., 1999, Distinctive patterns of PDGF-A, FGF-2, IGF-I, and TGF-beta1 gene expression during remyelination of experimentally-induced spinal cord demyelination. *Mol. Cell. Neurosci.* **14**: 153.

Hinks, G.L. and Franklin, R.J.M., 2000, Delayed changes in growth factor gene expression during slow remyelination in the CNS of aged rats. *Mol. Cell. Neurosci.* **16**: 542.

Ishibashi, T., Dakin, K.A., Stevens, B., Lee, P.R., Kozlov, S.V., Stewart, C.L. and Fields, R.D., 2006, Astrocytes promote myelination in response to electrical impulses. *Neuron* **49**: 823.

Jiang, F., Frederick, T.J. and Wood, T.L., 2001, IGF-I synergizes with FGF-2 to stimulate oligodendrocyte progenitor entry into the cell cycle. *Dev. Biol.* **232**: 414.

John, G.R., Shankar, S.L., Shafit-Zagardo, B., Massimi, A., Lee, S.C., Raine, C.S. and Brosnan, C.F., 2002, Multiple sclerosis: re-expression of a developmental pathway that restricts oligodendrocyte maturation. *Nat. Med.* **8**: 1115.

Karadottir, R., Cavelier, P., Bergersen, L.H. and Attwell, D., 2005, NMDA receptors are expressed in oligodendrocytes and activated in ischaemia. *Nature* **438**: 1162.

Karram, K., Chatterjee, N. and Trotter, J., 2005, NG2-expressing cells in the nervous system: role of the proteoglycan in migration and glial-neuron interaction. *J. Anat.* **207**: 735.

Keirstead, H.S. and Blakemore, W.F., 1997, Identification of post-mitotic oligodendrocytes incapable of remyelination within the demyelinated adult spinal cord. *J. Neuropathol. Exp. Neurol.* **56**: 1191.

Kohama, I., Lankford, K.L., Preiningerova, J., White, F.A., Vollmer, T.L. and Kocsis, J.D., 2001, Transplantation of cryopreserved adult human Schwann cells enhances axonal conduction in demyelinated spinal cord. *J. Neurosci.* **21**: 944.

Komoly, S., Hudson, L.D., Webster, H.D. and Bondy, C.A., 1992, Insulin-like growth factor I gene expression is induced in astrocytes during experimental demyelination. *Proc. Natl. Acad. Sci. USA* **89**: 1894.

Kondo, T. and Raff, M., 2000, Basic helix-loop-helix proteins and the timing of oligodendrocyte differentiation. *Development* **127**: 2989.

Kornek, B. and Lassmann, H., 2003, Neuropathology of multiple sclerosis – new concepts. *Brain Res. Bull.* **61**: 321.

Kornek, B., Storch, M.K., Weissert, R., Wallstroem, E., Stefferl, A., Olsson, T., Linington, C., Schmidbauer, M. and Lassmann, H., 2000, Multiple sclerosis and chronic autoimmune encephalomyelitis: a comparative quantitative study of axonal injury in active, inactive, and remyelinated lesions. *Am. J. Pathol.* **157**: 267.

Kotter, M.R., Li, W.W., Zhao, C. and Franklin, R.J.M., 2006, Myelin impairs CNS remyelination by inhibiting oligodendrocyte precursor cell differentiation. *J. Neurosci.* **26**: 328.

Lassmann, H., Bruck, W., Lucchinetti, C. and Rodriguez, M., 1997, Remyelination in multiple sclerosis. *Mult. Scler.* **3**: 133.

Levine, J.M. and Reynolds, R., 1999, Activation and proliferation of endogenous oligodendrocyte precursor cells during ethidium bromide-induced demyelination. *Exp. Neurol.* **160**: 333.

Levison, S.W. and Goldman, J.E., 1993, Both oligodendrocytes and astrocytes develop from progenitors in the subventricular zone of postnatal rat forebrain. *Neuron* **10**: 201.

Li, W.W., Setzu, A., Zhao, C. and Franklin, R.J.M., 2005, Minocycline-mediated inhibition of microglia activation impairs oligodendrocyte progenitor cell responses and remyelination in a non-immune model of demyelination. *J. Neuroimmunol.* **158**: 58.

Lucchinetti, C., Bruck, W., Parisi, J., Scheithauer, B., Rodriguez, M. and Lassmann, H., 2000, Heterogeneity of multiple sclerosis lesions: implications for the pathogenesis of demyelination. *Ann. Neurol.* **47**: 707.

Ludwin, S.K., 2006, The pathogenesis of multiple sclerosis: relating human pathology to experimental studies. *J. Neuropathol. Exp. Neurol.* **65**: 305.

Maeda, Y., Solanky, M., Menonna, J., Chapin, J., Li, W. and Dowling, P., 2001, Platelet-derived growth factor-alpha receptor-positive oligodendroglia are frequent in multiple sclerosis lesions. *Ann. Neurol.* **49**: 776.

Mallon, B.S., Shick, H.E., Kidd, G.J. and Macklin, W.B., 2002, Proteolipid promoter activity distinguishes two populations of NG2-positive cells throughout neonatal cortical development. *J. Neurosci.* **22**: 876.

Marcus, K., Schmidt, O., Schaefer, H., Hamacher, M., van Hall, A. and Meyer, H.E., 2004, Proteomics-application to the brain. *Int. Rev. Neurobiol.* **61**: 285.

Mason, J.L. and Goldman, J.E., 2002, A2B5+ and O4+ cycling progenitors in the adult forebrain white matter respond differentially to PDGF-AA, FGF-2, and IGF-1. *Mol. Cell. Neurosci.* **20**: 30.

Mason, J.L., Ye, P., Suzuki, K., D'Ercole, A.J. and Matsushima, G.K., 2000, Insulin-like growth factor-1 inhibits mature oligodendrocyte apoptosis during primary demyelination. *J. Neurosci.* **20**: 5703.

Mason, J.L., Suzuki, K., Chaplin, D.D. and Matsushima. G.K., 2001, Interleukin-1beta promotes repair of the CNS. *J. Neurosci.* **21**: 7046.

Mi, S., Lee, X., Shao, Z., Thill, G., Ji, B., Relton, J., Levesque, M., Allaire, N., Perrin, S., Sands, B., Crowell, T., Cate, R.L., McCoy, J.M. and Pepinsky, R.B., 2004, LINGO-1 is a component of the Nogo-66 receptor/p75 signaling complex. *Nat. Neurosci.* **7**: 221.

Miller, R.H., 1999, Contact with central nervous system myelin inhibits oligodendrocyte progenitor maturation. *Dev. Biol.* **216**: 359.

Morris, C.M. and Wilson, K.E., 2004, High throughput approaches in neuroscience. *Int. J. Dev. Neurosci.* **22**: 515.

Murtie, J.C., Zhou, Y.X., Le, T.Q., Vana, A.C. and Armstrong, R.C., 2005, PDGF and FGF2 pathways regulate distinct oligodendrocyte lineage responses in experimental demyelination with spontaneous remyelination. *Neurobiol. Dis.* **19**: 171.

O'Leary, M.T., Hinks, G.L., Charlton, H.M. and Franklin, R.J.M., 2002, Increasing local levels of IGF-I mRNA expression using adenoviral vectors does not alter oligodendrocyte remyelination in the CNS of aged rats. *Mol. Cell. Neurosci.* **19**: 32.

Olsen, I.M. and Ffrench-Constant, C., 2005, Dynamic regulation of integrin activation by intracellular and extracellular signals controls oligodendrocyte morphology. *BMC Biol.* **3**: 25.

Omlin, F.X., 1997, Optic disc and optic nerve of the blind cape mole-rat (Georychus capensis): a proposed model for naturally occurring reactive gliosis. *Brain Res. Bull.* **44**: 627.

Penderis, J., Shields, S.A. and Franklin, R.J.M., 2003a, Impaired remyelination and depletion of oligodendrocyte progenitors does not occur following repeated episodes of focal demyelination in the rat central nervous system. *Brain* **126**: 1382.

Penderis, J., Woodruff, R.H., Lakatos, A., Li, W.W., Dunning, M.D., Zhao, C., Marchionni, M. and Franklin, R.J.M., 2003b, Increasing local levels of neuregulin (glial growth factor-2) by direct infusion into areas of demyelination does not alter remyelination in the rat CNS. *Eur. J. Neurosci.* **18**: 2253.

Pluchino, S., Quattrini, A., Brambilla, E., Gritti, A., Salani, G., Dina, G., Galli, R., del Carro, U., Amadio, S., Bergami, A., Furlan, R., Comi, G., Vescovi, A.L. and Martino, G., 2003, Injection of adult neurospheres induces recovery in a chronic model of multiple sclerosis. *Nature* **422**: 688.

Prineas, J.W. and Connell, F., 1979, Remyelination in multiple sclerosis. *Ann. Neurol.* **5**: 22.

Prineas, J.W., Barnard, R.O., Kwon, E.E., Sharer, L.R. and Cho, E.S., 1993, Multiple sclerosis: remyelination of nascent lesions. *Ann. Neurol.* **33**: 137.

Raff, M.C., Miller, R.H. and Noble, M., 1983, A glial progenitor cell that develops in vitro into an astrocyte or an oligodendrocyte depending on culture medium. *Nature* **303**: 390.

Raine, C.S., 1997, The Norton Lecture: a review of the oligodendrocyte in the multiple sclerosis lesion. *J. Neuroimmunol.* **77**: 135.

Raine, C.S. and Wu, E., 1993, Multiple sclerosis: remyelination in acute lesions. *J. Neuropathol. Exp. Neurol.* **52**: 199.

Redwine, J.M. and Armstrong, R.C., 1998, In vivo proliferation of oligodendrocyte progenitors expressing PDGFalphaR during early remyelination. *J. Neurobiol.* **37**: 413.

Rodriguez, M., 2003, A function of myelin is to protect axons from subsequent injury: implications for deficits in multiple sclerosis. *Brain* **126**: 751.

Salter, M.G. and Fern, R., 2005, NMDA receptors are expressed in developing oligodendrocyte processes and mediate injury. *Nature* **438**: 1167.

Scolding, N., Franklin, R., Stevens, S., Heldin, C.H., Compston, A. and Newcombe, J., 1998, Oligodendrocyte progenitors are present in the normal adult human CNS and in the lesions of multiple sclerosis. *Brain* **121**: 2221.

Setzu, A., Ffrench-Constant, C. and Franklin, R.J.M., 2004, CNS axons retain their competence for myelination throughout life. *Glia* **45**: 307.

Sim, F.J., Hinks, G.L. and Franklin, R.J.M., 2000, The re-expression of the homeodomain transcription factor Gtx during remyelination of experimentally induced demyelinating lesions in young and old rat brain. *Neuroscience* **100**: 131.

Sim, F.J., Zhao, C., Penderis, J. and Franklin, R.J.M., 2002a, The age-related decrease in CNS remyelination efficiency is attributable to an impairment of both oligodendrocyte progenitor recruitment and differentiation. *J. Neurosci.* **22**: 2451.

Sobel, R.A., Chen, M., Maeda, A. and Hinojoza, J.R., 1995, Vitronectin and integrin vitronectin receptor localization in multiple sclerosis lesions. *J. Neuropathol. Exp. Neurol.* **54**: 202.

Spassky, N., Goujet-Zalc, C., Parmantier, E., Olivier, C., Martinez, S., Ivanova, A., Ikenaka, K., Macklin, W., Cerruti, I., Zalc, B. and Thomas, J.L., 1998, Multiple restricted origin of oligodendrocytes. *J. Neurosci.* **18**: 8331.

Spassky, N., Olivier, C., Perez-Villegas, E., Goujet-Zalc, C., Martinez, S., Thomas, J. and Zalc, B., 2000, Single or multiple oligodendroglial lineages: a controversy. *Glia* **29**: 143.

Stevens, B., Porta, S., Haak, L.L., Gallo, V. and Fields, R.D., 2002, Adenosine: a neuron-glial transmitter promoting myelination in the CNS in response to action potentials. *Neuron* **36**: 855.

Stidworthy, M.F., Genoud, S., Li, W.W., Leone, D.P., Mantei, N., Suter, U. and Franklin, R.J.M., 2004, Notch1 and Jagged1 are expressed after CNS demyelination, but are not a major rate-determining factor during remyelination. *Brain* **127**: 1928.

Tait, S., Gunn-Moore, F., Collinson, J.M., Huang, J., Lubetzki, C., Pedraza, L., Sherman, D.L., Colman, D.R. and Brophy, P.J., 2000, An oligodendrocyte cell adhesion molecule at the site of assembly of the paranodal axoglial junction. *J. Cell. Biol.* **150**: 657.

Talbott, J.F., Loy, D.N., Liu, Y., Qiu, M.S., Bunge, M.B., Rao, M.S. and Whittemore, S.R., 2005, Endogenous Nkx2.2+/Olig2+ oligodendrocyte precursor cells fail to remyelinate the demyelinated adult rat spinal cord in the absence of astrocytes. *Exp. Neurol.* **192**: 11.

Targett, M.P., Sussman, J., Scolding, N., O'Leary, M.T., Compston, D.A. and Blakemore, W.F., 1996, Failure to achieve remyelination of demyelinated rat axons following transplantation of glial cells obtained from the adult human brain. *Neuropathol. Appl. Neurobiol.* **22**: 199.

Tauber, H., Waehneldt, T.V. and Neuhoff, V., 1980, Myelination in rabbit optic nerves is accelerated by artificial eye opening. *Neurosci. Lett.* **16**: 235.

Totoiu, M.O., Nistor, G.I., Lane, T.E. and Keirstead, H.S., 2004, Remyelination, axonal sparing, and locomotor recovery following transplantation of glial-committed progenitor cells into the MHV model of multiple sclerosis. *Exp. Neurol.* **187**: 254.

Trapp, B.D., Peterson, J., Ransohoff, R.M., Rudick, R., Mork, S. and Bo, L., 1998, Axonal transection in the lesions of multiple sclerosis. *N. Engl. J. Med.* **338**: 278.

Vallstedt, A., Klos, J.M. and Ericson, J., 2005, Multiple dorsoventral origins of oligodendrocyte generation in the spinal cord and hindbrain. *Neuron* **45**: 55.

Viehover, A., Miller, R.H., Park, S.K., Fischbach, G. and Vartanian, T., 2001, Neuregulin: an oligodendrocyte growth factor absent in active multiple sclerosis lesions. *Dev. Neurosci.* **23**: 377.

Wang, S., Sdrulla, A.D., diSibio, G., Bush, G., Nofziger, D., Hicks, C., Weinmaster, G. and Barres, B.A., 1998, Notch receptor activation inhibits oligodendrocyte differentiation. *Neuron* **21**: 63.

Wang, S., Sdrulla, A., Johnson, J.E., Yokota, Y. and Barres, B.A., 2001, A role for the helix-loop-helix protein Id2 in the control of oligodendrocyte development. *Neuron* **29**: 603.

Wang, K.C., Kim, J.A., Sivasankaran, R., Segal, R. and He, Z., 2002, P75 interacts with the Nogo receptor as a coreceptor for Nogo, MAG and OMgp. *Nature* **420**: 74.

Waxman, S.G., 1998, Demyelinating diseases-new pathological insights, new therapeutic targets. *N. Engl. J. Med.* **338**: 323.
Wilson, H.C., Onischke, C. and Raine, C.S., 2003, Human oligodendrocyte precursor cells in vitro: phenotypic analysis and differential response to growth factors. *Glia* **44**: 153.
Wolswijk, G., 1998, Chronic stage multiple sclerosis lesions contain a relatively quiescent population of oligodendrocyte precursor cells. *J. Neurosci.* **18**: 601.
Wolswijk, G., 2000, Oligodendrocyte survival, loss and birth in lesions of chronic-stage multiple sclerosis. *Brain* **123**: 105.
Wolswijk, G. and Balesar, R., 2003, Changes in the expression and localization of the paranodal protein Caspr on axons in chronic multiple sclerosis. *Brain* **126**: 1638.
Wong, S.T., Henley, J.R., Kanning, K.C., Huang, K.H., Bothwell, M. and Poo, M.M., 2002, A p75(NTR) and Nogo receptor complex mediates repulsive signaling by myelin-associated glycoprotein. *Nat. Neurosci.* **5**: 1302.
Woodruff, R.H., Fruttiger, M., Richardson, W.D. and Franklin, R.J.M., 2004, Platelet-derived growth factor regulates oligodendrocyte progenitor numbers in adult CNS and their response following CNS demyelination. *Mol. Cell. Neurosci.* **25**: 252.
Wren, D., Wolswijk, G. and Noble, M., 1992, In vitro analysis of the origin and maintenance of O-2A adult progenitor cells. *J. Cell. Biol.* **116**: 167.
Ye, P., Carson, J. and D'Ercole, A.J., 1995, In vivo actions of insulin-like growth factor-I (IGF-I) on brain myelination: studies of IGF-I and IGF binding protein-1 (IGFBP-1) transgenic mice. *J. Neurosci.* **15**: 7344.
Zhang, S.C., Ge, B. and Duncan, I.D., 1999, Adult brain retains the potential to generate oligodendroglial progenitors with extensive myelination capacity. *Proc. Natl. Acad. Sci. USA* **96**: 4089.
Zhao, C., Fancy, S.P., Kotter, M.R., Li, W.W. and Franklin, R.J.M., 2005a, Mechanisms of CNS remyelination-the key to therapeutic advances. *J. Neurol. Sci.* **233**: 87.
Zhao, C., Fancy, S.P., Magy, L., Urwin, J.E. and Franklin, R.J.M., 2005b, Stem cells, progenitors and myelin repair. *J. Anat.* **207**: 251.
Zhao, C., Li, W.W. and Franklin, R.J.M., 2006, Differences in the early inflammatory responses to toxin-induced demyelination are associated with the age-related decline in CNS remyelination. *Neurobiol. Aging* **27**: 1298.

20

ADULT NEUROGENESIS IN NEURODEGENERATIVE DISEASES

Tomas Deierborg, Jia-Yi Li and Patrik Brundin

1. ABSTRACT

Neurogenesis occurs in the adult hippocampus and subventricular zone. It provides an exciting potential inroad to new treatments for slow neurodegenerative disorders. Changes in neurogenesis in human degenerative diseases may also teach us something about the underlying disease mechanisms. In the present chapter, we review data from clinical and experimental studies concerning neurogenesis in Alzheimer's, Huntington's and Parkinson's diseases. In brief, the clinical data have not clearly shown whether the rate of neurogenesis is changed by the pathogenesis in these three diseases. Whether neurogenesis occurs at all in the adult *substantia nigra* under baseline conditions remains controversial. Studies on animal models of the three neurodegenerative conditions provide a mixed picture, with disease/damage to the brain being associated both decreases and increases in neurogenesis in the different studied brain regions. In conclusion, this exciting research field is still in its infancy. It is too early to say whether neurogenesis in the adult brain can be targeted by novel therapies and be used to reduce functional deficits in slow neurodegenerative diseases.

2. INTRODUCTION

Neuroscientists have long been excited by the possibility that cellular plasticity could lead to functional recovery in the diseased or damaged brain. Until recently, the focus has been on mechanisms whereby remaining cells take over the functions of lost neurons, e.g. by collateral sprouting or structural and neurochemical changes in existing synapses. Alterations in the numbers and activity of glial cells have also been suggested to promote functional recovery in some situations. Little attention, however, has been paid to the possibility that newborn neurons could lead to improved neurological function in the compromised central nervous system. Over the past decade, the realization that neurogenesis takes place in the adult brain has come into focus, and with its speculation that new neurons could be taken over those lost to acute or in chronic diseases. The purpose of this book chapter is to review the field of adult neurogenesis in three slowly progressing neurological disorders: Alzheimer's

---

Tomas Deierborg – tomas.deierborg@med.lu.se
Neuronal Survival Unit, Wallenberg Neuroscience Center, Dep. Experimental Medical Sience, Lund University, BMC A10, 22184 Lund, Sweden

disease (AD), Huntington's disease (HD) and Parkinson's disease (PD). First, we provide a short overview of where neurogenesis occurs in the normal adult mammalian brain and the kind of stimuli that promote the process. Then we describe cell genesis in the three disorders. In each case, we begin by describing the main neuropathological features of the diseases. We then review findings concerning cell genesis in human post-mortem brain samples from patients with AD, HD and PD, and relate them to observations made in experimental animal models of the diseases. We critically evaluate whether the disease states affect cell genesis and if the newborn cells develop into neurons or other types of cells. Finally, for each disease, we briefly discuss whether experimental treatments could lead to increased neurogenesis in AD, HD and PD, and if this could be used to promote functional recovery.

3. BACKGROUND – NEUROGENESIS IN THE NORMAL ADULT BRAIN

The classical dogma stated that no new neurons are generated in the brains of adult mammals. In the early 1960s, a series of studies by Altman and coworkers suggested that neurogenesis occurs in the adult brain (Altman, 1962). They described that radioactively labelled thymidine could be incorporated into the DNA of neurons in the hippocampus in adult rats, indicating the neurons had undergone cell division. These findings were not followed up with great energy by other scientists with the exception of Michael Kaplan who, during the latter half of the 1970s and early 1980s, performed studies on dividing and differentiating cells in the adult rat brain. It was not until the mid-1990s that it became fully accepted that the mammalian adult brain has the potential to generate new neurons. Whilst the occurrence of adult neurogenesis is now considered a textbook fact, there is still debate regarding some of the locations where cell division can take place and concerning where the newborn cells migrate to and what precisely they differentiate into. In two brain regions in the adult mammalian brain neurogenesis definitely occurs: the subventricular zone (SVZ) within the ventricular walls and the subgranular zone of the dentate gyrus of the hippocampus, known to be important for memory and learning (Alvarez-Buylla and Lim, 2004). Neuroblasts generated in the rodent SVZ migrate through a path, the rostral migratory stream (RMS), to the olfactory bulb where they differentiate into interneurons (Hack et al., 2005). Neurons born in the subgranular zone migrate into the granular cell layer of the dentate gyrus and differentiate into granule neurons with functional connections to the pyramidal neurons in the CA3 region (Kempermann et al., 2004). In the adult human brain, so far neurogenesis has only been confirmed in the hippocampal dentate gyrus (Eriksson et al., 1998). Although neurogenesis has been demonstrated in the olfactory bulb of adult *non-human* primates (Kornack and Rakic, 2001), there is only indirect support (observation of cells coexpressing cell cycle proteins and immature neuronal markers) for neurogenesis in the human olfactory bulb (Bedard and Parent, 2004). It is clear that the adult human SVZ contains multipotent neural stem cells. However, they seem to differ from those of other mammals and do not form distinct chains of neuroblasts that migrate to the olfactory bulb or even migration along the SVZ (Sanai et al., 2004). A RMS in the human brain, if active, is likely to be comprised in relation to the rodent brain (Fig. 1).

The two neurogenic zones differ dramatically in the rodent brain in terms their capacity for cell proliferation. Thousands of new cells are born in the SVZ every day, and migrate several millimetres to the olfactory bulb where they are integrated into the normal circuitry. The newborn cells are dependent on sensory input for normal maturation and migration (Lledo and Saghatelyan, 2005) and have been shown to play a functional role in the responses to odours (Carlen et al., 2002). By contrast much fewer cells are generated daily in the hippocampus. They migrate less than 100 μm and differentiate into granule neurons. Despite the limited proliferation and migratory capacity of newborn cells in the hippocampus, neurogenesis in this brain region attracts particular scientific interest. As an illustration, in 2006, research reports about adult hippocampal neurogenesis outnumbered papers on olfactory bulb neurogenesis by a factor of more than three.

Hippocampus is recognized as "the gateway of memory" and has an important role in storage and recall of memories. The hippocampus is believed to process and evaluate information before long-term storage, involving specific neocortical areas, takes place. It consolidates declarative information (facts and events that consciously can be recalled) by relating things in time and space and is also important for spatial orientation and navigation (Holscher, 2003). Newborn granule neurons have been shown to integrate into the synaptic circuitry in the hippocampus of rodents. The rate of hippocampal neurogenesis decreases with age (Kuhn et al., 1996), an observation that is important in the context of neurodegenerative diseases that mostly strike in later stages of life. It has been suggested that neurogenesis in the hippocampus increases in response to performance of a learning and memory tasks (Gould et al., 1999). These data, however, have been challenged repeatedly (van Praag et al., 1999; Ambrogini et al., 2004; Van der Borght et al., 2005) and it is far from clear that there is a distinct causal relationship between adult hippocampal neurogenesis and cognitive function (Leuner et al., 2006).

4. ALZHEIMER'S DISEASE

AD affects elderly and up to one in 10 individuals over the age of 65 (Evans et al., 1989), which makes it the most common neurodegenerative disorder. Recent data predict that the prevalence of the disease will be doubled every 20 years when the population grows older (Ferri et al., 2005). The aetiology of AD is likely to be multifactorial and include both environmental and genetic factors. Initially, patients show memory problems, as well as spatial disorientation and personality changes. At the end stage, patients will have severe dementia, aphasia and immobility. AD is a progressive neurodegenerative disease. It is characterized by intracellular neurofibrillary tangles in neurons and accumulation of extra-cellular β-amyloid plaques. The plaque have a core of amyloid precursor protein (APP), complement factors, acute phase proteins and intercellular adhesion molecules surrounded by activated microglia, reactive astrocytes and dystrophic neurites (Rozemuller et al., 1990; Verbeek et al., 1996; Honda et al., 2000; Chaudhury et al., 2003). The cellular changes will consequently lead to synaptic dysfunction and neuronal loss (Gouras et al., 2005). Although the cause of the disease in most cases are unknown, genetic mutations in the genes coding for β-amyloid and its cleaving proteases, presenilin 1 and 2 (both function as γ-secretases), are known to lead to autosomal dominant AD (Tanzi and Bertram, 2005). The mutations alter the processing of β-amyloid and lead to plaque formation. Cholinergic neurons are most affected, and other neurons that produce the neurotransmitters γ-aminobutyric acid (GABA), norepinephrine, dopamine and serotonin are less disturbed (Lyness et al., 2003). Today acetylcholinesterase inhibitors are administered to patients. They improve the cholinergic transmission and reduce symptoms somewhat, especially in early disease, but do not affect progression of the pathology (Birks, 2006). The cellular pathology is manifested in regions essential for cognitive function such as the hippocampus, amygdala and the association cortices of the frontal and temporal lobes.

Degeneration in the hippocampus, particularly within the CA1 subregion (Csernansky et al., 2000), is a site of early neuropathology in AD (Murphy et al., 1993; Laakso et al., 1995). In contrast, the dentate gyrus is not strongly affected. The degeneration of pyramidal neurons in the CA1 correlates both with decreased hippocampal volume and reductions in memory function (Zarow et al., 2005).

At present, there is no effective neuroprotective treatment for AD (Pietrzik and Behl, 2005). Clinical trials based on vaccination against β-amyloid, which was a successful approach in animal models of AD, were halted in 2002 due to meningoencephalitis in 6% of the patients. Nonetheless, in some patients, cognitive stabilization and elimination of plaque was reported (Lemere et al., 2006). Nerve growth factor (NGF) is known to stimulate cholinergic function, improve memory and prevent cell degeneration in animal models. Administration of NGF was tested via pump (Eriksdotter Jonhagen et al., 1998) and recently, genetically modified cells producing NGF (Tuszynski et al., 2005) were transplanted into the brain, which was reported to improve cognition of the AD patient. Increased neurogenesis in the dentate gyrus has been discussed as a therapeutic target for AD (Tatebayashi et al., 2003). New neurons in the dentate gyrus are unlikely to restore the lost neurons in other parts of the hippocampal formation, but could conceivably improve function in a failing hippocampal circuitry. The hippocampal network architecture consists mainly of a tri-synaptic backbone. The flow of information in the hippocampus starts from the input from the enthorinal cortex to the dentate gyrus, where the granular cells connect to the CA3 pyramidal neurons via the mossy fibers. The Schaffer collaterals then connect the CA3 with the CA1 pyramidal neurons and finally the information passes through the subiculum back to the enthorinal cortex. In this circuitry, the connection between dentate gyrus and CA3, the mossy fiber tract, is recognized as the bottleneck of the circuitry. Increasing the output from the dentate gyrus into the CA3-CA1 circuitry by the generation of new granular cells could possibly strengthen the damaged circuitry. It could perhaps also provide additional trophic stimulation and inhibit neurodegeneration in AD. Thereby, one can speculate that new granule neurons might reduce the rate of decline of the hippocampal function and promote cognitive function in AD.

4.1. Is the Neurogenesis Changed in Patients with AD?

Cell cycle-dependent markers are upregulated in the hippocampus of patients with AD. This has led to the hypothesis that neuronal damage leads to re-entry of the neurons to the cell cycle (Nagy et al., 1997a, b; Nagy, 2000) that was reported to occur in brain following ischemia (Katchanov et al., 2001). It was suggested that post-mitotic neurons are unable to complete the cell cycle and instead undergo cell death (Yang et al., 2003). Kinases associated with cell-cycle regulation were proposed to phosphorylate tau instead, leading to tau hyperphosphorylation (Vincent et al., 1997; Busser et al., 1998). The increased levels of cell cycle-related proteins in AD brain suggest that there could be increased cell genesis and even that there are generation of neurons. One study has reported on expression of several markers of immature cells and, in particular, neurons in the hippocampal dentate gyrus of AD patients. Specifically, doublecortin, TUC-4 (turned-on-after-division/Ulip-1/CRMP-4), embryonic nerve cell adhesion molecule and neurogenic differentiation marker NeuroD (Jin et al., 2004b) were all upregulated. Doublecortin-positive cells were also found in the CA1 subregion of the AD patients, which is not the case in normal brains of humans and experimental animals. One can speculate that these phenomena are due to an elevated trophic stimulation of the AD hippocampus, in turn secondary to a response of the brain trying to handle the AD pathogenetic mechanisms. Although the study suggested that neurogenesis is increased in the hippocampus of AD patients, it has some important shortcomings. First, the study fails to demonstrate that the newly generated cells actually become mature neurons. Indeed, new cells expressing immature neuronal markers were described, but for these cells to be functional they must continue to mature and make synaptic connections. Second, the study illustrates that detailed cellular analysis of human post-mortem tissue is very difficult to achieve. To demonstrate that one and the same cell co-expresses two markers (e.g. one related to newborn cells and one related to neuronal fate), dual labelling immunohistochemistry is required. This still remains to be done in AD patient brain material and may be problematic to attain because of sub-optimal tissue fixation. Even if Western blot analysis can be used to verify the existence of the immature neuronal proteins, such as doublecortin and NeuroD, the physical confirmation of that the same cell express the different markers is essential. A third issue that complicates the analysis of neurogenesis in the AD brain is cell atrophy. The individual cells undergo atrophy before they die and the total volume of the hippocampus decreases in AD. Decreased total hippocampal volume may lead to relative increase (observed as higher density) of newborn neurons, which is not necessarily coupled to an absolute increase in cell genesis. Also, when mature cells undergo pathological atrophy, as a consequence, they can exhibit increased relative concentrations of proteins associated with immature cells, and therefore erroneously be classified as newly generated cells. Indeed, in the report from Jin et al., the cells expressing immature markers were described as having a "condensed morphology". This suggests that they either were "false-positive" (i.e. the concentration of the immature protein increased and was possible to detect with immunocytochemistry as the cell shrunk) or that they were undergoing apoptosis and would never develop into mature neurons (Kempermann, 2006). In summary, the study by Jin and coworkers presents tentative evidence for increased cell genesis in the brains of AD patients, but it needs to be interpreted with great caution. Before the results are confirmed in an independent study that addresses some of the difficult methodological issues at stake, the question of whether neurogenesis is increased in the hippocampus of AD patients should be considered unresolved.

4.2. Neurogenesis in Animal Models of AD

In the early 1990s, the first AD mouse models became available and presently there are several different genetically modified mouse strains that resemble AD. These mouse models display many behavioural and neuropathological features of human AD such as memory impairment, amyloid deposits, gliosis, synaptic alterations and signs of neurodegeneration (Higgins and Jacobsen, 2003). Studies on adult neurogenesis in AD mouse models have been performed in mice expressing mutant forms of APP or Presenilin, often with contradictory results. First, we describe studies on transgenic models that express different mutant forms of APP and thereafter experiments on mice expressing mutant presenilin.

4.3. Transgenic APP Mouse Models

Increased neurogenesis has been reported in both the dentate gyrus and the SVZ (Jin et al., 2004a) in a transgenic model expressing the Swedish and Indian APP mutations. A twofold increase in proliferation was observed in 3-month-old mice before neuronal loss and amyloid deposition were detected. In another study on APP transgenic mice, cell proliferation increased in the neocortex, but the newborn cells became microglia and not neurons (Bondolfi et al., 2002). Some cell cultures support the notion that AD-related pathology could increase neurogenesis. For example, the soluble secreted form of APP increases proliferation of adult rat neural stem cells (Caille et al., 2004) and embryonic neural stem cells (Ohsawa et al., 1999). Aggregated β-amyloid, but not soluble APP, has been reported to increase neuronal differentiation in neurosphere cultures of hippocampal neural stem cells, without affecting proliferation (Lopez-Toledano and Shelanski, 2004).

Contrary to these reports, the vast majority of studies demonstrate a reduction in neurogenesis and proliferation of neural stem cells in APP mouse and cell models of AD. In stem cell cultures from human embryonic cortex and rodent brain, β-amyloid has been reported to impair proliferation and neuronal differentiation, and promote apoptosis (Haughey et al., 2002). Proliferation and survival of neural progenitor cells are decreased in the dentate gyrus and SVZ in some mouse models where mutant APP is overexpressed (Sturchler-Pierrat et al., 1997; Haughey et al., 2002; Dong et al., 2004). In the study by Dong et al., these reductions are apparent at 3–9 months of age, and precede the formation of β-amyloid, impairments in hippocampal long-term potentiation and the appearance of behavioural deficits. A recent study, on yet another transgenic AD mouse model (Dodart et al., 2000), showed a 50% reduction in SGZ neurogenesis that developed with a time course similar to the development of the AD-like pathology and behavioural deficits (Donovan et al., 2006). In that particular study, there was an increased birth of immature neurons in the outer portion of the granular cell layer. A region that normally shows limited proliferative capacity, which they suggest as an explanation for the augmented neurogenesis reported in mouse models of AD.

Mutations in the gene for presenilin-1 (*PS-1*) are recognized as one of the most common cause of the early onset of the familial form of AD (St George-Hyslop, 2000). In the mouse, PS-1 is required for neural development and lack of the protein leads to premature differentiation of neural progenitor cells (Handler et al., 2000). Overexpression of mutant or wild-type forms of PS-1 inhibits the generation of neural progenitors in the adult hippocampus and overexpression of the wild-type, but not the mutant, PS-1 promotes survival and neuronal differentiation (Wen et al., 2002). Wang and collaborators studied a mouse strain where a mutation (M146V) is incorporated in the endogenous murine *PS-1* gene (Wang et al., 2004). They report impairment in a contextual fear-conditioning test, which is used to study hippocampus-dependent associative learning. This impairment was correlated with decreased neurogenesis in the dentate gyrus, but no changes were detected in synaptic plasticity in the CA1 or dentate gyrus using electrophysiological recordings. Another study used a transgenic mouse with the *PS-1* gene specifically knocked out in excitatory neurons of the adult forebrain (Feng et al., 2001). Although these mice did not display learning deficits, they lacked enrichment-induced increase in hippocampal neurogenesis, indicating that neurogenesis in their experimental set-up was not required to form memories. Interestingly, they suggest that hippocampal neurogenesis, on the other hand, is important in clearance of outdated hippocampal memories after they have been consolidated in cortical brain regions. This would make the hippocampus available for processing of new memories and promote normal memory function. There are opposing results that show an increase in cell proliferation in the dentate gyrus using a mouse strain with PS-1 mutation, A246E (Chevallier et al., 2005). The potential role of presenilins in adult neurogenesis may have serious implications for novel AD treatment strategies that are designed to inhibit the presenilin.

In summary, there is no consistent picture regarding the effects of AD-related mutations on neurogenesis in mouse models of AD. In the human condition, it is also not clear whether neurogenesis is altered, despite one claim that it is increased in patients with AD (Jin et al., 2004b). It is certain, however, that if endogenous neurogenesis is increased in sufferers from AD, it does not adequately compensate for cell loss and the cognitive decline. Interestingly, two classes of clinically available drugs for AD, i.e. acetylcholinesterase-inhibitors and an uncompetitive NMDA receptor antagonist (Jin et al., 2006), induce increases in hippocampal neurogenesis in normal rodents. It remains to be seen whether an enhancement of neurogenesis contributes to the mild beneficial effects of these drugs in AD.

5. HUNTINGTON'S DISEASE

HD is an autosomal dominant disorder with a prevalence of to around 1 in 10,000 (Harper, 1996). The disease comprises hyperkinetic involuntary motor symptoms, personality changes, cognitive and psychiatric symptoms. The disease is caused by an expansion of the CAG repeat (coding glutamine) in the gene encoding *huntingtin*. The mutant gene is located on the short arm of chromosome 4 and exhibits an expansion in the number of CAG trinucleotide repeats in the exon 1. Normal individuals have 35 or fewer CAG repeats in this locus, whereas *HD* gene carriers have 36 or more CAG repeats (Vonsattel and DiFiglia, 1998). The resulting abnormally long polyglutamine stretch causes the huntingtin to misfold and to acquire toxic properties. The age of onset of HD is inversely related to the number of CAG repeats. Individuals having around 40–55 CAG repeats typically develop symptoms around 35–45 years of age. If the CAG repeat expansion exceeds 70 repeats, the onset of disease can be juvenile. Patients typically live 15–20 years from the onset of the disease. They often die of aspiration pneumonia, urinary tract infection due to motor deficiency and general weakness. According to classical descriptions, the neuropathology in HD involves neuronal death and is focused on the basal ganglia and neocortex (Vonsattel and DiFiglia, 1998), but the hypothalamus is also clearly affected (Kremer et al., 1990, 1991; Petersen et al., 2005). The pathology is progressive and symptoms probably appear before there is extensive cell death. Mutant huntingtin accumulates in the cytoplasm and nucleus leading to the formation of insoluble aggregates/inclusions in the nucleus and the cytoplasm. These inclusions contain several other proteins, including components of the ubiquitin–proteasome pathway, chaperones, synaptic proteins and transcription factors (Ciechanover and Brundin, 2003; Landles and Bates, 2004; Qin et al., 2004). The inclusions themselves and their soluble precursors are both claimed to have negative impact on the neuronal machinery.

5.1. Is the Neurogenesis Altered in the Brain of HD Patients?

In HD patients, there is extensive neurodegeneration in the striatum that is adjacent to the SVZ. Therefore, the concept of restoring neurological functions through neurogenesis is particularly attractive in HD. An increase in neurogenesis was recently reported in the SVZ, adjacent to the caudate nucleus, in HD patients (Curtis et al., 2003, 2005). These claims were largely based on immunohistochemical stainings. The investigators reported a significant increase in cell proliferation in the SVZ based on increases in the cell cycle marker PCNA (proliferating cell nuclear antigen). These newborn cells were positive for certain neuronal markers, such as β-III tubulin (normally expressed by immature neurons) and neuropeptide

NPY. Some of the PCNA-labelled cells were also positive for the astrocyte marker, GFAP. These data indicate that increased neurogenesis and gliogenesis takes place in the brains of HD patients. Interestingly, the degrees of increased cell proliferation and neurogenesis were correlated to the severity of the disease (Curtis et al., 2003). Similar to the effect of AD on neurogenesis in the human brain, more clinical studies from HD patients are needed to confirm the existing reports that all come from the same laboratory. As discussed in the AD section above, more convincing immunohistochemical data are also required.

5.2. Neurogenesis in Animal Models of HD

In contrast to the human post-mortem studies of the SVZ region, reduced cell proliferation has been reported both in the hippocampus of R6/1 and R6/2 HD transgenic mice (Gil et al., 2004, 2005; Lazic et al., 2004). No evidence of changes in cell proliferation in the SVZ was found (Gil et al., 2005; Lazic et al., 2006). The reduction in hippocampal neurogenesis starts already at a pre-symptomatic stage (Gil et al., 2005), around three weeks of age in R6/2 mice, and progressively becomes more marked with increasing age. In these studies, the newborn hippocampal cells that have incorporated BrdU also express the mature neuronal marker NeuN or the immature marker doublecortin. The helix–loop–helix transcription factor NeuroD plays an important role in neural differentiation and survival in the hippocampus (Liu et al., 2000). In HD and related models, huntingtin-associated protein 1 (HAP1) is substantially reduced (Li et al., 2003), which has been suggested to inhibit the activity of NeuroD (Marcora et al., 2003). In turn, this may underlie the reduction of cell proliferation and neurogenesis in the dentate gyrus. If this is the case, it may also explain why there are no clear changes in neurogenesis in the SVZ of the R6/2, since here NeuroD does not have the same role. Reductions of hippocampal cell proliferation and neurogenesis may be features of HD and could contribute to the deficits of learning and memory seen in R6/2 HD transgenic mice. Therefore, enhancing hippocampal neurogenesis may have beneficial effects, even though its importance for memory and learning is under debate (Leuner et al., 2006). We tested the effects of asialoerythropoietin, a variant of the cytokine erythropoietin that is known to be neuroprotective and promote cell proliferation (Erbayrakar et al., 2003). After systemic administration, asialoerythropoetin had no effects on hippocampal neurogenesis in R6/2 mice (Gil et al., 2004). A recent study described increased proliferation in the SVZ and that new neurons were recruited into the striatum following subcutaneous injections of the growth factor FGF-2 into R6/2 mice. It was suggested that these newborn neurons migrated into the striatum and differentiated into striatal projection neurons expressing DARPP-32 (Jin et al., 2005). The R6/2 mice treated with FGF-2 displayed an expanded life span and improved motor function. It is still not clear whether this improvement is connected to reported increase in newborn cells.

It is important to point out that there are considerable inconsistencies regarding neurogenesis in HD, especially in two respects. One is the contradiction between results from studies on human post-mortem tissues and HD transgenic mice. Increased neurogenesis has been reported in the SVZ of HD patients (Curtis et al., 2003, 2005), but no change was observed in the SVZ of HD transgenic mice (Gil et al., 2005; Lazic et al., 2006). Why do results obtained in the animal model differ from the clinical data? Aside from the fact that the R6/2 mouse differs from HD in many respects at the level of molecular genetics (e.g. only exon 1 of human mutant huntingtin is expressed in the mouse), the inconsistencies may be explained at a more conceptual level. The progression of the pathology in HD mice is much more rapid (from weeks to months) compared to human HD patients (years to decades). In

HD patients, the reported increase cell proliferation in the SVZ is positively correlated to disease stage (Curtis et al., 2003). Possibly, it is a compensatory effect where new neurons are generated in the diseased brain in an attempt to restore the lost cells. Because there is relatively little cell death in the striatum of R6/2 mice, the same response may never develop. Another potential explanation is related to the techniques used to detect neurogenesis. All studies on human post-mortem HD tissues used immunohistochemistry for the cell cycle marker PCNA. It is abundant during cell division and down-regulates when the cell exits the mitotic phase. Therefore, PCNA immunohistochemistry is not ideal when studying cell differentiation and cannot readily be combined with neuronal markers, such as β-III tubulin and NeuN. In addition, adult cells can express PCNA as a consequence of stress or abnormal gene transcription. The studies performed in transgenic HD mouse models, on the other hand, have used BrdU to label newly generated cells. This technique makes impossible to ascertain whether the newborn cells differentiate into neurons. BrdU labelling, however, is also plagued by some shortcomings. Following death of a cell that has incorporated BrdU, it can be taken up by other cells and thereby give false positive results (Pearson, 2006). In view of the limitations of all methods used to label newborn cells in the brain, it would be valuable if multiple techniques were used in both human and mouse tissue. It is also desirable that the same team of investigators study both HD tissue and mouse models with the aim of performing direct comparisons.

So, will increased neurogenesis be a future therapeutic target for HD? The changes in neurogenesis in the striatum reported from the post-mortem studies in HD patients first need to be confirmed by more robust methods of analysis. If there is a true increase in neurogenesis, it is clearly not sufficient to compensate for the striatal cell death as the disease progresses. Possibly the newborn cells also die early due to the disease and could be stimulated to survive longer by, e.g. growth factor administration. Of course, one has to bear in mind that all the newborn cells will also carry the HD mutation. This is likely to eventually impair their function, albeit not until after several years when misfolded huntingtin has accumulated.

6. PARKINSON'S DISEASE

PD is a progressive neurodegenerative disorder characterized by tremor, difficulty with walking, initiation of movement and coordination. It affects around 1% of the population over 60 years of age (Fahn, 1989). Modern neuropathology shows that many brain areas are affected in PD, but most prominent is the loss of dopamine neurons in the *substantia nigra* in the midbrain. The progressive loss of nigral dopamine neurons leads to a deficit of dopamine in the target region of the nigrostriatal pathway, i.e. the striatum. The mainstay of current pharmacological therapy in PD is based on replacement of the lost dopamine. This is achieved through administration of the precursor of dopamine, levodopa, by directly acting on the remaining dopamine neurons or by augmentation of the remaining dopaminergic neurotransmission through inhibition of dopamine-degrading enzymes. Tremendous success with symptomatic relief has been reached by these approaches, but none of them leads to restoration or marked neuroprotection of the degenerating dopamine system. Also, eventually dopamine replacement treatments are afflicted with shortcomings. In advanced disease, the vast majority of patients experience disturbing fluctuations in response to the medication, which are referred to as the "on–off" effect. As the neurodegeneration progresses, the best response to the medication declines. Notwithstanding these shortcomings with pharmacological dopamine replacement, they are still important therapies and illustrate a fundamental principle: the nigrostriatal dopamine system is central to the symptomatology in PD. Therefore, much would be gained if it was possible to structurally restore these neurons in the brains of PD patients.

6.1. Does Neurogenesis Occur in the *Substantia Nigra* of Parkinson Patients?

Whether newborn neurons exist in the adult *substantia nigra* is controversial. Recently, Yoshimi and coworkers found polysialic acid (PSA)-like immunoreactivity in the human *substantia nigra* (Yoshimi et al., 2005). This molecule is expressed in young neurons. They further showed that PSA was increased in the *substantia nigra* following dopamine depletion in the rat and monkey brain and suggested that the same could occur in the PD brain. Taken together, the study does not provide any real evidence that new dopamine neurons are born in the adult *substantia nigra*, neither in humans nor in rodents. Changes in PSA-immunoreactivity may simply be a consequence of an ongoing disease process or secondary to the experimental lesion.

6.2. The Controversies of Ongoing Neurogenesis in the *Substantia Nigra* in Experimental Animals

In 2003, a study by Zhao and coworkers suggested that there is continuous replacement of dopamine neurons in the *substantia nigra* of adult mice (Zhao et al., 2003). They described how cells generated in the SVZ migrated to the *substantia nigra*, differentiated into neurons and grew axons to the striatum. Their data suggested that all dopamine neurons in *substantia nigra* could be regenerated during the life-time of a normal mouse. Moreover, they reported an increase in neurogenesis following partial destruction of the *substantia nigra* using systemic injections of the toxin MPTP. This report raised hopes for a new therapeutic approach for PD. Several other groups have also tried to find evidence for newborn dopamine neurons in the adult *substantia nigra* in rodents, but have failed (Kay and Blum, 2000; Lie et al., 2002; Cooper and Isacson, 2004; Frielingsdorf et al., 2004; Chen et al., 2005; Steiner et al., 2005). The exact reason for this discrepancy is not known. Zhao et al. (2003) employed unusual immunohistochemical markers for stem cells and very high concentrations of BrdU, an agent used to label dividing cells. Misinterpretation of labelled cells has also been suggested (Frielingsdorf et al., 2004; Mohapel and Brundin, 2004). The prevailing opinion is that no or very limited neurogenesis takes place in the *substantia nigra*, but it is difficult to definitely prove that a rare event never occurs. Possibly, in the future, completely novel treatments will be able to stimulate newborn cells to migrate to the adult *substantia nigra* and differentiate into dopaminergic neurons. There are reasons to argue in this direction. Thus, in rats with toxin-induced lesions of the nigrostriatal pathway, a population of actively dividing progenitors has been described in the *substantia nigra* (Lie et al., 2002). When left in the brain, they only matured into glial cells and no newborn neurons were detected. Interestingly, if these cells instead were harvested and transplanted into the neurogenic hippocampus they could differentiate into mature neurons (Lie et al., 2002). On the other hand, no neuronal differentiation could be found after transplantation into the *substantia nigra*, suggesting that this region is not permissive for neuronal differentiation. Other recent reports have reawakened the notion that new dopaminergic neurons can be generated in the *substantia nigra* of adult rodents (Van Kampen and Robertson, 2005; Yoshimi et al., 2005; Shan et al.,

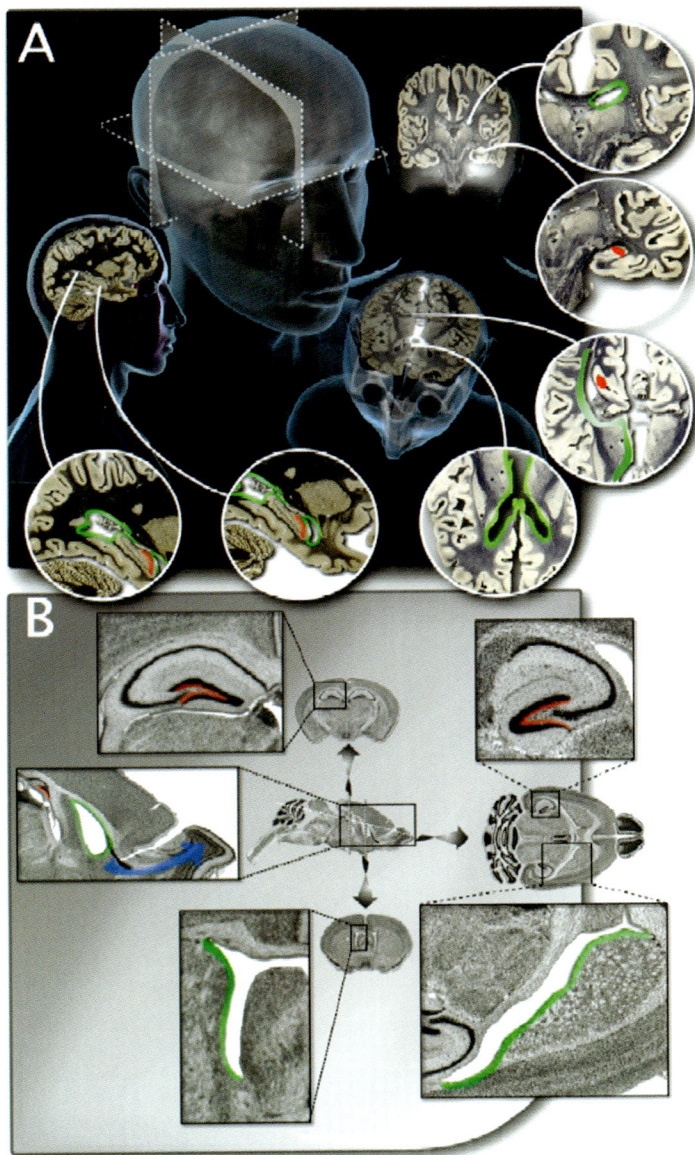

Fig. 1. The two neurogenic regions in the mammalian brain. The human brain (a) and the rodent brain (b) are visualized in coronal, sagital and horizontal views. Green colour represent the ventricular wall enclosing the germinative subventricular zone (SVZ). Red colour shows the neurogenic subgranular zone (SGZ) in the hippocampal dentate gyrus subregion. In the human brain, neurogenesis has so far only been properly characterized in the hippocampus

2006). Van Kampen and Robertson (2005) claim that there is an increase in newly generated nigral neurons when rats are treated with a dopamine D3 receptor ligand 7-OH-DPAT. Shan and collaborators argue that dopaminergic neurogenesis could arise from non-proliferative precursor. This question was addressed by using a transgenic mouse expressing LacZ reporter gene under an enhancer for the neural progenitor protein nestin. This labelling of progenitor cells was combined with BrdU labelling of newborn cells (Shan et al., 2006). They report a basal level of neurogenesis in the *substantia nigra* that increased following MPTP-lesion. However, no co-localization of TH and LacZ is shown and only between two and five cells positive for BrdU or LacZ per *substantia nigra* could be found. A very recent study has claimed that treatment with 7-OH-DPAT can increase neurogenesis in the damaged adult *substantia nigra* and even lead to recovery in tests of motor function (Van kampen and Eckman, 2006). However, the interpretation of reports of neurogenesis in the adult *substantia nigra* is often difficult. Typically, only very few newborn dopamine neurons are described and there are not always three-dimensional confocal images to confirm that cells were truly double labelled by the marker for proliferation and the label indicating that it is a dopamine neuron. In conclusion, the jury is still out concerning whether neurogenesis can occur in the *substantia nigra*.

7. ACKNOWLEDGMENTS

We wish to thank Denis Soulet for his help of making the figure. Our work was financed by the Swedish Research Council, the Swedish Society for Medicine, the Crafoord Foundation, the Hedlund Foundation, the Greta and Johan Kock Foundation.

8. REFERENCES

Altman, J., 1962, Are new neurons formed in the brains of adult mammals? *Science* **135**: 1127.
Alvarez-Buylla, A. and Lim, D.A., 2004, For the long run: maintaining germinal niches in the adult brain. *Neuron* **41**: 683.
Ambrogini, P., Orsini, L., Mancini, C., Ferri, P., Ciaroni, S. and Cuppini, R., 2004, Learning may reduce neurogenesis in adult rat dentate gyrus. *Neurosci. Lett.* **359**: 13.
Bedard, A. and Parent, A., 2004, Evidence of newly generated neurons in the human olfactory bulb. *Brain Res. Dev. Brain Res.* **151**: 159.
Birks, J., 2006, Cholinesterase inhibitors for Alzheimer's disease. *Cochrane Database Syst. Rev.* **1**: CD005593
Bondolfi, L., Calhoun, M., Ermini, F., Kuhn, H.G., Wiederhold, K.H., Walker, L., Staufenbiel, M. and Jucker, M., 2002, Amyloid-associated neuron loss and gliogenesis in the neocortex of amyloid precursor protein transgenic mice. *J. Neurosci.* **22**: 515.
Busser, J., Geldmacher, D.S. and Herrup, K., 1998, Ectopic cell cycle proteins predict the sites of neuronal cell death in Alzheimer's disease brain. *J. Neurosci.* **18**: 2801.
Caille, I., Allinquant, B., Dupont, E., Bouillot, C., Langer, A., Muller, U. and Prochiantz, A., 2004, Soluble form of amyloid precursor protein regulates proliferation of progenitors in the adult subventricular zone. *Development* **131**: 2173.
Carlen, M., Cassidy, R.M., Brismar, H., Smith, G.A., Enquist, L.W. and Frisen, J., 2002, Functional integration of adult-born neurons. *Curr. Biol.* **12**: 606.
Chaudhury, A.R., Gerecke, K.M., Wyss, J.M., Morgan, D.G., Gordon, M.N. and Carroll, S.L., 2003, Neuregulin-1 and erbB4 immunoreactivity is associated with neuritic plaques in Alzheimer disease brain and in a transgenic model of Alzheimer disease. *J. Neuropathol. Exp. Neurol.* **62**: 42.
Chen, Y., Ai, Y., Slevin, J.R., Maley, B.E. and Gash, D.M., 2005, Progenitor proliferation in the adult hippocampus and substantia nigra induced by glial cell line-derived neurotrophic factor. *Exp. Neurol.* **196**: 87.
Chevallier, N.L., Soriano, S., Kang, D.E., Masliah, E., Hu, G. and Koo, E.H., 2005, Perturbed neurogenesis in the adult hippocampus associated with presenilin-1 A246E mutation. *Am. J. Pathol.* **167**: 151.

Ciechanover, A. and Brundin, P., 2003, The ubiquitin proteasome system in neurodegenerative diseases: sometimes the chicken, sometimes the egg. *Neuron* **40**: 427.
Cooper, O. and Isacson, O., 2004, Intrastriatal transforming growth factor alpha delivery to a model of Parkinson's disease induces proliferation and migration of endogenous adult neural progenitor cells without differentiation into dopaminergic neurons. *J. Neurosci.* **24**: 8924.
Csernansky, J.G., Wang, L., Joshi, S., Miller, J.P., Gado, M., Kido, D., McKeel, D., Morris, J.C. and Miller, M.I., 2000, Early DAT is distinguished from aging by high-dimensional mapping of the hippocampus. Dementia of the Alzheimer type. *Neurology* **55**: 1636.
Curtis, M.A., Penney, E.B., Pearson, A.G., van Roon-Mom, W.M., Butterworth, N.J., Dragunow, M., Connor, B. and Faull, R.L., 2003, Increased cell proliferation and neurogenesis in the adult human Huntington's disease brain. *Proc. Natl Acad. Sci. USA* **100**: 9023.
Curtis, M.A., Penney, E.B., Pearson, J., Dragunow, M., Connor, B. and Faull, R.L., 2005, The distribution of progenitor cells in the subependymal layer of the lateral ventricle in the normal and Huntington's disease human brain. *Neuroscience* **132**: 777.
Dodart, J.C., Mathis, C., Saura, J., Bales, K.R., Paul, S.M. and Ungerer, A., 2000, Neuroanatomical abnormalities in behaviorally characterized APP(V717F) transgenic mice. *Neurobiol. Dis.* **7**: 71.
Dong, H., Goico, B., Martin, M., Csernansky, C.A., Bertchume, A. and Csernansky, J.G., 2004, Modulation of hippocampal cell proliferation, memory, and amyloid plaque deposition in APPsw (Tg2576) mutant mice by isolation stress. *Neuroscience* **127**: 601.
Donovan, M.H., Yazdani, U., Norris, R.D., Games, D., German, D.C. and Eisch, A.J., 2006, Decreased adult hippocampal neurogenesis in the PDAPP mouse model of Alzheimer's disease. *J. Comp. Neurol.* **495**: 70.
Erbayrahar, S., Grasso, G., Sfacteria, A., Xie, Q.W., Coleman, T., Kreilgaard, M., Torup, L., Sager, T., Erbayraktar, Z., Gokmen, N., Yilmaz, O., Ghezzi, P., Villa, P., Fratelli, M., Casagrande, S., Leist, M., Helboe, L., Gerwein, J., Christensen, S., Geist, M.A., Pedersen, L.O., Cerami-Hand, C., Wuerth, J.P., Cerami, A. and Brines, M., 2003, Asialoerythropoietin is a nonerythropoietic cytokine with broad neuroprotective activity in vivo. *Proc. Natl Acad. Sci. USA* **100**: 6741.
Eriksdotter Jonhagen, M., Nordberg, A., Amberla, K., Backman, L., Ebendal, T., Meyerson, B., Olson, L., Seiger, Shigeta, M., Theodorsson, E., Viitanen, M., Winblad, B. and Wahlund, L.O., 1998, Intracerebroventricular infusion of nerve growth factor in three patients with Alzheimer's disease. *Dement. Geriatr. Cogn. Disord.* **9**: 246.
Eriksson, P.S., Perfilieva, E., Bjork-Eriksson, T., Alborn, A.M., Nordborg, C., Peterson, D.A. and Gage. F.H., 1998, Neurogenesis in the adult human hippocampus. *Nat. Med.* **4**: 1313.
Evans, D.A., Funkenstein, H.H., Albert, M.S., Scherr, P.A., Cook, N.R., Chown, M.J., Hebert, L.E., Hennekens, C.H. and Taylor, J.O., 1989, Prevalence of Alzheimer's disease in a community population of older persons. Higher than previously reported. *JAMA* **262**: 2551.
Fahn, S., 1989, The history of parkinsonism. *Mov. Disord.* **4** (Suppl. 1): S2.
Feng, R., Rampon, C., Tang, Y.P., Shrom, D., Jin, J., Kyin, M., Sopher, B., Miller, M.W., Ware, C.B., Martin, G.M., Kim, S.H., Langdon, R.B., Sisodia, S.S. and Tsien, J.Z., 2001, Deficient neurogenesis in forebrain-specific presenilin-1 knockout mice is associated with reduced clearance of hippocampal memory traces. *Neuron* **32**: 911.
Ferri, C.P., Prince, M., Brayne, C., Brodaty, H., Fratiglioni, L,. Ganguli, M., Hall, K., Hasegawa, K., Hendrie, H., Huang, Y., Jorm, A., Mathers, C., Menezes, P.R., Rimmer, E. and Scazufca, M., 2005, Global prevalence of dementia: a Delphi consensus study. *Lancet* **366**: 2112.
Frielingsdorf, H., Schwarz, K., Brundin, P. and Mohapel, P., 2004, No evidence for new dopaminergic neurons in the adult mammalian substantia nigra. *Proc. Natl Acad. Sci. USA* **101**: 10177.
Gil, J.M., Leist, M., Popovic, N., Brundin, P. and Petersen, A., 2004, Asialoerythropoietin is not effective in the R6/2 line of Huntington's disease mice. *BMC Neurosci.* **5**: 17.
Gil, J.M., Mohapel, P., Araujo, I.M., Popovic, N., Li, J.Y., Brundin, P. and Petersen, A., 2005, Reduced hippocampal neurogenesis in R6/2 transgenic Huntington's disease mice. *Neurobiol. Dis.* **20**: 744.
Gould, E., Beylin, A., Tanapat, P., Reeves, A. and Shors, T.J., 1999, Learning enhances adult neurogenesis in the hippocampal formation. *Nat. Neurosci.* **2**: 260.
Gouras, G.K., Almeida, C.G. and Takahashi, R.H., 2005, Intraneuronal Abeta accumulation and origin of plaques in Alzheimer's disease. *Neurobiol. Aging* **26**: 1235.
Hack, M.A., Saghatelyan, A., de Chevigny, A., Pfeifer, A., Ashery-Padan, R., Lledo, P.M. and Gotz, M., 2005 Neuronal fate determinants of adult olfactory bulb neurogenesis. *Nat Neurosci* **8**: 865.
Handler, M., Yang, X. and Shen, J., 2000, Presenilin-1 regulates neuronal differentiation during neurogenesis. *Development* **127**: 2593.
Harper, P.S., 1996, *Huntington's Disease*. W.B. Saunders, London.

Haughey, N.J., Liu, D., Nath, A., Borchard, A.C. and Mattson, M.P., 2002, Disruption of neurogenesis in the subventricular zone of adult mice, and in human cortical neuronal precursor cells in culture, by amyloid β-peptide: implications for the pathogenesis of Alzheimer's disease. *Neuromol. Med.* **1**: 125.

Higgins, G.A. and Jacobsen, H., 2003, Transgenic mouse models of Alzheimer's disease: phenotype and application. *Behav. Pharmacol.* **14**: 419.

Holscher, C., 2003, Time, space and hippocampal functions. *Rev. Neurosci.* **14**: 253.

Honda, S., Itoh, F., Yoshimoto, M., Ohno, S., Hinoda, Y. and Imai, K., 2000, Association between complement regulatory protein factor H and AM34 antigen, detected in senile plaques. *J. Gerontol. A Biol. Sci. Med. Sci.* **55**: M265.

Jin, K., Galvan, V., Xie, L., Mao, X.O., Gorostiza, O.F., Bredesen, D.E. and Greenberg, D.A., 2004a, Enhanced neurogenesis in Alzheimer's disease transgenic (PDGF-APPSw,Ind) mice. *Proc. Natl Acad. Sci. USA* **101**: 13363.

Jin, K., Peel, A.L., Mao, X.O., Xie, L., Cottrell, B.A., Henshall, D.C. and Greenberg, D.A., 2004b, Increased hippocampal neurogenesis in Alzheimer's disease. *Proc. Natl Acad. Sci. USA* **101**: 343.

Jin, K., LaFevre-Bernt, M., Sun, Y., Chen, S., Gafni, J., Crippen, D., Logvinova, A., Ross, C.A., Greenberg, D.A. and Ellerby, L.M., 2005, FGF-2 promotes neurogenesis and neuroprotection and prolongs survival in a transgenic mouse model of Huntington's disease. *Proc. Natl Acad. Sci. USA* **102**: 18189.

Jin, K., Xie, L., Mao, X.O. and Greenberg, D.A., 2006, Alzheimer's disease drugs promote neurogenesis. *Brain Res.* **1085**: 183.

Katchanov, J., Harms, C., Gertz, K., Hauck, L., Waeber, C., Hirt, L., Priller, J., von Harsdorf, R., Bruck, W., Hortnagl, H., Dirnagl, U., Bhide, P.G. and Endres, M., 2001, Mild cerebral ischemia induces loss of cyclin-dependent kinase inhibitors and activation of cell cycle machinery before delayed neuronal cell death. *J. Neurosci.* **21**: 5045.

Kay, J.N. and Blum, M., 2000, Differential response of ventral midbrain and striatal progenitor cells to lesions of the nigrostriatal dopaminergic projection. *Dev. Neurosci.* **22**: 56.

Kempermann, G., 2006, Adult Neurogenesis, Stem Cells and Neuronal Development in the Adult Brain. Oxford University Press, New York.

Kempermann, G., Jessberger, S., Steiner, B. and Kronenberg, G., 2004, Milestones of neuronal development in the adult hippocampus. *Trends Neurosci.* **27**: 447.

Kornack, D.R. and Rakic, P., 2001, The generation, migration, and differentiation of olfactory neurons in the adult primate brain. *Proc. Natl Acad. Sci. USA* **98**: 4752.

Kremer, H.P., Roos, R.A., Dingjan, G., Marani, E. and Bots, G.T., 1990, Atrophy of the hypothalamic lateral tuberal nucleus in Huntington's disease. *J. Neuropathol. Exp. Neurol.* **49**: 371.

Kremer, H.P., Roos, R.A., Dingjan, G.M., Bots, G.T., Bruyn, G.W. and Hofman, M.A., 1991, The hypothalamic lateral tuberal nucleus and the characteristics of neuronal loss in Huntington's disease. *Neurosci. Lett.* **132**: 101.

Kuhn, H.G., Dickinson-Anson, H. and Gage, F.H., 1996, Neurogenesis in the dentate gyrus of the adult rat: age-related decrease of neuronal progenitor proliferation. *J. Neurosci.* **16**: 2027.

Laakso, M.P., Soininen, H., Partanen, K., Helkala, E.L., Hartikainen, P., Vainio, P., Hallikainen, M., Hanninen, T. and Riekkinen, P.J., Sr., 1995, Volumes of hippocampus, amygdala and frontal lobes in the MRI-based diagnosis of early Alzheimer's disease: correlation with memory functions. *J. Neural Transm. Park. Dis. Dement. Sect.* **9**: 73.

Landles, C. and Bates, G.P., 2004, Huntingtin and the molecular pathogenesis of Huntington's disease. *EMBO Rep.* **5**: 958.

Lazic, S.E., Grote, H., Armstrong, R.J., Blakemore, C., Hannan, A.J., van Dellen, A. and Barker, R.A., 2004, Decreased hippocampal cell proliferation in R6/1 Huntington's mice. *Neuroreport* **15**: 811.

Lazic, S.E., Grote, H.E., Blakemore, C., Hannan, A.J., van Dellen, A., Phillips, W. and Barker, R.A., 2006, Neurogenesis in the R6/1 transgenic mouse model of Huntington's disease: effects of environmental enrichment. *Eur. J. Neurosci.* **23**: 1829.

Lemere, C.A., Maier, M., Jiang, L., Peng, Y. and Seabrook, T.J., 2006, Amyloid-β immunotherapy for the prevention and treatment of Alzheimer disease: lessons from mice, monkeys, and humans. *Rejuvenation Res.* **9**: 77.

Leuner, B., Gould, E. and Shors, T.J., 2006, Is there a link between adult neurogenesis and learning? *Hippocampus* **16**: 216.

Li, S.H., Yu, Z.X., Li, C.L., Nguyen, H.P., Zhou, Y.X., Deng, C. and Li, X.J., 2003, Lack of huntingtin-associated protein-1 causes neuronal death resembling hypothalamic degeneration in Huntington's disease. *J. Neurosci.* **23**: 6956.

Lie, D.C., Dziewczapolski, G., Willhoite, A.R., Kaspar, B.K., Shults, C.W. and Gage, F.H., 2002, The adult substantia nigra contains progenitor cells with neurogenic potential. *J. Neurosci.* **22**: 6639.

Liu, M., Pleasure, S.J., Collins, A.E., Noebels, J.L., Naya, F.J., Tsai, M.J. and Lowenstein, D.H., 2000, Loss of BETA2/NeuroD leads to malformation of the dentate gyrus and epilepsy. *Proc. Natl Acad. Sci. USA* **97**: 865.

Lledo, P.M. and Saghatelyan, A., 2005, Integrating new neurons into the adult olfactory bulb: joining the network, life-death decisions, and the effects of sensory experience. *Trends Neurosci.* **28**: 248.

Lopez-Toledano, M.A. and Shelanski, M.L., 2004, Neurogenic effect of β-amyloid peptide in the development of neural stem cells. *J. Neurosci.* **24**: 5439.

Lyness, S.A., Zarow, C. and Chui, H.C., 2003, Neuron loss in key cholinergic and aminergic nuclei in Alzheimer disease: a meta-analysis. *Neurobiol. Aging* **24**: 1.

Marcora, E., Gowan, K. and Lee, J.E., 2003, Stimulation of NeuroD activity by huntingtin and huntingtin-associated proteins HAP1 and MLK2. *Proc. Natl Acad. Sci. USA* **100**: 9578.

Mohapel, P. and Brundin, P., 2004, Harnessing endogenous stem cells to treat neurodegenerative disorders of the basal ganglia. *Parkinsonism Relat. Disord.* **10**: 259.

Murphy, D.G., DeCarli, C.D., Daly, E., Gillette, J.A., McIntosh, A.R., Haxby, J.V., Teichberg, D., Schapiro, M.B., Rapoport, S.I. and Horwitz, B., 1993, Volumetric magnetic resonance imaging in men with dementia of the Alzheimer type: correlations with disease severity. *Biol. Psychiatry* **34**: 612.

Nagy, Z., 2000, Cell cycle regulatory failure in neurones: causes and consequences. *Neurobiol. Aging* **21**: 761.

Nagy, Z., Esiri, M.M., Cato, A.M. and Smith, A.D., 1997a, Cell cycle markers in the hippocampus in Alzheimer's disease. *Acta Neuropathol. (Berl.)* **94**: 6.

Nagy, Z., Esiri, M.M. and Smith, A.D., 1997b, Expression of cell division markers in the hippocampus in Alzheimer's disease and other neurodegenerative conditions. *Acta Neuropathol. (Berl.)* **93**: 294.

Ohsawa, I., Takamura, C., Morimoto, T., Ishiguro, M. and Kohsaka, S., 1999, Amino-terminal region of secreted form of amyloid precursor protein stimulates proliferation of neural stem cells. *Eur. J. Neurosci.* **11**:1907.

Pearson, H., 2006, Stem-cell tagging shows flaws. *Nature* **439**: 519.

Petersen, A., Gil, J., Maat-Schieman, M.L., Bjorkqvist, M., Tanila, H., Araujo, I.M., Smith, R., Popovic, N., Wierup, N., Norlen, P., Li, J.Y., Roos, R.A., Sundler, F., Mulder, H. and Brundin, P., 2005, Orexin loss in Huntington's disease. *Hum. Mol. Genet.* **14**: 39.

Pietrzik, C. and Behl, C., 2005, Concepts for the treatment of Alzheimer's disease: molecular mechanisms and clinical application. *Int. J. Exp. Pathol.* **86**: 173.

Qin, Z.H., Wang, Y., Sapp, E., Cuiffo, B., Wanker, E., Hayden, M.R., Kegel, K.B., Aronin, N. and DiFiglia, M., 2004, Huntingtin bodies sequester vesicle-associated proteins by a polyproline-dependent interaction. *J. Neurosci.* **24**: 269.

Rozemuller, J.M., Stam, F.C. and Eikelenboom, P., 1990, Acute phase proteins are present in amorphous plaques in the cerebral but not in the cerebellar cortex of patients with Alzheimer's disease. *Neurosci. Lett.* **119**: 75.

Sanai, N., Tramontin, A.D., Quinones-Hinojosa, A., Barbaro, N.M., Gupta, N., Kunwar, S., Lawton, M.T., McDermott, M.W., Parsa, A.T., Manuel-Garcia Verdugo, J., Berger, M.S. and Alvarez-Buylla, A., 2004, Unique astrocyte ribbon in adult human brain contains neural stem cells but lacks chain migration. *Nature* **427**: 740.

Shan, X., Chi, L., Bishop, M., Luo, C., Lien, L., Zhang, Z. and Liu, R., 2006, Enhanced de novo neurogenesis and dopaminergic neurogenesis in the substantia nigra of MPTP-induced Parkinson's disease-like mice. *Stem Cells* **24**: 1280.

Steiner, B., Winter, C., Hosman, K., Siebert, E., Kempermann, G. and Kupsch, A., 2005, Enriched environment induces cellular plasticity in the adult substantia nigra and improves motor behavior function in the 6-OHDA rat model of Parkinson's disease. *Exp. Neurol.* **199**: 291.

St George-Hyslop, P.H., 2000, Molecular genetics of Alzheimer's disease. *Biol. Psychiatry* **47**: 183.

Sturchler-Pierrat, C., Abramowski, D., Duke, M., Wiederhold, K.H., Mistl, C., Rothacher, S., Ledermann, B, Burki, K., Frey, P., Paganetti, P.A., Waridel, C., Calhoun, M.E., Jucker, M., Probst, A., Staufenbiel, M. and Sommer, B., 1997, Two amyloid precursor protein transgenic mouse models with Alzheimer disease-like pathology. *Proc. Natl Acad. Sci. USA* **94**: 13287.

Tanzi, R.E. and Bertram, L., 2005, Twenty years of the Alzheimer's disease amyloid hypothesis: a genetic perspective. *Cell* **120**: 545.

Tatebayashi, Y., Lee, M.H., Li, L., Iqbal, K. and Grundke-Iqbal, I., 2003, The dentate gyrus neurogenesis: a therapeutic target for Alzheimer's disease. *Acta Neuropathol. (Berl.)* **105**: 225.

Tuszynski, M.H., Thal, L., Pay, M., Salmon, D.P., U, H.S., Bakay, R., Patel, P., Blesch, A., Vahlsing, H.L., Ho, G., Tong, G., Potkin, S.G., Fallon, J., Hansen, L., Mufson, E.J., Kordower, J.H., Gall, C. and Conner, J., 2005, A phase 1 clinical trial of nerve growth factor gene therapy for Alzheimer disease. *Nat. Med.* **11**: 551.

Van der Borght, K., Wallinga, A.E., Luiten, P.G., Eggen, B.J. and Van der Zee, E.A., 2005, Morris water maze learning in two rat strains increases the expression of the polysialylated form of the neural cell adhesion molecule in the dentate gyrus but has no effect on hippocampal neurogenesis. *Behav. Neurosci.* **119**: 926.

Van kampen, J.M. and Eckman, C.B., 2006, Dopamine D3 receptor agonist delivery to a model of Parkinson's disease restores the nigrostriatal pathway and improves locomotor behaviour. *J. Neurosci.* **26**: 7272.

Van Kampen, J.M. and Robertson, H.A., 2005, A possible role for dopamine D3 receptor stimulation in the induction of neurogenesis in the adult rat substantia nigra. *Neuroscience* **136**: 381.

van Praag, H., Kempermann, G. and Gage, F.H., 1999, Running increases cell proliferation and neurogenesis in the adult mouse dentate gyrus. *Nat. Neurosci.* **2**: 266.

Verbeek, M.M., Otte-Holler, I., Wesseling, P., Ruiter, D.J. and de Waal, R.M., 1996, Differential expression of intercellular adhesion molecule-1 (ICAM-1) in the A β-containing lesions in brains of patients with dementia of the Alzheimer type. *Acta Neuropathol. (Berl.)* **91**: 608.

Vincent, I., Jicha, G., Rosado, M. and Dickson, D.W., 1997, Aberrant expression of mitotic cdc2/cyclin B1 kinase in degenerating neurons of Alzheimer's disease brain. *J. Neurosci.* **17**: 3588.

Vonsattel, J.P. and DiFiglia, M., 1998, Huntington disease. *J. Neuropathol. Exp. Neurol.* **57**: 369.

Wang, R., Dineley, K.T., Sweatt, J.D. and Zheng, H., 2004, Presenilin 1 familial Alzheimer's disease mutation leads to defective associative learning and impaired adult neurogenesis. *Neuroscience* **126**: 305.

Wen, P.H., Shao, X., Shao, Z., Hof, P.R., Wisniewski, T., Kelley, K., Friedrich, V.L., Jr., Ho, L., Pasinetti, G.M., Shioi, J., Robakis, N.K. and Elder, G.A., 2002, Overexpression of wild type but not an FAD mutant presenilin-1 promotes neurogenesis in the hippocampus of adult mice. *Neurobiol. Dis.* **10**: 8.

Yang, Y., Mufson, E.J. and Herrup, K., 2003, Neuronal cell death is preceded by cell cycle events at all stages of Alzheimer's disease. *J. Neurosci.* **23**: 2557.

Yoshimi, K., Ren, Y.R., Seki, T., Yamada, M., Ooizumi, H., Onodera, M., Saito, Y., Murayama, S., Okano, H., Mizuno, Y. and Mochizuki, H., 2005, Possibility for neurogenesis in substantia nigra of parkinsonian brain. *Ann. Neurol.* **58**: 31.

Zarow, C., Vinters, H.V., Ellis, W.G., Weiner, M.W., Mungas, D., White, L. and Chui, H.C., 2005, Correlates of hippocampal neuron number in Alzheimer's disease and ischemic vascular dementia. *Ann. Neurol.* **57**: 896.

Zhao, M., Momma, S., Delfani, K., Carlen, M., Cassidy, R.M., Johansson, C.B., Brismar, H., Shupliakov, O., Frisen, J. and Janson, A.M., 2003, Evidence for neurogenesis in the adult mammalian substantia nigra. *Proc. Natl Acad. Sci. USA* **100**: 7925.

21

ATP RECEPTORS IN THE PAIN SIGNALING: GLIAL CONTRIBUTION IN NEUROPATHIC PAIN

Kazuhide Inoue

1. ABSTRACT

There is abundant evidence that extracellular ATP has an important role in pain signaling at both the periphery and in the CNS. The focus of attention now is on the possibility that ATP and its receptor system might play important roles in chronic pathological pain states, particularly in neuropathic pain. Neuropathic pain is often a consequence of nerve injury through surgery, bone compression, diabetes, or infection. This type of pain can be so severe that even light touching can be intensely painful. Unfortunately, this state is generally resistant to currently available treatments. In this review, we summarize the role of ATP receptor $P2X_4$ and $P2X_7$ in spinal microglia in neuropathic pain. The activated microglia express $P2X_4$ after nerve injury, which can be stimulated by endogenous ATP, resulting in the release of brain-derived neurotrophic factor that is one of key molecules involving in neuropathic pain. The stimulation of microglial $P2X_7$ releases cytokines including TNF-α and IL-1β. Understanding the key roles of these ATP receptors may lead to new strategies for the management of intractable chronic pain.

2. INTRODUCTION

The first clue to a possibility that extracellular ATP might be an important substrate in the formation of pain (Bleehen et al., 1976; Bleehen and Keele, 1977; Chen et al., 1995; Lewis et al., 1995; Burnstock and Wood, 1996; Ralevic and Burnstock, 1998; Burnstock, 2000) was found about 30 years ago in clinical studies showing that ATP applied to blister bases (Bleehen et al., 1976; Bleehen and Keele, 1977) induced a pain sensation in humans. Significant advances in our understanding of mechanisms by which ATP causes pain have been made recently by the discovery of cell-surface receptors for detecting extracellular ATP and other nucleotides on sensory neurons (Chen et al., 1995; Lewis et al., 1995; Burnstock and Wood, 1996; Ravelic and Burstock, 1998; Burnstock, 2000). In a subset of primary afferent sensory neurons, ATP or its analogues produce electrophysiological and biological responses via ligand-gated ion-channel receptors, namely P2X receptors (P2XRs) (Cook

---

Kazuhide Inoue – kazuinouejp@ybb.ne.jp
Department of Molecular and System Pharmacology, Graduate School of Pharmaceutical Sciences, Kyushu University, 3-1-1 Maidashi, Higashi, Fukuoka 812-8582, Japan

et al., 1998; Li et al., 1999; Ueno et al., 1999; Tsuda et al., 2000; Chizh and Illes, 2001; Dunn et al., 2001), and G protein-coupled receptors, namely P2Y receptors (P2YRs) (Svichar et al., 1997; Koizumi et al., 2001; Tominaga et al., 2001; Molliver et al., 2002; Sanada et al., 2002; Moriyama et al., 2003) (Fig. 1). However, blocking P2XRs or P2YRs pharmacologically or suppressing their expression in sensory neurons or spinal cord had little effect on acute physiological pain evoked by heat or mechanical pressure in normal animals (Cockayne et al., 2000; Souslova et al., 2000; Jarvis et al., 2002), although inflammatory pain was attenuated (Hamilton et al., 1999; Cockayne et al., 2000). It seems likely that endogenous ATP and its receptors system may be more prominent in chronic pain states, particularly in neuropathic pain (Bland-Ward and Humphrey, 1997; Hamilton et al., 1999; Tsuda et al., 1999; Dorn et al., 2001; Barclay et al., 2002; Jarvis et al., 2002; Tsuda et al., 2002; Dai et al., 2004). Neuropathic pain, which often develops when nerves are damaged through surgery, cancer, bone compression, diabetes, or infection, is a type of pathological pain that does not resolve even when the overt tissue damage has healed (Woolf and Mannion, 1999; Woolf and Salter, 2000; Scholz and Woolf, 2002). In neuropathic pain state, even light contact with clothing can cause intense pain (tactile allodynia) and is often resistant to most current treatments including morphine. Accumulating evidence concerning how peripheral nerve injury causes neuropathic pain has suggested that molecular and cellular alterations in neuronal circuit between primary sensory neurons and spinal dorsal horn neurons after nerve injury have important role in the pain (Woolf and Mannion, 1999; Woolf and Salter, 2000; Scholz and Woolf, 2002). Several reports suggest that $P2X_3Rs$ (Tsuda et al., 1999; Dorn et al., 2001; Tsuzuki et al., 2001; Barclay et al., 2002; Kennedy et al., 2003) or $P2X_7$ (Casula et al., 2004; Chessell et al., 2005) have a role in neuropathic pain. And, we recently revealed that the $P2X_4R$ subtype in the spinal microglia is required for the expression of neurotrophic pain (Tsuda et al., 2003). More recently, we have reported that brain-derived neurotrophic factor (BDNF) released from microglia by the stimulation of $P2X_4$ causes the shift in neuronal anion gradient underlying neuropathic pain (Coull et al., 2005). Here we review the progress in the current understanding on the mechanisms of action of $P2X_4$ and $P2X_7$ receptors in the pathophysiology of neuropathic pain.

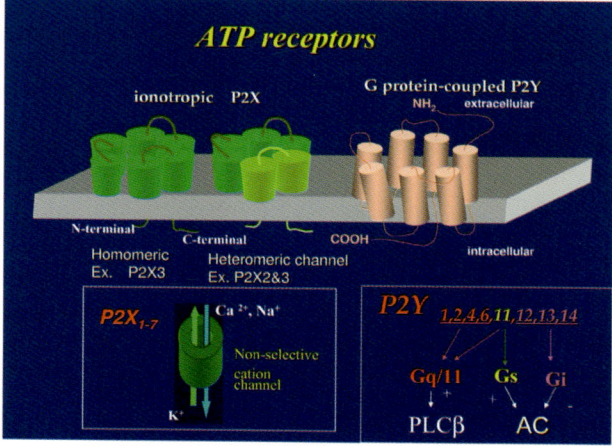

Fig. 1. ATP receptor subtypes

3. MICROGLIAL RESPONSE IN NERVE INJURY-INDUCED PAIN

Glial cells are classified into astrocytes, oligodendrocytes, and microglia and make-up over 70% of the total cell population in the CNS. Microglia are ubiquitously distributed in the CNS and represent a morphologically unique type of cell. In normal conditions, microglia are called as "resting microglia" and have a small soma bearing thin and branched processes (Kreutzberg, 1996; Stoll and Jander, 1999). Microglia are activated by neuronal injury, trauma, ischemia, infection, or neurological diseases and show stereotypic changes in morphology (amoeboid shape), gene expression, function, and number (Perry, 1994; Kreutzberg, 1996; Stoll and Jander, 1999). Such microglia are called as "activated microglia." The changes in expression of cell-surface molecule complement receptor 3, which is recognized by the antibody OX42, and by morphology, are widely used as the key diagnostic markers of activated microglia (Kreutzberg, 1996; Stoll and Jander, 1999).

A variety of animal models for studying neuropathic pain were developed on the clinical evidence that neuropathic pain results from damage to peripheral nerves in humans. In most animal models of neuropathic pain that have been intensively studied peripheral nerves are directly damaged (Wall et al., 1979; Bennett and Xie, 1988; Seltzer et al., 1990; Kim and Chung, 1992; Decosterd and Woolf, 2000). In such models, a dramatic change in microglia within the spinal dorsal horn has been reported after the nerve injury (Eriksson et al., 1993; Colburn et al., 1997; Coyle, 1998; Colburn et al., 1999; Stuesse et al., 2000). Spinal microglia become hypertrophic in their short and thick processes within the first 24 h after peripheral nerve injury (Eriksson et al., 1993). This is followed by a burst proliferation of microglia with a peak at around 2–3 days after the nerve injury (Gehrmann and Banati, 1995). Activated microglia exhibit upregulated OX42 labeling in the dorsal horn (Eriksson et al., 1993; Gehrmann and Banati, 1995; Liu et al., 1995; Coyle, 1998; Colburn et al., 1999; Stuesse et al., 2000), which starts to increase as early as 1 day after nerve injury and peaks at around 14 days (Coyle, 1998) (Fig. 2a). The time-course of OX42 upregulation in the dorsal horn correlated with that of the development of tactile allodynia (Coyle, 1998). It remained an open question whether spinal microglia play a causal role in neuropathic pain behavior though there have been many studies showing that activation of microglia in the dorsal horn is correlated with the development of pain hypersensitivity in a wide variety of nerve injury models (Eriksson et al., 1993; Gehrmann and Banati, 1995; Liu et al., 1995; Coyle, 1998; Colburn et al., 1999; Stuesse et al., 2000; Watkins et al., 2001).

4. THE ROLE OF P2X$_4$ ON MICROGLIA IN NERVE INJURY-INDUCED PAIN

The role of P2X$_4$ in the spinal cord in neuropathic pain first came from pharmacological investigations using the P2X antagonists TNP–ATP and PPADS (Tsuda et al., 2003). We found that the marked tactile allodynia that develops following the nerve injury was reversed by acutely administering TNP–ATP intrathecally (Fig. 2c) but was unaffected by administering PPADS. TNP–ATP had no effect on acute pain behavior in the uninjured state or on motor behavior. From the pharmacological profiles of TNP–ATP (blocking P2X$_4$ at high concentration) and PPADS (not blocking P2X$_4$), it was inferred that tactile allodynia depends upon P2X$_4$ in the spinal cord. The expression of P2X$_4$ protein, normally low in the naïve spinal cord, progressively increased in the days following nerve injury with a time-course parallel to that of the development of tactile allodynia (Fig. 2b). Double immunolabeling analysis using cell-specific markers demonstrated that microglia in the dorsal horn on the side of the nerve injury were intensely positive for $P2X_4$ protein. The cells expressing $P2X_4$ in the nerve-injured side of the dorsal horn were more numerous than under control conditions and showed high levels of OX42 labeling and morphological hypertrophy, all of which are characteristic markers of activated microglia. Moreover, it was found that reducing the expression of $P2X_4$ protein in spinal microglia by means of intrathecally administered antisense oligodeoxynucleotide targeting $P2X_4$ prevented the development of the nerve injury-induced tactile allodynia. Collectively, this evidence implies that activation of microglial $P2X_4$ is necessary for pain hypersensitivity following nerve injury.

The sufficiency of $P2X_4$ activation in microglia for the development of allodynia was demonstrated by intrathecal administration of activated microglia stimulated in vitro by ATP (Tsuda et al., 2003). In naïve rats, intrathecal administration of cultured microglia that were preincubated with ATP to activate $P2X_4$ on microglia produced tactile allodynia progressively over the 3–5 h following the administration. Microglia also express another subtype of P2X receptor, $P2X_7$, but this receptor subtype appears not to be involved because activation of $P2X_7$ requires a higher concentration (>1 mM) of ATP than used (50 µM) (Surprenant et al., 1996; Khakh et al., 2001) and because TNP–ATP, which does not affect $P2X_7$ (Virginio et al., 1998), prevents ATP from stimulating microglia to produce allodynia. These data indicate that $P2X_4$ stimulation of microglia is not only necessary for tactile allodynia, but is also sufficient to cause the allodynia.

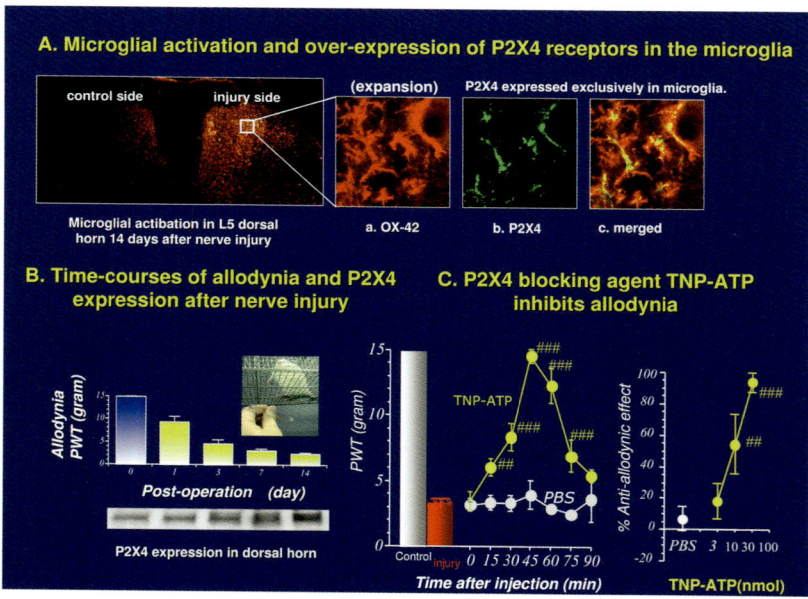

Fig. 2. Microglial activation and P2X4 receptors involved in neuropathic pain after nerve injury. (a) Double immunolabeling data: microglial activation and over-expression of $P2X_4$ receptors in the microglia. (b) Time-courses of allodynia and $P2X_4$ expression (Western blotting data) in L5 dorsal horn after nerve injury are similar. Tactile allodynia was reversed by acutely administering TNP–ATP (blocking $P2X_4$ at high concentration) intrathecally but was unaffected by administering PPADS (not blocking $P2X_4$). This data suggests that tactile allodynia depends upon $P2X_4$ in the spinal cord

5. BDNF RELEASE STIMULATED BY MICROGLIAL P2X$_4$ AND ALLODYNIA

The variety of biological effects produced in activated microglia by ATP in in vitro studies may provide hints toward clarifying the mechanisms by which microglia produce altered processing of information in the spinal cord dorsal horn (Fig. 3). It was recently reported that ATP-stimulated microglia signal to lamina I neurons, causing a collapse of their transmembrane anion gradient, and that BDNF is a crucial signaling molecule between microglia and neurons (Coull et al., 2005). It was considered that microglia may affect E_{anion} in LI neurons since it was already reported that the nerve injury-induced tactile allodynia depends on a depolarizing shift in the E_{anion} of spinal lamina I (LI) neurons in the dorsal spinal cord, resulting in converting the GABA$_A$-receptor- and glycine-receptor-mediated inhibition to excitation (Coull et al., 2003). To investigate this possibility, microglia were administered to the lumbar spinal level of naïve rats by an intrathecal catheter as described (Tsuda et al., 2003). Administering ATP-stimulated microglia caused a progressive tactile allodynia over the 5 h after injection. Voltage-clamp recording from LI neurons of slices prepared 5 h after intrathecal microglia administration revealed that E_{anion} in LI neurons from rats administered ATP-stimulated microglia was shifted to −61.6 from −68.3 mV (E_{anion} in spinal slices taken from normal rats). In addition, using current clamp recordings, GABA response switched from hyperpolarizing in control rats to depolarizing in rats treated with ATP-stimulated microglia. Activated microglia secrete various biologically active molecules, one of which, BDNF, was implicated in the hypersensitivity of dorsal horn neurons that follows sensitization and inflammation (Mannion et al., 1999; Thompson et al., 1999; Heppenstall and Lewin, 2001) and in anion gradient shifts in the hippocampus (Rivera et al., 2002).

Next question was whether BDNF could trigger shifts in pain hypersensitivity and in LI neuronal E_{anion} similar to those resulting from the application of ATP-stimulated microglia. To examine this, recombinant BDNF was administered intrathecally to normal rats. BDNF produced tactile allodynia comparable to that produced by ATP-stimulated microglia. E_{anion} of LI neurons in slices treated with BDNF were significantly less negative (>90 min, in vitro) than that of LI neurons from control slices. The proportion of neurons responding to GABA with a rise in [Ca^{2+}]i increased over time reaching 31% of neurons recorded between 80 and 120 min during perfusion with BDNF and in the presence of glutamate receptor blockers. The rise in [Ca^{2+}]i was prevented by the GABA$_A$ receptor blocker bicuculline, confirming that the effect was mediated by GABA$_A$ receptors. Thus, acute administration of BDNF in slices caused a depolarizing shift in E_{anion} and caused GABA to produce net excitation. Next, a BDNF-transducing recombinant adenovirus (adBDNF) (Gravel et al., 1997) was administered intrathecally to the rats to examine the effects of prolonged exposure to BDNF in vivo. A progressive tactile allodynia was observed over the 4 days after the treatment with adBDNF. E_{anion} in LI neurons from adBDNF-injected rats was significantly less negative than that in LI neurons from control rats. Thus, similar to acute administration of BDNF, sustained local release of BDNF caused the allodynia and a depolarizing shift in E_{anion}. Moreover, a function-blocking antibody against the TrκB receptor (anti-TrκB) and a BDNF-sequestering fusion protein (TrκB–Fc) acutely inhibited the allodynia and the shift of E_{anion} of LI neurons. These data indicate that endogenous BDNF is necessary to sustain both the tactile allodynia and the depolarizing shift in E_{anion} in LI neurons that result from nerve injury.

The administration of ATP-stimulated microglia with either anti-TrκB or TrκB-Fc did not result in tactile allodynia. After pretreatment of microglia with double-stranded short interfering RNA directed against BDNF (BDNF siRNA), the ATP-stimulated microglia injected intrathecally into normal rats did not cause the allodynia. Anti-TrκB and BDNF siRNA prevented the shift in E_{anion} induced by ATP-stimulated microglia. ATP stimulation caused release of BDNF from microglia in culture. Treating the cultures with the P2X receptor blocker TNP–ATP inhibited this effect of ATP. In addition, pretreatment of the microglia with BDNF siRNA prevented release of BDNF by ATP stimulation. E_{anion} of LI neurons was returned to normal value by bath-application of TNP–ATP to spinal slices taken from allodynic rats 2 weeks after nerve injury. Thus, $P2X_4$ receptor activation is necessary to sustain the depolarizing shift in E_{anion} in rats with nerve injury. These findings show that both the decrease in paw withdrawal threshold and the shift in E_{anion} in LI neurons caused by ATP-stimulated microglia through $P2X_4$ require BDNF–TrκB signaling and that the source of BDNF is the microglia themselves.

6. AN INVOLVEMENT OF OTHER P2X-TYPE RECEPTORS IN NEUROPATHIC PAIN: $P2X_7$ AND TUMOR NECROSIS FACTOR-α

Several cytokines such as IL-1β, IL-6, and tumor necrosis factor-α (TNF-α) in the dorsal horn are upregulated after nerve lesion (Sweitzer et al., 2001; Winkelstein et al., 2001; Raghavendra et al., 2003) and have been implicated in contributing to neuropathic pain (Robertson et al., 1997; DeLeo and Yezierski, 2001; Sweitzer et al., 2001; Watkins et al., 2001; Winkelstein et al., 2001; Raghavendra et al., 2003).

Recent evidences indicate the relationship between TNF-α and neuropathic pain (Wagner and Myers, 1996; Sommer and Schafers, 1998; Sommer et al., 1998; Sorkin and Doom, 2000; Schafers et al., 2002) and TNF-α released after injury is proposed as initiator of abnormal pain sensation (Wagner and Myers, 1996; Sommer and Schafers, 1998; Sommer et al., 1998; Sorkin and Doom, 2000). TNF-α is upregulated after nerve injury in both DRG (Schafers et al., 2002, 2003a) and spinal cord (Hashizume et al., 2000; Sweitzer et al., 2001; Winkelstein et al., 2001; Raghavendra et al., 2003). The inhibition of TNF-α reduces the hyperalgesia in neuropathic pain models (Sommer et al., 2001a, b). After peripheral nerve injury, DRG neurons robustly increase their expression of TNF-α (Schafers et al., 2002). Exogenous TNF-α applied to intact or compression-injured DRG induces sustained mechanical allodynia (Homma et al., 2002). However, the mechanisms by which TNF-α elicits pain behavior are still unclear. Previous studies suggest that TNF-α modulates neuronal activity in neurons (Sawada et al., 1990; Soliven and Albert, 1992; Sorkin et al., 1997; Furukawa and Mattson, 1998; Junger and Sorkin, 2000; Diem et al., 2001; Leem and Bove, 2002). Schafers et al. investigated responses of intact and nerve-injured DRG neurons to locally applied TNF-α using parallel in vivo and in vitro paradigms (Schafers et al., 2003a). In vivo, TNF-α (0.1–10 pg ml^{-1}) or vehicle was injected into L5 DRG in naïve rats and in rats that had received L5 and L6 spinal nerve ligation (SNL) immediately before injection. TNF-α elicits long-lasting allodynia in naïve rats. Subthreshold doses of TNF-α synergize with nerve injury to elicit faster onset of allodynia and spontaneous pain behavior in SNL rats. Pre-emptive treatment with etanercept, a TNF-α antagonist, reduces SNL-induced allodynia by almost 50%. Perfusion of TNF-α (100–1,000 pg ml^{-1}) to naïve DRG neuron evokes short-lasting discharges. In injured DRG, TNF-α elicits higher and longer lasting neuronal discharges in earlier onset at much lower concentrations. In naïve DRG adjacent to injured DRG, TNF-α also elicits high-frequency discharges at subthreshold concentrations. These findings suggest that injured and adjacent uninjured DRG neurons are sensitized to TNF-α after SNL, and sensitization to endogenous TNF-α may be essential for the development and maintenance of neuropathic pain.

Microglia are a major source of TNF-α. ATP potently stimulates the release of TNF-α following an increase in the TNF-α mRNA expression by activating MAPKs (Hide et al., 2000; Suzuki et al., 2004). The TNF-α release is maximally elicited by 1 mM ATP and also induced by a P2X$_7$ receptor agonist, BzATP, suggesting the involvement of P2X$_7$ receptor (Hide et al., 2000). ATP-induced TNF-α release is Ca^{2+}-dependent, and a sustained Ca^{2+} influx correlates with the TNF-α release. ATP-induced TNF-α release needs MAPKs activation. U0126, SP600125, and SB 203580, which inhibit MEK (MAPK kinase), JNK (c-Jun N-terminal kinase), and p38, respectively, all potently suppress the production of TNF-α protein in ATP-stimulated microglia, whereas the production of TNF-α mRNA is strongly inhibited by U0126 and SP600125 but not by SB203580. These findings suggest that a transcription of TNF-α mRNA is dependent on both of ERK and JNK but not on p38. SB203580 does not affect the increased levels of TNF-α mRNA but does prevent TNF-α mRNA from accumulating in the cytoplasm, suggesting that p38 plays an important role in the nucleocytoplasmic transport of TNF-α mRNA. The ATP-induced activation of JNK and p38, but not ERK are inhibited by brilliant blue G, a P2X$_7$ receptor blocker, and by genistein and 4-amino-5-(4-chlorophenyl)-7-(t-butyl)pyrazolo[3,4-D] pyrimidine, which are general and src-family-specific tyrosine kinase inhibitors, respectively. These findings indicate that an src family acts downstream of the P2X$_7$ receptor to activate JNK and p38 independently from channel action (Suzuki et al., 2004).

In cultured DRG neurons, exogenous TNF-α activates p38 MAPK (Pollock et al., 2002). Recently, p38 activation was shown to play a major role in the maintenance of pain (Ji et al., 2002; Jin et al., 2003; Schafers et al., 2003b; Tsuda et al., 2004). It is speculated that activation of the p38 cascade may represent a route correlating the development of pain after nerve injury. The question whether TNF-α activates the p38 cascade in vivo to trigger pain behavior after SNL was examined using etanercept (Schafers et al., 2003b). As the result, etanercept treatment starting 2 days before SNL attenuates mechanical allodynia. Interestingly, the treatment starting 1 or 7 days after SNL is ineffective. Similarly, intrathecal infusion of a p38 inhibitor (SB203580, 4 mg day^{-1}) is effective only when it is started before but not 7 days after SNL. In DRG, activated p38 is transiently elevated 5 h after SNL and returns to baseline by 1 day after SNL. Phosphorylated p38 is localized in small TNF-α-positive DRG neurons. In spinal cord, p38 is activated between 5 h and 3 day after SNL and returns to baseline level within 5 days. Pretreatment with etanercept blocks p38 activation only in DRG but not in spinal cord. These data indicate that phosphorylated p38 levels in spinal cord and DRG are transiently elevated after SNL treatment. In DRG, p38 activation is blocked by systemic TNF-α inhibition. Another report suggests the mechanism of TNF-α-induced pain in the line of the interaction with BDNF, which is thought to be a modulator of pain. Onda et al. investigated the effect of infliximab, a chimeric monoclonal antibody to TNF-α, on induction of BDNF using an experimental herniated nucleus pulposus (NP) model. Application of NP induces a marked increase of BDNF immunoreactivity in the DRG neurons and within the superficial layer in the dorsal horn compared with the sham group (Onda et al., 2004). Intraperitoneal injection with infliximab reduces the BDNF induction in both DRG and spinal cord.

7. INVOLVEMENT OF P2X$_7$ AND IL-1β IN NEUROPATHIC PAIN

Recently, accumulating evidences indicate the relationship between inflammatory cytokine IL-1β and neuropathic pain (Robertson et al., 1997; DeLeo and Yezierski, 2001; Sweitzer et al., 2001; Watkins et al., 2001; Winkelstein et al., 2001; Raghavendra et al., 2003). The expression of IL-1β is upregulated in the spinal cord of several rat mononeuropathy models (Sweitzer et al., 2001; Winkelstein et al., 2001; Raghavendra et al., 2003). Sweitzer et al. investigated whether blocking the action of central IL-1β and TNF-α attenuates mechanical allodynia in a gender-specific manner in a rodent L5 spinal nerve transection model of neuropathic pain with/without glial activation (Sweitzer et al., 2001). IL-1 receptor antagonist not alone but in combination with soluble TNF-α receptor decreases allodynia in a dose-dependent manner with remaining of glial activation. At days 3 and 7 post-transection, the level of IL-6, but not IL-1β, in the L5 spinal cord of animals receiving daily IL-1 receptor antagonist in combination with soluble TNF-α receptor is significantly less than that of control animals. These findings support a role for central IL-1β in the development and maintenance of neuropathic pain through induction of a pro-inflammatory cytokine cascade.

Di Virgilio's group first reported that extracellular ATP triggers IL-1β release from LPS-treated microglia (Ferrari et al., 1996, 1997; Sanz and Di Virgilio, 2000). They confirmed that ATP is a powerful stimulus for IL-1β release from LPS-treated microglia and that IL-1β release occurs much earlier than the leakage of cytoplasmic markers. Sanz and Di Virgilio examined the kinetics and mechanism of ATP-dependent IL-1β release from microglia (Sanz and Di Virgilio, 2000). The addition of extracellular ATP to LPS-primed microglia causes a burst of release of a large amount of processed IL-1β. ATP has no effect on the accumulation of intracellular pro-IL-1β in the absence of LPS. The optimal ATP concentration for IL-1β secretion is between 3 and 5 mM, but significant release can be observed at concentrations as low as 1 mM. At all ATP concentrations, the IL-1β release can be inhibited by increasing the extracellular K$^+$ concentration. The caspase inhibitors also inhibit the ATP-dependent IL-1β release. It was concluded that ATP triggers accelerated maturation and the release of intracellularly accumulated IL-1β by activating the IL-1β-converting enzyme/caspase 1 in mouse microglia. Extracellular ATP is the only endogenous compound known to cause a significant reduction in intracellular K$^+$ and consequent release of IL-1β (Perregaux and Gabel, 1994; Sanz and Di Virgilio, 2000). Substantial evidences suggest a key role of P2X$_7$ in the ATP-induced IL-1β release from LPS-primed microglia, i.e., (1) P2X$_7$ antagonist oATP inhibits the release from microglia (Ferrari et al., 1997) and (2) microglia lacking P2X$_7$ does not release IL-1β after ATP stimulation (Ferrari et al., 1996). Thus, it is suggested that an activation of P2X$_7$ by ATP induces movement of K$^+$, Na$^+$, and Ca^{2+} through cell membrane and provokes the release of IL-1β from microglia.

Several cytokines have been reported to alter synaptic transmission in the CNS including the spinal cord (Kerr et al., 1999; Vikman et al., 2003). The exogenous application of IL-1β enhances NMDA receptor-mediated Ca^{2+} responses via activating tyrosine protein kinase Src (Viviani et al., 2003) that is known to enhance NMDA receptor activity in dorsal horn neurons (Yu et al., 1997; Woolf and Salter, 2000). IL-1β also decreases GABA$_A$ receptor-mediated currents (Wang et al., 2000). Mechanical and thermal hyperalgesia was absent in both inflammatory and neuropathic pain models in mice with a disrupted P2X$_7$ gene, while normal nociceptive processing was preserved (Casula et al., 2004; Chessell et al., 2005), suggesting that the stimulation of $P2X_7$ receptor expressed by satellite and Schwann cells causes the release of IL-1β and upregulation of nerve growth factor, resulting in the pain. Thus, IL-1β released from activated microglia by activating $P2X_7$ receptors may also have modulatory effects on evoking neuropathic pain.

8. CONCLUSION

Almost all currently known drugs for neuropathic pain were developed to target neurons, and these drugs do not exhibit adequate therapeutic effects in patients with neuropathic pain (Watkins and Maier, 2003). Peripheral nerve injury leads to pathological changes in spinal microglia. The activated microglia express $P2X_4$ receptors, which can be stimulated by endogenous ATP, resulting in the release of BDNF, which is one of the key molecules involving in neuropathic pain (Fig. 3). Cytokines may be the other key molecules since there are many reports suggesting the relationship between cytokines and the pain. We expect that efforts to elucidate how P2X receptor signaling in microglia causes neuropathic pain will provide us both with exciting insights into pain mechanisms and with clues to developing new therapeutic agents that may fundamentally change the management of intractable pain.

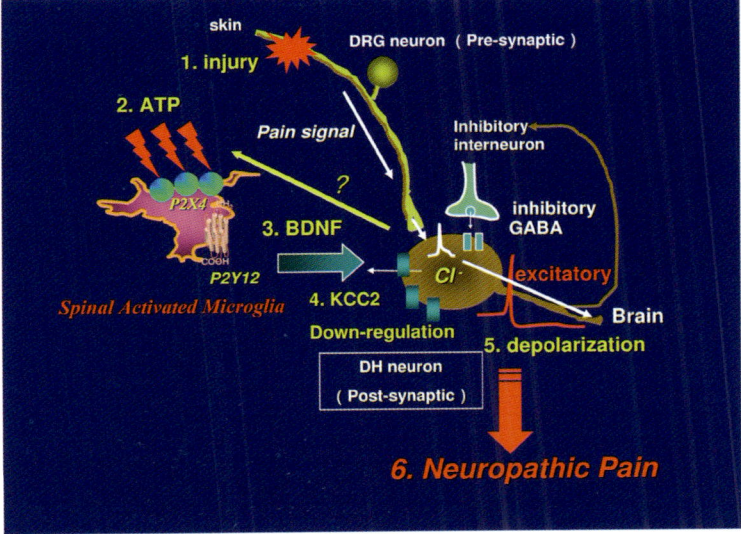

Fig. 3. Hypothesis: a mechanism of the expression of mechanical allodynia after nerve injury. In normal condition, pain impulse evoked by a stimulation of skin or other peripheral sites conducts and synaptically transmits to postsynaptic neuron. When the level of pain is too strong, an inhibitory interneuron releases GABA, resulting in an inhibition of pain. In pathophysiological condition, nerve injury causes release of many factors including the ATP that activates microglia and induces microglial chemotaxis through P2Y12. Microglia overexpresses $P2X_4$ receptors and the stimulation of $P2X_4$ evokes the release of BDNF. BDNF presumably downregulates the expression of KCC2, and causes internal Cl⁻ concentration increase and a depolarizing shift in E_{anion}. In this situation, GABA evokes depolarization in the postsynaptic neurons resulting in a neuropathic pain

9. REFERENCES

Barclay, J., Patel, S., Dorn, G., Wotherspoon, G., Moffatt, S., Eunson, L., Abdel'al, S., Natt, F., Hall, J., Winter, J., Bevan, S., Wishart, W., Fox, A. and Ganju, P., 2002, Functional downregulation of P2X3 receptor subunit in rat sensory neurons reveals a significant role in chronic neuropathic and inflammatory pain. *J. Neurosci.* **22**: 8139.
Bennett, G.J. and Xie, Y.K., 1988, A peripheral mononeuropathy in rat that produces disorders of pain sensation like those seen in man. *Pain* **33**: 87.
Bland-Ward, P.A. and Humphrey, P.P., 1997, Acute nociception mediated by hindpaw P2X receptor activation in the rat. *Br. J. Pharmacol.* **122**: 365.
Bleehen, T. and Keele, C.A., 1977, Observations on the algogenic actions of adenosine compounds on the human blister base preparation. *Pain* **3**: 367.
Bleehen, T., Hobbiger, F. and Keele, C.A., 1976, Identification of algogenic substances in human erythrocytes. *J. Physiol.* **262**: 131.
Burnstock, G., 2000, P2X receptors in sensory neurones. *Br. J. Anaesth.* **84**: 476.
Burnstock, G. and Wood, J.N., 1996, Purinergic receptors: their role in nociception and primary afferent neurotransmission. *Curr. Opin. Neurobiol.* **6**: 526.
Casula, M.A., Facer, P., Powell, A.J., Kinghorn, I.J., Plumpton, C., Tate, S.N., Bountra, C., Birch, R. and Anand, P., 2004, Expression of the sodium channel beta3 subunit in injured human sensory neurons. *Neuroreport* **15**: 1629.
Chen, C.C., Akopian, A.N., Sivilotti, L., Colquhoun, D., Burnstock, G. and Wood, J.N., 1995, A P2X purinoceptor expressed by a subset of sensory neurons. *Nature* **377**: 428.
Chessell, I.P., Hatcher, J.P., Bountra, C., Michel, A.D., Hughes, J.P., Green, P., Egerton, J., Murfin, M., Richardson, J., Peck, W.L., Grahames, C.B., Casula, M.A., Yiangou, Y., Birch, R., Anand, P. and Buell, G.N., 2005, Disruption of the P2X7 purinoceptor gene abolishes chronic inflammatory and neuropathic pain. *Pain* **114**: 386.
Chizh, B.A. and Illes, P., 2001, P2X receptors and nociception. *Pharmacol. Rev.* **53**: 553.
Cockayne, D.A., Hamilton, S.G., Zhu, Q.M., Dunn, P.M., Zhong, Y., Novakovic, S., Malmberg, A.B., Cain, G., Berson, A., Kassotakis, L., Hedley, L., Lachnit, W.G., Burnstock, G., McMahon, S.B. and Ford, A.P., 2000, Urinary bladder hyporeflexia and reduced pain-related behaviour in P2X3-deficient mice. *Nature* **407**: 1011.
Colburn, R.W., DeLeo, J.A., Rickman, A.J., Yeager, M.P., Kwon, P. and Hickey, W.F., 1997, Dissociation of microglial activation and neuropathic pain behaviors following peripheral nerve injury in the rat. *J. Neuroimmunol.* **79**: 163.
Colburn, R.W., Rickman, A.J. and DeLeo, J.A., 1999, The effect of site and type of nerve injury on spinal glial activation and neuropathic pain behavior. *Exp. Neurol.* **157**: 289.
Cook, S.P., Rodland, K.D. and McCleskey, E.W., 1998, A memory for extracellular Ca2+ by speeding recovery of P2X receptors from desensitization. *J. Neurosci.* **18**: 9238.
Coull, J.A., Boudreau, D., Bachand, K., Prescott, S.A., Nault, F., Sik, A., De Koninck, P. and De Koninck, Y., 2003, Trans-synaptic shift in anion gradient in spinal lamina I neurons as a mechanism of neuropathic pain. *Nature* **424**: 938.
Coull, J.A., Beggs, S., Boudreau, D., Boivin, D., Tsuda, M., Inoue, K., Gravel, C., Salter, M.W. and De Koninck, Y., 2005, BDNF from microglia causes the shift in neuronal anion gradient underlying neuropathic pain. *Nature* **438**: 1017.
Coyle, D.E., 1998, Partial peripheral nerve injury leads to activation of astroglia and microglia which parallels the development of allodynic behavior. *Glia* **23**: 75.
Dai, Y., Fukuoka, T., Wang, H., Yamanaka, H., Obata, K., Tokunaga, A. and Noguchi, K., 2004, Contribution of sensitized P2X receptors in inflamed tissue to the mechanical hypersensitivity revealed by phosphorylated ERK in DRG neurons. *Pain* **108**: 258.
Decosterd, I. and Woolf, C.J., 2000, Spared nerve injury: an animal model of persistent peripheral neuropathic pain. *Pain* **87**: 149.
DeLeo, J.A. and Yezierski, R.P., 2001, The role of neuroinflammation and neuroimmune activation in persistent pain. *Pain* **90**: 1.
Diem, R., Meyer, R., Weishaupt, J.H. and Bahr, M., 2001, Reduction of potassium currents and phosphatidylinositol 3-kinase-dependent AKT phosphorylation by tumor necrosis factor-(alpha) rescues axotomized retinal ganglion cells from retrograde cell death in vivo. *J. Neurosci.* **21**: 2058.
Dorn, G., Abdel'Al, S., Natt, F.J., Weiler, J., Hall, J., Meigel, I., Mosbacher, J. and Wishart, W., 2001, Specific inhibition of the rat ligand-gated ion channel P2X3 function via methoxyethoxy-modified phosphorothioated antisense oligonucleotides. *Antisense Nucleic Acid Drug Dev.* **11**: 165.
Dunn, P.M., Zhong, Y. and Burnstock, G., 2001, P2X receptors in peripheral neurons. *Prog. Neurobiol.* **65**: 107.

Eriksson, N.P., Persson, J.K., Svensson, M., Arvidsson, J., Molander, C. and Aldskogius, H., 1993, A quantitative analysis of the microglial cell reaction in central primary sensory projection territories following peripheral nerve injury in the adult rat. *Exp. Brain Res.* **96**: 19.

Ferrari, D., Villalba, M., Chiozzi, P., Falzoni, S., Ricciardi-Castagnoli, P. and Di Virgilio, F., 1996, Mouse microglial cells express a plasma membrane pore gated by extracellular ATP. *J. Immunol.* **156**: 1531.

Ferrari, D., Chiozzi, P., Falzoni, S. Hanau, S., and Di Virgilio, F., 1997, Purinergic modulation of interleukin-1 beta release from microglial cells stimulated with bacterial endotoxin. *J. Exp. Med.* **185**: 579.

Furukawa, K. and Mattson, M.P., 1998, The transcription factor NF-kappaB mediates increases in calcium currents and decreases in NMDA- and AMPA/kainate-induced currents induced by tumor necrosis factor-alpha in hippocampal neurons. *J. Neurochem.* **70**: 1876.

Gehrmann, J. and Banati, R.B., 1995, Microglial turnover in the injured CNS: activated microglia undergo delayed DNA fragmentation following peripheral nerve injury. *J. Neuropathol. Exp. Neurol.* **54**: 680.

Gravel, C., Gotz, R., Lorrain, A. and Sendtner, M., 1997, Adenoviral gene transfer of ciliary neurotrophic factor and brain-derived neurotrophic factor leads to long-term survival of axotomized motor neurons. *Nat. Med.* **3**: 765.

Hamilton, S.G., Wade, A. and McMahon, S.B., 1999, The effects of inflammation and inflammatory mediators on nociceptive behaviour induced by ATP analogues in the rat. *Br. J. Pharmacol.* **126**: 326.

Hashizume, H., DeLeo, J.A., Colburn, R.W. and Weinstein, J.N., 2000, Spinal glial activation and cytokine expression after lumbar root injury in the rat. *Spine* **25**: 1206.

Heppenstall, P.A. and Lewin, G.R., 2001, BDNF but not NT-4 is required for normal flexion reflex plasticity and function. *Proc. Natl. Acad. Sci. USA* **98**: 8107.

Hide, I., Tanaka, M., Inoue, A., Nakajima, K., Kohsaka, S., Inoue, K. and Nakata, Y., 2000, Extracellular ATP triggers tumor necrosis factor-alpha release from rat microglia. *J. Neurochem.* **75**: 965.

Homma, Y., Brull, S.J. and Zhang, J.M., 2002, A comparison of chronic pain behavior following local application of tumor necrosis factor alpha to the normal and mechanically compressed lumbar ganglia in the rat. *Pain* **95**: 239.

Jarvis, M.F., Burgard, E.C., McGaraughty, S., Honore, P., Lynch, K., Brennan, T.J., Subieta, A., Van Biesen, T., Cartmell, J., Bianchi, B., Niforatos, W., Kage, K., Yu, H., Mikusa, J., Wismer, C.T., Zhu, C.Z., Chu, K., Lee, C.H., Stewart, A.O., Polakowski, J., Cox, B.F., Kowaluk, E., Williams, M., Sullivan, J. and Faltynek, C., 2002, A-317491, a novel potent and selective non-nucleotide antagonist of P2X3 and P2X2/3 receptors, reduces chronic inflammatory and neuropathic pain in the rat. *Proc. Natl. Acad. Sci. USA* **99**: 17179.

Ji, R.R., Samad, T.A., Jin, S.X., Schmoll, R. and Woolf, C.J., 2002, p38 MAPK activation by NGF in primary sensory neurons after inflammation increases TRPV1 levels and maintains heat hyperalgesia. *Neuron* **36**: 57.

Jin, S.X., Zhuang, Z.Y., Woolf, C.J. and Ji, R.R., 2003, p38 mitogen-activated protein kinase is activated after a spinal nerve ligation in spinal cord microglia and dorsal root ganglion neurons and contributes to the generation of neuropathic pain. *J. Neurosci.* **23**: 4017.

Junger, H. and Sorkin, L.S., 2000, Nociceptive and inflammatory effects of subcutaneous TNFalpha. *Pain* **85**: 145.

Kennedy, C., Assis, T.S., Currie, A.J. and Rowan, E.G., 2003, Crossing the pain barrier: P2 receptors as targets for novel analgesics. *J. Physiol.* **553**: 683.

Kerr, B.J., Bradbury, E.J., Bennett, D.L., Trivedi, P.M., Dassan, P., French, J., Shelton, D.B., McMahon, S.B. and Thompson, S.W., 1999, Brain-derived neurotrophic factor modulates nociceptive sensory inputs and NMDA-evoked responses in the rat spinal cord. *J. Neurosci.* **19**: 5138.

Khakh, B.S., Burnstock, G., Kennedy, C., King, B.F., North, R.A., Seguela, P., Voigt, M. and Humphrey, P.P., 2001, International union of pharmacology. XXIV. Current status of the nomenclature and properties of P2X receptors and their subunits. *Pharmacol. Rev.* **53**: 107.

Kim, S.H. and Chung, J.M., 1992, An experimental model for peripheral neuropathy produced by segmental spinal nerve ligation in the rat. *Pain* **50**: 355.

Koizumi, S., Tsuda, M., Shigemoto, Y., Obama, T. and Inoue, K., 2001, Characterization of P2Y receptors in cultured rat dorsal root ganglion neurons. *Jpn. J. Pharmacol.* **85**: 149P.

Kreutzberg, G.W., 1996, Microglia: a sensor for pathological events in the CNS. *Trends Neurosci.* **19**: 312.

Leem, J.G. and Bove, G.M., 2002, Mid-axonal tumor necrosis factor-alpha induces ectopic activity in a subset of slowly conducting cutaneous and deep afferent neurons. *J. Pain* **3**: 45.

Lewis, C., Neidhart, S., Holy, C., North, R.A., Buell, G. and Surprenant, A., 1995, Coexpression of P2X2 and P2X3 receptor subunits can account for ATP- gated currents in sensory neurons. *Nature* **377**: 432.

Li, C., Peoples, R.W., Lanthorn, T.H., Li, Z.W. and Weight, F.F., 1999, Distinct ATP-activated currents in different types of neurons dissociated from rat dorsal root ganglion. *Neurosci. Lett.* **263**: 57.

Liu, L., Tornqvist, E., Mattsson, P., Eriksson, N.P., Persson, J.K., Morgan, B.P., Aldskogius, H. and Svensson, M., 1995, Complement and clusterin in the spinal cord dorsal horn and gracile nucleus following sciatic nerve injury in the adult rat. *Neuroscience* **68**: 167.

Mannion, R.J., Costigan, M., Decosterd, I., Amaya, F., Ma, Q.P., Holstege, J.C., Ji, R.R., Acheson, A., Lindsay, R.M., Wilkinson, G.A. and Woolf, C.J., 1999, Neurotrophins: peripherally and centrally acting modulators of tactile stimulus-induced inflammatory pain hypersensitivity. *Proc. Natl. Acad. Sci. USA* **96**: 9385.

Molliver, D.C., Cook, S.P., Carlsten, J.A., Wright, D.E. and McCleskey, E.W., 2002, ATP and UTP excite sensory neurons and induce CREB phosphorylation through the metabotropic receptor, P2Y2. *Eur. J. Neurosci.* **16**: 1850.

Moriyama, T., Iida, T., Kobayashi, K., Higashi, T., Fukuoka, T., Tsumura, H., Leon, C., Suzuki, N., Inoue, K., Gachet, C., Noguchi, K. and Tominaga, M., 2003, Possible involvement of P2Y2 metabotropic receptors in ATP-induced transient receptor potential vanilloid receptor 1-mediated thermal hypersensitivity. *J. Neurosci.* **23**: 6058.

Onda, A., Murata, Y., Rydevik, B., Larsson, K., Kikuchi, S. and Olmarker, K., 2004, Infliximab attenuates immunoreactivity of brain-derived neurotrophic factor in a rat model of herniated nucleus pulposus. *Spine* **29**: 1857.

Perregaux, D. and Gabel, C.A., 1994, Interleukin-1 beta maturation and release in response to ATP and nigericin. Evidence that potassium depletion mediated by these agents is a necessary and common feature of their activity. *J. Biol. Chem.* **269**: 15195.

Perry, V.H., 1994, Modulation of microglia phenotype. *Neuropathol. Appl. Neurobiol.* **20**: 177.

Pollock, J., McFarlane, S.M., Connell, M.C., Zehavi, U., Vandenabeele, P., MacEwan, D.J. and Scott, R.H., 2002, TNF-alpha receptors simultaneously activate Ca2+ mobilisation and stress kinases in cultured sensory neurones. *Neuropharmacology* **42**: 93.

Raghavendra, V., Tanga, F. and DeLeo, J.A., 2003, Inhibition of microglial activation attenuates the development but not existing hypersensitivity in a rat model of neuropathy. *J. Pharmacol. Exp. Ther.* **306**: 624.

Ralevic, V. and Burnstock, G., 1998, Receptors for purines and pyrimidines. *Pharmacol. Rev.* **50**: 413.

Rivera, C., Li, H., Thomas-Crusells, J., Lahtinen, H., Viitanen, T., Nanobashvili, A., Kokaia, Z., Airaksinen, M.S., Voipio, J., Kaila, K. and Saarma, M., 2002, BDNF-induced TrkB activation down-regulates the K^+-Cl^- cotransporter KCC2 and impairs neuronal Cl- extrusion. *J. Cell Biol.* **159**: 747.

Robertson, B., Xu, X.J., Hao, J.X., Wiesenfeld-Hallin, Z., Mhlanga, J., Grant, G. and Kristensson, K., 1997, Interferon-gamma receptors in nociceptive pathways: role in neuropathic pain-related behaviour. *Neuroreport* **8**: 1311.

Sanada, M., Yasuda, H., Omatsu-Kanbe, M., Sango, K., Isono, T., Matsuura, H. and Kikkawa, R., 2002, Increase in intracellular Ca(2+) and calcitonin gene-related peptide release through metabotropic P2Y receptors in rat dorsal root ganglion neurons. *Neuroscience* **111**: 413.

Sanz, J.M. and Di Virgilio, F., 2000, Kinetics and mechanism of ATP-dependent IL-1 beta release from microglial cells. *J. Immunol.* **164**: 4893.

Sawada, M., Hara, N. and Maeno, T., 1990, Extracellular tumor necrosis factor induces a decreased K+ conductance in an identified neuron of Aplysia kurodai. *Neurosci. Lett.* **115**: 219..

Schafers, M., Geis, C., Brors, D., Yaksh, T.L. and Sommer, C., 2002, Anterograde transport of tumor necrosis factor-alpha in the intact and injured rat sciatic nerve. *J. Neurosci.* **22**: 536.

Schafers, M., Lee, D.H., Brors, D., Yaksh, T.L. and Sorkin, L.S., 2003a, Increased sensitivity of injured and adjacent uninjured rat primary sensory neurons to exogenous tumor necrosis factor-alpha after spinal nerve ligation. *J. Neurosci.* **23**: 3028.

Schafers, M., Svensson, C.I., Sommer, C. and Sorkin, L.S., 2003b, Tumor necrosis factor-alpha induces mechanical allodynia after spinal nerve ligation by activation of p38 MAPK in primary sensory neurons. *J. Neurosci.* **23**: 2517.

Scholz, J. and Woolf, C.J., 2002, Can we conquer pain? *Nat. Neurosci.* **5** suppl.: 1062.

Seltzer, Z., Dubner, R. and Shir, Y., 1990, A novel behavioral model of neuropathic pain disorders produced in rats by partial sciatic nerve injury. *Pain* **43**: 205.

Soliven, B. and Albert, J., 1992, Tumor necrosis factor modulates Ca2+ currents in cultured sympathetic neurons. *J. Neurosci.* **12**: 2665.

Sommer, C. and Schafers, M., 1998, Painful mononeuropathy in C57BL/Wld mice with delayed wallerian degeneration: differential effects of cytokine production and nerve regeneration on thermal and mechanical hypersensitivity. *Brain Res.* **784**: 154.

Sommer, C., Marziniak, M. and Myers, R.R., 1998, The effect of thalidomide treatment on vascular pathology and hyperalgesia caused by chronic constriction injury of rat nerve. *Pain* **74**: 83.

Sommer, C., Schafers, M., Marziniak, M. and Toyka, K.V., 2001a, Etanercept reduces hyperalgesia in experimental painful neuropathy. *J. Peripher. Nerv. Syst.* **6**: 67.

Sommer, C., Lindenlaub, T., Teuteberg, P., Schafers, M., Hartung, T. and Toyka, K.V., 2001b, Anti-TNF-neutralizing antibodies reduce pain-related behavior in two different mouse models of painful mononeuropathy. *Brain Res.* **913**: 86.

Sorkin, L.S. and Doom, C.M., 2000, Epineurial application of TNF elicits an acute mechanical hyperalgesia in the awake rat. *J. Peripher. Nerv. Syst.* **5**: 96.

Sorkin, L.S., Xiao, W.H., Wagner, R. and Myers, R.R., 1997, Tumour necrosis factor-alpha induces ectopic activity in nociceptive primary afferent fibres. *Neuroscience* **81**: 255.

Souslova, V., Cesare, P., Ding, Y., Akopian, A.N., Stanfa, L., Suzuki, R., Carpenter, K., Dickenson, A., Boyce, S., Hill, R., Nebenuis-Oosthuizen, D., Smith, A.J., Kidd, E.J. and Wood, J.N., 2000, Warm-coding deficits and aberrant inflammatory pain in mice lacking P2X3 receptors. *Nature* **407**: 1015.

Stoll, G. and Jander, S., 1999, The role of microglia and macrophages in the pathophysiology of the CNS. *Prog. Neurobiol.* **58**: 233.

Stuesse, S.L., Cruce, W.L., Lovell, J.A., McBurney, D.L. and Crisp, T., 2000, Microglial proliferation in the spinal cord of aged rats with a sciatic nerve injury. *Neurosci. Lett.* **287**: 121.

Surprenant, A., Rassendren, F., Kawashima, E., North, R.A. and Buell, G., 1996, The cytolytic P2Z receptor for extracellular ATP identified as a P2X receptor (P2X7). *Science* **272**: 735.

Suzuki, T., Hide, I., Ido, K., Kohsaka, S., Inoue, K. and Nakata, Y., 2004, Production and release of neuroprotective tumor necrosis factor by P2X7 receptor-activated microglia. *J. Neurosci.* **24**: 1.

Svichar, N., Shmigol, A., Verkhratsky, A. and Kostyuk, P., 1997, ATP induces Ca2+ release from IP3-sensitive Ca2+ stores exclusively in large DRG neurones. *Neuroreport* **8**: 1555.

Sweitzer, S., Martin, D. and DeLeo, J.A., 2001, Intrathecal interleukin-1 receptor antagonist in combination with soluble tumor necrosis factor receptor exhibits an anti-allodynic action in a rat model of neuropathic pain. *Neuroscience* **103**: 529.

Thompson, S.W., Bennett, D.L., Kerr, B.J., Bradbury, E.J. and McMahon, S.B., 1999, Brain-derived neurotrophic factor is an endogenous modulator of nociceptive responses in the spinal cord. *Proc. Natl. Acad. Sci. USA* **96**: 7714.

Tominaga, M., Wada, M. and Masu, M., 2001, Potentiation of capsaicin receptor activity by metabotropic ATP receptors as a possible mechanism for ATP-evoked pain and hyperalgesia. *Proc. Natl. Acad. Sci. USA* **98**: 6951.

Tsuda, M., Ueno, S. and Inoue, K., 1999, In vivo pathway of thermal hyperalgesia by intrathecal administration of alpha,beta-methylene ATP in mouse spinal cord: involvement of the glutamate-NMDA receptor system. *Br. J. Pharmacol.* **127**: 449.

Tsuda, M., Koizumi, S., Kita, A., Shigemoto, Y., Ueno, S. and Inoue, K., 2000, Mechanical allodynia caused by intraplantar injection of P2X receptor agonist in rats: involvement of heteromeric P2X2/3 receptor signaling in capsaicin-insensitive primary afferent neurons. *J. Neurosci.* **20**: RC90.

Tsuda, M., Shigemoto-Mogami, Y., Ueno, S., Koizumi, S., Ueda, H., Iwanaga, T. and Inoue, K., 2002, Downregulation of P2X3 receptor-dependent sensory functions in A/J inbred mouse strain. *Eur. J. Neurosci.* **15**: 1444.

Tsuda, M., Shigemoto-Mogami, Y., Koizumi, S., Mizokoshi, A., Kohsaka, S., Salter, M.W. and Inoue, K., 2003, P2X4 receptors induced in spinal microglia gate tactile allodynia after nerve injury. *Nature* **424**: 778.

Tsuda, M., Mizokoshi, A., Shigemoto-Mogami, Y., Koizumi, S. and Inoue, K., 2004, Activation of p38 mitogen-activated protein kinase in spinal hyperactive microglia contributes to pain hypersensitivity following peripheral nerve injury. *Glia* **45**: 89.

Tsuzuki, K., Kondo, E., Fukuoka, T., Yi, D., Tsujino, H., Sakagami, M. and Noguchi, K., 2001, Differential regulation of P2X(3) mRNA expression by peripheral nerve injury in intact and injured neurons in the rat sensory ganglia. *Pain* **91**: 351.

Ueno, S., Tsuda, M., Iwanaga, T. and Inoue, K., 1999, Cell type-specific ATP-activated responses in rat dorsal root ganglion neurons. *Br. J. Pharmacol.* **126**: 429.

Vikman, K.S., Hill, R.H., Backstrom, E., Robertson, B. and Kristensson, K., 2003, Interferon-gamma induces characteristics of central sensitization in spinal dorsal horn neurons in vitro. *Pain* **106**: 241.

Virginio, C., Robertson, G., Surprenant, A. and North, R.A., 1998, Trinitrophenyl-substituted nucleotides are potent antagonists selective for P2X1, P2X3, and heteromeric P2X2/3 receptors. *Mol. Pharmacol.* **53**: 969.

Viviani, B., Bartesaghi, S., Gardoni, F., Vezzani, A., Behrens, M.M., Bartfai, T., Binaglia, M., Corsini, E., Di Luca, M., Galli, C.L. and Marinovich, M., 2003, Interleukin-1beta enhances NMDA receptor-mediated intracellular calcium increase through activation of the Src family of kinases. *J. Neurosci.* **23**: 8692.

Wagner, R. and Myers, R.R., 1996, Endoneurial injection of TNF-alpha produces neuropathic pain behaviors. *Neuroreport* **7**: 2897.

Wall, P.D., Devor, M., Inbal, R., Scadding, J.W., Schonfeld, D., Seltzer, Z. and Tomkiewicz, M.M., 1979, Autotomy following peripheral nerve lesions: experimental anaesthesia dolorosa. *Pain* **7**: 103.

Wang, S., Cheng, Q., Malik, S. and Yang, J., 2000, Interleukin-1beta inhibits gamma-aminobutyric acid type A (GABA(A)) receptor current in cultured hippocampal neurons. *J. Pharmacol. Exp. Ther.* **292**: 497.

Watkins, L.R. and Maier, S.F., 2003, GLIA: A novel drug discovery target for clinical pain. *Nat. Rev. Drug Discov.* **2**: 973.

Watkins, L.R., Milligan, E.D. and Maier, S.F., 2001, Spinal cord glia: new players in pain. *Pain* **93**: 201.
Winkelstein, B.A., Rutkowski, M.D., Sweitzer, S.M., Pahl, J.L. and DeLeo, J.A., 2001, Nerve injury proximal or distal to the DRG induces similar spinal glial activation and selective cytokine expression but differential behavioral responses to pharmacologic treatment. *J. Comp. Neurol.* **439**: 127.
Woolf, C.J. and Mannion, R.J., 1999, Neuropathic pain: aetiology, symptoms, mechanisms, and management. *Lancet* **353**: 1959.
Woolf, C.J. and Salter, M.W., 2000, Neuronal plasticity: increasing the gain in pain. *Science* **288**: 1765.
Yu, X.M., Askalan, R. and Keil, G.J. 2nd and Salter, M.W., 1997, NMDA channel regulation by channel-associated protein tyrosine kinase Src. *Science* **275**: 674.

22

DIABETIC RETINOPATHY, INFLAMMATION, AND PROTEASOME

António F. Ambrósio, Paulo Pereira and José Cunha-Vaz

1. ABSTRACT

Diabetic retinopathy is a leading cause of vision loss and blindness in adults in developed countries. Growing evidence indicates that a low grade and chronic inflammatory process may have a key role in the pathogenesis of diabetic retinopathy, even at the early stages of the disease. Increasing adhesion of leukocytes to retinal vessels, vascular permeability and production of inflammatory cytokines, and the appearance of reactive microglia in the retina, as well as new vessel formation and proliferation at the later stages, are common characteristics of diabetic retinopathy. The upregulation of adhesion molecules in both endothelial cells and leukocytes mediates leukostasis, which is well correlated with the breakdown of blood–retinal barrier. Vascular endothelial growth factor is considered the major player in the processes of retinal leakage and vascular proliferation, but protein kinase C has also a predominant role. Other growth factors and cytokines, such as interleukin-1β and TNF-α, have also increased expression in diabetic retinas, and therapies targeted to these molecules or its receptors are being considered. Increasing attention has also been given to microglial cells in the retina, since they become reactive early in the course of the disease and might be an important source of inflammatory mediators.

The ubiquitin–proteasome pathway plays a key role in a variety of biological processes and dysfunction of this pathway has been implicated in a number of diseases, including diabetic retinopathy and other complications associated with diabetes. There are multiple pathways through which hyperglycemia may lead to dysfunction of UPP, thus contributing to the pathophysiology of diabetic retinopathy. For example, hyperglycemia and associated oxidative stress may impair degradation and activity of transcription factors such as NFκB and hypoxia inducible factor-1 (HIF-1) which, in turn, may lead to increased vascularization of the retina, contributing to vision loss associated with diabetes. The importance of UPP in retinal tissue damage is illustrated by a number of in vitro and in vivo experiments. Moreover, therapeutic approaches targeted to specific components of the UPP in the retina may prove beneficial in controlling tissue damage associated with inflammation and with diabetic retinopathy.

José Cunha-Vaz – cunhavaz@aibili.pt
Center of Ophtalmology of Coimbra, IBILI, Faculty of Medicine, University of Coimbra, 3000-548 Coimbra, Portugal

2. INTRODUCTION

It is currently estimated that diabetes affects more than 200 million people around the world, mainly in western countries. The prevalence of type 2 diabetes (noninsulin-dependent) is increasing at an alarming rate, and therefore the number of diabetic patients will further increase over the next years.

The major complications of diabetes have been related with vascular alterations in the heart, kidney, legs, and the retina, indicating that vascular endothelial cells are particularly susceptible to hyperglycemia, at least in some organs and tissues.

Diabetic retinopathy is a leading cause of new cases of blindness in working age adults in the western countries (Cunha-Vaz, 2000). In recent years, important progresses have been made to better understand the pathogenesis of diabetic retinopathy, but the disease is still neither preventable nor curable, and the treatments available are not very effective. In many patients, the disease progresses to more severe forms, where neovascularization occurs (proliferative diabetic retinopathy) and laser photocoagulation therapy is needed. Even with available treatments, in many cases, retinopathy continues to progress to blindness.

Diabetic retinopathy has been considered a microvascular disease of the retina (Cunha-Vaz, 2001). The breakdown of blood–retinal barrier is a hallmark of diabetic retinopathy. In addition, diabetic retinopathy is also characterized by capillary occlusion, formation of microaneurysms and cotton-wool spots, lipid exudates, hemorrhages, macular edema, and proliferative diabetic retinopathy might also occur at the later stages of the disease. Visual impairment is associated with the later stages, and it is principally due to macular edema and the formation of new vessels on the surface of the retina. In addition to these morphological and physiological alterations, various mechanisms underlying the pathogenesis of diabetic retinopathy have been identified in diabetic humans and animal models: increased polyol pathway flux, increased formation of advanced glycation end products (AGEs), activation of protein kinase C (PKC), increased hexosamine pathway flux, increased production of free radicals, and activation of poly(ADP-ribose) polymerase (PARP) (Lorenzi and Gerhardinger, 2001; Brownlee, 2005; Leal et al., 2005c).

In the last decade, several important findings have shown that diabetic retinopathy has features of a neurodegenerative disease. Thus, in addition to endothelial cells and pericytes, other cells, such as neurons, Muller cells, astrocytes, and microglial cells are also affected by hyperglycemia (Lorenzi and Gerhardinger, 2001; Gardner et al., 2002; Barber, 2003), indicating that all cell types of the retina might be directly or indirectly affected by diabetes.

More than a decade ago, it was reported that interleukin (IL)-6, IL-1, IL-2, and tumor necrosis factor (TNF)-α are present in the vitreous of individuals with diabetic retinopathy, but not in any of the control vitreous (Franks et al., 1992), suggesting that diabetic retinopathy might be associated with chronic inflammation. Increasing evidence indicates that diabetic retinopathy has many features of a chronic and low-grade inflammatory disease, such as leukostasis, increased vascular permeability, increased IL-1β, and nitric oxide (NO) production, and reactive microglia (Gardner et al., 2002; Mohr, 2004). In addition, it has been shown that these inflammatory events occur early in the course of the disease (Carmo et al., 1999; Miyamoto et al., 1999; Rungger-Brandle et al., 2000; Joussen et al., 2001a).

Inflammatory mediators are usually implicated in the process of neovascularization in several ocular and retinal disorders. In the case of proliferative diabetic retinopathy, the process of neovascularization in the retina is a direct consequence of retinal ischemia, due to capillary occlusion, which triggers the release of high amounts of vasoproliferative factors, such as vascular endothelial growth factor (VEGF) and others (Shima et al., 1995; Ferrara and Gerber, 2001).

3. LEUKOCYTE ADHESION

Leukocytes are large cells that have the capacity to generate free radicals, cytotoxic products, and proteolytic enzymes. When activated, leukocytes can cause microvascular occlusion and vascular cell damage, which are early features of diabetic retinopathy. The first demonstration clearly showing that leukocytes, especially monocytes, cause capillary occlusions in diabetic retinas was made by Schroder et al. (1991), using alloxan-induced diabetic rats. An increase in the adhesion of leukocytes to rat retinal vessels was also demonstrated, in vivo, soon after diabetes induction with streptozotocin (Miyamoto et al., 1998, 1999), and this observation was temporally and spatially associated with retinal vascular leakage and nonperfusion.

The increase in retinal leukostasis is correlated with an increase in the expression of intercellular adhesion molecule-1 (ICAM-1) in diabetic retinas (Miyamoto et al., 1999; Joussen et al., 2001b). When neutralizing antibodies directed against ICAM-1 are administered into diabetic animals, both leukostasis and vascular leakage are prevented (Miyamoto et al., 1999; Joussen et al., 2001b). In addition to the increase in ICAM-1 expression, an increase in the expression of vascular cell adhesion molecule-1 (VCAM-1) was also reported in diabetic rat retinas (Bai et al., 2003). In human diabetic retinas, ICAM-1 immunoreactivity was shown to be significantly elevated in choroidal vasculature and retinal blood vessels (McLeod et al., 1995), but in another study with human subjects it was reported that the levels of vascular ICAM-1 in diabetic retinas are similar to ICAM-1 levels observed in control retinas (Hughes et al., 2004).

ICAM-1 is a member of the immunoglobulin superfamily of adhesion molecules. The ligands of ICAM-1 include the leukocyte $\beta 2$ integrins CD11a/CD18 and CD11b/CD18. The expression of the surface integrin subunits CD11a, CD11b, and CD18 is increased in neutrophils isolated from diabetic rats. The adhesion of neutrophils from diabetic animals to endothelial cell monolayers is significantly increased as well and pretreatment with antibodies against these integrins lowered leukocyte adhesion (Barouch et al., 2000). Retinal leukostasis is also decreased in the retinal microvasculature of diabetic animals treated with anti-CD18 fragments (Barouch et al., 2000; Joussen et al., 2001b) and in mice lacking the genes for CD18 and ICAM-1 (Joussen et al., 2004). These observations were correlated with lesser vascular leakage.

The adhesion of leukocytes to retinal endothelial cells also increases when these cells are pretreated with high concentrations of glucose. However, mannitol, which is normally used as osmotic control, also increases leukocyte adherence, indicating that the effects of high glucose may be due to hyperosmolarity (Bullard et al., 1994). More recently, Chen et al. (2003) suggested that dyslipidemia, but not hyperglycemia, increases the expression of ICAM-1 and VCAM-1 in human retinal vascular endothelial cells. However, endothelial cells were exposed to high glucose only for 24 h, which may be insufficient to observe some effects.

The apoptosis of retinal endothelial cells is a hallmark of diabetic retinopathy. The adhesion of leukocytes might have an important role in that process, since it was demonstrated that leukostasis is temporally and spatially associated with retinal endothelial cell injury and death, just one week after diabetes induction. The antibody-based neutralization of ICAM-1 and CD18 prevented retinal endothelial cell death (Joussen et al., 2001b). Endothelial cell death, a major cause of blood–retinal barrier breakdown in early diabetes, depends on a Fas–Fas ligand-dependent apoptosis mediated by leukocytes. Soon after diabetes induction, the expression of FasL and Fas is increased in rat neutrophils and in retinal vasculature, respectively. The inhibition of FasL reduces retinal vascular endothelial cell apoptosis and blood–retinal barrier breakdown, but does not decrease leukocyte adhesion to the retinal vasculature of diabetic animals. Also, neutrophils isolated from diabetic, but not control animals, induce endothelial cell apoptosis (Joussen et al., 2003).

A large body of evidence has shown that PKC, mainly β-isoform, plays a key role in the pathogenesis of diabetic retinopathy (Curtis and Scholfield, 2004). Oral administration of LY333531, a specific inhibitor of PKC-β, attenuates the increase in leukocyte adhesion in diabetic retinas (Nonaka et al., 2000; Abiko et al., 2003). Increased activity of the glycosylating enzyme (β)-1,6 acetylglucosaminyltransferase (core 2 GlcNAc-T) has been shown to be responsible for increased leukocyte adhesion to endothelial cells through PKC-β-dependent phosphorylation on serine/threonine residues of core 2 GlcNAc-T (Chibber et al., 2003), and these effects seem to be triggered by increased plasma levels of TNF-α (Ben-Mahmud et al., 2004). The use of etanercept, a soluble TNF-α receptor/Fc construct, has been shown to reduce leukocyte adhesion and suppress blood–retinal barrier breakdown (Joussen et al., 2002a).

AGEs have long been implicated in the pathogenesis of diabetic retinopathy (Stitt, 2003), and they also seem to play a role in chronic inflammatory processes, including complications of diabetes (Schmidt and Stern, 2000). Treatment of retinal microvascular endothelial cells with glycoaldehyde-modified albumin (AGE-Alb) increases the adhesion of leukocytes to endothelial cell monolayers. Also, mice infusion with AGE-Alb increases the expression of ICAM-1 mRNA and leukostasis in the retina. The breakdown of the blood–retinal barrier due to AGE-Alb was also demonstrated, and these effects seem to be mediated through an upregulation of NFκB (Moore et al., 2003). The increase in the adhesion of monocytes to bovine retinal endothelial cells (BRECs) induced by AGEs is mediated through VEGF-induced ICAM-1 expression and involves PKC-dependent signaling pathways (Mamputu and Renier, 2004).

VEGF is a permeabilizing and vasoproliferative factor, and it has been considered to have a major role in the development and progression of diabetic retinopathy (Caldwell et al., 2003). Intravitreal injections of VEGF increase protein and mRNA levels of ICAM-1 in retinal vessels (Lu et al., 1999). Joussen et al. (2002b) have shown that both ICAM-1 mRNA levels and leukocyte adhesion decrease in the retinas of diabetic animals treated with a highly specific VEGF-neutralizing Flt-Fc construct, VEGF TrapA(40), indicating that VEGF is an initiator of the early leukocyte adhesion in the retina. The VEGF isoform, VEGF(164), seems to be more potent than VEGF(120) at inducing retinal leukostasis and blood–retinal barrier breakdown mediated by ICAM-1. The specific blockade of endogenous VEGF(164) reduces leukostasis and blood–retinal barrier breakdown (Ishida et al., 2003).

Nitric oxide (NO) may also play a role in leukostasis. The expression of endothelial nitric oxide synthase (eNOS) is increased in diabetic retinas, and the inhibition of eNOS reduces leukocyte adhesion (Joussen et al., 2002b). In addition, recent evidences indicate that both neuronal NOS (nNOS) and inducible NOS (iNOS) are also involved in leukocyte adhesion and blood–retinal barrier breakdown (Leal et al., 2005a, b). The expression of eNOS, nNOS, and iNOS is increased in diabetic retinas, and treatment with L-NAME, an NOS inhibitor, decreases leukostasis and retinal leakage. Also, retinal leukostasis is decreased in iNOS knockout diabetic mice, and the in vitro adhesion of leukocytes to retinal endothelial cells exposed to a NO donor, NOC-18, is increased, probably due to the upregulation of ICAM-1 in endothelial cells (Fig. 1).

Treatment of diabetic rats with high dose of nonsteroidal anti-inflammatory drugs, such as aspirin and meloxicam, a cyclooxygenase 2 (COX-2) inhibitor, or intravitreal injection of dexamethasone, a corticosteroid, reduced the adhesion of leukocytes to retinal vessels, retinal ICAM-1 expression, and inhibited the blood–retinal barrier breakdown (Joussen et al., 2002a; Tamura et al., 2005), clearly indicating that anti-inflammatory strategies may be used to treat diabetic retinopathy.

Treatment with antioxidants, such as α-lipoic acid and D-α-tocopherol, and with a peroxynitrite decomposition catalyst, FP15, also reduces retinal leukostasis in diabetic rats (Abiko et al., 2003; Sugawara et al., 2004), suggesting that both oxidative and nitrosative stress contribute to the increase in adhesion molecules expression. Also, vitamins C and E were shown to inhibit the increase in ICAM-1 expression induced by AGEs in retinal endothelial cells (Mamputu and Renier, 2004).

In conclusion, a large body of evidence demonstrates that leukocyte activation and leukocyte adhesion to retinal vasculature have a major role in the pathogenesis of the early stages of diabetic retinopathy, being at least in part responsible for retinal vascular dysfunction and the breakdown of blood–retinal barrier.

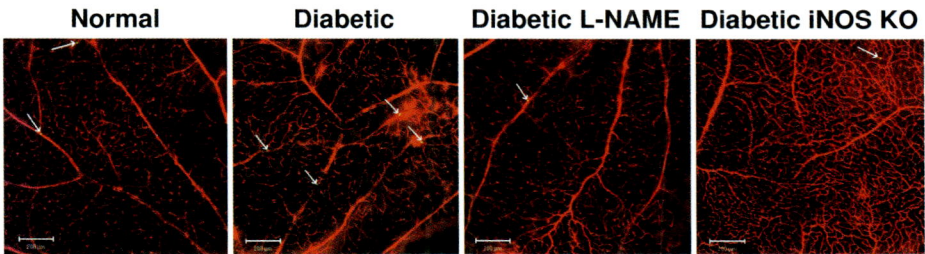

Fig. 1. Diabetes increases the adhesion of leukocytes to retinal vessels: involvement of NO. Leukocytes were isolated from the spleen of control mice, and then stained with Calcein-AM and injected, via tail vein, into control, diabetic, diabetic treated with L-NAME (an inhibitor of NOS isoforms), and diabetic iNOS knock-out mice. The figure shows representative images of adherent leukocytes in whole mounted retinas. Trapped leukocytes (*green*) are indicated by arrows. Evans blue (*red*), which was injected via the tail vein 10 min before animal sacrifice, outline the retinal vessels, allowing the detection of blood–retinal barrier breakdown

4. PERMEABILIZING FACTORS: THE IMPORTANCE OF VEGF

The existence of a blood–retinal barrier is essential to maintain retinal homeostasis. Actually, there are two retinal barriers, the inner blood–retinal barrier, which is confined to retinal vessels, and the outer blood–retinal barrier, located at retinal pigment epithelial cells (Cunha-Vaz, 1976). Both barriers depend on the presence of junctional structures between endothelial or epithelial cells, called tight junctions (TJ). These structures are composed of several transmembrane and cytosolic proteins, including occludin, claudins, zonula occludens, and others.

The breakdown of blood–retinal barrier is an early event in diabetic retinopathy. Both inner and outer blood–retinal barrier are affected by diabetes, but the retinal vasculature seems to be the primary site of leakage either in diabetic humans or experimental diabetic models (Cunha-Vaz et al., 1975; Vinores et al., 1993; Carmo et al., 1998). As mentioned before, VEGF is considered to have a major role in the pathogenesis of diabetic retinopathy, namely increasing vessel permeability at the earlier stages of the disease (Wilkinson-Berka, 2004). VEGF is increased in the retinas of diabetic subjects before the onset of retinopathy (Mathews et al., 1997), and in the retinas of diabetic animals, soon after the onset of diabetes (Hammes et al., 1998; Qaum et al., 2001). These observations have been correlated with increased retinal vascular permeability. The inhibition of VEGF bioactivity with the administration of specific VEGF-neutralizing soluble constructs significantly reduces the breakdown of blood–retinal barrier (Qaum et al., 2001; Saishin et al., 2003a). Also, intravitreal injection of VEGF triggers blood–retinal barrier breakdown and causes microangiopathy in primate retinas (Tolentino et al., 1996; Luna et al., 1997).

The increase in retinal permeability induced by intravitreal injection of VEGF is mediated by PKC, predominantly by the β-isoform. The effect of VEGF is suppressed by intravitreal or oral administration of LY333531, a specific inhibitor of PKC-β (Aiello et al., 1997). Retinal vascular permeability and retinal VEGF in diabetic rats decrease after intravitreal injection of GF109203X, a PKC inhibitor (Xu et al., 2004), and the inhibition of PKC prevents high glucose-induced increase in VEGF expression in vitro (Williams et al., 1997), suggesting that PKC also regulates VEGF expression.

Regarding the importance of TJ proteins in regulating blood–retinal barrier permeability, the breakdown of blood–retinal barrier may result from changes in the levels, expression, degradation, or distribution of TJ proteins. The content of occludin decreases in the retinas of diabetic rats and is redistributed in some retinal vessels of diabetic animals (Antonetti et al., 1998; Barber et al., 2000). Occludin might play a prominent role in this process, since the content of claudin-5, another transmembrane protein of TJ complex in the retina, is not affected by diabetes (Barber and Antonetti, 2003).

It has been demonstrated that VEGF increases retinal permeability by altering TJ protein complex. VEGF reduces the content and induces a rapid phosphorylation of occludin in cultured BRECs and in retinas, after intravitreal injection. VEGF also rapidly increases the phosphorylation of zonula occludens-1 (ZO-1) in tyrosine residues, indicating that the phosphorylation of TJ proteins might promote an increase in blood–retinal barrier permeability or signal a long-term decrease in tight junction proteins content (Antonetti et al., 1998, 1999).

Corticosteroids are a class of substances with anti-inflammatory properties, and it has been demonstrated that corticosteroids inhibit the expression of VEGF (Nauck et al., 1998). Treatment with triamcinolone acetonide blocks VEGF-induced blood–retinal barrier breakdown in rabbits (Edelman et al., 2005) and reduces increased VEGF expression in retinal pigment epithelial (RPE) cells exposed to oxidative stress (Matsuda et al., 2005). In addition, corticosteroids increase barrier properties. Hydrocortisone decreases the transport of water and solutes across BRECs monolayers, increases both mRNA and protein content of occludin and reduces occludin phosphorylation (Antonetti et al., 2002). Therefore, these anti-inflammatory drugs have a dual effect, since they inhibit VEGF action and regulate the expression, distribution, and phosphorylation of TJ proteins.

The increase in blood vessels permeability is a characteristic of inflammatory processes, and VEGF appears to be deeply involved in the process of blood–retinal barrier breakdown in diabetes. VEGF mediates leukocyte adhesion to retinal vessels, which in turn contribute to the increase in blood–retinal barrier permeability, and changes TJ expression and phosphorylation, mainly occludin, which has a key role in regulating paracellular permeability. Some anti-inflammatory drugs, such as corticosteroids, counteract the effect of VEGF, further supporting the concept that there is an inflammatory component in the pathogenesis of diabetic retinopathy, and therefore, the use of anti-inflammatory drugs.

In addition to VEGF, other growth factors can directly or indirectly, through VEGF, increase barrier permeability (Nauck et al., 1997). Platelet-derived growth factor (PDGF) increases the permeability to dextran and triggers the distribution of both occludin and ZO-1 from the cell border to the cytoplasm in MDCK cells (Harhaj et al., 2002), and insulin-like growth factor-1 (IGF-1) increases leakage in retinal vessels (Saishin et al., 2003b). However, these growth factors, including PDGF, IGF, and basic fibroblast growth factor (bFGF) are mainly implicated in retinal neovascularization in the later stages of diabetic retinopathy, in addition to VEGF, which is considered to have a predominant role in the process of neovascularization.

Histamine is also considered a permeabilizing factor, and it has been implicated in blood–retinal barrier breakdown in both experimental diabetes and diabetic patients (Enea et al., 1989; Gardner et al., 1995). The synthesis of histamine in the retinas of streptozotocin-induced diabetic rats is increased (Carroll et al., 1988) and inhibition of histamine receptors reduces the leakage of retinal vessels in diabetic rats and humans (Hollis et al., 1992; Gardner et al., 1995). It has been shown that histamine downregulates the expression of TJ proteins, particularly ZO-1, which might increase blood–retinal barrier permeability (Gardner, 1995). These evidences indicate that histamine may be also involved in the breakdown of blood–retinal barrier, but more studies are necessary to clarify its role.

5. PRODUCTION OF PROINFLAMMATORY MEDIATORS

The gene expression profile in the retinas of diabetic rats indicates that the identity of the genes that are upregulated is correlated with an inflammatory response (Joussen et al., 2001a). This retinal inflammatory process occurs very early upon the induction of diabetes. The expression of genes associated with inflammation was also reported to occur in Muller cells isolated from diabetic rat retinas (Gerhardinger et al., 2005).

The levels of IL-1β, a proinflammatory cytokine, are increased in the retinas of diabetic rats (Carmo et al., 1999; Kowluru and Odenbach, 2004a; Gerhardinger et al., 2005; Krady et al., 2005). The expression of IL-1β is also increased in BRECs exposed to high glucose (Kowluru and Odenbach, 2004b). The production of IL-1β is decreased in the retinas of diabetic animals treated with cyclosporin A, used as an anti-inflammatory drug, and this observation was correlated with a decrease in blood–retinal barrier permeability (Carmo et al., 2000). Treatment of diabetic rats with antioxidants also prevents the increase in retinal levels of IL-1β (Kowluru and Odenbach, 2004a).

The intravitreal administration of IL-1β induces the breakdown of the blood–retinal barrier, which appears to be mediated through recruited leukocytes and release of other vasoactive factors (Bamforth et al., 1997). Intravitreal injection of IL-1 also increases the number of acellular capillaries and apoptotic cells in retinal microvessels. The activation of the transcription factor NFκB and increased oxidative stress seem to mediate the effects of IL-1β (Kowluru and Odenbach, 2004a). Similar effects were found in cultured retinal endothelial cells. IL-1β activates NFκB, and increases the activity of caspase-3 and apoptosis. The presence of the interleukin-1 receptor antagonist (IL-1ra) decreases endothelial cell apoptosis (Kowluru and Odenbach, 2004b).

Caspase-1, formerly known as IL-1β-converting enzyme (ICE), is the enzyme responsible for the production of the proinflammatory cytokines IL-1β and IL-18. Mohr et al. (2002) have shown that the activity of several caspases, including caspase-1, is increased in the retina of diabetic mice in early diabetes. The activity of caspase-1 is also increased in the retinas obtained from human subjects with type 2 diabetes. Minocycline, a second-generation tetracycline derivative, which has been shown to exert neuroprotective effects in several neurodegenerative diseases, is able to prevent the activation of caspase-1 in the retina of diabetic mice, and therefore might prevent chronic inflammation and/or apoptosis (Mohr, 2004).

TNF-α, another proinflammatory cytokine, has been linked to proliferative diabetic retinopathy. However, it was demonstrated that serum levels of several inflammatory mediators, including TNF-α, are increased in patients with nonproliferative diabetic retinopathy, compared with healthy subjects, but the highest values were obtained in patients with proliferative diabetic retinopathy. The levels of TNF-α are also increased in the retinas of diabetic rats (Krady et al., 2005) and TNF-α increases the adhesion of leukocytes to retinal microvessels (Ben-Mahmud et al., 2004) and the permeability of blood–retinal barrier (Saishin et al., 2003b). The inhibition of TNF-α, with etanercept, inhibits leukostasis and prevents blood–retinal barrier breakdown (Joussen et al., 2002a), indicating that TNF-α is operative at the early stages of diabetic retinopathy.

As mentioned previously, NO, another inflammatory mediator, is involved in leukostasis and blood–retinal barrier breakdown, indicating that NO may also play an important role in the pathogenesis of diabetic retinopathy. NO is a product of the enzymatic conversion of arginine to citrulline by NOS. The activity of NOS and the uptake of L-arginine are enhanced in the retinas of streptozotocin (STZ)-induced diabetic rats, soon after the induction of diabetes (Carmo et al., 1999). The production of NO is increased in the retinas of diabetic rats and in retinal endothelial cells exposed to high glucose, and the increase in NO production is well correlated with the breakdown of blood–retinal barrier (Kowluru, 2001; Kowluru et al., 2001; Du et al., 2002). These findings are in agreement with the demonstration of increased expression of eNOS, nNOS, and iNOS in diabetic rat retinas (Carmo et al., 2000; Takeda et al., 2001; Leal et al., 2005a, b; Park et al., 2005). The inducible isoform of NOS seems to have a key role in the pathogenic process, in humans, since it was reported that iNOS is not expressed in the retinas of control subjects, but it is expressed in the retinas of patients with diabetes (Abu El-Asrar et al., 2001, 2004). The expression of iNOS decreases when diabetic rats are treated with cyclosporin A, an anti-inflammatory drug, and the increase in blood–retinal barrier permeability is also inhibited (Carmo et al., 2000; El-Remessy et al., 2003).

NO reacts rapidly with superoxide ions to form peroxynitrite, which may lead to the nitration of protein tyrosine residues and affect protein function. Therefore, the presence of nitrotyrosine on proteins can be used as a marker for peroxynitrite formation. It has been demonstrated that the levels of nitrotyrosine are increased in diabetic retinas and in retinal endothelial cells exposed to high glucose (Du et al., 2002; Kowluru et al., 2003), indicating that diabetes impairs retinal proteins function.

The expression of COX-2 and the production of prostaglandins are also increased in diabetic retinas and in retinal Muller cells cultured in high glucose, and cyclosporine A or high dose of aspirin inhibit these increases (Carmo et al., 2000; Du et al., 2002; Ayalasomayajula and Kompella, 2003). Also, celecobix, a selective COX-2 inhibitor, inhibits VEGF expression and vascular leakage in the retinas of diabetic rats (Ayalasomayajula and Kompella, 2003). As mentioned before, the inhibition of COX-2 decreases leukostasis and prevents blood–retinal barrier breakdown (Joussen et al., 2002a). These findings identify COX-2 as another inflammatory mediator that operates in the early stages of diabetic retinopathy. The increase in the production of cytotoxic prostaglandins via COX-2 may account for cell dysfunction and death in the retina.

These evidences point out that several inflammatory mediators, including IL-1β, TNF-α, nitric oxide, and prostaglandins, might play important roles in the pathophysiology and progression of diabetic retinopathy, and therefore they may serve as potential targets for the treatment of diabetic retinopathy. These molecules likely act together during the course of the disease, but the interactions between them remain to be elucidated. Some studies have proven that anti-inflammatory drugs have beneficial effects in animal models of diabetic retinopathy, but further studies are necessary to test the efficacy of anti-inflammatory drugs in the prevention and progression of diabetic retinopathy.

6. MICROGLIA ACTIVATION

Retinal microglial cells can be found in every layer of the human retina and may be activated upon retinal injury. In the retina of STZ-induced diabetic rats microglial cells appear activated, 4 weeks after diabetes induction. The density of microglial cells increases and they become hypertrophic (Rungger-Brandle et al., 2000; Zeng et al., 2000; Krady et al., 2005). Alterations in the morphology of microglia have been recently also observed in a mouse model of retinal complications in diabetes, the Ins2(Akita) mouse (Barber et al., 2005).

Several factors may account for microglial cell activation, including neural cell death, and the degree of microglia activation may itself influence the extent of retinal injury. Minocycline, an anti-inflammatory drug, represses the production of inflammatory cytokines by microglial cells and reduces the activity of caspase-3, an executor of apoptosis, in the retina of STZ-induced diabetic rats (Krady et al., 2005), suggesting that the inhibition of microglial activity may be an important strategy to treat diabetic retinopathy (Fig. 2).

7. OVERVIEW OF THE UBIQUITIN–PROTEASOME PATHWAY

The ubiquitin–proteasome pathway (UPP) is a major proteolytic pathway present in virtually all eukaryotic cells where it regulates vital biological and physiological processes. Cellular processes such as cell division, differentiation, signal transduction, quality control, and protein trafficking are all regulated to some extent by the UPP. The importance of UPP in normal physiology is tremendous and disruption of components of UPP has been implicated in a great variety of human diseases.

There are currently numerous forms of ubiquitin and ubiquitin-like proteins present in the cells. Ubiquitin was described as "Darwin's phosphate" (Hampton, FASEB meeting 2001,

Vermont, USA) and the total number of genes involved in ubiquitin and ubiquitin-like reactions is comparable to the number of genes involved in phosphorylation/dephosphorylation reactions (Lorick et al., 2005). Although this was initially recognized as a selective degradative pathway, more recent evidence showed that UPP has a number of nondegradative functions (Welchman et al., 2005) that have a significant role in numerous biological processes and that are implicated in a number of human diseases.

Many of the pathophysiological alterations associated with development and progression of diabetic retinopathy are under the control of the UPP in a direct or indirect manner. A major cause of sight loss associated with diabetic retinopathy is the neovascularization of the retina and the general inflammatory response that accompanies the progression of the disease. This section will focus on the regulatory effects of UPP in two major transcription factors that show altered activity in diabetic retinopathy: NFκB and HIF-1. Although the response and activity of these transcription factors is altered in diabetic retinopathy, the role of UPP in regulating these two transcription factors is diverse and inhibition of proteasome leads to quite different outcomes. This section will begin with an overview of the general biology of the UPP concluding with perspectives on putative therapies targeted to proteasome activity in diabetic retinopathy.

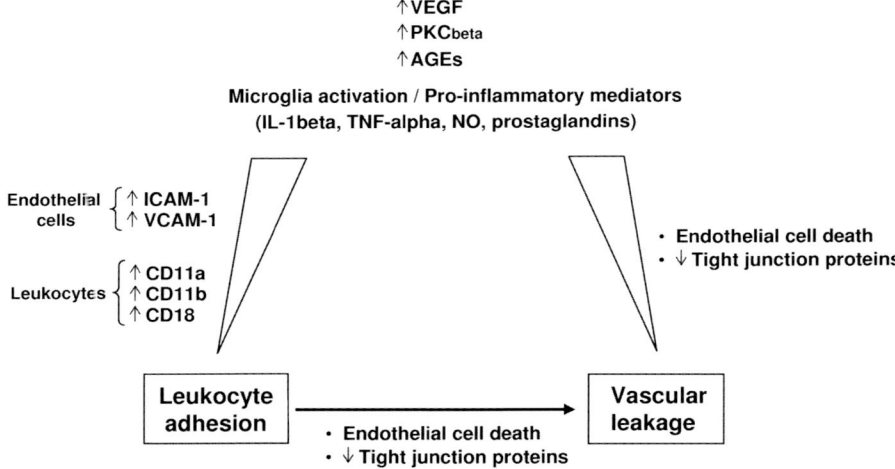

Fig. 2. Summary of the mechanisms and molecules that have been shown to be involved in the increase of leukocyte adhesion to retinal vessels and in blood–retinal barrier breakdown, in the early stages of diabetes

8. UBIQUITIN-DEPENDENT PROTEIN DEGRADATION

Degradation of a protein by the UPP involves two discrete and successive steps: first the substrate is tagged by covalent attachment of multiple ubiquitin molecules. In the second stage, the polyubiquitinylated protein is degraded by the 26S proteasome complex with the release of free ubiquitin that can be used in subsequent ubiquitinylation cycles. Removal of ubiquitin from substrates prior to degradation is also a regulated process that is carried out by deubiquitinylating enzymes. Ubiquitin is a small peptide of about 76 amino acids, which is highly conserved from yeast to humans. Indeed, there are only three amino acid substitutions between yeast and human ubiquitin (P19S, E24D, A28S). This makes ubiquitin the most highly conserved protein identified in eukaryotes so far (Ozkaynak et al., 1984). Conjugation of ubiquitin to substrates proceeds via a three-step mechanism. Initially, the ubiquitin activating enzyme (E1) activates ubiquitin in an ATP-dependent reaction to generate a high energy thiol ester intermediate. In the second step, one of a variety of ubiquitin conjugating enzymes (E2) transfers the activated ubiquitin moiety from the E1 to the substrate that is generally associated to one of a member of the ubiquitin ligase family (E3). In some cases, ubiquitin can be transferred from the E2 to the E3 and subsequently from the E3 to the substrates. This sequence of reactions is repeated until a chain of typically four or more ubiquitins is attached to the substrate. The most common and best-characterized fate of a polyubiquitinylated protein is its translocation to a large proteolytic complex, the 26S proteasome where it is degraded (Fig. 3). Although this is the canonical view of the system and indeed, the process responsible for the degradation of numerous cytosolic proteins, it should be emphasized that sometimes only one ubiquitin is attached to the protein in one or more amino acid residues. This results in the production of a monoubiquitinylated protein. Proteins can be monoubiquitinylated in different residues resulting in the formation of a multimonoubiquitinylated protein. In most cases the final destination of such proteins is not degradation but rather targeting to specific subcellular compartments, including the endocytic pathway.

The formation of a polyubiquitinylated protein is generally initiated by the formation of an isopeptide bond between the C-terminus of ubiquitin (G76) and the ϵ-NH2 group of an internal Lys residue in the substrate, to generate a covalent isopeptide bond. Less frequently, ubiquitin can be conjugated to the terminal amino-group of the substrate. The subsequent addition of other ubiquitin molecules involves frequently (but not always) the formation of a peptide bond between the G76 at the C-terminus of a newly activated ubiquitin and the Lys 48 of the previously attached ubiquitin molecule. This sequence of reactions leads to the formation of a polyubiquitin chain of varied size. The polyubiquitin chain is, then, recognized by the 26S proteasome and the substrate is degraded.

As the substrate is unfolded and enters the pore of the catalytic chamber of the 20S proteasome, deubiquitinylating enzymes present in the 19S caps of the proteasome remove polyubiquitin chains from the substrates, thus regenerating free ubiquitin.

The specificity and selectivity of the UPP was unclear for many years and some specific aspects of the system selectivity remain to be elucidated. It is currently accepted that the high selectivity of the system is primarily the result of the presence of specific signals in substrate proteins and the action of both ubiquitin ligases (E3s) and related ancillary proteins. In most instances, substrates are not constitutively or directly recognized by ligases. In some cases, the ligase must be activated by undergoing some post-translational modification to yield an active form that recognizes the substrates. In other instances, it is the substrate that must undergo specific modifications, which renders it a target for ubiquitinylation (Glickman and Ciechanover, 2002).

Ubiquitin ligases are therefore key players in determining the system specificity. The first E3s to be identified were very diverse and apparently unrelated. The first big family of E3s to be recognized presented a 350 amino acid residue sequence homologous to E-6AP (E6-associated protein) carboxy terminus and is currently known as HECT domain E3s

(Huibregtse et al., 1995). This domain contains a conserved Cys residue to which the ubiquitin moiety is transferred from E2s (Scheffner et al., 1995). The first ligase of the HECT domain family to be identified was the E6-AP. E6 is the oncoprotein of the high-risk human papillomaviruses (HPVs). The E6 oncoprotein binds to the cellular ubiquitin ligase E6-AP and targets the tumor suppressor protein p53 for rapid degradation (Scheffner et al., 1993) rendering cells more susceptible to tumor development. In the absence of the viral ancillary protein, E6-AP targets other cellular proteins for degradation without major noxious consequences to the cells. Mutations in E6-AP are implicated in diseases such as Angelman syndrome, a severe form of mental and motor retardation (Kishino et al., 1997). Other important members of the HECT domain family of ubiquitin ligases include NEDD4, Npi1/Rsp5, Smurf1, and Smurf2 (Glickman and Ciechanover, 2002).

Virtually all ubiquitin ligases, or E3s, that do not belong to the HECT domain belong to large family of RING (Really Interesting New Gene) finger proteins (Glickman and Ciechanover, 2002; Pickart, 2004). The mammalian RING finger domain family is very large and it is currently accepted that many of its members are, indeed, E3s. RING fingers consist of domains of approximately 70 amino acid residues containing a pattern of conserved Cys and His residues that bind zinc and stabilize the typical globular conformation of these proteins (Glickman and Ciechanover, 2002). Interestingly, it is the spacing of zinc ligands, rather than any primary sequence, that is conserved in the RING finger family, suggesting that RING finger are likely to function as structural elements in a complex, rather than presenting catalytic activity (Pickart, 2001). Indeed, it was shown that unlike HECT domain E3s, RING fingers proteins do not have a catalytic function in the ubiquitinylation process. Apparently, RING-fingers function as scaffolds that bring other proteins together, including E2s and substrates, thus facilitating ubiquitin transfer (Zheng et al., 2000). The RING finger E3s can form either single or multisubunit proteins of various sizes and composition. The multisubunit RING finger family of E3s is often subdivided into three major groups: APC (Anaphase Promoting Complex), SCF (Skp1-Cdc53-F-box protein), and VCB (Von-Hippel-Lindau tumor suppressor-Elongin C/B). The architecture and organization of E3s of the SCF and VCB group is similar in many ways. In both cases a central 100-residue RING finger protein called Rbx1 acts as a scaffold that strongly interacts with a subunit belonging to the cullin protein family (Cul1/Cdc53 for the SCF and Cul2 in VCB) (Pickart, 2001). Rbx1 also assists in recruiting the cognate E2s. The APC group of ligases has a central RING finger protein and displays a core ligase activity in conjunction with the ubiquitin conjugating enzyme Ubc4 (Pickart, 2001).

The single subunit members of the RING finger E3s family recognize the ubiquitynylation signals in their specific substrates through domains that are structurally distinct from the RING finger. For the multisubunit RING finger SCF the substrate recognition is delegated in a different protein of the complex. For example in the SCF E3s, substrate specific F-box proteins are recruited to SCF complexes through the adaptor protein Skp1 that recognizes the eponymous F-box motive (Feldman et al., 1997; Skowyra et al., 1997; Deshaies, 1999; Pickart, 2001). For the RING finger VCB E3, the substrate recognition is done by the pVHL, the product of the Von Hippel-Lindau tumor suppressor gene (Chau et al., 1989). The protein pVHL is recruited to the complex through interactions with the heterodimeric adaptor Elongin B/C (Pickart, 2001, for review). The existence of substrate binding (F-boxes), cullin and adaptor protein families (Skps, elongins) together with other data indicates that E3 specificity can be reprogrammed, among other factors, by changing the identity of the substrate recognition subunit (Pickart, 2004).

Examples of monomers or homodimers that contain both the RING finger domain and the substrate binding site in the same molecule include Mdm2 (that targets p53 for proteasome degradation) (Lorick et al., 1999; Boyd et al., 2000; Geyer et al., 2000), Ubr1/E3α (Kwon et al., 1998; Reiss and Hershko, 1990), and Parkin (Shimura et al., 2000). On the other hand, E3s of the APC family are involved in degradation of cell cycle regulators such as cyclins (Page and Hieter, 1999); the VCB complex is involved in degradation of HIF-1α (Maxwell et al., 1999; Kamura et al., 2000; Ivan et al., 2001; Jaakkola et al., 2001); the SCF family is usually involved in the degradation of signal and cell-cycle-induced phosphorylated proteins (Deshaies, 1999).

Recognition of protein substrates depends not only on the specificity of E3s and ancillary proteins but also on intrinsic signals present on the substrates that determine its half live and its susceptibility to ubiquitinylation and subsequent proteolysis. In the late 1980s, a set of short lived substrates of the UPP was identified, which contained specific amino acids in its N-terminus. This led to the idea that the stability of a protein was largely determined by the nature of its N-terminus amino acid (Bachmair et al., 1986; Pickart, 2004). This rule, for protein susceptibility to ubiquitin-dependent degradation, is still known as the N-end rule. Subsequent studies revealed that, for some substrates, there are elements that are both necessary and sufficient to determine its ubiquitinylation (Varshavsky, 1997; Pickart, 2004). These elements, which were eventually found in different locations in the protein structure, are called degrons and can consist of varied amino acid motifs or sequences.

Recognition by the N-end rule involves direct binding of the substrates to the ubiquitin ligase E3α/Ubr1 (Varshavsky et al., 2000). E3α/Ubr1 has two N-end rule recognition sites: one for substrates with basic amino acids at the NH_2 terminal and another one for substrates with hydrophobic residues at the NH_2 terminal (Kwon et al., 1998; Reiss et al., 1988). A third site in the ligase is involved in targeting non-N-end rule substrates.

One of the first amino acid sequences to be identified in substrates that render them prone to UPP-dependent degradation was the destruction box found in mitotic cyclins and many other substrates of the APC (Glotzer et al., 1991; Deshaies, 1999; Koepp et al., 1999; Page and Hieter, 1999). The destruction box is a short sequence RXALGXIXN, where the arginin and leucine residues are key determinants of specificity. This sequence, in the primary structure of the protein, appears to be sufficient to recruit the appropriate E3s and to promote degradation of substrates. However, it also appears that the signal conformation may play a role in the efficient binding of E3s to substrates (Pickart, 2001). Over the years, it also became apparent that phosphorylation is a common signal that targets proteins for degradation. Many of such proteins contain the so-called PEST regions. More recently, however, it became clear that although such proteins are often degraded by the ubiquitin system, the PEST regions provide phosphorylation sites to regulate the accessibility of the degradation signals.

Physiological recognition of degrons is not trivial and is subjected to a complex and highly coordinated sequence of events and interactions. Interactions between an E3 and a degron can be modulated by a variety of mechanisms, including post-transcriptional modifications that serve to link ubiquitinylation to other cellular events (Deshaies, 1999; Deshaies and Ferrell, 2001). For example, the cell cycle needs to be precisely regulated and this involves the phosphorylation of cyclin-dependent kinase (CDK) that is required to trigger ubiquitin-dependent degradation of CDK regulators. These regulators include (but are not limited to) the mammalian G1 cyclins D and E and the mammalian CDK inhibitor $p27^{KIP1}$ (Glickman and Ciechanover, 2002). Other regulatory proteins that need to be phosphorylated prior to ubiquitinylation include the transcriptional regulators IκBα and β-catenin (Kornitzer and Ciechanover, 2000) as well as the NFκB precursor p105 (Heissmeyer et al., 2001; Orian et al., 2000).

Another important example of the post-transcriptional modification of a protein, that is particularly relevant in the context of neovascularization associated with diabetic retinopathy, is the degradation of the HIF-1α. Indeed degradation of HIF-1α occurs continuously under normoxic conditions and requires prior oxygen-dependent hydroxylation of two specific proline residues that triggers recognition by a cullin-based E3 that has the Von Hippel Lindau (VHL) tumor suppressor protein as its specificity subunit (Ivan et al., 2001; Jaakkola et al., 2001).

9. DEGRADATION OF UBIQUITINYLATED PROTEINS

The second major step in the UPP consists on the recognition and degradation of ubiquitinylated substrates by the proteasome. The proteasome holoenzyme, also referred to as the 26S proteasome, is a 2.5-MDa complex made up of two copies. Each copy comprises, at least, 32 different subunits that are highly conserved among all eukaryotes. The overall structure can be divided into two major subcomplexes, the 20S complex or core particle (CP) that contains the proteolytic activity and a regulatory particle (RP) also referred to as 19S cap complex or PA700 in mammals and μ particle in *D. melanogaster* (Glickman et al., 1999; Gorbea et al., 1999; Pickart and VanDemark, 2000). The 20S particle is a cylindrical structure composed of four stacked heptameric rings. In yeast and higher eukaryotes, the rings are formed by 14 distinct subunits designated by α or β. The active sites reside within the β subunits, which provide the catalytic N-terminal threonine residues. However, only three of the seven different β subunits have free N-terminal threonine, which means that the proteasome will present a total of six active sites (Groll et al., 1997). The active sites are buried in a central chamber isolated from the external solvent. Access to the catalytic core is very restricted and substrates must be unfolded to fit to an axial pore of 13 Å in diameter (Pickart and VanDemark, 2000). The compartmentalization of the active sites inside a restricted chamber prevents indiscriminate degradation of proteins and confers some selectivity to the system. However, it is the 19S regulatory particle that confers both selectivity and specificity to the 26S proteasome activity. Biochemically, the 20S core presents only ATP-independent peptidase activity. However, the assembled 26S proteasome presents a variety of activities and functions including polyubiquitin chain recognition and binding, nucleotidase activity, isopeptidase activity, unfoldase, and endopeptidase activity (Groll et al., 1997; Glickman et al., 1999). Thus, binding of the 19S regulatory particle to the 20S proteasome confers selectivity and specificity of the proteasome toward ubiquitinylated proteins. Indeed, ubiquitinylated substrates need to be recognized, at a first stage, by specific protein subunits present in the 19S cap. Subsequently, the protein needs to be unfolded and fed to the catalytic chamber of the proteasome. As substrates are translocated through the central pore of the 20S core particle, ubiquitin is removed and recycled for reuse in other ubiquitinylation cycles. These and other activities are all located in specific subunits present at the 19S caps and many of them require ATP that is used up by the ATPases present in these regulatory subunits.

10. INFLAMMATION AND PROTEASOME

Proteasomes can be implicated in inflammatory response associated with diabetes in a number of ways. However, the most prominent role of proteasome in inflammation is, probably, due to its ability to regulate the signaling pathway dependent on NFκB. This nuclear factor is critical in the processing of a variety of proinflammatory signals, including the upregulation of numerous cytokines (interleukin-1, 6, and 8, and tumor necrosis factor), proangiogenic factors such as VEGF, inducible enzymes such as COX-2 and iNOS, cell adhesion molecules (intercellular adhesion molecule-1 and vascular cellular adhesion molecule-1), stress proteins, antiapoptotic factors (bcl-2, survinine), and immune system receptors (Baldwin, 1996). The activation of NFκB and the transcription of a number of NFκB-dependent genes have been well documented in diabetes, especially in the retinal vasculature of diabetic patients and in animal models of diabetes (Joussen et al., 2002b; Kowluru et al., 2003). Consistently, in vitro exposure of retinal endothelial cell and pericytes to high glucose has been shown to cause activation of NFκB (Romeo et al., 2002; Kowluru et al., 2003).

NFκB is a generic term for a dimeric transcription factor formed by the hetero or homodimerization of subunits belonging to a family of five members. Three of these proteins (p65, Re1B, and cRel) contain transactivation domains and two other proteins; p50 and p52 are expressed as the precursor proteins p105 (NFκB1) and p100 (NFκB2), respectively. These precursors require post-translational processing and do not contain transactivation domains. In fact, the ubiquitin-proteasome system is also involved in the limited processing of the precursor protein p105 to yield the active subunit p50. In quiescent cells, NFκB is located in the cytoplasm in an inactive form bound to a labile inhibitor molecule called IκBα, or other structurally related proteins (Hayden and Ghosh, 2004).

NFκB can be activated in response to a variety of different stimuli including viruses, growth factors, antigens, radiation, and chemotherapeutic drugs. However, the best studied mechanisms of activation of NFκB involve the response to proinflammatory stimuli such as TNF-α and the interleukin-1β (Karin and Ben-Neriah, 2000). The various pathways that lead to activation of NFκB appear to converge in the phosphorylation of IκB by a cascade of complexes that include TAK1 (transforming growth factor-β (TGFβ)-activated kinase-1) and the IκB Kinase complex (IKK). Activation of these kinases results in phosphorylation of IκB (Karin and Ben-Neriah, 2000). The phosphorylated motif docks with a specific ubiquitin ligase complex, resulting in conjugation of ubiquitin to IKBα (Carter et al., 2005). Polyubiquitinylated IkBα is then degraded by the 26S proteasome, NFκB is released and can translocate to the nucleus where it activates some of the genes described above (Maniatis, 1999; Karin and Ben-Neriah, 2000). IKK holoenzyme is composed of three subunits: α, β, and γ. It is the subunit β (IKKβ) that is responsible for IκB phosphorylation and KFκB activation in response to proinflammatory stimuli, whereas IKKα targets distinct cellular substrates in an alternative pathway for NFκB induction (Carter et al., 2005).

The molecular events that activate the kinases that phosphorylate IKK, targeting its degradation, are fairly complex and often involve different molecules that respond to different stimuli. For example, when TNF-α binds to TNF receptor-1, the receptor oligomerizes and associates with a complex that contains the ubiquitin ligase TRAF2 (TNF-receptor-associated factor-2). The RANK receptor associates with the ubiquitin ligase TRAF6. Working in association with the heterodimeric E2 UBC13-MMS2, TRAF2 attaches K63-linked polyubiquitin chains to the receptor interacting protein (RIP) (Meylan and

Tschopp, 2005). This triggers the TAK1 complex to bind to the membrane–receptor complex through the K63-linked polyubiquitin chains and results in IκB phosphorylation and its subsequent ubiquitinylation (Sun and Chen, 2004).

The switching off of this signaling cascade depends, among other factors, on deubiquitinylation. The importance of deubiquitinylation of proteins in switching off this signaling cascade is well illustrated by the human tumor condition cylindromatosis. Cylindromatosis is due to a mutation in the tumor suppressor CYLD, which cleaves K63-linked chains from TRAF2 (Wilkinson, 2003). More recently, the protein A20 has emerged as a dual-catalysis enzyme that downregulates the NFκB system (Wertz et al., 2004). A20 cleaves the K63-linked ubiquitin chain from RIP and adds K48-linked chains that results in proteasome degradation of a RIP.

There are number of compounds that were shown to interfere with the activity of NFκB in cells and these include antioxidants, glucocorticosteroids, nonsteroidal anti-inflammatory drugs, cytokines, peptides inhibiting the nuclear localization of NFκB, and proteasome inhibitors such as MG132, P-341, and PS-519.

Stabilization of NFκB, by proteasome inhibitors, for example, results in the blockade of NFκB activation and prevents the actions of NFκB within the cells (Palombella et al., 1994). An alternative method of controlling NFκB activity is via the modulation of phosphorylation of the DNA-binding subunits that can regulate the interactions between the transcription factor and regulatory proteins (Schmitz et al., 2001).

11. INFLAMMATION, RETINAL ISCHEMIA, AND PROTEASOME

Inflammation is a hallmark of early stages of diabetic retinopathy involving attachment to and transmigration of leukocytes through the retinal microvasculature (Adamis, 2002; Joussen et al., 2002a; Schroder et al., 1991). Proinflammatory cytokines, including TNF-α and IL-1β, are elevated in the extracellular matrix, endothelium, vessel walls, and vitreous of eyes of patients with proliferative diabetic retinopathy as well as in animal models of diabetes (Spranger et al., 1995; Limb et al., 1996, 1999a, b; Kowluru and Odenbach, 2004a). Consistently, inhibition of TNF-α and IL-1β signaling significantly reduced leukocyte adhesion and endothelial cell injury (Joussen et al., 2002a; Kowluru and Odenbach, 2004b).

Inflammation is further exacerbated by retinal ischemia that is frequently present in diabetes. A number of endothelial tissues including brain, heart, and retina upregulate expression of cell adhesion molecules following an ischemic episode that attract circulating leukocytes (Carroll et al., 2000). Once bound to the endothelium, these cells diapedese into the tissues and are responsible for much of the subsequently recorded damage. The results of proteasome inhibition by PS-519, in rodent models of cerebral ischemia, revealed attenuated expression of cell adhesion proteins, reduced invasion of leukocytes and, as a consequence, reduced brain tissue damage (Carroll et al., 2000; Phillips et al., 2000). Ischemic response is itself intrinsically regulated by the UPP. In addition to the effects of the activation of NFκB, the increased neovascularization associated with retinal ischemia is primarily due to the activation of the transcription factor HIF-1.

HIF-1 is an oxygen-dependent transcriptional activator, which plays crucial roles in the angiogenesis. HIF-1 induces the expression of more than 60 proteins including VEGF. HIF-1 consists of a constitutively expressed HIF-1β subunit and one of three inducible subunits: HIF-1α, HIF-2 α, and HIF-3 α. The stability of HIF-1 α is regulated in various ways.

Under nonhypoxic (normoxia) conditions, HIF-1α is subject to oxygen-dependent prolyl hydroxylation (Ivan et al., 2001; Jaakkola et al., 2001), which is required for binding of the von Hippel-Lindau tumor supressor protein (VHL), the recognition component of an ubiquitin–protein ligase, which targets HIF-1α for ubiquitin-dependent proteasomal degradation (Maxwell et al., 1999). HIF-1α contains two sites for hydroxylation, Pro402 and Pro564, within the oxygen-dependent degradation (ODD) domain and each site contains a conserved LXXLAP motif (Masson et al., 2001).

Ubiquitination of HIF-1α, as of most proteins, requires primarily formation of a polyubiquitin chain through lysine 48 of ubiquitin. Subsequently, the polyubiquitinylated HIF-1α needs to be translocated for the 26S proteasome for degradation. Under hypoxic conditions, oxygen becomes limiting for prolyl hydroxylase activity (Epstein et al., 2001) and ubiquitination of HIF-1α is inhibited (Sutter et al., 2000). As a result, HIF-1α accumulates, dimerizes with HIF-1β, and activates transcription of target genes, including VEGF.

VEGF expression is increased about 30-fold by hypoxia (Wise, 1956). However, it is still not clear whether hypoxia is the stimulus that leads to production of VEGF in diabetic retinopathy.

The mechanism for VEGF overexpression appears to be dependent on selective activation of various PKC isoforms, however, PKC-β seems to be especially important in VEGF signaling (Suzuma et al., 2002). In fact transcription of VEGF gene in response to hypoxia and/or hyperglycemia is dependent on activation of such specific kinases. The mechanisms through which hypoxia leads to increased production of VEGF and by consequence to neovascularization have been extensively studied over the last few years.

In further support of the hypothesis that increased production of VEGF is associated with diabetic retinopathy is the observation that increased levels of VEGF are present in the eyes of patients with diabetes even before the onset of detectable retinopathy.

Moreover, levels of VEGF correlate with new vessel formation in patients with diabetes (Aiello et al., 1994) and were further shown to be increased in the vitreous of eyes with neovascularization and diminish after panretinal photocoagulation (Aiello et al., 1994). In animal models local injection of VEGF causes neovascularization of the retina (Tolentino et al., 2002). Conversely strategies to block VEGF, such as those using antibodies that can bind to VEGF before it can activate its receptors or antisense oligonucleotides that inhibit VEGF mRNA, appear to prevent retinal neovascularization (Robinson et al., 1996; Tolentino et al., 2002).

It should be noted that VEGF regulates growth of new vessels not only in proliferative diabetic retinopathy but also during the normal physiological retinal vascular development. In both cases, formation of new vessels is derived from retinal ischemia; however, there is a critical difference in the direction of vessels growth. While in physiological vascularization, new vessels growth extend from the optic disc toward the peripheral avascular retina (Stone et al., 1996) (following the guidance of VEGF-expressing retinal astrocytes), in pathological neovascularization, the new vessels invade the vitreous cavity (Stone et al., 1995). This form of neovascularization leads to fibrovascular proliferation, resulting in vitreous hemorrhage and traction retinal detachment, which may seriously compromise vision. Parallel with the pathologic neovascularisation, the newly ischemic central areas of the retina are vascularized with normal-appearing blood vessels, a process called revascularization. The revascularization that accompanies diabetic retinopathy raises a problem in development of agent that target VEGF as a means of controlling pathological neovascularization. While preventing outgrowth of retinal vessels may be beneficial, the inhibition of revascularization is most likely harmful. A possible approach to limit the damage might be to block the specific isoforms of VEGF that contribute to pathological neovascularization (Ishida et al., 2003).

Despite its important role in angiogenesis, VEGF is probably not the only player that leads to neovascularization in diabetic retinopathy. Indeed, the switch from quiescent to active vessels involves not only an increase in inducers of neovascularization but also a decrease in concentration of negative regulators of angiogenesis such as pigment-derived factor (PEDF) (Dawson et al., 1999; Stellmach et al., 2001). PEDF is also responsible for the antiangiogenic activity of human vitreous and for excluding vessels from cornea (Dawson et al., 1999; Chader, 2001). PEDF is regulated by oxygen concentration and behaves in a manner opposite to VEGF, falling in concentration when oxygen is limited and rising when it is in good supply (Becerra, 1997; Dawson et al., 1999; Gao et al., 2001). It, thus, appears that the balance between the levels of VEGF and PEDF may be critical in preventing neovascularization. A number of agents that block VEGF or stimulate PEDF are currently being tested, including on clinical trials. For example, gene therapy may prove useful in increasing the levels of PEDF in the eye (Rasmussen et al., 2001; Frank, 2004).

12. PROTEASOME INHIBITORS AND THERAPEUTIC OPPORTUNITIES

The UPP is involved in such a variety of biological processes that the implication of this pathway in human diseases is not surprising. Consistently, inhibition of critical components of UPP is likely to have dramatic effects in cell and tissue function. For example, inhibition of ubiquitin activating enzyme, E1, is lethal. On the other hand deregulation of UPP was also demonstrated in many human diseases including cystic fibrosis, Liddle's syndrome, muscle wasting, neurodegenerative disorders such as CAG expansion diseases, Alzheimer's, and Parkinson's diseases. This would, in principle, suggest that the ability to manipulate the components of UPP could, in some cases, bring therapeutic benefits. Moreover, major human diseases such as inflammation and cancer, although not involving dysfunction of UPP, are often associated with deregulation of substrates of the UPP. This observation suggests that nonlethal inhibition of components of UPP, such as the proteasome, could provide an important therapeutic window for compounds such as proteasome inhibitors.

Over the last few years proteasome inhibitors assumed a particularly important role in biology and medicine given its potential use on the treatment of diseases such as cancer and inflammatory conditions, including ischemia reperfusion injury, rheumatoid arthritis, and asthma (Elliott et al., 2003).

Proteasomes belong to the class of threonine proteases and its proteolytic sites utilize N-terminal threonines as the active site nucleophils. Proteasome inhibitors belong to different chemical and pharmacological categories and act through different mechanisms. For example, MG132 (Z-Leu-Leu-Leucinal) belongs to a class of peptide aldehydes that were the first proteasome inhibitors to be developed (Vinitsky et al., 1992; Palombella et al., 1994; Rock et al., 1994). MG132 is probably the most widely used proteasome inhibitor in cell culture and in vitro studies. MG132 is a potent and selective proteasome inhibitor that is cell permeable (Elliott et al., 2003) and very useful in identifying physiological roles of the proteasome. However, MG132 also targets other proteases such as calpains and cathepsines and is unstable in aqueous solutions (Adams et al., 1998), which has limited its use in vivo. Boronate inhibitors of the proteasome are typically much more potent than their structurally analogous peptide aldehydes and include MG262 (analogous to MG132) and PS-341 (analogous to PS-402) (Elliott et al., 2003). Boronate inhibitors show high affinity and high specificity and are very selective inhibitors of the proteasome (Adams et al., 1998; Bogyo and Wang, 2002). In addition, these compounds are more stable in the circulation, which has encouraged their use in vivo (Elliott and Ross, 2001; Bogyo and Wang, 2002).

Bortezomib, formerly known as PS-341, is the clinically most advanced proteasome inhibitor, which reversibly and selectively inhibits the chymotrypsin active site of the proteasome β-subunits. In contrast to peptide aldehydes, which are removed from cells by the multidrug resistance carrier system, bortezomib is a poor substrate for this class of proteins (Adams, 2002).

A major and apparently successful use for proteasome inhibitors appears to be the treatment of several severe forms of cancer including multiple myeloma. Many forms of tumors result either from stabilization of oncoproteins that are under normal circumstances quickly removed from cells or by destabilization of tumor suppressors that need to be present in the cells (Goldberg and Rock, 2002; Adams and Kauffman, 2004). Many of such proteins are substrates for the UPP. For example, under normal physiological conditions, the UPP is responsible for degradation of N-myc, c-myc, C-Fos, c-Jun, Src, and viral protein E1A. Conversely, excessive degradation and/or destabilization of tumor suppressors such as p53 and p27 are also associated with increased malignancy (Ciechanover, 2006).

Proteasome inhibitors, such as bortezomib, can stabilize numerous cell cycle inhibitory proteins and cause cell cycle arrest and induce apoptosis, thus limiting tumor development (Hershko, 1997). In addition, tumor cells appear to be more sensitive to the proapoptotic effects of proteasome inhibitors than normal cells, which provide a therapeutic window for development of anticancer drugs.

The major biological effect of bortezomib appears to be inhibition of the activation of antiapoptotic transcription factor NFκB with subsequent inhibition of growth of tumor cells, induction of apoptosis, and inhibition of angiogenesis and cellular adhesion. Indeed, in cancer cells, there is often an increased expression of NFκB following chemotherapy or radiation. This renders tumor cells less sensitive to subsequent therapy. Therefore, inhibition of NFκB activity by proteasome inhibitors should make tumor cells more susceptible to chemotherapy and promote cancer cell apoptosis. Alternatively, proteasome inhibitors may inhibit degradation of abnormal proteins, which activates cell stress response and induces apoptosis.

Since neovascularization is a hallmark of proliferative diabetic retinopathy, it is possible to speculate that proteasome inhibitors may also prove useful in at least some forms of diabetic retinopathy. It should however be noted that there are currently no studies on the effect of proteasome inhibitors in the progression of diabetic retinopathy in humans. The outcomes of such studies are difficult to anticipate as inhibition of proteasome may have a variety of different, and possibly opposite, effects in characteristic features of diabetic retinopathy. For example, the proteasome is involved, among others processes, in the regulation of the transcription factors NFκB and HIF-1 in a quite different way. While proteasome inhibitor may prevent activation of NFκB by stabilizing its inhibitory protein IκB, an opposite effect is likely to be observed in forms of retinopathy associated with retinal ischemia. Indeed inhibition of the proteasome would prevent degradation of labile HIF-1α subunit, resulting in the activation of HIF-1 and in the transcription of target genes including VEGF.

It should, finally, be emphasized that the putative therapeutic effects of proteasome inhibitors in diabetic retinopathy, as in other diseases, are largely determined by the pharmacokinetics and bioavailability in target retinal cells as well as by the specific dose that is used and form of administration.

13. CONCLUSION

The evidences presented in this chapter clearly demonstrate that inflammation is an important component of the pathogenic process in diabetic retinopathy (Fig. 2). This inflammatory component occurs very early and persists during the course of the disease. Therefore, it can be assumed that diabetic retinopathy is associated with a situation of chronic and low level inflammation, suggesting that therapeutic strategies targeted to inflammatory molecules or components of the UPP (Fig. 3) may be useful to delay or prevent vision loss in patients with diabetic retinopathy.

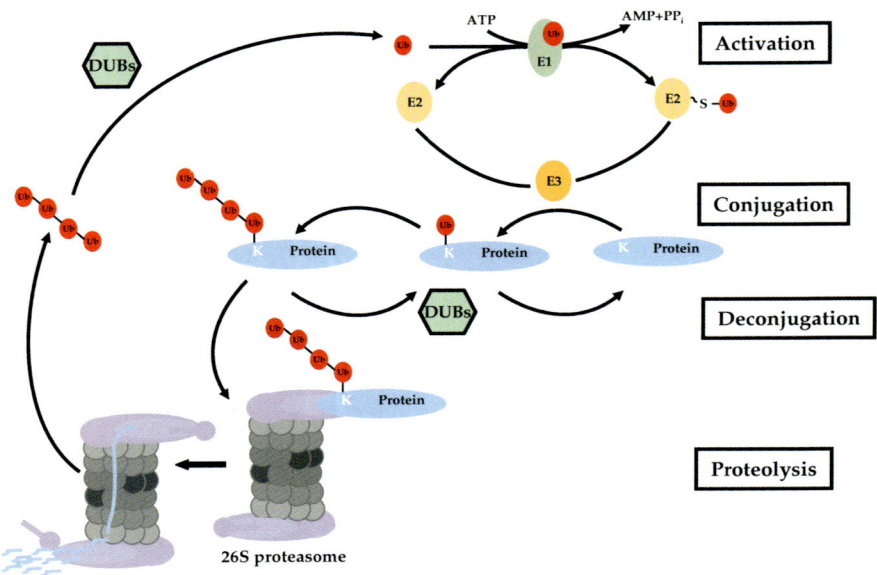

Fig. 3. The ubiquitin–proteasome pathway. (1) Activation of ubiquitin by the ubiquitin-activating enzyme, E1; (2) Transfer of activated ubiquitin from E1 to a member of the ubiquitin-conjugating enzymes, E2s; (3) Transfer of activated ubiquitin from E2 to a member of the ubiquitin–protein ligase family, E3; and (4) Conjugation of multiple ubiquitin molecules to the substrate. Ubiquitinated substrates are degraded by a large ~2,000-kDa protease called the 26S proteasome and the ubiquitin is recycled via the deubiquitinating enzymes

14. REFERENCES

Abiko, T., Abiko, A., Clermont, A.C., Shoelson, B., Horio, N., Takahashi, J., Adamis, A.P., King, G.L. and Bursell, S.E., 2003, Characterization of retinal leukostasis and hemodynamics in insulin resistance and diabetes: role of oxidants and protein kinase-C activation. *Diabetes* **52**: 829.

Abu El-Asrar, A.M., Desmet, S., Meersschaert, A., Dralands, L., Missotten, L. and Geboes, K., 2001, Expression of the inducible isoform of nitric oxide synthase in the retinas of human subjects with diabetes mellitus. *Am. J. Ophthalmol.* **132**: 551.

Abu El-Asrar, A.M., Meersschaert, A., Dralands, L., Missotten, L. and Geboes, K., 2004, Inducible nitric oxide synthase and vascular endothelial growth factor are colocalized in the retinas of human subjects with diabetes. *Eye* **18**: 306.

Adamis, A.P., 2002, Is diabetic retinopathy an inflammatory disease? *Br. J. Ophthalmol.* **86**: 363.

Adams, J., 2002, Development of the proteasome inhibitor PS-341. *Oncologist* **7**: 9.

Adams, J. and Kauffman, M., 2004, Development of the proteasome inhibitor Velcade (Bortezomib), *Cancer Invest.* **22**: 304.

Adams, J., Behnke, M., Chen, S., Cruickshank, A.A., Dick, L.R., Grenier, L., Klunder, J.M., Ma, Y.T., Plamondon, L. and Stein, R. L., 1998, Potent and selective inhibitors of the proteasome: dipeptidyl boronic acids. *Bioorg. Med. Chem. Lett.* **8**: 333.

Aiello, L.P., Avery, R.L., Arrigg, P.G., Keyt, B.A., Jampel, H.D., Shah, S.T., Pasquale, L.R., Thieme, H., Iwamoto, M.A., Park, J.E., et al., 1994, Vascular endothelial growth factor in ocular fluid of patients with diabetic retinopathy and other retinal disorders. *N. Engl. J. Med.* **331**: 1480.

Aiello, L.P., Bursell, S.E., Clermont, A., Duh, E., Ishii, H., Takagi, C., Mori, F., Ciulla, T.A., Ways, K., Jirousek, M., Smith, L.E. and King, G.L., 1997, Vascular endothelial growth factor-induced retinal permeability is mediated by protein kinase C in vivo and suppressed by an orally effective β-isoform-selective inhibitor. *Diabetes* **46**: 1473.

Antonetti, D.A., Barber, A.J., Khin, S., Lieth, E., Tarbell, J.M. and Gardner, T.W., 1998, Vascular permeability in experimental diabetes is associated with reduced endothelial occludin content: vascular endothelial growth factor decreases occludin in retinal endothelial cells. *Diabetes* **47**: 1953.

Antonetti, D.A., Barber, A.J., Hollinger, L.A., Wolpert, E.B. and Gardner, T.W., 1999, Vascular endothelial growth factor induces rapid phosphorylation of tight junction proteins occludin and zonula occluden. 1. A potential mechanism for vascular permeability in diabetic retinopathy and tumors. *J. Biol. Chem.* **274**: 23463.

Antonetti, D.A., Wolpert, E.B., DeMaio, L., Harhaj, N.S. and Scaduto, R.C., Jr., 2002, Hydrocortisone decreases retinal endothelial cell water and solute flux coincident with increased content and decreased phosphorylation of occludin. *J. Neurochem.* **80**: 667.

Ayalasomayajula, S.P. and Kompella, U.B., 2003, Celecoxib, a selective cyclooxygenase-2 inhibitor, inhibits retinal vascular endothelial growth factor expression and vascular leakage in a streptozotocin-induced diabetic rat model. *Eur. J. Pharmacol.* **458**: 283.

Bachmair, A., Finley, D. and Varshavsky, A., 1986, In vivo half-life of a protein is a function of its amino-terminal residue. *Science* **234**: 179.

Bai, N., Tang, S., Ma, J., Luo, Y. and Lin, S., 2003, Increased expression of intercellular adhesion molecule-1, vascular cellular adhesion molecule-1 and leukocyte common antigen in diabetic rat retina. *Yan Ke Xue Bao.* **19**: 176.

Baldwin, A.S., Jr., 1996, The NF-kappa B and I kappa B proteins: new discoveries and insights. *Annu. Rev. Immunol.* **14**: 649.

Bamforth, S.D., Lightman, S.L. and Greenwood, J., 1997, Interleukin-1 β-induced disruption of the retinal vascular barrier of the central nervous system is mediated through leukocyte recruitment and histamine. *Am. J. Pathol.* **150**: 329.

Barber, A.J., 2003, A new view of diabetic retinopathy: a neurodegenerative disease of the eye. *Prog. Neuropsychopharmacol. Biol. Psychiatry.* **27**: 283.

Barber, A.J. and Antonetti, D.A., 2003, Mapping the blood vessels with paracellular permeability in the retinas of diabetic rats. *Invest. Ophthalmol. Vis. Sci.* **44**: 5410.

Barber, A.J., Antonetti, D.A. and Gardner, T.W., 2000, Altered expression of retinal occludin and glial fibrillary acidic protein in experimental diabetes. *Invest. Ophthalmol. Vis. Sci.* **41**: 3561.

Barber, A.J., Antonetti, D.A., Kern, T.S., Reiter, C.E., Soans, R.S., Krady, J.K., Levison, S.W., Gardner, T.W. and Bronson, S.K., 2005, The Ins2Akita mouse as a model of early retinal complications in diabetes. *Invest. Ophthalmol. Vis. Sci.* **46**: 2210.

Barouch, F.C., Miyamoto, K., Allport, J.R., Fujita, K., Bursell, S.E., Aiello, L.P., Luscinskas, F.W. and Adamis, A. P., 2000, Integrin-mediated neutrophil adhesion and retinal leukostasis in diabetes. *Invest. Ophthalmol. Vis. Sci.* **41**: 1153.

Ben-Mahmud, B.M., Mann, G.E., Datti, A., Orlacchio, A., Kohner, E.M. and Chibber, R., 2004, Tumor necrosis factor-alpha in diabetic plasma increases the activity of core 2 GlcNAc-T and adherence of human leukocytes to retinal endothelial cells: significance of core 2 GlcNAc-T in diabetic retinopathy. *Diabetes* **53**: 2968.

Becerra, S.P., 1997, Structure-function studies on PEDF. A noninhibitory serpin with neurotrophic activity. *Adv. Exp. Med. Biol.* **425**: 223.

Bogyo, M. and Wang, E.W., 2002, Proteasome inhibitors: complex tools for a complex enzyme. *Curr. Top Microbiol. Immunol.* **268**: 185.

Boyd, S.D., Tsai, K.Y. and Jacks, T., 2000, An intact HDM2 RING-finger domain is required for nuclear exclusion of p53. *Nat. Cell Biol.* **2**: 563.

Brownlee, M., 2005, The pathobiology of diabetic complications: a unifying mechanism. *Diabetes* **54**: 1615.

Bullard, S.R., Hatchell, D.L., Cohen, H.J. and Rao, K.M., 1994, Increased adhesion of neutrophils to retinal vascular endothelial cells exposed to hyperosmolarity. *Exp. Eye Res.* **58**: 641.

Caldwell, R.B., Bartoli, M., Behzadian, M.A., El-Remessy, A.E., Al-Shabrawey, M., Platt, D.H. and Caldwell, R.W., 2003, Vascular endothelial growth factor and diabetic retinopathy: pathophysiological mechanisms and treatment perspectives. *Diabetes Metab. Res. Rev.* **19**: 442.

Carmo, A., Ramos, P., Reis, A., Proença, R. and Cunha-Vaz, J.G., 1998, Breakdown of the inner and outer blood retinal barrier in streptozotocin-induced diabetes. *Exp. Eye Res.* **67**: 569.

Carmo, A., Cunha-Vaz, J.G., Carvalho, A.P. and Lopes, M.C., 1999, L-arginine transport in retinas from streptozotocin diabetic rats: correlation with the level of IL-1β and NO synthase activity. *Vision Res.* **39**: 3817.

Carmo, A., Cunha-Vaz, J.G., Carvalho, A.P. and Lopes, M.C., 2000, Effect of cyclosporin-A on the blood–retinal barrier permeability in streptozotocin-induced diabetes. *Mediators Inflamm.* **9**: 243.

Carroll, W.J., Hollis, T.M. and Gardner, T.W., 1988, Retinal histamine synthesis is increased in experimental diabetes. *Invest. Ophthalmol. Vis. Sci.* **29**: 1201.

Carroll, J.E., Hess, D.C., Howard, E.F. and Hill, W.D., 2000, Is nuclear factor-kappaB a good treatment target in brain ischemia/reperfusion injury? *Neuroreport* **11**: R1.

Carter, R.S., Pennington, K.N., Arrate, P., Oltz, E.M. and Ballard, D.W., 2005, Site-specific monoubiquitination of IkappaB kinase IKKβ regulates its phosphorylation and persistent activation. *J. Biol. Chem.* **280**: 43272.

Chader, G.J., 2001, PEDF: raising both hopes and questions in controlling angiogenesis. *Proc. Natl. Acad. Sci. USA* **98**: 2122.

Chau, V., Tobias, J.W., Bachmair, A., Marriott, D., Ecker, D.J., Gonda, D.K. and Varshavsky, A., 1989, A multiubiquitin chain is confined to specific lysine in a targeted short-lived protein. *Science* **243**: 1576.

Chen, W., Jump, D.B., Grant, M.B., Esselman, W.J. and Busik, J.V. 2003, Dyslipidemia, but not hyperglycemia, induces inflammatory adhesion molecules in human retinal vascular endothelial cells. *Invest. Ophthalmol. Vis. Sci.* **44**: 5016.

Chibber, R., Ben-Mahmud, B.M., Mann, G.E., Zhang, J.J. and Kohner, E.M., 2003, Protein kinase C β2-dependent phosphorylation of core 2 GlcNAc-T promotes leukocyte-endothelial cell adhesion: a mechanism underlying capillary occlusion in diabetic retinopathy. *Diabetes* **52**: 1519.

Ciechanover, A., 2006, The ubiquitin proteolytic system: from a vague idea, through basic mechanisms and onto human diseases and drug targeting. *Neurology.* **66**: S7.

Cunha-Vaz, J.G., 1976, The blood–retinal barriers. *Doc. Ophthalmol.* **41**: 287.

Cunha-Vaz, J.G., 2000, Diabetic retinopathy: surrogate outcomes for drug development for diabetic retinopathy. *Ophthalmologica* **214**: 377.

Cunha-Vaz, J.G., 2001, Initial alterations in nonproliferative diabetic retinopathy. *Ophthalmologica* **215**: 7.

Cunha-Vaz, J., Faria de Abreu, J.R. and Campos, A.J., 1975, Early breakdown of the blood–retinal barrier in diabetes. *Br. J. Ophthalmol.* **59**: 649.

Curtis, T.M. and Scholfield, C.N., 2004, The role of lipids and protein kinase Cs in the pathogenesis of diabetic retinopathy. *Diabetes Metab. Res. Rev.* **20**: 28.

Dawson, D.W., Volpert, O.V., Gillis, P., Crawford, S.E., Xu, H., Benedict, W. and Bouck, N.P., 1999, Pigment epithelium-derived factor: a potent inhibitor of angiogenesis. *Science* **285**: 245.

Deshaies, R.J., 1999, SCF and Cullin/Ring H2-based ubiquitin ligases. *Annu. Rev. Cell Dev. Biol.* **15**: 435.

Deshaies, R.J. and Ferrel, J.E., Jr., 2001, Multisite phosphorylation and the countdown to S phase. *Cell* **107**: 819.

Du, Y., Smith, M.A., Miller, C.M. and Kern, T.S., 2002, Diabetes-induced nitrative stress in the retina, and correction by aminoguanidine. *J. Neurochem.* **80**: 771.

Edelman, J.L., Lutz, D. and Castro, M.R., 2005, Corticosteroids inhibit VEGF-induced vascular leakage in a rabbit model of blood–retinal and blood-aqueous barrier breakdown. *Exp. Eye Res.* **80**: 249.

Elliott, P.J. and Ross, J.S., 2001, The proteasome: a new target for novel drug therapies. *Am. J. Clin. Pathol.* **116**: 637.

Elliott, P.J., Zollner, T.M. and Boehncke, W.H., 2003, Proteasome inhibition: a new anti-inflammatory strategy. *J. Mol. Med.* **81**: 235.

El-Remessy, A.B, Behzadian, M.A., Abou-Mohamed, G., Franklin, T., Caldwell, R.W. and Caldwell, R.B., 2003, Experimental diabetes causes breakdown of the blood–retina barrier by a mechanism involving tyrosine nitration and increases in expression of vascular endothelial growth factor and urokinase plasminogen activator receptor. *Am. J. Pathol.* **162**: 1995.

Enea, N.A., Hollis, T.M., Kern, J.A. and Gardner, T.W., 1989, Histamine H1 receptors mediate increased blood–retinal barrier permeability in experimental diabetes. *Arch. Ophthalmol.* **107**: 270.

Epstein, A.C., Gleadle, J.M., McNeill, L.A., Hewitson, K.S., O'Rourke, J., Mole, D.R., Mukherji, M., Metzen, E., Wilson, M.I., Dhanda, A., et al., 2001, C. elegans EGL-9 and mammalian homologs define a family of dioxygenases that regulate HIF by prolyl hydroxylation. *Cell* **107**: 43.

Feldman, R.M., Correll, C.C., Kaplan, K.B. and Deshaies, R.J., 1997, A complex of Cdc4p, Skp1p, and Cdc53p/cullin catalyzes ubiquitination of the phosphorylated CDK inhibitor Sic1p. *Cell* **91**: 221.

Ferrara, N. and Gerber, H.P., 2001, The role of vascular endothelial growth factor in angiogenesis. *Acta Haematol.* **106**: 148.

Frank, R.N., 2004, Diabetic retinopathy. *N. Engl. J. Med.* **350**: 48.

Franks, W.A., Limb, G.A., Stanford, M.R., Ogilvie, J., Wolstencroft, R.A., Chignell, A.H. and Dumonde, D.C., 1992, Cytokines in human intraocular inflammation. *Curr. Eye Res.* **11**: 187.

Gao, G., Li, Y., Zhang, D., Gee, S., Crosson, C. and Ma, J., 2001, Unbalanced expression of VEGF and PEDF in ischemia-induced retinal neovascularization. *FEBS Lett.* **489**: 270.

Gardner, T.W., 1995, Histamine, ZO-1 and increased retinal–blood barrier permeability in diabetic retinopathy. *Trans. Am. Ophthalmol. Soc.* **93**: 583.

Gardner, T.W., Eller, A.W., Friberg, T.R., D'Antonio, J.A. and Hollis, T.M., 1995, Antihistamines reduce blood–retinal barrier permeability in type I (insulin-dependent) diabetic patients with nonproliferative retinopathy. A pilot study. *Retina* **15**: 134.

Gardner, T.W., Antonetti, D.A., Barber, A.J., LaNoue, K.F. and Levison, S.W., 2002, Diabetic retinopathy: more than meets the eye. *Surv. Ophthalmol.* **47**: 253.

Gerhardinger, C., Costa, M.B., Coulombe, M.C., Toth, I., Hoehn, T. and Grosu, P., 2005, Expression of acute-phase response proteins in retinal Muller cells in diabetes. *Invest. Ophthalmol. Vis. Sci.* **46**: 349.

Geyer, R.K., Yu, Z.K. and Maki, C.G., 2000, The MDM2 RING-finger domain is required to promote p53 nuclear export. *Nat .Cell Biol.* **2**: 569.

Glickman, M.H. and Ciechanover, A., 2002, The ubiquitin-proteasome proteolytic pathway: destruction for the sake of construction. *Physiol. Rev.* **82**: 373.

Glickman, M.H., Rubin, D.M., Fu, H., Larsen, C.N., Coux, O., Wefes, I., Pfeifer, G., Cjeka, Z., Vierstra, R., Baumeister, W., et al., 1999, Functional analysis of the proteasome regulatory particle. *Mol. Biol. Rep.* **26**: 21.

Glotzer, M., Murray, A.W. and Kirschner, M.W., 1991, Cyclin is degraded by the ubiquitin pathway. *Nature* **349**: 132.

Goldberg, A.L. and Rock, K., 2002, Not just research tools-proteasome inhibitors offer therapeutic promise. *Nat. Med.* **8**: 338.

Gorbea, C., Taillandier, D. and Rechsteiner, M.. 1999, Assembly of the regulatory complex of the 26S proteasome. *Mol. Biol. Rep.* **26**: 15.

Groll, M., Ditzel, L., Lowe, J., Stock, D., Bochtler, M., Bartunik, H.D. and Huber, R., 1997, Structure of 20S proteasome from yeast at 2.4 A resolution. *Nature* **386**: 463.

Hammes, H.P., Lin, J., Bretzel, R.G., Brownlee, M. and Breier, G., 1998, Upregulation of the vascular endothelial growth factor/vascular endothelial growth factor receptor system in experimental background diabetic retinopathy of the rat. *Diabetes* **47**: 401.

Harhaj, N.S., Barber, A.J. and Antonetti, D.A., 2002, Platelet-derived growth factor mediates tight junction redistribution and increases permeability in MDCK cells. *J. Cell Physiol.* **193**: 349.

Hayden, M.S. and Ghosh, S., 2004, Signaling to NF-kappaB. *Genes Dev.* **18**: 2195.

Heissmeyer, V., Krappmann, D., Hatada, E.N. and Scheidereit, C., 2001, Shared pathways of IkappaB kinase-induced SCF(βTrCP)-mediated ubiquitination and degradation for the NF-kappaB precursor p105 and IkappaBalpha. *Mol. Cell Biol.* **21**: 1024.

Hershko, A., 1997, Roles of ubiquitin-mediated proteolysis in cell cycle control. *Curr. Opin. Cell Biol.* **9**: 788.

Hollis, T.M., Sill, H.W., Butler, C., Campos, M.J. and Gardner, T.W., 1992, Astemizole reduces blood–retinal barrier leakage in experimental diabetes. *J. Diabetes Complications* **6**: 230.

Hughes, J.M., Brink, A., Witmer, A.N., Hanraads-de Riemer, M., Klaassen, I. and Schlingemann, R.O., 2004, Vascular leucocyte adhesion molecules unaltered in the human retina in diabetes. *Br. J. Ophthalmol.* **88**: 566.

Huibregtse, J.M., Scheffner, M., Beaudenon, S. and Howley, P.M., 1995, A family of proteins structurally and functionally related to the E6-AP ubiquitin-protein ligase. *Proc. Natl. Acad. Sci. USA* **92**: 2563.

Ishida, S., Usui, T., Yamashiro, K., Kaji, Y., Amano, S., Ogura, Y., Hida, T., Oguchi, Y., Ambati, J., Miller, J.W., et al., 2003, VEGF164-mediated inflammation is required for pathological, but not physiological, ischemia-induced retinal neovascularization. *J. Exp. Med.* **198**: 483.

Ivan, M., Kondo, K., Yang, H., Kim, W., Valiando, J., Ohh, M., Salic, A., Asara, J.M., Lane, W.S. and Kaelin, W.G., Jr., 2001, HIFalpha targeted for VHL-mediated destruction by proline hydroxylation: implications for O_2 sensing. *Science* **292**: 464.

Jaakkola, P., Mole, D.R., Tian, Y.M., Wilson, M.I., Gielbert, J., Gaskell, S.J., Kriegsheim, A., Hebestreit, H.F., Mukherji, M., Schofield, C.J., et al., 2001, Targeting of HIF-alpha to the von Hippel-Lindau ubiquitylation complex by O_2-regulated prolyl hydroxylation. *Science* **292**: 468.

Joussen, A.M., Huang, S., Poulaki, V., Camphausen, K., Beecken, W.D., Kirchhof, B. and Adamis, A.P., 2001a, In vivo retinal gene expression in early diabetes. *Invest. Ophthalmol. Vis. Sci.* **42**: 3047.

Joussen, A.M, Murata, T., Tsujikawa, A., Kirchhof, B., Bursell, S.E. and Adamis, A.P., 2001b, Leukocyte-mediated endothelial cell injury and death in the diabetic retina. *Am. J. Pathol.* **158**: 147.

Joussen, A.M., Poulaki, V., Mitsiades, N., Kirchhof, B., Koizumi, K., Dohmen, S. and Adamis, A.P., 2002a, Nonsteroidal anti-inflammatory drugs prevent early diabetic retinopathy via TNF-alpha suppression. *FASEB J.* **16**: 438.

Joussen, A.M., Poulaki, V., Qin, W., Kirchhof, B., Mitsiades, N., Wiegand, S.J., Rudge, J., Yancopoulos, G.D. and Adamis, A.P., 2002b, Retinal vascular endothelial growth factor induces intercellular adhesion molecule-1 and endothelial nitric oxide synthase expression and initiates early diabetic retinal leukocyte adhesion in vivo. *Am. J. Pathol.* **160**: 501.

Joussen, A.M., Poulaki, V., Mitsiades, N., Cai, W.Y., Suzuma, I., Pak, J., Ju, S.T., Rook, S.L., Esser, P., Mitsiades, C.S., Kirchhof, B., Adamis, A.P. and Aiello, L.P., 2003, Suppression of Fas-FasL-induced endothelial cell apoptosis prevents diabetic blood–retinal barrier breakdown in a model of streptozotocin-induced diabetes. *FASEB J.* **17**: 76.

Joussen, A.M., Poulaki, V., Le, M.L., Koizumi, K., Esser, C., Janicki, H., Schraermeyer, U., Kociok, N., Fauser, S., Kirchhof, B., Kern, T.S. and Adamis, A.P., 2004, A central role for inflammation in the pathogenesis of diabetic retinopathy. *FASEB J.* **18**: 1450.

Kamura, T., Sato, S., Iwai, K., Czyzyk-Krzeska, M., Conaway, R.C. and Conaway, J.W., 2000, Activation of HIF1alpha ubiquitination by a reconstituted von Hippel-Lindau (VHL) tumor suppressor complex. *Proc. Natl. Acad. Sci. USA* **97**: 10430.

Karin, M. and Ben-Neriah, Y., 2000, Phosphorylation meets ubiquitination: the control of NF-[kappa]B activity. *Annu. Rev. Immunol.* **18**: 621.

Kishino, T., Lalande, M. and Wagstaff, J., 1997, UBE3A/E6-AP mutations cause Angelman syndrome. *Nat. Genet.* **15**: 70.

Koepp, D.M., Harper, J.W. and Elledge, S.J., 1999, How the cyclin became a cyclin: regulated proteolysis in the cell cycle. *Cell* **97**: 431.

Kornitzer, D. and Ciechanover, A., 2000, Modes of regulation of ubiquitin-mediated protein degradation. *J. Cell Physiol.* **182**: 1.

Kowluru, R.A., 2001, Diabetes-induced elevations in retinal oxidative stress, protein kinase C and nitric oxide are interrelated. *Acta Diabetol.* **38**: 179.

Kowluru, R.A. and Odenbach, S., 2004a, Role of interleukin-1β in the development of retinopathy in rats: effect of antioxidants. *Invest. Ophthalmol. Vis. Sci.* **45**: 4161.

Kowluru, R.A. and Odenbach, S., 2004b, Role of interleukin-1β in the pathogenesis of diabetic retinopathy. *Br. J. Ophthalmol.* **88**: 1343.

Kowluru, R.A., Engerman, R.L., Case, G.L. and Kern, T.S., 2001, Retinal glutamate in diabetes and effect of antioxidants. *Neurochem. Int.* **38**: 385.

Kowluru, R.A., Koppolu, P., Chakrabarti, S. and Chen, S., 2003, Diabetes-induced activation of nuclear transcriptional factor in the retina, and its inhibition by antioxidants. *Free Radic. Res.* **37**: 1169.

Krady, J.K., Basu, A., Allen, C.M., Xu, Y., LaNoue, K.F., Gardner, T.W. and Levison, S.W., 2005, Minocycline reduces proinflammatory cytokine expression, microglial activation, and caspase-3 activation in a rodent model of diabetic retinopathy. *Diabetes* **54**: 1559.

Kwon, Y.T., Reiss, Y., Fried, V.A., Hershko, A., Yoon, J.K., Gonda, D.K., Sangan, P., Copeland, N.G., Jenkins, N.A. and Varshavsky, A., 1998, The mouse and human genes encoding the recognition component of the N-end rule pathway. *Proc. Natl. Acad. Sci. USA* **95**: 7898.

Leal, E.C., Manivannan, A., Aveleira, C., Serra, A., Castilho, A., Terasaki, T., Hosoya, K.-I., Cotter, M., Ambrosio, A. and Forrester, J.V., 2005a, Leukocyte adhesion and blood-retinal barrier (BRB) breakdown in diabetic retinopathy (DR): role of nitric oxide (NO). *IOVS* ARVO E-Abstract 423.

Leal, E.C., Manivannan, A., Cotter, M., Ambrosio, A.F. and Forrester, J.V., 2005b, Inducible nitric oxide synthase is involved in increased leukocyte adhesion to retinal vessels induced by diabetes. *Ophthalmic Res.* **37**.S1.05: 62.

Leal, E.C., Santiago, A.R. and Ambrosio, A.F., 2005c, Old and new drug targets in diabetic retinopathy: from biochemical changes to inflammation and neurodegeneration. *Curr. Drug Targets CNS Neurol. Disord.* **4**: 421.
Limb, G.A., Chignell, A.H., Green, W., LeRoy, F. and Dumonde, D.C., 1996, Distribution of TNF-alpha and its reactive vascular adhesion molecules in fibrovascular membranes of proliferative diabetic retinopathy. *Br. J. Ophthalmol.* **80**: 168.
Limb, G.A., Soomro, H., Janikoun, S., Hollifield, R.D. and Shilling, J., 1999a, Evidence for control of tumor necrosis factor-alpha (TNF-alpha) activity by TNF receptors in patients with proliferative diabetic retinopathy. *Clin. Exp. Immunol.* **115**: 409.
Limb, G.A., Webster, L., Soomro, H., Janikoun, S. and Shilling, J., 1999b, Platelet expression of tumor necrosis factor-alpha (TNF-alpha), TNF receptors and intercellular adhesion molecule-1 (ICAM-1) in patients with proliferative diabetic retinopathy. *Clin. Exp. Immunol.* **118**: 213.
Lorenzi, M. and Gerhardinger, C., 2001, Early cellular and molecular changes induced by diabetes in the retina. *Diabetologia.* **44**: 791.
Lorick, K.L., Jensen, J.P., Fang, S., Ong, A.M., Hatakeyama, S. and Weissman, A.M., 1999, RING fingers mediate ubiquitin-conjugating enzyme (E2)-dependent ubiquitination. *Proc. Natl. Acad. Sci. USA* **96**: 11364.
Lorick, K.L., Tsai, Y.-C., Yang, Y. and Weissman, A., 2005, In: *Protein Degradation,* vol. 1, R. J. Mayer, A. Ciechanover and M. Rechsteiner, (Eds.). Wiley-VHC, Weinheim pp. 44–101.
Lu, M., Perez, V.L., Ma, N., Miyamoto, K., Peng, H.B., Liao, J.K. and Adamis, A.P., 1999, VEGF increases retinal vascular ICAM-1 expression in vivo. *Invest. Ophthalmol. Vis. Sci.* **40**: 1808.
Luna, J.D., Chan, C.C., Derevjanik, N.L., Mahlow, J., Chiu, C., Peng, B., Tobe, T., Campochiaro, P.A. and Vinores, S.A., 1997, Blood–retinal barrier (BRB) breakdown in experimental autoimmune uveoretinitis: comparison with vascular endothelial growth factor, tumor necrosis factor alpha, and interleukin-1β-mediated breakdown. *J. Neurosci. Res.* **49**: 268.
Mamputu, J.C. and Renier, G., 2004, Advanced glycation end-products increase monocyte adhesion to retinal endothelial cells through vascular endothelial growth factor-induced ICAM-1 expression: inhibitory effect of antioxidants. *J. Leukoc. Biol.* **75**: 1062.
Maniatis, T., 1999, A ubiquitin ligase complex essential for the NF-kappaB, Wnt/Wingless, and Hedgehog signaling pathways. *Genes Dev.* **13**: 505.
Masson, N., Willam, C., Maxwell, P.H., Pugh, C.W. and Ratcliffe, P.J., 2001, Independent function of two destruction domains in hypoxia-inducible factor-alpha chains activated by prolyl hydroxylation. *EMBO J.* **20**: 5197.
Mathews, M.K., Merges, C., McLeod, D.S. and Lutty, G.A., 1997, Vascular endothelial growth factor and vascular permeability changes in human diabetic retinopathy. *Invest. Ophthalmol. Vis. Sci.* **38**: 2729.
Matsuda, S., Gomi, F., Oshima, Y., Tohyama, M. and Tano, Y., 2005, Vascular endothelial growth factor reduced and connective tissue growth factor induced by triamcinolone in ARPE19 cells under oxidative stress. *Invest. Ophthalmol. Vis. Sci.* **46**: 1062.
Maxwell, P.H., Wiesener, M.S., Chang, G.W., Clifford, S.C., Vaux, E.C., Cockman, M.E., Wykoff, C.C., Pugh, C.W., Maher, E.R. and Ratcliffe, P.J., 1999, The tumor suppressor protein VHL targets hypoxia-inducible factors for oxygen-dependent proteolysis. *Nature.* **399**: 271.
Meylan, E. and Tschopp, J., 2005, The RIP kinases: crucial integrators of cellular stress. *Trends Biochem. Sci.* **30**: 151.
McLeod, D.S., Lefer, D.J., Merges, C. and Lutty, G.A., 1995, Enhanced expression of intracellular adhesion molecule-1 and P-selectin in the diabetic human retina and choroids. *Am. J. Pathol.* **147**: 642.
Miyamoto, K., Hiroshiba, N., Tsujikawa, A. and Ogura, Y., 1998, In vivo demonstration of increased leukocyte entrapment in retinal microcirculation of diabetic rats. *Invest. Ophthalmol. Vis. Sci.* **39**: 2190.
Miyamoto, K., Khosrof, S., Bursell, S.-E., Rohan, R., Murata, T., Clermont, A., Aiello, L.P., Ogura, Y. and Adamis, A.P., 1999, Prevention of leukostasis and vascular leakage in streptozotocin-induced diabetic retinopathy via intercellular adhesion molecule-1 inhibition. *Proc. Natl. Acad. Sci. USA* **96**: 10836.
Mohr, S., 2004, Potential new strategies to prevent the development of diabetic retinopathy. *Expert. Opin. Investig. Drugs.* **13**: 189.
Mohr, S., Xi, X., Tang, J. and Kern, T.S., 2002, Caspase activation in retinas of diabetic and galactosemic mice and diabetic patients. *Diabetes.* **51**: 1172.
Moore, T.C, Moore, J.E., Kaji, Y., Frizzell, N., Usui, T., Poulaki, V., Campbell, I.L., Stitt, A.W., Gardiner, T.A., Archer, D.B. and Adamis, A.P., 2003, The role of advanced glycation end-products in retinal microvascular leukostasis. *Invest. Ophthalmol. Vis. Sci.* **44**: 4457.
Nauck, M., Roth, M., Tamm, M., Eickelberg, O., Wieland, H., Stulz, P. and Perruchoud, A.P., 1997, Induction of vascular endothelial growth factor by platelet-activating factor and platelet-derived growth factor is downregulated by corticosteroids. *Am. J. Respir. Cell Mol. Biol.* **16**: 398.

Nauck, M., Karakiulakis, G., Perruchoud, A.P., Papakonstantinou, E. and Roth, M., 1998, Corticosteroids inhibit the expression of the vascular endothelial growth factor gene in human vascular smooth muscle cells. *Eur. J. Pharmacol.* **341**: 309.

Nonaka, A., Kiryu, J., Tsujikawa, A., Yamashiro, K., Miyamoto, K., Nishiwaki, H., Honda, Y. and Ogura, Y., 2000, PKC-β inhibitor (LY333531) attenuates leukocyte entrapment in retinal microcirculation of diabetic rats. *Invest. Ophthalmol. Vis. Sci.* **41**: 2702.

Orian, A., Gonen, H., Bercovich, B., Fajerman, I., Eytan, E., Israel, A., Mercurio, F., Iwai, K., Schwartz, A.L. and Ciechanover, A., 2000, SCF(β)(-TrCP) ubiquitin ligase-mediated processing of NF-kappaB p105 requires phosphorylation of its C-terminus by IkappaB kinase. *EMBO J.* **19**: 2580.

Ozkaynak, E., Finley, D. and Varshavsky, A., 1984, The yeast ubiquitin gene: head-to-tail repeats encoding a polyubiquitin precursor protein. *Nature* **312**: 663.

Page, A.M. and Hieter, P., 1999, The anaphase-promoting complex: new subunits and regulators. *Annu. Rev. Biochem.* **68**: 583.

Palombella, V.J., Rando, O.J., Goldberg, A.L. and Maniatis, T., 1994, The ubiquitin-proteasome pathway is required for processing the NF-kappa B1 precursor protein and the activation of NF-kappa B. *Cell* **78**: 773.

Park, J.W., Park, S.J., Park, S.H., Kim, K.Y., Chung, J.W., Chun, M.H. and Oh, S.J., 2005, Upregulated expression of neuronal nitric oxide synthase in experimental diabetic retina. *Neurobiol. Dis.* (Epub ahead of print).

Phillips, J.B., Williams, A.J., Adams, J., Elliott, P.J. and Tortella, F.C., 2000, Proteasome inhibitor PS519 reduces infarction and attenuates leukocyte infiltration in a rat model of focal cerebral ischemia. *Stroke* **31**: 1686.

Pickart, C.M., 2001, Mechanisms underlying ubiquitination. *Annu. Rev. Biochem.* **70**: 503.

Pickart, C.M., 2004, Back to the future with ubiquitin, *Cell* **116**: 181.

Pickart, C.M. and VanDemark, A.P., 2000, Opening doors into the proteasome. *Nat. Struct. Biol.* **7**: 999.

Qaum, T., Xu, Q., Joussen, A.M., Clemens, M.W., Qin, W., Miyamoto, K., Hassessian, H., Wiegand, S.J., Rudge, J., Yancopoulos, G.D. and Adamis, A.P., 2001, VEGF-initiated blood–retinal barrier breakdown in early diabetes. *Invest. Ophthalmol. Vis. Sci.* **42**: 2408.

Rasmussen, H., Chu, K.W., Campochiaro, P., Gehlbach, P.L., Haller, J.A., Handa, J.T., Nguyen, Q.D. and Sung, J.U., 2001, Clinical protocol. An open-label, phase I, single administration, dose-escalation study of ADGVPEDF.11D (ADPEDF) in neovascular age-related macular degeneration (AMD). *Hum. Gene Ther.* **12**: 2029.

Reiss, Y. and Hershko, A., 1990, Affinity purification of ubiquitin-protein ligase on immobilized protein substrates. Evidence for the existence of separate NH2-terminal binding sites on a single enzyme. *J. Biol. Chem.* **265**: 3685.

Reiss, Y., Kaim, D. and Hershko, A., 1988, Specificity of binding of NH2-terminal residue of proteins to ubiquitin-protein ligase. Use of amino acid derivatives to characterize specific binding sites. *J. Biol. Chem.* **263**: 2693.

Robinson, G.S., Pierce, E.A., Rook, S.L., Foley, E., Webb, R. and Smith, L.E., 1996, Oligodeoxynucleotides inhibit retinal neovascularization in a murine model of proliferative retinopathy. *Proc. Natl. Acad. Sci. USA* **93**: 4851.

Rock, K.L., Gramm, C., Rothstein, L., Clark, K., Stein, R., Dick, L., Hwang, D. and Goldberg, A.L., 1994, Inhibitors of the proteasome block the degradation of most cell proteins and the generation of peptides presented on MHC class I molecules. *Cell* **78**: 761.

Romeo, G., Liu, W.H., Asnaghi, V., Kern, T.S. and Lorenzi, M., 2002, Activation of nuclear factor-kappaB induced by diabetes and high glucose regulates a proapoptotic program in retinal pericytes. *Diabetes* **51**: 2241.

Rungger-Brandle, E., Dosso, A.A. and Leuenberger, P.M., 2000, Glial reactivity, an early feature of diabetic retinopathy. *Invest. Ophthalmol. Vis. Sci.* **41**: 1971.

Saishin, Y., Saishin, Y., Takahashi, K., Lima e Silva, R., Hylton, D., Rudge, J.S., Wiegand, S.J. and Campochiaro, P.A., 2003a, VEGF-TRAP(R1R2) suppresses choroidal neovascularization and VEGF-induced breakdown of the blood–retinal barrier. *J. Cell Physiol.* **195**: 241.

Saishin, Y., Saishin, Y., Takahashi, K., Melia, M., Vinores, S.A. and Campochiaro, P.A., 2003b, Inhibition of protein kinase C decreases prostaglandin-induced breakdown of the blood–retinal barrier. *J. Cell Physiol.* **195**: 210.

Scheffner, M., Huibregtse, J.M., Vierstra, R.D. and Howley, P.M., 1993, The HPV-16 E6 and E6-AP complex functions as a ubiquitin-protein ligase in the ubiquitination of p53. *Cell* **75**: 495.

Scheffner, M., Nuber, U. and Huibregtse, J.M., 1995, Protein ubiquitination involving an E1-E2-E3 enzyme ubiquitin thioester cascade. *Nature* **373**: 81.

Schmidt, A.M. and Stern, D.M., 2000, RAGE: a new target for the prevention and treatment of the vascular and inflammatory complications of diabetes. *Trends Endocrinol. Metab.* **11**: 368.

Schmitz, M.L., Bacher, S. and Kracht, M., 2001, I kappa B-independent control of NF-kappa B activity by modulatory phosphorylations. *Trends Biochem. Sci.* **26**: 186.

Schroder, S., Palinski, W. and Schmid-Schonbein, G.W., 1991, Activated monocytes and granulocytes, capillary nonperfusion, and neovascularization in diabetic retinopathy. *Am. J. Pathol.* **139**: 81.

Shima, D.T., Adamis, A.P., Ferrara, N., Yeo, K.T., Yeo, T.K., Allende, R., Folkman, J. and D'Amore, P.A., 1995, Hypoxic induction of endothelial cell growth factors in retinal cells: identification and characterization of vascular endothelial growth factor (VEGF) as the mitogen. *Mol. Med.* 1: 182.

Shimura, H., Hattori, N., Kubo, S., Mizuno, Y., Asakawa, S., Minoshima, S., Shimizu, N., Iwai, K., Chiba, T., Tanaka, K., et al., 2000, Familial Parkinson disease gene product, parkin, is a ubiquitin-protein ligase. *Nat. Genet.* 25: 302.

Skowyra, D., Craig, K.L., Tyers, M., Elledge, S.J. and Harper, J.W., 1997, F-box proteins are receptors that recruit phosphorylated substrates to the SCF ubiquitin-ligase complex. *Cell* 91: 209.

Spranger, J., Meyer-Schwickerath, R., Klein, M., Schatz, H. and Pfeiffer, A., 1995, TNF-alpha level in the vitreous body. Increase in neovascular eye diseases and proliferative diabetic retinopathy. *Med. Klin.* (*Munich*) 90: 134.

Stellmach, V., Crawford, S.E., Zhou, W. and Bouck, N., 2001, Prevention of ischemia-induced retinopathy by the natural ocular antiangiogenic agent pigment epithelium-derived factor, *Proc. Natl. Acad. Sci. USA* 98: 2593.

Stitt, A.W., 2003, The role of advanced glycation in the pathogenesis of diabetic retinopathy. *Exp. Mol. Pathol.* 75: 95.

Stone, J., Itin, A., Alon, T., Pe'er, J., Gnessin, H., Chan-Ling, T. and Keshet, E., 1995, Development of retinal vasculature is mediated by hypoxia-induced vascular endothelial growth factor (VEGF) expression by neuroglia. *J. Neurosci.* 15: 4738.

Stone, J., Chan-Ling, T., Pe'er, J., Itin A., Gnessin, H. and Keshet, E., 1996, Roles of vascular endothelial growth factor and astrocyte degeneration in the genesis of retinopahty of prematurity. *Invest. Ophtalmol. Vis. Sci.* 37: 290.

Sugawara, R., Hikichi, T., Kitaya, N., Mori, F., Nagaoka, T., Yoshida, A. and Szabo, C., 2004, Peroxynitrite decomposition catalyst, FP15, and poly(ADP-ribose) polymerase inhibitor, PJ34, inhibit leukocyte entrapment in the retinal microcirculation of diabetic rats. *Curr. Eye Res.* 29: 11.

Sun, L. and Chen, Z.J., 2004, The novel functions of ubiquitination in signaling. *Curr. Opin. Cell Biol.* 16: 119.

Sutter, C.H., Laughner, E. and Semenza, G.L., 2000, Hypoxia-inducible factor 1-alpha protein expression is controlled by oxygen-regulated ubiquitination that is disrupted by deletions and missense mutations. *Proc. Natl. Acad. Sci. USA* 97: 4748.

Suzuma, K., Takahara, N., Suzuma, I., Isshiki, K., Ueki, K., Leitges, M., Aiello, L.P. and King, G.L., 2002, Characterization of protein kinase C β isoform's action on retinoblastoma protein phosphorylation, vascular endothelial growth factor-induced endothelial cell proliferation, and retinal neovascularization. *Proc. Natl. Acad. Sci. USA* 99: 721.

Takeda, M., Mori, F., Yoshida, A., Takamiya, A., Nakagomi, S., Sato, E. and Kiyama, H., 2001, Constitutive nitric oxide synthase is associated with retinal vascular permeability in early diabetic rats. *Diabetologia.* 44: 1043.

Tamura, H., Miyamoto, K., Kiryu, J., Miyahara, S., Katsuta, H., Hirose, F., Musashi, K. and Yoshimura, N., 2005, Intravitreal injection of corticosteroid attenuates leukostasis and vascular leakage in experimental diabetic retina. *Invest. Ophthalmol. Vis. Sci.* 46: 1440.

Tolentino, M.J., Miller, J.W., Gragoudas, E.S., Jakobiec, F.A., Flynn, E., Chatzistefanou, K., Ferrara, N. and Adamis, A.P., 1996, Intravitreous injections of vascular endothelial growth factor produce retinal ischemia and microangiopathy in an adult primate. *Ophthalmology* 103: 1820.

Tolentino, M.J., McLeod, D.S., Taomoto, M., Otsuji, T., Adamis, A.P. and Lutty, G.A., 2002, Pathologic features of vascular endothelial growth factor-induced retinopathy in the nonhuman primate. *Am. J. Ophthalmol.* 133: 373.

Varshavsky, A., 1997, The N-end rule pathway of protein degradation. *Genes Cells* 2: 13.

Varshavsky, A., Turner, G., Du, F. and Xie, Y., 2000, Felix Hoppe-Seyler Lecture 2000. The ubiquitin system and the N-end rule pathway. *Biol. Chem.* 381: 779.

Vinitsky, A., Michaud, C., Powers, J.C. and Orlowski, M., 1992, Inhibition of the chymotrypsin-like activity of the pituitary multicatalytic proteinase complex. *Biochemistry* 31: 9421.

Vinores, S.A., Van Niel, E., Swerdloff, J.L. and Campochiaro, P.A., 1993, Electron microscopic immunocytochemical demonstration of blood–retinal barrier breakdown in human diabetics and its association with aldose reductase in retinal vascular endothelium and retinal pigment epithelium. *Histochem. J.* 25: 648.

Welchman, R.L., Gordon, C. and Mayer, R.J., 2005, Ubiquitin and ubiquitin-like proteins as multifunctional signals. *Nat. Rev. Mol. Cell Biol.* 6: 599.

Wertz, I.E., O'Rourke, K.M., Zhou, H., Eby, M., Aravind, L., Seshagiri, S., Wu, P., Wiesmann, C., Baker, R., Boone, D.L., et al., 2004, Deubiquitination and ubiquitin ligase domains of A20 downregulate NF-kappaB signaling. *Nature* 430: 694.

Wilkinson, K.D., 2003, Signal transduction: aspirin, ubiquitin and cancer. *Nature* 424: 738.

Wilkinson-Berka, J.L., 2004, Vasoactive factors and diabetic retinopathy: vascular endothelial growth factor, cyclooxygenase-2 and nitric oxide. *Curr. Pharm. Des.* 10: 3331.

Williams, B., Gallacher, B., Patel, H. and Orme, C., 1997, Glucose-induced protein kinase C activation regulates vascular permeability factor mRNA expression and peptide production by human vascular smooth muscle cells in vitro. *Diabetes* 46: 1497.

Wise, G.N., 1956, Retinal neovascularization. *Trans. Am. Acad. Opththalmol. Soc.* 54: 729.

Xu, X., Zhu, Q., Xia, X., Zhang, S., Gu, Q. and Luo, D., 2004, Blood–retinal barrier breakdown induced by activation of protein kinase C via vascular endothelial growth factor in streptozotocin-induced diabetic rats. *Curr. Eye Res.* **28**: 251.

Zeng, X.X., Ng, Y.K. and Ling, E.A., 2000, Neuronal and microglial response in the retina of streptozotocin-induced diabetic rats. *Vis. Neurosci.* **17**: 463.

Zheng, N., Wang, P., Jeffrey, P.D. and Pavletich, N.P., 2000, Structure of a c-Cbl-UbcH7 complex: RING domain function in ubiquitin-protein ligases. *Cell* **102**: 533.

23

REJUVENATING NEURONS AND GLIA WITH MICROBIAL ENZYMES

Aubrey D.N.J. de Grey

1. ABSTRACT

All the major neurodegenerative diseases are characterised by the accumulation of proteinaceous aggregates within neurons. The commonest of such conditions, Alzheimer's disease, also features extracellular proteinaceous aggregates (amyloid). The role of these aggregates in the aetiology and progression of cognitive impairment is still unclear, but their absence in young adults suggests that their removal would at any rate not be harmful. However, no method for removing the intracellular aggregates in question has yet been developed. Moreover, the engulfment of amyloid by microglia as a result of immunisation may result in loss of microglial function if the amyloid then resists lysosomal digestion. A novel approach to eliminating intracellular aggregates in the brain (and elsewhere) was proposed by de Grey (2002) and has recently attracted enthusiastic support from all relevant specialities, which are unusually disparate. This approach is to isolate bacterial or fungal strains with the capacity to metabolise the recalcitrant aggregates, following which the genes encoding the enzymes responsible would be identified and modified for expression in mammalian cells and targeting to the appropriate subcellular compartment. Delivery could be either of the genes themselves, using somatic gene therapy, or of the enzymes they encode, which would be injected either directly into the brain or into the circulation in conjunction with agents to deliver them across the blood–brain barrier. Ambitious though this approach undoubtedly is, its potential for both the prevention and the treatment of the entire range of neurodegenerative diseases so far exceeds any alternative presently being explored that the case for pursuing it is strong.

2. INTRODUCTION

Arguably the most severe side-effect of modern medicine's success in enabling people to live to old age in unprecedented numbers is the prevalence of dementia in the elderly population. The most common age-related neurodegenerative condition, Alzheimer's disease (AD), affects huge numbers of individuals worldwide, placing immense strains on sufferers,

---

Aubrey D.N.J. de Grey – ag24@gen.cam.ac.uk
Department of Genetics, University of Cambridge, Downing Street, Cambridge CB2 3EH, UK

their loved ones and the medical systems of all industrialised nations. Although certain risk factors for AD have been identified, treatments are still elusive.

One of the most promising approaches for treating AD is to clear the extracellular aggregates with which it is associated, the senile plaques, by vaccination that stimulates microglia to internalise and degrade plaque material. This technique was spectacularly successful in a mouse model of AD (Schenk et al., 1999) and moved rapidly to clinical trials (Schenk, 2002). Though the first trials had to be terminated prematurely as a result of serious inflammatory side-effects in a small proportion of patients (Schenk, 2004), hope for this strategy remains widespread and a second round of Phase 1 trials is already underway using an improved immunisation protocol that is not expected to encounter the same problems (Ma et al., 2006).

However, it is quite likely that clearance of senile plaques will form only part of an effective therapy for AD. First, AD exhibits considerable cell loss in various brain regions, whose reversal will probably require stem cell or growth factor therapy to replace the lost cells; a protocol involving the introduction of NGF-secreting fibroblasts into the affected regions is at the clinical trial stage (Tuszynski et al., 2005). Second, AD sufferers accumulate a second type of proteinaceous aggregate, neurofibrillary tangles, in pyramidal neurons; being intracellular already, these are unlikely to be cleared by immunisation.

The latter problem, intracellular aggregates, is the topic of this chapter. Its importance extends beyond the specific case of neurofibrillary tangles for two reasons. First, the fate of senile plaque material following engulfment by microglia is uncertain; it may be degraded readily, but some reports indicate that degradation may be slow or partial (Paresce et al., 1997). Second, intraneuronal aggregates composed mainly of proteins distinct from the major constituent of tangles (hyperphosphorylated tau) are found in, and may contribute to the progression of, all the other major neurodegenerative diseases, including Parkinson's, Huntington's, progressive supranuclear palsy and Lewy body dementia. A means of clearing such aggregates thus has great potential for alleviating a range of conditions that are increaseingly prevalent throughout the industrialised world. However, no strategy for achieving this has yet been developed, even in animal models.

An intriguing feature of these aggregates is that their subcellular location varies. This is in contrast to the accumulation of undigested material in non-neuronal cell types, such as sterols in arterial macrophages or photoreceptor-derived material in the retina, which tend to be lysosomal. It might be expected that unwanted material would generally be targeted to the lysosome if its degradation elsewhere in the cell was unsuccessful, since the lysosome is the site of the cell's most powerful catabolic machinery. A well-supported hypothesis for the apparently non-lysosomal localisation of most neuronal proteinaceous aggregates is that targeting to the lysosome is impaired at an early stage in the disease, leading to the requirement to sequester unwanted material elsewhere; how this may work in the context of AD is summarised in Fig. 1. It may well be that different strategies are employed for the sequestration of different proteins – for example, the polyglutamine repeat diseases seem particularly to be associated with nuclear inclusions (Ross et al., 1999) – but the general theme seems to be that recalcitrant proteins, which might be toxic to the cell, can be rendered less so by simple aggregation, which lowers their exposed surface area. This hypothesis is reinforced by the observation that aggregates in some neurodegenerative diseases are often located near the microtubule organising centres, suggesting that they have been formed by active microtubule transport (Hoffner et al., 2002).

A corollary of the above model is that lysosomal dysfunction may be the key to all intracellular aggregates, irrespective of the aggregate's location. Indeed, as noted above, a cause independent of lysosomal function would seem paradoxical in view of the wide range of mechanisms by which material – especially proteins – can be targeted to the lysosome.

Moreover, it is difficult to envisage any mechanism whereby lysosomal function could progressively deteriorate with age other than the accumulation within the vacuolar apparatus of material that resists degradation. This gives the cause for optimism that a therapy that effectively cleared lysosomes of their problematic contents might be a powerful intervention against all neurodegenerative diseases. However, this still leaves us with the distinctly challenging task of developing such a therapy.

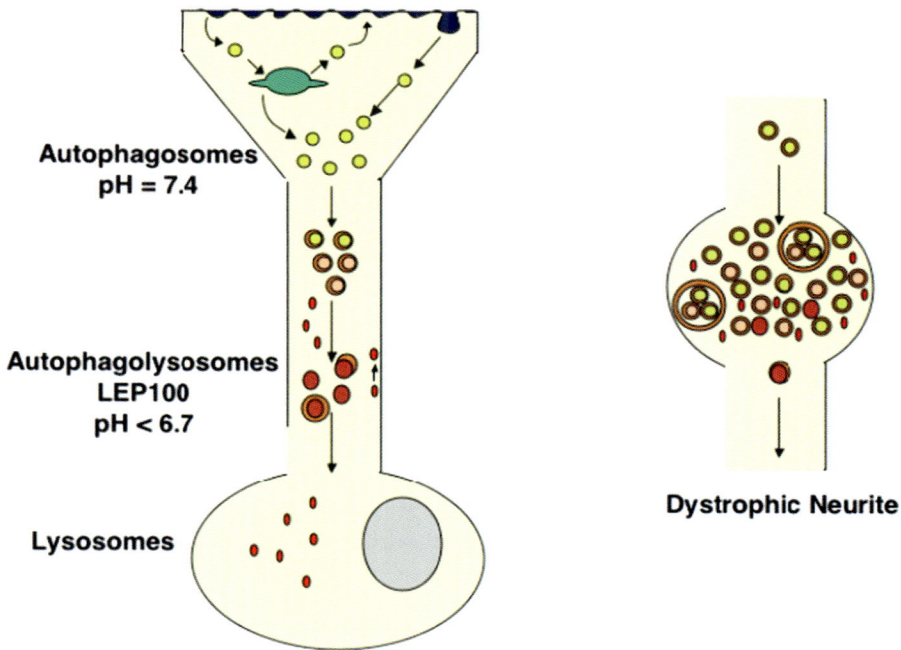

Fig. 1. Material accumulates in the neuronal cytoplasm as a result of failure to target it to the lysosome. Depicted here is the situation in Alzheimer's disease, in which fusion of autophagosomes with lysosomes is particularly affected

3. MICROBES – A BIOMEDICALLY NOVEL SOURCE OF CATABOLIC VERSATILITY

It was pointed out in 1952 that the removal of organic toxins from the environment might be facilitated by taking advantage of the selective pressures that such toxins impose on the microbial ecology in their vicinity (Gayle, 1952). Since most such toxins are energy-rich compounds, a bacterial or fungal strain that can degrade them should be able to survive in the contaminated environment in competition with other species that may grow more rapidly but only on more amenable materials. If so, the site might be remediable by simply isolating microbes from it, selecting for those that degrade the offending compound, expanding them in culture and returning them to the site in sufficient abundance that the toxin is eliminated. Since that time, this concept has been robustly confirmed and exploited, with the identification of bacteria that can metabolise such diverse compounds as TNT, dioxins and rubber (Wittich, 1998; Lewis et al., 2004; Rose and Steinbuchel, 2005). This field, termed bioremediation, has thus become a great commercial success.

A biomedical application for bioremediation technology hardly seems plausible, but just that was suggested by de Grey (2002). It is realised that environments enriched in human remains – graveyards, for instance – might, for the same reasons outlined above, harbour microbes capable of degrading the aggregates present in the human brain at death, and that the enzymes responsible for this catabolic function might be identifiable and adaptable for function within the mammalian cell, as illustrated in Fig. 2. This concept applies in principle to all recalcitrant compounds that accumulate in the body during life, whether in the brain or elsewhere; thus, it may also have applications to such conditions as atherosclerosis and macular degeneration, among others. It was thus appropriate that a group of eight distinguished experimentalists gathered in mid-2004 to discuss all aspects of this potentially ground-breaking approach to such a broad range of age-related diseases (de Grey et al., 2005). Enthusiasm for it has continued to grow, not least because there seem to be realistic approaches to overcoming all the obstacles that one can easily see it faces before it could become clinically relevant. Therefore, in the remainder of this paper some of those obstacles and their avoidance, with particular attention to the brain, will be discussed.

4. ISOLATION OF COMPETENT ENZYMES

The most established technique in bioremediation for identifying useful microbial strains is to select by starvation. In brief, bacteria and/or fungi isolated from the contaminated environment are cultured in the absence of any carbon source other than the target compound, so that those that are unable to degrade it are simply unable to grow and those that can degrade it thus proliferate (Warhurst and Fewson, 1994). Standard molecular biology techniques are then applicable to identify the relevant enzymes and the genes encoding them. This strategy necessarily requires one to know what compound one wants to degrade, and as has been explained above, in the case of neurodegenerative diseases this will typically be expected not to be the main constituent of the aggregates that define the disease, because those aggregates are not lysosomal in neurons (though they are perforce lysosomal in the case of senile plaques internalised by microglia).

At least in the case of AD, however, there is intriguing circumstantial evidence for the involvement of sterols. The most powerful of such evidence is that the genetic polymorphism most strongly correlated with susceptibility to AD is the apolipoprotein E gene (Roses, 1994), which occupies the same position of notoriety in regard to atherosclerosis. It is quite possible that sterol accumulation underlies dysfunction of the neuronal lysosome in other neurodegenerative diseases too, though a strong case for that hypothesis does not yet exist.

5. DELIVERY OF ENZYMES TO THE BRAIN

Even once enzymes with the function of interest are identified, they will be of little therapeutic value unless they can be targeted to the location of their desired target. The most established delivery technology for lysosomal storage diseases (congenital deficiencies for one of a number of lysosomal enzymes) is to manufacture recombinant enzyme in *E. coli* and inject it intravenously in quantities large enough that a therapeutic amount reaches the most severely affected cell type (Schiffmann and Brady, 2002). This approach is inadequate for diseases affecting the brain, however, because of the blood–brain barrier (BBB). Luckily, the importance of translocating therapeutic proteins across the BBB is such that considerable effort is currently underway to achieve this; a promising option is to exploit the system that naturally transports transferrin across the BBB (Pardridge, 2002).

Other options are also available, however. Gene therapy targeting the affected neurons would be effective (via intranasal delivery, for example, De Rosa et al., 2005) if the enzymes were engineered to be targeted to the lysosome after synthesis. This modification may not be trivial (though we must note that it will probably not even be necessary if the enzyme is initially identified in fungi rather than bacteria, since fungi have a vacuole that resembles the mammalian lysosome in both structure and function). A more straightforward approach may be cell therapy – to inject cells (typically fibroblasts) into the brain after engineering them to secrete the enzyme, which would then be taken up by the cells that need it. This approach has shown immense promise in stimulating hippocampal neurogenesis with NGF, a therapy that is currently in clinical trials (Tuszynski et al., 2005). The enzyme would still need to be modified for secretion, but that is likely to be simpler than lysosomal targeting since it involves N-terminal amino acid sequences, whereas lysosomal targeting involves either glycosylation (Ghosh et al., 2003) or internal sequences (Cuervo, 2004).

6. TOXICITY AND EFFICACY IN MAMMALIAN CELLS

The environment of an enzyme, especially the pH, greatly affects its function. This may be another reason to prioritise fungi as the source of promising enzymes, since their vacuole has a pH similar to the mammalian lysosome. However, the pH optimum of enzymes has been efficiently modified by in vitro evolution (Francis and Hansche, 1972) and this is another potential obstacle to which multiple solutions are plausible.

Toxicity within the mammalian cell is clearly a potential concern. However, it may be less severe than it seems at first sight, simply because enzymes with unusual substrates do not necessarily have an unusually broad range of substrates. Additionally, it may be relatively straightforward to emulate a trick that various mammalian lysosomal enzymes feature, which is to be synthesised as inactive proenzymes that are activated by cleavage of an N-terminal sequence only after arrival within the lysosome (Wittlin et al., 1999). Toxicity within the lysosome itself is unlikely, because the targets of these enzymes will be proteins, and all lysosomal proteins (including the catabolic enzymes themselves) are broken down rather rapidly as it is.

7. IMMUNE RESPONSE

Since these enzymes will be foreign, there is a concern that they will evoke an immune response leading to the elimination of the cells that express them – not the intention at all, to say the least. However, indications are that this should not be a serious problem. First, delivery by injection of recombinant enzyme would mean that the enzyme was only ever present in the vacuolar apparatus, not in the cytosol, and so should not be processed by the immunoproteasome and presented on MHC I complexes. This is thought to be the main reason why enzyme therapy for lysosomal storage diseases typically causes only a mild and temporary immune response, if any at all (Brooks et al., 2003). Second, delivery by cell or gene therapy faces this difficulty only to the same extent as does gene therapy for congenital deficiencies of vital proteins, since those proteins are just as foreign to the patient as a bacterial enzyme would be to all of us. Hence, one can be confident that this therapy can exploit the encouraging progress currently being made in tolerising the immune system to new proteins.

8. MULTIPLICITY OF REQUIRED ENZYMES

The material that this proposal seeks to eliminate consists, in all cases, of complex molecules – whose complete degradation is virtually certain to require several steps and thus several enzymes. Is it realistic to suppose that this capacity can be developed for clinical use?

Thankfully, this is a non-problem, because the precept of complete degradation is incorrect. The purpose that has driven the microbes to evolve this machinery is to extract energy, which would probably be achieved only by complete degradation to carbon dioxide or comparably simple molecules, whereas the biomedical requirement is only to transform the target compound to something that mammalian cells can already metabolise. This much less stringent requirement can be satisfied by providing only the first steps – with luck, only the first step – of the degradation pathway.

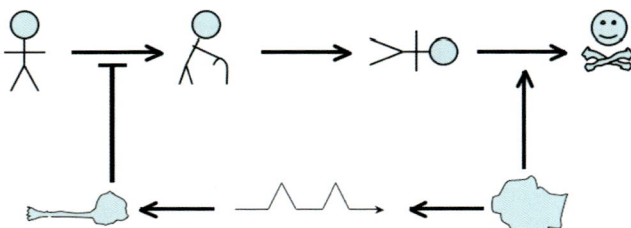

Fig. 2. The process of aging partly comprises the accumulation of initially inert but eventually pathogenic material, consisting mainly of energy-rich but degradation-resistant compounds. This accumulation, together with other aspects of aging, turns young people into old people and eventually into dead people. Thereafter, processes encoded in microorganisms turn them into decomposed people *devoid* of energy-rich compounds. These processes can thus, in principle, be incorporated into the mammalian catabolic arsenal, thereby retarding or even reversing the original accumulation

9. CONCLUSION

Eliminating the material that accumulates within neurons and in the extracellular space during the progression of neurodegenerative diseases is an immense challenge, but it is one that, on present evidence, may be an essential ingredient of any truly effective therapy. No strategy has previously been proposed for achieving this, and the absence of such a strategy has arguably limited ambitions concerning the cure of such diseases and focused attention unduly exclusively on prophylactic therapies, especially the early diagnosis of risk factors. While radical and ambitious, the approach described here may be a truly feasible way to exploit the ingenuity of evolution to solve this particularly intractable problem.

10. REFERENCES

Brooks, D.A., Kakavanos, R. and Hopwood, J.J., 2003, Significance of immune response to enzyme-replacement therapy for patients with a lysosomal storage disorder. *Trends Mol. Med.* **9**: 450.

Cuervo, A.M., 2004, Autophagy: in sickness and in health. *Trends Cell Biol.* **14**: 70.

de Grey, A.D.N.J., 2002, Bioremediation meets biomedicine: therapeutic translation of microbial catabolism to the lysosome. *Trends Biotechnol.* **20**: 452.

de Grey, A.D.N.J., Alvarez, P.J.J., Brady, R.O., Cuervo, A.M., Jerome, W.G., McCarty, P.L., Nixon, R.A., Rittmann, B.E. and Sparrow, J.R., 2005, Medical bioremediation: prospects for the application of microbial catabolic diversity to aging and several major age-related diseases. *Ageing Res. Rev.* **4**: 315.

De Rosa, R., Garcia, A.A., Braschi, C., Capsoni, S., Maffei, L., Berardi, N. and Cattaneo, A., 2005, Intranasal administration of nerve growth factor (NGF) rescues recognition memory deficits in AD11 anti-NGF transgenic mice. *Proc. Natl Acad. Sci. USA* **102**: 3811.

Francis, J.C. and Hansche, P.E., 1972, Directed evolution of metabolic pathways in microbial populations. I. Modification of the acid phosphatase pH optimum in *S. cerevisiae*. *Genetics* **70**: 59.

Gayle, E.F., 1952, *The Chemical Activities of Bacteria*. Academic Press, London.

Ghosh, P., Dahms, N.M. and Kornfeld, S., 2003, Mannose 6-phosphate receptors: new twists in the tale. *Nat. Rev. Mol. Cell Biol.* **4**: 202.

Hoffner, G., Kahlem, P. and Djian, P., 2002, Perinuclear localization of Huntington as a consequence of its binding to microtubules through an interaction with beta-tubulin: relevance to Huntington's disease. *J. Cell Sci.* **115**: 941.

Lewis, T.A., Newcombe, D.A. and Crawford, R.L., 2004, Bioremediation of soils contaminated with explosives. *J. Environ. Manage* **70**: 291.

Ma, Q.L., Lim, G.P., Harris-White, M.E., Yang, F., Ambegaokar, S.S., Ubeda, O.J., Glabe, C.G., Teter, B., Frautschy, S.A. and Cole, G.M., 2006, Antibodies against beta-amyloid reduce abeta oligomers, glycogen synthase kinase-3-beta activation and tau phosphorylation in vivo and in vitro. *J. Neurosci. Res.* **83**: 374.

Pardridge, W.M., 2002, Drug and gene delivery to the brain: the vascular route. *Neuron* **36**: 555.

Paresce, D.M., Chung, H. and Maxfield, F.R., 1997, Slow degradation of aggregates of the Alzheimer's disease amyloid beta-protein by microglial cells. *J. Biol. Chem.* **272**: 29390.

Rose, K. and Steinbuchel, A., 2005, Biodegradation of natural rubber and related compounds: recent insights into a hardly understood catabolic capability of microorganisms. *Appl. Environ. Microbiol.* **71**: 2803.

Roses, A.D., 1994, Apolipoprotein E affects the rate of Alzheimer's disease expression: beta-amyloid burden is a secondary consequence dependent on APOE genotype and duration of disease. *J. Neuropathol. Exp. Neurol.* **53**: 429.

Ross, C.A., Wood, J.D., Schilling, G., Peters, M.F., Nucifora, F.C., Cooper, J.K., Sharp, A.H., Margolis, R.L. and Borchelt, D.R., 1999, Polyglutamine pathogenesis. *Philos. Trans. R. Soc. Lond. B Biol. Sci.* **354**: 1005.

Schenk, D., 2002, Amyloid-β immunotherapy for Alzheimer's disease: the end of the beginning. *Nat. Rev. Neurosci.* **3**: 824.

Schenk, D., 2004, Hopes remain for an Alzheimer's vaccine. *Nature* **431**: 398.

Schenk, D., Barbour, R., Dunn, W., Gordon, G., Grajeda, H., Guido, T., Hu, K., Huang, J., Johnson-Wood, K., Khan, K., Kholodenko, D., Lee, M., Liao, Z., Lieberburg, I., Motter, R., Mutter, L., Soriano, F., Shopp, G., Vasquez, N., Vandevert, C., Walker, S., Wogulis, M., Yednock, T., Games, D. and Seubert, P., 1999, Immunization with amyloid-β attenuates Alzheimer-disease-like pathology in the PDAPP mouse. *Nature* **400**: 173.

Schiffmann, R. and Brady, R.O., 2002, New prospects for the treatment of lysosomal storage diseases. *Drugs* **62**: 733.

Tuszynski, M.H., Thal, L., Pay, M., Salmon, D.P., Sang, H., Bakay, R., Patel, P., Blesch, A., Vahlsing, H.L., Ho, G., Tong, G., Potkin, S.G., Fallon, J., Hansen, L., Mufson, E.J., Kordower, J.H., Gall, C. and Conner, J., 2005, A phase 1 clinical trial of nerve growth factor gene therapy for Alzheimer's disease. *Nat. Med.* **11**: 551.

Warhurst, A.M. and Fewson, C.A., 1994, Biotransformations catalyzed by the genus *Rhodococcus*. *Crit. Rev. Biotechnol.* **14**: 29.

Wittich, R.M., 1998, Degradation of dioxin-like compounds by microorganisms. *Appl. Microbiol. Biotechnol.* **49**: 489.

Wittlin, S., Rosel, J., Hofmann, F. and Stover, D.R., 1999, Mechanisms and kinetics of procathepsin D activation, *Eur. J. Biochem.* **265**: 384.

Abbreviations

6-OHDA - 6-hydroxydopamine
AD - Alzheimer's disease
AIDS - Acquired immunodeficiency syndrome
ALS - Amyotrophic lateral sclerosis
ANT - Adenine nucleotide translocator
APCs - Antigen presenting cells
APOE - Apolipoprotein E
APP - Amyloid precursor protein
ATP - Adenosine tri-phosphate
Bad - Bcl-2-associated death protein
BBB - Blood-brain barrier
BNDF - Brain-derived neurotrophic factor
cAMP - Cyclic adenosine monophosphate
CNTF - Ciliary neurotrophic factor
COX - Cyclooxigenase
CREB - cAMP response-element binding protein
CSF - Cerebrospinal fluid
CVOs - Circumventricular organs
DAG - Diacylglycerol
DD - Death domains
EAE - Autoimmune encephalomyelitis
EGF - Epidermal growth factor
ERK - Extracellular signal-related kinase
FAP - Familial amyloid polyneuropathy
FGF - Fibroblast growth factor
GABA - γ-aminobutyric acid
GDNF - Glial-derived neurotrophic factor
GFAP - Glial fibrillary acidic protein

HAD - HIV-1 associated dementia
HD - Huntington's disease
HIV - Human immunodeficiency virus
ICAM - Intercellular adhesion molecule
ICE - Interleukin-converting-enzyme
INF - Interferon
IGF - Insulin-like growth factor
IKK - IκB kinase
IL - Interleukin
IL-1 - Interleukin-1
IL-1R - IL-1 receptor
IL-1Ra - IL-1 receptor antagonist
IL-1RAcP - IL-1 receptor accessory protein
IRAK - IL-1 Receptor-associated kinase
JAKs - Janus kinases
JNK - c-jun N-terminal kinase
LIF - Leukemia-inhibitory factor
LPS - Lipopolysaccharide
LTP - Long-term potentiation
MAPK - Mitogen-activated protein kinase
MCP-1- Monocyte chemoattractant protein-1
MHC - Major histocompatibility complex
MIP - Macrophage inflammatory protein
MJD - Machado-Joseph disease
MPTP - 1-methyl-4-phenyl-1,2,3,6-tetrahydropyridine
MS - Multiple sclerosis
MT - Metallothioneins
mtDNA - Mitochondrial DNA

NCAM - Neural cell adhesion molecule
NGF - Nerve growth factor
NLS - Nuclear localization signal
NO - Nitric oxide
NOS - Nitric oxide synthase
NSAIDs - Non-steroidal anti-inflammatory drugs
NT-3 - Neurotrophin-3
NT-4/5 - Neurotrophin 4/5
OPCs - Oligodendrocyte precursor cells
$p75^{NTR}$ - p75 neurotrophin receptor
PAMPs - Pathogen associated molecular patterns
PARP - Poly (ADP-ribose) polymerase
PD - Parkinson's disease
PDGF - Platelet-derived growth factor
Pi3K - Phosphoinositide 3-kinase
PKC - Protein kinase C
PLC - Phospholipase C
RAGE - Advanced glycation end-products
RNS - Reactive nitrogen species
ROS - Reactive oxygen species
SOD - Superoxide dismutase
SVZ - Subventricular zone
TGF - Transforming growth factor
TIR - Toll/IL-1R motif
TLE - Temporal lobe epilepsy
TLR - Toll-like receptor
TNF - Tumor necrosis factor
TNFR - Tumor necrosis factor receptor
TRAF6 - Tumor necrosis factor receptor-associated factor 6
TREM - Triggering receptor expressed on myeloid cells
TTR - Transthyretin
VCAM-1 - Vascular adhesion molecule-1
VEGF - Vascular endothelial growth factor
ZnT - Zinc transporter

Index

Acetylcholine, 55, 68, 121, 254, 256, 260, 451
Acquired immunodeficiency syndrome (AIDS), 282–283, 288, 292–295
Adenine nucleotide translocator (ANT), 194, 197, 292
Adenosine, 53–56, 63, 65, 70, 145, 153, 246, 257, 435
    A1 receptor, 54–57
    A2 receptor, 64
Adenosine tri-phosphate (ATP), 19, 45, 53, 55, 68, 142–143, 175, 193–198, 205, 207, 210, 224, 228–230, 236, 270, 291–292, 320–322, 435, 461–469, 485, 488
Adhesion molecules, 5–8, 14, 40, 59, 92, 123, 180, 284–285, 287, 370, 434, 447, 477, 479, 489, 490
Advanced glycation end-products (RAGE), 247, 277–278, 313, 320
Aging, 88, 121–123, 150, 178, 199, 210, 231, 338, 346, 376, 382, 408, 431, 508
Allodynia, 142–143, 462–469
Amygdala, 22, 65, 83, 138, 224, 334, 338, 343, 347, 364–365, 368, 448
Amyloid, 61, 63, 124, 152–153, 157, 222, 234–235, 247–249, 269, 272–273, 275–278, 310–311, 313–320, 325, 334–335, 338, 340, 343–344, 404–408, 411, 414, 447–450
Amyloid precursor protein (APP), 234, 235, 249, 310, 316, 334–337, 339–345, 347, 350, 411, 447, 449–450

Amyotrophic Lateral Sclerosis (ALS), 67, 141, 157, 291, 296
Angiogenesis, 95, 287, 490, 492–493
Antiapoptotic, 18, 196–197, 493, 489
Antiepileptic, 116–117
Antigen, 5, 7–8, 37–38, 40, 48, 59–60, 62–63, 86, 99, 126, 157, 312–313, 451
Antigen presenting cells (APCs), 5, 7–8, 40, 63, 312
Anti-inflammatory, 8–10, 37, 41, 48, 58, 60, 62–64, 123, 127–131, 250, 252–256, 260–261, 323–324, 385, 415, 479–481, 490, 483
Antioxidant, 122, 174, 181, 183, 205, 227, 229, 234, 255, 260, 320, 322, 324, 347, 350
AP-1, 10, 14, 18, 433
Apolipoprotein E (APOE), 247, 253, 260 310, 316, 336, 506
Apoptosis, 5, 7, 15, 17–18, 20–21, 38, 84, 89–91, 147, 149–150, 152–153, 156–157, 182, 193, 196–199, 202–204, 208, 210, 228, 230, 285–286, 288–293, 295, 299, 319, 321–322, 326, 368, 398, 411–412, 436, 449–450, 477–478, 482–483, 493
Aspartate, 113, 345–346, 368, 398
Astrocytes, 3, 6–12, 14, 23, 45, 56–57, 59–62, 65–67, 83–87, 89, 91–93, 96–97, 111–115, 140–145, 148, 157, 158, 173–178, 181–182, 184–185, 195, 223, 235, 236–237, 248–250,

261, 284–288, 290, 292–294,
311–317, 319–321, 325, 349–350, 364,
368–371, 376, 378–379, 382–385, 429,
434–436, 438, 447, 463, 476, 491
Astrogliosis, 8, 67, 142, 180, 311
Ataxin-3, 402, 403, 406–416
Autoimmune, 59, 63, 123, 129, 142, 145, 379, 433
Autoimmune encephalomyelitis (EAE), 46, 129, 145
Axonal growth, 125, 173, 176, 180, 183

Bacteria, 4, 7, 124, 179, 414, 506–507
Basal ganglia, 55, 206, 451
Bax, 147, 151, 157, 196–197, 204, 208, 321, 368, 398
Bcl-2, 17, 147, 151, 153, 155, 182, 194–197, 208–209, 368, 398, 489
Bcl-2-associated death protein (Bad), 147, 153, 157, 197, 204, 321
Bcl-xl, 17, 147, 151, 182, 197, 208, 232, 398
Blood-brain barrier (BBB), 3, 5–9, 46, 59–60, 68, 150, 175, 252, 284, 294, 311, 318, 325, 368, 371, 414, 507
Bone marrow, 7, 37, 43, 58, 65, 68, 83, 94
Brain injury, 7, 41, 58–59, 62–63, 66, 82, 91, 95, 110, 141, 177, 221, 225, 233, 236–237, 317, 319, 336, 379
Brain parenchyma, 3, 5–7, 9, 59–60, 62, 66–69, 311
Brain repair, 61, 81–82, 89–93, 95–97, 180

CA3, 13, 64, 225, 233, 270, 363–367, 369–371, 376–377, 446, 448
CAG repeats, 392–394, 396, 451
Calcium (Ca^{2+}), 9, 11, 20–22, 54–55, 66, 112, 114, 138, 148–149, 153, 193–195, 197–200, 203, 207–209, 210, 221–222, 225–226, 230–233, 236–237, 287–288, 293–300, 323–324, 346, 350, 365, 267, 399, 469, 471–472
Calcium channel, 22, 54, 184, 222, 286, 295, 297, 396
Calcium homeostasis, 20, 193–194, 198, 200, 207, 395
Calpains, 155, 198, 208, 322, 367, 492

cAMP response-element binding protein (CREB), 139, 148, 153, 175, 399
Caspases, 15, 17, 155, 182, 196–199, 204, 208, 230, 277, 292, 295, 321–322, 363, 368, 396, 398, 411, 468, 483
Caspase-1 (Interleukin-converting enzyme, ICE), 12, 19, 124, 127, 129, 398, 411, 482
Caspase-2, 197, 208
Caspase-3, 15, 17, 151, 153, 157, 196–198, 204, 229, 236, 277, 292, 398, 482–483
Caspase-6, 15, 157, 196, 208, 368
Caspase-7, 153, 196, 204
Caspase-8, 15, 17, 196, 204, 208
Caspase-9, 157, 196–197, 204–205, 208–209
Catalase, 183, 255, 320–321, 347–348
Caudate-putamen, 174, 178–179, 395
CD11, 8, 39, 40, 42, 43, 45, 126, 313, 477
CD34, 37, 39–43, 48, 59
Cell cycle, 86, 89, 95, 149, 288, 384, 398, 433, 446, 448, 451, 453, 487, 493
Cell death, 3, 11, 14–15, 20–21, 38, 67, 122, 124, 126–127, 137–138, 142, 149–153, 155–157, 179, 183, 193, 196–199, 203–204, 206, 208–210, 229, 231, 237, 246, 271, 276 293, 295, 312, 319, 321–322, 345, 363, 367, 368, 371, 373, 384, 394–395, 412, 448, 451, 453, 478, 483
Cell debris, 4, 7–8, 12, 20, 59, 180, 313
Cell survival, 16, 18, 137–138, 146–147, 149, 153, 156, 176, 180, 183
Ceramide, 18, 149
Cerebellum, 86, 138, 143, 337, 364, 397, 409–410
Cerebrospinal fluid (CSF), 87, 157, 207, 234, 245, 283, 295, 319, 338–339, 350
Chaperone, 202, 342–343, 401, 406, 408, 417–418, 455
Chemokine, 3, 5–11, 14, 18, 38, 46–47, 91–92, 123, 130, 182, 247–249, 283–284, 286–291, 295–298, 311, 314–319, 351, 419
CCL2, 6, 46–47
CCL5, 46–47

Index 515

CCL19, 6
CCL17, 63
CCL21, 6, 46
CCR1, 296
CCR2, 92
CCR3, 284, 289–290, 296
CCR5, 284, 289–290, 296–297
CXCL10, 46, 63, 319
CXCL12, 91, 96
CXCR3, 46, 319
CXCR4, 11, 91, 284, 288–290, 297
Chemotaxis, 10, 19, 286, 315, 317, 473
Cholesterol, 15, 122, 205, 346, 349
Cholinergic, 122, 138, 143, 150, 152, 207, 256, 325, 451–452
Choroid plexus, 6, 44, 87
Ciliary neurotrophic factor (CNTF), 22, 90, 151
Circumventricular organs (CVOs), 6, 9
c-jun, 124, 129, 157, 204, 292, 467, 493
c-jun N-terminal kinase (JNK), 14–18, 124, 129–130, 149, 157, 292, 467
Cognitive, 81, 109–110, 121–122, 141, 206, 234, 245, 253, 255, 257–260, 282–284, 310, 316, 323, 325, 333–335, 337, 339, 364, 385, 392, 395, 447–448, 451, 503
Complement system, 61, 249, 318, 324
Convulsions, 65, 363–365
Cortex, 39, 45–46, 57, 65, 82–84, 86, 89–91, 93–94, 138, 143–144, 152, 155, 157–158, 180–181, 206–207, 209, 224, 250, 311, 315, 319–320, 322, 334, 343, 363–366, 370, 395, 397–398, 448, 450
Creutzfeldt-Jacob, 144
Cyclic adenosine monophosphate (cAMP), 11, 63, 144, 155–156, 399
Cyclooxygenase (COX), 21, 248–252, 260–261, 318–319, 323, 479, 483, 489
Cytochrome c, 17, 157, 193–199, 204, 230–231, 292, 321–322, 343, 368, 398
Cytokine, 9–10, 12, 14, 41, 45, 91, 123, 127–128, 130–131, 152, 182, 284, 290, 294–295, 316–317, 379, 385, 415, 434, 452, 468, 481–482

Death domains (DD), 15, 18
Dementia, 37, 48, 61, 110, 152, 245, 252–254, 257–261, 282–283, 285, 290, 310, 319, 334–336, 348, 395–396, 447, 504
   HIV-1 associated dementia (HAD), 282–285, 288–289, 291, 293–296, 298–299
Demyelination, 82, 89, 91, 94–95, 142, 368, 410, 428–438
Dendritic cells, 5–6, 40, 59, 63
Dentate gyrus, 13, 37, 39–42, 44, 46, 82, 233, 270, 364, 366, 370, 376, 378–380, 384–385, 446, 448–450, 452, 455
Diabetes, Diabetic, 57, 150, 209, 270, 395, 462, 476–484, 488–494
Diacylglycerol (DAG), 18, 148
DNA
   cDNA, 47, 178, 409–410
   Cell death, fragmentation, condensation, 152, 195–196, 199, 209, 229, 249, 323, 349, 365, 370, 407
   Cell division, 288, 379, 433, 446, 453, 483
   Mitochondrial DNA (mtDNA), 194, 202, 204, 207, 321–322
   Polymorphism, 272, 415
   Repair, 399, 402
   Transcription factors, 13–14, 139, 493
Dopamine, 55, 58, 65, 68, 87–88, 94, 96, 121, 151, 173–184, 201–205, 209, 397, 447, 453–454, 456
Doublecortin, 95, 378, 381, 384, 448–449, 452

Endothelial cells, 6, 7, 10, 11, 19, 41, 43, 46, 47, 290, 319, 475–478, 482, 489, 490
Endotoxin, 4
Energy, 112, 141, 198, 206, 210, 228, 230, 291–292, 321, 367, 371, 414, 446, 485, 505, 508
Entorhinal cortex, 46, 152, 224, 334, 363–366, 370
Ependymal cell, 83, 86, 99, 145, 376
Epidermal growth factor (EGF), 84, 85, 95, 97

Epilepsy, 10, 22, 57–58, 65, 81, 111, 116, 145, 222, 225–226, 231, 233, 291, 363–371, 376, 385, 395
Epileptogenesis, 364, 370, 384
Estrogen, 173, 182
Excitotoxic, 3, 20, 22, 111–113, 129, 144, 151, 155, 198, 221, 225, 227, 233, 236–237, 291–295, 366–368
Excitotoxicity, 3, 10–11, 19, 21, 55, 150, 153, 181, 195, 198, 208, 221, 226, 234–236, 283, 287, 289, 290–295, 297, 367, 368, 371, 415
Extracellular signal-related kinase (ERK), 18–19, 128, 146–149, 152–156, 175, 231, 277, 288, 467

Familial amyloidotic polyneuropathy (FAP), 271–277
Fibroblast growth factor (FGF), 85, 90–91, 94–95, 141, 178, 181, 382–383, 432–434, 452
Free radical, 122–123, 193, 205, 229, 285, 291, 295, 320, 346

GABA (γ-aminobutyric acid), 54–55, 68, 87, 94, 111–116, 142, 206, 233, 410, 447, 465, 469
  GABA receptor, 114–115, 117
  GABA transport, 111, 115–117, 233
Glial-derived neurotrophic factor (GDNF), 87, 151, 156, 173–185, 290
Glial fibrillary acidic protein (GFAP), 8, 85, 87, 96, 99, 126, 312, 369, 378–384, 398, 430, 452
Glial scar, 8, 92
Gliosis, 144, 270, 311, 334, 363, 371, 449
Glucocorticoid, 130, 209, 383
Glutamate, 19–23, 54–57, 68–69, 88, 111–117, 148, 150–151, 181–182, 195, 198, 208, 221–222, 226, 233–237, 250, 252, 281, 285–288, 290, 292–293, 295–297, 365–367, 371, 432, 435, 465
Glutamate receptor
  AMPA receptor, 19–22, 182, 222, 226, 237, 293, 365, 369, 436
  Kainate receptor, 21, 222, 237, 293, 365, 369

Metabotropic receptor, 19, 150, 208
NMDA receptor, 21, 58, 69, 121, 140, 151, 156, 195, 198, 209, 222, 297–298, 384, 451, 468
Glutamate release, 21, 55, 234
Glutamate transporter, 20, 23
Glutamate uptake, 57, 112–113, 288
Glutathione, 183, 195, 223, 255, 320–321, 347, 348, 350
Granular cell, 13, 446, 450
Growth factor, 6, 41, 66, 84, 87, 137, 141–151, 153, 156, 177, 209, 248, 299, 313, 316, 382, 433, 448, 452–453, 475, 477, 481, 489, 504

Hippocampus, 13, 39, 41, 43–44, 46–47, 57, 64, 82–83, 86–89, 92–93, 121–131, 138, 141, 143–144, 150, 152, 183, 198, 223–225, 311, 322, 334, 338, 347, 363–371, 375–377, 381, 382, 385, 445–452, 454, 455, 465
Human immunodeficiency virus (HIV), 37, 39, 45, 152, 281–299, 416
Huntingtin, 151, 206–210, 395, 413–414, 451–453
Hydrogen peroxide (H_2O_2), 195, 320, 347–350
6-hydroxydopamine (6-OHDA), 88, 175, 204
Hyperglycemia, 475–477, 491
Hypothalamus, 84, 95, 138, 410, 451
Hypoxia, 153, 196, 290, 365, 475, 491

IκB kinase (IKK), 13, 15–16, 489
Immune, 3–10, 12, 14, 17, 19, 37, 48, 53, 59–64, 66–70, 122–123, 126, 142, 145, 152, 174, 246, 281–286, 290, 293, 311–312, 322, 325, 370, 379, 427, 489, 508
  Adaptive, 5–6, 60, 62–63, 67, 70
  Innate, 4–6, 37, 48, 59–63, 67, 70, 284, 312, 318
Infection, 3–7, 45, 64, 68, 86, 125, 198, 281–287, 290–292, 293–295, 310, 318, 324, 451, 462–463
Inflammation, 3–13, 18–19, 53, 91–92, 94, 123, 128, 141, 198, 224, 237, 246, 248, 250, 252, 277, 311–313, 317–318,

323–324, 385, 415, 436–437, 465, 475–476, 481–482, 490–494
Inflammatory cytokines, 4–6, 8–9, 19, 37, 41, 62–63, 121, 123, 127, 129–131, 179–180, 290, 293, 350, 385, 436, 483
Insulin-like growth factor (IGF), 209, 289, 433, 481
Integrin, 40, 180, 183, 313, 432, 434, 477
Intercellular adhesion molecule 1 (ICAM-1), 6, 180, 477, 489
Interferon (INF), 60, 62, 320
Interleukin-1 (IL-1), 4–10, 12–14, 16, 23, 45, 123, 127–129, 173, 180–182, 185, 248–249, 316–317, 324, 350, 468, 476–481
IL-1 receptor (IL-1R), 5, 12–13, 123, 129, 180
IL-1 receptor accessory protein (IL-1RAcP), 12–13, 16
IL-1 receptor antagonist (IL-1Ra), 8, 10, 12, 22, 486
IL-1 receptor-associated kinase (IRAK), 13–14, 129
Interleukin-2 (IL-2), 62, 126, 316, 476
Interleukin-3 (IL-3), 316
Interleukin-4 (IL-4), 60, 62, 91, 121, 127–131, 317
Interleukin-6 (IL-6), 4, 6, 45, 92, 123–125, 248, 251, 317, 323, 350, 379, 468
Interleukin-8 (IL-8), 350
Interleukin-10 (IL-10), 8, 10, 60, 62, 121, 127–131, 317
Interleukin-12 (IL-12), 63, 96, 126
Interleukin-13 (IL-13), 317
Interleukin-15 (IL-15), 316
Interleukin-16 (IL-16), 284
Interleukin-18 (IL-18), 123–124, 482
Interneurons, 85–87, 151, 208, 225, 363–366, 370, 378, 446
Ion channel, 11, 291–292, 296, 367
Ischemia, 10, 19, 21, 180, 198, 222, 224–227, 229–233, 237, 291, 297–298, 365, 370, 448, 463, 476, 490–493

Janus kinases (JAKs), 127–128
Josephin, 400–403, 407–408

Kinases, 13–14, 21, 124, 127, 137, 147–148, 203, 277, 335, 339–340, 448, 489, 491
Kindling, 22, 65
Knock-out, 20–21, 416, 479

Leukemia-inhibitory factor (LIF), 90, 385
Leukocyte, 11, 40, 46, 64, 285, 313, 477–479, 481, 484, 490
Lipid peroxidation, 122, 198, 292, 321, 347, 367
Lipopolysaccharide (LPS), 4, 9, 19, 65, 67–68, 124–125, 127–129, 179, 182, 468
Long-term potentiation (LTP), 21, 122, 124–127, 129–131, 139, 296
Lymphocyte, 3, 5–6
  B, 5–6
  T, 5–7, 60, 62, 64, 123, 126, 128, 142, 145, 379
Lysosomes, 199, 209, 315, 505

Machado-Joseph disease (MJD), 270, 391, 394, 396–397, 403, 407, 409, 411–415
Macrophages, 4–7, 9–10, 14, 37–38, 40, 44, 60, 63, 67, 84, 123, 125–126, 142, 145, 180, 246, 250, 257, 282–290, 293–295, 350, 436, 438, 504
Macrophage inflammatory protein (MIP), 6, 8, 17, 248, 286–287, 295
Major histocompatibility complex (MHC), 5, 7–8, 40, 59, 247, 312–313, 369–370, 508
Memantine, 254, 291, 293, 295–298
Memory, 21, 82–83, 121, 152, 254, 257–259, 310, 319, 323, 333, 337, 364, 375, 381–382, 385, 446–450, 452
Metallothioneins (MT), 222–224, 227, 231–233, 237, 339
Microglia, 3–12, 14, 19–20, 37–45, 48, 58–62, 64–68, 92, 121, 123, 125–128, 130–131, 140–145, 151, 157, 173–174, 179–182, 184–185, 246–249, 251, 269, 285–287, 289, 290, 292, 293–294, 311, 313–320, 323–325, 349–350, 363, 369–371, 379, 384–385, 436, 438, 447, 461–469, 476, 483, 503–504, 506

Activated, 8, 19–21, 38–41, 59, 61, 64–65, 92, 125–126, 130–131, 141, 143, 173, 179–180, 247, 283–286, 311–313, 315, 317, 319, 320, 350, 371, 379, 465, 467, 469, 477
Resting, 38, 40, 42, 180, 463
Survival, 45
Microgliosis, 38, 40–41, 43, 45, 67, 179
Minocycline, 92, 126, 179, 209, 482–483
Mitochondria
  Amyloid, 311, 320–325
  Apoptosis, 18, 157, 230, 269–270, 321–322, 326, 368, 398, 411–412
  Dysfunction, 122, 199–210
  Huntington's disease, 193, 206–210
  Zinc, 221–222, 225–233, 236–237
Mitochondrial DNA (mtDNA), 122, 202, 204, 207, 321–322
Mitochondrial membrane potential, 194, 228, 292, 321
Mitochondrial respiratory chain, 193–195, 199, 207, 320
Mitogen-activated protein kinase (MAPK), 13–15, 17, 124, 128, 147–148, 156, 183, 292–295, 299, 467
Monocyte, 7, 46, 91, 284, 313–317
Monocyte chemoattractant protein-1 (MCP-1), 91, 248, 284, 294, 313, 317
Mossy fibers, 224–225, 365–366, 371, 448
MPTP (1-methyl-4-phenyl-1, 2, 3, 6-tetrahydropyridine), 58, 65-67, 88, 175, 177–179, 182, 201, 204–205, 321, 454–456
Myelin, 62, 94, 142, 152, 229, 249, 284, 427–438
Myeloid cells, 45, 48, 61, 65

Natural killer cells, 10, 84, 126
Necrosis, 18, 23, 150, 193, 196–199, 228, 235, 291–292, 321, 363, 368
Nestin, 96, 288–289, 378–379, 384, 456
Neural cell adhesion molecule (NCAM), 99, 175, 434
Neural stem cell, 81, 84, 287, 288, 382, 446, 450
Neuroblast, 82–84, 86–87, 89, 91, 95, 378, 385, 446

Neuroblastoma, 152, 181, 204, 410
Neurodegeneration, 3, 11, 21, 56, 57–59, 62, 65–66, 110, 111, 115, 141, 174, 179, 182, 193, 199, 200, 202, 208, 249, 270–71, 273, 276–277, 283, 285, 309–310, 312, 317, 363, 366–367, 369, 370–371, 392, 394, 397, 399, 404, 410, 413, 415, 448–449, 451, 453
Neurodegenerative diseases, 4, 7, 20, 25, 89, 109–110, 122, 125, 141, 178, 193, 199–200, 206, 210, 269, 299, 318, 391, 403, 447, 482, 504–506, 509
  Alzheimer's disease (AD), 20, 58, 61–63, 89, 123, 125, 141, 144, 151–152, 157, 202, 222, 234, 246, 252, 254, 260–261, 291, 296, 298–299, 309, 333–334, 336, 338, 349, 411, 447, 503, 505
  Huntington's disease (HD), 89, 94–95, 97, 141, 144, 151, 193, 199, 206, 296, 391, 395–396, 446, 451
  Multiple sclerosis (MS), 37, 62, 89, 94, 97, 141–142, 144–145, 152, 157, 224, 291, 427–428
  Parkinson's disease (PD), 58, 61–62, 65, 67, 81, 94–95, 123, 141, 151–152, 174, 201, 205, 270, 296, 446, 453
Neurofibrillary tangles, 245, 249, 311, 335, 339, 447
Neuroinflammation, 3, 7, 53, 56, 58–70, 123, 126, 130, 182, 245–247, 281–282, 309, 311–313, 316, 385
Neurogenesis, 61, 81–83, 87–93, 95–96, 290, 377–381, 383–385, 449–456
Neurons
  Death, 3, 8, 11, 14–15, 20–22, 38, 67, 92, 94, 122, 124, 137–138, 141–142, 149–153, 155–157, 175–176, 179–180, 182–183, 193, 195–198, 200, 203–204, 206–210, 222, 225, 228–231, 235, 245–246, 252, 258, 273, 278–279, 287–298, 301, 312–315, 318, 320–321, 323–325, 347, 365, 367–370, 373, 386, 398–399, 401–402, 411–412, 416–417, 452, 455, 457, 487

Survival, 3, 8, 10–11, 18, 21, 88, 90, 97, 137–138, 140, 142, 144, 146–147, 149, 151–153, 155–157, 173–176, 179–180, 182–184, 290, 318, 328, 381, 383, 387, 399, 402, 454, 456
Neuropathic pain, 142, 270, 296, 461–463, 466–469
Neuropeptide Y (NPY), 382–383, 452
Neuroprotection, Neuroprotective, 8, 12, 20–22, 41, 55–58, 60–63, 65–67, 70, 121, 129, 137, 142, 146, 150–151, 153–155, 173–175, 178, 180–182, 184–185, 200, 224, 226–227, 229, 230, 232, 234, 245–247, 249, 252, 256–258, 261–262, 283, 289, 298–301, 317, 325–326, 417, 452, 456–457, 486
Neurosphere, 82, 84–87, 450
Neurotoxic, Neurotoxicity, 8, 12, 20–22, 41, 55, 61, 63, 65, 67, 157, 180, 182, 184–185, 226, 229, 231, 233, 247–248, 252, 255, 284, 287–290, 292–293, 295–297, 299–300, 312, 316–317, 328, 338, 352, 407–408, 414, 419
Neurotransmitter release, 11, 112, 114
Neurotrophins, 7–8, 60–61, 90, 137–138, 140–142, 144–147, 149–153, 155–157, 292, 318, 386
  Brain-derived neurotrophic factor (BNDF), 7, 90, 93, 95, 137–145, 149–154, 156, 178, 180–181, 292, 318, 399, 465–466, 469–471, 473
  Nerve growth factor (NGF), 7, 22, 137–142, 144–145, 148–153, 156–157, 177–178, 181, 452, 508, 511
  Neurotrophin-3 (NT-3), 7, 137–138, 142, 150–153, 156, 318
  Neurotrophin 4/5 (NT-4/5), 90, 137–138, 151, 156
  p75 neurotrophin receptors (p75^{NTR}), 137–138, 140–144, 149, 156–157, 440
NFκB, 9–10, 13–15, 18–21, 147, 149, 479, 482, 486, 488, 492–495, 497–498
Nigrostriatal, 173–185, 397, 401, 453–454

Nitric oxide (NO), 7–9, 14, 20, 22, 92, 141, 145, 198, 223, 229–230, 248, 252, 277, 285–286, 290–292, 295, 297, 312, 319, 350, 476, 483
Nitric oxide synthase (NOS), 14, 141, 195, 198, 229, 252, 277, 291, 319, 478
Non-steroidal anti-inflammatory drugs (NSAIDs), 58, 61, 245, 250–254, 319, 323
Noradrenaline, 68, 88, 122, 293
Nuclear localization signal (NLS), 13, 398

Olfactory bulb, 82–83, 85–88, 90–91, 95, 337–338, 376, 378, 446–447
Oligodendrocytes, 6–7, 10–12, 82–84, 89, 91, 94, 97, 140, 142, 144–145, 152, 156–157, 229, 235, 290, 349, 368, 376, 427, 429, 430–431, 434–435, 463
Oxidative stress, 141, 175, 182–183, 195, 197, 200–203, 206, 221, 223, 227, 229–230, 235–237, 250, 271, 277–278, 309, 319–322, 324, 333, 337, 339, 346, 348, 415, 475, 480, 482

p38 MAPK, 14–15, 125, 128, 156, 292, 295, 299, 467
p53, 96, 157, 204, 290, 486, 487, 493
Pathogen, 4–5, 7–8, 12, 59, 123, 127, 276, 318
Pathogen associated molecular patterns (PAMPs), 4
Pathogenesis, 58, 123, 193, 200–203, 206, 208, 234, 283, 285, 295, 310, 321, 337, 338, 350, 381, 391, 394, 398–399, 403–407, 411–413, 415–416, 428, 432, 436, 445, 475, 476, 482–486
Pathogenic, 151, 200, 273, 275–277, 309–310, 315, 319, 336, 391–393, 397, 403, 405, 408–410, 413–414, 416, 482, 494, 508
Perivascular, 6–8, 10, 40, 44–45, 283, 290, 368
Peroxynitrite (ONOO⁻), 21, 158, 195, 229–230, 286, 292, 297, 319, 479, 482

Phagocytic, 4, 8–9, 38, 48, 61, 173, 180, 184, 247, 312–313, 315, 350, 370
Phagocytosis, 8, 37–38, 48, 180, 312, 315, 325
Phosphoinositide 3-kinase (Pi3K), 18, 146, 175, 183, 199, 295
Phospholipase C (PLC), 18, 146, 148, 175
Phospholipids, 11, 146
Plaque, 222, 235, 247–248, 311, 315–317, 334–335, 338–339, 344–346, 349–350, 427–428, 431, 447–448, 504
Platelet-derived growth factor (PDGF), 432, 481
Poly (ADP-ribose) polymerase (PARP), 198, 204, 229, 231, 480
Polyglutamine, 206, 208–209, 391–400, 402–416, 451, 504
Presenilin, 234, 334, 336, 342, 447, 449–450
Presynaptic, 8, 112–115, 144, 291, 295, 397
Prion, 309–310, 405, 410
Progenitor cell, Precursor cell, 11, 40, 44, 82–84, 86, 88, 90–91, 92, 94, 96–97, 290–291, 378, 380–382, 384–387, 454, 458
  Glial, 83, 383
  Microglial, 38, 41, 44, 313
  Neuronal, 11, 81, 84, 88–90, 92–93, 96, 378–379, 381, 383, 385, 392, 394–395, 454
  Oligodendrocyte (OPCs), 91, 94, 429–431, 433–435
Prostaglandins, 9, 14, 22, 59, 248–250, 294, 314, 483
Protease, 125, 139, 179, 196, 248, 283, 294, 334, 341, 344, 384, 399–403, 409, 413, 494–498
Proteasome, 13, 199–200, 203, 396, 401–402, 404–408, 455, 479, 487–489, 491–498, 512
Protein aggregation, 200–201, 269, 394, 407–408, 414
Protein kinase C, 18, 140, 475–476
Protein misfolding, 202–203, 271, 278, 403, 405, 413
Psychiatric disorder, 139, 260

Purinergic receptors
  P2X, 45, 461, 463–464, 469
  P2Y, 45, 462
Pyramidal cell, 13, 39, 93, 225, 363–364, 365–366, 370–371

Reactive nitrogen species (RNS), 319
Reactive oxygen species (ROS), 3, 8, 122–123, 179, 193, 195, 197–198, 202, 204, 221, 222, 228–231, 233, 235–236, 247–248, 250, 255, 294–295, 297, 311, 314–316, 320–326, 341–342, 348–349, 352, 370
Remyelination, 91, 94, 145, 427–438
Retina, 21, 141–143, 150–151, 198, 291, 396, 409, 475–477, 482–484, 490–491, 504

Schwann cell, 142, 144, 156–157, 183, 275, 288, 469
Secretase-gamma, 157, 234–235, 252, 261, 312, 324–325, 336–337, 344–345, 451
Seizures, 19–22, 90, 92, 111, 115, 141, 157, 291, 363–368, 370–371, 383–385
  Kainate-induced, 21, 366
Serotonin, 55, 88, 447
Signaling, 3–4, 9, 11–15, 18–19, 22, 41, 45, 63, 68, 90, 95, 123–124, 127–131, 137–138, 140, 142–143, 145–150, 152–153, 155–157, 174, 178, 182–184, 199, 221–223, 230–232, 236, 273, 279, 283, 287–288, 293, 321, 324, 326, 384, 386, 415, 435–436, 438–441, 465, 469–470, 473, 493–495
Signal transducers and activators of transcription (STAT), 128
Sphingolipid, 18, 149
Sphingomyelinase, 18, 149
Spinal cord, 60, 64, 86, 92, 94, 138, 142 145, 157, 180, 197, 226, 291, 299, 340, 394, 397, 427, 433, 462–468
Sprouting, 173–174, 176–177, 180, 182–184, 370–371, 445
Status epilepticus, 233–234, 363–369, 371, 383

Stem cells, 38, 48, 81–88, 93–94, 96–97, 123, 287–288, 375, 376–379, 382, 385, 430, 446, 450, 454
Striatum, 57, 61, 82, 89, 91, 93–95, 143, 151, 173, 184–185, 201, 207–208, 397, 451–454
Stroke, 46, 58, 62, 64, 81, 224, 246, 260, 291, 296, 299, 427
Subgranular zone, 270, 376–377, 380, 446, 455
Substantia nigra, 83, 89, 94–95, 138, 151, 173, 175–185, 201, 204–205, 397–398, 407, 410–411, 445, 453–454, 456
Subventricular zone (SVZ), 81–99, 376, 430, 446, 455
Superoxide, 21, 141, 175, 179, 195, 229, 287–288, 322, 348, 487
Superoxide dismutase (SOD), 141, 180, 183, 195, 320–321, 341, 347
Synaptic plasticity, 54–55, 122, 130, 143–144, 148, 450
Synaptic transmission, 11, 54, 57, 146, 344, 468
Synuclein-alpha, 151, 201–204

Temporal lobe epilepsy (TLE), 64–65, 145, 363–364
Thalamus, 84, 95, 144, 233, 364–365, 368, 397
Toll/IL-1R motif (TIR), 5, 12–13
Toll-like receptor (TLR), 4–5, 13
Transcription factor, 13, 18, 20, 92, 124, 139, 141, 147–148, 153, 175, 224, 277, 292, 392, 404, 429, 433–434, 451–452, 475, 482, 484, 490, 493
Transforming growth factor (TGF), 8, 10, 41, 45, 60, 62, 92–93, 95–96, 248, 289, 313, 316, 370, 379, 489
Transthyretin (TTR), 269–278
Trauma, 19, 41, 58, 64, 81, 90, 95, 125, 157, 197, 225–226, 233, 336, 365, 463

Triggering receptor expressed on myeloid cells (TREM), 48, 61
TrkB receptors, 90, 93, 138, 142, 144–145, 148–149, 151–152
Tumor necrosis factor (TNF), 4, 6–10, 12–15, 18–22, 41, 45, 48, 60–64, 66, 91–92, 125–126, 137, 141, 149, 152, 156, 180, 248–249, 257, 284, 286, 290, 293–294, 316–317, 320, 323, 368, 436, 461, 466–468, 475–476, 478, 482–483, 489–490
Tumor necrosis factor receptor (TNFR)
TNFR1, 14–15, 18–21, 141
TNFR2, 14–15, 19–21
Tumor necrosis factor receptor-associated factor 6 (TRAF6), 13, 16, 149, 157, 494
Tyrosine hydroxylase, 61, 95, 176, 205, 410–411
Tyrosine kinase, 21, 137, 148, 156, 174, 467

Ubiquitin, 13, 18, 157, 199, 201, 203, 311, 313, 395–396, 398–400, 402–408, 410, 413–415, 417, 455, 479, 487–496, 498

Vascular adhesion molecule-1 (VCAM-1), 17, 284, 487
Vascular endothelial growth factor (VEGF), 6, 90–91, 95–96, 475, 477–481, 483, 489–493
Ventricles, 82, 84–86, 94–95, 97, 247
Virus, 7, 45, 62, 129, 152, 282, 283–284, 287, 289, 294–295

White matter, 40, 82, 89, 94, 145, 150, 260, 376, 428–429

Zinc, 141, 157, 221, 224, 233–235, 336–346, 349, 429, 433, 486
Zinc Transporter (ZnT), 224, 339, 341, 344, 346

Printed in Singapore

© 2007 Springer Science+Business Media, LLC

This electronic component package is protected by federal copyright law and international treaty. If you wish to return this book and the electronic component package to Springer Science+Business Media, LLC, do not open the disc envelope or remove it from the book. Springer Science+Business Media, LLC, will not accept any returns if the package has been opened and/or separated from the book. The copyright holder retains title to and ownership of the package. U.S. copyright law prohibits you from making any copy of the entire electronic component package for any reason without the written permission of Springer Science+Business Media, LLC, except that you may download and copy the files from the electronic component package for your own research, teaching, and personal communications use. Commercial use without the written consent of Springer Science+Business Media, LLC, is strictly prohibited. Springer Science+Business Media, LLC, or its designee has the right to audit your computer and electronic components usage to determine whether any unauthorized copies of this package have been made.

Springer Science+Business Media, LLC, or the author(s) makes no warranty or representation, either express or implied, with respect to this electronic component package or book, including their quality, merchantability, or fitness for a particular purpose. In no event will Springer Science+Business Media, LLC, or the author(s) be liable for direct, indirect, special, incidental, or consequential damages arising out of the use or inability to use the electronic component package or book, even if Springer Science+Business Media, LLC, or the author(s) has been advised of the possibility of such damages.